Springer Tracts in Advanced Robotics

Volume 147

The Springer Tracts in Advanced Robotics (STAR) publish new developments and advances in the fields of robotics research, rapidly and informally but with a high quality. The intent is to cover all the technical contents, applications, and multidisciplinary aspects of robotics, embedded in the fields of Mechanical Engineering, Computer Science, Electrical Engineering, Mechatronics, Control, and Life Sciences, as well as the methodologies behind them. Within the scope of the series are monographs, lecture notes, selected contributions from specialized conferences and workshops, as well as selected PhD theses.

Special offer: For all clients with a print standing order we offer free access to the electronic volumes of the Series published in the current year.

Indexed by SCOPUS, DBLP, EI Compendex, zbMATH, SCImago.

All books published in the series are submitted for consideration in Web of Science.

Peter Corke · Witold Jachimczyk · Remo Pillat

Robotics, Vision and Control

Fundamental Algorithms in MATLAB®

3rd edition 2023

 Springer

Peter Corke
Queensland University of Technology
Brisbane, QLD, Australia

Witold Jachimczyk
MathWorks
Natick, MA, USA

Remo Pillat
MathWorks
Natick, MA, USA

ISSN 1610-7438   ISSN 1610-742X (electronic)
Springer Tracts in Advanced Robotics
ISBN 978-3-031-07261-1   ISBN 978-3-031-07262-8 (eBook)
https://doi.org/10.1007/978-3-031-07262-8

This Springer imprint is published by the registered company Springer Nature Switzerland AG
The registered company address is: Gewerbestrasse 11, 6330 Cham, Switzerland

*To my family Phillipa, Lucy and Madeline for their indulgence
and support;
my parents Margaret and David for kindling my curiosity;
and to Richard (Lou) Paul who planted the seed that became
this book. – PC*

*To my wife Halina, my daughter Julia and my son Alec for their
encouragement and support;
my mom Janina and my uncle Alfred for inspiring me to be
an engineer. – WJ*

*To my beloved wife, Kristi, for her enduring patience and tolerance
for new projects;
to my children Tobin and Evabelle for giving up some play time;
and to my parents Lothar and Birgit for allowing me to pursue
my dreams. – RP*

Foreword

Once upon a time, a very thick document of a dissertation from a faraway land came to me for evaluation. *Visual robot control* was the thesis theme and *Peter Corke* was its author. Here, I am reminded of an excerpt of my comments, which reads, *this is a masterful document, a quality of thesis one would like all of one's students to strive for, knowing very few could attain – very well considered and executed.*

The connection between robotics and vision has been, for over three decades, the central thread of Peter Corke's productive investigations and successful developments and implementations. In this third edition of the book on *Robotics, Vision and Control* he is joined by Witold Jachimczyk and Remo Pillat from MathWorks, the publishers of MATLAB. In its melding of theory and application, this third edition has considerably benefited from the authors' unique and diverse experience in academia, commercial software development, and real-world applications across robotics, computer vision and self-driving vehicles.

There have been numerous textbooks in robotics and vision, but few have reached the level of integration, analysis, dissection, and practical illustrations evidenced in this book. The discussion is thorough, the narrative is remarkably informative and accessible, and the overall impression is of a significant contribution for researchers and future investigators in our field. Most every element that could be considered as relevant to the task seems to have been analyzed and incorporated, and the effective use of toolbox software echoes this thoroughness.

The reader is taken on a realistic walk through the fundamentals of mobile robots, navigation, localization, manipulator-arm kinematics, dynamics, and joint-level control, as well as camera modeling, image processing, feature extraction, and multi-view geometry. These areas are finally brought together through extensive discussion of visual servo system and large-scale real-world examples. In the process, the authors provide insights into how complex problems can be decomposed and solved using powerful numerical tools and effective software.

The *Springer Tracts in Advanced Robotics (STAR)* is devoted to bringing to the research community the latest advances in the robotics field on the basis of their significance and quality. Through a wide and timely dissemination of critical research developments in robotics, our objective with this series is to promote more exchanges and collaborations among the researchers in the community and contribute to further advancements in this rapidly growing field.

The authors bring a great addition to our STAR series with an authoritative book, reaching across fields, thoughtfully conceived and brilliantly accomplished.

Oussama Khatib
Stanford, California
March 2023

Preface

» *Tell me and I will forget.*
Show me and I will remember.
Involve me and I will understand.
– Chinese proverb

Simple things should be simple,
complex things should be possible.
– Alan Kay

These are exciting times for robotics. Since the first edition of this book was published over ten years ago we have seen great progress: the actuality of self-driving cars on public roads, multiple robots on Mars (including one that flies), robotic asteroid and comet sampling, the rise of robot-enabled businesses like Amazon, and the DARPA Subterranean Challenge where teams of ground and aerial robots autonomously mapped underground spaces. We have witnessed the drone revolution – flying machines that were once the domain of the aerospace giants can now be bought for just tens of dollars. All of this has been powered by the ongoing improvement in computer power and tremendous advances in low-cost inertial sensors and cameras – driven largely by consumer demand for better mobile phones and gaming experiences. It's getting easier for individuals to create robots – 3D printing is now very affordable, the Robot Operating System (ROS) is capable and widely used, and powerful hobby technologies such as the Arduino, Raspberry Pi, and Dynamixel servo motors are available at low cost. This in turn has driven the growth of the global maker community, and empowered individuals working at home, and enabled small startups to do what would once have been done by major corporations.

Robots are machines which acquire data, process it, and take action based on it. The data comes from a variety of sensors that measure, for example, the velocity of a wheel, the angle of a robot arm's joint, or the intensities of millions of pixels that comprise an image of the world. For many robotic applications the amount of data that needs to be processed, in real-time, is massive. For a vision sensor it can be on the order of tens to hundreds of megabytes per second. Progress in robots and machine vision has been, and continues to be, driven by more effective ways to process sensory data.

One axis of progress has been driven by the relentless increase in affordable computational power. ▶ Moore's law predicts that the number of transistors on a chip will double every two years, and this enables ever-increasing amounts of memory, and parallel processing with multiple cores and graphical processing units (GPUs). Concomitantly, the size of transistors has shrunk and clock speed has increased.

The other axis of progress is algorithmic, exploiting this abundance of computation and memory to solve robotics problems. Over decades, the research community has developed many solutions for important problems in perception, localization, planning, and control. However, for any particular problem there is a wide choice of algorithms, and each of them may have several implementations. These will be written in a variety of languages, with a variety of API styles and conventions, and with variable code quality, documentation, support, and licence conditions. This is a significant challenge for robotics today, and "cobbling together" disparate pieces of software has become an essential skill for roboticists. The ROS framework ▶ has helped by standardizing interfaces and allowing common functions to be composed to solve a particular problem. Nevertheless, the software side of robotics is still harder and more time-consuming than it should be. This unfortunate complexity, and the sheer range of choice, presents a very real barrier to somebody new entering the field.

When the first author started in robotics and vision in the mid 1980s, the IBM PC had been recently released – it had a 4.77 MHz 16-bit microprocessor and 16 kbytes (expandable to 256 k) of memory. In the 1990s it took a rack full of custom electronics to process video data in real time.

See ▶ https://ros.org.

VME rack
- 68030 CPU @ 25 MHz
  + 16 MB RAM + VxWorks
- Datacube pixel
  processing @ 10 Mpix/s

Robot's eye view

Camera strobe control

Multibus i/o rack
- motor current
- wrist force + torque

Analog sensor filters

Unimation robot
controller

◻ Fig. 1 Once upon a time a lot of equipment was needed to do vision-based robot control. The first author with a large rack full of real-time image processing and robot control equipment and a PUMA 560 robot (1992). Over the intervening 30 years, the number of transistors on a chip has increased by a factor of $2^{30/2} \approx 30{,}000$ according to Moore's law

》 the software tools used in this book aim to reduce complexity for the reader

The software tools that are used in this book aim to reduce complexity for the reader by providing a coherent and complete set of functionality. We use MATLAB®, a popular mathematical and engineering computing environment and associated licensed toolboxes that provide functionality for robotics and computer vision. This makes common algorithms tangible and accessible. You can read much of the code, you can apply it to your own problems, or you can extend it. It gives you a "leg up" as you begin your journey into robotics.

》 allow the user to work with real problems, not just trivial examples

This book uses that software to illustrate each topic, and this has a number of benefits. Firstly, the software allows the reader to work with real problems, not just trivial examples. For real robots, those with more than two links, or real images with millions of pixels the computation required is beyond unaided human ability. Secondly, these software tools help us gain insight which can otherwise get lost in the complexity. We can rapidly and easily experiment, play *what if* games, and depict the results graphically using the powerful 2D and 3D graphical display tools of MATLAB.

》 a cohesive narrative that covers robotics and computer vision – both separately and together

The book takes a conversational approach, weaving text, mathematics, and lines of code into a cohesive narrative that covers robotics and computer vision – separately, and together as *robotic vision*. It shows how complex problems can be decomposed and solved using just a few simple lines of code. More formally this is an inductive learning approach, going from specific and concrete examples to the more general.

》 show how complex problems can be decomposed and solved

The topics covered in this book are guided by real problems faced by practitioners of robotics, computer vision, and robotic vision. Consider the book as a grand tasting menu and we hope that by the end of this book you will share our enthusiasm for these topics.

》 consider the book as a grand tasting menu

We were particularly motivated to present a solid introduction to computer vision for roboticists. The treatment of vision in robotics textbooks tends to concentrate on simple binary vision techniques. In this book we will cover a broad range of topics including color vision, advanced segmentation techniques, image warping, image retrieval, stereo vision, pose estimation, deep learning, bundle adjustment, and visual odometry. We also cover nonperspective imaging using fisheye lenses, catadioptric optics, and the emerging area of light-field cameras. These topics are growing in importance for robotics but are not commonly covered. Vision is a powerful sensing modality, and roboticists should have a solid grounding in modern fundamentals. The last part of the book shows how to apply vision and robotics concepts to solving real-world automated driving and UAV applications.

This book is unlike other textbooks, and deliberately so. While there are already a number of excellent textbooks that cover robotics and computer vision separately and in depth, there are few that cover both in an integrated fashion. Achieving such integration was a principal goal of this book.

» software is a first-class citizen in this book

Software is a tangible instantiation of the algorithms described – it can be read and it can be pulled apart, modified and put back together again – software is a first-class citizen in this book. There are a number of classic books that use software in an illustrative fashion and have influenced this approach, for example *LaTeX: A document preparation system* (Lamport 1994), *Numerical Recipes in C* (Press et al. 2007), *The Little Lisper* (Friedman et al. 1987), and *Structure and Interpretation of Classical Mechanics* (Sussman et al. 2001). In this book, over 2000 lines of code across over 1600 examples illustrate how the software can be used and generally provide *instant gratification* in just a couple of lines of MATLAB code.

» instant gratification in just a couple of lines of MATLAB code

Thirdly, building the book around MATLAB and the associated toolboxes that we are able to tackle more realistic and more complex problems than other books.

» this book provides a complementary approach

A key motivation of this book is to make robotics, vision, and control algorithms *accessible* to a wide audience. The mathematics that underpins robotics is inescapable, but the theoretical complexity has been minimized and the book assumes no more than an undergraduate-engineering level of mathematical knowledge. The software-based examples help to ground the abstract concepts and make them tangible. This approach is complementary to the many other excellent textbooks that cover these same topics but which take a stronger, and more traditional, theoretical approach. This book is best read in conjunction with those other texts, and the end of each chapter has a section on further reading that provides pointers to relevant textbooks and key papers.

The fields of robotics and computer vision are underpinned by theories developed by mathematicians, scientists, and engineers over many hundreds of years. Some of their names have become adjectives like Coriolis, Gaussian, Laplacian, or Cartesian; nouns like Jacobian, or units like Newton and Coulomb. They are interesting characters from a distant era when science was a hobby and their day jobs were as doctors, alchemists, gamblers, astrologers, philosophers, or mercenaries. To know whose shoulders we are standing on, the book includes small vignettes about the lives of some of these people – a smattering of history as a backstory.

Many people have helped with critical comments over previous editions – this edition is the better for their input and we thank: Paul Newman, Daniela Rus, Cédric Pradalier, Tim Barfoot, Dmitry Bratanov, Duncan Campbell, Donald Dansereau, Tom Drummond, Malcolm Good, Peter Kujala, Obadiah Lam, Jörn

Malzahn, Felipe Nascimento Martins, Ajay Pandey, Dan Richards, Sareh Shirazi, Surya Singh, Ryan Smith, Ben Talbot, Dorian Tsai, Ben Upcroft, François Chaumette, Donald Dansereau, Kevin Lynch, Robert Mahony, and Frank Park. Thanks also to all those who have submitted errata. We are grateful to our colleagues who have provided detailed and insightful feedback on the latest chapter drafts: Christina Kazantzidou (who helped to polish words and mathematical notation), Tobias Fischer, Will Browne, Jesse Haviland, Feras Dayoub, Dorian Tsai, Alessandro De Luca, Renaud Detry, Steve Eddins, Qu Cao, Labhansh Atriwal, Birju Patel, Hannes Daepp, Karsh Tharyani, Jianxin Sun, Brian Fanous, Cameron Stabile, Akshai Manchana, and Zheng Dong.

We have tried hard to eliminate errors but inevitably some will remain. Please contact us on ▶ https://github.com/petercorke/RVC3-MATLAB with issues or suggestions for improvements and extensions.

We thank our respective employers, Queensland University of Technology and MathWorks®, for their support of this project. Over all editions, this book has enjoyed strong support from the MathWorks book program, and from the publisher. At Springer-Nature, Thomas Ditzinger has supported this project since before the first edition, and Wilma McHugh and Ellen Blasig have assisted with this edition. Special thanks also to Yvonne Schlatter and the team at le-tex for their wonderful support with typesetting.

Notes on the Third Edition

The first edition (2011), the second edition as a single volume (2017) and then as a two-volume set (2022).

The first two editions of this book ◄ were based on MATLAB® in conjunction with open-source toolboxes that are now thirty years old – that's a long time for any piece of software. Much has happened in the last decade that motivates a change to the software foundations of the book, and that has led to *two* third editions:

- The version you are reading, rewritten with colleagues from MathWorks, is based on MATLAB, and state-of-the-art toolboxes developed by MathWorks including: Robotics System Toolbox™, Navigation Toolbox™, Computer Vision Toolbox™, and Image Processing Toolbox™.
  To run the examples in this book you require appropriate software licenses and details are given in ▶ App. A.
- The alternative version, is based on Python which is a popular open-source language with strong third-party support. The old MATLAB-based toolboxes have been ported to Python (Corke 2021).

In addition to changing the software underpinnings of the book, this third edition also provides an opportunity to fix errors, improve mathematical notation, and clarify the narrative throughout. Chapters 2 and 7 have been extensively rewritten. This edition also includes new topics such as graph-based path planning, Dubins and Reeds-Shepp paths, branched robots, URDF models, collision checking, task-space control, deep learning for object detection and semantic segmentation, fiducial markers, and point clouds. Chapter 16, previously advanced visual servoing, has been replaced with large-scale application examples that showcase advanced features of many MathWorks toolboxes.

Peter Corke
Witold Jachimczyk
Remo Pillat
Brisbane, Australia and Boston, USA
July 2022

Contents

1	**Introduction**	1
1.1	**A Brief History of Robots**	2
1.2	**Types of Robots**	4
1.3	**Definition of a Robot**	8
1.4	**Robotic Vision**	9
1.5	**Ethical Considerations**	11
1.6	**About the Book**	12
1.6.1	MATLAB and the Book	13
1.6.2	Notation, Conventions, and Organization	13
1.6.3	Audience and Prerequisites	15
1.6.4	Learning with the Book	15
1.6.5	Teaching with the Book	16
1.6.6	Outline	17
1.6.7	Further Reading	18

I Foundations

2	**Representing Position and Orientation**	21
2.1	**Foundations**	22
2.1.1	Relative Pose	22
2.1.2	Coordinate Frames	24
2.1.3	Pose Graphs	27
2.1.4	Summary	28
2.2	**Working in Two Dimensions (2D)**	30
2.2.1	Orientation in Two Dimensions	31
2.2.2	Pose in Two Dimensions	34
2.3	**Working in Three Dimensions (3D)**	42
2.3.1	Orientation in Three Dimensions	43
2.3.2	Pose in Three Dimensions	60
2.4	**Advanced Topics**	66
2.4.1	Pre- and Post-Multiplication of Transforms	66
2.4.2	Active and Passive Transformations	67
2.4.3	Direction Cosine Matrix	68
2.4.4	Efficiency of Representation	68
2.4.5	Distance Between Orientations	69
2.4.6	Normalization	70
2.4.7	Understanding the Exponential Mapping	71
2.4.8	More About Twists	73
2.4.9	Configuration Space	77
2.5	**MATLAB Classes for Pose and Rotation**	79
2.6	**Wrapping Up**	82
2.6.1	Further Reading	83
2.6.2	Exercises	85
3	**Time and Motion**	87
3.1	**Time-Varying Pose**	88
3.1.1	Rate of Change of Orientation	88
3.1.2	Rate of Change of Pose	89
3.1.3	Transforming Spatial Velocities	90
3.1.4	Incremental Rotation	91
3.1.5	Incremental Rigid-Body Motion	94

3.2	**Accelerating Bodies and Reference Frames**	95
3.2.1	Dynamics of Moving Bodies	95
3.2.2	Transforming Forces and Torques	96
3.2.3	Inertial Reference Frame	97
3.3	**Creating Time-Varying Pose**	98
3.3.1	Smooth One-Dimensional Trajectories	99
3.3.2	Multi-Axis Trajectories	101
3.3.3	Multi-Segment Trajectories	102
3.3.4	Interpolation of Orientation in 3D	104
3.3.5	Cartesian Motion in 3D	105
3.4	**Application: Inertial Navigation**	107
3.4.1	Gyroscopes	108
3.4.2	Accelerometers	111
3.4.3	Magnetometers	114
3.4.4	Inertial Sensor Fusion	118
3.5	**Wrapping Up**	121
3.5.1	Further Reading	121
3.5.2	Resources	122
3.5.3	Exercises	122

II   Mobile Robotics

4	**Mobile Robot Vehicles**	127
4.1	**Wheeled Mobile Robots**	129
4.1.1	Car-Like Vehicle	130
4.1.2	Differentially-Steered Vehicle	142
4.1.3	Omnidirectional Vehicle	144
4.2	**Aerial Robots**	147
4.3	**Advanced Topics**	154
4.3.1	Nonholonomic and Underactuated Systems	154
4.4	**Wrapping Up**	157
4.4.1	Further Reading	157
4.4.2	Exercises	159
5	**Navigation**	161
5.1	**Introduction to Reactive Navigation**	163
5.1.1	Braitenberg Vehicles	164
5.1.2	Simple Automata	166
5.2	**Introduction to Map-Based Navigation**	168
5.3	**Planning with a Graph-Based Map**	170
5.3.1	Breadth-First Search	173
5.3.2	Uniform-Cost Search	174
5.3.3	A* Search	177
5.3.4	Minimum-Time Path Planning	179
5.3.5	Wrapping Up	179
5.4	**Planning with an Occupancy Grid Map**	180
5.4.1	Distance Transform	180
5.4.2	D*	185
5.5	**Planning with Roadmaps**	187
5.5.1	Introduction to Roadmap Methods	187
5.5.2	Probabilistic Roadmap Method (PRM)	189
5.6	**Planning Drivable Paths**	192
5.6.1	Dubins Path Planner	193
5.6.2	Reeds-Shepp Path Planner	194
5.6.3	Lattice Planner	197

5.6.4   Clothoids 200
5.6.5   Planning in Configuration Space (RRT) 203
5.7   **Advanced Topics** 207
5.7.1   A* vs Dijkstra Search 207
5.7.2   Converting Grid Maps to Graphs 207
5.7.3   Converting Between Graphs and Matrices 208
5.7.4   Local and Global Planning 208
5.8   **Wrapping Up** 209
5.8.1   Further Reading 209
5.8.2   Resources 211
5.8.3   Exercises 211

6   **Localization and Mapping** 213
6.1   **Dead Reckoning Using Odometry** 218
6.1.1   Modeling the Robot 218
6.1.2   Estimating Pose 221
6.2   **Landmark Maps** 226
6.2.1   Localizing in a Landmark Map 226
6.2.2   Creating a Landmark Map 232
6.2.3   EKF SLAM 234
6.2.4   Sequential Monte-Carlo Localization 239
6.2.5   Rao-Blackwellized SLAM 244
6.3   **Occupancy Grid Maps** 244
6.3.1   Application: Lidar 245
6.3.2   Lidar-Based Odometry 246
6.3.3   Lidar-Based Map Building 249
6.3.4   Lidar-Based Localization 251
6.3.5   Simulating Lidar Sensors 255
6.4   **Pose-Graph SLAM** 256
6.4.1   Pose-Graph Landmark SLAM 261
6.4.2   Pose-Graph Lidar SLAM 263
6.5   **Wrapping Up** 266
6.5.1   Further Reading 268
6.5.2   Exercises 271

III   **Robot Manipulators**

7   **Robot Arm Kinematics** 275
7.1   **Forward Kinematics** 277
7.1.1   Forward Kinematics from a Pose Graph 278
7.1.2   Forward Kinematics as a Chain of Robot Links 283
7.1.3   Branched Robots 291
7.1.4   Unified Robot Description Format (URDF) 293
7.1.5   Denavit-Hartenberg Parameters 295
7.2   **Inverse Kinematics** 297
7.2.1   2-Dimensional (Planar) Robotic Arms 298
7.2.2   3-Dimensional Robotic Arms 301
7.2.3   Underactuated Manipulator 306
7.2.4   Overactuated (Redundant) Manipulator 308
7.3   **Trajectories** 308
7.3.1   Joint-Space Motion 308
7.3.2   Cartesian Motion 311
7.3.3   Kinematics in Simulink 312
7.3.4   Motion Through a Singularity 313

7.4	**Applications**	315
7.4.1	Writing on a Surface	315
7.4.2	A 4-Legged Walking Robot	316
7.5	**Advanced Topics**	320
7.5.1	Creating the Kinematic Model for a Robot	320
7.5.2	Modified Denavit-Hartenberg Parameters	321
7.5.3	Products of Exponentials	323
7.5.4	Collision Checking	324
7.6	**Wrapping Up**	325
7.6.1	Further Reading	326
7.6.2	Exercises	327
8	**Manipulator Velocity**	329
8.1	**Manipulator Jacobian**	330
8.1.1	Jacobian in the World Coordinate Frame	330
8.1.2	Jacobian in the End Effector Coordinate Frame	333
8.1.3	Analytical Jacobian	333
8.2	**Application: Resolved-Rate Motion Control**	335
8.3	**Jacobian Condition and Manipulability**	338
8.3.1	Jacobian Singularities	338
8.3.2	Velocity Ellipsoid and Manipulability	339
8.3.3	Dealing with Jacobian Singularity	342
8.3.4	Dealing with a Non-Square Jacobian	342
8.4	**Force Relationships**	346
8.4.1	Transforming Wrenches to Joint Space	347
8.4.2	Force Ellipsoids	348
8.5	**Numerical Inverse Kinematics**	348
8.6	**Advanced Topics**	350
8.6.1	Computing the Manipulator Jacobian Using Twists	350
8.6.2	Manipulability, Scaling, and Units	350
8.7	**Wrapping Up**	351
8.7.1	Further Reading	351
8.7.2	Exercises	352
9	**Dynamics and Control**	355
9.1	**Independent Joint Control**	356
9.1.1	Actuators	356
9.1.2	Friction	358
9.1.3	Link Mass	359
9.1.4	Gearbox	360
9.1.5	Modeling the Robot Joint	361
9.1.6	Velocity Control Loop	362
9.1.7	Position Control Loop	367
9.1.8	Summary	369
9.2	**Rigid-Body Equations of Motion**	369
9.2.1	Gravity Term	372
9.2.2	Inertia (Mass) Matrix	373
9.2.3	Coriolis and Centripetal Matrix	375
9.2.4	Effect of Payload	375
9.2.5	Base Wrench	376
9.2.6	Dynamic Manipulability	377
9.3	**Forward Dynamics**	378
9.4	**Rigid-Body Dynamics Compensation**	380
9.4.1	Feedforward Control	381
9.4.2	Computed-Torque Control	382

9.5   **Task-Space Dynamics and Control** 382
9.6   **Application** 387
9.6.1   Series-Elastic Actuator (SEA) 387
9.7   **Wrapping Up** 389
9.7.1   Further Reading 389
9.7.2   Exercises 391

IV   Computer Vision

10   **Light and Color** 395
10.1   **Spectral Representation of Light** 396
10.1.1 Absorption 399
10.1.2 Reflectance 400
10.1.3 Luminance 400
10.2   **Color** 401
10.2.1 The Human Eye 402
10.2.2 Camera Sensor 404
10.2.3 Measuring Color 405
10.2.4 Reproducing Colors 407
10.2.5 Chromaticity Coordinates 410
10.2.6 Color Names 414
10.2.7 Other Color and Chromaticity Spaces 415
10.2.8 Transforming Between Different Primaries 417
10.2.9 What Is White? 419
10.3   **Advanced Topics** 420
10.3.1 Color Temperature 420
10.3.2 Color Constancy 421
10.3.3 White Balancing 422
10.3.4 Color Change Due to Absorption 423
10.3.5 Dichromatic Reflection 424
10.3.6 Gamma 425
10.4   **Application: Color Images** 427
10.4.1 Comparing Color Spaces 427
10.4.2 Shadow Removal 428
10.5   **Wrapping Up** 430
10.5.1 Further Reading 431
10.5.2 Data Sources 432
10.5.3 Exercises 433

11   **Images and Image Processing** 435
11.1   **Obtaining an Image** 436
11.1.1 Images from Files 436
11.1.2 Images from an Attached Camera 441
11.1.3 Images from a Video File 441
11.1.4 Images from the Web 444
11.1.5 Images from Code 445
11.2   **Image Histograms** 447
11.3   **Monadic Operations** 448
11.4   **Dyadic Operations** 451
11.4.1 Application: Chroma Keying 452
11.4.2 Application: Motion detection 453
11.5   **Spatial Operations** 455
11.5.1 Linear Spatial Filtering 456
11.5.2 Template Matching 468
11.5.3 Nonlinear Operations 473

11.6	**Mathematical Morphology**	474
11.6.1	Noise Removal	477
11.6.2	Boundary Detection	479
11.6.3	Hit or Miss Transform	479
11.6.4	Distance Transform	481
11.7	**Shape Changing**	482
11.7.1	Cropping	482
11.7.2	Image Resizing	483
11.7.3	Image Pyramids	484
11.7.4	Image Warping	485
11.8	**Wrapping Up**	488
11.8.1	Further Reading	489
11.8.2	Sources of Image Data	489
11.8.3	Exercises	490
12	**Image Feature Extraction**	493
12.1	**Region Features**	496
12.1.1	Pixel Classification	496
12.1.2	Representation: Distinguishing Multiple Objects	506
12.1.3	Description	510
12.1.4	Object Detection Using Deep Learning	519
12.1.5	Summary	520
12.2	**Line Features**	521
12.2.1	Summary	525
12.3	**Point Features**	525
12.3.1	Classical Corner Detectors	526
12.3.2	Scale-Space Feature Detectors	531
12.4	**Applications**	534
12.4.1	Character Recognition	534
12.4.2	Image Retrieval	535
12.5	**Wrapping Up**	539
12.5.1	MATLAB Notes	539
12.5.2	Further Reading	539
12.5.3	Exercises	541
13	**Image Formation**	543
13.1	**Perspective Camera**	544
13.1.1	Perspective Projection	544
13.1.2	Modeling a Perspective Camera	548
13.1.3	Discrete Image Plane	549
13.1.4	Camera Matrix	551
13.1.5	Projecting Points	553
13.1.6	Lens Distortion	556
13.2	**Camera Calibration**	558
13.2.1	Calibrating with a 3D Target	558
13.2.2	Decomposing the Camera Calibration Matrix	560
13.2.3	Pose Estimation with a Calibrated Camera	562
13.2.4	Camera Calibration Tools	563
13.3	**Wide Field-of-View Cameras**	566
13.3.1	Fisheye Lens Camera	566
13.3.2	Catadioptric Camera	569
13.3.3	Spherical Camera	572
13.4	**Unified Imaging Model**	574
13.4.1	Mapping Wide-Angle Images to the Sphere	575
13.4.2	Mapping from the Sphere to a Perspective Image	577

13.5   **Novel Cameras**   578
13.5.1   Multi-Camera Arrays   578
13.5.2   Light-Field Cameras   579
13.6   **Applications**   581
13.6.1   Fiducial Markers   581
13.6.2   Planar Homography   583
13.7   **Advanced Topics**   584
13.7.1   Projecting 3D Lines and Quadrics   584
13.7.2   Nonperspective Cameras   586
13.8   **Wrapping Up**   587
13.8.1   Further Reading and Resources   588
13.8.2   Exercises   590

14   **Using Multiple Images**   593
14.1   **Point Feaature Correspondence**   595
14.2   **Geometry of Multiple Views**   599
14.2.1   The Fundamental Matrix   601
14.2.2   The Essential Matrix   603
14.2.3   Estimating the Fundamental Matrix from Real Image Data   604
14.2.4   Planar Homography   609
14.3   **Sparse Stereo**   614
14.3.1   3D Triangulation   614
14.3.2   Bundle Adjustment (advanced)   618
14.4   **Dense Stereo**   623
14.4.1   Stereo Failure Modes   627
14.4.2   Refinement and Reconstruction   629
14.4.3   Stereo Image Rectification   631
14.5   **Anaglyphs**   634
14.6   **Other Depth Sensing Technologies**   635
14.6.1   Depth from Structured Light   635
14.6.2   Depth from Time-Of-Flight   636
14.7   **Point Clouds**   637
14.7.1   Fitting Geometric Objects into a Point Cloud   639
14.7.2   Matching Two Sets of Points   641
14.8   **Applications**   644
14.8.1   Perspective Correction   644
14.8.2   Image Mosaicing   647
14.8.3   Visual Odometry   650
14.9   **Wrapping Up**   654
14.9.1   Further Reading   654
14.9.2   Resources   658
14.9.3   Exercises   659

V   Robotics, Vision & Control

15   **Vision-Based Control**   665
15.1   **Position-Based Visual Servoing**   668
15.2   **Image-Based Visual Servoing**   670
15.2.1   Camera and Image Motion   671
15.2.2   Controlling Feature Motion   678
15.2.3   Estimating Feature Depth   681
15.2.4   Performance Issues   683
15.3   **Wrapping Up**   686
15.3.1   Further Reading   687
15.3.2   Exercises   690

16   **Real-World Applications** 693
16.1   **Lane and Vehicle Detection with a Monocular Camera** 694
16.2   **Highway Lane Change Planner and Controller** 697
16.3   **UAV Package Delivery** 699
16.4   **Pick-and-Place Workflow in Gazebo Using Point-Cloud Processing and RRT**
       Path Planning 701

       Supplementary Information 705
       A Software Installation 706
       B Linear Algebra 708
       C Geometry 716
       D Lie Groups and Algebras 732
       E Linearization, Jacobians, and Hessians 737
       F Solving Systems of Equations 742
       G Gaussian Random Variables 751
       H Kalman Filter 754
       I Graphs 760
       J Peak Finding 763
       References 767
       Index of People 783
       Index of Functions, Classes, and Methods 785
       Index of Apps 793
       Index of Models 795
       General Index 797

Nomenclature

The notation used in robotics and computer vision varies considerably across books and research papers. The symbols used in this book, and their units where appropriate, are listed below. Some symbols have multiple meanings and their context must be used to disambiguate them.

Notation	Description
$\boldsymbol{v}$	a vector
$\hat{\boldsymbol{v}}$	a unit vector parallel to $\boldsymbol{v}$
$\tilde{\boldsymbol{v}}$	homogeneous representation of the vector $\boldsymbol{v}$
v_i	i^{th} element of vector $\boldsymbol{v}$
$\tilde{v}_i$	i^{th} element of the homogeneous vector $\tilde{\boldsymbol{v}}$
v_x, v_y, v_z	elements of the coordinate vector $\boldsymbol{v}$
$\mathbf{A}$	a matrix
$a_{i,j}$	the element of $\mathbf{A}$ at row i and column j
$\mathbf{0}_{m \times n}$	an $m \times n$ matrix of zeros
$\mathbf{1}_{m \times n}$	an $m \times n$ identity matrix
$\check{q}$	a quaternion, $\check{q} \in \mathbb{H}$
$\mathring{q}$	a unit quaternion, $\mathring{q} \in S^3$
$f(x)$	a function of x
$F_x(x)$	the first derivative of $f(x)$, $\partial f / \partial x$
$F_{xy}(x, y)$	the second derivative of $f(x, y)$, $\partial^2 f / \partial x \partial y$
$\{F\}$	coordinate frame F

Vectors are generally lower-case Roman or Greek letters, while matrices are generally upper-case Roman or Greek letters. Various decorators are applied to symbols:

Decorators	Description
x^*	desired value of x
x^+	predicted value of x
$x^\#$	measured, or observed, value of x
$\hat{x}$	estimated value of x, also a unit vector as above
$\bar{x}$	mean of x or relative value
$x\langle k \rangle$	k^{th} element of a time series

where x could be a scalar, vector, or matrix.

Symbol	Description	Unit
B	viscous friction coefficient	$\mathrm{N\,m\,s\,rad^{-1}}$
B	magnetic flux density	T
$\mathbf{C}$	camera matrix, $\mathbf{C} \in \mathbb{R}^{3\times4}$	
$\mathbf{C}(\boldsymbol{q}, \dot{\boldsymbol{q}})$	manipulator centripetal and Coriolis term	$\mathrm{kg\,m^2\,s^{-1}}$
C	configuration space of a robot with N joints: $C \subset \mathbb{R}^N$	
e	mathematical constant, base of natural logarithms, $\mathrm{e} = 2.71828\ldots$	
E	illuminance (in lux)	lx
f	focal length	m
f	force	N
$\boldsymbol{f}(\dot{\boldsymbol{q}})$	manipulator friction torque	Nm
g	gravitation acceleration, see ◨ Fig. 3.11	$\mathrm{ms^{-2}}$
$\boldsymbol{g}(\boldsymbol{q})$	manipulator gravity term	Nm
$\mathbb{H}$	the set of all quaternions (H for Hamilton)	
J	inertia	$\mathrm{kg\,m^2}$
$\mathbf{J}$	inertia tensor, $\mathbf{J} \in \mathbb{R}^{3\times3}$	$\mathrm{kg\,m^2}$
$\mathbf{J}$	Jacobian matrix	
$^A\mathbf{J}_B$	Jacobian transforming velocities in frame {B} to frame {A}	
k, K	constant	
$\mathbf{K}$	camera intrinsic matrix, $\mathbf{K} \in \mathbb{R}^{3\times3}$	
K_i	amplifier gain (transconductance)	$\mathrm{A\,V^{-1}}$
K_m	motor torque constant	$\mathrm{Nm\,A^{-1}}$
L	luminance (in nit)	nt
m	mass	kg
$\mathbf{M}(\boldsymbol{q})$	manipulator inertia tensor	$\mathrm{kg\,m^2}$
$\mathbb{N}$	the set of natural numbers $\{1, 2, 3, \ldots\}$	
$\mathbb{N}_0$	the set of natural numbers including zero $\{0, 1, 2, \ldots\}$	
$N(\mu, \sigma^2)$	a normal (Gaussian) distribution with mean μ and standard deviation σ	
p	an image plane point	
P	a world point	
$\mathbf{p}$	coordinate vector of an image plane point, $\mathbf{p} \in \mathbb{R}^2$	
$\mathbf{P}$	coordinate vector of a world point, $\mathbf{P} \in \mathbb{R}^3$	
$\mathbb{P}^2$	projective space of all 2D points, elements are a 3-tuple	
$\mathbb{P}^3$	projective space of all 3D points, elements are a 4-tuple	
$\boldsymbol{q}$	generalized coordinates, robot configuration $\boldsymbol{q} \in C$	m, rad
$\boldsymbol{Q}$	generalized force $\boldsymbol{Q} \in \mathbb{R}^N$	N, Nm
$\mathbf{R}$	an orthonormal rotation matrix, $\mathbf{R} \in \mathbf{SO}(2)$ or $\mathbf{SO}(3)$	
$\mathbb{R}$	set of real numbers	
$\mathbb{R}_{>0}$	set of positive real numbers	

Symbol	Description	Unit
$\mathbb{R}_{\geq 0}$	set of non-negative real numbers	
$\mathbb{R}^2$	set of all 2D points, elements are a 2-tuple	
$\mathbb{R}^3$	set of all 3D points, elements are a 3-tuple	
s	distance along a path or trajectory, $s \in [0, 1]$	
s	Laplace transform operator	
$\mathbf{S}^1$	set of points on the unit circle $\sim$ set of angles $[0, 2\pi)$	
$\mathbf{S}^n$	unit sphere embedded in $\mathbb{R}^{n+1}$	
$\mathbf{se}(n)$	Lie algebra for $\mathbf{SE}(n)$, a vector space of $\mathbb{R}^{(n+1) \times (n+1)}$ augmented skew-symmetric matrices	
$\mathbf{so}(n)$	Lie algebra for $\mathbf{SO}(n)$, a vector space of $\mathbb{R}^{n \times n}$ skew-symmetric matrices	
$\mathbf{SE}(n)$	special Euclidean group of matrices, $\mathbf{T} \in \mathbf{SE}(n)$, $\mathbf{T} \subset \mathbb{R}^{(n+1) \times (n+1)}$, represents pose in n dimensions, aka homogeneous transformation matrix, aka rigid-body transformation	
$\mathbf{SO}(n)$	special orthogonal group of matrices, $\mathbf{R} \in \mathbf{SO}(n)$, $\mathbf{R} \subset \mathbb{R}^{n \times n}$, $\mathbf{R}^\top = \mathbf{R}^{-1}$, $\det(\mathbf{R}) = 1$, represents orientation in n dimensions, aka rotation matrix	
S	twist in 2 or 3 dimensions, $S \in \mathbb{R}^3$ or $\mathbb{R}^6$	
t	time	s
$\mathcal{T}$	task space of robot, $\mathcal{T} \subset \mathbf{SE}(3)$	K
T	sample interval	s
T	temperature	K
T	optical transmission	m^{-1}
$\mathbf{T}$	a homogeneous transformation matrix $\mathbf{T} \in \mathbf{SE}(2)$ or $\mathbf{SE}(3)$	
$^A\mathbf{T}_B$	a homogeneous transformation matrix representing the pose of frame $\{B\}$ with respect to frame $\{A\}$. If A is not given then assumed relative to world coordinate frame $\{0\}$. Note that $^A\mathbf{T}_B = (^B\mathbf{T}_A)^{-1}$	
u, v	coordinates of point on a camera's image plane	pixels
u_0, v_0	coordinates of a camera's principal point	pixels
$\bar{u}, \bar{v}$	normalized image plane coordinates, relative to the principal point	pixels
v	velocity	m s^{-1}
$\mathbf{v}$	velocity vector	m s^{-1}
$\mathbf{w}$	wrench, a vector of forces and moments $(m_x, m_y, m_z, f_x, f_y, f_z)$ $\in \mathbb{R}^6$	N, Nm
$\hat{\mathbf{x}}, \hat{\mathbf{y}}, \hat{\mathbf{z}}$	unit vectors aligned with the x-, y- and z-axes, 3D basis vectors	
X, Y, Z	Cartesian coordinates of a point	
$\bar{x}, \bar{y}$	retinal image-plane coordinates	m
$\mathbb{Z}$	set of all integers $\{\ldots, -2, -1, 0, 1, 2, \ldots\}$	
$\varnothing$	null motion	
α	roll angle	rad
β	pitch angle	rad

Symbol	Description	Unit
δ	an increment	
ϵ	an error or residual, ideally zero	
$\boldsymbol{\Gamma}$	3-angle representation of rotation, $\boldsymbol{\Gamma} \in \mathbb{R}^3$	rad
$\boldsymbol{\Gamma}$	body torque, $\boldsymbol{\Gamma} \in \mathbb{R}^3$	Nm
θ	an angle, first Euler angle, yaw angle, heading angle	rad
λ	wavelength	m
λ	an eigenvalue	
$\boldsymbol{\nu}$	innovation	
$\boldsymbol{\nu}$	spatial velocity, $(\omega_x, \omega_y, \omega_z, v_x, v_y, v_z) \in \mathbb{R}^6$	m s^{-1} rad s^{-1}
$\boldsymbol{\xi}$	abstract representation of pose (pronounced ksigh)	
$^A\boldsymbol{\xi}_B$	abstract representation of relative pose, frame {B} with respect to frame {A} or rigid-body motion from frame {A} to {B}	
$\boldsymbol{\xi}^{t_i}(d)$	abstract pose that is pure translation of d along axis $i \in \{x, y, z\}$	
$\boldsymbol{\xi}^{r_i}(\theta)$	abstract pose that is pure rotation a pure rotation of θ about axis $i \in \{x, y, z\}$	
π	mathematical constant $\pi = 3.14159\ldots$	
$\boldsymbol{\pi}$	a plane	
$\rho_{\mathrm{w}}, \rho_{\mathrm{h}}$	pixel width and height	m
σ	standard deviation	
σ	robot joint type, $\sigma = \mathrm{R}$ for revolute and $\sigma = \mathrm{P}$ for prismatic	
$\boldsymbol{\Sigma}$	element of the Lie algebra of $\mathbf{SE}(3)$, $\boldsymbol{\Sigma} = [\boldsymbol{v}] \in \mathbf{se}(3), \boldsymbol{v} \in \mathbb{R}^6$	
τ	torque	N m
τ_{C}	Coulomb friction torque	N m
ϕ	angle, second Euler angle, pitch angle	rad
ψ	third Euler angle, roll angle, steered-wheel angle for mobile robot	rad
ω	rotational rate	rad s^{-1}
$\boldsymbol{\omega}$	angular velocity vector	rad s^{-1}
ϖ	rotational speed of a motor shaft, wheel or propeller	rad s^{-1}
$\boldsymbol{\Omega}$	element of the Lie algebra of $\mathbf{SO}(3)$, $\boldsymbol{\Omega} = [\boldsymbol{v}]_\times \in \mathbf{so}(3), \boldsymbol{v} \in \mathbb{R}^3$	
$\boldsymbol{v}_1 \cdot \boldsymbol{v}_2$	dot, or inner, product, also $\boldsymbol{v}_1^\top \boldsymbol{v}_2 \colon \mathbb{R}^n \times \mathbb{R}^n \mapsto \mathbb{R}$	
$\|\cdot\|$	norm, or length, of vector, also $\sqrt{\boldsymbol{v} \cdot \boldsymbol{v}} \colon \mathbb{R}^n \mapsto \mathbb{R}_{\geq 0}$	
$\boldsymbol{v}_1 \times \boldsymbol{v}_2$	cross, or vector, product: $\mathbb{R}^3 \times \mathbb{R}^3 \mapsto \mathbb{R}^3$	
$\tilde{\cdot}$	$\mathbb{R}^n \mapsto \mathbb{P}^n$	
$\epsilon(\cdot)$	$\mathbb{P}^n \mapsto \mathbb{R}^n$	
$\mathbf{A}^{-1}$	inverse of $\mathbf{A}$: $\mathbb{R}^{n \times n} \mapsto \mathbb{R}^{n \times n}$	
$\mathbf{A}^+$	pseudo-inverse of $\mathbf{A}$: $\mathbb{R}^{n \times m} \mapsto \mathbb{R}^{m \times n}$	
$\mathbf{A}^*$	adjugate of $\mathbf{A}$, also $\det(\mathbf{A})\mathbf{A}^{-1}$: $\mathbb{R}^{n \times n} \mapsto \mathbb{R}^{n \times n}$	
$\mathbf{A}^\top$	transpose of $\mathbf{A}$: $\mathbb{R}^{n \times m} \mapsto \mathbb{R}^{m \times n}$	
$\mathbf{A}^{-\top}$	transpose of inverse of $\mathbf{A}$, $(\mathbf{A}^\top)^{-1} \equiv (\mathbf{A}^{-1})^\top$: $\mathbb{R}^{n \times n} \mapsto \mathbb{R}^{n \times n}$	

Operator	Description
$\oplus$	composition of abstract pose: ${}^x\xi_y \oplus {}^y\xi_z \mapsto {}^x\xi_z$
$\ominus$	composition of abstract pose with inverse: ${}^x\xi_y \ominus {}^z\xi_y \equiv {}^x\xi_y \oplus {}^y\xi_z \mapsto {}^x\xi_z$
$\ominus$	inverse of abstract pose: $\ominus\, {}^x\xi_y \equiv {}^y\xi_x$
$\cdot$	transform a point (coordinate vector) by abstract relative pose: ${}^x\xi_y \cdot {}^y p \mapsto {}^x p$
$\Delta(\cdot)$	maps incremental relative pose to differential motion: $\mathbf{SE}(3) \mapsto \mathbb{R}^6$
$\Delta^{-1}(\cdot)$	maps differential motion to incremental relative pose: $\mathbb{R}^6 \mapsto \mathbf{SE}(3)$
$[\cdot]_t$	translational component of pose: $\mathbf{SE}(n) \mapsto \mathbb{R}^n$
$[\cdot]_R$	rotational component of pose: $\mathbf{SE}(n) \mapsto \mathbf{SO}(n)$
$[\cdot]_{xy\theta}$	2D pose to configuration: $\mathbf{SE}(2) \mapsto \mathbb{R}^2 \times \mathbf{S}^1$
$[\cdot]_\times$	skew-symmetric matrix: $\mathbb{R} \mapsto \mathbf{so}(2),\ \mathbb{R}^3 \mapsto \mathbf{so}(3)$
$[\cdot]$	augmented skew-symmetric matrix: $\mathbb{R}^3 \mapsto \mathbf{se}(2),\ \mathbb{R}^6 \mapsto \mathbf{se}(3)$
$\vee_\times(\cdot)$	*unpack* skew-symmetric matrix: $\mathbf{so}(2) \mapsto \mathbb{R},\ \mathbf{so}(3) \mapsto \mathbb{R}^3$
$\vee(\cdot)$	*unpack* augmented skew-symmetric matrix: $\mathbf{se}(2) \mapsto \mathbb{R}^3,\ \mathbf{se}(3) \mapsto \mathbb{R}^6$
$\mathrm{Ad}(\cdot)$	adjoint representation: $\mathbf{SE}(3) \mapsto \mathbb{R}^{6\times 6}$
$\mathrm{ad}(\cdot)$	logarithm of adjoint representation: $\mathbf{SE}(3) \mapsto \mathbb{R}^{6\times 6}$
$\circ$	quaternion (Hamiltonian) multiplication: $\mathbb{H} \times \mathbb{H} \mapsto \mathbb{H}$
$\check{v}$	pure quaternion: $\mathbb{R}^3 \mapsto \mathbb{H}$
$\sim$	equivalence of representation
$\simeq$	homogeneous coordinate equivalence
$\ominus$	smallest angular difference between two angles: $\mathbb{S}^1 \times \mathbb{S}^1 \mapsto [-\pi, \pi)$
$\mathcal{K}(\cdot)$	forward kinematics: $C \mapsto \mathcal{T}$
$\mathcal{K}^{-1}(\cdot)$	inverse kinematics: $\mathcal{T} \mapsto C$
$\mathcal{D}^{-1}(\cdot)$	manipulator inverse dynamics function: $C, \mathbb{R}^N, \mathbb{R}^N \mapsto \mathbb{R}^N$
$\mathcal{P}(\cdot)$	camera projection function: $\mathbb{R}^3 \mapsto \mathbb{R}^2$
$*$	image convolution
$\otimes$	image correlation
$\equiv$	colormetric equivalence
$\oplus$	morphological dilation
$\ominus$	morphological erosion
$\circ$	morphological opening
$\bullet$	morphological closing
$[a, b]$	interval a to b inclusive
(a, b)	interval a to b exclusive, not including a or b
$[a, b)$	interval a to b, not including b
$(a, b]$	interval a to b, not including a

Toolbox Conventions

- A Cartesian point, a coordinate vector, is mathematically described as a column vector in this book, but the MATLAB toolboxes generally expect it as row vector.
- A set of points is expressed as a matrix with rows representing the coordinate vectors of individual points.
- Time series data is generally expressed as a matrix with rows representing time steps.
- A rectangular region is frequently represented by a 4-vector: the coordinate of its top-left corner, and its width and height $[x \; y \; w \; h]$.
- A robot configuration, a set of joint angles, is expressed as a row vector.
- A MATLAB matrix has subscripts (i, j) which represent row and column indices respectively. Image coordinates are written (u, v), so an image represented by a matrix `I` is indexed as `I(v,u)`.
- The indices of MATLAB matrices start from one.
- Matrices with three or more dimensions are frequently used:
  - A series of poses or orientations has 3 dimensions where the third index is the sequence index, and `A(:,:,k)` is an $\mathbf{SE}(n)$ or $\mathbf{SO}(n)$ matrix.
  - A color image has 3 dimensions: row, column, color plane.
  - A grayscale image sequence has 3 dimensions: row, column, index.
  - A color image sequence has 4 dimensions: row, column, color plane, index.

Common Abbreviations

1D – 1-dimensional

2D – 2-dimensional

3D – 3-dimensional

CoM – Center of mass

DoF – Degrees of freedom

n-**tuple** – A group of n numbers, it can represent a point or a vector

Introduction

Contents

1.1   A Brief History of Robots – 2

1.2   Types of Robots – 4

1.3   Definition of a Robot – 8

1.4   Robotic Vision – 9

1.5   Ethical Considerations – 11

1.6   About the Book – 12

© The Author(s), under exclusive license to Springer Nature Switzerland AG 2023
P. Corke, W. Jachimczyk, R. Pillat, *Robotics, Vision and Control*, Springer Tracts in Advanced Robotics 147,
https://doi.org/10.1007/978-3-031-07262-8_1

1

1.1 A Brief History of Robots

The word *robot* means different things to different people. Science fiction books and movies have strongly influenced what many people expect a robot to be or what it can do – sadly the practice of robotics is far behind this popular conception. The word *robot* is also emotive and some people are genuinely fearful of a future with robots, while others are concerned about what robots mean for jobs, privacy, or warfare. One thing is certain though – robotics will be an important technology in this century. Products such as vacuum cleaning robots have already been with us for over two decades, and self-driving cars are now on the roads with us. These are the vanguard of a wave of smart machines that will appear in our homes and workplaces in the near to medium future.

In the eighteenth century, Europeans were fascinated by automata such as Vaucanson's duck shown in ◘ Fig. 1.1a. These machines, complex by the standards of the day, demonstrated what then seemed *life-like* behavior. The duck used a cam mechanism to sequence its movements and Vaucanson went on to explore the mechanization of silk weaving. Jacquard extended these ideas and developed a loom, shown in ◘ Fig. 1.1b, that was essentially a programmable weaving machine. The pattern to be woven was encoded as a series of holes on punched cards. ◄ This machine has many hallmarks of a modern robot: it performed a physical task and was reprogrammable.

This in turn influenced Sir Charles Babbage in his quest to mechanize computation, which in turn influenced Countess Ada Lovelace to formalize computation and create the first algorithm.

The term *robot* first appeared in a 1920 Czech science fiction play "Rossum's Universal Robots" by Karel Čapek (pronounced chapek). The word *roboti* was suggested by his brother Josef, and in the Czech language has the same linguistic roots

◘ **Fig. 1.1** Early programmable machines. **a** Vaucanson's duck (1739) was an automaton that could flap its wings, eat grain, and defecate. It was driven by a clockwork mechanism and executed a single program; **b** the Jacquard loom (1801) was a reprogrammable machine and the program was held on punched cards (Image by George P. Landow from ► https://www.victorianweb.org)

◘ **Fig. 1.2** Universal automation. **a** A plan view of the machine from Devol's patent; **b** the first Unimation robot – the Unimate – working at a General Motors factory (Image courtesy of George C. Devol)

Fig. 1.3 Manufacturing robots, technological descendants of the Unimate shown in ◘ Fig. 1.2. **a** A modern six-axis robot designed for high accuracy and throughput (Image courtesy ABB robotics); **b** Universal Robots collaborative robot can work safely with a human co-worker (© 2022 Universal Robots A/S. All Rights Reserved. This image has been used with permission from Universal Robots A/S)

Excurse 1.1: George C. Devol

Devol (1912–2011) was a prolific American inventor, born in Louisville, Kentucky. In 1932, he founded United Cinephone Corp. which manufactured phonograph arms and amplifiers, registration controls for printing presses and packaging machines. In 1954, he applied for US patent 2,988,237 for Programmed Article Transfer which introduced the concept of Universal Automation or "Unimation". Specifically, it described a track-mounted polar-coordinate arm mechanism with a gripper and a programmable controller – the precursor of all modern robots.

In 2011 he was inducted into the National Inventors Hall of Fame. (Image courtesy of George C. Devol)

as servitude and slavery. The robots in the play were artificial people or androids and, as in so many robot stories that follow this one, the robots rebel and it ends badly for humanity. Isaac Asimov's robot series, comprising many books and short stories written between 1950 and 1985, explored issues of human and robot interaction, and morality. The robots in these stories are equipped with "positronic

Excurse 1.2: Unimation Inc.

Devol sought financing to develop his unimation technology and at a cocktail party in 1954 he met Joseph Engelberger who was then an engineer with Manning, Maxwell and Moore. In 1956 they jointly established Unimation (1956–1982), the first robotics company, in Danbury, Connecticut. The company was acquired by Consolidated Diesel Corp. (Condec) and became Unimate Inc. a division of Condec. Their first robot went to work in 1961 at a General Motors die-casting plant in New Jersey. In 1968 they licensed technology to Kawasaki Heavy Industries which produced the first Japanese industrial robot. Engelberger served as chief executive until it was acquired by Westinghouse in 1982. People and technologies from this company have gone on to be very influential across the whole field of robotics.

Excurse 1.3: Joseph F. Engelberger

Engelberger (1925–2015) was an American engineer and entrepreneur who is often referred to as the "Father of Robotics". He received his B.S. and M.S. degrees in physics from Columbia University, in 1946 and 1949, respectively. Engelberger was a tireless promoter of robotics. In 1966, he appeared on *The Tonight Show Starring Johnny Carson* with a Unimate robot which poured a beer, putted a golf ball, and directed the band. He promoted robotics heavily in Japan, which led to strong investment and development of robotic technology in that country.

Engelberger served as chief executive of Unimation Inc. until 1982, and in 1984 founded Transitions Research Corporation which became HelpMate Robotics Inc., an early entrant in the hospital service robot sector. He was elected to the National Academy of Engineering, received the Beckman Award and the Japan Prize, and wrote two books: *Robotics in Practice* (1980) and *Robotics in Service* (1989). Each year the Association for Advancing Automation (▶ https://www.automate.org) presents an award in his honor to "persons who have contributed outstandingly to the furtherance of the science and practice of robotics."

brains" in which the "Three Laws of Robotics" are encoded. These stories have influenced subsequent books and movies, which in turn have shaped the public perception of what robots are. The mid-twentieth century also saw the advent of the field of *cybernetics* – an uncommon term today but then an exciting science at the frontiers of understanding life and creating intelligent machines.

The first patent for what we would now consider a robot manipulator was filed in 1954 by George C. Devol and issued in 1961. The device comprised a mechanical arm with a gripper that was mounted on a track and the sequence of motions was encoded as magnetic patterns stored on a rotating drum. The first robotics company, Unimation, was founded by Devol and Joseph Engelberger in 1956 and their first industrial robot, shown in ◙ Fig. 1.2b, was installed in 1961. The original vision of Devol and Engelberger for robotic automation has subsequently become a reality. Many millions of robot manipulators, such as shown in ◙ Fig. 1.3, have been built and put to work at tasks such as welding, painting, machine loading and unloading, assembly, sorting, packaging, and palletizing. The use of robots has led to increased productivity and improved product quality. Today, many products that we buy have been assembled or handled by a robot.

1.2 Types of Robots

The *first generation* of robots were fixed in place and could not move about the factory. By contrast, *mobile robots* shown in ◙ Figs. 1.4 and 1.5 can move through the world using various forms of mobility. They can locomote over the ground using wheels or legs, fly through the air using fixed wings or multiple rotors, move through the water, or sail over it.

An alternative taxonomy is based on the function that the robot performs. *Manufacturing* robots operate in factories and are the technological descendants of the first generation robots created by Unimation Inc. *Service robots* provide services to people such as cleaning, personal care, medical rehabilitation or fetching and

◘ **Fig. 1.4** Non-land-based mobile robots. **a** Small autonomous underwater vehicle (Todd Walsh © 2013 MBARI); **b** Global Hawk uncrewed aerial vehicle (UAV) (Image courtesy of NASA)

◘ **Fig. 1.5** Mobile robots. **a** Perseverance rover on Mars, self portrait. The mast contains many cameras including stereo camera pairs from which the robot can compute the 3-dimensional structure of its environment (Image courtesy of NASA/JPL-Caltech/MSSS); **b** Savioke Relay delivery robot (Image courtesy Savioke); **c** self-driving car (Image courtesy Dept. Information Engineering, Oxford Univ.); **d** Boston Dynamics Spot® legged robot (Image courtesy Dorian Tsai)

carrying as shown in ◘ Fig. 1.5b. *Field robots*, such as those shown in ◘ Fig. 1.4, work outdoors on tasks such as environmental monitoring, agriculture, mining, construction, and forestry. *Humanoid robots* such as shown in ◘ Fig. 1.6 have the physical form of a human being – they are both mobile robots and service robots.

1

Excurse 1.4: Rossum's Universal Robots (RUR)

This 1920 play by Karel Čapek introduced the word "robot" to the world. In the introductory scene Helena Glory is visiting Harry Domin, the director general of Rossum's Universal Robots, and his robotic secretary Sulla.

Domin – Sulla, let Miss Glory have a look at you.

Helena – (stands and offers her hand) Pleased to meet you. It must be very hard for you out here, cut off from the rest of the world [the factory is on an island]

Sulla – I do not know the rest of the world Miss Glory. Please sit down.

Helena – (sits) Where are you from?

Sulla – From here, the factory

Helena – Oh, you were born here.

Sulla – Yes I was made here.

Helena – (startled) What?

Domin – (laughing) Sulla isn't a person, Miss Glory, she's a *robot*.

Helena – Oh, please forgive me . . .

The full play can be found at ► https://www.gutenberg.org/ebooks/59112. One hundred years later, it is prescient and still thought provoking. (Image courtesy of Library of Congress, item 96524672)

◘ **Fig. 1.6** Humanoid robots. **a** Honda's Asimo humanoid robot (Image courtesy Honda Motor Co. Japan); **b** Hubo robot that won the DARPA Robotics Challenge in 2015 (Image courtesy KAIST, Korea)

A *manufacturing robot* is typically an arm-type manipulator on a fixed base, such as shown in ◘ Fig. 1.3a, that performs repetitive tasks within a local work cell. Parts are presented to the robot in an orderly fashion which maximizes the advantage of the robot's high speed and precision. High-speed robots are hazardous and safety is achieved by excluding people from robotic work places – typically the robot is placed inside a cage. In contrast, collaborative robots such as shown in ◘ Fig. 1.3b, are human safe – they operate at low speed and stop moving when they encounter an obstruction.

Excurse 1.5: Telerobots

The Manhattan Project in World War 2 developed the first nuclear weapons and this required handling of radioactive material. Remotely controlled arms were developed by Ray Goertz at Argonne National Laboratory to exploit the manual dexterity of human operators, while keeping them away from the hazards of the material they were handling. The operators viewed the work space through thick lead-glass windows or via a television link and manipulated the leader arm (on the left). The follower arm (on the right) followed the motion, and forces felt by the follower arm were reflected back to the leader arm, allowing the operator to feel weight and interference force. Telerobotics is still important today for many tasks where people cannot work but which are too complex for a machine to perform by itself, for instance the underwater robots that surveyed the wreck of the Titanic. (Image courtesy of Argonne National Laboratory)

Field and service robots face specific and significant challenges. The first challenge is that the robot must operate and move in a complex, cluttered and changing environment. For example, a delivery robot in a hospital must operate despite crowds of people and a time-varying configuration of parked carts and trolleys. A Mars rover, as shown in ◘ Fig. 1.5a, must navigate rocks and small craters despite not having an accurate local map in advance of its travel. Robotic, or self-driving, cars, such as shown in ◘ Fig. 1.5c, must follow roads, avoid obstacles, and obey traffic signals and the rules of the road. The second challenge for these types of robots is that they must operate safely in the presence of people. The hospital delivery robot operates amongst people, the robotic car contains people, and a robotic surgical device operates *inside* people.

Telerobots are robot-like machines that are remotely controlled by a human operator. Perhaps the earliest example was a radio-controlled boat demonstrated by Nikola Tesla in 1898 and which he called a teleautomaton. Such machines were an important precursor to robots, and are still important today for tasks conducted in environments where people cannot work, but which are too complex for a machine to perform by itself. For example, the "underwater robots" that surveyed the wreck of the Titanic were technically remotely operated vehicles (ROVs). A modern surgical robot as shown in ◘ Fig. 1.7 is also teleoperated – the motion of the small tools inside the patient are remotely controlled by the surgeon. The patient benefits because the procedure is carried out using much smaller incisions than the old-fashioned approach where the surgeon works inside the body with their hands.

The various Mars rovers autonomously navigate the surface of Mars but human operators provide the high-level goals. That is, the operators tell the robot where to go and the robot itself determines the details of the route. Local decision making on Mars is essential given that the communications delay is several minutes. Some telerobots are hybrids, and the control task is shared or traded with a human operator. In traded control, the control function is passed back and forth between the human operator and the computer. For example, an aircraft pilot can pass control to an autopilot and take control back. In shared control, the control function is performed by the human operator and the computer working together. For example, an autonomous passenger car might have the computer keeping the car safely in the lane while the human driver controls speed.

1

Excurse 1.6: Cybernetics, Artificial Intelligence and Robotics

Cybernetics flourished as a research field from the 1930s until the 1960s and was fueled by a heady mix of new ideas and results from neurology, control theory, and information theory. Research in neurology had shown that the brain was an electrical network of neurons. Harold Black, Henrik Bode and Harry Nyquist at Bell Labs were researching negative feedback and the stability of electrical networks, Claude Shannon's information theory described digital signals, and Alan Turing was exploring the fundamentals of computation. Walter Pitts and Warren McCulloch proposed an artificial neuron in 1943 and showed how it might perform simple logical functions. In 1951 Marvin Minsky built SNARC (from a B24 autopilot and comprising 3000 vacuum tubes) which was perhaps the first neural-network-based learning machine as his graduate project. William Grey Walter's robotic tortoises showed lifelike behavior (see ▶ Sect. 5.1). Maybe an electronic brain could be built!

An important early book was Norbert Wiener's *Cybernetics or Control and Communication in the Animal and the Machine* (Wiener 1965). A characteristic of a cybernetic system is the use of feedback which is common in engineering and biological systems. The ideas were later applied to evolutionary biology, psychology, and economics.

In 1956, a watershed conference was hosted by John McCarthy at Dartmouth College and attended by Minsky, Shannon, Herbert Simon, Allen Newell and others. This meeting defined the term artificial intelligence (AI) as we know it today, with an emphasis on digital computers and symbolic manipulation and led to new research in robotics, vision, natural language, semantics and reasoning. McCarthy and Minsky formed the AI group at MIT, and McCarthy left in 1962 to form the Stanford AI Laboratory. Minsky focused on artificially simple "blocks world". Simon, and his student Newell, were influential in AI research at Carnegie-Mellon University from which the Robotics Institute was spawned in 1979. These AI groups were to be very influential in the development of robotics and computer vision in the USA. Societies and publications focusing on cybernetics are still active today.

◨ Fig. 1.7 The working end of a surgical robot, multiple tools work inside the patient but pass through a single small incision (Image © 2015 Intuitive Surgical, Inc)

1.3   Definition of a Robot

» *I can't define a robot, but I know one when I see one*
    – Joseph Engelberger

So what is a robot? There are many definitions and not all of them are particularly helpful. A definition that will serve us well in this book is

» *a goal-oriented machine that can sense, plan, and act.*

A robot *senses* its environment and uses that information, together with a goal, to *plan* some *action*. The action might be to move the tool of a robot manipulator arm to grasp an object, or it might be to drive a mobile robot to some place.

Sensing is critical to robots. *Proprioceptive* sensors measure the state of the robot itself: the angle of the joints on a robot arm, the number of wheel revolutions on a mobile robot, or the current drawn by an electric motor. *Exteroceptive* sensors measure the state of the world with respect to the robot. The sensor might be a simple bump sensor on a robot vacuum cleaner to detect collision. It might be a GPS receiver that measures distances to an orbiting satellite constellation, or a compass that measures the direction of the Earth's magnetic field vector relative to the robot. It might also be an active sensor that emits acoustic, optical or radio pulses in order to measure the distance to points in the world based on the time taken for a reflection to return to the sensor.

1.4 Robotic Vision

Another way to sense the world is to capture and interpret patterns of ambient light reflected from the scene. This is what our eyes and brain do, giving us the sense of vision. Our own experience is that vision is a very effective sensor for most things that we do, including recognition, navigation, obstacle avoidance and manipulation. We are not alone in this, and almost all animal species use eyes — in fact evolution has *invented* the eye many times over. ◻ Fig. 1.8 shows some of the diversity of eyes found in nature.

It is interesting to note that even very simple animals, bees for example, with brains comprising just 10^6 neurons (compared to our 10^{11}) are able to perform complex and life-critical tasks such as finding food and returning it to the hive using only vision. For more complex animals such as ourselves, the benefits of vision outweigh the very high biological cost of owning an eye: the complex eye itself, muscles to move it, eyelids and tear ducts to protect it, and a large visual cortex (one third of our brain) to process the data it produces.

◻ **Fig. 1.8** The diversity of eyes. **a** Robber fly, Holocephala fusca; **b** jumping spider, Phidippus putnami (Images **a** and **b** courtesy Thomas Shahan, ▶ https://thomasshahan.com); **c** scallop (Image courtesy of Sönke Johnsen), each of the small blue spheres is an eye; **d** human eye

◻ Fig. 1.9   A cluster of cameras on an outdoor mobile robot: forward looking stereo pair, side looking wide-angle camera, overhead panoramic camera mirror (Image courtesy of CSIRO)

The sense of vision has long been of interest to robotics researchers – cameras can mimic the function of an eye and create images, and computer vision algorithms can extract meaning from the images. Combined, they could create powerful vision-based competencies for robots such as recognizing and manipulating objects and navigating within the world. For example, a soccer playing robot could determine the coordinate of a round red object in the scene, while a drone could estimate its motion in the world based on how the world appears to move relative to the drone. ◻ Fig. 1.9 shows a robot with a number of different types of cameras to enable outdoor navigation.

Technological developments have made it increasingly feasible for robots to use cameras as eyes and computers as brains, especially with recent advancements in deep learning. For much of the history of computer vision, dating back to the 1960s, electronic cameras were cumbersome and expensive, and computer power was inadequate. Today we have CMOS cameras developed for cell phones that cost just a few dollars each, and personal computers come standard with massive parallel computing power. New algorithms, cheap sensors, and plentiful computing power make vision a practical sensor today, and that is a strong focus of this book.

An important limitation of a camera, or an eye, is that the 3-dimensional structure of the scene is lost in the resulting 2-dimensional image. Despite this, humans are particularly good at inferring the 3-dimensional nature of a scene using a number of visual cues. One approach, used by humans and robots, is stereo vision where information from two eyes is combined to estimate the 3-dimensional structure of the scene. The Mars rover shown in ◻ Fig. 1.5a has a stereo camera on its mast, and the robot in ◻ Fig. 1.9 has a stereo camera on its turret.

If the robot's environment is unchanging it could make do with an accurate map and do away with sensing the state of the world, apart from determining its own position. Imagine driving a car with the front window completely covered over and

> **Excurse 1.7: Evolution of the Eye**
>
> It is believed that all animal eyes share a common ancestor in a proto-eye that evolved 540 million years ago. However major evolutionary advances seem to have occurred in just the last few million years. The very earliest eyes, called eyespots, were simple patches of photoreceptor protein in single-celled animals. Multi-celled animals evolved multicellular eyespots which could sense the brightness of light but not its direction. Gradually, the eyespot evolved into a shallow cup shape which gave a limited ability to discriminate directional brightness according to which cells were illuminated. The pit deepened, the opening became smaller, and the number of photoreceptor cells increased, forming a pin-hole camera that was capable of distinguishing shapes. Next came an overgrowth of transparent cells to protect the eyespot which led to a filled eye chamber, and eventually the eye as we know it today. The lensed eye has evolved independently seven different times across species. Nature has evolved ten quite distinct eye designs including those shown in ◘ Fig. 1.8.

just looking at the GPS navigation system. If you had the road to yourself, you could probably drive successfully from A to B although it might be quite stressful. However, if there were other cars, pedestrians, traffic signals or roadworks this approach would not work. To deal with this realistic scenario you need to look outwards – to sense the world and plan your actions accordingly. For humans this is easy, done without conscious thought, but it is not yet easy to program a machine to do the same – this is the challenge of *robotic vision*.

1.5 Ethical Considerations

A number of ethical issues arise from the advent of robotics. Perhaps the greatest concern to the wider public is "robots taking jobs from people". Already today, artificial intelligence systems are encroaching on many information handling tasks including image analysis, decision making, credit assessment, and most recently, answering questions and writing text. People fear that robots will soon encroach on physical tasks, but currently the skill of robots for everyday tasks is poor, reliability and speed is low, and the cost is high. However, it is highly likely that, over time, these challenges will be overcome – we cannot shy away from the fact that many jobs now done by people could, in the future, be performed by robots.

The issue of robots and jobs, even today, is complex. Clearly there are jobs which people should not do, for example working in unhealthy or hazardous environments. There are many low-skilled jobs where human labor is increasingly hard to source, for instance in jobs like fruit picking. In many developed countries people no longer aspire to hard physical outdoor work in remote locations. What are the alternatives if people don't want to do the work? Consider again the fruit picking example, and in the absence of available human labor – do we stop eating fruit? do we dramatically raise the wages of fruit pickers (and increase the cost of fruit)? do we import fruit from other places (and increase the cost of fruit as well as its environmental footprint)? or do we use robots to pick locally grown fruit?

In areas like manufacturing, particularly car manufacturing, the adoption of robotic automation has been critical in raising productivity which has allowed that industry to be economically viable in high-wage countries like Europe, Japan and the USA. Without robots, these industries could not exist; they would not employ any people, not pay any taxes, and not consume products and services from other parts of the economy. Automated industry might employ fewer people, but it still makes an important contribution to society. Rather than taking jobs, we could argue

1

that robotics and automation has helped to keep manufacturing industries viable in high-labor cost countries. How do we balance the good of the society with the good of the individual?

There are other issues besides jobs. Consider self-driving cars. We are surprisingly accepting of human-driven cars even though they kill more than one million people every year, yet many are uncomfortable with the idea of self-driving cars even though they will dramatically reduce this loss of life. We worry about who to blame if a robotic car makes a mistake while the carnage caused by human drivers continues. Similar concerns are raised when talking about robotic healthcare and surgery – human surgeons are not perfect but robots are seemingly held to a much higher account. There is a lot of talk about using robots to look after elderly people, but does this detract from their quality of life by removing human contact, conversation and companionship? Should we use robots to look after our children, and even teach them? What do we think of armies of robots fighting and killing human beings?

Robotic fruit picking, cars, health care, elder care, and child care might bring economic benefits to our society but is it the right thing to do? Is it a direction that we want our society to go? Once again, how do we balance the good of the society with the good of the individual? These are deep ethical questions that cannot and should not be decided by roboticists alone. But neither should roboticists ignore them. We must not sleepwalk into a technological future just "because we can". It is our responsibility to help shape the future in a deliberate way and ensure that it is good for all. This is an important discussion for all of society, and roboticists have a responsibility to be active participants in this debate.

1.6 About the Book

This is a book about robotics and computer vision – separately, and together as robotic vision. These are big topics and the combined coverage is necessarily broad. The goals of the book are:
- to provide a broad and solid base of understanding through theory and the use of examples to make abstract concepts tangible;
- to tackle more complex problems than other more specialized textbooks by virtue of the powerful numerical tools and software that underpins it;
- to provide instant gratification by solving complex problems with relatively little code;
- to complement the many excellent texts in robotics and computer vision;
- to encourage intuition through hands-on numerical experimentation.

The approach used is to present background, theory, and examples in an integrated fashion. Code and examples are first-class citizens in this book and are not relegated to the end of the chapter or an associated web site. The examples are woven into the discussion like this

```
>> R = rotm2d(0.3)
R =
   0.9553  -0.2955
   0.2955   0.9553
```

where the code illuminates the topic being discussed and generally results in a crisp numerical result or a graph in a figure that is then discussed. The examples illustrate how to use the software tools and that knowledge can then be applied to other problems.

1.6.1 MATLAB and the Book

» *To do good work, one must first have good tools.*
   – Chinese proverb

The computational foundation of this book is MATLAB®, a software package developed by MathWorks®. MATLAB is an interactive mathematical software environment that makes linear algebra, data analysis, and high-quality graphics a breeze. It also supports a programming language which allows the creation of complex algorithms. MATLAB is a popular package and one that is very likely to be familiar to engineering students as well as researchers. It can run on the desktop or in a web browser or be compiled into efficient C/C++/CUDA/HDL code for embedded systems. MATLAB is proprietary software and must be licensed: many universities provide licenses for students and academics; and many companies in the technology sector, large and small, provide licenses for their staff.

A strength of MATLAB is that its functionality can be extended using *toolboxes* which are collections of functions targeted at particular topics. While many MATLAB toolboxes are open source, this book requires a number of licensed MATLAB toolboxes, and the details are given in ▶ App. A.

This book provides examples of how to use many toolbox functions in the context of solving specific problems, but it is not a reference manual. Comprehensive documentation of all toolbox functions is available through the MATLAB built-in documentation browser or online at ▶ https://mathworks.com/help/matlab.

1.6.2 Notation, Conventions, and Organization

The mathematical notation used in the book is summarized in the ▶ Nomenclature section. Since the coverage of the book is broad there are just not enough good symbols to go around, and unavoidably some symbols have different meanings in different parts of the book.

Excurse 1.8: MATLAB

The MATLAB software we use today has a long history. It starts with the LINPACK and EISPACK projects, run by the Argonne National Laboratory in the 1970s, to produce high-quality, tested and portable mathematical software for super-computers. LINPACK is a collection of routines for linear algebra and EISPACK is a library of numerical algorithms for computing eigenvalues and eigenvectors of matrices. These packages were written in Fortran which was then the language of choice for large-scale numerical problems. LAPACK is the intellectual descendant of those early packages and is the foundation of most linear algebra code today.

Cleve Moler, then at the University of New Mexico, was part of the LINPACK and EISPACK projects. He wrote the first version of MATLAB in the late 1970s to allow interactive use of LINPACK and EISPACK for problem solving without having to write and compile Fortran code. MATLAB quickly spread to other universities and found a strong audience within the applied mathematics and engineering community. In 1984 Cleve Moler and Jack Little founded The MathWorks Inc. which exploited the newly released IBM PC – the first widely available desktop computer.

Cleve Moler received his bachelor's degree from Caltech in 1961, and a Ph.D. from Stanford University. He was a professor of mathematics and computer science at universities including University of Michigan, Stanford University, and the University of New Mexico. He has served as president of the Society for Industrial and Applied Mathematics (SIAM) and was elected to the National Academy of Engineering in 1997.

See also ▶ https://sn.pub/9FRP5R which includes a video of Cleve Moler and also ▶ http://history.siam.org/pdfs2/Moler_final.pdf.

1

There is a lot of MATLAB code in the book and this is indicated in fixed-width font such as

```
>> 6 * 7
ans =
42
```

The MATLAB command prompt is >> and what follows is the command issued to MATLAB by the user. Subsequent lines, without the prompt, are MATLAB's response. Long commands require continuation lines which also begin with >>. ◄ All functions, classes, and methods mentioned in the text or in code segments are cross-referenced and have their own index at the end of the book, allowing you to find different ways that particular functions can be used.

The code examples conform to the conventions for MATLAB example coding ◄ and rely on recent MATLAB language extensions: strings which are delimited by double quotation marks (introduced in R 2016b); and `name=value` syntax for passing arguments to functions (introduced in R 2021a), for example, `plot(x,y,LineWidth=2)` instead of the old-style `plot(x,y,"LineWidth",2)`.

Various boxes are used to organize and differentiate parts of the content.

> These highlight boxes indicate content that is particularly important.

🛈 These warning boxes highlight points that are often traps for those starting out.

As authors, we have to balance the tension between completeness, clarity, and conciseness. For this reason a lot of detail has been pushed into side notes ◄ and information boxes, and on a first reading, these can be skipped. Some chapters have an *Advanced Topics* section at the end that can also be skipped on a first reading. For clarity, references are cited sparingly in the text but are included at the end of each chapter. However if you are trying to understand a particular algorithm and apply it to your own problem, then understanding the details and nuances can be important and the sidenotes, references and *Advanced Topics* will be helpful.

Each chapter ends with a *Wrapping Up* section that summarizes the important lessons from the chapter, discusses some suggested further reading, and provides some exercises. The *Further Reading* subsection discusses prior work and provides extensive references to prior work and more complete description of the algorithms. *Resources* provides links to relevant online code and datasets. *Exercises* extend the concepts discussed within the chapter and are generally related to specific code examples discussed in the chapter. The exercises vary in difficulty from straightforward extension of the code examples to more challenging problems.

Many of the code examples include a plot command which generates a figure. The figure that is included in the book is generally embellished with axis labels, grid, legends and adjustments to line thickness. The embellishment requires many extra lines of code which would clutter the examples, so the book's GitHub repository, see ▶ App. A, provides the detailed MATLAB scripts that generate all of the published figures.

Finally, the book contains links to online resources which can be simply accessed through a web browser. Some of these have short URLs of the form

> **Excurse 1.9: Information**
>
> These information boxes provide technical, historical, or biographical information that augment the main text.

Marginal notes (left column):

Code blocks with the prompt characters can be copy and pasted into the MATLAB console and will be interpreted correctly.

With the exception that we prefer the object-oriented dot syntax to functional syntax for object methods.

They are placed as marginal notes near the corresponding marker.

chapter2.mlx

▶ sn.pub/KI1xtA

`sn.pub/XXXXXX` (clickable in the e-book version) and QR codes, which are placed in the margin. Each chapter has a link to a MATLAB live script running on MATLAB® Online™, allowing you to run that chapter's code in a browser window (even on a tablet) with zero installation on your own computer. Some figures depict three-dimensional scenes and the included two-dimensional image do not do them justice – the links connect to web-based interactive 3D viewers of those scenes.

1.6.3 Audience and Prerequisites

The book is intended primarily for third or fourth year engineering undergraduate students, Masters students and first year Ph.D. students. For undergraduates the book will serve as a companion text for a robotics, mechatronics, or computer vision course or support a major project in robotics or vision. Students should study Part I and the appendices for foundational concepts, and then the relevant part of the book: mobile robotics, arm robots, computer vision or robotic vision. The code in the book shows how to solve common problems, and the exercises at the end of each chapter provide additional problems beyond the worked examples in the book.

For students commencing graduate study in robotics, and who have previously studied engineering or computer science, the book will help fill the gaps between what you learned as an undergraduate, and what will be required to underpin your deeper study of robotics and computer vision. The book's working code base can help bootstrap your research, enable you to get started quickly, and work productively on your own problems and ideas.

For those who are no longer students, the researcher or industry practitioner, the book will serve as a useful companion for your own reference to a wide range of topics in robotics and computer vision, as well as a handbook and guide for the Toolboxes.

The book assumes undergraduate-level knowledge of linear algebra (matrices, vectors, eigenvalues), basic set theory, basic graph theory, probability, calculus, dynamics (forces, torques, inertia) and control theory. The appendices provide a concise reiteration of key concepts. Computer science students may be unfamiliar with concepts in ▶ Chaps. 4 and 9 such as the Laplace transform, transfer functions, linear control (proportional control, proportional-derivative control, proportional-integral control) and block diagram notation. That material could be skimmed over on a first reading, and Albertos and Mareels (2010) may be a useful introduction to some of these topics. The book also assumes the reader is familiar with using, and programming in, MATLAB and also familiar with MATLAB object-oriented programming techniques (which is conceptually similar to C++, Java or Python). Familiarity with Simulink®, the graphical block-diagram modeling tool integrated with MATLAB will be helpful but not essential.

1.6.4 Learning with the Book

The best way to learn is by doing. Although the book shows the MATLAB commands and the response, there is something special about doing it for yourself. Consider the book as an invitation to tinker. By running the code examples yourself you can look at the results in ways that you prefer, plot the results in a different way, or try the algorithm on different data or with different parameters. The paper edition of the book is especially designed to stay open which enables you to type in commands as you read. You can also look at the online documentation for the

Toolbox functions, discover additional features and options, and experiment with those.

Most of the code examples are quite short so typing them in to MATLAB is not too onerous, however there are a lot of them – more than 1600 code examples. The book's GitHub repository, see ▶ App. A, provides the code for each chapter as a MATLAB live script, which allows the code to be executed cell by cell, or it can be cut and pasted into your own scripts. If you have a MATLAB® Online™ account and a license for the required products, then clicking on the link at the start of a chapter will take you to a MATLAB session in your browser with the appropriate live script ready to run in the Live Editor environment.

The Robot Academy (▶ https://robotacademy.net.au) is a free online resource with over 200 short video lessons that cover much, but not all, of the content of this book. Many lessons use code to illustrate concepts, and those examples use the older open-source MATLAB toolboxes, which can be found at ▶ https://petercorke.com.

1.6.5 Teaching with the Book

This book can be used to support courses in robotics, mechatronics and computer vision. All courses should include the introduction to position, orientation, pose, pose composition, and coordinate frames that is discussed in ▶ Chap. 2. For a mobile robotics or computer vision course it is sufficient to teach only the 2-dimensional case. For robotic manipulators or multi-view geometry the 2-D and 3-dimensional cases should be taught.

All code and most figures in this book are available from the book's GitHub respository – you are free to use them with attribution. Code examples in this book are available as MATLAB live scripts and could be used as the basis for demonstrations in lectures or tutorials. Line drawings are provided as PDF files and MATLAB-generated figures are provided as scripts that will recreate the figures. See ▶ App. A for details.

The exercises at the end of each chapter can be used as the basis of assignments, or as examples to be worked in class or in tutorials. Most of the questions are rather open ended in order to encourage exploration and discovery of the effects of parameters and the limits of performance of algorithms. This exploration should be supported by discussion and debate about performance measures and what *best* means. True understanding of algorithms involves an appreciation of the effects of parameters, how algorithms fail and under what circumstances.

The teaching approach could also be inverted, by diving head first into a particular problem and then teaching the appropriate prerequisite material. Suitable problems could be chosen from the *Application* sections of chapters (see *applications* in the main index), or from any of the exercises. Particularly challenging exercises are so marked.

If you wanted to consider a flipped learning approach then the Robot Academy (▶ https://robotacademy.net.au) could be used in conjunction with your class. Students would watch the video lessons and undertake some formative assessment out of the classroom, and you could use classroom time to work through problem sets.

For graduate-level teaching the papers and textbooks mentioned in the *Further Reading* could form the basis of a student's reading list. They could also serve as candidate papers for a reading group or journal club.

1.6.6 Outline

We promised a book with instant gratification but before we can get started in robotics there are some fundamental concepts that we absolutely need to understand, and understand well. Part I introduces the concepts of pose and coordinate frames – how we represent the position and orientation of a robot, a camera or the objects that the robot needs to work with. We discuss how motion between two poses can be *decomposed* into a sequence of elementary translations and rotations, and how elementary motions can be *composed* into more complex motions. ▶ Chap. 2 discusses how pose can be represented in a computer, and ▶ Chap. 3 discusses the relationship between velocity and the derivative of pose, generating a sequence of poses that smoothly follow some path in space and time, and estimating motion from sensors.

With these formalities out of the way we move on to the first main event – robots. There are two important classes of robot: mobile robots and manipulator arms and these are covered in Parts II and III respectively. ▶

Part II begins, in ▶ Chap. 4, with motion models for several types of wheeled vehicles and a multi-rotor flying vehicle. Various control laws are discussed for wheeled vehicles such as moving to a point, following a path, and moving to a specific pose. ▶ Chap. 5 is concerned with navigation, that is, how a robot finds a path between points A and B in the world. Two important cases, with and without a map, are discussed. Most navigation techniques require knowledge of the robot's position and ▶ Chap. 6 discusses various approaches to this problem based on dead-reckoning, or landmark observation, and a map. We also show how a robot can make a map, and even determine its location while simultaneously mapping an unknown region – the SLAM problem.

Part III is concerned with arm-type robots, or more precisely serial-link manipulators. Manipulator arms are used for tasks such as assembly, welding, material handling and even surgery. ▶ Chap. 7 introduces the topic of kinematics which relates the angles of the robot's joints to the 3-dimensional pose of the robot's tool. Techniques to generate smooth paths for the tool are discussed and two examples show how an arm-robot can draw a letter on a surface, and how multiple arms (acting as legs) can be used to create a model for a simple walking robot. ▶ Chap. 8 discusses the relationships between the rates of change of joint angles and tool pose. It introduces the Jacobian matrix and concepts such as singularities, manipulability, null-space motion, and resolved-rate motion control. It also discusses under- and overactuated robots and the general numerical solution to inverse kinematics. ▶ Chap. 9 introduces the design of joint control systems, the dynamic equations of motion for a serial-link manipulator, and the relationship between joint forces and joint motion. It discusses important topics such as variation in inertia, the effect of payload, flexible transmissions, independent-joint versus nonlinear control strategies, and task-space control.

Computer vision is a large field concerned with processing images in order to enhance them for human benefit, interpret the contents of the scene or create a 3D model corresponding to the scene. Part IV is concerned with machine vision, a subset of computer vision, and defined here as the extraction of numerical features from images to provide input for control of a robot. The discussion starts in ▶ Chap. 10 with the fundamentals of light, illumination and color. ▶ Chap. 11 discusses *image processing* which is a domain of 2-dimensional signal processing that transforms one image into another image. The discussion starts with acquiring real-world images and then covers various arithmetic and logical operations that can be performed on images. We then introduce spatial operators such as convolution, segmentation, morphological filtering and finally image shape and size changing. These operations underpin the discussion in ▶ Chap. 12 which describe how numerical features are extracted from images. The features describe homo-

Although robot arms came first chronologically, mobile robotics is mostly a 2-dimensional problem and easier to understand than the 3-dimensional arm-robot case.

1

geneous regions (blobs), lines or distinct points in the scene and are the basis for vision-based robot control. The application of deep-learning approaches for object detection is also introduced. ▶ Chap. 13 describes the geometric model of perspective image creation using lenses and discusses topics such as camera calibration and pose estimation. We introduce nonperspective imaging using wide-angle lenses and mirror systems, camera arrays and light-field cameras. ▶ Chap. 14 is concerned with estimating the underlying three-dimensional geometry of a scene using classical methods such as structured lighting and also combining features found in different views of the same scene to provide information about the geometry and the spatial relationship between the camera views which is encoded in fundamental, essential and homography matrices. This leads to the topics of stereo vision, bundle adjustment and structure from motion, and applications including perspective correction, mosaicing, image retrieval and visual odometry.

Part V is about robotic vision, the combination of robotics and computer vision, and pulls together concepts introduced in the earlier parts of the book. ▶ Chap. 15 introduces vision-based control or visual servoing, how information from the camera's view can be used to control the pose of the camera using approaches known as position-based and image-based visual servoing. ▶ Chap. 16 covers a number of complex application examples concerned with highway driving, drone navigation and robot manipulator pick and place.

This is a big book but any one of the parts can be read standalone, with more or less frequent visits to the required earlier material. ▶ Chap. 2 is the only essential reading. Parts II, III or IV could be used respectively for an introduction to mobile robots, arm robots or computer vision class. An alternative approach, following the instant gratification theme, is to jump straight into any chapter and start exploring – visiting the earlier material as required.

1.6.7 Further Reading

The Handbook of Robotics (Siciliano and Khatib 2016) provides encyclopedic coverage of the field of robotics today, covering theory, technology and the different types of robot such as telerobots, service robots, field robots, aerial robots, underwater robots and so on. The classic work by Sheridan (2003) discusses the spectrum of autonomy from remote control, through shared and traded control, to full autonomy.

A comprehensive coverage of computer vision is the book by Szeliski (2022). Fascinating reading about eyes in nature can be found in Srinivasan and Venkatesh (1997), Land and Nilsson (2002), Ings (2008), Frisby and Stone (2010), and Stone (2012). A solid introduction to artificial intelligence is the text by Russell and Norvig (2020).

A number of very readable books discuss the future impacts of robotics and artificial intelligence on society, for example Ford (2015), Brynjolfsson and McAfee (2014), and Bostrom (2016). The YouTube video Grey (2014) makes some powerful points about the future of work and is always a great discussion starter.

Foundations

Contents

Chapter 2   Representing Position and Orientation – 21

Chapter 3   Time and Motion – 87

Representing Position and Orientation

Contents

2.1   Foundations – 22

2.2   Working in Two Dimensions (2D) – 30

2.3   Working in Three Dimensions (3D) – 42

2.4   Advanced Topics – 66

2.5   MATLAB Classes for Pose and Rotation – 79

2.6   Wrapping Up – 82

© The Author(s), under exclusive license to Springer Nature Switzerland AG 2023
P. Corke, W. Jachimczyk, R. Pillat, *Robotics, Vision and Control*, Springer Tracts in Advanced Robotics 147,
https://doi.org/10.1007/978-3-031-07262-8_2

Orientation is often referred to as *attitude*.

We are familiar with numbers for counting and measuring, and tuples of numbers (coordinates) to describe the position of points in 2- or 3-dimensions. In robotics, we are particularly interested in where *objects* are in the world – objects such as mobile robots, the links and tool of a robotic manipulator arm, cameras or work pieces. ► Sect. 2.1 introduces foundational concepts and then, in ► Sect. 2.2 and 2.3 respectively, we apply them to the 2-dimensional and 3-dimensional scenarios that we encounter in robotics. ► Sect. 2.4 covers some advanced topics that could be omitted on a first reading, and ► Sect. 2.5 has additional details about the MATLAB® toolboxes we will use throughout the rest of the book.

2.1 Foundations

We start by introducing important general concepts that underpin our ability to describe where objects are in the world, both graphically and algebraically.

2.1.1 Relative Pose

For the 2D and 3D examples in ◼ Fig. 2.1 we see red and blue versions of an object that have a different position and orientation. The combination of position and orientation ◄ of an object is its *pose*, so the red and blue versions of the object have different poses.

The object's initial pose, shown in blue, has been *transformed* into the pose shown in red. The object's shape has not changed, so we call this a *rigid-body transformation*. It is natural to think of this transformation as a motion – the object has moved in some complex way. Its position has changed (it has *translated*) and its orientation has changed (it has *rotated*).

We will use the symbol ξ (pronounced ksigh) to denote motion and depict it graphically as a thick curved arrow. Importantly, as shown in ◼ Fig. 2.2, the motion is always defined with respect to an initial pose.

Next, we will consider the very simple 2D example shown in ◼ Fig. 2.3. We have elaborated our notation to make it clear where the motion is from and to. The leading superscript indicates the initial pose and the trailing subscript indicates the resulting pose. Therefore $^x\xi_y$ means the motion from pose x to pose y.

Motions can be *joined up* to create arbitrary motions – a process called *composition* or *compounding*. We can write this algebraically as

$$^0\xi_2 = {}^0\xi_1 \oplus {}^1\xi_2 \tag{2.1}$$

which says that the motion $^0\xi_2$ is the same as the motion $^0\xi_1$ followed by the motion $^1\xi_2$. We use the special symbol $\oplus$ to remind ourselves that we are *adding*

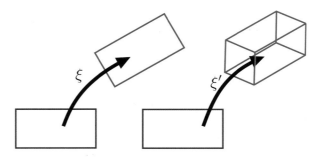

◼ **Fig. 2.1** Rigid motion of a shape **a** rectangular object in 2D; **b** cuboid object in 3D, the initial blue view is a "side on" view of a 3D cuboid

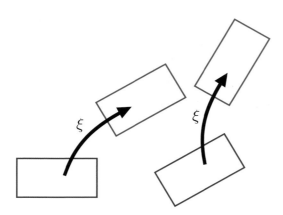

Fig. 2.2 The motion is always defined relative to the starting pose shown in blue. $\boldsymbol{\xi}$ is the same in both cases

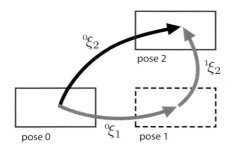

Fig. 2.3 The motion from pose 0 to pose 2 is equivalent to the motion from pose 0 to pose 1 and then to pose 2

motions not regular numbers. The motions are consecutive, and this means that the pose names on each side of the $\oplus$ must be the same. Although **Fig. 2.3** only shows translations, the motions that we compose can include rotations. For the 3D case, there is a richer set of motions to choose from – there are more directions to translate in, and more axes to rotate about.

It is important to note that the order of the motions is important. For example, if we walk forward 1 m and then turn clockwise by 90°, we have a different pose than if we turned 90° clockwise and then walked forward 1 m. Composition of motions is not commutative.

The corollary is that any motion can be *decomposed* into a number of smaller or simpler motions. For the 2D case in **Fig. 2.1**, one possible decomposition of $\boldsymbol{\xi}$ is a horizontal translation, followed by a vertical translation followed by a rotation about its center.

For any motion there is an *inverse* or *opposite* motion – the motion that gets us back to where we started. For example, for a translation of 2 m to the right, the inverse motion is a translation of 2 m to the left, while for a rotation of 30° clockwise, the inverse is a rotation of 30° counter-clockwise. In general, for a motion from pose X to pose Y the inverse motion is from pose Y to pose X, that is

$$^{X}\boldsymbol{\xi}_{Y} \oplus {}^{Y}\boldsymbol{\xi}_{X} = {}^{X}\boldsymbol{\xi}_{X} = \varnothing \tag{2.2}$$

where $\varnothing$ is the *null motion* and means no motion.

We will introduce another operator $\ominus$ that turns any motion into its inverse

$$^{Y}\boldsymbol{\xi}_{X} \equiv \ominus {}^{X}\boldsymbol{\xi}_{Y} ,$$

and now we can write (2.2) as

$$^{X}\boldsymbol{\xi}_{Y} \ominus {}^{X}\boldsymbol{\xi}_{Y} = \varnothing .$$

2

While this looks a lot like subtraction, we use the special symbol $\ominus$ to remind ourselves that we are *subtracting* motions not regular numbers. It follows, that adding or subtracting a *null motion* to a motion leaves the motion unchanged, that is

$$^X\xi_Y \oplus \varnothing = {}^X\xi_Y, \ {}^X\xi_Y \ominus \varnothing = {}^X\xi_Y$$

For the example in ◨ Fig. 2.3, we could also write

$$^0\xi_1 = {}^0\xi_2 \oplus {}^2\xi_1$$

and while there is no arrow in ◨ Fig. 2.3 for $^2\xi_1$, it would simply be the reverse of the arrow shown for $^1\xi_2$. Since $^2\xi_1 = \ominus\, {}^1\xi_2$, we can write

$$^0\xi_1 = {}^0\xi_2 \ominus {}^1\xi_2$$

which is very similar to (2.1) except that we have effectively *subtracted* $^1\xi_2$ from both sides – to be very precise, we have subtracted it from the right of *both* sides of the equation since order is critically important. Taking this one step at a time

$$^0\xi_2 = {}^0\xi_1 \oplus {}^1\xi_2$$
$$^0\xi_2 \ominus {}^1\xi_2 = {}^0\xi_1 \oplus {}^1\xi_2 \ominus {}^1\xi_2$$
$$^0\xi_2 \ominus {}^1\xi_2 = {}^0\xi_1 \oplus \cancel{{}^1\xi_2 \ominus {}^1\xi_2}$$
$$^0\xi_2 \ominus {}^1\xi_2 = {}^0\xi_1 \oplus \varnothing$$
$$= {}^0\xi_1 \ .$$

Alternatively, just as we do in regular algebra, we could do this in a single step by "*taking $^1\xi_2$ across to the other side*" and "*negating*" it

$$\overset{\oplus\ \to\ \ominus}{^0\xi_2 = {}^0\xi_1 \underset{\frown}{\left(\oplus\ {}^1\xi_2\right)}}$$

At the outset, we talked about the pose of the objects in ◨ Fig. 2.1. There is no absolute way to describe the pose of an object, we can only describe its pose with respect to some other pose, and we introduce a *reference pose* for this purpose. Any pose is always a *relative pose*, described by the motion required to get there from the reference pose.

2.1.2 Coordinate Frames

To describe relative pose, we need to describe its two components: translation and rotation. To achieve this, we rigidly attach a right-handed coordinate frame to the object, as shown in ◨ Fig. 2.4. Coordinate frames are a familiar concept from mathematics and comprise two or three orthogonal axes which intersect at a point called the *origin*.

The position and orientation of the coordinate frame is sufficient to fully describe the position and orientation of the object it is attached to. Each frame is named, in this case {A} or {B}, and the name is also reflected in the subscripts on the axis labels. Translation is the distance between the origins of the two coordinate frames, and rotation can be described by the set of angles between corresponding axes of the coordinate frames. Together, these represent the change in pose, the relative pose, or the motion described by ξ.

The coordinate frame associated with the reference pose is denoted by {0} and called the *reference frame*, the *world frame* and denoted by {W}, or the *global frame*. If the leading superscript is omitted, the reference frame is assumed. The

► sn.pub/SGAYt6

■ **Fig. 2.4** An object with attached coordinate frame shown in two different poses. The axes of 3D coordinate frames are frequently colored red, green and blue for the x-, y- and z-axes respectively. The reference frame is shown in black

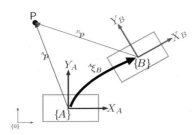

■ **Fig. 2.5** The point P can be described by coordinate vectors expressed in frame {A} or {B}. The pose of {B} relative to {A} is $^{A}\xi_{B}$

frame associated with a moving body, such as a vehicle, is often called the *body frame* or *body-fixed frame* and denoted by {B}. Any point on the moving body is defined by a constant coordinate vector with respect to the body frame.

■ Fig. 2.5 shows that a point P can be described by two different coordinate vectors: $^{A}\boldsymbol{p}$ with respect to frame {A}, or $^{B}\boldsymbol{p}$ with respect to frame {B}. The *basis* of the coordinate frames are unit frames in the directions of the frame's axes. These two coordinate vectors are related by

$$^{A}\boldsymbol{p} = {}^{A}\boldsymbol{\xi}_{B} \cdot {}^{B}\boldsymbol{p}$$

where the $\cdot$ operator *transforms* the coordinate vector, resulting in a new coordinate vector that describes the same point but with respect to the coordinate frame resulting from the motion $^{A}\boldsymbol{\xi}_{B}$.

We can think of the right-hand side as a motion from {A} to {B} and then to P. We indicate this *composition* of a motion and a vector with the $\cdot$ operator to distinguish it from proper motion composition where we use the $\oplus$ operator.

Coordinate frames are an extremely useful way to think about real-world robotics problems such as shown in ■ Fig. 2.6. We have attached coordinate frames to the key objects in the problem and, next, we will look at the relationships between the poses that they represent.

We have informally developed a useful set of rules for dealing with motions. We can compose motions

$$^{X}\boldsymbol{\xi}_{Z} = {}^{X}\boldsymbol{\xi}_{Y} \oplus {}^{Y}\boldsymbol{\xi}_{Z} \tag{2.3}$$

which we read as the motion from X to Z is equivalent to a motion from X to Y and then from Y to Z. The pose names on either side of the $\oplus$ operator must match.

There is an operator for inverse motion

$$^{Y}\boldsymbol{\xi}_{X} = \ominus {}^{X}\boldsymbol{\xi}_{Y} \tag{2.4}$$

■ **Fig. 2.6** Examples of coordinate frames for a real-world scenario. Frames include the mobile robot {M}, arm base {B}, arm end effector {E}, camera {C}, workpiece {P} and the reference frame {0} (Image by John Skinner and Dorian Tsai)

and a null motion denoted by $\varnothing$ such that

$$
{}^{X}\boldsymbol{\xi}_{X} = \varnothing, \quad {}^{X}\boldsymbol{\xi}_{Y} \ominus {}^{X}\boldsymbol{\xi}_{Y} = \varnothing, \quad \ominus {}^{X}\boldsymbol{\xi}_{Y} \oplus {}^{X}\boldsymbol{\xi}_{Y} = \varnothing,
$$
$$
{}^{X}\boldsymbol{\xi}_{Y} \oplus \varnothing = {}^{X}\boldsymbol{\xi}_{Y}, \quad {}^{X}\boldsymbol{\xi}_{Y} \ominus \varnothing = {}^{X}\boldsymbol{\xi}_{Y}, \quad \varnothing \oplus {}^{X}\boldsymbol{\xi}_{Y} = {}^{X}\boldsymbol{\xi}_{Y}
\tag{2.5}
$$

A motion transforms the coordinate vector of a point to reflect the change in coordinate frame due to that motion

$$
{}^{X}\boldsymbol{p} = {}^{X}\boldsymbol{\xi}_{Y} \cdot {}^{Y}\boldsymbol{p} \ .
\tag{2.6}
$$

While this looks like regular algebra it is important to remember that $\boldsymbol{\xi}$ is not a number, it is a relative pose or motion. In particular, the order of operations matters when rotations are involved. We cannot assume commutativity as we do with regular numbers. Formally, $\boldsymbol{\xi}$ belongs to a non-abelian group with the operator $\oplus$, an inverse $\ominus$ and an identity element $\varnothing$.

In the literature, it is common to use a symbol like $*$, $\cdot$ or nothing instead of $\oplus$, and X^{-1} to indicate inverse instead of $\ominus X$.

Excurse 2.1: The Reference Frame Rule

In relative pose composition, we can check that we have our reference frames correct by ensuring that the subscript and superscript on each side of the $\oplus$ operator are matched. We can then *cancel out* the intermediate subscripts and superscripts, leaving just the outermost subscript and superscript.

Excurse 2.2: Groups

Groups are a useful and important mathematical concept and are directly related to our discussion about motion and relative pose. Formally, a *group* is an abstract system represented by an ordered pair $G = (G, \diamond)$ which comprises a set of objects G equipped with a single *binary operator* denoted by $\diamond$ which combines any two elements of the group to form a third element, also belonging to the group. A group satisfies the following axioms:

Closure	$a \diamond b \in G$	$\forall a, b \in G$
Associativity	$(a \diamond b) \diamond c = a \diamond (b \diamond c)$	$\forall a, b, c \in G$
Identity element	$a \diamond e = e \diamond a = a$	$\forall a \in G, \exists e \in G$
Inverse element	$a \diamond a^{-1} = a^{-1} \diamond a = e$	$\forall a \in G, \exists a^{-1} \in G$

A group does *not* satisfy the axiom of commutativity which means that the result of applying the group operation depends on the order in which they are written – an abelian group is one that does satisfy the axiom of commutativity.

For the group of relative motions, the group operator is composition. The axioms state that the result of composing two motions is another motion, that the order of motions being composed is significant, and that there exists an inverse motion, and an identity (null) motion. In this book, for the group of relative motions, elements are denoted by ξ, the group operator is $\oplus$, inverse is $\ominus$, and the identity element is $\varnothing$.

We will shortly introduce mathematical objects to represent relative motion, for example, rotation matrices, homogeneous transformation matrices, quaternions, and twists – these all form groups under the operation of composition.

2.1.3 Pose Graphs

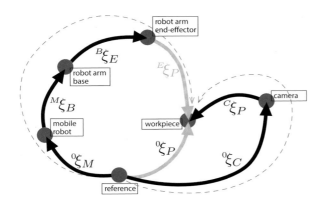 Fig. 2.7 is a pose graph – a directed graph ▶ which comprises vertices (blue circles) representing poses and edges (arrows) representing relative poses or motions. This is a clear, but abstract, representation of the spatial relationships present in the problem shown in ◼ Fig. 2.6. Pose graphs can be used for 2D or 3D problems – that just depends on how we *implement* ξ.

The black arrows represent known relative poses, and the gray arrows are unknown relative poses that we wish to determine. In order for the robot to grasp the workpiece, we need to know its pose relative to the robot's end effector, that is, $^E\xi_P$. We start by looking for a pair of equivalent paths – two different paths that have the same start and end pose, one of which includes the unknown. We choose the paths shown as red dashed lines in ◼ Fig. 2.7. Each path has the same start and end pose, they are equivalent motions, so we can equate them

$$^0\xi_M \oplus {}^M\xi_B \oplus {}^B\xi_E \oplus \underline{{}^E\xi_P} = {}^0\xi_C \oplus {}^C\xi_P \; .$$

See ▶ App. I for more details about graphs.

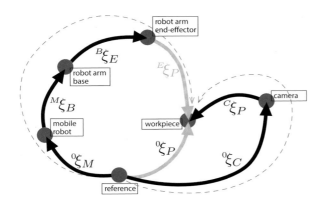

◼ **Fig. 2.7** A pose graph for the scenario shown in ◼ Fig. 2.6. Blue circles are vertices of the graph and represent poses, while the arrows are edges of the graph and represent relative poses or motions. The dashed arrows represent the same relative pose, but achieved by different paths through the pose graph

2

Now we can perform the algebraic operations discussed earlier and manipulate the expression to isolate the unknown motion

$$\underline{{}^E\boldsymbol{\xi}_P} = \ominus\,{}^B\boldsymbol{\xi}_E \ominus\,{}^M\boldsymbol{\xi}_B \ominus\,{}^0\boldsymbol{\xi}_M \oplus\,{}^0\boldsymbol{\xi}_C \oplus\,{}^C\boldsymbol{\xi}_P$$

It is easy to write such an expression by inspection. To determine ${}^X\boldsymbol{\xi}_Y$, find any path through the pose graph from {X} to {Y} and write down the relative poses of the edges in a left to right order – if we traverse the edge in the direction of its arrow, precede it with the $\oplus$ operator, otherwise use $\ominus$.

2.1.4 Summary

🔲 Fig. 2.8 provides a graphical summary of the key concepts we have just covered. The key points to remember are:

- The position and orientation of an object is referred to as its pose.
- A motion, denoted by $\boldsymbol{\xi}$, causes a change in pose – it is a relative pose defined with respect to the initial pose.
- There is no absolute pose, a pose is always relative to a reference pose.
- We can perform algebraic manipulation of expressions written in terms of relative poses using the operators $\oplus$ and $\ominus$, and the concept of a null motion $\varnothing$.
- We can represent a set of poses, with known relative poses, as a pose graph.
- To quantify pose, we rigidly attach a coordinate frame to an object. The origin of that frame is the object's position, and the directions of the frame's axes describe its orientation.
- The constituent points of a rigid object are described by constant coordinate vectors relative to its coordinate frame.
- Any point can be described by a coordinate vector with respect to any coordinate frame. A coordinate vector can be transformed between frames by applying the relative pose of those frames to the vector using the · operator.

The details of what $\boldsymbol{\xi}$ *really is* have been ignored up to this point. This was a deliberate strategy to have us think generally about the pose of objects and motions between poses, rather than immediately diving into the details of implementation. The true nature of $\boldsymbol{\xi}$ depends on the problem. Is it a 2D or 3D problem? Does it involve translations, rotations, or both? The implementation of $\boldsymbol{\xi}$ could be a vector, a matrix or something more exotic like a twist, a unit quaternion or a unit dual quaternion.

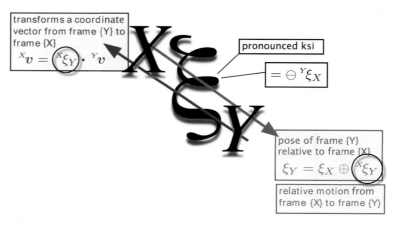

🔲 **Fig. 2.8** Everything you need to know about relative pose

Excurse 2.3: Euclid of Alexandria

Euclid (325–265 BCE) was a Greek mathematician, who was born and lived in Alexandria, Egypt, and is considered the "father of geometry", His great work *Elements*, comprising 13 books, captured and systematized much early knowledge about geometry and numbers. It deduces the properties of planar and solid geometric shapes from a set of 5 axioms and 5 postulates.

Elements is probably the most successful book in the history of mathematics. It describes plane geometry and is the basis for most people's first introduction to geometry and formal proof, and is the basis of what we now call Euclidean geometry. Euclidean distance is simply the distance between two points on a plane. Euclid also wrote *Optics* which describes geometric vision and perspective.

Excurse 2.4: Euclidean versus Cartesian Geometry

Euclidean geometry is concerned with points and lines in the Euclidean plane (2D) or Euclidean space (3D). It is based entirely on a set of axioms and makes no use of arithmetic. Descartes added a coordinate system (2D or 3D) and was then able to describe points, lines and other curves in terms of algebraic equations. The study of such equations is called analytic geometry and is the basis of all modern geometry. The Cartesian plane (or space) is the Euclidean plane (or space) with all its axioms and postulates *plus* the extra facilities afforded by the added coordinate system. The term Euclidean geometry is often used to mean that Euclid's fifth postulate (parallel lines never intersect) holds, which is the case for a planar surface but not for a curved surface.

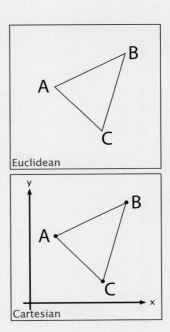

In the rest of this chapter, we will explore a number of concrete representations of rotation, translation and pose, and ways to convert between them. To make it real, we introduce software tools for MATLAB that can create and manipulate these mathematical objects, and will support everything we do in the rest of the book.

Excurse 2.5: René Descartes

Descartes (1596–1650) was a French philosopher, mathematician and part-time mercenary. He is famous for the philosophical statement "*Cogito, ergo sum*" or "*I am thinking, therefore I exist*" or "*I think, therefore I am*". He was a sickly child and developed a life-long habit of lying in bed and thinking until late morning. A possibly apocryphal story is that during one such morning he was watching a fly walk across the ceiling and realized that he could describe its position in terms of its distance from the two edges of the ceiling. This is the basis of the *Cartesian* coordinate system and modern (analytic) geometry, which he described in his 1637 book *La Géométrie*. For the first time, mathematics and geometry were connected, and modern calculus was built on this foundation by Newton and Leibniz. Living in Sweden, at the invitation of Queen Christina, he was obliged to rise at 5 A.M., breaking his lifetime habit – he caught pneumonia and died. His remains were later moved to Paris, and are now lost apart from his skull which is in the Musée de l'Homme. After his death, the Roman Catholic Church placed his works on the Index of Prohibited Books – the Index was not abolished until 1966.

2.2   Working in Two Dimensions (2D)

The relative orientation of the x- and y-axes obey the right-hand rule. For the 2D case the y-axis is obtained by rotating the x-axis counter-clockwise by 90°.

A 2-dimensional world, or plane, is familiar to us from high-school geometry. We use a right-handed ◄ Cartesian coordinate system or coordinate frame with orthogonal axes, denoted x and y, and typically drawn with the x-axis pointing to the right and the y-axis pointing upwards. The point of intersection is called the origin.

The basis vectors are unit vectors parallel to the axes and are denoted by $\hat{x}$ and $\hat{y}$. A point is represented by its x- and y-coordinates (x, y) or as a coordinate vector from the origin to the point

$$\boldsymbol{p} = x\hat{\boldsymbol{x}} + y\hat{\boldsymbol{y}} \tag{2.7}$$

which is a linear combination of the basis vectors.

◘ Fig. 2.9 shows a red coordinate frame {B} that we wish to describe with respect to the blue frame {A}. We see clearly that the origin of {B} has been displaced, and the frame rotated counter-clockwise. We will consider the problem in two parts: pure rotation and then rotation plus translation.

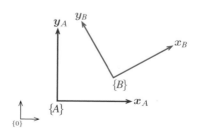

◘ **Fig. 2.9** Two 2D coordinate frames {A} and {B}, defined with respect to the reference frame {0}. {B} is rotated and translated with respect to {A}

2.2.1 Orientation in Two Dimensions

2.2.1.1 2D Rotation Matrix

◩ Fig. 2.10 shows two coordinate frames {A} and {B} with a common origin but different orientation. Frame {B} is obtained by rotating frame {A} by θ in the positive (counter-clockwise) direction about the origin.

Frame {A} is described by the directions of its axes, and these are parallel to the basis vectors $\hat{\boldsymbol{x}}_A$ and $\hat{\boldsymbol{y}}_A$ which are defined with respect to the reference frame {0}. In this 2D example, the basis vectors have two elements which we write as column vectors and stack side-by-side to form a 2×2 matrix $(\hat{\boldsymbol{x}}_A \quad \hat{\boldsymbol{y}}_A)$. This matrix *completely* describes frame {A}.

Similarly, frame {B} is completely described by its basis vectors $\hat{\boldsymbol{x}}_B$ and $\hat{\boldsymbol{y}}_B$ and these can be expressed in terms of the basis vectors of frame {A}

$$\hat{\boldsymbol{x}}_B = \hat{\boldsymbol{x}}_A \cos\theta + \hat{\boldsymbol{y}}_A \sin\theta$$
$$\hat{\boldsymbol{y}}_B = -\hat{\boldsymbol{x}}_A \sin\theta + \hat{\boldsymbol{y}}_A \cos\theta$$

or in matrix form as

$$(\hat{\boldsymbol{x}}_B \quad \hat{\boldsymbol{y}}_B) = (\hat{\boldsymbol{x}}_A \quad \hat{\boldsymbol{y}}_A) \underbrace{\begin{pmatrix} \cos\theta & -\sin\theta \\ \sin\theta & \cos\theta \end{pmatrix}}_{^A\mathbf{R}_B(\theta)} \tag{2.8}$$

where

$$^A\mathbf{R}_B(\theta) = \begin{pmatrix} \cos\theta & -\sin\theta \\ \sin\theta & \cos\theta \end{pmatrix}$$

is a special type of matrix called a *rotation matrix* that transforms frame {A}, described by $(\hat{\boldsymbol{x}}_A \quad \hat{\boldsymbol{y}}_A)$, into frame {B} described by $(\hat{\boldsymbol{x}}_B \quad \hat{\boldsymbol{y}}_B)$. Rotation matrices have some special properties:

- The columns are the basis vectors that define the axes of the rotated 2D coordinate frame, and therefore have unit length and are orthogonal.
- It is an orthogonal (also called orthonormal) matrix ▶ and therefore its inverse is the same as its transpose, that is, $\mathbf{R}^{-1} = \mathbf{R}^\top$.
- The matrix-vector product $\mathbf{R}\boldsymbol{v}$ preserves the length and relative orientation of vectors $\boldsymbol{v}$ and therefore its determinant is $+1$.
- It is a member of the Special Orthogonal (SO) group of dimension 2 which we write as $\mathbf{R} \in \mathbf{SO}(2) \subset \mathbb{R}^{2\times2}$. Being a group under the operation of matrix multiplication means that the product of any two matrices belongs to the group, as does its inverse.

▶ App. B is a refresher on vectors, matrices and linear algebra.

◩ Fig. 2.5 shows a point and two coordinate frames, {A} and {B}. The point can be described by the coordinate vector $\left(^A p_x, {}^A p_y \right)^\top$ with respect to frame {A} or the

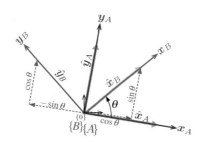

◩ **Fig. 2.10** Rotated coordinate frames in 2D. Basis vectors are shown as thick arrows

coordinate vector $\left(\,^{B}p_x,\ ^{B}p_y\right)^{\top}$ with respect to frame {B}. From (2.7), we can write a coordinate vector as a linear combination of the basis vectors of the reference frame which in matrix form is

$$
^{A}\boldsymbol{p} = \begin{pmatrix} \hat{\boldsymbol{x}}_A & \hat{\boldsymbol{y}}_A \end{pmatrix} \begin{pmatrix} ^{A}p_x \\ ^{A}p_y \end{pmatrix}, \tag{2.9}
$$

$$
^{B}\boldsymbol{p} = \begin{pmatrix} \hat{\boldsymbol{x}}_B & \hat{\boldsymbol{y}}_B \end{pmatrix} \begin{pmatrix} ^{B}p_x \\ ^{B}p_y \end{pmatrix}. \tag{2.10}
$$

Substituting (2.8) into (2.10) yields

$$
^{A}\boldsymbol{p} = \begin{pmatrix} \hat{\boldsymbol{x}}_A & \hat{\boldsymbol{y}}_A \end{pmatrix} {}^{A}\mathbf{R}_B(\theta) \begin{pmatrix} ^{B}p_x \\ ^{B}p_y \end{pmatrix} \tag{2.11}
$$

and we have changed the left-hand side's reference frame, since it is now expressed in terms of the basis vectors of {A} – we write $^{A}\boldsymbol{p}$ instead of $^{B}\boldsymbol{p}$. Equating the two definitions of $^{A}\boldsymbol{p}$ in (2.9) and (2.11), we can write

$$
\begin{pmatrix} \hat{\boldsymbol{x}}_A & \hat{\boldsymbol{y}}_A \end{pmatrix} \underline{\begin{pmatrix} ^{A}p_x \\ ^{A}p_y \end{pmatrix}} = \begin{pmatrix} \hat{\boldsymbol{x}}_A & \hat{\boldsymbol{y}}_A \end{pmatrix} \underline{{}^{A}\mathbf{R}_B(\theta) \begin{pmatrix} ^{B}p_x \\ ^{B}p_y \end{pmatrix}} \tag{2.12}
$$

and then, equating the underlined coefficients, leads to

$$
\begin{pmatrix} ^{A}p_x \\ ^{A}p_y \end{pmatrix} = {}^{A}\mathbf{R}_B(\theta) \begin{pmatrix} ^{B}p_x \\ ^{B}p_y \end{pmatrix} \tag{2.13}
$$

which shows that the rotation matrix $^{A}\mathbf{R}_B(\theta)$ transforms a coordinate vector from frame {B} to {A}.

A rotation matrix has all the characteristics of relative pose $\boldsymbol{\xi}$ described by (2.3) to (2.6).

ξ as an SO(2) Matrix

For the case of pure rotation in 2D, $\boldsymbol{\xi}$ can be implemented by a rotation matrix $\mathbf{R} \in \mathbf{SO}(2)$. Its implementation is:

composition	$\boldsymbol{\xi}_1 \oplus \boldsymbol{\xi}_2$	$\mapsto \mathbf{R}_1\mathbf{R}_2$, matrix multiplication
inverse	$\ominus\boldsymbol{\xi}$	$\mapsto \mathbf{R}^{-1} = \mathbf{R}^{\top}$, matrix transpose
identity	$\varnothing$	$\mapsto \mathbf{R}(0) = \mathbf{1}_{2\times2}$, identity matrix
vector-transform	$\boldsymbol{\xi} \cdot \boldsymbol{v}$	$\mapsto \mathbf{R}\boldsymbol{v}$, matrix-vector product

Composition is commutative, that is, $\mathbf{R}_1\mathbf{R}_2 = \mathbf{R}_2\mathbf{R}_1$, and $\mathbf{R}(-\theta) = \mathbf{R}^{\top}(\theta)$.

To make this tangible, we will create an **SO(2)** rotation matrix using MATLAB

```
>> R = rotm2d(0.3)
R =
    0.9553   -0.2955
    0.2955    0.9553
```

where the angle is specified in radians. The orientation represented by a rotation matrix can be visualized as a coordinate frame

```
>> plottform2d(R);
```

We can observe some of the properties just mentioned, for example, the determinant is equal to one

```
>> det(R)
ans =
    1.0000
```

and that the product of two rotation matrices is also a rotation matrix

```
>> det(R*R)
ans =
    1.0000
```

The functions also support symbolic mathematics, ▶ for example

You will need to have the Symbolic Math Toolbox® installed.

```
>> syms theta real
>> R = rotm2d(theta)
R =
[cos(theta), -sin(theta)]
[sin(theta),  cos(theta)]
>> simplify(R * R)
ans =
[cos(2*theta), -sin(2*theta)]
[sin(2*theta),  cos(2*theta)]
>> det(R)
ans =
cos(theta)^2 + sin(theta)^2
>> simplify(ans)
ans =
1
```

2.2.1.2 Matrix Exponential for Rotation

There is a fascinating, and very useful, connection between a rotation matrix and the exponential of a skew-symmetric matrix. We can easily demonstrate this by considering, again, a pure rotation of 0.3 radians expressed as a rotation matrix

```
>> R = rotm2d(0.3);
```

Excurse 2.6: 2D Skew-Symmetric Matrix

In 2 dimensions, the skew- or anti-symmetric matrix is

$$[\omega]_\times = \begin{pmatrix} 0 & -\omega \\ \omega & 0 \end{pmatrix} \in \mathbf{so}(2) \tag{2.14}$$

which has a distinct structure with a zero diagonal and only one unique value $\omega \in \mathbb{R}$, and $[\omega]_\times^\top = -[\omega]_\times$. The vector space of 2D skew-symmetric matrices is denoted $\mathbf{so}(2)$ and is the Lie algebra of $\mathbf{SO}(2)$. The $[\cdot]_\times$ operator is implemented by

```
>> X = vec2skew(2)
X =
     0    -2
     2     0
```

and the inverse operator $\vee_\times(\cdot)$ by

```
>> skew2vec(X)
ans =
     2
```

logm is different to the function log which computes the logarithm of each element of the matrix. A logarithm can be computed using a power series with a matrix argument, rather than a scalar. The logarithm of a matrix is not unique and logm computes the principal logarithm.

We can take the matrix logarithm using the function logm ◄

```
>> L = logm(R)
L =
         0   -0.3000
    0.3000         0
```

and the result is a simple matrix with just two nonzero elements – they have a magnitude of 0.3 which is the rotation angle. This matrix is an example of a 2 × 2 skew-symmetric matrix. It has only one unique element

```
>> S = skew2vec(L)
S =
    0.3000
```

which is the rotation angle. This is our first encounter with Lie (pronounced lee) group theory, and we will continue to explore this topic in the rest of this chapter.

Exponentiating the logarithm of the rotation matrix L, using the matrix exponential function expm, ◄ yields the original rotation matrix

expm is different to the function exp which computes the exponential of each element of the matrix. The matrix exponential can be computed using a power series with a matrix argument, rather than a scalar: $\exp m(\mathbf{A}) = \mathbf{1} + \mathbf{A} + \mathbf{A}^2/2! + \mathbf{A}^3/3! + \cdots$

```
>> expm(L)
ans =
    0.9553   -0.2955
    0.2955    0.9553
```

The exponential of a skew-symmetric matrix is always a rotation matrix with all the special properties outlined earlier. The skew-symmetric matrix can be reconstructed from the single element of s, so we can also write

```
>> expm(vec2skew(S))
ans =
    0.9553   -0.2955
    0.2955    0.9553
```

In general, we can write

$$\mathbf{R} = e^{[\theta]_\times} \in \mathbf{SO}(2)$$

where θ is the rotation angle, and $[\cdot]_\times : \mathbb{R} \mapsto \mathbf{so}(2) \subset \mathbb{R}^{2\times2}$ is a mapping from a scalar to a skew-symmetric matrix.

2.2.2 Pose in Two Dimensions

To describe the relative pose of the frames shown in ◘ Fig. 2.9, we need to account for the translation between the origins of the frames as well as the rotation. The frames are shown in more detail in ◘ Fig. 2.11.

◘ **Fig. 2.11** Rotated and translated coordinate frames where the axes of frame {A′} are parallel to the axes of frame {A}

2.2.2.1 2D Homogeneous Transformation Matrix

The first step is to transform the coordinate vector ${}^B\boldsymbol{p} = ({}^Bx, \, {}^By)$, with respect to frame $\{B\}$, to ${}^{A'}\boldsymbol{p} = ({}^{A'}x, \, {}^{A'}y)$ with respect to frame $\{A'\}$ using the rotation matrix ${}^{A'}\mathbf{R}_B(\theta)$ which is a function of the orientation θ. Since frames $\{A'\}$ and $\{A\}$ are parallel, the coordinate vector ${}^A\boldsymbol{p}$ is obtained by adding ${}^A\boldsymbol{t}_B = (t_x, t_y)^\top$ to ${}^{A'}\boldsymbol{p}$

$$\begin{pmatrix} {}^Ax \\ {}^Ay \end{pmatrix} = \begin{pmatrix} {}^{A'}x \\ {}^{A'}y \end{pmatrix} + \begin{pmatrix} t_x \\ t_y \end{pmatrix} \tag{2.15}$$

$$= \begin{pmatrix} \cos\theta & -\sin\theta \\ \sin\theta & \cos\theta \end{pmatrix} \begin{pmatrix} {}^Bx \\ {}^By \end{pmatrix} + \begin{pmatrix} t_x \\ t_y \end{pmatrix} \tag{2.16}$$

$$= \begin{pmatrix} \cos\theta & -\sin\theta & t_x \\ \sin\theta & \cos\theta & t_y \end{pmatrix} \begin{pmatrix} {}^Bx \\ {}^By \\ 1 \end{pmatrix} \tag{2.17}$$

or more compactly as

$$\begin{pmatrix} {}^Ax \\ {}^Ay \\ 1 \end{pmatrix} = \begin{pmatrix} {}^A\mathbf{R}_B(\theta) & {}^A\boldsymbol{t}_B \\ \boldsymbol{0}_{1\times2} & 1 \end{pmatrix} \begin{pmatrix} {}^Bx \\ {}^By \\ 1 \end{pmatrix} \tag{2.18}$$

where ${}^A\boldsymbol{t}_B$ is the translation of the origin of frame $\{B\}$ with respect to frame $\{A\}$, and ${}^A\mathbf{R}_B(\theta)$ is the orientation of frame $\{B\}$ with respect to frame $\{A\}$.

> If we consider the homogeneous transformation as a relative pose or rigid-body motion, this corresponds to the coordinate frame being first translated by ${}^A\boldsymbol{t}_B$ with respect to frame $\{A\}$, and *then* rotated by ${}^A\mathbf{R}_B(\theta)$.

The coordinate vectors for point P are now expressed in *homogeneous form* and we denote that with a tilde. Now we can write

$${}^A\tilde{\boldsymbol{p}} = \begin{pmatrix} {}^A\mathbf{R}_B(\theta) & {}^A\boldsymbol{t}_B \\ \boldsymbol{0}_{1\times2} & 1 \end{pmatrix} {}^B\tilde{\boldsymbol{p}}$$

$$= {}^A\mathbf{T}_B \, {}^B\tilde{\boldsymbol{p}}$$

Excurse 2.7: Homogeneous Vectors

A vector $\boldsymbol{p} = (x, y)$ is written in homogeneous form as $\tilde{\boldsymbol{p}} = (x_1, x_2, x_3) \in \mathbb{P}^2$ where $\mathbb{P}^2$ is the 2-dimensional projective space, and the tilde indicates the vector is homogeneous.

Homogeneous vectors have the important property that $\tilde{\boldsymbol{p}}$ is equivalent to $\lambda\tilde{\boldsymbol{p}}$ for all $\lambda \neq 0$ which we write as $\tilde{\boldsymbol{p}} \simeq \lambda\tilde{\boldsymbol{p}}$. That is, $\tilde{\boldsymbol{p}}$ represents the same point in the plane irrespective of the overall scaling factor. Homogeneous representation is also used in computer vision which we discuss in Part IV. Additional details are provided in ▶ App. C.2.

To convert a point to homogeneous form we typically append an element equal to one, for the 2D case this is $\tilde{\boldsymbol{p}} = (x, y, 1)$. The dimension of the vector has been increased by one, and a point on a plane is now represented by a 3-vector. The Euclidean or nonhomogeneous coordinates are related by $x = x_1/x_3$, $y = x_2/x_3$ and $x_3 \neq 0$.

and $^A\mathbf{T}_B$ is referred to as a *homogeneous transformation* – it transforms homogeneous vectors. The matrix has a very specific structure and belongs to the Special Euclidean (SE) group of dimension 2 which we write as $\mathbf{T} \in \mathbf{SE}(2) \subset \mathbb{R}^{3\times3}$.

The matrix $^A\mathbf{T}_B$ represents translation and orientation or relative pose and has all the characteristics of relative pose $\boldsymbol{\xi}$ described by (2.3) to (2.6).

$\boldsymbol{\xi}$ as an SE(2) Matrix

For the case of rotation and translation in 2D, $\boldsymbol{\xi}$ can be implemented by a homogeneous transformation matrix $\mathbf{T} \in \mathbf{SE}(2)$ which is sometimes written as an ordered pair $(\mathbf{R}, \boldsymbol{t}) \in \mathbf{SO}(2) \times \mathbb{R}^2$. The implementation is:

$$\text{composition} \quad \boldsymbol{\xi}_1 \oplus \boldsymbol{\xi}_2 \quad \mapsto \mathbf{T}_1\mathbf{T}_2 = \begin{pmatrix} \mathbf{R}_1\mathbf{R}_2 & \boldsymbol{t}_1 + \mathbf{R}_1\boldsymbol{t}_2 \\ \mathbf{0}_{1\times2} & 1 \end{pmatrix}, \text{matrix}$$

multiplication

$$\text{inverse} \quad \ominus\boldsymbol{\xi} \quad \mapsto \mathbf{T}^{-1} = \begin{pmatrix} \mathbf{R}^\top & -\mathbf{R}^\top\boldsymbol{t} \\ \mathbf{0}_{1\times2} & 1 \end{pmatrix}, \text{matrix inverse}$$

$$\text{identity} \quad \varnothing \quad \mapsto \mathbf{1}_{3\times3}, \text{identity matrix}$$

$$\text{vector-transform} \quad \boldsymbol{\xi} \cdot \boldsymbol{v} \quad \mapsto \epsilon(\mathbf{T}\tilde{\boldsymbol{v}}), \text{matrix-vector product}$$

where $\tilde{\cdot} : \mathbb{R}^2 \mapsto \mathbb{P}^2$ and $\epsilon(\cdot) : \mathbb{P}^2 \mapsto \mathbb{R}^2$. Composition is not commutative, that is, $\mathbf{T}_1\mathbf{T}_2 \neq \mathbf{T}_2\mathbf{T}_1$.

To recap, the rotation matrix for a rotation of 0.3 rad is

```
>> rotm2d(0.3)
ans =
    0.9553   -0.2955
    0.2955    0.9553
```

which can be composed with other rotation matrices, or used to transform coordinate vectors. The homogeneous transformation matrix for a rotation of 0.3 rad is

```
>> tformr2d(0.3)
ans =
    0.9553   -0.2955         0
    0.2955    0.9553         0
         0         0    1.0000
```

and we see the rotation matrix in the top-left corner, and zeros and a one on the bottom row. The top two elements of the right-hand column are zero indicating zero translation. This matrix represents relative pose, it can be composed with other relative poses, or used to transform *homogeneous* coordinate vectors.

Next, we will compose two relative poses: a translation of $(1, 2)$ followed by a rotation of 30°

```
>> TA = trvec2tform([1 2])*tformr2d(deg2rad(30))
TA =
    0.8660   -0.5000    1.0000
    0.5000    0.8660    2.0000
         0         0    1.0000
```

For describing rotations, there are functions to create a rotation matrix (`rotm2d`) or a homogeneous transformation with zero translation (`tformr2d`).

The function `trvec2tform` creates a 2D relative pose with a finite translation but zero rotation, while `tformr2d` creates a relative pose with a finite rotation but zero translation. ◄ We can plot a coordinate frame representing this pose, relative to the reference frame, by

```
>> axis([0 5 0 5]);  % new plot with both axes from 0 to 5
>> plottform2d(TA,frame="A",color="b");
```

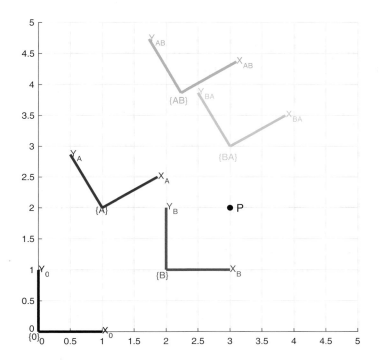

■ **Fig. 2.12**   Coordinate frames drawn using the function `plottform2d`

The options to `plottform` specify that the label for the frame is {A} and it is colored blue and this is shown in ■ Fig. 2.12. For completeness, we will add the reference frame to the plot

```
>> T0 = trvec2tform([0 0]);
>> plottform2d(T0, frame="0",color="k");   % reference frame
```

We create another relative pose which is a displacement of $(2, 1)$ and zero rotation

```
>> TB = trvec2tform([2 1])
TB =
      1      0      2
      0      1      1
      0      0      1
```

which we plot in red

```
>> plottform2d(TB,frame="B",color="r");
```

Now we can compose the two relative poses

```
>> TAB = TA*TB
TAB =
    0.8660   -0.5000    2.2321
    0.5000    0.8660    3.8660
         0         0    1.0000
```

and plot the result as a green coordinate frame

```
>> plottform2d(TAB,frame="AB",color="g");
```

We see that the displacement of $(2, 1)$ has been applied with respect to frame {A}. It is important to note that our final displacement is not $(3, 3)$ because the displacement is with respect to the rotated coordinate frame. The noncommutativity of composition is clearly demonstrated by reversing the order of multiplication

```
>> TBA = TB*TA;
>> plottform2d(TBA,frame="BA",color="c");
```

and we see that frame {BA} is different to frame {AB}.

2

Now we define a point (3,2) relative to the world frame

```
>> P = [3;2];  % column vector
```

which is a column vector, and add it to the plot

```
>> plotpoint(P',"ko",label="P");
```

To determine the coordinate of the point with respect to {A} we write

$$^{0}\boldsymbol{p} = {}^{0}\boldsymbol{\xi}_{A} \cdot {}^{A}\boldsymbol{p}$$

and then rearrange as

$$^{A}\boldsymbol{p} = (\ominus\,{}^{0}\boldsymbol{\xi}_{A}) \cdot {}^{0}\boldsymbol{p} \ .$$

Substituting numerical values, we obtain

```
>> inv(TA)*[P;1]
ans =
    1.7321
   -1.0000
    1.0000
```

where we first converted the Euclidean point coordinates to homogeneous form by appending a one. The result is also in homogeneous form and has a negative y-coordinate in frame {A}. We can convert this to nonhomogeneous form by

```
>> h2e(ans')
ans =
    1.7321   -1.0000
```

More concisely, we can write

```
>> homtrans(inv(TA),P')
ans =
    1.7321   -1.0000
```

which handles conversion of the coordinate vectors to and from homogeneous form. The functions `plotpoint`, `h2e`, `e2h` and `homtrans` accept and return points as row vectors – this is a common convention across MATLAB toolboxes.

2.2.2.2 Rotating a Coordinate Frame

The pose of a coordinate frame is fully described by an **SE**(2) matrix and here we will explore rotation of coordinate frames. First, we create and plot a reference coordinate frame {0} and a target frame {X}

```
>> axis([-5 4 -1 5]);
>> T0 = trvec2tform([0 0]);
>> plottform2d(T0,frame="0",color="k");  % reference frame
>> TX = trvec2tform([2 3]);
>> plottform2d(TX,frame="X",color="b");  % frame {X}
```

Next, we create an **SE**(2) matrix representing a rotation of 2 radians (nearly 120°)

```
>> TR = tformr2d(2);
```

and plot the effect of the two possible orders of composition

```
>> plottform2d(TR*TX,framelabel="RX",color="g");
>> plottform2d(TX*TR,framelabel="XR",color="g");
```

The results are shown as green coordinate frames in ◨ Fig. 2.13. We see that the frame {RX} has been rotated about the origin, while frame {XR} has been rotated about the origin of {X}.

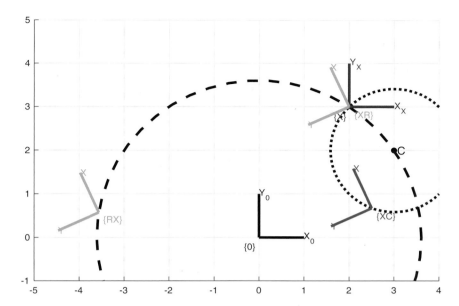

◻ Fig. 2.13 The frame {X} is rotated by 2 radians about {0} to give frame {RX}, about {X} to give {XR}, and about point C to give frame {XC}

What if we wished to rotate a coordinate frame about some arbitrary point C? The first step is to define the coordinate of the point and display it

```
>> C = [3 2];
>> plotpoint(C,"ko",label="C");
```

and then compute a transform to rotate about point C

```
>> TC = trvec2tform(C)*TR*trvec2tform(-C)
TC =
    -0.4161    -0.9093     6.0670
     0.9093    -0.4161     0.1044
          0          0     1.0000
```

We apply it to frame {X}

```
>> plottform2d(TC*TX,framelabel="XC",color="r");
```

and see that the red frame has been rotated about point C.

To understand how this works we can unpack what happens when we apply TC to TX. Since $\mathbf{T}_C$ premultiplies $\mathbf{T}_X$, the first transform that is applied to {X} is the right-most one, `trvec2tform(-C)`, which performs an origin shift that places C at the origin of the reference frame. Then we apply $\mathbf{T}_R$, a pure rotation which rotates the shifted version of {X} about the origin, which is where C now is – therefore {X} is rotated about C. Finally, we apply `trvec2tform(C)` which is the inverse origin shift, that places C back at its original position and {X} to its final pose. This transformation is an example of a conjugation but its construction was somewhat elaborate. Premultiplication of matrices is discussed in ▶ Sect. 2.4.1. A more intuitive way to achieve this same result is by using twists which we introduce in ▶ Sect. 2.2.2.4.

2.2.2.3 Matrix Exponential for Pose

We can take the logarithm of the **SE**(2) matrix TC from the previous example

```
>> L = logm(TC)
L =
          0    -2.0000     4.0000
     2.0000          0    -6.0000
          0          0          0
```

2

Excurse 2.8: 2D Augmented Skew-Symmetric Matrix

In 2 dimensions, the augmented skew-symmetric matrix corresponding to the vector $S = (\omega, v_x, v_y)$ is

$$[S] = \left(\begin{array}{cc|c} 0 & -\omega & v_x \\ \omega & 0 & v_y \\ \hline 0 & 0 & 0 \end{array} \right) \in \mathbf{se}(2) \tag{2.19}$$

which has a distinct structure with a zero diagonal and bottom row, and a skew-symmetric matrix in the top-left corner. The vector space of 2D augmented skew-symmetric matrices is denoted $\mathbf{se}(2)$ and is the Lie algebra of $\mathbf{SE}(2)$. The $[\cdot]$ operator is implemented by

```
>> X = vec2skewa([1 2 3])
X =
        0      -1       2
        1       0       3
        0       0       0
```

and the inverse operator $\vee(\cdot)$ by

```
>> skewa2vec(X)
ans =
        1       2       3
```

---

and the result is an augmented skew-symmetric matrix: the upper-left corner is a 2×2 skew-symmetric matrix: the upper right column is a 2-vector, and the bottom row is zero. The three unique elements can be unpacked

```
>> S = skewa2vec(L)
S =
    2.0000    4.0000    -6.0000
```

and the first element is the rotation angle.

Exponentiating `L` yields the original $\mathbf{SE}(2)$ matrix, and `L` can be reconstructed from just the three elements of `s`

```
>> expm(vec2skewa(S))
ans =
   -0.4161    -0.9093     6.0670
    0.9093    -0.4161     0.1044
         0          0     1.0000
```

In general, we can write

$$\mathbf{T} = e^{[S]} \in \mathbf{SE}(2)$$

where $S \in \mathbb{R}^3$ and $[\cdot] : \mathbb{R}^3 \mapsto \mathbf{se}(2) \subset \mathbb{R}^{3 \times 3}$. The vector S is a *twist vector* which we will discuss next.

2.2.2.4 2D Twists

In ▶ Sect. 2.2.2.2, we transformed a coordinate frame by rotating it about a specified point. The corollary is that, given any two frames, we can find a rotational center and rotation angle that will *rotate* the first frame into the second. This is the key concept behind what is called a twist.

A rotational, or revolute, twist about the point specified by the coordinate `c` is created by

```
>> S = Twist2d.UnitRevolute(C)
S =
( 1; 2  -3 )
```

where the class method `UnitRevolute` constructs a rotational twist. The result is a `Twist2d` object that encapsulates a 2D twist vector $(\omega, \boldsymbol{v}) \in \mathbb{R}^3$ comprising a scalar ω and a *moment* $\boldsymbol{v} \in \mathbb{R}^2$. This particular twist is a *unit twist* that describes a rotation of 1 rad about the point `c`.

For the example in ▶ Sect. 2.2.2.2, we require a rotation of 2 rad about the point C, so we scale the unit twist and exponentiate it

```
>> expm(vec2skewa(2*S.compact))
ans =
    -0.4161    -0.9093    6.0670
     0.9093    -0.4161    0.1044
          0          0    1.0000
```

where `S.compact` is the twist vector as a MATLAB row vector. The `Twist2d` object has a shorthand method for this

```
>> S.tform(2)
ans =
    -0.4161    -0.9093    6.0670
     0.9093    -0.4161    0.1044
          0          0    1.0000
```

This has the same value as the transformation computed in the previous section, ▶ but more concisely specified in terms of the center and magnitude of rotation. The center of rotation, called the pole, is encoded in the twist

```
>> S.pole
ans =
     3     2
```

For the case of pure translational motion, the rotational center is at infinity. A translational, or prismatic, unit twist is therefore specified only by the direction of motion. For example, motion in the y-direction is created by

```
>> S = Twist2d.UnitPrismatic([0 1])
S =
( 0; 0  1 )
```

which represents a displacement of 1 in the y-direction. To create an **SE**(2) transformation for a specific translation, we scale and exponentiate it

```
>> S.tform(2)
ans =
     1     0     0
     0     1     2
     0     0     1
```

and the result has a zero rotation and a translation of 2 in the y-direction.

For an arbitrary 2D homogeneous transformation

```
>> T = trvec2tform([3 4])*tformr2d(0.5)
T =
    0.8776    -0.4794    3.0000
    0.4794     0.8776    4.0000
         0          0    1.0000
```

the twist is

```
>> S = Twist2d(T)
S =
( 0.5; 3.9372  3.1663 )
```

which describes a rotation of

```
>> S.w
ans =
    0.5000
```

The related method `exp` returns the same result but as an `se2` object, this is discussed further in ▶ Sect. 2.5.

about the point

```
>> S.pole
ans =
    -3.1663    3.9372
```

and exponentiating this twist

```
>> S.tform()
ans =
    0.8776    -0.4794    3.0000
    0.4794     0.8776    4.0000
         0          0    1.0000
```

yields the original **SE**(2) transform – completely described by the three elements of the twist vector.

ξ as a 2D Twist

For the case of rotation and translation in 2D, ξ can be implemented by a twist $S \in \mathbb{R}^3$. The implementation is:

composition	$\xi_1 \oplus \xi_2$	$\mapsto \log\left(e^{[S_1]}e^{[S_2]}\right)$, product of exponential
inverse	$\ominus \xi$	$\mapsto -S$, negation
identity	$\varnothing$	$\mapsto \mathbf{0}_{1\times 3}$, zero vector
vector-transform	$\xi \cdot v$	$\mapsto \epsilon(e^{[S]}\tilde{v})$, matrix-vector product using homogeneous vectors

where $\tilde{\cdot} : \mathbb{R}^2 \mapsto \mathbb{P}^2$ and $\epsilon(\cdot) : \mathbb{P}^2 \mapsto \mathbb{R}^2$. Composition is not commutative, that is, $e^{[S_1]}e^{[S_2]} \neq e^{[S_2]}e^{[S_1]}$. Note that log and exp have efficient closed form, rather than transcendental, solutions which makes composition relatively inexpensive.

2.3 Working in Three Dimensions (3D)

These basis vectors are sometimes denoted e_1, e_2, e_3.

In all these identities, the symbols from left to right (across the equals sign) are a cyclic rotation of the sequence $\ldots z, x, y, z, x, y, \ldots$.

The 3-dimensional case is an extension of the 2-dimensional case discussed in ► Sect. 2.2. We add an extra coordinate axis, denoted by z, that is orthogonal to both the x- and y-axes. The direction of the z-axis obeys the *right-hand rule* and forms a *right-handed coordinate frame*. The basis vectors ◄ are unit vectors parallel to the axes, denoted by $\hat{x}$, $\hat{y}$ and $\hat{z}$ and related such that ◄

$$\hat{z} = \hat{x} \times \hat{y}, \quad \hat{x} = \hat{y} \times \hat{z}, \quad \hat{y} = \hat{z} \times \hat{x} .$$

(2.20)

Excurse 2.9: Right-Hand Rule

A right-handed coordinate frame is defined by the first three fingers of your right hand which indicate the relative directions of the x-, y- and z-axes respectively.

▶ sn.pub/iIjZHE

◼ **Fig. 2.14** Two 3D coordinate frames {A} and {B} defined with respect to the reference frame. {B} is rotated and translated with respect to {A}. There is an ambiguity when viewing 3D coordinate frames on a page. Your brain may flip between two interpretations, with the origin either toward or away from you – knowing the right-hand rule resolves the ambiguity

A point P is represented by its x-, y- and z-coordinates (x, y, z) or as a coordinate vector from the origin to the point

$$\boldsymbol{p} = x\hat{\boldsymbol{x}} + y\hat{\boldsymbol{y}} + z\hat{\boldsymbol{z}}$$

which is a linear combination of the basis vectors.

◼ Fig. 2.14 shows two 3-dimensional coordinate frames and we wish to describe the red frame {B} with respect to the blue frame {A}. We can see clearly that the origin of {B} has been displaced by the 3D vector $^A\boldsymbol{t}_B$ and then rotated in some complex fashion. We will again consider the problem in two parts: pure rotation and then rotation plus translation. Rotation is surprisingly complex for the 3-dimensional case and the next section will explore some of the many ways to describe it.

2.3.1 Orientation in Three Dimensions

The two frames shown in ◼ Fig. 2.14 clearly have different orientations, and we want some way to describe the orientation of one with respect to the other. We can imagine picking up frame {A} in our hand and rotating it until it looked just like frame {B}.

We start by considering rotation about a single coordinate frame axis. ◼ Fig. 2.15 shows a right-handed coordinate frame, and that same frame after it has been rotated by various angles about different axes.

Rotation in 3D has some subtleties which are illustrated in ◼ Fig. 2.16, where a sequence of two rotations are applied in different orders. We see that the final orientation depends on the order in which the rotations are applied – this is a confounding characteristic of the 3-dimensional world. It is important to remember:

❗ In three dimensions, rotation is not commutative – the result depends on the order in which rotations are applied.

Mathematicians have developed many ways to represent rotation and we will introduce those that are most commonly encountered in robotics: rotation matrices, Euler and Cardan angles, rotation axis and angle, exponential coordinates, and unit quaternions. All can be represented as MATLAB matrices or vectors, or as MATLAB classes. The toolboxes provides many ways to create and convert between these representations.

2

Excurse 2.10: Rotation About a Vector
Wrap your right hand around the vector with your thumb
(your x-finger) in the direction of the arrow. The curl of your
fingers indicates the direction of increasing angle.

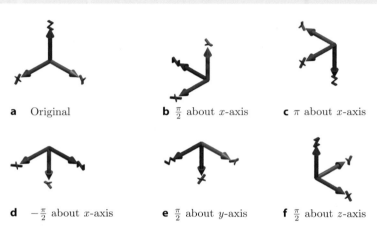

a Original   **b** $\frac{\pi}{2}$ about x-axis   **c** π about x-axis

d $-\frac{\pi}{2}$ about x-axis   **e** $\frac{\pi}{2}$ about y-axis   **f** $\frac{\pi}{2}$ about z-axis

☐ **Fig. 2.15** Rotation of a 3D coordinate frame. **a** The original coordinate frame, **b–f** frame **a** after
various rotations, in units of radians, as indicated

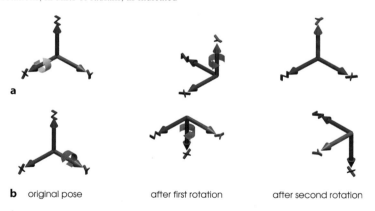

a

b original pose   after first rotation   after second rotation

☐ **Fig. 2.16** Example showing the noncommutativity of rotation. In the top row the coordinate frame
is rotated by $\frac{\pi}{2}$ about the x-axis and then $\frac{\pi}{2}$ about the y-axis. In the bottom row the order of rotations is
reversed. The results are clearly different

2.3.1.1 **3D Rotation Matrix**

Just as for the 2-dimensional case, we can represent the orientation of a coordinate
frame by its basis vectors expressed in terms of the reference coordinate frame.
Each basis vector has three elements and they form the columns of a 3×3 *orthogonal matrix* $^A\mathbf{R}_B$

$$\begin{pmatrix} ^Ax \\ ^Ay \\ ^Az \end{pmatrix} = {}^A\mathbf{R}_B \begin{pmatrix} ^Bx \\ ^By \\ ^Bz \end{pmatrix}$$

(2.21)

▶ sn.pub/j7PTLU

which transforms a coordinate vector defined with respect to frame {B} to a coordinate vector with respect to frame {A}. A 3-dimensional rotation matrix **R** has the same special properties as its 2D counterpart:

- The columns are the basis vectors that define the axes of the rotated 3D coordinate frame, and therefore have unit length and are orthogonal.
- It is orthogonal (also called orthonormal) ▸ and therefore its inverse is the same as its transpose, that is, $\mathbf{R}^{-1} = \mathbf{R}^{\top}$.
- The matrix-vector product $\mathbf{R}v$ preserves the length and relative orientation of vectors v and therefore its determinant is $+1$.
- It is a member of the Special Orthogonal (SO) group of dimension 3 which we write as $\mathbf{R} \in \mathbf{SO}(3) \subset \mathbb{R}^{3\times3}$. Being a group under the operation of matrix multiplication means that the product of any two matrices belongs to the group, as does its inverse.

See ▸ App. B for a refresher on vectors, matrices and linear algebra.

The rotation matrices that correspond to a coordinate frame rotation of θ about the x-, y- and z-axes are

$$\mathbf{R}_x(\theta) = \begin{pmatrix} 1 & 0 & 0 \\ 0 & \cos\theta & -\sin\theta \\ 0 & \sin\theta & \cos\theta \end{pmatrix}$$

$$\mathbf{R}_y(\theta) = \begin{pmatrix} \cos\theta & 0 & \sin\theta \\ 0 & 1 & 0 \\ -\sin\theta & 0 & \cos\theta \end{pmatrix}$$

$$\mathbf{R}_z(\theta) = \begin{pmatrix} \cos\theta & -\sin\theta & 0 \\ \sin\theta & \cos\theta & 0 \\ 0 & 0 & 1 \end{pmatrix}$$

There are functions to compute these elementary rotation matrices, for example $\mathbf{R}_x(\frac{\pi}{2})$ is

```
>> R = rotmx(pi/2)
R =
    1    0    0
    0    0   -1
    0    1    0
```

and the functions `rotmy` and `rotmz` compute $\mathbf{R}_y(\theta)$ and $\mathbf{R}_z(\theta)$ respectively.

The orientation represented by a rotation matrix can be visualized as a coordinate frame – rotated with respect to the reference coordinate frame

```
>> plottform(R);
```

which is shown in ◘ Fig. 2.17a. The interpretation as a motion or relative pose, can be made explicit by visualizing the rotation matrix as an animation ▸

```
>> animtform(R)
```

which shows the reference frame moving to the specified relative pose. If you have a pair of anaglyph stereo glasses you can see this in more vivid 3D by either of

```
>> plottform(R,anaglyph=true)
>> animtform(R,anaglyph=true);
```

2

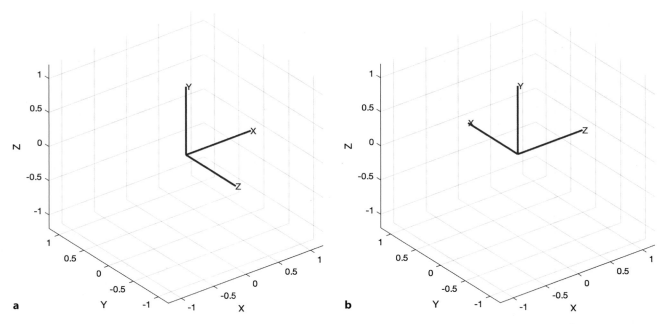

□ **Fig. 2.17**   3D coordinate frames displayed using `plottform`. **a** Reference frame rotated by $\frac{\pi}{2}$ about the *x*-axis; **b** frame from **a** rotated by $\frac{\pi}{2}$ about the *y*-axis

Now we will apply another rotation, this time about the *y*-axis resulting from the first rotation

```
>> R = rotmx(pi/2)*rotmy(pi/2)
R =
      0      0      1
      1      0      0
      0      1      0
>> plottform(R)
```

which gives the frame shown in □ Fig. 2.17b and the *x*-axis now points in the direction of the world *y*-axis. This frame is the same as the rightmost frame in □ Fig. 2.16a.

The noncommutativity of rotation is clearly shown by reversing the order of the rotations

```
>> rotmy(pi/2)*rotmx(pi/2)
ans =
      0      1      0
      0      0     -1
     -1      0      0
```

which has a very different value, and this orientation is shown in □ Fig. 2.16b.

ξ as an SO(3) Matrix

For the case of pure rotation in 3D, ξ can be implemented by a rotation matrix $\mathbf{R} \in \mathbf{SO}(3)$. The implementation is:

composition	$\xi_1 \oplus \xi_2$	$\mapsto \mathbf{R}_1\mathbf{R}_2$, matrix multiplication
inverse	$\ominus\xi$	$\mapsto \mathbf{R}^{-1} = \mathbf{R}^\top$, matrix transpose
identity	$\varnothing$	$\mapsto \mathbf{1}_{3\times3}$, identity matrix
vector-transform	$\xi \cdot v$	$\mapsto \mathbf{R}v$, matrix-vector product

Composition is not commutative, that is, $\mathbf{R}_1\mathbf{R}_2 \neq \mathbf{R}_2\mathbf{R}_1$.

Excurse 2.11: Reading a Rotation Matrix

The columns from left to right tell us the directions of the new frame's axes in terms of the current coordinate frame.

In this case, the new frame has its x-axis in the old x-direction $(1, 0, 0)$, its y-axis in the old z-direction $(0, 0, 1)$, and the new z-axis in the old negative y-direction $(0, -1, 0)$. The x-axis was unchanged, since this is the axis around which the rotation occurred. The rows are the converse and represent the current frame axes in terms of the new frame's axes.

2.3.1.2 Three-Angle Representations

» Euler's rotation theorem:

Any two independent orthonormal coordinate frames can be related by a sequence of rotations (not more than three) about coordinate axes, where no two successive rotations may be about the same axis.

– Kuipers (1999)

Euler's rotation theorem means that a rotation between any two coordinate frames in 3D can be represented by a sequence of three rotation angles each associated with a particular coordinate frame axis. The rotations are applied consecutively: the first rotates the world frame {0} around some axis to create a new coordinate frame {1}; then a rotation about some axis of {1} results in frame {2}; and finally a rotation about some axis of {2} results in frame {3}.

There are a total of twelve unique rotation sequences. Six involve repetition, but not successive, of rotations about one particular axis: XYX, XZX, YXY, YZY, ZXZ, or ZYZ. Another six are characterized by rotations about all three axes: XYZ, XZY, YZX, YXZ, ZXY, or ZYX.

⊘ "Euler angles" is an ambiguous term

It is common practice to refer to all possible three-angle representations as Euler angles but this is insufficiently precise since there are twelve rotation sequences. The particular angle sequence needs to be specified, but it is often an implicit convention within a particular technological field.

In mechanical dynamics, the ZYZ sequence is commonly used

$$\mathbf{R}(\phi, \theta, \psi) = \mathbf{R}_z(\phi)\,\mathbf{R}_y(\theta)\,\mathbf{R}_z(\psi) \tag{2.22}$$

and the Euler angles are written as the 3-vector $\boldsymbol{\Gamma} = (\phi, \theta, \psi) \in (S^1)^3$. ▶ To compute the equivalent rotation matrix for $\boldsymbol{\Gamma} = (0.1, -0.2, 0.3)$, we write

```
>> R = rotmz(0.1)*rotmy(-0.2)*rotmz(0.3)
R =
    0.9021   -0.3836   -0.1977
    0.3875    0.9216   -0.0198
    0.1898   -0.0587    0.9801
```

Refer to the nomenclature section. $(S^1)^3$ or $S^1 \times S^1 \times S^1$ denotes a 3-tuple of angles, whereas S^3 denote an angle in 4D space.

2

Excurse 2.12: Leonhard Euler
Euler (1707–1783) was a Swiss mathematician and physicist who dominated eighteenth century mathematics. He was a student of Johann Bernoulli and applied new mathematical techniques such as calculus to many problems in mechanics and optics. He developed the functional notation, $y = f(x)$, and in robotics we use his rotation theorem and his equations of motion in rotational dynamics.

He was prolific and his collected works fill 75 volumes. Almost half of this was produced during the last seventeen years of his life when he was completely blind.

or more conveniently

```
>> R = eul2rotm([0.1 -0.2 0.3],"ZYZ")
R =
    0.9021   -0.3836   -0.1977
    0.3875    0.9216   -0.0198
    0.1898   -0.0587    0.9801
```

The inverse problem is finding the ZYZ-Euler angles that correspond to a given rotation matrix

```
>> gamma = rotm2eul(R,"ZYZ")
gamma =
    0.1000   -0.2000    0.3000
```

If θ is positive, the rotation matrix is

```
>> R = eul2rotm([0.1 0.2 0.3],"ZYZ")
R =
    0.9021   -0.3836    0.1977
    0.3875    0.9216    0.0198
   -0.1898    0.0587    0.9801
```

and the inverse function

```
>> gamma = rotm2eul(R,"ZYZ")
gamma =
   -3.0416   -0.2000   -2.8416
```

returns a set of different ZYZ-Euler angles where the sign of θ has changed, and the values of ϕ and ψ have been offset by $-\pi$. However, the corresponding rotation matrix

```
>> eul2rotm(gamma,"ZYZ")
ans =
    0.9021   -0.3836    0.1977
    0.3875    0.9216    0.0198
   -0.1898    0.0587    0.9801
```

is the same. This means that there are two different sets of ZYZ-Euler angles that generate the same rotation matrix – the mapping from a rotation matrix to ZYZ-Euler angles is not unique. The function *always* returns a negative value for θ.

For the case where $\theta = 0$

```
>> R = eul2rotm([0.1 0 0.3],"ZYZ")
R =
    0.9211   -0.3894        0
    0.3894    0.9211        0
         0         0   1.0000
```

the inverse function returns

```
>> rotm2eul(R,"ZYZ")
ans =
         0         0    0.4000
```

which is again quite different, but these ZYZ-Euler angles will generate the same rotation matrix. The explanation is that if $\theta = 0$ then $\mathbf{R}_y = \mathbf{1}_{3\times 3}$ and (2.22) becomes

$$\mathbf{R} = \mathbf{R}_z(\phi)\mathbf{R}_z(\psi) = \mathbf{R}_z(\phi + \psi)$$

which is a function of the sum $\phi + \psi$. The inverse operation can determine this sum and arbitrarily split it between ϕ and ψ – by convention we choose $\phi = 0$. The case $\theta = 0$ is a *singularity* and will be discussed in more detail in the next section.

The other widely used convention are the Cardan angles: roll, pitch and yaw which we denote as α, β and γ respectively. ▶

ⓘ Roll-pitch-yaw angle ambiguity

Confusingly, there are two different roll-pitch-yaw sequences in common use: ZYX or XYZ, depending on whether the topic is mobile robots or robot manipulators respectively.

When describing the attitude of vehicles such as ships, aircraft and cars, the convention is that the x-axis of the body frame points in the forward direction and its z-axis points either up or down. ▶ We start with the world reference frame and in order:

- rotate about the world z-axis by the yaw angle, γ, so that the x-axis points in the direction of travel, then
- rotate about the y-axis of the frame above by the pitch angle, β, which sets the angle of the vehicle's longitudinal axis relative to the horizontal plane (pitching the nose up or nose down), and then finally
- rotate about the x-axis of the frame above by the roll angle, α, so that the vehicle's body rolls about its longitudinal axis.

which leads to the ZYX angle sequence

$$\mathbf{R}(\alpha, \beta, \gamma) = \mathbf{R}_z(\gamma)\,\mathbf{R}_y(\beta)\,\mathbf{R}_x(\alpha) \tag{2.23}$$

and the roll, pitch and yaw angles are written as the 3-vector $\boldsymbol{\Gamma} = (\gamma, \beta, \alpha) \in (\mathbf{S}^1)^3$. For example

```
>> R = eul2rotm([0.3 0.2 0.1],"ZYX")
R =
    0.9363   -0.2751    0.2184
    0.2896    0.9564   -0.0370
   -0.1987    0.0978    0.9752
```

where the arguments are given in the order yaw, pitch and roll.

Roll-pitch-yaw angles are also known as Tait-Bryan angles, after Peter Tait a Scottish physicist and quaternion supporter, and George Bryan a pioneering Welsh aerodynamicist. They are also known as nautical angles. For aeronautical applications, the angles are called bank, attitude and heading respectively.

By convention z is up for ground vehicles and down for aircraft.

2

Excurse 2.13: Gerolamo Cardano

Cardano (1501–1576) was an Italian Renaissance mathematician, physician, astrologer, and gambler. He was born in Pavia, Italy, the illegitimate child of a mathematically gifted lawyer. He studied medicine at the University of Padua and later was the first to describe typhoid fever. He partly supported himself through gambling and his book about games of chance *Liber de ludo aleae* contains the first systematic treatment of probability as well as effective cheating methods. His family life was problematic: his eldest son was executed for poisoning his wife, and his daughter was a prostitute who died from syphilis (about which he wrote a treatise). He computed and published the horoscope of Jesus, was accused of heresy, and spent time in prison until he abjured and gave up his professorship.

He published the solutions to the cubic and quartic equations in his book *Ars magna* in 1545, and also invented the combination lock, the gimbal consisting of three concentric rings allowing a compass or gyroscope to rotate freely (see ▪ Fig. 2.19), and the Cardan shaft with universal joints – the drive shaft used in motor vehicles today.

⊘ Roll-Pitch-Yaw Angles are in the Opposite Order

The order of the angles passed to `eul2rotm` is yaw, pitch and roll. This is the opposite to the way we refer to these angles as "roll, pitch and yaw angles" but it is the order they appear in (2.23). In other editions of this book the vector is reversed (α, β, γ).

The inverse is

```
>> gamma = rotm2eul(R,"ZYX")
gamma =
    0.3000    0.2000    0.1000
```

When describing the orientation of a robot gripper, as shown in ▪ Fig. 2.20, the convention is that its coordinate frame has the z-axis pointing forward and the y-axis is parallel to a line between the finger tips. This leads to the XYZ angle sequence

$$\mathbf{R}(\alpha, \beta, \gamma) = \mathbf{R}_x(\gamma)\,\mathbf{R}_y(\beta)\,\mathbf{R}_z(\alpha) \tag{2.24}$$

for example

```
>> R = eul2rotm([0.3 0.2 0.1],"XYZ")
R =
    0.9752   -0.0978    0.1987
    0.1538    0.9447   -0.2896
   -0.1593    0.3130    0.9363
```

and the inverse is

```
>> gamma = rotm2eul(R,"XYZ")
gamma =
    0.3000    0.2000    0.1000
```

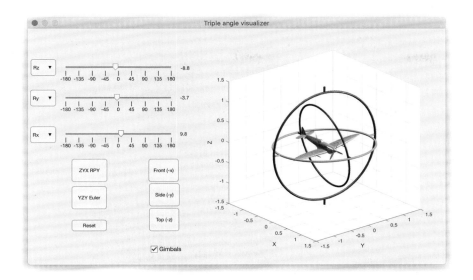

◻ Fig. 2.18 The `tripleangle` app allows you to interactively explore any triple angle sequence, the ZYX sequence is shown, and the angles can be adjusted to see how they change the orientation of a body in 3D space

The roll-pitch-yaw sequence allows all angles to have arbitrary sign and it has a singularity when $\beta = \pm\frac{\pi}{2}$.

The RVC Toolbox includes an interactive App

```
>> tripleangleApp
```

that allows you to experiment with Euler angles or roll-pitch-yaw angles and visualize their effect on the orientation of an object, as shown in ◻ Fig. 2.18.

2.3.1.3 Singularities and Gimbal Lock

A fundamental problem with all the three-angle representations just described is singularity. This is also known as gimbal lock, a geeky term made famous in the movie Apollo 13. The term is related to mechanical gimbal systems.

One example is the mechanical gyroscope, shown in ◻ Fig. 2.19, that was used for spacecraft navigation. The innermost assembly is the *stable member* which has three orthogonal gyroscopes that hold it at a constant orientation with respect to the universe. It is mechanically connected to the spacecraft via a gimbal mechanism which allows the spacecraft to rotate around the stable platform without exerting any torque on it. The attitude of the spacecraft is determined directly by measuring the angles of the gimbal axes with respect to the stable platform – in this design, the gimbals form a Cardanian YZX sequence giving yaw-roll-pitch angles. ▶ The orientation of the spacecraft's body-fixed frame {B} with respect to the stable platform frame {S} is

$$^S\mathbf{R}_B = \mathbf{R}_y(\beta)\mathbf{R}_z(\alpha)\mathbf{R}_x(\gamma) \ . \tag{2.25}$$

Consider the situation when the rotation angle of the middle gimbal (α, roll about the spacecraft's z-axis) is 90° – the axes of the inner (β) and outer (γ) gimbals are aligned and they share the *same* rotation axis. Instead of the original three rotational axes, since two are parallel, there are now only two effective rotational axes – we say that one degree of freedom has been lost. ▶ Substituting the identity ▶

$$\mathbf{R}_y(\theta)\,\mathbf{R}_z\left(\tfrac{\pi}{2}\right) \equiv \mathbf{R}_z\left(\tfrac{\pi}{2}\right)\mathbf{R}_x(\theta)$$

for the first two terms of (2.25), leads to

$$^S\mathbf{R}_B = \mathbf{R}_z(\tfrac{\pi}{2})\,\mathbf{R}_x(\beta)\,\mathbf{R}_x(\gamma) = \mathbf{R}_z(\tfrac{\pi}{2})\,\mathbf{R}_x(\beta + \gamma)$$

"The LM (Lunar Module) Body coordinate system is right-handed, with the +X axis pointing up through the thrust axis, the +Y axis pointing right when facing forward which is along the +Z axis. The rotational transformation matrix is constructed by a 2-3-1 [YZX] Euler sequence, that is: Pitch about Y, then Roll about Z and, finally, Yaw about X. Positive rotations are pitch up, roll right, yaw left." (Hoag 1963).

Operationally, this was a significant limiting factor with this particular gyroscope (Hoag 1963) and could have been alleviated by adding a fourth gimbal, as was used on other spacecraft. It was omitted on the Lunar Module for reasons of weight and space.

Rotations obey the cyclic rotation rules

$$\mathbf{R}_x(\tfrac{\pi}{2})\mathbf{R}_y(\theta)\mathbf{R}_x^\top(\tfrac{\pi}{2}) \equiv \mathbf{R}_z(\theta)$$
$$\mathbf{R}_y(\tfrac{\pi}{2})\mathbf{R}_z(\theta)\mathbf{R}_y^\top(\tfrac{\pi}{2}) \equiv \mathbf{R}_x(\theta)$$
$$\mathbf{R}_z(\tfrac{\pi}{2})\mathbf{R}_x(\theta)\mathbf{R}_z^\top(\tfrac{\pi}{2}) \equiv \mathbf{R}_y(\theta)$$

and anti-cyclic rotation rules

$$\mathbf{R}_y^\top(\tfrac{\pi}{2})\mathbf{R}_x(\theta)\mathbf{R}_y(\tfrac{\pi}{2}) \equiv \mathbf{R}_z(\theta)$$
$$\mathbf{R}_z^\top(\tfrac{\pi}{2})\mathbf{R}_y(\theta)\mathbf{R}_z(\tfrac{\pi}{2}) \equiv \mathbf{R}_x(\theta)$$
$$\mathbf{R}_x^\top(\tfrac{\pi}{2})\mathbf{R}_z(\theta)\mathbf{R}_x(\tfrac{\pi}{2}) \equiv \mathbf{R}_y(\theta)$$

2

■ **Fig. 2.19** Schematic of Apollo Lunar Module (LM) inertial measurement unit (IMU). The vehicle's coordinate system has the *x*-axis pointing up through the thrust axis, the *z*-axis forward, and the *y*-axis pointing right. Starting at the stable platform {S} and working outwards toward the spacecraft's body frame {B}, the rotation angle sequence is *YZX*. The components labeled X_g, Y_g and Z_g are the *x*-, *y*- and *z*-axis gyroscopes and those labeled X_a, Y_a and Z_a are the *x*-, *y*- and *z*-axis accelerometers (redrawn from Apollo Operations Handbook, LMA790-3-LM)

which is unable to represent any rotation about the *y*-axis. This is dangerous because any spacecraft rotation about the *y*-axis would also rotate the stable element and thus ruin its precise alignment with the stars: hence the anxiety onboard Apollo 13.

The loss of a degree of freedom means that mathematically we cannot invert the transformation, we can only establish a linear relationship between two of the angles. In this case, the best we can do is to determine the sum of the pitch and yaw angles. We observed a similar phenomenon with the ZYZ-Euler angle singularity in the previous section.

For an aircraft or submarine, normal operation involves a pitch angle around zero, and a singularity only occurs for impossible orientations such as nose straight up or straight down.

All three-angle representations of orientation, whether Eulerian or Cardanian, suffer the problem of gimbal lock when two axes become aligned. For ZYZ-Euler angles, this occurs when $\theta = k\pi$, $k \in \mathbb{Z}$; and for roll-pitch-yaw angles when pitch $\beta = \pm(2k+1)\frac{\pi}{2}$. The best we can do is carefully choose the angle sequence and co-ordinate system to ensure that the singularity is relegated to an orientation outside the normal operating envelope of the system. ◄

Singularities are an unfortunate consequence of using a minimal representation – in this case, using just three parameters to represent orientation. To eliminate this problem, we need to adopt different representations of orientation. Many in the Apollo LM team would have preferred a four gimbal system and the clue to success, as we shall see shortly in ▶ Sect. 2.3.1.7, is to introduce a fourth parameter.

2.3.1.4 Two-Vector Representation

For arm-type robots, it is useful to consider a coordinate frame {E} attached to the end effector, as shown in ◘ Fig. 2.20. By convention, the axis of the tool is associated with the z-axis and is called the *approach vector* and denoted $\hat{a} = (a_x, a_y, a_z)$. For some applications, it is more convenient to specify the approach vector than to specify ZYZ-Euler or roll-pitch-yaw angles.

However, specifying the direction of the z-axis is insufficient to describe the coordinate frame – we also need to specify the direction of the x- or y-axes. For robot grippers, this is given by the *orientation vector*, $\hat{o} = (o_x, o_y, o_z)$ which is parallel to the gripper's y-axis and is typically a vector between the finger tips. These two unit vectors are sufficient to completely define the rotation matrix

$$
\mathbf{R} = \begin{pmatrix} n_x & o_x & a_x \\ n_y & o_y & a_y \\ n_z & o_z & a_z \end{pmatrix} \tag{2.26}
$$

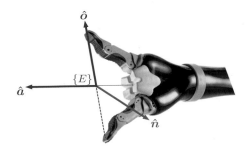

◘ **Fig. 2.20** Robot end-effector coordinate system defines the pose in terms of an *approach* vector $\hat{a}$ and an *orientation* vector $\hat{o}$ parallel to a line joining the finger tips. The $\hat{n}$, $\hat{o}$ and $\hat{a}$ vectors correspond to the x-, y- and z-axes respectively of the end-effector coordinate frame (Image of robot gripper courtesy of Kinova Robotics)

2

since the remaining column, the normal vector $\hat{\boldsymbol{n}} = (n_x, n_y, n_z)$, can be computed using (2.20) as $\hat{\boldsymbol{n}} = \hat{\boldsymbol{o}} \times \hat{\boldsymbol{a}}$. Consider an example where the robot's gripper is facing downward toward a work table

```
>> a = [0 0 -1];
```

and its finger tips lie on a line parallel to the vector $(1, 1, 0)$

```
>> o = [1 1 0];
```

In this case, o is not a unit vector but oa2rotm will unitize it.

The rotation matrix is ◄

```
>> R = oa2rotm(o,a)
R =
   -0.7071      0.7071           0
    0.7071      0.7071           0
         0           0     -1.0000
```

Any two nonparallel vectors are sufficient to define a coordinate frame. Even if the two vectors $\hat{\boldsymbol{a}}$ and $\hat{\boldsymbol{o}}$ are not orthogonal, they still define a plane, and the computed $\hat{\boldsymbol{n}}$ is normal to that plane. In this case, we need to compute a new value for $\hat{\boldsymbol{o}}' = \hat{\boldsymbol{a}} \times \hat{\boldsymbol{n}}$ which lies in the plane but is orthogonal to each of $\hat{\boldsymbol{a}}$ and $\hat{\boldsymbol{n}}$.

There are many other situations where we know the orientation in terms of two vectors. For a camera, we might use the optical axis (by convention the z-axis) as $\hat{\boldsymbol{a}}$, and the right-side of the camera (by convention the x-axis) as $\hat{\boldsymbol{n}}$. For a mobile robot, we might use the gravitational acceleration vector measured with accelerometers (by convention the z-axis) as $\hat{\boldsymbol{a}}$, and the heading direction measured with an electronic compass (by convention the x-axis) as $\hat{\boldsymbol{n}}$.

2.3.1.5 Rotation About an Arbitrary Vector

Two coordinate frames of arbitrary orientation are related by a *single* rotation about some axis in space. For the example rotation used earlier

```
>> R = eul2rotm([0.1 0.2 0.3]);
```

we can determine such an angle and axis by

```
>> rotm2axang(R)
ans =
    0.7900      0.5834      0.1886      0.3655
```

where the first three elements are a unit vector $\hat{\boldsymbol{v}}$ parallel to the axis around which the rotation occurs, and the last element is the rotation angle θ about that axis. Note that this is not unique since a rotation of $-\theta$ about the vector $-\boldsymbol{v}$ results in the same orientation as a rotation of θ about the vector $\boldsymbol{v}$ – this is referred to as a double mapping.

This information is encoded in the eigenvalues and eigenvectors of $\mathbf{R}$

```
>> [x,e] = eig(R)
x =
  -0.7900 + 0.0000i   -0.1073 - 0.4200i   -0.1073 + 0.4200i
  -0.5834 + 0.0000i   -0.0792 + 0.5688i   -0.0792 - 0.5688i
  -0.1886 + 0.0000i    0.6944 + 0.0000i    0.6944 + 0.0000i

e =
   1.0000 + 0.0000i    0.0000 + 0.0000i    0.0000 + 0.0000i
   0.0000 + 0.0000i    0.9339 + 0.3574i    0.0000 + 0.0000i
   0.0000 + 0.0000i    0.0000 + 0.0000i    0.9339 - 0.3574i
```

The matrices e and x are complex, and MATLAB denotes the imaginary part with $i = \sqrt{-1}$. Some elements are real – they have a zero imaginary part.

where the eigenvalues are returned on the diagonal of the matrix e and the associated eigenvectors are the corresponding columns of x. ◄

From the definition of eigenvalues and eigenvectors, we recall that

$$\mathbf{R}\boldsymbol{v} = \lambda\boldsymbol{v}$$

Excurse 2.15: Olinde Rodrigues
Rodrigues (1795–1850) was a French banker and mathematician who wrote extensively on politics, social reform and banking. He received his doctorate in mathematics in 1816 from the University of Paris, for work on his first well-known formula which is related to Legendre polynomials. His eponymous rotation formula was published in 1840 and is perhaps the first time the representation of a 3D rotation as a scalar and a vector was articulated. His formula was invented earlier by Gauss but not published. He is buried in the Pere-Lachaise cemetery in Paris.

where v is the eigenvector corresponding to the eigenvalue λ. For the case $\lambda = 1$, we can write

$$\mathbf{R}v = v$$

which implies that the corresponding eigenvector v is *unchanged* by the rotation. There is only one such vector, and that is the vector *about which* the rotation occurs. In the example above, the first eigenvalue is equal to one, so the rotation axis v is the first column of x.

A rotation matrix will always have one real eigenvalue at $\lambda = 1$ and in general a complex pair $\lambda = \cos\theta \pm j\sin\theta$, where θ is the rotation angle. The angle of rotation in this case is ▶

```
>> theta = angle(e(2,2))
theta =
    0.3655
```

The inverse problem, converting from angle and vector to a rotation matrix, is achieved using Rodrigues' rotation formula

The rotation angle is also related to the trace of R as discussed in ▶ App. B.2.1 but is limited to solutions in quadrants 1 and 2 only. The angle function finds a solution in any quadrant.

$$\mathbf{R} = \mathbf{1}_{3\times3} + \sin\theta[\hat{\boldsymbol{v}}]_\times + (1 - \cos\theta)[\hat{\boldsymbol{v}}]_\times^2 \qquad (2.27)$$

where $[\hat{\boldsymbol{v}}]_\times$ is a skew-symmetric matrix as discussed in the next section. We can use this formula to determine the rotation matrix for a rotation of 0.3 rad about the x-axis

```
>> R = axang2rotm([1 0 0 0.3])
R =
    1.0000         0         0
         0    0.9553   -0.2955
         0    0.2955    0.9553
```

which is equivalent to rotmx(0.3).

The Euler-Rodrigues formula is a variation that directly transforms a coordinate vector x

$$x' = \mathbf{R}(\theta, \hat{v})x = x + 2s(\omega \times x) + 2(\omega \times (\omega \times x))$$

where $s = \cos\frac{\theta}{2}$ and $\omega = \hat{v}\sin\frac{\theta}{2}$. Collectively, $(s, \omega) \in \mathbb{R}^4$ are known as the Euler parameters of the rotation. ◄

The angle-axis representation is parameterized by four numbers: one for the angle of rotation, and three for the unit vector defining the axis. We can pack these four numbers into a single vector $v \in \mathbb{R}^3$ – the direction of the rotational axis is given by the unit vector $\hat{v}$ and the rotation angle is some function of the vector magnitude $\|v\|$. ◄ This leads to some common 3-parameter representations such as:

- $\hat{v}\theta$, the Euler vector or exponential coordinates,
- $\hat{v}\tan\frac{\theta}{2}$, the Rodrigues' vector,
- $\hat{v}\sin\frac{\theta}{2}$, the unit quaternion vector component, or
- $\hat{v}\tan\theta$.

These forms are minimal and efficient in terms of data storage but the direction $\hat{v}$ is not defined when $\theta = 0$. That case corresponds to zero rotation and needs to be detected and dealt with explicitly.

2.3.1.6 Matrix Exponential for Rotation

Consider an x-axis rotation expressed as a rotation matrix

```
>> R = rotmx(0.3)
R =
    1.0000         0         0
         0    0.9553   -0.2955
         0    0.2955    0.9553
```

As we did for the 2-dimensional case, we can take the logarithm using the function `logm` ◄

```
>> L = logm(R)
L =
         0         0         0
         0         0   -0.3000
         0    0.3000         0
```

and the result is a sparse matrix with two elements that have a magnitude of 0.3, which is the rotation angle. This matrix has a zero diagonal and is another example of a skew-symmetric matrix, in this case 3×3.

Unpacking the skew-symmetric matrix gives

```
>> S = skew2vec(L)
S =
    0.3000         0         0
```

and we find the original rotation angle is in the first element, corresponding to the x-axis about which the rotation occurred. This is a computational alternative to the eigenvector approach introduced in ► Sect. 2.3.1.5.

Exponentiating the logarithm of the rotation matrix L using the matrix exponential function `expm`, ◄ yields the original rotation matrix

```
>> expm(L)
ans =
    1.0000         0         0
         0    0.9553   -0.2955
         0    0.2955    0.9553
```

The exponential of a skew-symmetric matrix is always a rotation matrix, with all the special properties outlined earlier. The skew-symmetric matrix can be re-

This is actually a unit quaternion, which is introduced in ► Sect. 2.3.1.7.

A 3D unit vector contains redundant information and can be described using just two numbers, the unit-norm constraint means that any element can be computed from the other two. Another way to think about this is to consider a unit sphere, then all possible unit vectors from the center can be described by the latitude and longitude of the point at which they touch the surface of the sphere.

`logm` is different to the function `log` which computes the logarithm of each element of the matrix. A logarithm can be computed using a power series with a matrix argument, rather than a scalar. The logarithm of a matrix is not unique and `logm` computes the principal logarithm.

`expm` is different to the function `exp` which computes the exponential of each element of the matrix. The matrix exponential can be computed using a power series with a matrix, rather than scalar, argument: $\text{expm}(\mathbf{A}) = \mathbf{1} + \mathbf{A} + \mathbf{A}^2/2! + \mathbf{A}^3/3! + \cdots$

Excurse 2.16: 3D Skew-Symmetric Matrix

In three dimensions, the skew- or anti-symmetric matrix has the form

$$
[\boldsymbol{\omega}]_\times = \begin{pmatrix} 0 & -\omega_z & \omega_y \\ \omega_z & 0 & -\omega_x \\ -\omega_y & \omega_x & 0 \end{pmatrix} \in \mathbf{so}(3) \tag{2.28}
$$

which has a distinct structure with a zero diagonal and only three unique elements $\boldsymbol{\omega} \in \mathbb{R}^3$, and $[\boldsymbol{\omega}]_\times^\top = -[\boldsymbol{\omega}]_\times$. The vector space of 3D skew-symmetric matrices is denoted $\mathbf{so}(3)$ and is the Lie algebra of $\mathbf{SO}(3)$. The vector cross product can be written as a matrix-vector product $\boldsymbol{v}_1 \times \boldsymbol{v}_2 = [\boldsymbol{v}_1]_\times \boldsymbol{v}_2$. The $[\cdot]_\times$ operator is implemented by

```
>> X = vec2skew([1 2 3])
X =
      0    -3     2
      3     0    -1
     -2     1     0
```

and the inverse operation, the $\vee_\times(\cdot)$ operator by

```
>> skew2vec(X)
ans =
      1     2     3
```

Both functions work for the 3D case, shown here, and the 2D case where the vector is a 1-vector.

---

constructed from the 3-vector `s`, so we can also write

```
>> expm(vec2skew(S))
ans =
    1.0000         0         0
         0    0.9553   -0.2955
         0    0.2955    0.9553
```

In fact, the function

```
>> R = rotmx(0.3);
```

is equivalent to

```
>> R = expm(0.3*vec2skew([1 0 0]));
```

where we have specified the rotation in terms of a rotation angle and a rotation axis (as a unit vector).

In general, we can write

$$
\mathbf{R} = e^{\theta[\hat{\boldsymbol{\omega}}]_\times} \in \mathbf{SO}(3)
$$

where θ is the rotation angle, $\hat{\boldsymbol{\omega}}$ is a unit vector parallel to the rotation axis, and the notation $[\cdot]_\times : \mathbb{R}^3 \mapsto \mathbf{so}(3) \subset \mathbb{R}^{3\times3}$ indicates a mapping from a vector to a skew-symmetric matrix. Since $\theta[\hat{\boldsymbol{\omega}}]_\times = [\theta\hat{\boldsymbol{\omega}}]_\times$, we can completely describe the rotation by the vector $\theta\hat{\boldsymbol{\omega}} \in \mathbb{R}^3$ – a rotational representation known as exponential coordinates. Rodrigues' rotation formula (2.27) is a computationally efficient means of computing the matrix exponential for the special case where the argument is a skew-symmetric matrix. ▶

However, the MATLAB function `expm` is faster than a MATLAB implementation of (2.27).

2.3.1.7 Unit Quaternions

» *Quaternions came from Hamilton after his really good work had been done; and, though beautifully ingenious, have been an unmixed evil to those who have touched them in any way, including Clark Maxwell.*
– Lord Kelvin, 1892

Unit quaternions are widely used today for robotics, computer vision, computer graphics and aerospace navigation systems. The quaternion was invented nearly 200 years ago as an extension of the complex number – a hypercomplex number – with a real part and three complex parts

$$\breve{q} = s + ui + vj + wk \in \mathbb{H} \tag{2.29}$$

where the orthogonal complex numbers i, j and k are defined such that

$$i^2 = j^2 = k^2 = ijk = -1 \; . \tag{2.30}$$

Today, it is more common to consider a quaternion as an ordered pair $\breve{q} = (s, \boldsymbol{v})$, sometimes written as $s + \boldsymbol{v}$, where $s \in \mathbb{R}$ is the scalar part and $\boldsymbol{v} = (u, v, w) \in \mathbb{R}^3$ is the vector part.

Quatenions form a vector space $\breve{q} \in \mathbb{H}$ and therefore allow operations such as addition and subtraction, performed element-wise, and multiplication by a scalar. Quaternions also allow conjugation

$$\breve{q}^* = s - ui - vj - wk \in \mathbb{H} \; , \tag{2.31}$$

and quaternion multiplication

$$\breve{q}_q \circ \breve{q}_2 = \underbrace{(s_1 s_2 - \boldsymbol{v}_1 \cdot \boldsymbol{v}_2)}_{\text{real}} + \underbrace{(s_1 \boldsymbol{v}_2 + s_2 \boldsymbol{v}_1 + \boldsymbol{v}_1 \times \boldsymbol{v}_2)}_{\text{vector}} \in \mathbb{H} \tag{2.32}$$

which is known as the quaternion or Hamilton product and is not commutative. ◄ The inner product is a scalar

$$\breve{q}_1 \cdot \breve{q}_2 = s_1 s_2 + u_1 u_2 + v_1 v_2 + w_1 w_2 \in \mathbb{R} \tag{2.33}$$

and the magnitude of a quaternion is

$$\|\breve{q}\| = \sqrt{\breve{q} \cdot \breve{q}} = \sqrt{s^2 + u^2 + v^2 + w^2} \in \mathbb{R} \; . \tag{2.34}$$

A pure quaternion is one whose scalar part is zero.

To represent rotations, we use unit quaternions, quaternions with unit magnitude $\|\breve{q}\| = 1$ and denoted by $\mathring{q}$. ◄ They can be considered as a rotation of θ about the unit vector $\hat{\boldsymbol{v}}$ which are related to the components of the unit quaternion by

$$\mathring{q} = \underbrace{\cos\tfrac{\theta}{2}}_{\text{real}} + \underbrace{\sin\tfrac{\theta}{2}(\hat{v}_x i + \hat{v}_y j + \hat{v}_z k)}_{\text{vector}} \in S^3 \tag{2.35}$$

which has similarities to the angle-axis representation of ► Sect. 2.3.1.5. It also has a double mapping which means that $\mathring{q}$ and $-\mathring{q}$ represent the same orientation.

In MATLAB, both quaternions and unit quaternions are implemented by the `quaternion` class. The constructor converts a passed argument such as a rotation matrix to a unit quaternion, for example ◄

```
>> q = quaternion(rotmx(0.3),"rotmat","point")
q =
  quaternion
    0.98877 +  0.14944i +       0j +       0k
```

The class supports a number of standard operators and methods, for example, quaternion multiplication ◄ is achieved by the overloaded multiplication operator

```
>> q = q*q;
```

If we write the quaternion as a 4-vector (s, u, v, w), then multiplication can be expressed as a matrix-vector product where

$$\breve{q} \circ \breve{q}' = \begin{pmatrix} s & -u & -v & -w \\ u & s & w & -v \\ v & w & s & u \\ w & v & -u & s \end{pmatrix} \begin{pmatrix} s' \\ u' \\ v' \\ w' \end{pmatrix}$$

The four elements of the unit quaternion form a vector and are known as the Euler parameters of the rotation.

The meaning of the argument `"point"` is described in ► Sect. 2.4.2.

Compounding two orthonormal rotation matrices requires 27 multiplications and 18 additions. The quaternion form requires 16 multiplications and 12 additions. This saving can be particularly important for embedded systems.

Excurse 2.17: Sir William Rowan Hamilton

Hamilton (1805–1865) was an Irish mathematician, physicist, and astronomer. He was a child prodigy with a gift for languages and by age thirteen knew classical and modern European languages as well as Persian, Arabic, Hindustani, Sanskrit, and Malay. Hamilton taught himself mathematics at age 17, and discovered an error in Laplace's Celestial Mechanics. He spent his life at Trinity College, Dublin, and was appointed Professor of Astronomy and Royal Astronomer of Ireland while still an undergraduate. In addition to quaternions, he contributed to the development of optics, dynamics, and algebra. He also wrote poetry and corresponded with Wordsworth who advised him to devote his energy to mathematics.

According to legend, the key quaternion equation, (2.30), occurred to Hamilton in 1843 while walking along the Royal Canal in Dublin with his wife, and this is commemorated by a plaque on Broome bridge:

» Here as he walked by on the 16th of October 1843 Sir William Rowan Hamilton in a flash of genius discovered the fundamental formula for quaternion

multiplication $i^2 = j^2 = k^2 = ijk = -1$ & cut it on a stone of this bridge.

His original carving is no longer visible, but the bridge is a pilgrimage site for mathematicians and physicists.

and inversion, the conjugate of a unit quaternion, is

```
>> q.conj
ans =
  quaternion
     0.95534 - 0.29552i +        0j +        0k
```

Multiplying a unit quaternion by its inverse, yields the identity quaternion

```
>> q*q.conj()
ans =
  quaternion
     1 + 0i + 0j + 0k
```

which represents a null rotation.

The unit quaternion $\mathring{q}$ can be converted to a rotation matrix by ►

```
>> q.rotmat("point")
ans =
     1.0000          0          0
          0     0.8253    -0.5646
          0     0.5646     0.8253
```

The meaning of the argument `"point"` is described in ► Sect. 2.4.2.

and its components can be converted to a vector by

```
>> q.compact()
ans =
     0.9553    0.2955          0          0
```

A coordinate vector, as a row vector, can be rotated by a quaternion

```
>> q.rotatepoint([0 1 0])
ans =
          0    0.8253     0.5646
```

The `quaternion` class can form 1- or 2-dimensional arrays, and has many methods and operators which are described fully in the documentation.

2

ξ as a Unit Quaternion

For the case of pure rotation in 3D, $\boldsymbol{\xi}$ can be implemented with a unit quaternion $\overset{\circ}{q} \in S^3$. The implementation is:

composition	$\boldsymbol{\xi}_1 \oplus \boldsymbol{\xi}_2$	$\mapsto \overset{\circ}{q}_1 \circ \overset{\circ}{q}_2$, Hamilton product
inverse	$\ominus \boldsymbol{\xi}$	$\mapsto \overset{\circ}{q}^* = s - \boldsymbol{v}$, quaternion conjugation
identity	$\varnothing$	$\mapsto 1 + \mathbf{0}$
vector-transform	$\boldsymbol{\xi} \cdot \boldsymbol{v}$	$\mapsto \overset{\circ}{q} \circ \overset{\vee}{v} \circ \overset{\circ}{q}^*$, where $\overset{\vee}{v} = 0 + \boldsymbol{v}$ is a pure quaternion

Composition is not commutative, that is, $\overset{\circ}{q}_1 \circ \overset{\circ}{q}_2 \neq \overset{\circ}{q}_2 \circ \overset{\circ}{q}_1$.

It is possible to store only the vector part of a unit quaternion, since the scalar part can be recovered by $s = \pm(1 - u^2 - v^2 - w^2)^{1/2}$. The sign ambiguity can be resolved by ensuring that the scalar part is positive, negating the quaternion if required, before taking the vector part. This 3-vector form of a unit quaternion is frequently used for optimization problems such as posegraph relaxation or bundle adjustment – it has minimum dimensionality and is singularity free.

❶ Beware the left-handed quaternion

Hamilton defined the complex numbers such that $ijk = -1$ but in the aerospace community it is common to use quaternions where $ijk = 1$. These are known as JPL, flipped or left-handed quaternions. This can be confusing, and you need to understand the convention adopted by any software libraries or packages you might use. This book and the toolbox software adopt Hamilton's convention.

❶ Quaternion element ordering

We have described quaternions with the components implicitly ordered as s, u, v, w, and the quaternion object constructor and the `compact` object function consider the quaternion as a 4-element row vector with this ordering. Some software tools, notably ROS, consider the quaternion as a vector ordered as u, v, w, s.

2.3.2 Pose in Three Dimensions

To describe the relative pose of the frames shown in ◘ Fig. 2.14, we need to account for the translation between the origins of the frames, as well as the rotation. We will combine particular methods of representing rotation with representations of translation, to create tangible representations of relative pose in 3D.

2.3.2.1 Homogeneous Transformation Matrix

The derivation for the 3D homogeneous transformation matrix is similar to the 2D case of (2.18) but extended to account for the z-dimension

$$\begin{pmatrix} {}^A x \\ {}^A y \\ {}^A z \\ 1 \end{pmatrix} = \begin{pmatrix} {}^A \mathbf{R}_B & {}^A \boldsymbol{t}_B \\ \mathbf{0}_{1 \times 3} & 1 \end{pmatrix} \begin{pmatrix} {}^B x \\ {}^B y \\ {}^B z \\ 1 \end{pmatrix}$$

where ${}^A \boldsymbol{t}_B \in \mathbb{R}^3$, as shown in ◘ Fig. 2.14, is a vector defining the origin of frame {B} with respect to frame {A}, and ${}^A \mathbf{R}_B \in \mathbf{SO}(3)$ is the rotation matrix which describes the orientation of frame {B} with respect to frame {A}. If points are represented by homogeneous coordinate vectors, then

$$
\begin{aligned}
{}^{A}\tilde{\boldsymbol{p}} &= \begin{pmatrix} {}^{A}\mathbf{R}_B & {}^{A}\boldsymbol{t}_B \\ \mathbf{0}_{1\times3} & 1 \end{pmatrix} {}^{B}\tilde{\boldsymbol{p}} \\
&= {}^{A}\mathbf{T}_B\,{}^{B}\tilde{\boldsymbol{p}}
\end{aligned}
\tag{2.36}
$$

where ${}^{A}\mathbf{T}_B$ is a 4×4 homogeneous transformation matrix which is commonly called a "transform". This matrix has a very specific structure and belongs to the Special Euclidean (SE) group of dimension 3 which we write as $\mathbf{T} \in \mathbf{SE}(3) \subset \mathbb{R}^{4\times4}$.

The 4×4 homogeneous transformation matrix is very commonly used in robotics, computer graphics and computer vision. It is supported by the toolboxes and will be used throughout this book as a concrete representation of $\boldsymbol{\xi}$ for 3-dimensional relative pose.

The toolboxes have many functions to create 3D homogeneous transformations. We can demonstrate composition of transforms by

```
>> T = trvec2tform([2 0 0])*tformrx(pi/2)*trvec2tform([0 1 0])
T =
     1      0      0      2
     0      0     -1      0
     0      1      0      1
     0      0      0      1
```

The function `trvec2tform` creates a relative pose with a finite translation but zero rotation, while `tformrx` creates a relative pose corresponding to a rotation of $\frac{\pi}{2}$ about the x-axis with zero translation. ▶ We can think of this expression as representing a walk along the x-axis for 2 units, *then* a rotation by 90° about the x-axis, *then* a walk of 1 unit along the new y-axis. The result, as shown in the last column of the resulting matrix is a translation of 2 units along the original x-axis and 1 unit along the original z-axis. We can plot the corresponding coordinate frame by

For describing rotation, there are functions that create a rotation matrix (eg. `rotmx`, `eul2rotm`) or a homogeneous transformation with zero translation (eg. `tformrx`, `eul2tform`.

```
>> plottform(T);
```

The rotation matrix component of `T` is

```
>> tform2rotm(T)
ans =
     1      0      0
     0      0     -1
     0      1      0
```

and the translation component is a row vector

```
>> tform2trvec(T)
ans =
     2      0      1
```

2

> **ξ as an SE(3) Matrix**
>
> For the case of rotation and translation in 3D, $\boldsymbol{\xi}$ can be implemented by a homogeneous transformation matrix $\mathbf{T} \in \mathbf{SE}(3)$ which is sometimes written as an ordered pair $(\mathbf{R}, \boldsymbol{t}) \in \mathbf{SO}(3) \times \mathbb{R}^3$. The implementation is:
>
> composition $\qquad \boldsymbol{\xi}_1 \oplus \boldsymbol{\xi}_2 \qquad \mapsto \mathbf{T}_1\mathbf{T}_2 = \begin{pmatrix} \mathbf{R}_1\mathbf{R}_2 & \boldsymbol{t}_1 + \mathbf{R}_1\boldsymbol{t}_2 \\ \mathbf{0}_{1\times3} & 1 \end{pmatrix},$ matrix multiplication
>
> inverse $\qquad \ominus\boldsymbol{\xi} \qquad \mapsto \mathbf{T}^{-1} = \begin{pmatrix} \mathbf{R}^\top & -\mathbf{R}^\top\boldsymbol{t} \\ \mathbf{0}_{1\times3} & 1 \end{pmatrix},$ matrix inverse
>
> identity $\qquad \varnothing \qquad\qquad \mapsto \mathbf{1}_{4\times4},$ identity matrix
>
> vector-transform $\quad \boldsymbol{\xi} \cdot \boldsymbol{v} \qquad \mapsto \epsilon(\mathbf{T}\tilde{\boldsymbol{v}}),$ matrix-vector product using homogeneous vectors
>
> where $\tilde{\cdot} : \mathbb{R}^3 \mapsto \mathbb{P}^3$ and $\epsilon(\cdot) : \mathbb{P}^3 \mapsto \mathbb{R}^3$. Composition is not commutative, that is, $\mathbf{T}_1\mathbf{T}_2 \neq \mathbf{T}_2\mathbf{T}_1$.

2.3.2.2 Matrix Exponential for Pose

Consider the $\mathbf{SE}(3)$ matrix

```
>> T = trvec2tform([2 3 4])*tformrx(0.3)
T =
    1.0000         0         0    2.0000
         0    0.9553   -0.2955    3.0000
         0    0.2955    0.9553    4.0000
         0         0         0    1.0000
```

and its logarithm

```
>> L = logm(T)
L =
         0         0         0    2.0000
         0         0   -0.3000    3.5775
         0    0.3000         0    3.5200
         0         0         0         0
```

is a sparse matrix and the rotation magnitude, 0.3, is clearly evident. The structure of this matrix is an augmented skew-symmetric matrix: the upper-left corner is a 3×3 skew-symmetric matrix, the upper right column is a 3-vector, and the bottom row is zero. We can unpack the six unique elements by

```
>> S = skewa2vec(L)
S =
    0.3000         0         0    2.0000    3.5775    3.5200
```

and the first three elements describe rotation, and the value 0.3 is in the first "slot" indicating a rotation about the first axis – the x-axis. Exponentiating the logarithm matrix yields the original $\mathbf{SE}(3)$ matrix, and the logarithm can be reconstructed from just the six elements of s

```
>> expm(vec2skewa(S))
ans =
    1.0000         0         0    2.0000
         0    0.9553   -0.2955    3.0000
         0    0.2955    0.9553    4.0000
         0         0         0    1.0000
```

In general, we can write

$$\mathbf{T} = \mathrm{e}^{[S]} \in \mathbf{SE}(3)$$

> **Excurse 2.19: 3D Augmented Skew-Symmetric Matrix**
>
> In 3 dimensions, the augmented skew-symmetric matrix, corresponding to the vector $S = (\omega_x, \omega_y, \omega_z, v_x, v_y, v_z)$, is
>
> $$[S] = \left(\begin{array}{ccc|c} 0 & -\omega_z & \omega_y & v_x \\ \omega_z & 0 & -\omega_x & v_y \\ -\omega_y & \omega_x & 0 & v_z \\ \hline 0 & 0 & 0 & 0 \end{array} \right) \in \mathbf{se}(3) \qquad (2.37)$$
>
> which has a distinct structure with a zero diagonal and bottom row, and a skew-symmetric matrix in the top-left corner. $\mathbf{se}(3)$ is the vector space of 3D augmented skew-symmetric matrices, and the Lie algebra of $\mathbf{SE}(3)$. The $[\cdot]$ operator is implemented by
>
> ```
> >> X = vec2skewa([1 2 3 4 5 6])
> X =
>         0      -3       2       4
>         3       0      -1       5
>        -2       1       0       6
>         0       0       0       0
> ```
>
> and the inverse operator $\vee(\cdot)$ by
>
> ```
> >> skewa2vec(X)
> ans =
>         1       2       3       4       5       6
> ```

where $S \in \mathbb{R}^6$ and $[\cdot] : \mathbb{R}^6 \mapsto \mathbf{se}(3) \subset \mathbb{R}^{4\times4}$. The vector S is a *twist vector* as we will discuss next.

2.3.2.3 3D Twists

In 3D, a twist is equivalent to screw or helicoidal motion – motion about and along some line in space as shown in ◘ Fig. 2.25. Rigid-body motion between any two poses can be represented by such screw motion. For pure translation, the rotation is about a screw axis at infinity.

We represent a screw as a twist vector $(\boldsymbol{\omega}, \boldsymbol{v}) \in \mathbb{R}^6$ comprising a vector $\boldsymbol{\omega} \in \mathbb{R}^3$ parallel to the screw axis and a moment $\boldsymbol{v} \in \mathbb{R}^3$. The moment encodes a point lying on the twist axis, as well as the screw pitch which is the ratio of the distance along the screw axis to the rotation about the screw axis.

Consider the example of a rotation about the x-axis which has a screw axis parallel to the x-axis and passing through the origin

```
>> S = Twist.UnitRevolute([1 0 0],[0 0 0])
S =
( 1  0  0; 0  0  0 )
```

This particular twist is a *unit twist* that describes a rotation of 1 rad about the x-axis.

To create an $\mathbf{SE}(3)$ transformation for a specific rotation, we scale it and exponentiate it

```
>> expm(0.3*S.skewa());
```

where the `skewa` method is the twist vector as an augmented skew-symmetric matrix, or more compactly

```
>> S.tform(0.3)
ans =
    1.0000         0         0         0
         0    0.9553   -0.2955         0
```

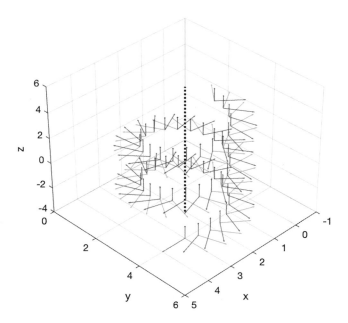

Fig. 2.21 A coordinate frame displayed for different values of θ about a screw parallel to the z-axis and passing through the point $(2, 3, 2)$. The x-, y- and z-axes are indicated by red, green and blue lines respectively

```
      0     0.2955    0.9553         0
      0          0         0    1.0000
```

The related method `exp` returns the same result but as an `se3` object, this is discussed further in ▶ Sect. 2.5.

which is identical to the value ◄ we would obtain using `tformrx(0.3)`, but the twist allows us to express rotation about an axis passing through any point, not just the origin.

To illustrate the underlying screw motion, we will progressively rotate a coordinate frame about a screw. We define a screw parallel to the z-axis that passes through the point $(2, 3, 2)$ and has a pitch of 0.5

```
>> S = Twist.UnitRevolute([0 0 1],[2 3 2],0.5);
```

and the coordinate frame to be rotated is described by an **SE**(3) matrix

```
>> X = trvec2tform([3 4 -4]);
```

For values of θ in the range 0 to 15 rad, we evaluate the twist for each value of θ, apply it to the frame {X} and plot the result

```
>> clf; hold on
>> view(3)
>> for theta = [0:0.3:15]
>>    plottform(S.tform(theta)*X,style="rgb",LineWidth=2)
>> end
```

The plot in ■ Fig. 2.21 clearly shows the screw motion in the successive poses of the frame {X} as it is rotated about the screw axis

```
>> L = S.line();
>> L.plot("k:",LineWidth=2);
```

The 3D line is represented by a `Plucker` object in terms of Plücker coordinates which are covered in ▶ App. C.1.2.2.

which is shown in the plot as a black dotted line. ◄

A translational, or prismatic, unit twist in the y-direction is created by

```
>> S = Twist.UnitPrismatic([0 1 0])
S =
( 0  0  0; 0  1  0 )
```

and describes a displacement of 1 in the specified direction. To create an **SE**(3) transformation for a specific translation, we scale and exponentiate the unit twist

```
>> S.tform(2)
ans =
        1        0        0        0
        0        1        0        2
        0        0        1        0
        0        0        0        1
```

which has a zero rotation and a translation of 2 in the y-direction.

We can also convert an arbitrary **SE**(3) matrix to a twist. For the transform from ► Sect. 2.3.2.2, the twist vector is

```
>> T = trvec2tform([1 2 3])*eul2tform([0.3 0.4 0.5]);
>> S = Twist(T)
S =
( 0.42969  0.46143  0.19003; 0.53084  2.5512  2.7225 )
```

and we see that the twist vector is the unique elements of the logarithm of the **SE**(3) matrix. In this case, the screw is parallel to

```
>> S.w
ans =
    0.4297    0.4614    0.1900
```

passes through the point

```
>> S.pole
ans =
    1.1714    -1.6232    1.2927
```

has a pitch of

```
>> S.pitch
ans =
    1.9226
```

and the rigid-body motion requires a rotation about the screw of

```
>> S.theta
ans =
    0.6585
```

radians, which is about 1/10th of a turn.

ξ as a 3D Twist

For the case of rotation and translation in 3D, ξ can be implemented by a twist $S \in \mathbb{R}^6$. The implementation is:

composition	$\xi_1 \oplus \xi_2$	$\mapsto \log\left(e^{[S_1]}e^{[S_2]}\right)$, product of exponential
inverse	$\ominus \xi$	$\mapsto -S$, negation
identity	$\varnothing$	$\mapsto \mathbf{0}_{1 \times 6}$, zero vector
vector-transform	$\xi \cdot v$	$\mapsto \epsilon(e^{[S]}\tilde{v})$, matrix-vector product using homogeneous vectors

where $\tilde{\cdot} : \mathbb{R}^3 \mapsto \mathbb{P}^3$ and $\epsilon(\cdot) : \mathbb{P}^3 \mapsto \mathbb{R}^3$. Composition is not commutative, that is, $e^{[S_1]}e^{[S_2]} \neq e^{[S_2]}e^{[S_1]}$. Note that log and exp have efficient closed form, rather than transcendental, solutions which makes composition relatively inexpensive.

2.3.2.4 Vector-Quaternion Pair

Another way to represent pose in 3D is to combine a translation vector and a unit quaternion. It represents pose using just 7 numbers, is easy to compound, and is singularity free.

2

ξ as a Vector-Quaternion Pair

For the case of rotation and translation in 3D, ξ can be implemented by a vector and a unit quaternion written as an ordered pair $(t, \mathring{q}) \in \mathbb{R}^3 \times S^3$. The implementation is:

composition	$\xi_1 \oplus \xi_2$	$\mapsto (t_1 + \mathring{q}_1 \cdot t_2, \mathring{q}_1 \circ \mathring{q}_2)$
inverse	$\ominus \xi$	$\mapsto (-\mathring{q}^* \cdot t, \mathring{q}^*)$
identity	$\varnothing$	$\mapsto (\mathbf{0}_3, 1 + \mathbf{0})$
vector-transform	$\xi \cdot v$	$\mapsto \mathring{q} \cdot v + t$

Composition is not commutative, that is, $\xi_1 \xi_2 \neq \xi_2 \xi_1$. The vector transformation operator $\cdot : S^3 \times \mathbb{R}^3 \mapsto \mathbb{R}^3$ can be found in ▶ Sect. 2.3.1.7.

2.3.2.5 Unit Dual Quaternion

An early approach to representing pose with quaternions was Hamilton's biquaternion where the quaternion coefficients were themselves complex numbers. Later, William Clifford ◀ developed the dual number which is defined as an ordered pair $d = (x, y)$ that can be written as $d = x + y\varepsilon$, where $\varepsilon^2 = 0$. Clifford created a quaternion dual number with $x, y \in \mathbb{H}$ which he also called a bi-quaternion but is today called a dual quaternion.

William Clifford (1845–1879) was an English mathematician and geometer. He made contributions to geometric algebra, and Clifford Algebra is named in his honor.

ξ as a Unit Dual Quaternion

For the case of rotation and translation in 3D, ξ can be implemented by a unit dual quaternion $(\mathring{q}_r, \check{q}_d) \in S^3 \times \mathbb{H}$ which comprises a unit quaternion $\mathring{q}_r$ (the real part) representing rotation, and a quaternion $\check{q}_d = \frac{1}{2}\check{t} \circ \mathring{q}_r$ (the dual part) where $\check{t}$ is a pure quaternion representing translation. The implementation is:

composition	$\xi_1 \oplus \xi_2$	$\mapsto (\mathring{q}_{r_1} \circ \mathring{q}_{r_2}, \mathring{q}_{r_1} \circ \check{q}_{d_2} + \check{q}_{d_1} \circ \mathring{q}_{r_2})$
inverse	$\ominus \xi$	$\mapsto (\mathring{q}_r, \check{q}_d)^* = (\mathring{q}_r^*, \check{q}_d^*)$, conjugation
identity	$\varnothing$	$\mapsto (1 + \mathbf{0}, 1 + \mathbf{0})$
vector-transform	$\xi \cdot v$	$\mapsto (\mathring{q}_r, \check{q}_d) \circ (1 + \mathbf{0}, \check{v}) \circ (\mathring{q}_r, \check{q}_d)^*$, where the middle term is a pure dual quaternion.

Composition is not commutative, that is, $\xi_1 \xi_2 \neq \xi_2 \xi_1$. Note that the Hamilton product of a non-unit quaternion $\check{p}$ and a unit quaternion $\mathring{q}$ is a non-unit quaternion.

This representation is quite compact, requiring just 8 numbers; it is easy to compound using a special multiplication table; and it is easy to normalize to eliminate the effect of finite precision arithmetic. However, it has no real useful computational advantage over matrix methods.

2.4 Advanced Topics

2.4.1 Pre- and Post-Multiplication of Transforms

Many texts make a distinction between *pre-multiplying* and *post-multiplying* relative transformations. For the case of a series of rotation matrices, the former rotates with respect to the world reference frame, whereas the latter rotates with respect to the current or rotated frame. The difference is best understood using the pose

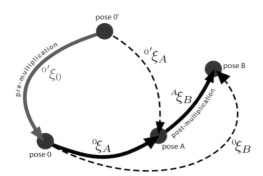

■ **Fig. 2.22** Pre- and post-multiplication shown as a pose graph

graph shown in ■ Fig. 2.22. The pose {A} is defined with respect to the reference pose {0}.

Post-multiplication should be familiar by now, and is the motion from {A} to {B} expressed in frame {A} – the rotated frame. The result, $^0\boldsymbol{\xi}_B$, is the motion from {0} to {A} and then to {B}.

Pre-multiplication is shown by the red arrow. It requires that we introduce a new pose for the start of the arrow which we will call {0'}. We can compose these motions – the super- and subscripts match – as $^{0'}\boldsymbol{\xi}_0 \oplus {}^0\boldsymbol{\xi}_A = {}^{0'}\boldsymbol{\xi}_A$. Effectively this is a change in the reference frame – from {0} to {0'}. The transformation is therefore with respect to the reference frame, albeit a new one.

A three-angle rotation sequence such as ZYX roll-pitch-yaw angles from (2.23)

$$^0\mathbf{R}_B = \mathbf{R}_z(\gamma)\,\mathbf{R}_y(\beta)\,\mathbf{R}_x(\alpha)$$

can be interpreted in two different, but equivalent, ways. Firstly, using pre-multiplication, we can consider this right to left as consecutive rotations about the reference frame by $\mathbf{R}_x(\alpha)$, then by $\mathbf{R}_y(\beta)$ and then by $\mathbf{R}_z(\gamma)$ – these are referred to as extrinsic rotations. Alternatively, from left to right as sequential rotations, each with respect to the previous frame, by $\mathbf{R}_z(\gamma)$, then by $\mathbf{R}_y(\beta)$, and then by $\mathbf{R}_x(\alpha)$ – these are referred to as intrinsic rotations.

2.4.2 Active and Passive Transformations

Many sources, particularly from physics, distinguish between transforming the co-ordinate vector of a point $\boldsymbol{p}$ by *transforming the point* (active, alibi or intrinsic transformation)

$$\tilde{\boldsymbol{q}} = \mathbf{T}^a\,\tilde{\boldsymbol{p}}$$

and *transforming the reference frame* (passive, extrinsic or alias transformation)

$$\tilde{\boldsymbol{q}} = \mathbf{T}^p\,\tilde{\boldsymbol{p}}\ .$$

This becomes less confusing if we rewrite using our superscript and subscript notation as shown in ■ Fig. 2.23.

The active transformation, ■ Fig. 2.23a, moves the point from the blue position to the red position. The point has the same relative position with respect to the frame – the frame moves and the point moves with it. The coordinate vector of the moved point with respect to the original coordinate frame is

$$^A\tilde{\boldsymbol{p}} = {}^A\mathbf{T}_B\,{}^B\tilde{\boldsymbol{p}}\ .$$

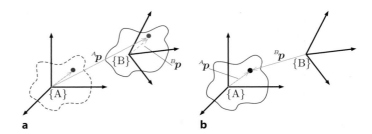

Fig. 2.23 **a** Active transformation of the point from its initial position (blue) to a final position (red); **b** passive transformation of the point

For the passive transformation, ◻ Fig. 2.23b, the point does not move but the coordinate frame does. The coordinate vector of the point with respect to the moved coordinate frame is

$$^B\tilde{p} = {}^B\mathbf{T}_A\,{}^A\tilde{p} \; .$$

We can clearly see that the active and passive transformations are inverses.

> This book takes an active transformation approach.

In this book, we try to avoid making the distinction between active and passive transformations. ◀ It is simpler to consider that points are fixed to objects and defined by coordinate vectors relative to the object's coordinate frame. We use our established notation to transform coordinate vectors between frames: superscript and subscript to denote frames, and a mechanism to describe motion between frames.

🛇 Somes MATLAB classes, for example `quaternion`, do make this distinction. In order for matrix premultiplication of a coordinate vector by a rotation matrix to give the same result as the `rotatepoint` method of a `quaternion` object, we must use the option `"point"` (active transformation) when constructing the `quaternion` from a rotation matrix.

2.4.3 Direction Cosine Matrix

Consider two coordinate frames {A} and {B}. The rotation matrix $^A\mathbf{R}_B$ can be written as

$$^A\mathbf{R}_B = \begin{pmatrix} \cos(\hat{x}_A, \hat{x}_B) & \cos(\hat{x}_A, \hat{y}_B) & \cos(\hat{x}_A, \hat{z}_B) \\ \cos(\hat{y}_A, \hat{x}_B) & \cos(\hat{y}_A, \hat{y}_B) & \cos(\hat{y}_A, \hat{z}_B) \\ \cos(\hat{z}_A, \hat{x}_B) & \cos(\hat{z}_A, \hat{y}_B) & \cos(\hat{z}_A, \hat{z}_B) \end{pmatrix} \tag{2.38}$$

where $\cos(\hat{u}_A, \hat{v}_B)$, $u, v \in \{x, y, z\}$ is the cosine of the angle between the unit vector $\hat{u}_A$ in {A}, and the unit vector $\hat{v}_B$ in {B}. Since $\hat{u}_A$ and $\hat{v}_B$ are unit vectors $\cos(\hat{u}_A, \hat{v}_B) = \hat{u}_A \cdot \hat{v}_B$. This matrix is called the direction cosine matrix (DCM) and is frequently used in aerospace applications.

2.4.4 Efficiency of Representation

Many of the representations discussed in this chapter have more parameters than strictly necessary. For instance, an **SO**(2) matrix is 2×2 and has four elements that represent a single rotation angle. This redundancy can be explained in terms of constraints: the columns are each unit-length ($\|c_1\| = \|c_2\| = 1$), which provides two constraints; and the columns are orthogonal ($c_1 \cdot c_2 = 0$), which adds

☐ **Table 2.1** Number of parameters required for different pose representations. The † variant does not store the constant bottom row of the matrix

Type	N	N_{min}	N_{min}/N
SO(2) matrix	4	1	25%
SE(2) matrix	9	3	33%
SE(2) as 2×3 matrix †	6	3	50%
2D twist vector	3	3	**100%**
SO(3) matrix	9	3	33%
Unit quaternion	4	3	75%
Unit quaternion vector part	3	3	**100%**
Euler parameters	3	3	**100%**
SE(3) matrix	16	6	38%
SE(3) as 3×4 matrix †	12	6	50%
3D twist vector	6	6	**100%**
Vector + unit quaternion	7	6	86%
Unit dual quaternion	8	6	75%

another constraint. Four matrix elements with three constraints is effectively one independent value.

☐ Table 2.1 shows the number of parameters N for various representations, as well as the minimum number of parameters N_{min} required. The matrix representations are the least efficient, they have the lowest N_{min}/N, which has a real cost in terms of computer memory. However, this is offset by the simplicity of composition which is simply matrix multiplication and that can be executed very efficiently on modern computer hardware. If space is an issue, then we can use simple tricks like not storing the bottom row of matrices which is equivalent to representing the homogeneous transformation matrix as an ordered pair $(\mathbf{R}, \boldsymbol{t})$ which is common in computer vision.

2.4.5 Distance Between Orientations

The distance between two Euclidean points is the length of the line between them, that is, $\|\boldsymbol{p}_1 - \boldsymbol{p}_2\|$, $\boldsymbol{p}_i \in \mathbb{R}^n$. The *distance* between two orientations is more complex and, unlike translational distance, it has an upper bound. Some common metrics for rotation matrices and quaternions are

$$d(\mathbf{R}_1, \mathbf{R}_2) = \|\mathbf{1}_{3\times3} - \mathbf{R}_1\mathbf{R}_2^\top\| \in [0, 2] \tag{2.39}$$

$$d(\mathbf{R}_1, \mathbf{R}_2) = \|\log \mathbf{R}_1\mathbf{R}_2^\top\| \in [0, \pi] \tag{2.40}$$

$$d(\mathring{q}_1, \mathring{q}_2) = 1 - |\mathring{q}_1 \cdot \mathring{q}_2| \in [0, 1] \tag{2.41}$$

$$d(\mathring{q}_1, \mathring{q}_2) = \cos^{-1} |\mathring{q}_1 \cdot \mathring{q}_2| \in [0, \pi/2] \tag{2.42}$$

$$d(\mathring{q}_1, \mathring{q}_2) = 2\tan^{-1} \frac{\|\mathring{q}_1 - \mathring{q}_2\|}{\|\mathring{q}_1 + \mathring{q}_2\|} \in [0, \pi/2] \tag{2.43}$$

$$d(\mathring{q}_1, \mathring{q}_2) = 2\cos^{-1} |\mathring{q}_1 \cdot \mathring{q}_2| \in [0, \pi] \tag{2.44}$$

where the · operator for unit quaternions is the inner product given by (2.33). These measures are all proper distance metrics which means that $d(x, x) = 0$, $d(x, y) =$

$d(y, x)$, and the triangle inequality holds $d(x, z) \leq d(x, y) + d(y, z)$. For a metric, $d(x, y) = 0$ implies that $x = y$ but for unit quaternions this is not true since, due to the double mapping, it is also true that $d(\mathring{q}, -\mathring{q}) = 0$. This problem can be avoided by ensuring that unit quaternions always have a non-negative scalar part. (2.42) and (2.43) are equivalent but the latter is numerically more robust.

The `dist` method of the `quaternion` class computes the metric (2.44). For example

```
>> q1 = quaternion(rotmx(pi/2),"rotmat","point");
>> q2 = quaternion(rotmz(-pi/2),"rotmat","point");
>> q1.dist(q2)
ans =
    2.0944
```

2.4.6 Normalization

The IEEE-754 standard for double precision (64-bit) floating point arithmetic has around 16 decimal digits of precision.

Floating-point arithmetic has finite precision ◄ and each operation introduces some very small error. Consecutive operations will accumulate that error. A rotation matrix has, by definition, a determinant of one

```
>> R = eye(3);
>> det(R)-1
ans =
    0
```

but if we repeatedly multiply by a valid rotation matrix

```
>> for i = 1:100
>>    R = R*eul2rotm([0.2 0.3 0.4]);
>> end
>> det(R) - 1
ans =
    3.7748e-15
```

This may vary between CPU types and software.

the result has a small error ◄ – the determinant is no longer equal to one and the matrix is no longer a proper orthonormal rotation matrix. To fix this, we need to normalize the matrix, a process which enforces the constraints on the columns c_i of the rotation matrix $\mathbf{R} = [c_1, c_2, c_3]$.

We assume that the third column has the correct direction

$$c_3' = c_3$$

then the first column is made orthogonal to the last two

$$c_1' = c_2 \times c_3' \ .$$

However, the last two columns may not have been orthogonal so

$$c_2' = c_3' \times c_1' \ .$$

Finally, the columns are all normalized to unit magnitude

$$c_i'' = \frac{c_i'}{\|c_i'\|}, \ i = 1, 2, 3$$

Normalization is implemented by

```
>> R = tformnorm(R);
```

and the determinant is now much closer to one ▶

```
>> det(R)-1
ans =
  -1.1102e-16
```

This error is now almost at the limit of IEEE-754 standard 64-bit (double precision) floating point arithmetic which is 2.2204×10^{-16} and given by the MATLAB variable eps.

Normalization can also be applied to an **SE**(3) matrix, in which case only the **SO**(3) rotation submatrix is normalized.

Arithmetic issues also occur when multiplying unit quaternions – the norm, or magnitude, of the unit quaternion diverges from one. However, this is much easier to fix since normalizing the quaternion simply involves dividing all elements by the norm

$$\mathring{q}' = \frac{\mathring{q}}{\|\mathring{q}\|}$$

which is implemented by

```
>> q = q.normalize();
```

Normalization does not need to be performed after every multiplication since it is an expensive operation. However it is advisable for situations like the example above, where one transform is being repeatedly updated.

2.4.7 Understanding the Exponential Mapping

In this chapter, we have glimpsed some connection between rotation matrices, homogeneous transformation matrices, skew-symmetric matrices, matrix logarithms and matrix exponentiation. The roots of this connection are in the mathematics of Lie groups which are covered in text books on algebraic topology. This is advanced mathematics and many people starting out in robotics will find their content somewhat inaccessible. An introduction to the essentials of this topic is given in ▶ App. D. In this section, we will use an intuitive approach, based on undergraduate engineering mathematics, to shed some light on the connection.

◼ Fig. 2.24 shows a point P, defined by a coordinate vector $\boldsymbol{p}$, being rotated about an axis. The axis is parallel to the vector $\boldsymbol{\omega}$ whose magnitude $\|\boldsymbol{\omega}\|$ specifies the rate of rotation about the axis. ▶ We wish to rotate the point by an angle θ about this axis and the velocity of the point is known from mechanics to be

Such a vector is an angular velocity vector which will be properly introduced in the next chapter.

$$\dot{\boldsymbol{p}} = \boldsymbol{\omega} \times \boldsymbol{p}$$

and we can replace the cross product with a skew-symmetric matrix-vector product

$$\dot{\boldsymbol{p}} = [\boldsymbol{\omega}]_\times \boldsymbol{p} \ . \tag{2.45}$$

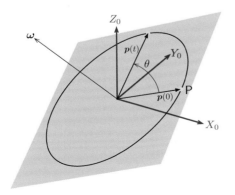

◼ **Fig. 2.24** The point P follows a circular path in the plane normal to the axis $\boldsymbol{\omega}$

2

We can find the solution to this first-order differential equation by analogy to the simple scalar case

$$\dot{x} = ax$$

whose solution is

$$x(t) = e^{at} x(0)$$

which implies that the solution to (2.45) is

$$p(t) = e^{[\omega]_\times t} p(0) \ .$$

If $\|\omega\| = 1$, then after t seconds, the coordinate vector will have rotated by t radians. We require a rotation by θ, so we can set $t = \theta$ to give

$$p(\theta) = e^{[\hat{\omega}]_\times \theta} p(0)$$

which describes the vector $p(0)$ being rotated to $p(\theta)$. A matrix that rotates a coordinate vector is a rotation matrix, and this implies that our matrix exponential is a rotation matrix

$$\mathbf{R}(\theta, \ \hat{\omega}) = e^{[\hat{\omega}]_\times \theta} \in \mathbf{SO}(3) \ .$$

Now consider the more general case of rotational and translational motion. We can write

$$\dot{p} = [\omega]_\times p + v$$

and, rearranging into matrix form

$$\begin{pmatrix} \dot{p} \\ 0 \end{pmatrix} = \begin{pmatrix} [\omega]_\times & v \\ 0 & 0 \end{pmatrix} \begin{pmatrix} p \\ 1 \end{pmatrix}$$

and introducing homogeneous coordinates, this becomes

$$\dot{\tilde{p}} = \begin{pmatrix} [\omega]_\times & v \\ 0 & 0 \end{pmatrix} \tilde{p}$$
$$= \mathbf{\Sigma}(\omega, v) \tilde{p}$$

where $\mathbf{\Sigma}(\omega, v)$ is a 4×4 augmented skew-symmetric matrix. Again, by analogy with the scalar case, we can write the solution as

$$\tilde{p}(\theta) = e^{\mathbf{\Sigma}(\omega, v)\theta} \tilde{p}(0) \ .$$

A matrix that rotates and translates a point in homogeneous coordinates is a homogeneous transformation matrix, and this implies that our matrix exponential is a homogeneous transformation matrix

$$\mathbf{T}(\theta, \hat{\omega}, v) = e^{\mathbf{\Sigma}(\omega, v)\ \theta} = e^{\begin{pmatrix} [\hat{\omega}]_\times & v \\ 0 & 0 \end{pmatrix} \theta} \in \mathbf{SE}(3)$$

where $[\hat{\omega}]_\times \theta$ defines the magnitude and axis of rotation and $v\theta$ is the translation.

2.4.8 More About Twists

» Chasles' theorem:

Any displacement of a body in space can be accomplished by means of a rotation of the body about a unique line in space accompanied by a translation of the body parallel to that line.

In this chapter, we have introduced and demonstrated twists in 2D and 3D. Here, we will define them more formally and discuss the relationship between twists and homogeneous transformation matrices via the exponential mapping.

The motion described by Chasles' theorem is that of a body, rigidly attached to a nut, rotating about a screw as illustrated in ◘ Fig. 2.25. As the nut rotates, a frame attached to the nut rotates and translates in space. Simplistically, this is the essence of screw theory and its mathematics was developed by Sir Robert Ball in the late 19th century for the analysis of mechanisms.

The general displacement of a rigid body in 3D can be represented by a twist vector

$$S = (\omega, v) \in \mathbb{R}^6$$

where $\omega \in \mathbb{R}^3$ is the direction of the screw axis, and $v \in \mathbb{R}^3$ is referred to as the moment and encodes the position of the screw axis in space and the pitch of the screw.

🛈 **Twist ordering convention**

There is no firm convention for the order of the elements in the twist vector. This book always uses $S = (\omega, v)$ but some other sources, including other editions of this book use $S = (v, \omega)$.

▶ sn.pub/97Lsi6

◘ **Fig. 2.25** Conceptual depiction of a screw. The blue coordinate frame is attached to a nut by a rigid rod. As the nut rotates around the screw, the pose of the frame changes as shown in red. The corollary is that, given any two frames, we can determine a screw axis, pitch and amount of rotation that will transform one frame into the other (the screw thread shown is illustrative only, and does not reflect the pitch of the screw)

2

We begin with unit twists. For the case of pure rotation, the unit twist parallel to the unit vector $\hat{\boldsymbol{a}}$ and passing through the point Q defined by coordinate vector $\boldsymbol{q}$ is

$$\hat{\boldsymbol{S}} = (\hat{\boldsymbol{a}}, \boldsymbol{q} \times \hat{\boldsymbol{a}})$$

and corresponds to unit rotation, 1 radian, about the screw axis.

The screw pitch is the ratio of the distance along the screw axis to the rotation about the axis. For pure rotation, the pitch is zero. For the more general case, where the screw axis is parallel to the vector $\hat{\boldsymbol{a}}$ and passes through the point defined by $\boldsymbol{q}$, and has a pitch of p, the unit twist is

$$\hat{\boldsymbol{S}} = (\hat{\boldsymbol{a}}, \boldsymbol{q} \times \hat{\boldsymbol{a}} + p\hat{\boldsymbol{a}})$$

and the screw pitch is

$$p = \hat{\boldsymbol{w}}^\top \boldsymbol{v} \ .$$

At one extreme, we have a pure rotation, where the screw pitch is zero. At the other extreme, we have a pure translation, which is equivalent to an infinite pitch. For translation parallel to the vector $\hat{\boldsymbol{a}}$, the unit twist is

$$\hat{\boldsymbol{S}} = (0, \hat{\boldsymbol{a}})$$

and corresponds to unit motion along the screw axis.

A unit twist describes the motion induced by a rotation of 1 rad about the screw axis. A unit rotational twist has $\|\boldsymbol{\omega}\| = 1$, while a unit translational twist has $\boldsymbol{\omega} = 0$ and $\|\boldsymbol{v}\| = 1$. We can also consider a unit twist as defining a family of motions

$$\boldsymbol{S} = \theta \hat{\boldsymbol{S}}$$

where $\hat{\boldsymbol{S}}$ completely defines the screw, and the scalar parameter θ describes the amount of rotation about the screw.

A twist vector can be written in a non-compact form as an augmented skew-symmetric matrix, which for the 3D case is

$$[\boldsymbol{S}] = \left(\begin{array}{ccc|c} 0 & -\omega_z & \omega_y & v_x \\ \omega_z & 0 & -\omega_x & v_y \\ -\omega_y & \omega_x & 0 & v_z \\ \hline 0 & 0 & 0 & 0 \end{array} \right) \in \mathbf{se}(3) \ .$$

The matrix belongs to the vector space $\mathbf{se}(3)$, the Lie algebra of $\mathbf{SE}(3)$, and is the *generator* of a rigid-body displacement. The displacement as an $\mathbf{SE}(3)$ matrix is obtained by the exponential mapping

$$\mathbf{T} = e^{[\boldsymbol{S}]} \in \mathbf{SE}(3) \ .$$

If the motion is expressed in terms of a unit twist $\theta \hat{\boldsymbol{S}}$ we can write

$$\mathbf{T}(\theta, \hat{\boldsymbol{S}}) = e^{\theta[\hat{\boldsymbol{S}}]} \in \mathbf{SE}(3)$$

and the matrix exponential has an efficient closed-form

$$\mathbf{T}(\theta, \hat{\boldsymbol{S}}) = \begin{pmatrix} \mathbf{R}(\theta, \hat{\boldsymbol{\omega}}) & \left(\theta \mathbf{1}_{3\times 3} + (1 - \cos \theta)[\hat{\boldsymbol{\omega}}]_\times + (\theta - \sin \theta)[\hat{\boldsymbol{\omega}}]_\times^2 \right) \hat{\boldsymbol{v}} \\ \mathbf{0}_{1\times 3} & 1 \end{pmatrix} \quad (2.46)$$

where $\mathbf{R}(\theta, \hat{\boldsymbol{\omega}})$ can be computed cheaply using Rodrigues' rotation formula (2.27).

The RVC Toolbox implements twists as classes: `Twist` for the 3D case and `Twist2d` for the 2D case – and they have many properties and methods. A 3D rotational unit twist is created by

```
>> S = Twist.UnitRevolute([1 0 0],[0 0 0])
S =
( 1  0  0; 0  0  0 )
```

which describes unit rotation about the x-axis. The twist vector, and its components, can be ▶

In the equations above ω and v are column vectors, but the `Twist` implementation stores them as row vectors.

```
>> S.compact()
ans =
     1     0     0     0     0     0
>> S.v
ans =
     0     0     0
>> S.w
ans =
     1     0     0
```

We can display the Lie algebra of the twist

```
>> S.skewa()
ans =
     0     0     0     0
     0     0    -1     0
     0     1     0     0
     0     0     0     0
```

which is an **se**(3) matrix. If we multiply this by 0.3 and exponentiate it

```
>> expm(0.3*S.skewa)
ans =
    1.0000         0         0         0
         0    0.9553   -0.2955         0
         0    0.2955    0.9553         0
         0         0         0    1.0000
```

the result is the same as `tformrx(0.3)`. The class provides a shorthand way to achieve this result

```
>> S.tform(0.3)
ans =
    1.0000         0         0         0
         0    0.9553   -0.2955         0
         0    0.2955    0.9553         0
         0         0         0    1.0000
```

The class supports composition using the overloaded multiplication operator

```
>> S2 = S*S
S2 =
( 2  0  0; 0  0  0 )
>> S2.printline(orient="axang", unit="rad")
t = (0, 0, 0), R = (2rad | 1, 0, 0)
```

and the result in this case is a twist of two units, or 2 rad, about the x-axis.

The `line` method returns the screw's line of action as a `Plucker` line object that represents a line in 3-dimensional space expressed in Plücker coordinates

```
>> line = S.line()
{ -0   -0   -0; 1   0   0}
```

which can be plotted by

```
>> line.plot("k:",LineWidth=2);
```

Any rigid-body motion described by $\mathbf{T} \in \mathbf{SE}(n)$ can also be described by a twist

$$\boldsymbol{S} = \vee(\log \mathbf{T}) \ .$$

The screw is defined by the unit twist

$$\hat{\boldsymbol{S}} = \boldsymbol{S}/\theta$$

and the motion parameter is

$$\theta = \begin{cases} \|\boldsymbol{\omega}\|, & \text{if } \|\boldsymbol{\omega}\| > 0 \\ \|\boldsymbol{v}\|, & \text{if } \|\boldsymbol{\omega}\| = 0 \end{cases} \ .$$

Consider the example

```
>> T = trvec2tform([1 2 3])*eul2tform([0.3 0.4 0.5]);
>> S = Twist(T)
S =
( 0.42969  0.46143  0.19003; 0.53084  2.5512  2.7225 )
```

The unit twist is

```
>> S/S.theta
ans =
( 0.6525  0.70069  0.28857; 0.8061  3.8741  4.1342 )
```

which is the same result as

```
>> S.unit();
```

Scaling the twist by zero is the null motion expressed as an $\mathbf{SE}(3)$ matrix

```
>> S.tform(0)
ans =
         1         0         0         0
         0         1         0         0
         0         0         1         0
         0         0         0         1
```

and scaling by one is the original rigid-body displacement

```
>> S.tform(1)
ans =
     0.8799    -0.0810     0.4682     1.0000
     0.2722     0.8936    -0.3570     2.0000
    -0.3894     0.4416     0.8083     3.0000
          0          0          0     1.0000
```

If we took just half the required rotation

```
>> S.tform(0.5)
ans =
     0.9692    -0.0687     0.2367     0.3733
     0.1179     0.9727    -0.2001     1.1386
    -0.2165     0.2218     0.9508     1.4498
          0          0          0     1.0000
```

we would have an intermediate displacement. This is not a linear interpolation but rather a *helicoidal* interpolation.

> 🛈 **The proper way to compound twists**
>
> Since twists are the logarithms of rigid-body transformations, there is a temptation to compound the transformations by *adding* the twists rather than multiplying $\mathbf{SE}(3)$ matrices. This works of course for real numbers, if $x = \log X$ and $y = \log Y$, then
>
> $$Z = XY = e^x e^y = e^{x+y}$$
>
> but for the matrix case this would *only be true* if the matrices commute, and rotation matrices do not, therefore
>
> $$Z = XY = e^x e^y \neq e^{x+y} \quad \text{if} \quad x, y \in \mathbf{so}(n) \quad \text{or} \quad \mathbf{se}(n) .$$
>
> The bottom line is that there is no shortcut to compounding poses, we must compute $z = \log(e^x e^y)$ not $z = x + y$. Fortunately, these matrices have a special structure and efficient means to compute the logarithm and exponential as shown in (2.46).

2.4.9 Configuration Space

So far, we have considered the pose of objects in terms of the position and orientation of a coordinate frame affixed to them. For an arm-type robot, we might affix a coordinate frame to its end effector, while for a mobile robot or UAV we might affix a frame to its body – its body-fixed frame. This is sufficient to describe the state of the robot in the familiar 2D or 3D Euclidean space which is referred to as the task space or operational space, since it is where the robot performs tasks or operates. However, it says nothing about the joints and the shape of the arm that positions the end effector.

An alternative way of thinking about this comes from classical mechanics and is referred to as the *configuration* of a system. The configuration is the smallest set of parameters, called generalized coordinates, that are required to fully describe the position of *every* particle in the system. This is not as daunting as it may appear since real-world objects are generally rigid, and, within each of these, the particles have a constant coordinate vector with respect to the object's coordinate frame.

If the *system* is a train moving along a track, then all the particles comprising the train move together and we need only a single generalized coordinate q, the distance along the track from some datum, to describe their location. A robot arm with a fixed base and two rigid links, connected by two rotational joints has a configuration that is completely described by two generalized coordinates – the two joint angles (q_1, q_2). The generalized coordinates can, as their name implies, represent displacements or rotations.

2

Excurse 2.21: Sir Robert Ball

Ball (1840–1913) was an Irish astronomer born in Dublin. He became Professor of Applied Mathematics at the Royal College of Science in Dublin in 1867, and in 1874 became Royal Astronomer of Ireland and Andrews Professor of Astronomy at the University of Dublin. In 1892, he was appointed Lowndean Professor of Astronomy and Geometry at Cambridge University and became director of the Cambridge Observatory. He was a Fellow of the Royal Society and in 1900 became the first president of the Quaternion Society.

He is best known for his contributions to the science of kinematics described in his treatise *The Theory of Screws* (1876), but he also published *A Treatise on Spherical Astronomy* (1908) and a number of popular articles on astronomy. He is buried at the Parish of the Ascension Burial Ground in Cambridge.

That is, there are no holonomic constraints on the system.

The number of independent ◄ generalized coordinates N is known as the number of degrees of freedom of the system. Any configuration of the system is represented by a point q in its N-dimensional configuration space, or C-space, denoted by C and $q \in C$. We can also say that $\dim C = N$. For the train example $C \subset \mathbb{R}$, which says that the displacement is a bounded real number. For the 2-joint robot, the generalized coordinates are both angles, so $C \subset \mathbf{S}^1 \times \mathbf{S}^1$.

Consider again the train moving along its rails. We might wish to describe the train in terms of its position on a plane which we refer to as its task space – in this case, the task space is $\mathcal{T} \subset \mathbb{R}^2$. We might also be interested in the latitude and longitude of the train, in which case the task space would be $\mathcal{T} \subset \mathbf{S}^2$. We could also choose the task space to be our everyday 3D world where $\mathcal{T} \subset \mathbb{R}^3 \times \mathbf{S}^3$ which accounts for height changes as the train moves up and down hills and its orientation changes as it moves around curves and vertical gradients. Any point in the configuration space can be mapped to a point in the task space $q \in C \mapsto \tau \in \mathcal{T}$ and the mapping depends on the particular task space that we choose. However, in general, not all points in the task space can be mapped to the configuration space – some points in the task space are not accessible. While every point along the rail line can be mapped to a point in the task space, most points in the task space will not map to a point on the rail line. The train is constrained by its fixed rails to move in a subset of the task space. If the dimension of the task space exceeds the dimension of the configuration space, $\dim \mathcal{T} > \dim C$, the system can only access a lower-dimensional subspace of the entire task space.

The simple 2-joint robot arm can access a subset of points in a plane, so a useful task space might be $\mathcal{T} \subset \mathbb{R}^2$. The dimension of the task space equals the dimension of the configuration space, $\dim \mathcal{T} = \dim C$, and this means that the mapping between task and configuration spaces is bi-directional but it is not necessarily unique – for this type of robot, in general, two different configurations map to a single point in task space. Points in the task space beyond the physical reach of the robot are not mapped to the configuration space. If we chose a task space with more dimensions such as 2D or 3D pose, then $\dim \mathcal{T} > \dim C$ and the robot would only be able to access points within a subset of that space.

Now consider a snake-robot arm, such as shown in ◘ Fig. 8.10, with 20 joints and $C \subset (\mathbf{S}^1)^{20}$ and $\dim \mathcal{T} < \dim C$. In this case, an infinite number of configurations in a $20 - 6 = 14$-dimensional subspace of the 20-dimensional configuration space will map to the same point in task space. This means that in addition to the task of positioning the robot's end effector, we can *simultaneously* perform motion in the configuration subspace to control the shape of the arm to avoid obstacles in the environment. Such a robot is referred to as overactuated or redundant and this topic is covered in ▶ Sect. 8.3.4.2.

The airframe of a quadrotor, such as shown in ◨ Fig. 4.23d, is a single rigid-body whose configuration is completely described by six generalized coordinates, its position and orientation in 3D space $C \subset \mathbb{R}^3 \times (\mathbf{S}^1)^3$ where the orientation is expressed in some three-angle representation. For such a robot the most logical task space would be $\mathbb{R}^3 \times \mathbf{S}^3$ which is equivalent to the configuration space and $\dim \mathcal{T} = \dim C$. However, the quadrotor has only four actuators which means it cannot *directly* access all the points in its configuration space and hence its task space. Such a robot is referred to as an underactuated system and we will revisit this important topic in ▶ Sect. 4.2.

2.5 MATLAB Classes for Pose and Rotation

MATLAB has strong support for multi-dimensional matrices, and matrices can be used to represent rotation matrices, homogeneous transformation matrices, quaternions and twists. However, quaternions and twists were introduced earlier in this chapter as MATLAB objects rather than matrices. Classes bring a number of advantages:

- They improve code readability and can be used in an almost identical fashion to the native matrix types. For example, binary operations on the objects can be implemented by overloading the standard MATLAB operators allowing us to write `obj1*obj2` rather than `objmul(obj1, obj2)`.
- They provide type safety. MATLAB allows us to add two **SO**(3) matrices even though that is not a defined group operation. A class could raise an error if that operation was attempted.
- Functions that operate on the type can be made methods of the object rather than global functions, and this avoids polluting MATLAB's global namespace.
- We can easily create an *n*-dimensional array of objects that can be indexed in the normal MATLAB way. This is useful to represent a set of related poses or orientations, for example, the poses of all the link frames in a robot manipulator arm or poses along a trajectory.

Given the advantages of classes we would like to use them for rotation matrices and rigid-body transformations. MATLAB supports this, but for historical reasons there are two different sets of objects:

- The Robotics System Toolbox™ provides classes called `so2`, `so3`, `se2` and `se3` respectively for rotation matrices **SO**(2) and **SO**(3), and rigid-body transformation matrices **SE**(2) and **SE**(3). Many functions that expect a rotation matrix or rigid-body transformation will accept a native MATLAB matrix or an equivalent object. ▶

- The Computer Vision™ and Image Processing™ Toolboxes provide a `rigidtform3d` class which represents an **SE**(3) matrix and is similar to the `se3` class but with some notable differences summarized in ◨ Tab. 2.2. These toolboxes also provide related classes for projective (`projtform2d`), similarity (`simtform2d`) and affine (`affinetform3d`) transformations.

Note that these MATLAB classes have lower-case names. The `se3` class represents an **SE**(3) matrix not an **se**(3) matrix.

We will illustrate the difference between native matrices and these classes using an example of an **SE**(3) matrix with a specific rotation and translation. To create a native matrix we compound **SE**(3) matrices representing the translation and the rotation

```
>> T = trvec2tform([1 2 3])*tformrx(0.3)
T =
    1.0000         0         0    1.0000
```

2

■ **Table 2.2** Key operations of the `se3` and `rigidtform3d` classes. Methods are indicated by `()` and properties without. Notes: † when transforming points, the points are row vectors or $n \times 3$ matrices with one row per point; ‡ this is a function, not a method

Operation	se3	rigidtform3d
Translation as a native 1×3 matrix	`.trvec()`	`.Translation`
Rotation as a native 3×3 matrix	`.rotm()`	`.R`
Transformation as a native 4×4 matrix	`.tform()`	`.A`
Composition	`*`	not supported
Inverse	`.inv()`	`.invert()`
Transform point †	`.transform()`	`.transformPointsForward()`
Plot frame	`plotTransforms(T)` ‡	not supported

```
        0    0.9553   -0.2955    2.0000
        0    0.2955    0.9553    3.0000
        0         0         0    1.0000
>> whos T
  Name        Size              Bytes  Class      Attributes
  T           4x4                 128  double
```

To create an `se3` instance we pass the rotation matrix and translation vector to the constructor

```
>> T = se3(rotmx(0.3),[1 2 3])
T =
  se3
   1.0000         0         0    1.0000
        0    0.9553   -0.2955    2.0000
        0    0.2955    0.9553    3.0000
        0         0         0    1.0000
>> whos T
  Name        Size              Bytes  Class      Attributes
  T           1x1                 136  se3
```

The visual appearance of the displayed value of the `se3` object is the same as the standard MATLAB representation of a 4×4 matrix. The class encapsulates the **SE**(3) matrix as a native MATLAB matrix

```
>> T.tform()
ans =
   1.0000         0         0    1.0000
        0    0.9553   -0.2955    2.0000
        0    0.2955    0.9553    3.0000
        0         0         0    1.0000
>> whos ans
  Name        Size              Bytes  Class      Attributes
  ans         4x4                 128  double
```

and we can also extract the rotational and translational components as native MATLAB matrices

```
>> T.rotm()
ans =
   1.0000         0         0
        0    0.9553   -0.2955
        0    0.2955    0.9553
>> T.trvec()
ans =
     1     2     3
```

We can plot the coordinate frame represented by this pose

```
>> plotTransforms(T,FrameLabel="A",FrameColor="b");
```

or using the RVC Toolbox function

```
>> plottform(T,frame="A",color="b");
```

which can plot a frame given an **SO**(3) or **SE**(3) matrix, or an `so3`, `quaternion`, `se3`, `rigidtform3d` or `Twist` object. `animtform` is similar but can animate a coordinate frame. We can display a homogeneous transformation in a concise single-line format using the RVC Toolbox function

```
>> printtform(T);
     T: t = (1, 2, 3), RPY/zyx = (0.3, 0, 0) rad
```

and we see that the roll angle (about the x-axis since we are using the ZYX convention) is 0.3 rad. Various options allow for control over how orientation is represented.

We can transform a 3D point by

```
>> P = [4 5 6];
>> T.transform(P)
ans =
    5.0000    5.0036   10.2096
```

which handles the details of converting the point between Euclidean and homogeneous forms. The `transform` method assumes that the points are provided as row vectors which is the convention across many MATLAB toolboxes. If our data was arranged with points as column vector we could write

```
>> T.transform(P',IsCol=true)
ans =
    5.0000
    5.0036
   10.2096
```

The `rigidtform3d` object behaves similarly

```
>> T = rigidtform3d(rotmx(0.3),[1 2 3]) % create rigidtform3d object
T =
  rigidtform3d with properties:
    Dimensionality: 3
               R: [3×3 double]
     Translation: [1 2 3]
               A: [4×4 double]
>> whos T
  Name      Size              Bytes  Class          Attributes
  T         1x1                  97  rigidtform3d
>> T.A
ans =
    1.0000         0         0    1.0000
         0    0.9553   -0.2955    2.0000
         0    0.2955    0.9553    3.0000
         0         0         0    1.0000
>> T.R
ans =
    1.0000         0         0
         0    0.9553   -0.2955
         0    0.2955    0.9553
>> T.Translation
ans =
     1     2     3
>> T.transformPointsForward(P)
ans =
    5.0000    5.0036   10.2096
```

For the rest of this book, we will use the classes `se3` and `rigidtform3d` in preference to native matrices for rigid-body transformations.

Finally, it is worthwhile noting that MATLAB is unusual amongst object-oriented languages in the way it blurs the distinction between methods and functions. For example

```
>> q = quaternion(rotmx(0.3),"rotmat","point")
q =
  quaternion
     0.98877 +  0.14944i +        0j +        0k
```

is a quaternion object, and its conjugate is obtained by a method using either function call syntax or the dot syntax common in other object-oriented languages

```
>> conj(q)
ans =
  quaternion
     0.98877 -  0.14944i +        0j +        0k
>> q.conj()
ans =
  quaternion
     0.98877 -  0.14944i +        0j +        0k
```

For the first case, if the object has a method with the same name as the function, then that will be invoked, otherwise the object will be passed to a global function of that name. For the latter case, the empty parentheses are not even required

```
>> q.conj
ans =
  quaternion
     0.98877 -  0.14944i +        0j +        0k
```

In this book we use dot notation to distinguish methods from functions, and we use parentheses to distinguish a method call from an object property value.

2.6   Wrapping Up

In this chapter, we introduced relative pose which describes the position and orientation of one object with respect to another. We can think of relative pose as the motion of a rigid body. We discussed how such motions can be composed and decomposed, developed some algebraic rules for manipulating poses, and showed how sets of related poses can be represented as a pose graph. It is important to remember that composition is noncommutative – the order in which relative poses are applied is significant.

To quantify the pose of an object, we affix a 2D or 3D coordinate frame to it. Relative pose is the distance between the origins of the frames and the relative angles between the axes of the coordinate frames. Points within the object are represented by constant coordinate vectors relative to a coordinate frame.

We have discussed a variety of mathematical objects to tangibly represent pose. We have used orthonormal rotation matrices for the 2- and 3-dimensional case to represent orientation and shown how it can rotate a point's coordinate vector from one coordinate frame to another. Its extension, the homogeneous transformation matrix, can be used to represent both orientation and translation, and we have shown how it can rotate and translate a point expressed as a homogeneous coordinate vector from one frame to another. Rotation in 3 dimensions has subtlety and complexity and we have looked at various parameterizations such as ZYZ-Euler angles, roll-pitch-yaw angles, and unit quaternions. Touching on Lie group theory, we showed that rotation matrices, from the group $\mathbf{SO}(2)$ or $\mathbf{SO}(3)$, are the result of exponentiating skew-symmetric generator matrices. Similarly, homogeneous transformation matrices, from the group $\mathbf{SE}(2)$ or $\mathbf{SE}(3)$, are the result of exponentiating augmented skew-symmetric generator matrices. We have also introduced twists as a concise way of describing relative pose in terms of rotation around a screw axis, a

◘ Fig. 2.26 Physical coordinate frames are very helpful when thinking about robotics problems. For information about how build your own coordinate frames visit ▶ https://petercorke.com/resources/3d-frame. There are frames you can 2D print and fold, or 3D print (Image by Dorian Tsai)

notion that comes to us from screw theory, and these twists are the unique elements of the generator matrices.

A simple graphical summary of key concepts was given in ◘ Fig. 2.8. There are two important lessons from this chapter. The first is that there are *many* mathematical objects that can be used to represent pose. There is no right or wrong – each has strengths and weaknesses, and we typically choose the representation to suit the problem at hand. Sometimes we wish for a vectorial representation, perhaps for interpolation, in which case (x, y, θ) or $(x, y, z, \boldsymbol{\Gamma})$ might be appropriate, where $\boldsymbol{\Gamma}$ is a 3-angle sequence, but this representation cannot be easily compounded. Sometimes we may only need to describe 3D rotation in which case $\boldsymbol{\Gamma}$ or $\overset{\circ}{q}$ is appropriate.

The second lesson is that coordinate frames are your friend. The essential first step in many vision and robotics problems is to assign coordinate frames to all objects of interest as shown in ◘ Fig. 2.6. Then add the important relative poses, known and unknown, in a pose graph, write down equations for the loops and solve for the unknowns. ◘ Fig. 2.26 shows how to build your own coordinate frames – you can pick them up and rotate them to make these ideas tangible. Don't be shy, embrace the coordinate frame.

We now have solid foundations for moving forward. The notation has been defined and illustrated, and we have started our hands-on work with MATLAB and the toolboxes. The next chapter discusses motion and coordinate frames that change with time, and after that we are ready to move on and discuss robots.

2.6.1 Further Reading

The treatment in this chapter is a hybrid mathematical and graphical approach that covers the 2D and 3D cases by means of abstract representations and operators which are later made tangible. The standard robotics textbooks such as Kelly (2013), Siciliano et al. (2009), Spong et al. (2006), Craig (2005), and Paul (1981) all introduce homogeneous transformation matrices for the 3-dimensional case but differ in their approach. These books also provide good discussion of the other representations, such as angle-vector and three-angle vectors. Spong et al. (2006, § 2.5.1) have a good discussion of singularities. Siegwart et al. (2011) explicitly cover the 2D case in the context of mobile robotics.

The book by Lynch and Park (2017) is a comprehensive introduction to twists and screws as well as covering the standard matrix approaches. Solà et al. (2018) presents the essentials of Lie-group theory for robotics. Algebraic topology is a challenging topic but Selig (2005) provides a readable way into the subject. Grassia (1998) provides an introduction to exponential mapping for rotations.

Quaternions are discussed in Kelly (2013) and briefly in Siciliano et al. (2009). The books by Vince (2011) and Kuipers (1999) are readable and comprehensive introductions to quaternions. Quaternion interpolation is widely used in computer graphics and animation and the classic paper by Shoemake (1985) is very readable introduction to this topic. Solà (2017) provides a comprehensive discussion of quaternion concepts, including JPL quaternions, and Kenwright (2012) describes the fundamentals of quaternions and dual quaternions. Dual quaternions, and a library with MATLAB bindings, are presented Adorno and Marhinho (2021). The first publication about quaternions for robotics is probably Taylor (1979), and followed up in subsequent work by Funda et al. (1990). Terzakis et al. (2018) describe and compare a number of common rotational representations, while Huynh (2009) introduces and compares a number of distance metrics for 3D rotations.

You will encounter a wide variety of different notation for rotations and transformations in textbooks and research articles. This book uses $^{A}\mathbf{T}_{B}$ to denote an **SE**(3) representation of a rigid-body motion from frame {A} to frame {B}. A common alternative notation is $\mathbf{T}_{B}^{A}$ or even $^{A}_{B}\mathbf{T}$. To describe points, this book uses $^{A}\boldsymbol{p}_{B}$ to denote a coordinate vector from the origin of frame {A} to the point B whereas others use $\boldsymbol{p}_{B}^{A}$, or even $^{C}\boldsymbol{p}_{B}^{A}$ to denote a vector from the origin of frame {A} to the point B but with respect to coordinate frame {C}. Twists can be written as either $(\boldsymbol{\omega}, \boldsymbol{v})$ as in this book, or as $(\boldsymbol{v}, \boldsymbol{\omega})$.

■ ■ **Tools and resources.**

This chapter has introduced MATLAB functionality for creating and manipulating objects such as rotation matrices, homogeneous transform matrices, quaternions and twists. There are numerous other packages that perform similar functions, but have different design objectives. Manif is a Lie theory library written in C++, ▶ https://artivis.github.io/manif. Sophus is a C++ implementation of Lie groups for 2D and 3D geometric problems, ▶ https://github.com/strasdat/Sophus. tf2 is part of the ROS ecosystem and keeps track of multiple coordinate frames and maintains the relationship between them in a tree structure which is similar to the pose graph introduced in this chapter. tf2 can manipulate vectors, quaternions and homogeneous transformations, and read and write those types as ROS geometry messages, ▶ http://wiki.ros.org/tf2. The ROS Toolbox for MATLAB® provides an interface to the transform tree capability, ▶ https://sn.pub/DCfQyx.

■ ■ **Historical and general.**

Quaternions had a tempestuous beginning. Hamilton and his supporters, including Peter Tait, were vigorous in defending Hamilton's precedence in inventing quaternions and promoted quaternions as an alternative to vectors. Vectors are familiar to us today, but at that time were just being developed by Josiah Gibbs. The paper by Altmann (1989) is an interesting description on this tussle of ideas.

Rodrigues developed his eponymous formula in 1840 although Gauss discovered it in 1819 but, as usual, did not publish it. It was published in 1900. The article by Pujol (2012) revisits the history of mathematical ideas for representing rotation. Quaternions have even been woven into fiction (Pynchon 2006).

2.6.2 Exercises

1. Create a 2D rotation matrix. Visualize the rotation using `plottform2d`. Use it to transform a vector. Invert it and multiply it by the original matrix; what is the result? Reverse the order of multiplication; what is the result? What is the determinant of the matrix and its inverse?
2. Build a coordinate frame as shown in ◘ Fig. 2.26 and reproduce the 3D rotation shown in ◘ Fig. 2.15 and 2.16.
3. Create a 3D rotation matrix. Visualize the rotation using `plottform` or `animtform`. Use it to transform a vector. Invert it and multiply it by the original matrix; what is the result? Reverse the order of multiplication; what is the result? What is the determinant of the matrix and its inverse?
4. Explore the many options associated with `plottform`.
5. Animate a rotating cube:
    a) Write a function to plot the edges of a cube centered at the origin.
    b) Modify the function to accept an argument which is a homogeneous transformation that is applied to the cube vertices before plotting.
    c) Animate rotation about the x-axis.
    d) Animate rotation about all axes.
6. Create a vector-quaternion class to describe pose and which supports composition, inverse and point transformation.
7. Compute the matrix exponential using the power series. How many terms are required to match the result shown to standard 64-bit floating point precision?
8. Generate the sequence of plots shown in ◘ Fig. 2.16.
9. Write the function `na2rotm` which is analogous to `oa2rotm` but creates a rotation matrix from the normal and approach vectors.
10. For the 3-dimensional rotation about the vector $[2, 3, 4]$ by 0.5 rad compute an **SO**(3) rotation matrix using: the matrix exponential functions `expm` and Rodrigues' rotation formula (code this yourself), and the function `axang2rotm`. Compute the equivalent unit quaternion.
11. Create two different rotation matrices, in 2D or 3D, representing frames {A} and {B}. Determine the rotation matrix $^A\mathbf{R}_B$ and $^B\mathbf{R}_A$. Express these as a rotation axis and angle, and compare the results. Express these as a twist.
12. Create a 2D or 3D homogeneous transformation matrix. Visualize the rigid-body displacement using `animtform2d` or `animtform`. Use it to transform a vector. Invert it and multiply it by the original matrix, what is the result? Reverse the order of multiplication; what happens?
13. Create three symbolic variables to represent roll, pitch and yaw angles, then use these to compute a rotation matrix (from first principles since `eul2rotm` does not accept symbolic values). You may want to symbolically simplify the result. Use this to transform a unit vector in the z-direction. Looking at the elements of the rotation matrix, devise an algorithm to determine the roll, pitch and yaw angles. Hint: find the pitch angle first.
14. Experiment with the `tripleangle` application in the RVC Toolbox. Explore roll, pitch and yaw motions about the nominal attitude and at singularities.
15. Using the composition rule for **SE**(3) matrices from ▶ Sect. 2.3.2.1, show that $\mathbf{TT}^{-1} = \mathbf{1}$.
16. Is the inverse of a homogeneous transformation matrix equal to its transpose?
17. In ▶ Sect. 2.2.2.2, we rotated a frame about an arbitrary point. Derive the expression for computing TC that was given.
18. Explore the effect of negative roll, pitch or yaw angles. Does transforming from RPY angles to a rotation matrix then back to RPY angles give a different result to the starting value as it does for ZYZ-Euler angles?
19. Show that $e^x e^y \neq e^{x+y}$ for the case of matrices. Hint: expand the first few terms of the exponential series.

2

20. A camera has its z-axis parallel to the vector $(0, 1, 0)$ in the world frame, and its y-axis parallel to the vector $(0, 0, -1)$. What is the attitude of the camera with respect to the world frame expressed as a rotation matrix and as a unit quaternion?

21. Pick a random **SE**(3) matrix, and plot it and the reference frame. Compute and display the screw axis that rotates the reference frame to the chosen frame. Hint: convert the chosen frame to a twist.

Time and Motion

» *The only reason for time is so that everything doesn't happen at once.*
— Albert Einstein

Contents

3.1   Time-Varying Pose – 88

3.2   Accelerating Bodies and Reference Frames – 95

3.3   Creating Time-Varying Pose – 98

3.4   Application: Inertial Navigation – 107

3.5   Wrapping Up – 121

P. Corke, W. Jachimczyk, R. Pillat, *Robotics, Vision and Control*, Springer Tracts in Advanced Robotics 147,
https://doi.org/10.1007/978-3-031-07262-8_3

3

chapter3.mlx

► sn.pub/7AGvQP

The previous chapter was concerned with describing the pose of objects in two or three-dimensional space. This chapter extends those concepts to poses that change as a function of time. ► Sect. 3.1 introduces the derivative of time-varying position, orientation and pose and relates these to concepts from classical mechanics such as velocity and angular velocity. We also cover discrete-time approximations to the derivatives which are useful for computer implementation of algorithms such as inertial navigation. ► Sect. 3.2 is a brief introduction to dynamics, the motion of objects under the influence of forces and torques. We also discuss the important difference between inertial and noninertial reference frames.

► Sect. 3.3 discusses how to generate a temporal sequence of poses, a trajectory, that smoothly changes from an initial pose to a final pose. This could describe the path followed by a robot gripper moving to grasp an object, or the flight path of an aerial robot. ► Sect. 3.4 brings many of these topics together for the important application of inertial navigation. We introduce three common types of inertial sensor and learn how to use their measurements to update the estimate of pose for a moving object such as a robot.

3.1 Time-Varying Pose

This section considers poses that change with time, and we discuss how to describe the rate of change of 3D pose which has both a translational and rotational velocity component. The translational velocity is straightforward: it is the rate of change of the position of the origin of the coordinate frame. Rotational velocity is a little more complex.

3.1.1 Rate of Change of Orientation

We consider the time-varying orientation of the body frame {B} with respect to a fixed frame {A}. There are many ways to represent orientation but the rotation matrix and exponential form, introduced in ► Sect. 2.3.1.6, is convenient

$$^A\mathbf{R}_B(t) = e^{\theta(t)\left[^A\hat{\boldsymbol{\omega}}_B(t)\right]_\times} \in \mathbf{SO}(3)$$

where the rotation is described by a rotational axis $^A\hat{\boldsymbol{\omega}}_B(t)$ defined with respect to frame {A} and a rotational angle $\theta(t)$, and where $[\cdot]_\times$ is a skew-symmetric, or anti-symmetric, matrix.

At an instant in time t, we will assume that the axis has a fixed direction and the frame is rotating around that axis. The derivative with respect to time is

$$^A\dot{\mathbf{R}}_B(t) = \dot{\theta}(t)\left[^A\hat{\boldsymbol{\omega}}_B\right]_\times e^{\theta(t)\left[^A\hat{\boldsymbol{\omega}}_B\right]_\times} \in \mathbb{R}^{3\times3}$$
$$= \dot{\theta}(t)\left[^A\hat{\boldsymbol{\omega}}_B\right]_\times {}^A\mathbf{R}_B(t)$$

which is a general 3×3 matrix, not a rotation matrix. We can write this succinctly as

$$^A\dot{\mathbf{R}}_B = \left[^A\boldsymbol{\omega}_B\right]_\times {}^A\mathbf{R}_B \in \mathbb{R}^{3\times3} \tag{3.1}$$

where $^A\boldsymbol{\omega}_B = \dot{\theta}\,^A\hat{\boldsymbol{\omega}}_B$ is the *angular velocity* of frame {B} with respect to frame {A}. This is a vector quantity $^A\boldsymbol{\omega}_B = (\omega_x, \omega_y, \omega_z)$ that defines the *instantaneous* axis and rate of rotation. The unit vector $^A\hat{\boldsymbol{\omega}}_B$ is parallel to the axis about which the coordinate frame is rotating at a particular instant of time, and the magnitude $\| {}^A\boldsymbol{\omega}_B \|$ is the rate of rotation $\dot{\theta}$ about that axis.

Consider now that angular velocity is measured in the moving frame {B}, for example, it is measured by gyroscope sensors onboard a moving vehicle. We know that

$${}^A\boldsymbol{\omega}_B = {}^A\mathbf{R}_B \, {}^B\boldsymbol{\omega}_B$$

and, using the identity $[\mathbf{R}\boldsymbol{v}]_\times = \mathbf{R}[\boldsymbol{v}]_\times\mathbf{R}^\top$, it follows that

$${}^A\dot{\mathbf{R}}_B = {}^A\mathbf{R}_B\left[{}^B\boldsymbol{\omega}_B\right]_\times \in \mathbb{R}^{3\times3} \tag{3.2}$$

and we see that the order of the rotation matrix and the skew-symmetric matrix have been swapped.

The derivatives of a unit quaternion, the quaternion equivalent of (3.1) and (3.2), are

$${}^A\dot{\mathring{q}}_B = \tfrac{1}{2}\,{}^A\breve{\omega}_B \circ {}^A\mathring{q}_B = \tfrac{1}{2}\,{}^A\mathring{q}_B \circ {}^B\breve{\omega}_B \in \mathbb{H} \tag{3.3}$$

which is a regular quaternion, not a unit quaternion, and $\breve{\omega}$ is a pure quaternion formed from the angular velocity vector.

3.1.2 Rate of Change of Pose

The derivative of pose can be determined by expressing pose as an **SE**(3) matrix

$${}^A\mathbf{T}_B = \begin{pmatrix} {}^A\mathbf{R}_B & {}^A\boldsymbol{t}_B \\ \mathbf{0}_{1\times3} & 1 \end{pmatrix} \in \mathbf{SE}(3)$$

and taking the derivative with respect to time, then substituting (3.1) gives

$${}^A\dot{\mathbf{T}}_B = \begin{pmatrix} {}^A\dot{\mathbf{R}}_B & {}^A\dot{\boldsymbol{t}}_B \\ \mathbf{0}_{1\times3} & 0 \end{pmatrix} = \begin{pmatrix} \left[{}^A\boldsymbol{\omega}_B\right]_\times {}^A\mathbf{R}_B & {}^A\dot{\boldsymbol{t}}_B \\ \mathbf{0}_{1\times3} & 0 \end{pmatrix} \in \mathbb{R}^{4\times4}\ . \tag{3.4}$$

The rate of change can be described in terms of the current orientation ${}^A\mathbf{R}_B$ and *two* velocities. The linear, or translational, velocity ${}^A\boldsymbol{v}_B = {}^A\dot{\boldsymbol{t}}_B \in \mathbb{R}^3$ is the velocity of the origin of {B} with respect to {A}. The angular velocity ${}^A\boldsymbol{\omega}_B \in \mathbb{R}^3$ has already been introduced. We can combine these two velocity vectors to create the spatial velocity vector

$${}^A\boldsymbol{\nu}_B = \left({}^A\boldsymbol{\omega}_B,\ {}^A\boldsymbol{v}_B \right) \in \mathbb{R}^6$$

which is the instantaneous velocity of frame {B} with respect to {A}.

Every point in the body has the same angular velocity, but the translational velocity of a point depends on its position within the body. It is common to place {B} at the body's center of mass.

3

There is no firm convention for the order of the velocities in the spatial velocity vector. This book always uses $v = (\omega, v)$ but some other sources, including other editions of this book use $v = (v, \omega)$.

🛈 Spatial velocity convention

There is no firm convention for the order of the velocities in the spatial velocity vector. This book always uses $v = (\omega, \ v)$ but some other sources, including other editions of this book use $v = (v, \ \omega)$.

3.1.3 Transforming Spatial Velocities

� Fig. 3.1a shows two fixed frames and a moving coordinate frame. An observer on fixed frame {B} observes an object moving with a spatial velocity $^B v$ with respect to frame {B}. For an observer on frame {A}, that frame's spatial velocity is

$$
^A v = \begin{pmatrix} ^A\mathbf{R}_B & \mathbf{0}_{3\times 3} \\ \mathbf{0}_{3\times 3} & ^A\mathbf{R}_B \end{pmatrix} \, ^B v = \, ^A\mathbf{J}_B(\,^A\mathbf{T}_B) \, ^B v \tag{3.5}
$$

where $^A\mathbf{J}_B(\cdot)$ is a Jacobian or interaction matrix which is a function of the relative orientation of the two frames, and independent of the translation between them. We

We use `aTb` to denote $^A\mathbf{T}_B$ in the code in a readable and valid MATLAB® way.

can investigate this numerically, ◄ mirroring the setup of ◘ Fig. 3.1a

```
>> aTb = se3(eul2rotm([-pi/2 0 pi/2]),[-2 0 0]);
```

If the spatial velocity in frame {B} is

```
>> bV = [1 2 3 4 5 6]';
```

then the spatial velocity in frame {A} is

```
>> aJb = velxform(aTb);
>> size(aJb)
ans =
     6     6
>> aV = aJb*bV;
>> aV'   % transpose for display
ans =
   -3.0000   -1.0000    2.0000   -6.0000   -4.0000    5.0000
```

where `velxform` has returned a 6 × 6 matrix. We see that the velocities have been transposed, and sometimes negated due to the different orientation of the frames. The x-axis translational and rotational velocities in frame {B} have been mapped to the $-y$-axis translational and rotational velocities in frame {A}, the y-axis velocities have been mapped to the z-axis, and the z-axis velocities have been mapped to the $-x$-axis.

► sn.pub/DPIahV

► sn.pub/Ga8A7O

◘ **Fig. 3.1** Spatial velocity can be expressed with respect to frame {A} or frame {B}. **a** Both frames are stationary and the object frame is moving with velocity v; **b** both frames are moving together

Next, we consider the case where the two frames are rigidly attached, and moving together, as shown in ◼ Fig. 3.1b. There is an observer in each frame who is able to estimate their velocity with respect to their own frame. The velocities they observe are related by

$$
{}^A\boldsymbol{\nu} = \begin{pmatrix} {}^A\mathbf{R}_B & \mathbf{0}_{3\times 3} \\ \left[{}^A\boldsymbol{t}_B \right]_\times {}^A\mathbf{R}_B & {}^A\mathbf{R}_B \end{pmatrix} {}^B\boldsymbol{\nu} = \mathrm{Ad}({}^A\mathbf{T}_B) \; {}^B\boldsymbol{\nu}
$$

which uses the adjoint matrix of the relative pose and this does depend on the translation between the frames. For example, using the relative pose from the previous example, and with frame {B} experiencing pure translation, the velocity experienced by the observer in frame {A}

```
>> aV = tform2adjoint(aTb)*[0 0 0 1 2 3]';
>> aV'  % transpose for display
ans =
         0         0         0   -3.0000   -1.0000    2.0000
```

is also pure translation, but the axes have been transposed due to the relative orientation of the two frames. If the velocity in frame B is pure angular velocity about the x-axis

```
>> aV = tform2adjoint(aTb)*[1 0 0 0 0 0]';
>> aV'  % transpose for display
ans =
    0.0000   -1.0000         0         0         0    2.0000
```

then the observer in frame {A} measures that same angular velocity magnitude, but now transposed to be about the $-y$-axis. There is also a translational velocity in the z-direction which is due to the origin of {A} following a circular path around the x-axis of frame B. If we combine these translational and rotational velocities in frame {B}, then the spatial velocity in frame {A} will be

```
>> aV = tform2adjoint(aTb)*[1 0 0 1 2 3]';
>> aV'  % transpose for display
ans =
    0.0000   -1.0000         0   -3.0000   -1.0000    4.0000
```

❗ Twist confusion

Many sources (including ROS) refer to spatial velocity, as defined above, as a *twist*. This is not to be confused with the twists introduced in ▶ Sect. 2.2.2.4 and 2.3.2.3. The textbook Lynch and Park (2017) uses the term *velocity twist* $\bar{V}$ which has important differences to the spatial velocity introduced above (the bar is introduced here to clearly differentiate the notation). The velocity twist of a body-fixed frame {B} is ${}^B\bar{V} = ({}^B\bar{\omega}, {}^B\bar{v})$ which has a rotational and translational velocity component, but ${}^B\bar{v}$ is the body-frame velocity of an imaginary point rigidly attached to the body but located at the world frame origin. The body- and world-frame velocity twists are related by the adjoint matrix rather than (3.5). Confusingly, the older book by Murray et al. (1994) refers to the velocity twist as spatial velocity.

3.1.4 Incremental Rotation

In robotics, there are many applications where we need to integrate angular velocity, measured by sensors in the body frame, in order to estimate the orientation of the body frame. This is a key operation in inertial navigation systems which we will discuss in greater detail in ▶ Sect. 3.4. If we assume that the rotational axis is constant over the sample interval δ_t, then the change in orientation over the interval can be expressed in angle-axis form where angle is $\|\boldsymbol{\omega}\delta_t\|$. The orientation update is

$$\mathbf{R}_{B\langle t+\delta_t\rangle} = \mathbf{R}_{B\langle t\rangle}\, e^{\delta_t[\boldsymbol{\omega}]_\times} \tag{3.6}$$

and involves the matrix exponential which is expensive to compute.

We introduced $\dot{\mathbf{R}}$ in ► Sect. 3.1.1 which describes the rate of change of rotation matrix elements with time. We can write this as a first-order approximation to the derivative ◄

The only valid operators for the group **SO**(n) are composition ⊕ and inverse ⊖, so the result of subtraction cannot belong to the group. The result is a 3×3 matrix of element-wise differences. Groups are introduced in ► Sect. 2.1.2 and ► App. D.

$$\dot{\mathbf{R}} \approx \frac{\mathbf{R}_{\langle t+\delta_t\rangle} - \mathbf{R}_{\langle t\rangle}}{\delta_t} \in \mathbb{R}^{3\times3} \tag{3.7}$$

where $\mathbf{R}_{\langle t\rangle}$ is the rotation matrix at time t and δ_t is an infinitesimal time step. Consider an object whose body frames {B} at two consecutive time steps $\mathbf{R}_{B\langle t\rangle}$ and $\mathbf{R}_{B\langle t+\delta_t\rangle}$ are related

$$\mathbf{R}_{B\langle t+\delta_t\rangle} = \mathbf{R}_{B\langle t\rangle}{}^{B\langle t\rangle}\mathbf{R}_{B\langle t+\delta_t\rangle} = \mathbf{R}_{B\langle t\rangle}\,\mathbf{R}^{\Delta}$$

by a small rotation $\mathbf{R}^{\Delta} \in \mathbf{SO}(3)$ expressed in the first body frame. We substitute (3.2) into (3.7) and rearrange to obtain

$$\mathbf{R}^{\Delta} \approx \delta_t\left[{}^{B}\boldsymbol{\omega}\right]_\times + \mathbf{1}_{3\times3} \tag{3.8}$$

This is the first two terms of the Rodrigues' rotation formula (2.27) when $\theta = \delta_t\omega$.

which says that a small rotation can be approximated by the sum of a skew-symmetric matrix and an identity matrix. ◄ We can see this structure clearly for a small rotation about the x-axis

```
>> rotmx(0.001)
ans =
    1.0000         0         0
         0    1.0000   -0.0010
         0    0.0010    1.0000
```

Rearranging (3.8) allows us to approximate the angular velocity vector

$$\boldsymbol{\omega} \approx \frac{1}{\delta_t}\vee_\times\left(\mathbf{R}_{B\langle t\rangle}^{\top}\,\mathbf{R}_{B\langle t+\delta_t\rangle} - \mathbf{1}_{3\times3}\right)$$

from two consecutive rotation matrices where the operator $\vee_\times(\cdot)$ unpacks the unique elements of a skew-symmetric matrix into a vector. The exact value can be found by $\vee_\times \log(\mathbf{R}_{B\langle t\rangle}^{\top}\mathbf{R}_{B\langle t+\delta_t\rangle})$.

Returning now to the rotation integration problem, we can substitute (3.2) into (3.7) and rearrange as

$$\mathbf{R}_{B\langle t+\delta_t\rangle} \approx \mathbf{R}_{B\langle t\rangle}(\mathbf{1}_{3\times3} + \delta_t[\boldsymbol{\omega}]_\times) \tag{3.9}$$

which is cheap to compute and involves no trigonometric operations, but is an approximation. We can numerically explore the tradeoffs in this approximation

```
>> Rexact = eye(3); Rapprox = eye(3);  % null rotation
>> w = [1 0 0]';    % rotation of 1rad/s about x-axis
>> dt = 0.01;       % time step
>> tic
>> for i = 1:100    % exact integration over 100 time steps
>>    Rexact = Rexact*expm(vec2skew(w*dt));   % update by composition
>> end
>> toc  % display the execution time
```

```
>> tic
>> for i = 1:100       % approximate integration over 100 time steps
>>    Rapprox = Rapprox + Rapprox*vec2skew(w*dt); % update by addition
>> end
>> toc  % display the execution time
Elapsed time is 0.003184 seconds.
Elapsed time is 0.000856 seconds.
```

The approximate solution is four times faster, ▶ but the repeated non-group addition operations will result in an improper rotation matrix. We can check this by examining its determinant

This is a very crude way to measure code execution time, see also `timeit`. Results will vary with different computers.

```
>> det(Rapprox)-1
ans =
    0.0100
```

which is significantly different to +1. Even the exact solution has an improper rotation matrix

```
>> det(Rexact)-1
ans =
  -2.5535e-15
```

though the error is very small, and is due to accumulated finite-precision arithmetic error as discussed in ▶ Sect. 2.4.6.

We can normalize the rotation matrices in both cases using `tformnorm`, and then check the overall result which should be a rotation around the x-axis by 1 radian. We can check this conveniently if we convert the rotation matrices to axis-angle form, and to do this we use the function `rotm2axang` which returns a vector $(\hat{\boldsymbol{v}}, \theta) \in \mathbb{R}^4$ containing the rotational axis and rotation angle

```
>> rotm2axang(tformnorm(Rexact))
ans =
      1     0     0     1
>> rotm2axang(tformnorm(Rapprox))
ans =
    1.0000         0         0    1.0000
```

We see that the approximate solution is good to the default printing precision of five significant figures. The sample time used in this example is rather high, but was chosen for the purpose of illustration.

We can also approximate the unit quaternion derivative by a first-order difference

$$\dot{\mathring{q}} \approx \frac{\mathring{q}\langle k+1\rangle - \mathring{q}\langle k\rangle}{\delta_t} \in \mathbb{H}$$

which combined with (3.3) gives us the approximation

$$\mathring{q}\langle k+1\rangle \approx \mathring{q}\langle k\rangle + \tfrac{\delta_t}{2}\mathring{q}\langle k\rangle \circ \breve{\boldsymbol{\omega}} \tag{3.10}$$

where $\breve{\boldsymbol{\omega}}$ is a pure quaternion. This is even cheaper to compute than the rotation matrix approach, and while the result will not be a proper unit quaternion it can be normalized cheaply as discussed in ▶ Sect. 2.4.6.

❶ Addition is not a group operator for rotation matrices

In this section we have updated rotation matrices by addition, but addition is not a group operation for **SO**(3). The resulting matrix will not belong to **SO**(3) – its determinant will not equal one and its columns will not be orthogonal unit vectors.

3

Similarly, we have updated unit quaternions by addition which is not a group operation for $\mathbf{S}^3$. The result will not be a proper unit quaternion belonging to $\mathbf{S}^3$ – its magnitude will not be one.

However, if the values we add are *small*, then this problem is minor, and can be largely corrected by *normalization* as discussed in ▶ Sect. 2.4.6. We can ensure that those values are small by ensuring that δ_t is small which implies a high sample rate. For inertial navigation systems operating on low-end computing hardware, there is a tradeoff between sample rate and accuracy when using approximate update methods.

3.1.5 Incremental Rigid-Body Motion

Consider two poses $\boldsymbol{\xi}_1$ and $\boldsymbol{\xi}_2$ which differ infinitesimally and are related by

$$\boldsymbol{\xi}_2 = \boldsymbol{\xi}_1 \oplus \boldsymbol{\xi}^\Delta$$

where $\boldsymbol{\xi}^\Delta = \ominus \boldsymbol{\xi}_1 \oplus \boldsymbol{\xi}_2$. If $\boldsymbol{\xi}$ is represented by an $\mathbf{SE}(3)$ matrix we can write

$$\mathbf{T}^\Delta = \mathbf{T}_1^{-1}\mathbf{T}_2 = \begin{pmatrix} \mathbf{R}^\Delta & \boldsymbol{t}^\Delta \\ \mathbf{0}_{1\times3} & 1 \end{pmatrix}$$

then $\boldsymbol{t}^\Delta \in \mathbb{R}^3$ is an incremental displacement. $\mathbf{R}^\Delta \in \mathbf{SO}(3)$ is an incremental rotation matrix which, by (3.8), will be an identity matrix plus a skew-symmetric matrix which has only three unique elements $\boldsymbol{r}^\Delta = \vee_\times(\mathbf{R}_\Delta - \mathbf{1}_{3\times3})$. The incremental rigid-body motion can therefore be described by just six parameters

This is useful in optimization procedures that seek to minimize the error between two poses: we can choose the cost function $e = \|\Delta(\xi_1, \xi_2)\|$ which is equal to zero when $\xi_1 \equiv \xi_2$. This is very approximate when the poses are significantly different, but becomes ever more accurate as $\xi_1 \to \xi_2$.

$$\Delta(\boldsymbol{\xi}_1, \boldsymbol{\xi}_2) \mapsto \boldsymbol{\delta} = (\boldsymbol{r}^\Delta, \boldsymbol{t}^\Delta) \in \mathbb{R}^6 \qquad (3.11)$$

which is a *spatial displacement*. ◀ A body with constant spatial velocity $\boldsymbol{v}$ for δ_t seconds undergoes a spatial displacement of $\boldsymbol{\delta} = \delta_t \boldsymbol{v}$. This is an approximation to the more expensive operation $\vee(\log \mathbf{T}^\Delta)$.

The inverse of this operator is

$$\Delta^{-1}(\boldsymbol{\delta}) \mapsto \boldsymbol{\xi}^\Delta \qquad (3.12)$$

Excurse 3.1: Sir Isaac Newton
Newton (1642–1727) was an English natural philosopher and alchemist. He was Lucasian professor of mathematics at Cambridge, Master of the Royal Mint, and the thirteenth president of the Royal Society. His achievements include the three laws of motion, the mathematics of gravitational attraction and the motion of celestial objects, the theory of light and color (see ▶ Exc. 10.1), and building the first reflecting telescope.

Many of these results were published in 1687 in his great 3-volume work *The Philosophiae Naturalis Principia Mathematica* (Mathematical principles of natural philosophy). In 1704 he published *Opticks*, which was a study of the nature of light and color and the phenomena of diffraction. The SI unit of force is named in his honor. He is buried in Westminster Abbey, London.

and if $\boldsymbol{\xi}$ is represented by an $\mathbf{SE}(3)$ matrix then

$$\mathbf{T}_\Delta = \begin{pmatrix} \left[r^\Delta \right]_\times + \mathbf{1}_{3\times 3} & t^\Delta \\ \mathbf{0}_{1\times 3} & 1 \end{pmatrix}$$

which is an approximation to the more expensive operation $e^{[\delta]}$.

The spatial displacement operator and its inverse are implemented by the RVC Toolbox functions `tform2delta` and `delta2tform` ▶ respectively. These functions are computationally cheap compared to the exact operations using `logm` and `expm`, but do assume that the displacements are infinitesimal – they become increasingly inaccurate with larger displacement magnitude.

> The function `delta2se` is similar to `delta2tform` but returns an `se3` object rather than a native matrix.

3.2 Accelerating Bodies and Reference Frames

So far we have considered only the first derivative, the velocity of a coordinate frame. However, all motion is ultimately caused by a force or a torque which leads to acceleration and this is the domain of dynamics.

3.2.1 Dynamics of Moving Bodies

For translational motion, Newton's second law describes, in the inertial frame $\{0\}$, the acceleration of a particle with position x and mass m

$$m\,{}^0\ddot{x}_B = {}^0 f_B \tag{3.13}$$

due to the applied force ${}^0 f_B$.

Rotational motion in three dimensions is described by Euler's equations of motion which relates the angular acceleration of the body in the body frame

$$ {}^B\mathbf{J}_B\,{}^B\dot{\boldsymbol{\omega}}_B + {}^B\boldsymbol{\omega}_B \times \left({}^B\mathbf{J}_B\,{}^B\boldsymbol{\omega}_B \right) = {}^B\boldsymbol{\tau}_B \tag{3.14}$$

to the applied torque or moment ${}^B\boldsymbol{\tau}_B$ and a positive-definite rotational inertia tensor ${}^B\mathbf{J}_B \in \mathbb{R}^{3\times 3}$. Nonzero angular acceleration implies that angular velocity, the axis and angle of rotation, evolves over time. ▶

We will simulate the rotational motion of an object moving in space. We define the inertia tensor as

```
>> J = [2 -1 0; -1 4 0; 0 0 3];
```

which is positive definite ▶ and the nonzero off-diagonal terms will cause it to *tumble*. The initial conditions for orientation and angular velocity are

```
>> orientation = quaternion([1 0 0 0]); % null rotation
>> w = 0.2*[1 2 2]';  % initial angular velocity
```

The simulation is implemented as an animation loop which computes angular acceleration by (3.14), and uses rectangular integration to update the angular velocity and orientation.

```
>> dt = 0.05;  % time step
>> h = plottform(se3);  % create coordinate frame and return handle
```

> In the absence of torque, the angular velocity a body is not necessarily constant, for example, the angular velocity of a satellite tumbling in space is not constant. This is quite different to the linear velocity case where, in the absence of force, velocity remains constant. For rotational motion, it is angular momentum $h = \mathbf{J}\boldsymbol{\omega}$ in the inertial frame that is constant.

> Off-diagonal elements of a positive-definite matrix can be negative, it is the eigenvalues that must be positive.

3

> **Excurse 3.2: Rotational Inertia Tensor**
>
> The rotational inertia of a body that moves in three dimensions is described by a rank-2 tensor whose elements depend on the choice of the coordinate frame. An inertia tensor can be written and used like a 3×3 matrix.
>
> $$\mathbf{J} = \begin{pmatrix} J_{xx} & J_{xy} & J_{xz} \\ J_{xy} & J_{yy} & J_{yz} \\ J_{xz} & J_{yz} & J_{zz} \end{pmatrix}$$
>
> which is symmetric and positive definite, that is, it has positive eigenvalues. The eigenvalues also satisfy the triangle inequality: the sum of any two eigenvalues is always greater than, or equal to, the other eigenvalue.
>
> The diagonal elements of $\mathbf{J}$ are the positive moments of inertia, and the off-diagonal elements are products of inertia. Only six of these nine elements are unique: three moments and three products of inertia. The products of inertia are all zero if the object's mass distribution is symmetrical with respect to the coordinate frame.

```
>> for t = 0:dt:10
>>    wd = -inv(J)*(cross(w,J*w)); % angular acceleration by (3.14)
>>    w = w + wd*dt;
>>    orientation = orientation*quaternion(w'*dt,"rotvec"); % update
>>    plotform(orientation,handle=h)  % update displayed frame
>>    pause(dt)
>> end
```

The constructor `quaternion(..., "rotvec")` creates a new `quaternion` from a 3-element rotation vector or Euler vector. At line 2 `plotform` creates a graphical coordinate frame and this is updated, using the returned graphics handle, at line 7 so as to create an animation effect.

3.2.2 Transforming Forces and Torques

The translational force $\boldsymbol{f} = (f_x, f_y, f_z)$ and rotational torque, or moment, $\boldsymbol{m} = (m_x, m_y, m_z)$ applied to a body can be combined into a 6-vector that is called a wrench $\boldsymbol{w} = (m_x, m_y, m_z, f_x, f_y, f_z) \in \mathbb{R}^6$.

A wrench $^B\boldsymbol{w}$ is defined with respect to the coordinate frame {B} and applied at the origin of that frame. For coordinate frame {A} attached to the same body, the wrench $^A\boldsymbol{w}$ is equivalent if it causes the same motion of the body when applied to the origin of frame {A}. The wrenches are related by

$$^A\boldsymbol{w} = \begin{pmatrix} ^B\mathbf{R}_A & \mathbf{0}_{3\times3} \\ \left[^B\boldsymbol{t}_A \right]_\times {}^B\mathbf{R}_A & {}^B\mathbf{R}_A \end{pmatrix}^\top {}^B\boldsymbol{w} = \mathrm{Ad}^\top\!\left(^B\mathbf{T}_A \right) {}^B\boldsymbol{w} \tag{3.15}$$

which is similar to the spatial velocity transform of (3.5) but uses the transpose of the adjoint of the *inverse* relative pose.

Continuing the example from ◘ Fig. 3.1b, we define a wrench with respect to frame {B} that exerts force along each axis

```
>> bW = [0 0 0 1 2 3]';
```

The equivalent wrench in frame {A} would be

```
>> aW = tform2adjoint(inv(aTb))'*bW;
>> aW'  % transpose for display
ans =
        0    4.0000    2.0000   -3.0000   -1.0000    2.0000
```

which has the same force components as applied at {B}, but transposed and negated to reflect the different orientations of the frames. The forces applied to the origin of {B} will also exert a moment on the body, so the equivalent wrench at {A} must include this: a torque of 4 Nm about the y-axis and 2 Nm about the z-axis. Note that there is no moment about the x-axis of frame {A}.

> 🛈 **Wrench convention**
> There is no firm convention for the order of the forces and torques in the wrench vector. This book always uses $w = (m, \ f)$ but other sources, including other editions of this book, use $w = (f, \ m)$.

3.2.3 Inertial Reference Frame

In robotics it is important to distinguish between an *inertial reference frame* and a non-inertial reference frame. An inertial reference frame is crisply defined as:

» *a reference frame that is not accelerating or rotating.*

Consider a particle P at rest with respect to a stationary reference frame {0}. Frame {B} is moving with constant velocity ${}^0v_B$ relative to frame {0}. From the perspective of frame {B}, the particle would be moving at constant velocity, in fact ${}^Bv_P = -{}^0v_B$. The particle is not accelerating and obeys Newton's first law:

» *in the absence of an applied force, a particle moves at a constant velocity*

and therefore frame {B} is an inertial reference frame.

Now consider that frame {B} is accelerating, at a constant acceleration ${}^0a_B$ with respect to {0}. From the perspective of frame {B}, the particle appears to be accelerating without an applied force, in fact ${}^Ba_P = -{}^0a_B$. This violates Newton's first law and an observer in frame {B} would have to invoke some intangible force to explain what they observe. We call such a force a fictitious, apparent, pseudo, inertial or d'Alembert force – they only exist in an accelerating or noninertial reference frame. This accelerating frame {B} is *not* an inertial reference frame.

Gravity could be considered to be an intangible force since it causes objects to accelerate with respect to a frame that is not accelerating. However, in Newtonian mechanics, gravity is considered a real body force mg and a free object will accelerate relative to the inertial frame. ▶

An everyday example of a noninertial reference frame is an accelerating car or airplane. Inside an accelerating vehicle we observe fictitious forces pushing objects, including ourselves, around in a way that is not explained by Newton's first law.

For a rotating reference frame, things are more complex still. Imagine two people standing on a rotating turntable, and throwing a ball back and forth. They would observe that the ball follows a curved path in space, and they would also have to invoke an intangible force to explain what they observe.

If the reference frame {B} is rotating with angular velocity ω about its origin, then Newton's second law (3.13) becomes

Einstein's equivalence principle is that *"we assume the complete physical equivalence of a gravitational field and a corresponding acceleration of the reference system"* – we are unable to distinguish between gravity and being on a rocket accelerating at 1 g far from the gravitational influence of any celestial object.

$$m\left({}^B\dot{v} + \underbrace{\omega \times (\omega \times {}^Bp)}_{\text{centripetal}} + \underbrace{2\omega \times {}^Bv}_{\text{Coriolis}} + \underbrace{\dot{\omega} \times {}^Bp}_{\text{Euler}} \right) = {}^0f$$

3

Coriolis (1792–1843) was a French mathematician, mechanical engineer and scientist. Born in Paris, in 1816 he became a tutor at the École Polytechnique where he carried out experiments on friction and hydraulics and later became a professor at the École des Ponts and Chaussées (School of Bridges and Roads). He extended ideas about kinetic energy and work to rotating systems and in 1835 wrote the famous paper *Sur les équations du mouvement relatif des systèmes de corps* (On the equations of relative motion of a system of bodies) which dealt with the transfer of energy in rotating systems such as waterwheels. In the late 19th century, his ideas were used by the meteorological community to incorporate effects due to the Earth's rotation. He is buried in Paris's Montparnasse Cemetery.

with three *new* acceleration terms. Centripetal acceleration always acts inward toward the origin. If the point is moving, then Coriolis acceleration will be normal to its velocity. If the rotational velocity varies with time, then Euler acceleration will be normal to the position vector. The centripetal term can be moved to the right-hand side, in which case it becomes a fictitious outward centrifugal force. This complexity is symptomatic of being in a noninertial reference frame, and another definition of an inertial frame is:

» *one in which the physical laws hold good in their simplest form*
  – Albert Einstein: The foundation of the general theory of relativity.

In robotics, the term inertial frame and world coordinate frame tend to be used loosely and interchangeably to indicate a frame fixed to some point on the Earth. This is to distinguish it from the body frame attached to the robot or vehicle. The surface of the Earth is an approximation of an inertial reference frame – the effect of the Earth's rotation is a finite acceleration less than $0.03\,\mathrm{m\,s^{-2}}$ due to centripetal acceleration. From the perspective of an Earth-bound observer, a moving body will experience Coriolis acceleration. ◄ These effects are small, dependent on latitude, and typically ignored.

Coriolis acceleration is significant for large-scale weather systems and therefore meteorological prediction, but is below the sensitivity of low-cost sensors such as those found in smart phones.

3.3 Creating Time-Varying Pose

In robotics, we often need to generate a pose that varies smoothly with time from ξ_0 to ξ_1, for example, the desired motion of a robot end effector or a drone. We require position and orientation to vary smoothly with time. In simple terms, this is moving smoothly along a path that is itself smooth. Smoothness in this context means that its first few temporal derivatives of position and orientation are continuous. Typically, velocity and acceleration are required to be continuous and sometimes also the derivative of acceleration or jerk.

The first problem is to construct a smooth path from ξ_0 to ξ_1 which could be defined by some function $\xi(s)$, $s \in [0, 1]$. The second problem is to construct a smooth trajectory, that is smooth motion along the path which requires that $s(t)$ is a smooth function of time.

We start by discussing how to generate smooth trajectories in one dimension. We then extend that to the multi-dimensional case and then to piecewise-linear trajectories that visit a number of intermediate points without stopping.

3.3.1 Smooth One-Dimensional Trajectories

We start our discussion with a scalar function of time that has a specified initial and final value. An obvious candidate for such a smooth function is a polynomial function of time. Polynomials are simple to compute and can easily provide the required smoothness and boundary conditions. A quintic (fifth-order) polynomial is commonly used

$$q(t) = At^5 + Bt^4 + Ct^3 + Dt^2 + Et + F \tag{3.16}$$

where time $t \in [0, T]$. The first- and second-derivatives are also polynomials

$$\dot{q}(t) = 5At^4 + 4Bt^3 + 3Ct^2 + 2Dt + E \tag{3.17}$$

$$\ddot{q}(t) = 20At^3 + 12Bt^2 + 6Ct + 2D \tag{3.18}$$

and therefore smooth. The third-derivative, jerk, will be a quadratic.

The trajectory has defined boundary conditions for position (q_0, q_T), velocity $(\dot{q}_0, \dot{q}_T)$ and acceleration $(\ddot{q}_0, \ddot{q}_T)$. Writing (3.16) to (3.18) for the boundary conditions $t = 0$ and $t = T$ gives six equations which we can write in matrix form as

$$
\begin{pmatrix} q_0 \\ q_T \\ \dot{q}_0 \\ \dot{q}_T \\ \ddot{q}_0 \\ \ddot{q}_T \end{pmatrix} = \begin{pmatrix} 0 & 0 & 0 & 0 & 0 & 1 \\ T^5 & T^4 & T^3 & T^2 & T & 1 \\ 0 & 0 & 0 & 0 & 1 & 0 \\ 5T^4 & 4T^3 & 3T^2 & 2T & 1 & 0 \\ 0 & 0 & 0 & 2 & 0 & 0 \\ 20T^3 & 12T^2 & 6T & 2 & 0 & 0 \end{pmatrix} \begin{pmatrix} A \\ B \\ C \\ D \\ E \\ F \end{pmatrix}.
$$

The matrix is square ▶ and, if $T \neq 0$, we can invert it to solve for the coefficient vector (A, B, C, D, E, F).

We can generate the trajectory described by (3.16)

```
>> t = linspace(0,1,50); % 0 to 1 in 50 steps
>> [q,qd,qdd] = quinticpolytraj([0 1],[0 1],t);
```

where the arguments are respectively a vector containing the start and end positions, a vector containing the start and end time, and a vector of times at which to sample the trajectory. The outputs q, qd and qdd are respectively the trajectory, velocity and acceleration – each a 1×50 row vector. We can plot these values

```
>> stackedplot(t,[q' qd' qdd'])
```

as shown in ◘ Fig. 3.2a. We observe that the initial and final velocity and acceleration are all zero – the default value.

The initial and final velocities can be set to nonzero values, for example, an initial velocity of 10 and a final velocity of 0

```
>> [q2,qd2,qdd2] = quinticpolytraj([0 1],[0 1],t, ...
>>    VelocityBoundaryCondition=[10 0]);
```

creates the trajectory shown in ◘ Fig. 3.2b. This illustrates an important problem with polynomials – nonzero initial velocity causes the polynomial to overshoot the terminal value, in this example, peaking at over 2 on a trajectory from 0 to 1.

Another problem with polynomials, a very practical one, can be seen in the middle graph of ◘ Fig. 3.2a. The velocity peaks when $t = 0.5\,$s which means that,

This is the reason for choice of quintic polynomial. It has six coefficients that enable it to meet the six boundary conditions on initial and final position, velocity and acceleration.

3

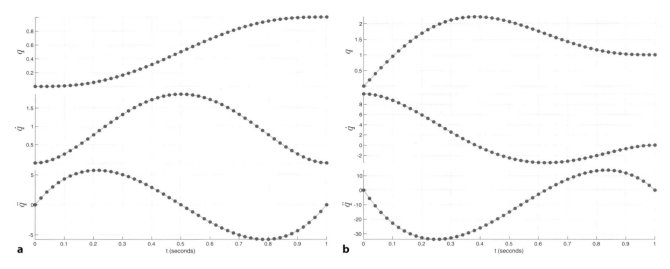

Fig. 3.2 Quintic polynomial trajectory. From top to bottom is position, velocity and acceleration versus time step. **a** With zero initial and final velocity, **b** initial velocity of 10 and a final velocity of 0. The discrete-time points are indicated by dots

for most of the time, the velocity is far less than the maximum. The mean velocity

```
>> mean(qd)/max(qd)
ans =
    0.5231
```

is only 52% of the peak so we are not using the motor as fully as we could. A real robot joint has a well-defined maximum speed and, for minimum-time motion, we want to be operating at that maximum for as much of the time as possible. We would like the velocity curve to be *flatter* on top.

A well-known alternative is a trapezoidal hybrid trajectory

```
>> [q,qd,qdd] = trapveltraj([0 1],50);
>> stackedplot(t,[q' qd' qdd'])
```

where the arguments are the start and end position, and the number of time steps. The trajectory, plotted in a similar way, is shown in ◻ Fig. 3.3a.

The velocity trajectory has a trapezoidal shape, hence the name, comprising three linear segments. The corresponding segments of the position trajectory are

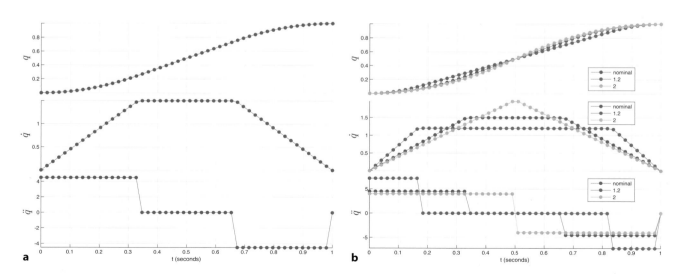

Fig. 3.3 Trapezoidal trajectory: **a** trajectory for default linear segment velocity; **b** trajectories for specified linear segment velocities

a straight line (constant velocity segment) with parabolic blends. The term blend refers to a trajectory segment that smoothly joins linear segments. This type of trajectory is commonly used in industrial motor drives. It is continuous in position and velocity, but not in acceleration.

The function `trapveltraj` has *chosen* the velocity of the linear segment to be

```
>> max(qd)
ans =
    1.5000
```

but this can be overridden by specifying an additional argument

```
>> [q2,qd2,qdd2] = trapveltraj([0 1],50,EndTime=1,PeakVelocity=1.2);
>> [q3,qd3,qdd3] = trapveltraj([0 1],50,EndTime=1,PeakVelocity=2);
```

The trajectories for these different cases are overlaid in ◘ Fig. 3.3b. We see that as the velocity of the linear segment increases, its duration decreases and ultimately its duration would be zero. In fact, the velocity cannot be chosen arbitrarily – too high or too low a value for the maximum velocity will result in an infeasible trajectory and the function will return an error. ▶

The system has one design degree of freedom. There are six degrees of freedom (blend time, three parabolic coefficients and two linear coefficients) and five constraints (total time, initial and final position and velocity).

3.3.2 Multi-Axis Trajectories

Most useful robots have more than one axis of motion and it is quite straightforward to extend the smooth scalar trajectory to the vector case. In terms of configuration space (▶ Sect. 2.4.9), these axes of motion correspond to the dimensions of the robot's configuration space – to its degrees of freedom. We represent the robot's configuration as a vector $q \in \mathbb{R}^N$ where N is the number of degrees of freedom. The configuration of a 3-joint robot would be its joint coordinates $q = (q_1, q_2, q_3)$. The configuration vector of wheeled mobile robot might be its position $q = (x, y)$ or its position and heading angle $q = (x, y, \theta)$. For a 3-dimensional body that had an orientation in **SO**(3) we would use roll-pitch-yaw angles $q = (\alpha, \beta, \gamma)$, or for a pose in **SE**(3) we would use $q = (x, y, z, \alpha, \beta, \gamma)$. ▶ In all these cases, we would require smooth multi-dimensional motion from an initial configuration vector to a final configuration vector.

Any 3-parameter representation could be used: roll-pitch-yaw angles, Euler angles or exponential coordinates.

Consider a 2-axis xy-type robot moving from configuration $(0, 2)$ to $(1, -1)$ in 50 steps with a trapezoidal profile. This is achieved by

```
>> q0 = [0 2]; qf = [1 -1];
>> q = trapveltraj([q0' qf'],50,EndTime=1);
```

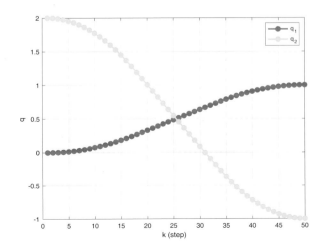

◘ **Fig. 3.4** Multi-axis motion. q_1 varies from $0 \rightarrow 1$ and q_2 varies from $2 \rightarrow -1$

which results in a 2×50 matrix `q` with one column per time step and one row per axis. The first argument is a 2-column matrix that specified the initial and final configurations. The trajectory

```
>> plot(q')
```

is shown in ◘ Fig. 3.4.

If we wished to create a trajectory for 3-dimensional pose, we could convert an `se3` pose instance `T` to a 6-vector by a command like

```
q = [T.trvec() rotm2eul(T.rotm())]
```

though as we shall see later, interpolation of 3-angle representations has some complexities.

3.3.3 Multi-Segment Trajectories

In robotics applications, there is often a need to move smoothly along a path through one or more intermediate or *via* points or *waypoints* without stopping. This might be to avoid obstacles in the workplace, or to perform a task that involves following a piecewise-linear trajectory such as welding a seam or applying a bead of sealant in a manufacturing application.

To formalize the problem, consider that the trajectory is defined by M configurations $q_k, k = 1, \ldots, M$ so there will be $M - 1$ motion segments. As in the previous section, $q_k \in \mathbb{R}^N$ is a configuration vector. The robot starts from q_1 at rest and finishes at q_M at rest, and passes through, and stops at, the intermediate configurations.

Consider again a 2-axis xy-type robot following a path defined by the corners of a rotated square. The trajectory can be generated by

```
>> cornerPoints = [-1 1; 1 1; 1 -1; -1 -1; -1 1];
>> R = so2(deg2rad(30),"theta");
>> via = R.transform(cornerPoints)';
>> [q,qd,qdd,t] = trapveltraj(via,100,EndTime=5);
```

The first argument is a matrix whose columns are the initial configuration, a number of via points, and the final configuration. The remaining arguments are the total number of sample points, and the time per motion segment. The function returns a 2×100 matrix with one column per time step for position, velocity and acceleration. The values in each column correspond to the axis positions. The time is given by the corresponding column of `t`. If we want the robot to slow down before the via points, we can specify a particular acceleration time

```
>> q2 = trapveltraj(via,100,EndTime=5,AccelTime=2);
```

To see the path of the robot we can plot q_2 against q_1

```
>> plot(q(1,:),q(2,:),".-")
```

which is shown in ◘ Fig. 3.5a.

The configuration variables, as a function of time, are shown in ◘ Fig. 3.5b and the linear segments and blends can be clearly seen. This function also accepts optional parameters for peak velocity, acceleration and acceleration time.

Keep in mind that this function simply interpolates configuration represented as a vector. In this example, the vector was assumed to be Cartesian coordinates. For rotation, we could consider applying this function to Euler or roll-pitch-yaw angles but this is not an ideal way to interpolate orientation as we will discuss in the next section.

For a continuous trajectory that visits all the via points without stopping we surrender the ability to move along straight line segments. Such a trajectory, that

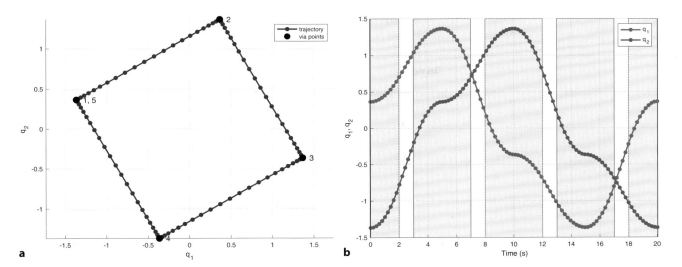

Fig. 3.5 Multi-segment multi-axis trajectory that stops at via points: **a** configuration of robot (tool position); **b** configuration versus time with segment blends indicated by gray boxes and via point indicated by solid lines. In all cases, acceleration time is 2 seconds

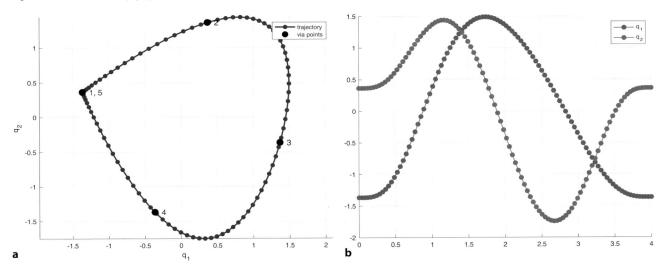

Fig. 3.6 Multi-segment multi-axis trajectory that moves through via points: **a** configuration of robot (tool position); **b** configuration versus time

minimizes jerk, is given by

```
>> [q,qd,qdd] = minjerkpolytraj(via,[1 2 3 4 5],500);
>> plot(q(1,:),q(2,:),".-")
```

and is shown in ◘ Fig. 3.6. The second argument is the arrival time at the corresponding via point.

If we want the axes to obey velocity and acceleration limits we can use a constrained time-optimal trajectory

```
>> vel_lim = [-1 1; -1 1]; accel_lim= [-2 2; -2 2];
>> [q,qd,qdd] = contopptraj(via,vel_lim,accel_lim,numsamples=100);
>> plot(q(1,:),q(2,:),".-")
```

where the rows of `vel_lim` and `accel_lim` correspond to axes and are the minimum and maximum values of velocity or acceleration respectively.

3

3.3.4 **Interpolation of Orientation in 3D**

In robotics, we often need to interpolate orientation, for example, we require the end effector of a robot to smoothly change from orientation ξ_0 to ξ_1 in $\mathbf{S}^3$. We require some function $\xi(s) = \sigma(\xi_0, \xi_1, s)$ where $s \in [0, 1]$, with boundary conditions $\sigma(\xi_0, \xi_1, 0) = \xi_0$ and $\sigma(\xi_0, \xi_1, 1) = \xi_1$ and where $\sigma(\xi_0, \xi_1, s)$ varies *smoothly* for intermediate values of s. How we implement this depends very much on our concrete representation of ξ.

A workable and commonly used approach is to consider a 3-parameter representation such as Euler or roll-pitch-yaw angles, $\Gamma \in (\mathbf{S}^1)^3$, or exponential coordinates and use linear interpolation

$$\sigma(\Gamma_0, \Gamma_1, s) = (1 - s)\Gamma_0 + s\Gamma_1, s \in [0, 1]$$

and converting the interpolated vector back to a rotation matrix. For example, we define two orientations

```
>> rpy0 = [-1 -1 0]; rpy1 = [1 1 0];
```

and create a trajectory between them with 50 uniform steps

```
>> rpy = quinticpolytraj([rpy0' rpy1'],[0 1],linspace(0,1,50));
```

and `rpy` is a 50×3 matrix of roll-pitch-yaw angles. This is most easily visualized as an animation, so we will first convert those angles to a sequence of $\mathbf{SO}(3)$ matrices ◄ and then animate them

> A matrix with three dimensions `A(i,j,k)` where `A(:,:,k)` is the k'th **SO**(3) matrix in the sequence.

```
>> animtform(so3(rpy',"eul"))
```

For large orientation change we see that the axis, around which the coordinate frame rotates, changes along the trajectory. The motion, while smooth, can look a bit uncoordinated. There will be problems if either ξ_0 or ξ_1 is close to a singularity in the particular vector representation of orientation.

Interpolation of unit quaternions is only a little more complex than for 3-angle vectors and results in a rotation about a *fixed* axis in space. We first find the two equivalent unit quaternions

```
>> q0 = quaternion(eul2rotm(rpy0),"rotmat","point");
>> q1 = quaternion(eul2rotm(rpy1),"rotmat","point");
```

and then interpolate them in 50 uniform steps

```
>> q = q0.slerp(q1,linspace(0,1,50));
>> whos q
  Name       Size              Bytes  Class         Attributes
  q          1x50               1600  quaternion
```

which results in a row vector of 50 quaternion objects which we can animate by

```
>> animtform(q)
```

Quaternion interpolation is achieved using spherical linear interpolation (*slerp*) in which the unit quaternions follow a great circle ◄ path on a 4-dimensional hypersphere. The result in three dimensions is rotation about a fixed axis in space.

> A great circle on a sphere is the intersection of the sphere and a plane that passes through its center. On Earth, the equator and all lines of longitude are great circles. Ships and aircraft prefer to follow great circles because they represent the shortest path between two points on the surface of a sphere.

Several alternatives are supported. For example, to obtain angular velocity and acceleration as well we can use

```
>> tpts = [0 5]; tvec = [0:0.01:5];
>> [q,w,a] = rottraj(q0,q1,tpts,tvec);
```

where the arguments can be `quaternion` objects, `so3` objects or $\mathbf{SO}(3)$ matrices, `w` is angular velocity at each time step and `a` is angular acceleration at each time step. Interpolation is performed using quaternion slerp interpolation. This is constant velocity motion with a discontinuity at the start and end points. For continuous motion we can interpolate using trapezoidal or cubic time scaling, and the latter can be achieved by

```
>> [s,sd,sdd] = cubicpolytraj([0 1],tpts,tvec);
>> [q,w,a] = rottraj(q0,q1,tpts,tvec,TimeScaling=[s;sd;sdd]);
```

> 🛑 **We cannot linearly interpolate a rotation matrix**
>
> If pose is represented by an orthogonal rotation matrix $\mathbf{R} \in \mathbf{SO}(3)$, we *cannot* use linear interpolation $\sigma(\mathbf{R}_0, \mathbf{R}_1, s) = (1 - s)\mathbf{R}_0 + s\mathbf{R}_1$. Scalar multiplication and addition are not valid operations for the group of $\mathbf{SO}(3)$ matrices, and that means that the resulting $\mathbf{R}$ would not be an $\mathbf{SO}(3)$ matrix – the column norm and inter-column orthogonality constraints would be violated.

3.3.4.1 Direction of Rotation

When moving between two points on a circle, we can travel clockwise or counterclockwise – the result is the same but the distance traveled may be different. The same choices exist when we move on a great circle. In this example we animate a rotation about the z-axis, from an angle of -2 radians to $+1$ radians

```
>> q1 = quaternion(rotmz(-2),"rotmat","point");
>> q2 = quaternion(rotmz(1),"rotmat","point");
>> animtform(q1.slerp(q2,linspace(0,1,50)))
```

which is a rotation of 3 radians counter-clockwise. If we require a rotation of 4 radians

```
>> q2 = quaternion(rotmz(2),"rotmat","point");
>> animtform(q1.slerp(q2,linspace(0,1,50)))
```

the motion is now clockwise through $2\pi - 4 = 2.28$ radians. The quaternion slerp interpolation method always takes the shortest path along the great circle.

3.3.5 Cartesian Motion in 3D

Another common requirement in robotics is to find a smooth path between two 3D poses in $\mathbb{R}^3 \times \mathbf{S}^3$ which involves change in position as well as in orientation. In robotics, this is often referred to as Cartesian motion.

We represent the initial and final poses as $\mathbf{SE}(3)$ matrices

```
>> T0 = se3(eul2rotm([1.5 0 0]),[0.4 0.2 0]);
>> T1 = se3(eul2rotm([-pi/2 0 -pi/2]),[-0.4 -0.2 0.3]);
```

The se3 object has a method interp that interpolates between two poses for normalized distance $s \in [0, 1]$ along the path, for example the midway pose between T0 and T1 is

```
>> T0.interp(T1,0.5)
ans =
  se3
    0.7239   -0.4622    0.5122         0
   -0.0256    0.7239    0.6894         0
   -0.6894   -0.5122    0.5122    0.1500
         0         0         0    1.0000
```

where the object is the initial pose, the arguments to the interp method are the final pose and the normalized distance, and the result is an se3 object. The translational component is linearly interpolated, while rotation is interpolated using unit quaternion spherical linear interpolation as introduced above. A trajectory between the two poses in 50 uniform steps is created by

```
>> tpts = [0 1]; tvec = linspace(tpts(1),tpts(2),50);
>> Ts = transformtraj(T0,T1,tpts,tvec);
```

3

and the result is a row vector of `se3` objects

```
>> whos Ts
  Name      Size              Bytes  Class     Attributes
  Ts        1x501             68136  se3
```

representing the pose at each time step, and once again the easiest way to visualize this is by animation

```
>> animtform(Ts)
```

which shows the coordinate frame moving and rotating from pose `T0` to pose `T1`. The **SE**(3) value at the mid-point on the path is

```
>> Ts(251)
ans =
  se3
     0.7239   -0.4622    0.5122         0
    -0.0256    0.7239    0.6894         0
    -0.6894   -0.5122    0.5122    0.1500
          0         0         0    1.0000
```

and is the same as the result given above.

The translational part of this trajectory is obtained by

```
>> P = Ts.trvec();
>> size(P)
ans =
   501     3
```

which is a matrix where the rows represent position at consecutive time steps. This is plotted

```
>> plot(P);
```

in ◘ Fig. 3.7 along with the orientation expressed in roll-pitch-yaw angles

```
>> plot(rotm2eul(Ts.rotm()));
```

We see that the position coordinates vary linearly with time, but that the roll-pitch-yaw angles are nonlinear with time and that has two causes. Firstly, roll-

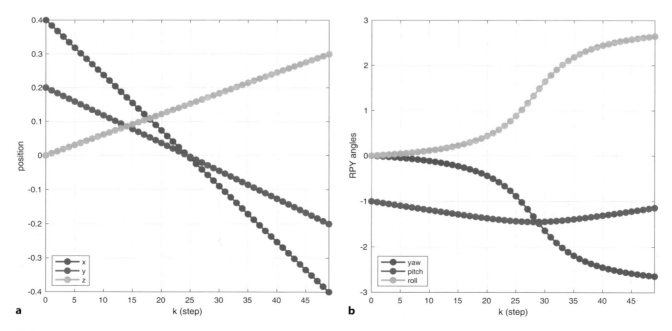

◘ **Fig. 3.7** Cartesian motion. **a** Cartesian position versus time; **b** roll-pitch-yaw angles versus time

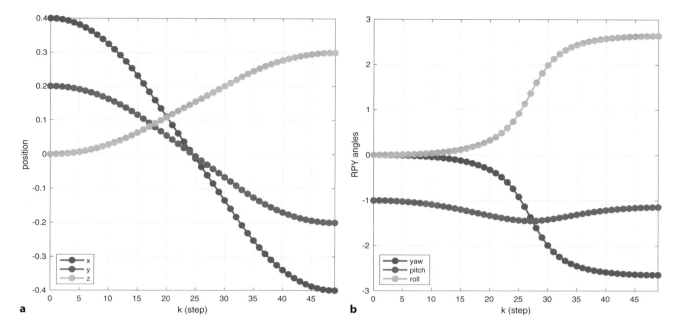

◻ Fig. 3.8 Cartesian motion with trapezoidal path distance profile. **a** Cartesian position versus time; **b** roll-pitch-yaw angles versus time

pitch-yaw angles are a nonlinear transformation of the linearly-varying quaternion orientation. Secondly, this particular trajectory passes very close to the roll-pitch-yaw singularity, ► at around steps 24 and 25, and a symptom of this is the rapid rate of change of roll-pitch-yaw angles around this point. The coordinate frame is not rotating faster at this point – you can verify that in the animation – the rotational parameters are changing very quickly, and this is a consequence of the particular representation.

Covered in ► Sect. 2.3.1.3.

However, the motion has a velocity and acceleration *discontinuity* at the first and last points. While the path is smooth in space, the distance s along the path is not smooth in time. Speed along the path jumps from zero to some finite value and then drops to zero at the end – there is no initial acceleration or final deceleration. The scalar functions `quinticpolytraj`, `minjerkpolytraj`, and `trapveltraj` discussed earlier can be used to generate the interpolation variable s so that motion *along* the path is smooth. We can pass a vector of normalized distances along the path as the second argument to `interp`

```
>> Ts = transformtraj(T0,T1,tpts,trapveltraj(tpts,50));
```

The path is unchanged, but the coordinate frame now accelerates to a constant speed along the path and decelerates at the end, resulting in the smoother trajectories shown in ◻ Fig. 3.8.

3.4 Application: Inertial Navigation

An inertial navigation system or INS is a self contained unit that estimates its velocity, orientation and position by measuring accelerations and angular velocities and integrating them over time. Importantly, it has no external inputs such as radio signals from satellites. This makes it well suited to applications such as submarine, spacecraft and missile guidance where it is not possible to communicate with radio navigation aids, or which must be immune to radio jamming. These particular applications drove development of the technology during the cold war and space

3

Fig. 3.9 Inertial sensors. **a** SPIRE (Space Inertial Reference Equipment) from 1953 was 1.5 m in diameter and weighed 1200 kg; **b** A modern inertial navigation system the LORD Micro-Strain 3DM-GX4-25 has triaxial gyroscopes, accelerometers and magnetometer, a pressure altimeter, is only 36×24×11 mm and weighs 16 g (Image courtesy of LORD MicroStrain); **c** 9 Degrees of Freedom IMU Breakout (LSM9DS1-SEN-13284 from Spark-Fun Electronics), the chip itself is only 3.5 × 3 mm

race of the 1950s and 1960s. Those early systems were large, see ◘ Fig. 3.9a, extremely expensive and the technical details were national secrets. Today, INSs are considerably cheaper and smaller as shown in ◘ Fig. 3.9b; the sensor chips shown in ◘ Fig. 3.9c can cost as little as a few dollars and they are built into every smart phone.

An INS estimates its pose with respect to an inertial reference frame which is typically denoted as {0}. ◄ The frame typically has its z-axis upward or downward and the x- and y-axes establish a locally horizontal tangent plane. Two common conventions have the x-, y- and z-axes respectively parallel to north-east-down (NED) or east-north-up (ENU) directions. The coordinate frame {B} is attached to the moving vehicle or robot and is known as the body- or body-fixed frame.

As discussed in ► Sect. 3.2.3, the Earth's surface is not an inertial reference frame but, for most robots with nonmilitary grade sensors, this is a valid assumption.

3.4.1 Gyroscopes

Any sensor that measures the rate of change of orientation is known, for historical reasons, as a gyroscope.

3.4.1.1 How Gyroscopes Work

The term gyroscope conjures up an image of a childhood toy – a spinning disk in a round frame that can balance on the end of a pencil. Gyroscopes are confounding devices – you try to turn them one way but they resist and turn (precess) in a different direction. This unruly behavior is described by a simplified version of (3.14)

$$\tau = \boldsymbol{\omega} \times \boldsymbol{h} \tag{3.19}$$

where $\boldsymbol{h}$ is the angular momentum of the gyroscope, a vector parallel to the rotor's axis of spin and with magnitude $\|\boldsymbol{h}\| = J\varpi$, where J is the rotor's inertia and ϖ its rotational speed. It is the cross product in (3.19) that makes the gyroscope move in a contrary way.

If no torque is applied to the gyroscope, its angular momentum remains constant in the inertial reference frame which implies that the axis will maintain a *constant direction* in that frame. Two gyroscopes with orthogonal axes form a stable platform that will maintain a *constant orientation* with respect to the inertial

◨ **Fig. 3.10** Gyroscope in strapdown configuration. Angular velocity ω induces a torque τ which can be sensed as forces at the bearings shown in red

reference frame – fixed with respect to the universe. This was the principle of many early spacecraft navigation systems such as that shown in ◨ Fig. 2.19 – the gimbals allowed the spacecraft to rotate about the stable platform, and the relative orientation was determined by measuring the gimbal angles. ▶

Alternatively, we can fix the gyroscope to the vehicle in the strapdown configuration as shown in ◨ Fig. 3.10. If the vehicle rotates with an angular velocity ω, the attached gyroscope will *precess* and exert an orthogonal torque τ which can be measured. ▶ If the magnitude of h is high, then this kind of sensor is very sensitive – a very small angular velocity leads to an easily measurable torque.

Over the last few decades, this rotating disk technology has been eclipsed by sensors based on optical principles such as the ring-laser gyroscope (RLG) and the fiber-optic gyroscope (FOG). These are highly accurate sensors but expensive and bulky. The low-cost sensors used in smart phones and drones are based on micro-electro-mechanical systems (MEMS) fabricated on silicon chips. Details of the designs vary but all contain a mass which vibrates at many kHz in a plane. Rotation about an axis normal to the plane causes an orthogonal displacement within the plane that is measured capacitively.

Gyroscopic angular velocity sensors measure rotation about a single axis. Typically, three gyroscopes are packaged together and arranged so that their sensitive axes are orthogonal. The three outputs of such a triaxial gyroscope are the components of the angular velocity vector ${}^{B}\omega_{B}^{\#}$ measured in the body frame {B}, and we introduce the notation $x^{\#}$ to explicitly indicate a value of x measured by a sensor.

Interestingly, nature has invented gyroscopic sensors. All vertebrates have angular velocity sensors as part of their vestibular system. In each inner ear, we have three semi-circular canals – fluid-filled organs that measure angular velocity. They are arranged orthogonally, just like a triaxial gyroscope, with two measurement axes in a vertical plane and one diagonally across the head.

3.4.1.2 Estimating Orientation

If the orientation of the sensor frame is initially ξ_B, then the evolution of estimated orientation can be written in discrete-time form as

$$\hat{\xi}_{B\langle k+1\rangle} \leftarrow \hat{\xi}_{B\langle k\rangle} \oplus {}^{B\langle k\rangle}\xi_{B\langle k+1\rangle} \tag{3.20}$$

where we use the hat notation to explicitly indicate an estimate of orientation, and $k \in \mathbb{N}_0$ is the index of the time step. ${}^{B\langle k\rangle}\xi_{B\langle k+1\rangle}$ is the incremental rotation over the time step which can be computed from measured angular velocity.

The engineering challenge was to create a mechanism that allowed the vehicle to rotate around the stable platform without exerting any torque on the gyroscopes. This required exquisitely engineered low-friction gimbals and bearing systems.

Typically by strain gauges attached to the bearings of the rotor shaft.

3

It is common to assume that $^B\boldsymbol{\omega}_B$ is constant over the time interval δ_t. In terms of $\mathbf{SO}(3)$ rotation matrices, the orientation update is

$$\hat{\mathbf{R}}_{B\langle k+1\rangle} \leftarrow \hat{\mathbf{R}}_{B\langle k\rangle}\, e^{\delta_t \left[^B\boldsymbol{\omega}_B^\#\right]_\times} \tag{3.21}$$

and the result should be periodically normalized to eliminate accumulated numerical error, as discussed in ▶ Sect. 2.4.6.

We will demonstrate this integration using unit quaternions and simulated angular velocity data for a tumbling body. The RVC Toolbox function

```
>> TrueMotion = imudata();
```

returns a structure which describe the true motion of the body. The rows of the matrix `TrueMotion.omega` represent consecutive body-frame angular velocity measurements with corresponding times given by elements of the vector `TrueMotion.t`. We choose the initial orientation to be the null rotation

```
>> orientation(1) = quaternion([1 0 0 0]);  % identity quaternion
```

and then for each time step we update the orientation and keep the orientation history in a vector of quaternions

```
>> for k = 1:size(TrueMotion.omega,1)-1
>>    orientation(k+1) = orientation(k) * ...
>>      quaternion(TrueMotion.omega(k,:)*TrueMotion.dt,"rotvec");
>> end
>> whos orientation
Name            Size            Bytes  Class        Attributes
  orientation     1x401            12832  quaternion
```

The orientation is updated by a unit quaternion, returned by the `"rotvec"` constructor, that corresponds to a rotation angle and axis given by the magnitude and direction of its argument. We can animate the changing orientation of the body frame

```
>> animtform(orientation)
```

or view the roll-pitch-yaw angles as a function of time

```
>> stackedplot(orientation.euler("zyx","point"))
```

3.4.2 Accelerometers

Accelerometers are sensors that measure acceleration. Even when not moving, they sense the acceleration due to gravity which defines the direction we know as *downward*. Gravitational acceleration is a function of our distance from the Earth's center, and the material in the Earth beneath us. The Earth is not a perfect sphere ▶ and points in the equatorial region are further from the center. Gravitational acceleration can be approximated by

$$g \approx 9.780327\left(1 + 0.0053024 \sin^2 \theta - 0.0000058 \sin^2 2\theta\right) - 0.000003086h$$

where θ is the angle of latitude and h is height above sea level. A map of gravity showing the effect of latitude and topography is shown in ◻ Fig. 3.11.

The technical term is an oblate spheroid, it bulges out at the equator because of centrifugal acceleration due to the Earth's rotation. The equatorial diameter is around 40 km greater than the polar diameter.

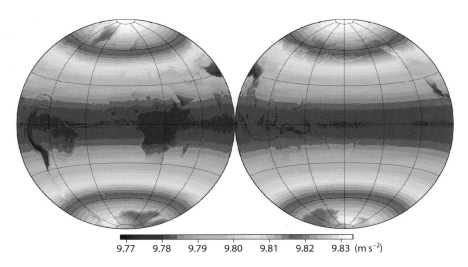

9.77   9.78   9.79   9.80   9.81   9.82   9.83 (m s⁻²)

◻ **Fig. 3.11** Variation in Earth's gravitational acceleration shows the imprint of continents and mountain ranges. The hemispheres shown are centered on the prime (left) and anti (right) meridian respectively (from Hirt et al. 2013)

Excurse 3.5: Charles Stark (Doc) Draper

Draper (1901–1987) was an American scientist and engineer, often referred to as "the father of inertial navigation." Born in Windsor, Missouri, he studied at the University of Missouri then Stanford where he earned a B.A. in psychology in 1922, then at MIT an S.B. in electro-chemical engineering and an S.M. and Sc.D. in physics in 1928 and 1938 respectively. He started teaching while at MIT and became a full professor in aeronautical engineering in 1939. He was the founder and director of the MIT Instrumentation Laboratory which made important contributions to the theory and practice of inertial navigation to meet the needs of the cold war and the space program.

Draper was named one of Time magazine's Men of the Year in 1961 and inducted to the National Inventors Hall of Fame in 1981. The Instrumentation lab was renamed Charles Stark Draper Laboratory (CSDL) in his honor. (Image courtesy of The Charles Stark Draper Laboratory Inc.)

3

3.4.2.1 How Accelerometers Work

An accelerometer is conceptually a very simple device comprising a mass, known as the proof mass, supported by a spring as shown in ◨ Fig. 3.12. In the inertial reference frame, Newton's second law for the proof mass is

$$m\ddot{x}_m = F_s - mg \qquad\qquad (3.22)$$

and, for a spring with natural length l_0, the relationship between force and extension d is

$$F_s = kd \ .$$

The various displacements are related by

$$x_b - (l_0 + d) = x_m$$

and taking the double derivative then substituting (3.22) gives

$$\ddot{x}_b - \ddot{d} = \frac{1}{m}kd - g \ .$$

The quantity we wish to measure is the acceleration of the accelerometer $a = \ddot{x}_b$ and the relative displacement of the proof mass ◀

We assume that $\ddot{d} = 0$ in steady state. Typically, there is a frictional element to damp out oscillation of the proof mass. This adds a term $-B\dot{x}_m$ to the right-hand side of (3.22).

$$d = \frac{m}{k}(a + g)$$

is linearly related to that acceleration. In an accelerometer, the displacement is measured and scaled by k/m so that the output of the sensor is

$$a^\# = a + g \ .$$

If this accelerometer is stationary, then $a = 0$ yet the measured acceleration would be $a^\# = 0 + g = g$ in the upward direction. This is because our model has included the Newtonian gravity force $m\boldsymbol{g}$, as discussed in ▶ Sect. 3.2.3. Accelerometer output is sometimes referred to as specific, inertial or proper acceleration.

◨ **Fig. 3.12** The essential elements of an accelerometer and notation

❗ Accelerometer Readings are Nonintuitive

It is quite unintuitive that a stationary accelerometer indicates an upward acceleration of $1\,g$ since it is clearly not accelerating. Intuition would suggest that, if anything, the acceleration should be in the downward direction which is how the device would accelerate if dropped. Adding to the confusion, some smart phone sensor apps incorrectly report positive acceleration in the *downward* direction when the phone is stationary.

Accelerometers measure acceleration along a single axis. Typically, three accelerometers are packaged together and arranged so that their sensitive axes are orthogonal. The three outputs of such a triaxial accelerometer are the components of the acceleration vector $^B a_B^{\#}$ measured in the body frame {B}.

Nature has also invented the accelerometer. All vertebrates have acceleration sensors called ampullae as part of their vestibular system. We have two in each inner ear to help us maintain balance: ▶ the saccule measures vertical acceleration, and the utricle measures front-to-back acceleration. The proof mass in the ampullae is a collection of calcium carbonate crystals called otoliths, literally ear stones, on a gelatinous substrate which serves as the spring and damper. Hair cells embedded in the substrate measure the displacement of the otoliths due to acceleration.

Inconsistency between motion sensed in our ears and motion perceived by our eyes is the root cause of motion sickness.

3.4.2.2 Estimating Pose and Body Acceleration

❑ Fig. 3.13 shows frame {0} with its z-axis vertically upward, and the acceleration is

$$^0 a_B = \begin{pmatrix} 0 \\ 0 \\ g \end{pmatrix}$$

where g is the local gravitational acceleration from ❑ Fig. 3.11. In a body-fixed frame {B} at an arbitrary orientation expressed in terms of ZYX roll-pitch-yaw angles ▶

We could use any 3-angle sequence.

$$^0\boldsymbol{\xi}_B = \boldsymbol{\xi}^{r_z}(\gamma) \oplus \boldsymbol{\xi}^{r_y}(\beta) \oplus \boldsymbol{\xi}^{r_x}(\alpha)$$

the gravitational acceleration will be

$$^B a_B = \left(\ominus\, ^0\boldsymbol{\xi}_B \right) \cdot\, ^0 a_B = \begin{pmatrix} -g\sin\beta \\ g\cos\beta\sin\alpha \\ g\cos\beta\cos\alpha \end{pmatrix}. \tag{3.23}$$

The *measured* acceleration vector from the sensor in frame {B} is

$$^B a_B^{\#} = (a_x^{\#}, a_y^{\#}, a_z^{\#})^\top$$

❑ Fig. 3.13 Gravitational acceleration is defined as the z-axis of the fixed frame {0} and is measured by an accelerometer in the rotated body frame {B}

3

and equating this with (3.23) we can solve for the roll and pitch angles

$$\sin \hat{\beta} = \frac{-a_x^\#}{g}$$

$$\tan \hat{\alpha} = \frac{a_y^\#}{a_z^\#}, \quad \hat{\beta} \neq \pm\frac{\pi}{2} \tag{3.24}$$

These angles are sufficient to determine whether a phone, tablet or camera is in portrait or landscape orientation.

and we use the hat notation to indicate that these are estimates of the angles. ◄ Notice that there is no solution for the yaw angle and in fact γ does not even appear in (3.23). The gravity vector is parallel to the vertical axis and rotating around that axis, yaw rotation, will not change the measured value at all. ◄ Also note that $\hat{\beta}$ depends on the local value of gravitational acceleration whose variation is shown in ◘ Fig. 3.11.

Another way to consider this is that we are essentially measuring the direction of the gravity vector with respect to the frame {B} and a vector provides only two unique *pieces* of directional information, since one component of a unit vector can always be written in terms of the other two.

🛆 **Accelerometers measure gravity *and* body motion**

We have made a very strong assumption that the measured acceleration $^B a_B^\#$ is only due to gravity. On a real robot, the sensor will experience additional acceleration as the robot moves and this will introduce an error in the estimated orientation.

Frequently, we want to estimate the motion of the vehicle in the inertial frame, and the total measured acceleration in {0} is due to gravity *and* motion

$$^0 a^\# = {}^0 g + {}^0 a_B \ .$$

We observe acceleration in the body frame so the acceleration in the world frame is

$$^0 \hat{a}_B = {}^0 \hat{\mathbf{R}}_B \, {}^B a_B^\# - {}^0 g \tag{3.25}$$

The first assumption is a strong one and is problematic in practice. Any error in the rotation matrix results in incorrect cancellation of the gravity component of $a^\#$ which leads to an error in the estimated body acceleration.

and we assume that $^0 \hat{\mathbf{R}}_B$ and g are both known. ◄ Integrating acceleration with respect to time

$$^0 \hat{v}_B(t) = \int_0^t {}^0 \hat{a}_B(t) \, dt \tag{3.26}$$

gives the velocity of the body frame, and integrating again

$$^0 \hat{p}_B(t) = \int_0^t {}^0 \hat{v}_B(t) \, dt \tag{3.27}$$

gives its position. We can assume vehicle acceleration is zero and estimate orientation, or assume orientation and estimate vehicle acceleration. We cannot estimate both since there are more unknowns than measurements.

3.4.3 **Magnetometers**

The Earth is a massive but weak magnet. The poles of this geomagnet are the Earth's north and south magnetic poles which are constantly moving and located quite some distance from the planet's rotational axis. At any point on the planet, the magnetic flux lines can be considered a vector m whose magnitude and direction can be accurately predicted and mapped as shown in ◘ Fig. 3.14.

◘ Fig. 3.14 A predicted model of the Earth magnetic field parameters for 2015. **a** Magnetic field flux density (nT); **b** magnetic declination (degrees); **c** magnetic inclination (degrees). Magnetic poles indicated by *asterisk* (maps by NOAA/NGDC and CIRES ► http://ngdc.noaa.gov/geomag/WMM, published Dec 2014)

Fig. 3.15 A compass needle aligns with the horizontal component of the Earth's magnetic field vector which is offset from true north by the declination angle D. The magnetic field vector has a dip or inclination angle of I with respect to horizontal

The direction of the Earth's north rotational pole, where the rotational axis intersects the surface of the northern hemisphere.

In the Northern hemisphere, inclination is positive, that is, the vector points into the ground.

By comparison, a modern MRI machine has a magnetic field strength of 4–8 T.

◻ Fig. 3.15 show the vector's direction in terms of two angles: declination and inclination. A horizontal projection of the vector m points in the direction of magnetic north and the declination angle D is measured from true north ◄ clockwise to that projection. The inclination or dip angle I of the vector is measured in a vertical plane downward ◄ from horizontal to m. The length of the vector, the magnetic flux density, is measured by a magnetometer in units of Tesla (T) and for the Earth this varies from 25–65 μT ◄ as shown in ◻ Fig. 3.14a.

3.4.3.1 How Magnetometers Work

The key element of most modern magnetometers is a Hall-effect sensor, a semiconductor device which produces a voltage proportional to the magnetic flux density in a direction normal to the current flow. Typically, three Hall-effect sensors are packaged together, and arranged so that their sensitive axes are orthogonal. The

Excurse 3.6: Edwin Hall

Hall (1855–1938) was an American physicist born in Maine. His Ph.D. research in physics at the Johns Hopkins University in 1880 discovered that a magnetic field exerts a force on a current in a conductor. He passed current through thin gold leaf and, in the presence of a magnetic field normal to the leaf, was able to measure a very small potential difference between the sides of the leaf. This is now known as the Hall effect. While it was then known that a magnetic field exerted a force on a current carrying conductor, it was believed the force acted on the conductor not the current itself – electrons were yet to be discovered. He was appointed as professor of physics at Harvard in 1895 where he worked on thermoelectric effects.

three outputs of such a triaxial magnetometer are the components of the Earth's magnetic flux density vector ${}^{B}\boldsymbol{m}^{\#}$ measured in the body frame {B}.

Yet again nature leads, and creatures from bacteria to turtles and birds are known to sense magnetic fields. The effect is particularly well known in pigeons and there is debate about whether or not humans have this sense. The actual biological sensing mechanism has not yet been discovered.

3.4.3.2 Estimating Heading

◘ Fig. 3.16 shows an inertial coordinate frame {0} with its z-axis vertically upward and its x-axis in the horizontal plane and pointing toward *magnetic* north. The magnetic field vector therefore lies in the xz-plane

$$
{}^{0}\boldsymbol{m} = B \begin{pmatrix} \cos I \\ 0 \\ \sin I \end{pmatrix}
$$

where B is the magnetic flux density and I the inclination angle which are both known from ◘ Fig. 3.14. In a body-fixed frame {B} at an arbitrary orientation expressed in terms of ZYX roll-pitch-yaw angles ▶

We could use any 3-angle sequence.

$$
{}^{0}\boldsymbol{\xi}_{B} = \boldsymbol{\xi}^{r_z}(\gamma) \oplus \boldsymbol{\xi}^{r_y}(\beta) \oplus \boldsymbol{\xi}^{r_x}(\alpha)
$$

the magnetic flux density will be

$$
\begin{aligned}
{}^{B}\boldsymbol{m} &= \left(\ominus\, {}^{0}\boldsymbol{\xi}_{B}\right) \cdot {}^{0}\boldsymbol{m} \\
&= \begin{pmatrix} B \sin I \sin \alpha \sin \gamma + \cos \alpha \cos \gamma \sin \beta + B \cos I \cos \beta \cos \gamma \\ B \cos I \cos \beta \sin \gamma - B \sin I \cos \gamma \sin \alpha - \cos \alpha \sin \beta \sin \gamma \\ B \sin I \cos \alpha \cos \beta - B \cos I \sin \beta \end{pmatrix} .
\end{aligned} \quad (3.28)
$$

The measured magnetic flux density vector from the sensor in frame {B} is

$$
{}^{B}\boldsymbol{m}^{\#} = (m_x^{\#}, m_y^{\#}, m_z^{\#})^{\top}
$$

and equating this with (3.28) we can solve for the yaw angle

$$
\tan \hat{\gamma} = \frac{\cos \beta \left(m_z^{\#} \sin \alpha - m_y^{\#} \cos \alpha \right)}{m_x^{\#} + B \sin I \sin \beta}
$$

which is the magnetic heading and related to the true heading by

$$
{}^{\mathrm{tn}}\hat{\gamma} = \hat{\gamma} - D .
$$

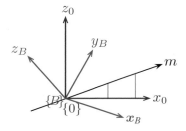

◘ **Fig. 3.16** The reference frame {0} has its x-axis aligned with the horizontal projection of $\boldsymbol{m}$ and its z-axis vertical. Magnetic field is measured by a magnetometer in the rotated body frame {B}

These can be calibrated out but the process requires that the sensor is physically rotated. The compass app in your phone might sometimes ask you to do that.

This assumes that we know the local declination angle as well as the roll and pitch angles α and β. The latter could be estimated from acceleration measurements using (3.24), and many triaxial Hall-effect sensor chips also include a triaxial accelerometer for just this purpose.

Magnetometers are great in theory but problematic in practice. Firstly, our modern world is full of magnets and electromagnets. Buildings contain electrical wiring and robots themselves are full of electric motors, batteries and electronics. These all add to, or overwhelm, the local geomagnetic field. Secondly, many objects in our world contain ferromagnetic materials such as the reinforcing steel in buildings or the steel bodies of cars, ships or robots. These distort the geomagnetic field leading to local changes in its direction. These effects are referred to respectively as hard- and soft-iron distortion of the magnetic field. ◄

3.4.4 Inertial Sensor Fusion

An inertial navigation system uses the sensors we have just discussed to determine the pose of a vehicle – its position and its orientation. Early inertial navigation systems, such as shown in ◘ Fig. 2.19, used mechanical gimbals to keep the accelerometers at a constant orientation with respect to the stars using a gyro-stabilized platform. The acceleration measured on this platform is by definition referred to the inertial frame, and simply needs to be integrated to obtain the velocity of the platform, and integrated again to obtain its position. In order to achieve accurate position estimates over periods of hours or days, the gimbals and gyroscopes had to be of extremely high quality so that the stable platform did not drift. The acceleration sensors also needed to be extremely accurate.

Increasingly, these sensor packages also include a barometric pressure sensor to measure changes in altitude.

The modern strapdown inertial measurement configuration uses no gimbals. The angular velocity, acceleration and magnetic field sensors are rigidly attached to the vehicle. The collection of inertial sensors is referred to as an inertial measurement unit or IMU. A 6-axis IMU comprises triaxial gyroscopes and accelerometers while a 9-axis IMU comprises triaxial gyroscopes, accelerometers and magnetometers. ◄ A system that only determines orientation is called an attitude and heading reference system or AHRS.

Some sensors may not be perfectly linear and a sensor datasheet will characterize this.

Some sensors also exhibit cross-sensitivity. They may give a weak response to a signal in an orthogonal direction or from a different mode, quite commonly low-cost gyroscopes respond to vibration and acceleration as well as rotation.

The sensors we use, particularly the low-cost ones in smart phones and drones, are far from perfect. The output of any sensor $x^{\#}$ – gyroscope, accelerometer or magnetometer – is a corrupted version of the true value x and a common model is

$$x^{\#} = sx + b + \varepsilon$$

where s is the scale factor, b is the offset or bias, and ε is random error or "noise". s is usually specified by the manufacturer to some tolerance, perhaps $\pm 1\%$, and for a particular sensor this can be determined by some calibration procedure. ◄ Bias b is ideally equal to zero but will vary from device to device. Bias that varies over time is often called sensor drift. Scale factor and bias are typically both a function of temperature. ◄

The effect of an orientation error is dangerous on something like a quadrotor. For example, if the estimated pitch angle is too high then the vehicle control system will pitch down by the same amount to keep the craft "level", and this will cause it to accelerate forward.

In practice, bias is the biggest problem because it varies with time and temperature and has a very deleterious effect on the estimated position and orientation. Consider a positive bias on the output of a gyroscopic sensor – the output is higher than it should be. At each time step in (3.20), the incremental rotation will be bigger than it should be, which means that the orientation error will grow linearly with time. ◄

If we use (3.25) to estimate the vehicle's acceleration, then the error in orientation means that the measured gravitational acceleration is incorrectly canceled out and will be indistinguishable from *actual* vehicle acceleration. This offset in acceleration becomes a linear-time error in velocity and a quadratic-time error in position. Given that the orientation error is already linear in time, we end up with

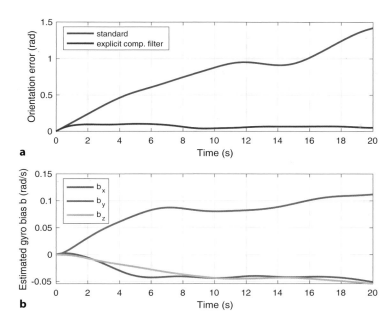

■ **Fig. 3.17** **a** Effect of gyroscope bias on simple gyro integration and explicit complementary filter; **b** estimated gyroscope bias from the explicit complementary filter

a cubic-time error in position, and this is ignoring the effects of accelerometer bias. Sensor bias is problematic! A rule of thumb is that gyroscopes with bias stability of $0.01 \deg\,h^{-1}$ will lead to position error growing at a rate of $1\,nmi\,h^{-1}$ ($1.85\,km\,h^{-1}$). Military-grade systems have very impressive stability, for missiles it is less than $0.000\,02 \deg\,h^{-1}$. This is in stark contrast to consumer grade devices which are in the range of 0.01–0.2 deg per *second*.

To see the effect of bias on the estimated orientation, we will use the RVC Toolbox function

```
>> [TrueMotion,IMU] = imudata();
```

which returns two structures that are respectively the true motion of a rotating body and simulated IMU data from that body. ▶ The latter contains "measured" gyroscope, accelerometer and magnetometer data which include a fixed but unknown bias. ▶ The data is organized as rows of the matrices `IMU.gyro`, `IMU.accel` and `IMU.magno` respectively, with one row per time step, and the corresponding times are given by elements of the vector `IMU.t`. We repeat the example from ▶ Sect. 3.4.1.2, but now with sensor bias

```
>> orientation(1) = quaternion([1 0 0 0]);  % identity quaternion
>> for k = 1:size(IMU.gyro,1)-1
>>    orientation(k+1) = orientation(k) ...
>>                       *quaternion(IMU.gyro(k,:)*IMU.dt,"rotvec");
>> end
```

The angular error ▶ between the estimated and true orientation

```
>> plot(IMU.t,dist(orientation,TrueMotion.orientation),"r");
```

is shown as the red line in ■ Fig. 3.17a. We can clearly see growth in angular error over time due to bias on the sensor signals.

If we knew the bias, we could subtract it from the sensor measurement before integration. A simple way of estimating bias is to leave the IMU stationary for a few seconds and compute the average value of all the sensor outputs. ▶ This is really only valid over a short time period because the bias is not constant.

An alternative approach would be to use `imuSensor` which is a MATLAB system object that models an inertial sensor and provides sophisticated error models.

Not completely unknown, you can look at the source code of `imudata.m`.

As discussed in ▶ Sect. 2.4.5.

Many hobby drones do this just before they take off.

Our brain has an online mechanism to cancel out the bias in our vestibular gyroscopes. It uses the recent average rotation as the bias, based on the reasonable assumption that we do not undergo prolonged rotation. If we do, then that angular velocity becomes the new normal so that when we stop rotating we perceive rotation in the opposite direction. We call this dizziness.

A more sophisticated approach is to estimate the bias online ◄ but, to do this, we need to combine information from different sensors – an approach known as sensor fusion. We rely on the fact that different sensors have complementary characteristics. Bias on angular rate sensors causes the orientation estimate error to grow with time, but for accelerometers it will only cause an orientation offset. Accelerometers respond to translational motion while good gyroscopes do not. Magnetometers provide partial information about roll, pitch and yaw, are immune to acceleration, but do respond to stray magnetic fields and other distortions. There are many ways to achieve this kind of fusion. A common approach is to use an estimation tool called an extended Kalman filter described in ► App. H. Given a full nonlinear mathematical model that relates the sensor signals and their biases to the vehicle pose and knowledge about the noise (uncertainty) on the sensor signals, the filter gives an optimal estimate of the pose and bias that best explain the sensor signals.

Here we will consider a simple, but effective, alternative called the explicit complementary filter. The rotation update step is performed using an augmented version of (3.21)

$$ {}^B\boldsymbol{\xi}^\Delta_{\langle k \rangle} = e^{\delta_t \left[{}^B\boldsymbol{\omega}^\#_{B\langle k \rangle} - \hat{\boldsymbol{b}}_{\langle k \rangle} + k_P \boldsymbol{\sigma}_{R\langle k \rangle} \right]_\times} \; . \tag{3.29} $$

The key differences are that the estimated bias $\hat{\boldsymbol{b}}$ is subtracted from the sensor measurement and a term based on the orientation error $\boldsymbol{\sigma}_R$ is added. The estimated bias changes with time according to

$$ \hat{\boldsymbol{b}}_{\langle k+1 \rangle} \leftarrow \hat{\boldsymbol{b}}_{\langle k \rangle} - k_I \boldsymbol{\sigma}_{R\langle k \rangle} \tag{3.30} $$

and also depends on the orientation error $\boldsymbol{\sigma}_R$. $k_P > 0$ and $k_I > 0$ are both well-chosen constants. The orientation error is derived from N *vector measurements* ${}^B\boldsymbol{v}^\#_i$

$$ \boldsymbol{\sigma}_{R\langle k \rangle} = \sum_{i=1}^N k_i \, {}^B\boldsymbol{v}^\#_{i\langle k \rangle} \times {}^B\boldsymbol{v}_{i\langle k \rangle} $$

where

$$ {}^B\boldsymbol{v}_{i\langle k \rangle} = \left(\ominus \, {}^0\hat{\boldsymbol{\xi}}_{B\langle k \rangle} \right) \cdot {}^0\boldsymbol{v}_i $$

is a vector signal, known in the inertial frame (for example gravitational acceleration), and rotated into the body-fixed frame by the inverse of the estimated orientation ${}^0\hat{\boldsymbol{\xi}}_B$. Any error in direction between these vectors will yield a nonzero cross-product which is the axis around which to rotate one vector into the other. The filter uses this difference – the innovation – to improve the orientation estimate by feeding it back into (3.29). This filter allows an unlimited number of body-frame vectorial measurements ${}^B\boldsymbol{v}_i$ to be fused together, for example, we could include magnetic field or any other kind of direction data such as the altitude and azimuth of visual landmarks, stars or planets.

Now we implement the explicit complementary filter. It has just a few extra lines of code compared to the example above

```
>> kI = 0.2; kP = 1;
>> bias = zeros(size(IMU.gyro,1),3);
>> orientation_ECF(1) = quaternion([1 0 0 0]);
>> for k = 1:size(IMU.gyro,1)-1
>>    invq = conj(orientation_ECF(k));  % unit quaternion inverse
>>    sigmaR = cross(IMU.accel(k,:), ...
>>            invq.rotatepoint(TrueMotion.g0))+ ...
>>            cross(IMU.magno(k,:),invq.rotatepoint(TrueMotion.B0));
>>    wp = IMU.gyro(k,:) - bias(k,:) + kP*sigmaR;
>>    orientation_ECF(k+1) = orientation_ECF(k) ...
>>                           *quaternion(wp*IMU.dt,"rotvec");
>>    bias(k+1,:) = bias(k,:) - kI*sigmaR*IMU.dt;
>> end
```

and plot the angular difference between the estimated and true orientation

```
>> plot(TrueMotion.t,dist(orientation_ECF, ...
>>    TrueMotion.orientation),"b");
```

as the blue line in ◨ Fig. 3.17a.

Bringing together information from multiple sensors has checked the growth in orientation error, despite all the sensors having a bias. The estimated gyroscope bias is shown in ◨ Fig. 3.17b and we can see the bias estimates converging on their true value.

3.5 Wrapping Up

This chapter has discussed the important issue of time-varying pose, from several different perspectives. The first was from the perspective of differential calculus, and we showed that the temporal derivative of a rotation matrix or a quaternion is a function of the angular velocity of the body. The skew-symmetric matrix appears in the rotation matrix case, and we should no longer be surprised about this given its intimate connection to rotation via Lie group theory as discussed in the previous chapter. We then looked at a finite-time difference as an approximation to the derivative and showed how this leads to computationally cheap methods to update rotation matrices and quaternions given knowledge of angular velocity. We also discussed the dynamics of moving bodies that translate and rotate under the influence of forces and torques, inertial and noninertial reference frames and the notion of fictitious forces.

The second perspective was creating time-varying motion – a sequence of poses, a trajectory, that a robot can follow. An important characteristic of a trajectory is that it is smooth with respect to time – position and orientation have a continuous first, and possibly higher, derivative. We started by discussing how to generate smooth trajectories in one dimension and then extended that to the multi-dimensional case and then to piecewise-linear trajectories that visit a number of intermediate points. Smoothly-varying rotation was achieved by interpolating roll-pitch-yaw angles and unit quaternions.

The third perspective was to use measured angular velocity, acceleration and magnetic field data to estimate orientation and position. We used our knowledge to investigate the important problem of inertial navigation, estimating the pose of a moving body given imperfect measurements from sensors on the moving body.

3.5.1 Further Reading

The earliest work on manipulator Cartesian trajectory generation was by Paul (1972, 1979) and Taylor (1979). The multi-segment trajectory is discussed by Paul (1979, 1981) and the concept of segment transitions or blends is discussed by Lloyd and Hayward (1991). These early papers, and others, are included in the compilation on Robot Motion (Brady et al. 1982). Polynomial and trapezoidal trajectories are described in detail by Spong et al. (2006) and multi-segment trajectories are covered at length in Siciliano et al. (2009) and Craig (2005).

There is a lot of literature related to the theory and practice of inertial navigation systems. The thesis of Achtelik (2014) describes a sophisticated extended Kalman filter for estimating the pose, velocity and sensor bias for a small aerial robot. The explicit complementary filter used in this chapter is described by Hua et al. (2014). The book by Groves (2013) covers inertial and terrestrial radio and satellite navigation and has a good coverage of Kalman filter state estimation techniques. Titterton and Weston (2005) provides a clear and concise description of the principles underlying inertial navigation with a focus on the sensors themselves but is perhaps a little dated with respect to modern low-cost sensors. Data sheets on many low-cost inertial and magnetic field sensing chips can be found at ► https://www.sparkfun.com in the Sensors category.

The book *Digital Apollo* (Mindell 2008) is a very readable story of the development of the inertial navigation system for the Apollo Moon landings. The article by Corke et al. (2007) describes the principles of inertial sensors and the functionally equivalent sensors located in the inner ear of mammals that play a key role in maintaining balance.

3.5.2 Resources

A detailed worked example for trapezoidal trajectories can be be found at ► https://sn.pub/cxekTH.

3.5.3 Exercises

1. Express the incremental rotation $\mathbf{R}^\triangle$ as an exponential series and verify (3.9).
2. Derive the unit quaternion update equation (3.10).
3. Derive the expression for fictitious forces in a rotating reference frame from ► Sect. 3.2.3.
4. Make a simulation that includes a particle moving at constant velocity and a rotating reference frame. Plot the position of the particle in the inertial and the rotating reference frame and observe how the motion in the latter changes as a function of the particle's translational velocity and the reference frame angular velocity.
5. Compute the centripetal acceleration due to the Earth's orbital rotation about the Sun. How does this compare to that due to the Earth's axial rotation?
6. For the rotational integration example (► Sect. 3.4.1.2):
    a) experiment by changing the sample interval and adding rotation matrix normalization at each step, or every Nth step.
    b) use `tic` and `toc` to measure the execution time of the approximate solution with normalization, and the exact solution using the matrix exponential.
    c) Redo the example using quaternions, and experiment by changing the sample interval and adding normalization at each step, or every Nth step.
7. Redo the quaternion-based angular velocity integration (► Sect. 3.4.1.2) using rotation matrices.
8. At your location, determine the magnitude and direction of centripetal acceleration you would experience. If you drove at $100\,\mathrm{km\,h^{-1}}$ due east what is the magnitude and direction of the Coriolis acceleration you would experience? What about at $100\,\mathrm{km\,h^{-1}}$ due north? The vertical component is called the Eötvös effect, how much lighter does it make you?
9. For a `quinticpolytraj` trajectory from 0 to 1 in 50 steps explore the effects of different initial and final velocities, both positive and negative. Under what circumstances does the quintic polynomial overshoot and why?
10. For a `trapveltraj` trajectory from 0 to 1 in 50 steps explore the effects of specifying the velocity for the constant velocity segment. What are the minimum and maximum bounds possible?
11. For a trajectory from 0 to 1 in T seconds with a maximum possible velocity of 0.5, compare the total trajectory times required for each of the `quinticpolytraj` and `trapveltraj` trajectories.
12. Use the `animate` method to compare rotational interpolation using quaternions, Euler angles and roll-pitch-yaw angles. Hint: use the quaternion `slerp` method and `trapveltraj`.

13. Develop a method to quantitatively compare the performance of the different orientation interpolation methods. Hint: plot the locus followed by $\hat{z}$ on a unit sphere.

14. Repeat the example of ◻ Fig. 3.7 for the case where:
    a) the interpolation does *not* pass through a singularity. Hint – change the start or goal pitch angle. What happens?
    b) the final orientation is at a singularity. What happens?

15. Convert an **SE**(3) matrix to a twist and exponentiate the twist with values in the interval [0,1]. How does this form of interpolation compare with others covered in this chapter?

16. For the multisegment example (▶ Sect. 3.3.3)
    a) Investigate the effect of increasing the acceleration time. Plot total time as a function of acceleration time.
    b) The function has an additional output argument that provides parameters for each segment, investigate these.
    c) Investigate other trajectory generator function that support via points, such as `cubicpolytraj` and `bsplinepolytraj`

17. There are a number of iOS and Android apps that display and record sensor data from the device's gyros, accelerometers and magnetometers. Alternatively you could use an inertial sensor shield for a RaspberryPi or Arduino and write your own logging software. You could also use MATLAB, see ▶ https://sn.pub/d2R3yh. Run one of these and explore how the sensor signals change with orientation and movement. If the app supports logging, what acceleration is recorded when you throw the phone into the air?

18. Consider a gyroscope with a 20 mm diameter steel rotor that is 4 mm thick and rotating at 10,000 rpm. What is the magnitude of h? For an angular velocity of 5 deg s^{-1}, what is the generated torque?

19. Using (3.19), can you explain how a toy gyroscope is able to balance on a single point with its spin axis horizontal? What holds it upright?

20. An accelerometer has fallen off the table. Assuming no air resistance and no change in orientation, what value does it return as it falls?

21. Implement the algorithm to determine roll and pitch angles from accelerometer measurements.
    a) Devise an algorithm to determine if it is in portrait or landscape orientation.
    b) Create a trajectory for the accelerometer using `quinticpolytraj` to generate motion in either the x- or y-direction. What effect does the acceleration along the path have on the estimated angle?
    c) Calculate the orientation using quaternions rather than roll-pitch-yaw angles.

22. Determine the Euler angles as a function of the measured acceleration. If you have the Symbolic Math Toolbox™ you could use that to help with the algebra.

23. You are in an aircraft flying at 30,000 feet over your current location. How much lighter are you?

24. Determine the magnetic field strength, declination and inclination at your location. Visit the website ▶ http://www.ngdc.noaa.gov/geomag-web.

25. Using the sensor reading app from above, orient the phone so that the magnetic field vector has only a z-axis component. Where is the magnetic field vector with respect to your phone?

26. Using the sensor reading app from above, log some inertial sensor data from a phone while moving it around. Use that data to estimate the changing orientation or full pose of the phone. Can you do this in real time?

27. Experiment with varying the parameters of the explicit complementary filter (▶ Sect. 3.4.4). Change the bias or add Gaussian noise to the simulated sensor readings.

Mobile Robotics

Contents

Chapter 4   Mobile Robot Vehicles – 127

Chapter 5   Navigation – 161

Chapter 6   Localization and Mapping – 213

Mobile robots are a more recent technology than robot manipulator arms, which we cover in Part III, but since the most common mobile robots operate in two dimensions, they allow for a conceptually more gentle introduction.

Today, there is intense interest in mobile robots – the next frontier in robotics – with advances in self-driving cars, planetary rovers, humanoids and drones, and new application areas such as warehouse logistics and aerial taxis. The first commercial applications of mobile robots came in the 1980s when automated guided vehicles (AGVs) were developed for transporting material around factories and these have since become a mature technology. Those early free-ranging mobile wheeled vehicles typically used fixed infrastructure for guidance, for example, a painted line on the floor, a buried cable that emits a radio-frequency signal, or wall-mounted fiducial markers. The last decade has seen significant achievements in mobile robots that can operate without navigational infrastructure.

The chapters in this part of the book cover the fundamentals of mobile robotics. ▶ Chap. 4 discusses the motion and control of two exemplar robot platforms: wheeled vehicles that operate on a planar surface, and aerial robots that move in 3-dimensional space – specifically quadrotor aerial robots. ▶ Chap. 5 is concerned with navigation, how a robot finds its way to a goal in an environment that contains obstacles using direct sensory information or a map. Most navigation strategies require knowledge of the robot's position, and this is the topic of ▶ Chap. 6 which examines techniques such as dead reckoning, and the use of maps combined with observations of landmarks. We also show how a robot can make a map, and even determine its location while simultaneously navigating an unknown region. To achieve all this, we will leverage our knowledge from ▶ Sect. 2.2 about representing position, orientation and pose in two dimensions.

Mobile Robot Vehicles

Contents

4.1   Wheeled Mobile Robots – 129

4.2   Aerial Robots – 147

4.3   Advanced Topics – 154

4.4   Wrapping Up – 157

4

Fig. 4.1   Some mobile ground robots: **a** The Roomba® robotic vacuum cleaner, 2008 (Image courtesy iRobot®). **b** Boss, Tartan Racing team's autonomous car that won the DARPA Urban Challenge, 2007 (© Carnegie-Mellon University)

Fig. 4.2   Some mobile air and water robots: **a** Yamaha RMAX® helicopter with 3 m blade diameter (Image by Sanjiv Singh); **b** ScanEagle, fixed-wing robotic aircraft (Image courtesy of Insitu); **c** Ranger-Bot, a 6-thruster underwater robot (2020) (Image courtesy Biopixel); **d** Autonomous Surface Vehicle (Image by Matthew Dunbabin)

Figs. 4.1 to 4.3 show examples of mobile robots that are diverse in form and function. Mobile robots are not just limited to operations on the ground, and Fig. 4.2 shows examples of aerial robots known as uncrewed aerial vehicles (UAVs), underwater robots known as autonomous underwater vehicles (AUVs), and robotic boats which are known as autonomous or uncrewed surface vehicles (ASVs or USVs). Collectively these are often referred to as autonomous uncrewed systems (AUSs). Field robotic systems such as trucks in mines, container transport vehicles in shipping ports, and self-driving tractors for broad-acre agriculture are shown in Fig. 4.3. While some of these are very specialized or research prototypes, commercial field robots are available for a growing range of applications.

The diversity we see is in the robotic platform – the robot's physical embodiment, its means of locomotion, and its sensors. The robot's sensors are clearly visible in Figs. 4.1b, 4.2a, d, and 4.3a. Internally, these robots would have strong similarity in their sensor-processing pipeline and navigation algorithms. The robotic vacuum cleaner of Fig. 4.1a has few sensors and uses reactive strategies to clean the floor. By contrast, the 2007-era self-driving vehicle shown in

Fig. 4.3 a Exploration: Mars Science Laboratory (MSL) rover, known as Curiosity, undergoing testing (Image courtesy NASA/Frankie Martin); **b** Logistics: an automated straddle carrier that moves containers at the Port of Brisbane, 2006 (Image courtesy of Port of Brisbane Pty Ltd); **c** Mining: autonomous haul truck (Copyright © 2015 Rio Tinto); **d** Agriculture: QUT's AgBot-2 broad-acre weeding robot (Image courtesy Owen Bawden)

Fig. 4.1b displays a multitude of sensors that provide the vehicle with awareness of its surroundings.

This chapter discusses how a robot platform moves, that is, how its pose changes with time as a function of its control inputs. There are many different types of robot platform but, in this chapter, we will consider only four important exemplars. ▶ Sect. 4.1 covers three different types of wheeled vehicles that operate in a 2-dimensional world. They can be propelled forward or backward and their heading direction is controlled by some steering mechanism. ▶ Sect. 4.2 describes a quadrotor, an aerial vehicle, which is an example of a robot that moves in 3-dimensional space. Quadrotors, or "drones", are becoming increasingly popular as a robot platform since they are low-cost and can be easily modeled and controlled. ▶ Sect. 4.3 revisits the concept of configuration space and dives more deeply into important issues of underactuation and nonholonomy.

4.1 Wheeled Mobile Robots

Wheeled locomotion is one of humanity's great innovations. The wheel was invented around 3000 BCE and the two-wheeled cart around 2000 BCE. Today, four-wheeled vehicles are ubiquitous, and the total automobile population of the planet is well over one billion. ▶ The effectiveness of cars, and our familiarity with them, makes them a natural choice for robot platforms that move across the ground.

We know from our everyday experience with cars that there are limitations on how they move. It is not possible to drive sideways, but with some practice we can learn to follow a path that results in the vehicle being to one side of its initial position – this is parallel parking. Neither can a car rotate on the spot, but we can follow a path that results in the vehicle being at the same position but rotated by $180°$ – a three-point turn. The necessity to perform such maneuvers is the hallmark of a system that is nonholonomic – an important concept, which is discussed further in ▶ Sect. 4.3. Despite these minor limitations, the car is the simplest and most effec-

There are also around one billion bicycles, but robotic bicycles are not common. While balancing is not too hard to automate, stopping and starting requires additional mechanical complexity.

4

FIG. 124.

From Sharp 1896

Often, perhaps incorrectly, called the Ackermann model.

tive means of moving in a planar world that we have yet found. The car's motion model and the challenges it raises for control will be discussed in ▶ Sect. 4.1.1.

In ▶ Sect. 4.1.2 we will introduce differentially-steered vehicles, which are mechanically simpler than cars and do not have steered wheels. This is a common configuration for small mobile robots and for larger machines such as bulldozers. ▶ Sect. 4.1.3 introduces novel types of wheels that *are* capable of omnidirectional motion, and then models a vehicle based on these wheels.

4.1.1 Car-Like Vehicle

Cars with steerable wheels are a very effective class of vehicles, and the archetype for most ground robots such as those shown in ◘ Fig. 4.3. In this section we will create a model for a car-like vehicle, and develop controllers that can drive the car to a point, along a line, follow an arbitrary path, and finally, drive to a specific pose.

A commonly used model for the low-speed behavior of a four-wheeled car-like vehicle is the kinematic bicycle model ◀ shown in ◘ Fig. 4.4. The bicycle has a rear wheel fixed to the body and the plane of the front wheel rotates about the vertical axis to steer the vehicle.

The pose of the vehicle is represented by its body coordinate frame {B} shown in ◘ Fig. 4.4, which has its x-axis in the vehicle's forward direction and its origin at the center of the rear axle. The orientation of its x-axis with respect to the world frame is the vehicle's heading angle (yaw angle or simply *heading*). The *configuration* of the vehicle is represented by the generalized coordinates $q = (x, y, \theta) \in C$ where $C \subset \mathbb{R}^2 \times \mathbf{S}^1$ is the configuration space. We assume that the velocity of each wheel is in the plane of the wheel, and that the wheels roll without skidding or slipping sideways

$$^B\boldsymbol{v} = (v, 0) \ . \tag{4.1}$$

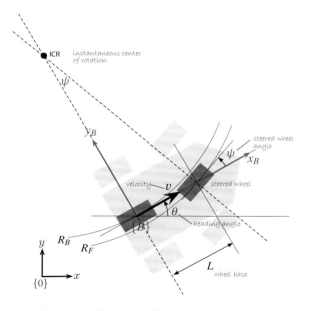

◘ **Fig. 4.4** Bicycle model of a car-like vehicle. The 4-wheeled car is shown in light gray, and the equivalent two-wheeled bicycle model in dark gray. The vehicle's body frame is shown in red, and the world coordinate frame in black. The steered wheel angle is ψ with respect to the body frame, and the velocity of the back wheel, in the body's x-direction, is v. The two wheel axes are extended as dashed lines and intersect at the instantaneous center of rotation (ICR). The back and front wheels follow circular arcs around the ICR of radius R_B and R_F respectively

The dashed lines show the direction along which the wheels cannot move, the lines of no motion, and these intersect at a point known as the Instantaneous Center of Rotation (ICR). The reference point of the vehicle, the origin of frame {B}, thus follows a circular path and its angular velocity is

$$\dot{\theta} = \frac{v}{R_B} \tag{4.2}$$

and by simple geometry the turning radius is

$$R_B = \frac{L}{\tan \psi} \tag{4.3}$$

where L is the length of the vehicle or the *wheel base*. As we would expect, the turning radius increases with vehicle length – a bus has a bigger turning radius than a car. The angle of the steered wheel, ψ, is typically limited mechanically and its maximum value dictates the minimum value of R_B.

In a car, we use the *steering wheel* to control the angle of the steered wheels on the road, and these two angles are linearly related by the steering ratio. ▶ Curves on roads are circular arcs or clothoids so that road following is easy for a driver – requiring constant or smoothly varying steering-wheel angle. Since $R_F > R_B$, the front wheel must follow a longer path and therefore rotate more quickly than the back wheel. Dubins and Reeds-Shepp paths consist of constant radius circular arcs

Steering ratio is the ratio of the steering-wheel angle to the steered-wheel angle. For cars, it is typically in the range 12 : 1 to 20 : 1. Some cars have "variable ratio" steering where the ratio is higher (less sensitive) around the zero angle.

4

and straight-line segments. See ▶ Chap. 5 how these paths can be used for path planning of car-like vehicles.

When a four-wheeled vehicle goes around a corner, the two steered wheels follow circular paths of slightly different radii and therefore the angles of the steered wheels ψ_L and ψ_R should be slightly different. This is achieved by the commonly used Ackermann steering mechanism that results in lower wear and tear on the tires. The driven wheels must rotate at different speeds on corners, so a differential gearbox is required between the motor and the driven wheels.

The velocity of the robot in the world frame is described by

$$
\begin{aligned}
\dot{x} &= v \cos \theta \\
\dot{y} &= v \sin \theta \\
\dot{\theta} &= \frac{v}{L} \tan \psi
\end{aligned}
\tag{4.4}
$$

which are the equations of motion, or the motion model. This is a kinematic model since it describes the velocities of the vehicle but not the forces or torques that cause the velocity. The rate of change of heading $\dot{\theta}$ is referred to as turn rate, heading rate, or yaw rate and can be measured by a gyroscope. It can also be deduced from the angular velocity of the wheels on the left- and right-hand sides of the vehicle which follow arcs of different radii, and therefore rotate at different speeds.

Equation (4.4) captures some other important characteristics of a car-like vehicle. When $v = 0$, then $\dot{\theta} = 0$. Therefore, it is not possible to change the vehicle's orientation when it is not moving. As we know from driving, we must be moving to turn. When the steering angle $\psi = \frac{\pi}{2}$, the front wheel is orthogonal to the back wheel, the vehicle cannot move forward, and the model enters an undefined region.

Expressing (4.1) in the world coordinate frame, the vehicle's velocity in its lateral or body-frame y-direction is

$$
\dot{y} \cos \theta - \dot{x} \sin \theta \equiv 0
\tag{4.5}
$$

The model also includes a maximum velocity limit, a velocity rate limiter to model finite acceleration, and a limiter on the steered-wheel angle to model the finite range of the steered wheel. These can be accessed by double clicking the `Bicycle Kinematic Model` block in Simulink.

which is called a nonholonomic constraint and will be discussed further in ▶ Sect. 4.3.1.

We can explore the behavior of the vehicle for simple steering input using the Simulink® model

```
>> sl_lanechange
```

shown in ❑ Fig. 4.5 which uses the `Bicycle Kinematic Model` block that implements (4.4). ◀ The velocity input is a constant, and the steered wheel angle is a finite positive pulse followed by a negative pulse. Running the model simulates the motion of the vehicle and adds a new variable `out` to the workspace

Besides the inputs and outputs, you can log arbitrary signals in the model by left-clicking a signal line in the model and selecting "Log Signal" in the toolstrip. See the "Running Simulink Models" box for more detail. Signals logged through this mechanism display a distinct icon above the signal line.

```
>> out = sim("sl_lanechange")
```

❑ **Fig. 4.5** Simulink model `sl_lanechange` that demonstrates a simple lane-changing maneuver. The pulse generator drives the angle of the steered wheel left, then right. The vehicle has a default wheel base $L = 1$ ◀

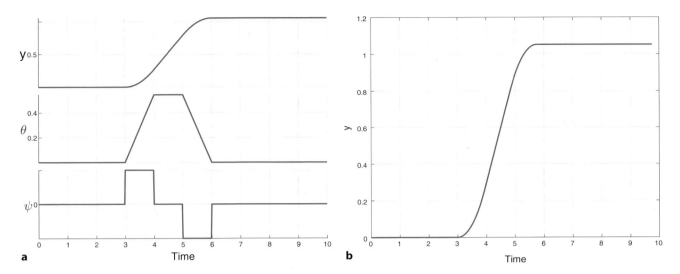

◘ Fig. 4.6 Simple lane changing maneuver. **a** Vehicle response as a function of time, **b** Motion in the xy-plane, the vehicle moves in the positive x-direction

sl_lanechange

▶ sn.pub/6Vy1rz

```
out =
  Simulink.SimulationOutput:
            logsout: [1x1 Simulink.SimulationData.Dataset]
                  t: [504x1 double]
                  y: [504x4 double]
  SimulationMetadata: [1x1 Simulink.SimulationMetadata]
        ErrorMessage: [0x0 char]
```

from which we can retrieve the simulation time and other variables

```
>> t = out.get("t");
>> q = out.get("y");
```

The configuration of the robot is plotted against time with the `stackedplot` function

```
>> stackedplot(t,q, ...
>>   DisplayLabels=["x","y","theta","psi"], ...
>>   GridVisible=1,XLabel="Time")
```

as shown in ◘ Fig. 4.6a, and the result in the xy-plane

```
>> plot(q(:,1),q(:,2))
```

shown in ◘ Fig. 4.6b, demonstrating a simple *lane-changing* trajectory.

4.1.1.1 Driving to a Point

Consider the problem of moving toward a goal point (x^*, y^*) on the xy-plane. We will control the robot's velocity to be proportional to its distance from the goal

$$v^* = K_v \sqrt{(x^* - x)^2 + (y^* - y)^2}$$

and to steer toward the goal, which is at the vehicle-relative angle in the world frame of ▶

$$\theta^* = \tan^{-1} \frac{y^* - y}{x^* - x} \qquad (4.6)$$

radians. A simple proportional controller

$$\psi = K_h(\theta^* \ominus \theta), \quad K_h > 0 \qquad (4.7)$$

This angle is in the interval $[-\pi, \pi)$ and is computed using the `atan2` function. See ▶ Exc. 4.4 for more details.

4

Excurse 4.3: Running Simulink Models

To run the Simulink model called model, we first load it

```
>> model
```

and a new window is created that displays the model. The simulation can be started by pressing the Run button on the toolbar of the model's window. The model can also be run directly from the MATLAB® command line

```
>> sim("model")
```

Many RVC Toolbox models create additional figures to display robot animations or graphs as they run.

All models in this chapter have the simulation data export option set to create a SimulationOutput object. All the signals feeding into output ports are concatenated, in port number order, to form a row vector and these are stacked to form a matrix y with one row per timestep. The corresponding time values form a vector t. These variables are packaged in a Simulink.SimulationOutput object which is written to the workspace variable out or returned if the simulation is invoked from MATLAB

```
>> r = sim("model")
```

Displaying r or out lists the variables that it contains, and their value is obtained using the find method, for example

```
>> t = r.find("t");
```

Besides the inputs and outputs, you can also log arbitrary signals in the model. Simply left-click a signal line in the model and select "Log Signal" in the toolstrip. By default, data from this logging will be stored in out.logsout as a Dataset object .

Log
Signals

The advantage of the "Log Signal" workflow is that it is easy to visualize the simulation output with the Simulation Data Inspector. You can analyze a single simulation run and easily compare the outcomes of multiple runs by calling plot on the simulation output object

```
>> r.plot;
```

An example of the Simulation Data Inspector visualization after simulating the sl_lanechange model is shown below.

The function angdiff computes the difference between two angles and returns a difference in the interval $[-\pi, \pi)$. The function wrapToPi ensures that an angle is in the interval $[-\pi, \pi)$. This is also the shortest distance around the circle, as discussed in ▶ Sect. 3.3.4.1. Both functions are also available as blocks in the RVC Toolbox Simulink library roblocks.

is sufficient – it turns the steered wheel toward the target. We use the operator ⊖ since $\theta^*, \theta \in \mathbf{S}^1$ are angles, not real numbers, and the result is always in the interval $[-\pi, \pi)$. ◄

A Simulink model

```
>> sl_drivepoint
```

is shown in ◘ Fig. 4.7. We specify a goal coordinate

```
>> xg = [5 5];
```

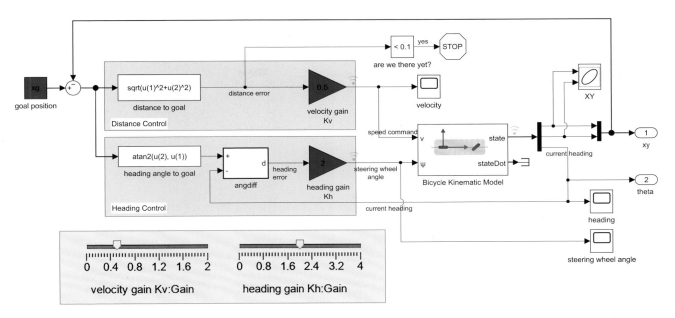

◻ Fig. 4.7 `sl_drivepoint`, the Simulink model that drives the vehicle to a point. Red blocks have parameters that you can adjust to investigate the effect on performance. The velocity and heading controller gains can also be changed during the simulation through the sliders on the bottom of the model

and an initial vehicle configuration

```
>> x0 = [8 5 pi/2];
```

and then simulate the motion

```
>> r = sim("sl_drivepoint");
```

The variable `r` is an object that contains the simulation results from which we extract the configuration as a function of time

```
>> q = r.find("y");
```

sl_drivepoint

▶ sn.pub/zJjaVo

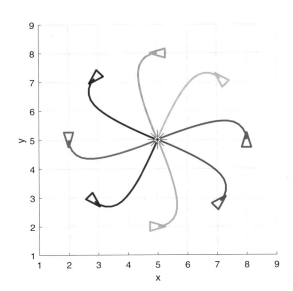

◻ Fig. 4.8 Vehicle motion to a point from different initial configurations, using the model `sl_drivepoint` shown in ◻ Fig. 4.7. The goal (x^*, y^*) is $(5, 5)$, and the controller parameters are $K_v = 0.5$ and $K_h = 2$

Excurse 4.4: atan vs atan2

In robotics, we frequently need to find θ given an expression for one or more of $\sin\theta$, $\cos\theta$, or $\tan\theta$. The inverse functions are not one-to-one and have ranges of $[-\frac{\pi}{2}, \frac{\pi}{2}]$, $[0, \pi]$ and $[-\frac{\pi}{2}, \frac{\pi}{2}]$ respectively which are all subsets of the full possible range of θ which is $[-\pi, \pi)$.

In MATLAB, these inverse functions are implemented by `asin`, `acos`, `atan`, and `atan2`. The figure shows the trigonometric quadrants associated with the sign of the argument to the inverse functions: blue for positive and red for negative.

Consider the example for θ in quadrant 3 (Q3), where $\sin\theta = s = \frac{-1}{\sqrt{2}}$ and $\cos\theta = c = \frac{-1}{\sqrt{2}}$. Then `asin(s)` $\rightarrow -\pi/4$, `acos(c)` $\rightarrow 3\pi/4$ and `atan(s/c)` $\rightarrow \pi/4$, three valid but quite different results. For the `atan(s/c)` case, the quotient of the two negative numbers is positive, placing the angle in quadrant 1. In contrast, the `atan2` function is passed the numerator and denominator separately and takes the individual signs into account to determine the angle in any quadrant. In this case, `atan2(s, c)` $\rightarrow -3\pi/4$, which is in quadrant 3.

The vehicle's path on the xy-plane is

```
>> plot(q(:,1),q(:,2));
```

which is shown in ■ Fig. 4.8 for several starting configurations. In each case the vehicle has moved forward and turned onto a path toward the goal point. The final part of each path approaches a straight line, and the final orientation therefore depends on the starting point.

4.1.1.2 Driving Along a Line

2-dimensional lines in homogeneous form are discussed in ▶ App. C.2.1

Another useful task for a mobile robot is to follow a line on the xy-plane ◀ defined by $ax + by + c = 0$. We achieve this using two steering controllers. The first controller

$$\psi_d = K_d d, \quad K_d > 0$$

turns the robot toward the line to minimize d, the robot's normal distance from the line

$$d = \frac{(a, b, c) \cdot (x, y, 1)}{\sqrt{a^2 + b^2}}$$

where $d > 0$ if the robot is to the right of the line. The second controller adjusts the heading angle, or orientation, of the vehicle to be parallel to the line

$$\theta^* = \tan^{-1}\frac{a}{-b}$$

using the proportional controller

$$\psi_h = K_h(\theta^* \ominus \theta), \quad K_h > 0 .$$

The combined control law

$$\psi = \psi_d + \psi_h = K_d d + K_h(\theta^* \ominus \theta)$$

turns the steering wheel to drive the robot toward the line and move along it.

■ Fig. 4.9 The Simulink model `sl_driveline` drives a vehicle along a line. The line parameters (a, b, c) are set in the workspace variable L. Red blocks have parameters that you can adjust to investigate the effect on performance

The Simulink model

```
>> sl_driveline
```

sn.pub/nMLJiJ

is shown in ■ Fig. 4.9. We specify the target line as a 3-vector (a, b, c)

```
>> L = [1 -2 4];
```

and an initial configuration

```
>> x0 = [8 5 pi/2];
```

and then simulate the motion

```
>> r = sim("sl_driveline");
```

The vehicle's path for several different starting configurations is shown in ■ Fig. 4.10.

4.1.1.3 Driving Along a Path

Instead of following a straight line, we might wish to follow an arbitrary path on the xy-plane. The path might come from one of the path planners that are introduced in ► Chap. 5.

A simple and effective algorithm for path following is *pure pursuit* in which we move at constant velocity and *chase* a point that is a constant distance ahead on the path. The path is defined by a set of N points $\boldsymbol{p}_i, i = 1, \ldots, N$ and the robot's current position is $\boldsymbol{p}_r$. The first step is to find the index of the point on the path that is closest to the robot

$$j = \arg \min_{i=1}^{N} \| \boldsymbol{p}_r - \boldsymbol{p}_i \| \,.$$

Next, we find the first point along the path that is at least a distance d ahead of that point

$$k = \arg \min_{i=j+1}^{N} | \| \boldsymbol{p}_i - \boldsymbol{p}_j \| - d |$$

4

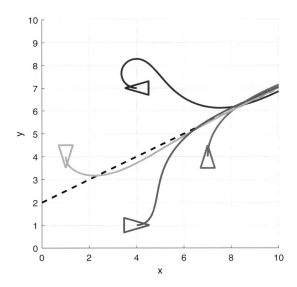

A simplified version of the pure pursuit algorithm that steers straight to the target point is commonly called "follow the carrot" in the literature.

sl_pursuit

► sn.pub/fabiQS

and then the robot drives toward the point p_k. In terms of the old fable about using a carrot to lead a donkey, what we are doing is always keeping the carrot a distance d in front of the donkey and moving it from side to side to steer the donkey. ◄

This problem is now the same as the one we tackled in ► Sect. 4.1.1.1, moving to a point, except that this time the point is moving. Robot speed is constant so only heading control, (4.6) and (4.7), is required.

The Simulink model shown in Fig. 4.11 defines a piecewise-linear path defined by several waypoints. It can be simulated by

```
>> sl_pursuit
>> r = sim("sl_pursuit");
```

and the results are shown in Fig. 4.12a. The robot starts at the origin but catches up to, and follows, the moving goal. Fig. 4.12b shows how the speed converges on a steady state value when following at the desired distance. Note the slow down

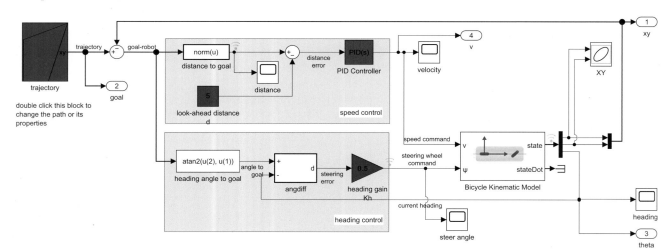

Fig. 4.11 The Simulink model `sl_pursuit` drives the vehicle along a piecewise-linear trajectory. Red blocks have parameters that you can adjust to investigate the effect on performance

◻ Fig. 4.12 Pure pursuit motion along a path defined by waypoints, using the simulation model `sl_pursuit` shown in ◻ Fig. 4.11. **a** Path of the robot in the xy-plane. The red dashed line is the path to be followed and the blue line is the path followed by the robot, which starts at the origin. The look-ahead distance $d = 5$, and the heading control gain is $K_h = 0.5$; **b** The speed of the robot versus time

at the end of each segment as the robot *shortcuts* across the corner. Adjusting the look-ahead distance will influence the controller behavior. Reducing the distance ensures that the vehicle follows the path more closely, but the controller will become more unstable at higher speeds. Picking a larger distance leads to the shortcut behavior around corners (since the target carrot point is already past the corner), but provides better control stability.

This section showed how to implement the pure pursuit algorithm from scratch. A ready-to-use implementation of the algorithm is available through the `controllerPurePursuit` object and a corresponding Simulink block. It takes a 2D trajectory and implements the heading control that is shown in ◻ Fig. 4.11, but assumes a fixed speed. To achieve a slowdown behavior when getting closer to the goal, we can implement a speed controller as shown in ◻ Fig. 4.11 or something similar. An in-depth example

```
>> openExample("nav/" + ...
>>    "PathFollowingWithObstacleAvoidanceInSimulinkExample");
```

demonstrates how the Simulink block is used for path following and obstacle avoidance.

4.1.1.4 Driving to a Configuration

The final control problem we discuss is driving to a specific configuration (x^*, y^*, θ^*). The controller of ◻ Fig. 4.7 can drive the robot to a goal position, but the final orientation depends on the starting position. For tasks like parking, the orientation of the vehicle is important.

To control the final orientation, we first rewrite (4.4) in matrix form

$$
\begin{pmatrix} \dot{x} \\ \dot{y} \\ \dot{\theta} \end{pmatrix} = \begin{pmatrix} \cos\theta & 0 \\ \sin\theta & 0 \\ 0 & 1 \end{pmatrix} \begin{pmatrix} v \\ \omega \end{pmatrix}
$$

where the inputs to the vehicle model are the speed v and the turning rate ω which can be achieved by choosing the steered-wheel angle as ▶

$$
\psi = \tan^{-1} \frac{\omega L}{v}
$$

▶ sn.pub/R5dwZw

We have effectively converted the bicycle kinematic model to a unicycle model which we discuss in
▶ Sect. 4.1.2.

Fig. 4.13 Polar coordinate notation for the bicycle model vehicle moving toward a goal configuration: ρ is the distance to the goal, β is the angle of the goal vector with respect to the world frame, and α is the angle of the goal vector with respect to the vehicle frame

We then transform the equations into polar coordinate form using the notation shown in ◘ Fig. 4.13 and apply a change of variables

$$\rho = \sqrt{\Delta_x^2 + \Delta_y^2}$$

$$\alpha = \tan^{-1}\frac{\Delta_y}{\Delta_x} - \theta$$

$$\beta = -\theta - \alpha$$

which results in

$$\begin{pmatrix} \dot\rho \\ \dot\alpha \\ \dot\beta \end{pmatrix} = \begin{pmatrix} -\cos\alpha & 0 \\ \dfrac{\sin\alpha}{\rho} & -1 \\ -\dfrac{\sin\alpha}{\rho} & 0 \end{pmatrix} \begin{pmatrix} v \\ \omega \end{pmatrix}, \quad \text{if } \alpha \in \left(-\frac{\pi}{2}, \frac{\pi}{2}\right] \tag{4.8}$$

and assumes that the goal frame {G} is in front of the vehicle. The linear control law

$$v = K_\rho\rho$$

$$\omega = K_\alpha\alpha + K_\beta\beta$$

drives the robot to a unique equilibrium at $(\rho, \alpha, \beta) = (0, 0, 0)$. The intuition behind this controller is that the terms $K^\rho\rho$ and $K_\alpha\alpha$ drive the robot along a line toward {G} while the term $K_\beta\beta$ rotates the line so that $\beta \to 0$. The closed-loop system

$$\begin{pmatrix} \dot\rho \\ \dot\alpha \\ \dot\beta \end{pmatrix} = \begin{pmatrix} -K_\rho\rho\cos\alpha \\ K_\rho\sin\alpha - K_\alpha\alpha - K_\beta\beta \\ -K_\rho\sin\alpha \end{pmatrix}$$

is stable so long as

$$K_\rho > 0, \quad K_\beta < 0, \quad K_\alpha - K_\rho > 0 \ .$$

◻ Fig. 4.14 The Simulink model `sl_drivepose` drives the vehicle to a specified configuration. The initial and final configurations are set by the workspace variables `x0` and `xf`, respectively. Red blocks have parameters that you can adjust to investigate the effect on performance

To implement this in practice, the distance and bearing angle to the goal (ρ, α) could be measured by a camera or lidar, and the angle β could be derived from α and vehicle heading θ as measured by a compass.

For the case where the goal is behind the robot, that is $\alpha \notin (-\frac{\pi}{2}, \frac{\pi}{2}]$, we reverse the vehicle by negating v and ψ in the control law. This negation depends on the initial value of α and remains constant throughout the motion.

So far, we have described a *regulator* that drives the vehicle to the configuration $(0, 0, 0)$. To move the robot to an arbitrary configuration (x^*, y^*, θ^*), we perform a change of coordinates

$$x' = x - x^*, \quad y' = y - y^*, \quad \theta' = \theta, \quad \beta = \beta' + \theta^* \ .$$

This configuration controller is implemented by the Simulink model

```
>> sl_drivepose
```

shown in ◻ Fig. 4.14 and the transformation from bicycle to unicycle kinematics is clearly shown, mapping angular velocity ω to steered-wheel angle ψ. We specify a goal configuration

```
>> xg = [5 5 pi/2];
```

sl_drivepose

▶ sn.pub/g9hFYb

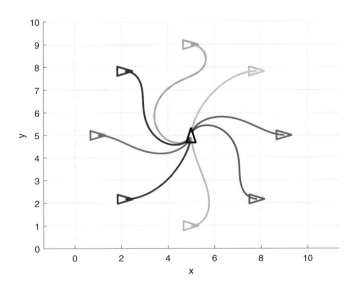

◻ Fig. 4.15 Vehicle motion to a configuration from different initial configurations, using the simulation model `sl_drivepoase` shown in ◻ Fig. 4.14. The goal configuration $(5, 5, \frac{\pi}{2})$ is shown in black, and the controller parameters are $K_\rho = 1$, $K_\alpha = 5$ and $K_\beta = -2$. Note that in some cases the robot has *backed into* the goal configuration

4

■ **Fig. 4.16**　Clearpath Robotics ™ Husky ™ robot with differential steering (Image by Tim Barfoot)

The controller is based on the bicycle model but the Simulink block `Bicycle Kinematic Model` has additional hard nonlinearities including steered-wheel angle limits and velocity rate limiting. If those limits are violated, the configuration controller may fail.

FIG 171.

From Sharp 1896

and an initial configuration

```
>> x0 = [9 5 0];
```

and then simulate the motion

```
>> r = sim("sl_drivepose");
```

As before, the simulation results are stored in `r` and can be plotted

```
>> q = r.find("y");
>> plot(q(:,1),q(:,2));
```

to show the vehicle's path in the xy-plane. The vehicle's path for several starting configurations is shown in ■ Fig. 4.15. The vehicle moves forward or backward, and takes a smooth path to the goal configuration. ◄

4.1.2　Differentially-Steered Vehicle

A car-like vehicle has steerable wheels that are mechanically complex. Differential steering eliminates this complexity and steers by independently controlling the speed of the wheels on each side of the vehicle – if the speeds are not equal, the vehicle will turn. Very simple differentially-steered robots have two driven wheels and a front and/or back castor to provide stability. Larger differentially-steered vehicles such as the one shown in ■ Fig. 4.16 employ a pair of wheels on each side, with each pair sharing a drive motor via some mechanical transmission. Very large differentially-steered vehicles such as bulldozers and tanks sometimes employ caterpillar tracks instead of wheels.

A commonly used model for the low-speed behavior of a differentially-steered vehicle is the kinematic unicycle model shown in ■ Fig. 4.17. The unicycle has a notional single wheel located at the centroid of the wheels.

The pose of the vehicle is represented by the body coordinate frame {B} shown in ■ Fig. 4.17, with its x-axis in the vehicle's forward direction and its origin at the centroid of the unicycle wheel. The configuration of the vehicle is represented by the generalized coordinates $q = (x, y, \theta) \in C$ where $C \subset \mathbb{R}^2 \times \mathbf{S}^1$ is the configuration space. The vehicle's velocity is v in the vehicle's x-direction, and we again assume

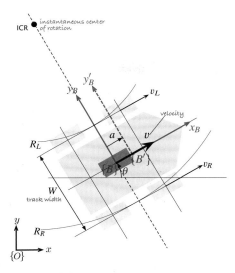

◻ Fig. 4.17 Unicycle model of a 4-wheeled differential-drive vehicle. The 4-wheeled vehicle is shown in light gray, and the equivalent 1-wheeled unicycle model in dark gray. The vehicle's body frame is shown in red, and the world coordinate frame in black. The left and right wheels follow circular arcs around the Instantaneous Center of Rotation (ICR) of radius R_L and R_R respectively. We can use the alternative body frame {B′} for trajectory tracking control

that the wheels roll without slipping sideways

$$^B\boldsymbol{v} = (v, 0) \ .$$

The vehicle follows a curved path centered on the Instantaneous Center of Rotation (ICR). The left-hand wheels roll at a speed of v_L and move along an arc of radius R_L. Similarly, the right-hand wheels roll at a speed of v_R and move along an arc of radius R_R. The angular velocity of {B} is

$$\dot{\theta} = \frac{v_L}{R_L} = \frac{v_R}{R_R}$$

and since $R_R = R_L + W$, we can write the turn rate

$$\dot{\theta} = \frac{v_R - v_L}{W} \tag{4.9}$$

in terms of the differential velocity and wheel separation W. The equations of motion are therefore

$$\dot{x} = v\cos\theta$$
$$\dot{y} = v\sin\theta$$
$$\dot{\theta} = \frac{v_\Delta}{W} \tag{4.10}$$

where $v = \frac{1}{2}(v_R + v_L)$ and $v_\Delta = v_R - v_L$ are the average and differential velocities respectively. For a desired speed v and turn rate $\dot{\theta}$, the wheel speeds are

$$v_R = v + \frac{W}{2}\dot{\theta}$$
$$v_L = v - \frac{W}{2}\dot{\theta} \tag{4.11}$$

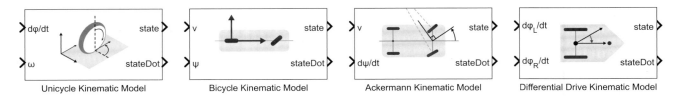

Fig. 4.18 The kinematic models in `robotmobilelib`

Depending on the control inputs of our robot, we can either directly control the wheel speeds, v_R and v_L (most common for differentially-steered vehicles), or control the velocity v and turn rate $\dot{\theta}$.

There are similarities and differences to the bicycle model of (4.4). The turn rate for this vehicle is directly proportional to v_Δ and is independent of speed – the vehicle can turn even when not moving forward. For the 4-wheeled case shown in ◘ Fig. 4.17 the axes of the wheels do not intersect the ICR, so when the vehicle is turning, the wheel velocity vectors v_L and v_R are not tangential to the path – there is a component in the lateral direction that violates the no-slip constraint. This causes skidding or scuffing ◄ which is extreme when the vehicle is turning on the spot – hence differential steering is also called skid steering. Similar to the car-like vehicle, we can write an expression for velocity in the vehicle's y-direction expressed in the world coordinate frame

$$\dot{y}\cos\theta - \dot{x}\sin\theta \equiv 0 \tag{4.12}$$

which is the nonholonomic constraint. It is important to note that the ability to turn on the spot does not make the vehicle holonomic and is fundamentally different to the ability to move in an arbitrary direction which we will discuss next.

If we move the vehicle's reference frame to $\{B'\}$ as shown in ◘ Fig. 4.17 and ignore orientation, we can rewrite (4.10) in matrix form as

$$\begin{pmatrix} \dot{x} \\ \dot{y} \end{pmatrix} = \begin{pmatrix} \cos\theta & -a\sin\theta \\ \sin\theta & a\cos\theta \end{pmatrix} \begin{pmatrix} v \\ \omega \end{pmatrix}$$

and if $a \neq 0$, this can be be inverted

$$\begin{pmatrix} v \\ \omega \end{pmatrix} = \begin{pmatrix} \cos\theta & \sin\theta \\ -\frac{1}{a}\sin\theta & \frac{1}{a}\cos\theta \end{pmatrix} \begin{pmatrix} \dot{x} \\ \dot{y} \end{pmatrix} \tag{4.13}$$

to give the required forward speed and turn rate to achieve an arbitrary velocity $(\dot{x}, \dot{y})$ for the origin of frame $\{B'\}$.

The Simulink block library `robotmobilelib` contains several blocks (see ◘ Fig. 4.18) that implement the kinematic models discussed in the previous sections. The `Unicycle Kinematic Model` implements the unicycle (with velocity v and turn rate $\dot{\theta}$ inputs) and the `Differential Drive Kinematic Model` supports left and right wheel speed inputs, v_R and v_L. All blocks have the same output. Each of these blocks has a corresponding MATLAB class as well. For example, the unicycle model is implemented in `unicycleKinematics`.

4.1.3 Omnidirectional Vehicle

The vehicles we have discussed so far have a constraint on lateral motion, the nonholonomic constraint, which necessitates complex maneuvers in order to achieve some goal configurations. Alternative wheel designs, such as shown in ◘ Fig. 4.19, remove this constraint and allow omnidirectional motion. Even more radical is the spherical wheel shown in ◘ Fig. 4.20.

For indoor applications, this can destroy carpet.

Fig. 4.19 Two types of omnidirectional wheels; note the different roller orientation. **a** Allows the wheel to *roll* sideways (courtesy VEX Robotics); **b** allows the wheel to *drive* sideways (courtesy of Nexus Robotics)

Fig. 4.20 The Rezero ballbot developed at ETH Zürich (Image by Péter Fankhauser)

In this section, we will discuss the mecanum or *Swedish* wheel ▶ shown in ◨ Fig. 4.19b and schematically in ◨ Fig. 4.21. It comprises a number of rollers set around the circumference of the wheel with their axes at an angle of α relative to the axle of the wheel. The dark roller is the one on the bottom of the wheel and currently in contact with the ground. The rollers have a barrel shape so that only one point on the roller is in contact with the ground at any time.

◨ Fig. 4.21 shows the wheel coordinate frame $\{W\}$ with its x-axis pointing in the direction of wheel motion. Rotation of the wheel will cause forward velocity of $R\varpi\hat{x}_w$ where R is the wheel radius and ϖ is the wheel's rotational rate. However,

Mecanum was a Swedish company where the wheel was invented by Bengt Ilon in 1973. It is described in US patent 3,876,255 (expired).

4

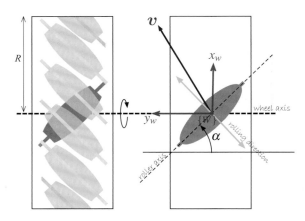

☐ **Fig. 4.21**   Schematic of a mecanum wheel in plan view. The rollers are shown in gray and the dark gray roller is in contact with the ground. The green arrow indicates the rolling direction

because the roller is free to roll in the direction indicated by the green line, normal to the roller's axis, there is potentially arbitrary velocity in that direction. A desired velocity $\boldsymbol{v}$ can be resolved into two components, one parallel to the direction of wheel motion and one parallel to the rolling direction

$$
\boldsymbol{v} = \underbrace{v_w \hat{\boldsymbol{x}}_w}_{\text{driven}} + \underbrace{v_r(\cos\alpha\,\hat{\boldsymbol{x}}_w + \sin\alpha\,\hat{\boldsymbol{y}}_w)}_{\text{rolling}}
$$
$$
= (v_w + v_r\cos\alpha)\hat{\boldsymbol{x}}_w + v_r\sin\alpha\,\hat{\boldsymbol{y}}_w \tag{4.14}
$$

where v_w is the speed due to wheel rotation and v_r is the rolling speed. Expressing $\boldsymbol{v} = v_x\hat{\boldsymbol{x}}_w + v_y\hat{\boldsymbol{y}}_w$ in component form allows us to solve for the rolling speed $v_r = v_y/\sin\alpha$ and substituting this into the first term we can solve for the required wheel velocity

$$
v_w = v_x - v_y\cot\alpha \; . \tag{4.15}
$$

The required wheel rotation rate is then $\varpi = v_w/R$. If $\alpha = 0$ then v_w is undefined since the roller axes are parallel to the wheel axis and the wheel can provide no traction. If $\alpha = \frac{\pi}{2}$ as in ☐ Fig. 4.19a, the wheel allows sideways rolling but not sideways driving since there is zero coupling between v_w to v_y.

A single mecanum wheel does not allow any control in the rolling direction but for three or more mecanum wheels, suitably arranged, the motion in the rolling direction of any one wheel will be driven by the other wheels. A vehicle with four mecanum wheels is shown in ☐ Fig. 4.22. Its pose is represented by the body

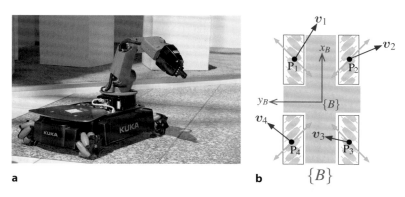

a   b

☐ **Fig. 4.22   a** KUKA youBot®, which has four mecanum wheels (Image courtesy youBot Store); **b** schematic of a vehicle with four mecanum wheels in the youBot configuration frame {B} with its x-axis in the vehicle's forward direction and its origin at the centroid of the four wheels. The configuration of the vehicle is represented by the generalized coordinates $\boldsymbol{q} = (x, y, \theta) \in C$ where $C \subset \mathbb{R}^2 \times \mathbf{S}^1$. The rolling axes of the wheels are orthogonal, and when the wheels are not rotating, the vehicle cannot roll in any direction or rotate – it is effectively locked.

The four wheel-contact points indicated by black dots have coordinate vectors $^B\boldsymbol{p}_i, i = 1, 2, 3, 4$. For a desired body velocity $^B\boldsymbol{v}_B$ and turn rate $^B\omega_B$, the velocity at each wheel-contact point is

$$^B\boldsymbol{v}_i = {}^B\boldsymbol{v}_B + {}^B\omega_B\hat{\boldsymbol{z}}_B \times {}^B\hat{\boldsymbol{p}}_i$$

and we then apply (4.14) and (4.15) to determine wheel rotational rates ϖ_i, while noting that α has the opposite sign for wheels 1 and 3 in (4.14).

Mecanum-wheeled robots exist, for example the KUKA youBot in ◘ Fig. 4.22a, but they are not common. The wheels themselves are more expensive than conventional wheels, and their additional complexity means more points of wear and failure. Mecanum wheels work best on a hard smooth floor, making them well-suited to indoor applications such as warehouse logistics. Conversely, they are ill-suited to outdoor applications where the ground might be soft or uneven, and the rollers could become fouled with debris. To create an omnidirectional base, the minimum requirement is three mecanum wheels placed at the vertices of an equilateral triangle and with their rolling axes intersecting at the center.

4.2 Aerial Robots

» *In order to fly, all one must do is simply miss the ground.*
– Douglas Adams

Aerial robots or uncrewed aerial vehicles (UAV) are becoming increasingly common and span a great range of sizes and shapes as shown in ◘ Fig. 4.23. Applications include military operations, surveillance, meteorological observation, robotics research, commercial photography and, increasingly, hobbyist and personal use. A growing class of flying machines are known as micro air vehicles or MAVs, which are smaller than 15 cm in all dimensions. Fixed-wing UAVs are similar in principle to passenger aircraft with wings to provide lift, a propeller or jet to provide forward thrust, and control surfaces for maneuvering. Rotorcraft UAVs have a variety of configurations that include the conventional *helicopter* design with a main and tail rotor, a *coax* with counter-rotating coaxial rotors, and *quadrotors*. Rotorcraft UAVs have the advantage of being able to take off vertically and to hover.

Aerial robots differ from ground robots in some important ways. Firstly, they have 6 degrees of freedom and their configuration $\boldsymbol{q} \in C$ where $C \subset \mathbb{R}^3 \times (\mathbf{S}^1)^3$ is the 3-dimensional configuration space. Secondly, they are actuated by forces. Their motion model is expressed in terms of forces, torques, and accelerations rather than velocities as was the case for the ground vehicle models – we will use a dynamic rather than a kinematic model. Underwater robots have many similarities to aerial robots and can be considered as vehicles that *fly through water*, and there are underwater equivalents to fixed-wing aircraft and rotorcraft. The principal differences underwater are an upward buoyancy force, drag forces that are much more significant than in air, and added mass.

In this section, we will create a model for a quadrotor aerial vehicle such as shown in ◘ Fig. 4.23d. Quadrotors are now widely available, both as commercial products and as open-source projects. Compared to fixed-wing aircraft, they are highly maneuverable and can be flown safely indoors, which makes them well suited for laboratory or hobbyist use. Compared to conventional helicopters, with a large main rotor and tail rotor, the quadrotor is easier to fly, does not have the complex swash plate mechanism, and is easier to model and control.

4

■ **Fig. 4.23**   Aerial robots. **a** Global Hawk uncrewed aerial vehicle (UAV) (Image courtesy of NASA); **b** a micro air vehicle (MAV) (Image courtesy of AeroVironment, Inc.); **c** Ingenuity on Mars, uses two counter-rotating coaxial propellers (Image courtesy of NASA); **d** a quadrotor that has four rotors and a block of sensing and control electronics in the middle (Image courtesy of 3D Robotics)

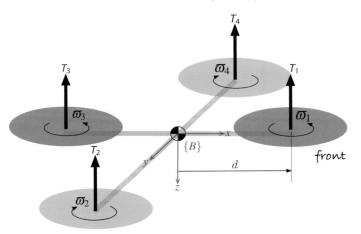

■ **Fig. 4.24**   Quadrotor notation showing the four rotors, their thrust vectors, and directions of rotation. The body frame {B} is attached to the vehicle and has its origin at the vehicle's center of mass. Rotors 1 and 3, shown in blue, rotate counterclockwise (viewed from above) while rotors 2 and 4, shown in pink, rotate clockwise

The notation for the quadrotor model is shown in ■ Fig. 4.24. The body coordinate frame {B} has its z-axis downward following the aerospace convention. The quadrotor has four rotors, labeled 1 to 4, mounted at the end of each cross arm. Hex- and octo-rotors are also popular, with the extra rotors providing greater payload lift capability. The approach described here can be generalized to N rotors, where N is even.

The rotors are driven by electric motors powered by electronic speed controllers. Some low-cost quadrotors use small motors and reduction gearing to achieve sufficient torque. The rotor speed is ϖ_i and the thrust is an upward vector

$$T_i = b\varpi_i^2, \quad i \in \{1, 2, 3, 4\} \tag{4.16}$$

Excurse 4.5: Rotor Flapping

The propeller blades on a rotor craft have fascinating dynamics. When flying into the wind, the blade tip coming forward experiences greater lift while the receding blade has less lift. This is equivalent to a torque about an axis pointing into the wind and the rotor blades behave like a gyroscope (see ▶ Sect. 3.4.1.1) so the net effect is that the plane of the rotor disk pitches up by an amount proportional to the apparent or net wind speed, countered by the blade's bending stiffness and the change in lift as a function of blade bending. The pitched rotor disk causes a component of the thrust vector to retard the vehicle's forward motion, and this velocity-dependent force acts like a friction force. This is known as blade flapping and is an important characteristic of blades on all types of rotorcrafts.

in the vehicle's $-z$-direction, where $b > 0$ is the lift constant that depends on air density, the cube of the radius of the *rotor disk*, ▶ the number of rotor blades, and the chord length of the blade. ▶

The translational dynamics of the vehicle in world coordinates are given by Newton's second law

The disk described by the tips of the rotor blades. Its radius is the rotor blade length.

When flying close to the ground a rotorcraft experiences *ground effect* – increased lift due to a cushion of air trapped beneath it. For helicopters this effect occurs for heights up to 2–3 rotor radii, while for multirotors it occurs for heights up to 5–6 rotor radii (Sharf et al. 2014).

$$
m\dot{\boldsymbol{v}} = \begin{pmatrix} 0 \\ 0 \\ mg \end{pmatrix} - {}^{0}\mathbf{R}_B \begin{pmatrix} 0 \\ 0 \\ T \end{pmatrix} - B\boldsymbol{v}
\tag{4.17}
$$

where $\boldsymbol{v}$ is the velocity of the vehicle's center of mass in the world frame, g is gravitational acceleration, m is the total mass of the vehicle, B is aerodynamic friction, and T is the total upward thrust

$$
T = \sum_{i=1}^{N} T_i \; .
\tag{4.18}
$$

The first term in (4.17) is the weight force due to gravity which acts downward in the world frame, the second term is the total thrust in the vehicle frame rotated into the world coordinate frame, and the third term is aerodynamic drag.

Pairwise differences in rotor thrusts cause the vehicle to rotate. The torque about the vehicle's x-axis, the *rolling* torque, is generated by the moments

$$
\tau_x = d\,T_4 - d\,T_2
$$

where d is the distance from the rotor axis to the center of mass. We can write this in terms of rotor speeds by substituting (4.16)

$$
\tau_x = db\big(\varpi_4^2 - \varpi_2^2\big)
\tag{4.19}
$$

and similarly for the y-axis, the *pitching* torque is

$$
\tau_y = db\big(\varpi_1^2 - \varpi_3^2\big) \; .
\tag{4.20}
$$

The torque applied to each propeller by the motor is opposed by aerodynamic drag

$$
Q_i = k\varpi_i^2
$$

where $k > 0$ depends on the same factors as b. This torque exerts a reaction torque on the airframe which acts to rotate the airframe about the propeller shaft in the

4

opposite direction to its rotation. The total reaction torque about the z-axis is

$$
\begin{aligned}
\tau_z &= Q_1 - Q_2 + Q_3 - Q_4 \\
&= k\left(\varpi_1^2 + \varpi_3^2 - \varpi_2^2 - \varpi_4^2\right)
\end{aligned}
\tag{4.21}
$$

where the different signs are due to the different rotation directions of the rotors. A yaw torque can be created simply by appropriate coordinated control of all four rotor speeds.

The total torque applied to the airframe according to (4.19) to (4.21) is $\tau = (\tau_x, \tau_y, \tau_z)^\top$ and the rotational acceleration is given by Euler's equation of motion from (3.14)

$$
\mathbf{J}\dot{\boldsymbol{\omega}} = -\boldsymbol{\omega} \times \mathbf{J}\boldsymbol{\omega} + \boldsymbol{\tau}
\tag{4.22}
$$

where $\mathbf{J}$ is the 3×3 inertia tensor of the vehicle and $\boldsymbol{\omega}$ is the angular velocity vector.

The motion of the quadrotor is obtained by integrating the forward dynamics equations (4.17) and (4.22). The forces and moments on the airframe, given by (4.18) to (4.21), can be written in matrix form

$$
\begin{pmatrix} \boldsymbol{\tau} \\ T \end{pmatrix} = \begin{pmatrix} 0 & -db & 0 & db \\ db & 0 & -db & 0 \\ k & -k & k & -k \\ -b & -b & -b & -b \end{pmatrix} \begin{pmatrix} \varpi_1^2 \\ \varpi_2^2 \\ \varpi_3^2 \\ \varpi_4^2 \end{pmatrix} = \mathbf{M} \begin{pmatrix} \varpi_1^2 \\ \varpi_2^2 \\ \varpi_3^2 \\ \varpi_4^2 \end{pmatrix}
\tag{4.23}
$$

and are functions of the squared rotor speeds. The mixing matrix $\mathbf{M}$ is constant, and full rank since $b, k, d > 0$ and can be inverted

$$
\begin{pmatrix} \varpi_1^2 \\ \varpi_2^2 \\ \varpi_3^2 \\ \varpi_4^2 \end{pmatrix} = \mathbf{M}^{-1} \begin{pmatrix} \tau_x \\ \tau_y \\ \tau_z \\ T \end{pmatrix}
\tag{4.24}
$$

to solve for the rotor speeds required to apply a specified moment $\boldsymbol{\tau}$ and thrust T to the airframe.

To control the vehicle, we will employ a nested control structure that we describe for pitch and x-translational motion. The innermost loop uses a proportional and derivative controller ◄ to compute the required pitching torque on the airframe

> The rotational dynamics has a second-order transfer function of $\beta(s)/\tau_y(s) = 1/(Js^2 + Bs)$ where J is rotational inertia, B is aerodynamic damping which is generally quite small, and s is the Laplace operator. To regulate a second-order system requires a proportional-derivative (PD) controller, as the rate of change as well as the value of the signal needs to be taken into account.

$$
\tau_y^* = K_{\tau,p}\left(\beta^* - \hat{\beta}\right) + K_{\tau,d}\left(\dot{\beta}^* - \dot{\hat{\beta}}\right)
\tag{4.25}
$$

based on the error between the desired and estimated pitch angle, as well as their rates of change. ◄ The gains $K_{\tau,p}$ and $K_{\tau,d}$ are determined by classical control design approaches based on an approximate dynamic model and then tuned to achieve good performance. The estimated vehicle pitch angle $\hat{\beta}$ would come from an inertial navigation system as discussed in ▶ Sect. 3.4 and $\dot{\hat{\beta}}$ would be derived from gyroscopic sensors. The required rotor speeds are then determined using (4.24).

> The term $\dot{\beta}^*$ is commonly ignored.

Consider a coordinate frame {B′} attached to the vehicle and with the same origin as {B} but with its x- and y-axes in the horizontal plane and parallel to the ground. The thrust vector is parallel to the $-z$-axis of frame {B} and pitching the nose down, rotating about the y-axis by β, generates a force

$$
{}^{B'}\!f = \boldsymbol{\xi}^{r_y}(\beta) \cdot \begin{pmatrix} 0 \\ 0 \\ -T \end{pmatrix} = \begin{pmatrix} -T\sin\beta \\ 0 \\ -T\cos\beta \end{pmatrix},
$$

which has a component

$$^{B'}\!f_x = -T\sin\beta \approx -T\beta$$

that accelerates the vehicle in the $-^{B'}\!x$-direction, and we have assumed that β is small. We can control the velocity in this direction with a proportional control law

$$^{B'}\!f_x^* = m K_f\left(^{B'}\!v_x^* - ^{B'}\!\hat{v}_x\right)$$

where $K_f > 0$ is a gain. ▶ Combining these two equations, we obtain the desired pitch angle

$$\beta^* \approx -\frac{m}{T}K_f\left(^{B'}\!v_x^* - ^{B'}\!\hat{v}_x\right) \tag{4.26}$$

required to achieve the desired forward velocity. Using (4.25), we compute the required pitching torque, and then, using (4.24), the required rotor speeds. For a vehicle in vertical equilibrium, the total thrust equals the weight force, so $m/T \approx 1/g$.

The estimated vehicle velocity $^{B}\!\hat{v}_x$ could come from an inertial navigation system as discussed in ▶ Sect. 3.4 or a GPS receiver. If the desired position of the vehicle in the xy-plane of the world frame is $^0\!p \in \mathbb{R}^2$, then the desired velocity is given by the proportional control law

$$^0\!v^* = K_p\left(^0\!p^* - ^0\!\hat{p}\right) \tag{4.27}$$

based on the error between the desired and actual position. The desired velocity in the xy-plane of frame $\{B'\}$ is

$$^{B'}\!v = \ominus\,{}^0\!\boldsymbol{\xi}^r_{B'}(\gamma)\!\cdot$$

which is a function of the yaw angle γ

$$\begin{pmatrix} ^{B'}\!v_x \\ ^{B'}\!v_y \end{pmatrix} = \begin{pmatrix} \cos\gamma & -\sin\gamma \\ \sin\gamma & \cos\gamma \end{pmatrix}^{\!\top}\begin{pmatrix} v_x \\ v_y \end{pmatrix}$$

◘ Fig. 4.25 shows a Simulink model of the complete control system for a quadrotor ▶ which can be loaded and displayed by

```
>> sl_quadrotor
```

The negative sign is because a positive pitch creates a thrust in the $-x$-direction. For the roll axis, there is no negative sign, positive roll angle leads to positive force and velocity in the y-direction.

This model is hierarchical and organized in terms of subsystems. Click the down arrow on a subsystem (can be seen onscreen but not in the figure) to reveal the detail. Double-click on the subsystem box to modify its parameters.

sl_quadrotor

▶ sn.pub/2wIDIm

◘ **Fig. 4.25** The Simulink model `sl_quadrotor` for a closed-loop simulation of a quadrotor. A Simulink bus is used for the 12-element state vector x output by the `Quadrotor` block. x contains the quadrotor state: position, orientation, velocity, and orientation rate. The vehicle lifts off and flies in a circle at constant altitude while yawing about its own center. To reduce the number of lines in the diagram we have used `Goto` and `From` blocks to transmit and receive the state vector

4

Working our way left to right and starting at the top, we have the desired position of the quadrotor in world coordinates. The position error is rotated from the world frame to the body frame and becomes the desired velocity. The velocity controller implements (4.26) and its equivalent for the roll axis, and outputs the desired pitch and roll angles of the quadrotor. The attitude controller is a proportional-derivative controller that determines the appropriate pitch and roll torques to achieve these angles based on feedback of current attitude and attitude rate. ◄ The yaw control block determines the error in heading angle and implements a proportional-derivative controller to compute the required yaw torque that is achieved by speeding up one pair of rotors and slowing the other pair.

Altitude is controlled by a proportional-derivative controller

$$T = K_p(z^* - \hat{z}) + K_d\left(\dot{z}^* - \dot{\hat{z}}\right) + T_w$$

Using the coordinate conventions shown in ◘ Fig. 4.24, x-direction motion requires a negative pitch angle, and y-direction motion requires a positive roll angle, so the gains have different signs for the roll and pitch loops.

which determines the average rotor speed. $T_w = mg$ is the weight of the vehicle and this is an example offeedforward control – used here to counter the effect of gravity that is otherwise a constant disturbance to the altitude control loop. The alternatives to feedforward control would be to have very high value of K_p for the altitude loop which might lead to instability, or a proportional-integral-derivative (PID) controller that might require a long time for the integral term to remove the steady-state error, and then lead to overshoot. We will revisit gravity compensation in ▶ Chap. 9 in the context of robot manipulator arms.

The control mixer block combines the three torque demands and the vertical thrust demand and implements (4.24) to determine the appropriate rotor speeds. Rotor speed limits are applied here. These are input to the quadrotor block ◄ which implements the forward dynamics, integrating (4.17) to (4.22) to determine the position, orientation, velocity, and orientation rate. The output of this block is the 12-element state vector $\boldsymbol{x} = (\,^0\boldsymbol{p},\,^0\boldsymbol{\Gamma},\,^B\dot{\boldsymbol{p}},\,^B\dot{\boldsymbol{\Gamma}})$. As is common in aerospace applications, we represent orientation $\boldsymbol{\Gamma} \in (\mathbf{S}^1)^3$ and orientation rate $\dot{\boldsymbol{\Gamma}} \in \mathbb{R}^3$ in terms of ZYX roll-pitch-yaw angles. Note that position and attitude are in the world frame, while the rates are expressed in the body frame.

This block can simulate an N-rotor vehicle, the default is $N = 4$.

The parameters of a specific quadrotor can be loaded with

```
>> mdl_quadrotor
```

which creates a structure called `quadrotor` in the workspace, and its elements are the various dynamic properties of the quadrotor. The simulation can be run using the Simulink toolstrip or from the MATLAB command line

```
>> r = sim("sl_quadrotor");
```

Opening the model using
`>> sl_quadrotor` automatically loads the default quadrotor model. This is done by the `PreLoadFcn` callback set from the model's properties
`Modeling → Model Settings →`
`Model Properties → Callbacks`
`→ PreLoadFcn`.

and it displays an animation in a separate window. ◄ The vehicle lifts off and flies around a circle while spinning slowly about its own z-axis. A snapshot is shown in ◘ Fig. 4.26. The simulation writes the results from each timestep into a matrix in the workspace

```
>> size(r.result)
ans =
        2355            16
```

which has one row per timestep, and each row contains the state vector (elements 1–12) and the commanded rotor speeds ω_i (elements 13–16). We can plot x and y versus time using

```
>> plot(r.t,r.result(:,1:2))
```

To recap on control of the quadrotor:

- A position error leads to a required translational velocity to correct the error.
- To achieve that velocity, we adjust the pitch and roll of the vehicle so that a component of the vehicle's thrust acts in the horizontal plane and generates a force to accelerate it.

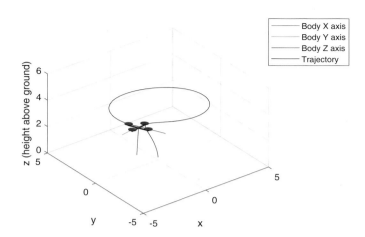

◘ Fig. 4.26 One frame from the quadrotor model `sl_quadrotor` shown in ◘ Fig. 4.25. The rotor disks are shown as circles and their orientation includes the effect of blade flapping

- To achieve the required pitch and roll angles, we apply moments to the airframe by means of differential-propeller thrust which, in turn, requires rotor speed control.
- When the vehicle is performing this motion, the total thrust must be increased so that the vertical component still balances gravity.
- As the vehicle approaches its goal, the airframe must be rotated in the opposite direction so that a component of thrust decelerates the motion.

This indirection from translational motion to rotational motion is a consequence of the vehicle being underactuated – we have just four rotor speeds to adjust, but the vehicle's configuration space is 6-dimensional. In the configuration space, we cannot move in the x- or y-directions, but we can move in the pitch- or roll-direction which *leads* to motion in the x- or y-directions. The cost of underactuation is once again a maneuver. The pitch and roll angles are a means to achieve translational control but cannot be independently set if we wish to hold a constant position.

For application such as waypoint following and motion planning, modeling the complete quadrotor dynamics and controllers (as shown in ◘ Fig. 4.25) might be overkill. During the rapid prototyping stage, we would like to quickly experiment

`sl_quadrotor_guidance`

▶ sn.pub/QMbkWT

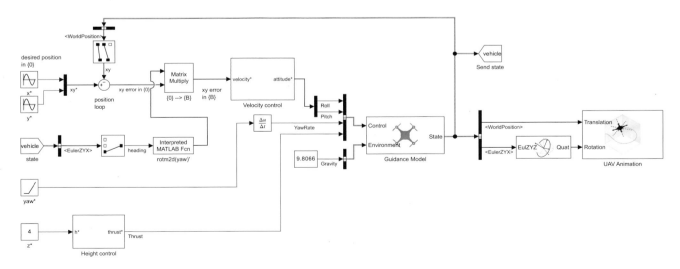

◘ Fig. 4.27 The Simulink model `sl_quadrotor_guidance` which simplifies the model shown in ◘ Fig. 4.25 by replacing the attitude controllers, control mixer, and dynamics simulation with a single `Guidance Model` block

4

and tune parameters, so a simpler model can help in reducing the cycle time. Ideally, we want to use a model that is less complex but captures the basic behavior of the UAV. In later verification stages, we can switch to a more accurate model, which will take more effort to build. The model in ◻ Fig. 4.27 uses the `Guidance Model` block to accomplish this simplification. The `Guidance Model` block approximates the behavior of the closed-loop system of an autopilot controller and a second-order quadrotor kinematic model. Additional approximations of the closed-loop system are described in Mellinger (2012). When running this model, we will notice that it simulates much faster (on the order of 4–5× faster) than the model shown in ◻ Fig. 4.25 but produces essentially the same trajectory. We can easily tune the behavior of the simulated UAV by changing the parameters of the `Guidance Model` block.

4.3 Advanced Topics

4.3.1 Nonholonomic and Underactuated Systems

We introduced the notion of configuration space in ▶ Sect. 2.4.9, and it is useful to revisit it now that we have discussed several different types of mobile robot platforms. Common vehicles – as diverse as cars, hovercrafts, ships, and aircraft – are all able to move forward effectively but are unable to instantaneously move sideways. This is a very sensible tradeoff that simplifies design and caters to the motion we most commonly require of the vehicle. Lateral motion for occasional tasks such as parking a car, docking a ship, or landing an aircraft are possible, albeit with some complex maneuvering but humans can learn this skill.

The configuration space of a train was introduced in ▶ Sect. 2.4.9. To fully describe all its constituent particles, we need to specify just one generalized coordinate: its distance q along the track from some datum. It has one degree of freedom, and its configuration space is $C \subset \mathbb{R}$. The train has one motor, or actuator, to move it along the track and this is equal to the number of degrees of freedom. We say the train is fully actuated.

Consider a hovercraft that moves over a planar surface. To describe its configuration, we need to specify three generalized coordinates: its position in the xy-plane and its heading angle. It has three degrees of freedom, and its configuration space is $C \subset \mathbb{R}^2 \times \mathbf{S}^1$. This hovercraft has two propellers whose axes are parallel but not collinear. The sum of their thrusts provides a forward force and the difference in thrusts generates a yawing torque for steering. The number of actuators, two, is less than its degrees of freedom $\dim C = 3$, and we call this an underactuated system. This imposes significant limitations on the way in which it can move. At any point in time, we can control the forward (parallel to the thrust vectors) acceleration and the rotational acceleration of the hovercraft, but there is zero sideways (or lateral) acceleration since it cannot generate any lateral thrust. Nevertheless, with some clever maneuvering, like with a car, the hovercraft can follow a path that will take it to a place to one side of where it started. In the hovercraft's 3-dimensional configuration space, this means that at any point there are certain directions in which *acceleration* is not possible. We can reach points in those directions but not directly, only by following some circuitous path.

All aerial and underwater vehicles have a configuration that is completely described by six generalized coordinates – their position and orientation in 3D space. The configuration space is $C \subset \mathbb{R}^3 \times (\mathbf{S}^1)^3$ where the orientation is expressed in some three-angle representation – since $\dim C = 6$, the vehicles have six degrees of freedom. A quadrotor has four actuators, four thrust-generating propellers, and this is fewer than its degrees of freedom making it underactuated. Controlling the four propellers causes motion in the up/down, roll, pitch, and yaw directions of

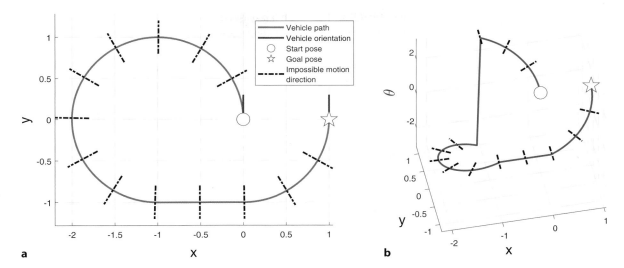

◻ Fig. 4.28 Sideways motion of a car-like vehicle. The car cannot move directly from the circle to the star due to the nonholonomic constraint. **a** Motion in the plane and the black line segments indicate the directions in which motion is not possible; **b** the same motion shown in configuration space, note that the heading angle has wrapped around

the configuration space, but not in the forward/backward or left/right directions. To access those degrees of freedom, it is necessary to perform a *maneuver*: pitch down so that the thrust vector provides a horizontal force component, accelerate forward, pitch up so that the thrust vector provides a horizontal force component to decelerate, and then level out.

For a helicopter, only four of the six degrees of freedom are practically useful: up/down, forward/backward, left/right, and yaw. Therefore, a helicopter requires a minimum of four actuators: the main rotor generates a thrust vector whose magnitude is controlled by the collective pitch and whose direction is controlled by the lateral and longitudinal cyclic pitch. The tail rotor provides a yawing moment. This leaves two degrees of freedom unactuated, roll and pitch angles, but clever design ensures that gravity actuates them and keeps them close to zero like the keel of a boat – without gravity a helicopter cannot work. A fixed-wing aircraft moves forward very efficiently and also has four actuators: engine thrust provides acceleration in the forward direction and the ailerons, elevator and rudder exert respectively roll, pitch, and yaw moments on the aircraft. ▶ To access the missing degrees of freedom such as up/down and left/right translation, the aircraft must pitch or yaw while moving forward.

The advantage of underactuation is having fewer actuators. In practice, this means real savings in terms of cost, complexity, and weight. The consequence is that, at any point in its configuration space, there are certain directions in which the vehicle cannot move. Fully-actuated designs are possible but not common, for example, the RangerBot underwater robot shown in ◻ Fig. 4.2c has six degrees of freedom and six actuators. These can exert an arbitrary force and torque on the vehicle, allowing it to accelerate in any direction or about any axis.

A 4-wheeled car has many similarities to the hovercraft discussed above. It moves over a planar surface and its configuration is fully described by its position in the xy-plane and a rotation angle. It has three degrees of freedom and its configuration space is $C \subset \mathbb{R}^2 \times \mathbf{S}^1$. A car has two actuators, one to move forwards or backwards and one to change the heading direction. A car, like a hovercraft, is underactuated. We know from our experience with cars that we cannot move directly in certain directions and sometimes need to perform a maneuver to reach our goal as shown in ◻ Fig. 4.28.

Some low-cost hobby aircraft have no rudder and rely only on ailerons to bank and turn the aircraft. Even cheaper hobby aircraft have no elevator and rely on motor speed to control height.

4

◻ Table 4.1 Summary of configuration space characteristics for various robots. A nonholonomic system is underactuated and/or has a rolling constraint

	dim C	Degrees of freedom	Number of actuators	Actuation	Rolling constraints	Holonomic
Train	1	1	1	Full		✓
2-joint robot arm	2	2	2	Full		✓
6-joint robot arm	6	6	6	Full		✓
10-joint robot arm	10	10	10	Over		✓
Hovercraft	3	3	2	Under		
Car	3	2	2	Under	✓	
Helicopter	6	6	4	Under		
Fixed-wing aircraft	6	6	4	Under		
RangerBot	6	6	6	Full		✓

A differentially- or skid-steered vehicle, such as a tank, is also underactuated – it has only two actuators, one for each track. While this type of vehicle can turn on the spot, it cannot move sideways. To do that it has to turn, proceed, stop, then turn – this need to maneuver is the clear signature of an underactuated system.

We might often wish for an ability to drive our car sideways, but the standard wheel provides real benefit when cornering – lateral friction between the wheels and the road provides, for free, the centripetal force which would otherwise require an extra actuator to provide. The hovercraft has many similarities to a car but we can push a hovercraft sideways – we cannot do that with a car. This lateral friction is a distinguishing feature of the car.

The hovercraft, aerial, and underwater vehicles are controlled by forces, so in this case the constraints are on vehicle acceleration in configuration space, not velocity.

The inability to slip sideways is a constraint, the *rolling* constraint, on the velocity ◄ of the vehicle just as underactuation is. A vehicle with one or more *velocity constraints*, due to underactuation or a rolling constraint, is referred to as a nonholonomic system. A key characteristic of these systems is that they cannot always move *directly* from one configuration to another. Sometimes, they must take a longer indirect path which we recognize as a maneuver. A car has a velocity constraint due to its wheels and is also underactuated.

For example, fixing the end of a 10-joint robot arm introduces six holonomic constraints (position and orientation) so the arm would have only 4 degrees of freedom.

In contrast, a holonomic constraint restricts the possible configurations that the system can achieve – it can be expressed as an equation written in terms of the configuration variables. ◄ A nonholonomic constraint, such as (4.5) and (4.12), is one that restricts the *velocity* (or acceleration) of a system in configuration space – it can only be expressed in terms of the derivatives of the configuration variables. ◄ The nonholonomic constraint does not restrict the possible configurations the system can achieve, but it does preclude instantaneous velocity or acceleration in certain directions as shown by the black line segments in ◻ Fig. 4.28.

The constraint cannot be integrated to a constraint in terms of configuration variables, so such systems are also known as nonintegrable systems.

In control theoretic terms, Brockett's theorem states that nonholonomic systems are controllable, but they cannot be stabilized to a desired state using a differentiable, or even continuous, pure state-feedback controller. A time-varying or nonlinear control strategy is required which means that the robot follows some generally nonlinear path. For example, the controller (4.8) introduces a discontinuity at $\rho = 0$. One exception is an underactuated system moving in a 3-dimensional space within a force field, for example, a gravity field – gravity acts like an additional actuator and makes the system linearly controllable (but not holonomic), as we showed for the quadrotor example in ► Sect. 4.2.

Mobility parameters for the various robots that we have discussed here, and earlier in ► Sect. 2.4.9, are tabulated in ◻ Tab. 4.1. We will discuss under- and over-actuation in the context of arm robots in ► Sect. 8.3.4.

4.4 Wrapping Up

In this chapter, we have created and discussed models and controllers for several common, but quite different, robot platforms. We first discussed wheeled robots. For car-like vehicles, we developed a kinematic model that we used to develop several different controllers in order that the platform could perform useful tasks such as driving to a point, driving along a line, following a path, or driving to a configuration. We then discussed differentially-steered vehicles on which many robots are based, and omnidirectional robots based on novel wheel types. Then we discussed a simple but common aerial vehicle, the quadrotor, and developed a dynamic model and a hierarchical control system that allowed the quadrotor to fly a circuit. This hierarchical or nested control approach is described in more detail in ▶ Sect. 9.1.7 in the context of robot arms.

We also extended our earlier discussion about configuration space to include the velocity constraints due to underactuation and rolling constraints from wheels.

The next chapters will discuss how to plan paths for robots through complex environments that contain obstacles and then how to determine the location of a robot.

4.4.1 Further Reading

Comprehensive modeling of mobile ground robots is provided in the book by Siegwart et al. (2011). In addition to the models covered here, it presents in-depth discussion of a variety of wheel configurations with different combinations of driven wheels, steered wheels, and passive castors. The book by Kelly (2013) also covers vehicle modeling and control. Both books also provide a good introduction to perception, localization, and navigation, which we will discuss in the coming chapters.

The paper by Martins et al. (2008) discusses kinematics, dynamics, and control of differentially-steered robots. Astolfi (1999) develops pose control for a car-like vehicle. The Handbook of Robotics (Siciliano and Khatib 2016, part E) covers modeling and control of various vehicle types including aerial and underwater. The theory of helicopters with an emphasis on robotics is provided by Mettler (2003), but the definitive reference for helicopter dynamics is the very large book by Prouty (2002). The book by Antonelli (2014) provides comprehensive coverage of modeling and control of underwater robots.

Some of the earliest papers on quadrotor modeling and control are by Pounds, Mahony and colleagues (Hamel et al. 2002; Pounds et al. 2004, 2006). The thesis by Pounds (2007) presents comprehensive aerodynamic modeling of a quadrotor with a particular focus on blade flapping, a phenomenon well known in conventional helicopters but largely ignored for quadrotors. A tutorial introduction to the control of multi-rotor aerial robots is given by Mahony, Kumar, and Corke (2012). For a comprehensive treatment of fixed-wing dynamic models, low-level autopilot design, state estimation, and high-level path planning, see the book by Beard and McLain (2012). The `Guidance Model` block that we introduced in ▶ Sect. 4.2 approximates the behavior of the closed-loop system of an autopilot controller and a second-order quadrotor kinematic model. This and other approximations of the closed-loop system are described in Mellinger (2012).

Mobile ground robots are now a mature technology for transporting parts around manufacturing plants. The research frontier is now for vehicles that operate autonomously in outdoor environments (Siciliano and Khatib 2016, part F). Research into the automation of passenger cars has been ongoing since the 1980s, and increasing autonomous driving capability is becoming available in new cars.

4

■ ■ Historical and interesting

The Navlab project at Carnegie-Mellon University started in 1984 and a series of autonomous vehicles, Navlabs, were built and a large body of research has resulted. All vehicles made strong use of computer vision for navigation. In 1995, the supervised autonomous Navlab 5 made a 3000-mile journey, dubbed "No Hands Across America" (Pomerleau and Jochem 1995, 1996). The vehicle steered itself 98% of the time largely by visual sensing of the white lines at the edge of the road.

In Europe, Ernst Dickmanns and his team at Universität der Bundeswehr München demonstrated autonomous control of vehicles. In 1988, the VaMoRs system, a 5-tonne Mercedes-Benz van, could drive itself at speeds over $90\,\mathrm{kmh^{-1}}$ (Dickmanns and Graefe 1988b; Dickmanns and Zapp 1987; Dickmanns 2007). The European Prometheus Project ran from 1987 to 1995 and, in 1994, the robot vehicles VaMP and VITA-2 drove more than $1000\,\mathrm{km}$ on a Paris multi-lane highway in standard heavy traffic at speeds up to $130\,\mathrm{kmh^{-1}}$. They demonstrated autonomous driving in free lanes, convoy driving, automatic tracking of other vehicles, and lane changes with autonomous passing of other cars. In 1995, an autonomous S-Class Mercedes-Benz made a $1600\,\mathrm{km}$ trip from Munich to Copenhagen and back. On the German Autobahn, speeds exceeded $175\,\mathrm{kmh^{-1}}$ and the vehicle executed traffic maneuvers such as overtaking. The mean time between human interventions was $9\,\mathrm{km}$ and it drove up to $158\,\mathrm{km}$ without any human intervention. The UK part of the project demonstrated autonomous driving of an XJ6 Jaguar with vision (Matthews et al. 1995) and radar-based sensing for lane keeping and collision avoidance. In the USA in the 2000s, DARPA ran a series of Grand Challenges for autonomous cars. The 2005 desert and 2007 urban challenges are comprehensively described in compilations of papers from the various teams in Buehler et al. (2007, 2010). More recent demonstrations of self-driving vehicles are a journey along the fabled silk road described by Bertozzi et al. (2011) and a classic road trip through Germany by Ziegler et al. (2014).

Ackermann's magazine can be found online at ▶ https://smithandgosling. wordpress.com/2009/12/02/ackermanns-repository-of-arts, and the carriage steering mechanism is published in the March and April issues of 1818 (pages 162 and 234). King-Hele (2002) provides a comprehensive discussion about the prior work on steering geometry and Darwin's earlier invention.

■ ■ Toolbox Notes

In addition to the Simulink blocks used in this chapter, the RVC Toolbox provides MATLAB classes that implement the kinematic equations. For example, we can create a vehicle model with steered-wheel angle and speed limits

```
>> veh = BicycleVehicle(MaxSpeed=1,MaxSteeringAngle=deg2rad(30));
>> veh.q
ans =
     0     0     0
```

which has an initial configuration of $(0, 0, 0)$. We can apply (v, γ) and after one time step the configuration is

```
>> veh.step(0.3,0.2)
ans =
    0.0300       0.0061
>> veh.q
ans =
    0.0300          0     0.0061
```

We can evaluate (4.4) for a particular configuration and set of control inputs (v, γ)

```
>> deriv = veh.derivative(0,veh.q,[0.3 0.2]);
deriv =
      0.3000     0.0018     0.0608
```

We will use these MATLAB classes for localization and mapping in ▶ Chap. 6.

4.4.2 **Exercises**

1. For a 4-wheeled vehicle with $L = 2$ m and track width between wheel centers of 1.5 m
   a) What steering wheel angle is needed for a turn rate of $10\,°s^{-1}$ at a forward speed of $20\,km\,h^{-1}$?
   b) Compute the difference in angle for the left and right steered-wheels, in an Ackermann steering system, when driving around arcs of radius 10, 50, and 100 m.
   c) If the vehicle is moving at $80\,km\,h^{-1}$, compute the difference in back wheel rotation rates for curves of radius 10, 50, and 100 m.
2. Write an expression for turn rate in terms of the angular rotation rate of the two back wheels. Investigate the effect of errors in wheel radius and vehicle width.
3. Consider a car and bus with $L = 4$ and $L = 12$ m respectively. To follow an arc with radius of 10, 20, and 50 m, determine the respective steered-wheel angles.
4. For several steered-wheel angles in the range -45 to $45°$ and a velocity of $2\,ms^{-1}$, overlay plots of the vehicle's trajectory in the xy-plane.
5. Implement the $\ominus$ operator used in ▶ Sect. 4.1.1.1 and check against the code for `angdiff`.
6. Driving to a point (▶ Sect. 4.1.1.1), plot x, y, and θ against time.
7. Pure pursuit example (▶ Sect. 4.1.1.3)
   a) Investigate what happens if you vary the look-ahead distance.
   b) Modify the pure pursuit example so that the robot follows a slalom course.
   c) Modify the pure pursuit example to follow a target moving back and forth along a line.
8. Driving to a configuration (▶ Sect. 4.1.1.4)
   a) Repeat the example with a different initial orientation.
   b) Implement a parallel-parking maneuver. Is the resulting path practical?
   c) Experiment with different controller parameters.
9. Create a GUI interface with a simple steering wheel and velocity control and use this to create a very simple driving simulator. Alternatively, interface a gaming steering wheel and pedal to MATLAB.
10. Adapt the various controllers in ▶ Sect. 4.1.1 to the differentially-steered robot.
11. Derive (4.9) from the preceding equation.
12. For constant forward velocity, plot v_L and v_R as a function of ICR position along the y-axis. Under what conditions do v_L and v_R have a different sign?
13. Using Simulink, implement a controller using (4.13) that moves a unicycle robot in its y-direction. How does the robot's orientation change as it moves?
14. Sketch the design for a robot with three mecanum wheels. Ensure that it cannot roll freely and that it can drive in any direction. Write code to convert from desired vehicle translational and rotational velocity to wheel rotation rates.
15. For the 4-wheeled omnidirectional robot of ▶ Sect. 4.1.3, write an algorithm that will allow it to move in a circle of radius 0.5 m around a point with its nose always pointed towards the center of the circle.
16. Aerial Robots (▶ Sect. 4.2)
    a) At equilibrium, compute the speed of all the propellers.
    b) Experiment with different control gains. What happens if you reduce the the gains applied to rates of change of quantities?
    c) Remove the gravity feedforward constant and experiment with large altitude gain or a PI controller.
    d) When the vehicle has nonzero roll and pitch angles, the magnitude of the vertical thrust is reduced, and the vehicle will slowly descend. Add compensation to the vertical thrust to correct this.
    e) Simulate the quadrotor flying inverted, that is, its z-axis is pointing upwards.

f) Program a ballistic motion. Have the quadrotor take off at 45° to horizontal, then remove all thrust.

g) Program a smooth landing. What does smooth mean?

h) Program a barrel roll maneuver. Have the quadrotor fly horizontally in its x-direction and then increase the roll angle from 0 to 2π.

i) Program a flip maneuver. Have the quadrotor fly horizontally in its x-direction and then increase the pitch angle from 0 to 2π.

j) Repeat the exercise but for a 6-rotor configuration.

k) Use the function `mstraj` to create a trajectory through ten via points $(X_i, Y_i, Z_i, \theta_y)$ and modify the controller of ◼ Fig. 4.25 for smooth pursuit of this trajectory.

l) Create a MATLAB GUI interface with a simple joystick control and use this to create a very simple flying simulator. Alternatively, interface a gaming joystick to MATLAB.

Navigation

» Navigation: *the process of directing a vehicle so as to reach the intended destination*
  – IEEE Standard 172-1983

Contents

5.1   **Introduction to Reactive Navigation** – 163

5.2   **Introduction to Map-Based Navigation** – 168

5.3   **Planning with a Graph-Based Map** – 170

5.4   **Planning with an Occupancy Grid Map** – 180

5.5   **Planning with Roadmaps** – 187

5.6   **Planning Drivable Paths** – 192

5.7   **Advanced Topics** – 207

5.8   **Wrapping Up** – 209

© The Author(s), under exclusive license to Springer Nature Switzerland AG 2023
P. Corke, W. Jachimczyk, R. Pillat, *Robotics, Vision and Control*, Springer Tracts in Advanced Robotics 147,
https://doi.org/10.1007/978-3-031-07262-8_5

5

Fig. 5.1 **a** Elsie the robotic tortoise developed in 1940 by William Grey Walter. Burden Institute Bristol (1948). Now in the collection of the Smithsonian Institution but not on display (Image courtesy Reuben Hoggett collection); **b** Shakey. SRI International (1968). It is now in the Computer Museum in Mountain View (Image courtesy SRI International)

chapter5.mlx

▶ sn.pub/276wTJ

Robot navigation is the problem of guiding a robot toward a goal. The goal might be specified in terms of some feature in the environment, for instance, moving toward a light, or in terms of some geometric coordinate or map reference.

The human approach to navigation is to make maps and erect signposts, and at first glance it seems obvious that robots should operate the same way. However, many robotic tasks can be achieved without any map at all, using an approach referred to as *reactive navigation* – sensing the world and reacting to what is sensed. For example, the early robotic tortoise Elsie, shown in ◘ Fig. 5.1a, *reacted* to her environment and could seek out a light source and move around obstacles without having any explicit plan, or any knowledge of the position of herself or the light.

Fig. 5.2 Time lapse photograph of an early iRobot® Roomba® robot cleaning a room (Image by Chris Bartlett)

Today, tens of millions of robotic vacuum cleaners are cleaning floors and many of them do so without using any map of the rooms in which they work. Instead, they do the job by making random moves and sensing only that they have made contact with an obstacle as shown in ◙ Fig. 5.2.

An alternative approach was embodied in the 1960s robot Shakey, shown in ◙ Fig. 5.1b, which was capable of 3D perception. Shakey created a map of its environment and then reasoned about the map to plan a path to its destination. This is a more human-style *map-based navigation* like that used today in the autopilots of aircraft and ships, self-driving cars and mobile robots. The problem of using a map to find a path to a goal is known as path planning or motion planning. This approach supports more complex tasks but is itself more complex. It requires having a map of the environment and knowing the robot's position with respect to the map at all times – both of these are non-trivial problems that we introduce in ▶ Chap. 6.

Reactive and map-based approaches occupy opposite ends of a spectrum for mobile robot navigation. Reactive systems can be fast and simple, since sensation is connected directly to action. This means there is no need for resources to acquire and hold a representation of the world, nor any capability to reason about that representation. In nature, such strategies are used by simple organisms such as insects. Systems that make maps and reason about them require more resources but are capable of performing more complex tasks. In nature, such strategies are used by more complex creatures such as mammals.

The remainder of this chapter discusses the reactive and map-based approaches to robot navigation with a focus on wheeled robots operating in a planar environment. Reactive navigation is covered in ▶ Sect. 5.1 and map-based motion planning approaches are introduced in ▶ Sect. 5.2.

5.1 Introduction to Reactive Navigation

Surprisingly complex tasks can be performed by a robot even if it has no map and no "idea" about where it is. The first generation of Roomba robotic vacuum cleaners used only random motion and information from contact sensors to perform a complex cleaning task, as shown in ◙ Fig. 5.2. Insects such as ants and bees

---

Excurse 5.1: William Grey Walter

Walter (1910–1977) was a neurophysiologist and pioneering cyberneticist born in Kansas City, Missouri. He studied at King's College, Cambridge and, unable to obtain a research fellowship at Cambridge, he undertook neurophysiological research at hospitals in London and, from 1939, at the Burden Neurological Institute in Bristol. He developed electro-encephalographic brain topography which used multiple electrodes on the scalp and a triangulation algorithm to determine the amplitude and location of brain activity.

Walter was influential in the emerging field of cybernetics. He built robots to study how complex reflex behavior could arise from neural interconnections. His tortoise Elsie (of the species *Machina Speculatrix*) was built in 1948. It was a three-wheeled robot capable of phototaxis that could also find its way to a recharging station. A second-generation tortoise (from 1951) is in the collection of the Smithsonian Institution. Walter published popular articles in Scientific American (1950 and 1951) and a book *The Living Brain* (1953). He was

badly injured in a car accident in 1970 from which he never fully recovered. (Image courtesy Reuben Hoggett collection)

5

gather food and return it to their nest based on input from their senses, they have far too few neurons to create any kind of mental map of the world and plan paths through it. Even single-celled organisms such as flagellate protozoa exhibit goal-seeking behaviors. In this case, we need to temporarily modify our earlier definition of a robot to be:

» a goal-oriented machine that can sense, ~~plan~~ and act.

Walter's robotic tortoise shown in ◘ Fig. 5.1 demonstrated that it could move toward a light source. It was seen to exhibit "life-like behavior" and was an important result in the then emerging scientific field of Cybernetics. This might seem like an exaggeration from the technologically primitive 1940s but this kind of behavior, known as phototaxis, ◂ is exhibited by simple organisms.

More generally, a *taxis* is the response of an organism to a stimulus gradient.

5.1.1 **Braitenberg Vehicles**

A very simple class of goal-achieving robots are known as Braitenberg vehicles and are characterized by direct connection between sensors and motors. They have no explicit internal representation of the environment in which they operate and they make no explicit plans. ◂

This is a fine philosophical point, the plan could be considered implicit in the details of the connections between the motors and sensors.

Consider the problem of a robot, moving in two dimensions, that is seeking the local maximum of a scalar field – the field could be light intensity or the concentration of some chemical. ◂ The Simulink® model

This is similar to the problem of moving to a point discussed in ► Sect. 4.1.1.1.

```
>> sl_braitenberg
```

shown in ◘ Fig. 5.3 achieves this using velocity and steering inputs derived directly from the sensors. ◂

This is similar to Braitenberg's Vehicle 4a.

To ascend the gradient, we need to estimate the gradient direction at the robot's location and this requires at least two measurements of the field. ◂ In this example, we use a common trick from nature, bilateral sensing, where two sensors are arranged symmetrically on each side of the robot's body. The sensors are modeled

We can make the measurements simultaneously using two spatially separated sensors or from one sensor over time as the robot moves.

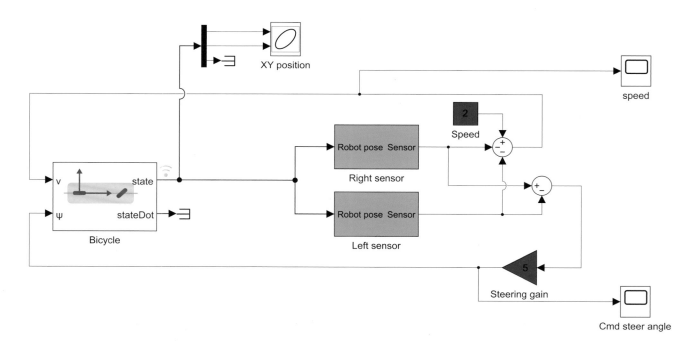

◘ **Fig. 5.3** The Simulink model `sl_braitenberg` drives a vehicle toward the maximum of a scalar field, which is modeled inside the `Right sensor` and `Left sensor` blocks. The vehicle plus controller is an example of a Braitenberg vehicle

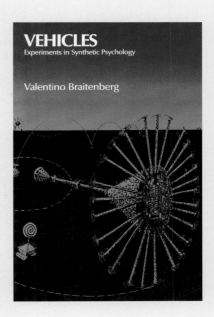

by the green sensor blocks in Fig. 5.3 and are parameterized by the position of the sensor with respect to the robot's body, and the sensing function. The sensors are positioned at ± 2 units in the vehicle's lateral or y-direction.

The field to be sensed is a simple inverse-square field defined by

```
1 function sensor = sensorfield(x, y)
2     xc = 60; yc = 90;
3     sensor = 200./((x-xc).^2 + (y-yc).^2 + 200);
4 end
```

sl_braitenberg

▶ sn.pub/cZb8Zy

that returns the sensor value $s(x, y) \in [0, 1]$ which is a function of the sensor's position in the plane. This particular function has a peak value at the coordinate $(60, 90)$.

The vehicle speed is

$$v = 2 - s_R - s_L \qquad (5.1)$$

where s_R and s_L are the right and left sensor readings respectively. At the goal, where $s_R = s_L = 1$, the velocity becomes zero.

Steering angle is based on the difference between the sensor readings

$$\psi = K_s(s_R - s_L) \qquad (5.2)$$

and K_s is the steering gain. When the field is equal in the left- and right-hand sensors, the robot moves straight ahead. ▶

We run the simulation from the Simulink toolstrip or the command line

```
>> sim("sl_braitenberg");
```

and the path of the robot is shown in Fig. 5.4. We see that the robot moves toward the maximum of the scalar field – turning toward the goal and slowing down as it approaches – asymptotically achieving the goal position. The initial configuration, wheel base, and steering angle limits can all be changed through the parameters of the `Bicycle` block. Integrating the state (x, y, θ) of the bicycle model is naturally expressed in Simulink. For more advanced use cases and custom integration behavior, see the `bicycleKinematics` function.

This particular sensor-action control law results in a specific robotic *behavior*. An obstacle would block this robot since its only behavior is to steer toward the

Similar strategies are used by moths whose two antennae are exquisitely sensitive odor detectors. A male moth uses the differential signal to "steer" toward a pheromone-emitting female.

5

□ **Fig. 5.4** Path of the Braitenberg vehicle using the model `sl_braitenberg` shown in □ Fig. 5.3. The vehicle moves toward the maximum of a 2D scalar field, whose magnitude is color coded

goal, but an additional behavior could be added to handle this case and drive around an obstacle. We could add another behavior to search randomly for the source if none was visible. Walter's tortoise had four behaviors, and switching was based on light level and a touch sensor.

Multiple behaviors and the ability to switch between them leads to an approach known as behavior-based robotics. The subsumption architecture was proposed as a means to formalize the interaction between different behaviors. Complex, some might say *intelligent-looking*, behaviors can be manifested by such systems. However, as more behaviors are added, the complexity of the system grows rapidly and interactions between behaviors become more complex to express and debug. Ultimately, the advantage of simplicity in not using a map becomes a limiting factor in achieving efficient and taskable behavior.

5.1.2   Simple Automata

Another class of reactive robots are known as *bugs* – simple automata that perform goal seeking in the presence of obstacles. Many *bug* algorithms have been proposed and they share the ability to sense when they are in proximity to an obstacle. In this respect, they are similar to the Braitenberg class vehicle, but the *bug* includes a state machine and other logic in between the sensor and the motors. The automata have memory, which the earlier Braitenberg vehicle lacked. ◄ In this section, we will investigate a specific *bug* algorithm known as *bug2*.

Braitenberg's book describes a series of increasingly complex vehicles, some of which incorporate memory. However, the term *Braitenberg vehicle* has become associated with the simplest vehicles he described.

We start by loading a model of a house

```
>> load house
>> whos floorplan
  Name         Size          Bytes    Class     Attributes
  floorplan   397x596       1892896   double
```

which defines a matrix variable `floorplan` in the workspace whose elements are zero or one, representing free space or obstacle respectively. The matrix is displayed as an image in □ Fig. 5.5. This matrix is an example of an occupancy grid map (specifically, a *binary* occupancy grid map, since each cell can only have two possible values), which will be properly introduced in ▶ Sect. 5.2. Tools to generate such maps are discussed later in ▶ Exc. 5.7. For now, we can simply convert this matrix into a `binaryOccupancyMap`, which is a convenience object for storing such maps. ◄

The `floorplan` matrix is in image coordinates. Image coordinate (1,1) is in the upper-left corner, whereas typical map coordinates have (1,1) in the lower-left corner. To account for this change, we use the `flipud` function to flip the matrix about the horizontal axis before passing it to the `binaryOccupancyMap`.

```
>> floorMap = binaryOccupancyMap(flipud(floorplan));
>> figure; floorMap.show
```

◻ Fig. 5.5 Drivable space is white, obstacles are red, and named places are indicated by hollow black stars. Approximate scale is 45 mm per cell. The start location is a solid blue circle and the goal is a solid blue star. The path taken by the bug2 algorithm is shown by a green line. The black dashed line is the m-line that connects the start and goal

The `load` command also loaded a list of named places within the house, as elements of a structure ▶

The structure has many elements and the output has been cropped for inclusion in this book.

```
>> places
places =
  struct with fields:
      kitchen: [320 190]
       garage: [500 150]
          br1: [50 220]
          ...
```

For the *bug* algorithm, we make several assumptions:
- the robot operates in a grid world and occupies one grid cell
- the robot is capable of omnidirectional motion and can move to any of its eight neighboring grid cells
- the robot can determine its position on the plane. This is a nontrivial problem that will be discussed in detail in ▶ Chap. 6
- the robot can only sense its goal and whether adjacent cells are occupied using a simulated proximity sensor

We create an instance of the `Bug2` class

```
>> bug = Bug2(floorMap);
```

and pass in the occupancy grid map which is used to generate the sensory inputs for the simulated robot. We can display the robot's environment by

```
>> bug.plot
```

and the simulation is run by

```
>> bug.run(places.br3,places.kitchen,animate=true);
```

whose arguments are the start and goal positions (x, y) of the robot within the house. The method displays an animation of the robot moving toward the goal and the path is shown as a series of green dots in ◻ Fig. 5.5.

The strategy of the *bug2* algorithm is quite simple. The first thing it does is to compute a straight line from the start position to the goal – the m-line – and attempts to drive along it. If it encounters an obstacle, it turns right and continues

5

It could be argued that the m-line represents an explicit plan. Thus, *bug* algorithms occupy a position somewhere between Braitenberg vehicles and map-based planning systems in the spectrum of approaches to navigation.

until it encounters a point on the m-line that is closer to the goal than when it departed from the m-line. ◄

The `run` method returns the robot's path

```
>> p = bug.run(places.br3,places.kitchen);
```

as a matrix with one row per time step, and in this case the path comprises 1300 points.

```
>> whos p
  Name      Size        Bytes  Class      Attributes
  p         1300x2      20800  double
```

In this example, the *bug2* algorithm has reached the goal but it has taken a very suboptimal route, traversing the inside of a wardrobe, behind doors, and visiting two bathrooms. For this example, it might have been quicker to turn left rather than right at the first obstacle, but that strategy might give a worse outcome somewhere else. Many variants of the *bug* algorithm have been developed – some may improve performance for one type of environment but will generally show worse performance in others. The robot is fundamentally limited by not using a map – without the "big picture" it is doomed to take paths that are locally, rather than globally, optimal.

5.2 Introduction to Map-Based Navigation

The key to achieving the *best* path between points A and B, as we know from everyday life, is to use a map. Typically, *best* means the shortest distance but it may also include some penalty term or cost related to traversability which is how easy the terrain is to drive over – it might be quicker to travel further, but faster, over better roads. A more sophisticated planner might also consider the size of the robot, the kinematics and dynamics of the vehicle, and avoid paths that involve turns that are tighter than the vehicle can execute. Recalling our earlier definition of a robot as a

» goal-oriented machine that can sense, ⃞plan⃞ and act,

the remaining sections of this chapter concentrate on path planning. We will discuss two types of maps that can be conveniently represented within a computer to solve navigation problems: mathematical graphs and occupancy grid maps.

The key elements of a graph, shown in ◼ Fig. 5.6a, are nodes or vertices (the dots) and edges (the lines) that respectively represent places and paths between places. Graphs can be directed, where the edges are arrows that indicate a direction of travel; or undirected, where the edges are without arrows and travel is possible in either direction. For mapping, we typically consider that the nodes are placed or *embedded* in a Cartesian coordinate system at the position of the places that they represent – this is called an embedded graph. The edges *represent* drivable routes ◄

The edges of the graph are generally not the actual route between nodes, they simply represent the fact that there is a drivable path between them.

between pairs of nodes and have an associated cost to traverse – typically distance or time, but it could also be road roughness. Edges could have other attributes relevant to planning such as a speed limit, toll fees, or the number of coffee shops.

Non-embedded graphs are also common. Subway maps, for example, do not necessarily place the stations at their geographic coordinates, rather they are placed so as to make the map easier to understand. Topological graphs, often used in robotics, comprise nodes, representing places, which can be recognized even if their position in the world is not known, and paths that link those places. For example, a robot may be able to recognize that a place is a kitchen, and that by going through a particular doorway it arrives at a place it can recognize as a living room. This is sufficient to create two nodes and one edge of the topological graph, even though the nodes have no associated Cartesian coordinates.

a

b

Fig. 5.6 Two common map representations. **a** A mathematical graph comprising nodes (places) and edges (routes); **b** occupancy grid map comprising an array of square cells that have two values shown as red or white to describe whether the cell is occupied or empty

Excurse 5.3: Graph

A graph is an abstract representation of a set of objects connected by links typically denoted as an ordered pair (N, E). The objects, N, are called nodes or vertices, and the links, E, that connect some pairs of nodes are called edges or arcs. Edges can be directed (arrows) or undirected, as in the case shown here. Edges can have an associated weight or cost associated with moving from one of its nodes to the other. A sequence of edges from one node to another is a path. Graphs can be used to represent transport or communications networks and even social relationships, and the associated branch of mathematics is graph theory. See ▶ App. I for more details on graph data structures.

Excurse 5.4: Humans and Maps

Animals, particularly mammals, have a powerful ability to navigate from one place to another. This ability includes recognition of familiar places or landmarks, planning ahead, reasoning about changes, and adapting the route if required. The hippocampus structure in the vertebrate brain holds spatial memory and plays a key role in navigation.

Humans are mammals with a unique ability for written and graphical communications, so it is not surprising that recording and sharing navigation information in the form of maps is an ancient activity.

The oldest surviving map is carved on a mammoth tusk and is over 25,000 years old. Archaeologists believe it may have depicted the landscape around the village of Pavlov in the Czech Republic, showing hunting areas and the Thaya river. Today, maps are ubiquitous, and we rely heavily on them for our daily life.

5

Excurse 5.5: The Origin of Graph Theory

FIGURE 98. *Geographic Map: The Königsberg Bridges.*

The origins of graph theory can be traced back to the Königsberg bridge problem. This city, formerly in Germany but now in Russia and known as Kaliningrad, included an island in the river Pregel and it was linked to the land on either side of the river by seven bridges (of which only three stand today).

A topic of debate was whether it was possible to walk through the city and cross each bridge just once. In 1736, Leonhard Euler proved that it was *not possible*, and his solution was the beginning of graph theory.

Universal Transverse Mercator (UTM) is a mapping system that maps the Earth's sphere as sixty north-south strips or zones numbered 1 to 60, each 6° wide. Each zone is divided into bands lettered C to X, each 8° high. The zone and band define a grid zone, which is represented as a planar grid with its coordinate frame origin at the center.

49L	50L	51L	52L	53L	54L	55L	56L
49K	50K	51K	52K	53K	54K	55K	56K
49J	50J	51J	52J	53J	54J	55J	56J
49H	50H	51H	52H	53H	54H	55H	56H
49G	50G	51G	52G	53G	54G	55G	56G

◨ Fig. 5.6b shows a grid-based map. Here, the environment is a grid of cells, typically square, and each cell holds information about the traversability of that cell. The simplest case is called a binary occupancy grid map, and a cell has two possible values: one indicates the cell is occupied and we cannot move into it – it is an obstacle; or zero if it is unoccupied. The size of the cell depends on the application. The memory required to hold the binary occupancy grid map increases with the spatial area represented, and inversely with the cell size. For modern computers, this representation is feasible even for very large areas.

5.3   Planning with a Graph-Based Map

We will work with embedded graphs where the position of the nodes in the world is known. For this example, we consider the towns and cities shown in ◨ Fig. 5.7, and we import a datafile

```
>> data = jsondecode(fileread("queensland.json"))
data =
  struct with fields:
    places: [1×1 struct]
    routes: [29×1 struct]
```

where `data` is a structure containing two items. Firstly, `data.places` is a structure of place names, where each place has an associated latitude, longitude, and UTM map coordinates in units of kilometers. ◄

These places are cities and towns in the state of Queensland, Australia.

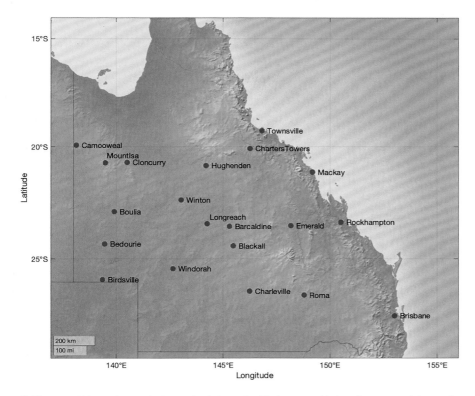

Fig. 5.7 Cities and towns in Queensland, Australia. Their geographic locations are used for graph-based planning

We can easily plot these places, along with their names

```
>> figure
>> geobasemap colorterrain
>> hold on
>> for p = string(fieldnames(data.places))'
>>   place = data.places.(p);
>>   geoplot(place.lat,place.lon,".b",MarkerSize=20)
>>   text(place.lat,place.lon+0.25,p)
>> end
```

on a physical map in ◨ Fig. 5.7.

Secondly, `data.routes` is a structure array, where each structure defines a graph edge: the name of its start and end place, the route length, and the driving speed. For example,

```
>> data.routes(1)
ans =
  struct with fields:
      start: 'Camooweal'
        end: 'Mount Isa'
   distance: 188
      speed: 100
```

We will use the `UGraph` object to create a graph representation of the data. The first step is to create an undirected graph object and then add nodes and edges from the data we just loaded

```
>> g = UGraph;
>> for name = string(fieldnames(data.places))'
>>   place = data.places.(name);
>>   g.add_node(place.utm,name=place.name);
>> end
>> for route = data.routes'
```

5

```
>>    g.add_edge(route.start,route.end,route.distance);
>> end
```

The `UGraph` object has many methods, for example, we can plot the graph we just created

```
>> figure; g.plot;
```

which produces a plot similar to ◼ Fig. 5.6a. The graph object has properties that report the number of nodes and edges.

```
>> g.n
ans =
  20
>> g.ne
ans =
  29
```

All methods that take a node input can either be called with the node name (more convenient) or with the node ID (slightly faster). For example, we can retrieve the neighbors of the "Brisbane" node, which also has node ID 1

```
>> g.neighbors("Brisbane")
ans =
  1×2 string array
    "Roma"     "Rockhampton"
>> nodeID = g.lookup("Brisbane")
nodeID =
  1
>> g.neighbors(nodeID)
ans =
  17    20
```

The degree of a node is the number of neighbors it is connected to

```
>> g.degree("Brisbane")
ans =
  2
```

and the average degree of the entire graph is

```
>> mean(g.degree)
ans =
  2.9000
```

The average degree indicates how well-connected a graph is. A higher value means that there are, on average, more paths between nodes and hence more options for travel.

This graph is connected and has a single connected component

```
>> g.nc
ans =
  1
```

which means there is a path between any pair of nodes. A disconnected or disjoint graph has multiple components – we can find a path within a component but not between them.

A list of the edge IDs from the Brisbane node is given by

```
>> e = g.edges("Brisbane")
e =
  1    2
```

The edge IDs are not very informative, so we can print more information about these edges

```
>> g.edgeinfo(e)
[Brisbane] -- [Roma], cost=482
[Brisbane] -- [Rockhampton], cost=682
```

5.3.1 Breadth-First Search

Now, we are ready to plan a path, and will consider the specific example of a journey from Hughenden to Brisbane. A simple-minded solution would be to randomly choose one of the neighboring nodes, move there, and repeat the process. We might eventually reach our goal – but there is no guarantee of that – and it is extremely unlikely that the path would be optimal. A more systematic approach is required, and we are fortunate this problem has been well-studied – it is the well-known graph search problem.

Starting at Hughenden, there are just four roads we can take to get to a neighboring place. From one of those places, say Winton, we can visit in turn each of its neighbors. However, one of those neighbors is Hughenden, our starting point, and backtracking cannot lead to an optimal path. We need some means of keeping track of where we have been. The frontier (or open set) contains all the nodes that are scheduled for exploration, and the explored set (or closed set) contains all the nodes that we have already explored.

We will introduce three algorithms, starting with breadth-first search (BFS)

```
>> [pathNodes,pathLength] = g.path_BFS("Hughenden","Brisbane")
pathNodes =
  1×5 string array
    "Hughenden"  "Barcaldine"  "Emerald"  "Rockhampton"  "Brisbane"

pathLength =
        1759
```

which has returned two values: a list of nodes defining the path from start to goal, and the total path length in kilometers. We can visualize the path by first displaying the plot, then overlaying the path

```
>> gPlot = g.plot;
>> g.highlight_path(gPlot,pathNodes,labels=true)
```

which is shown in ◘ Fig. 5.8a. The algorithm is sequential, and it is instructive to look at its step-by-step actions ▸

This method prints out many lines. The output has been cropped for inclusion in this book.

```
>> g.path_BFS("Hughenden","Brisbane",verbose=true)
FRONTIER: Hughenden
EXPLORED:
  expand Hughenden
```

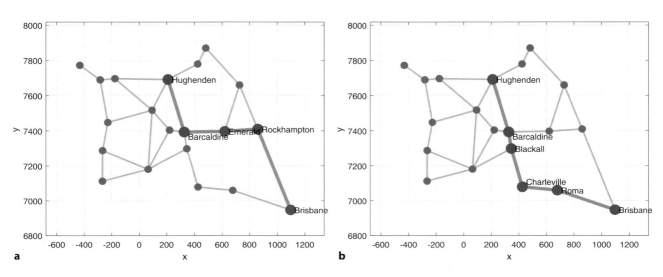

◘ **Fig. 5.8** Path planning results from Hughenden to Brisbane. **a** Breadth-first search (BFS) with a path length of 1759 km; **b** Uniform-cost search (UCS) and A* search with a path length of 1659 km

5

```
         add Cloncurry to the frontier
         add Charters Towers to the frontier
         add Winton to the frontier
         add Barcaldine to the frontier
         move Hughenden to the explored list

FRONTIER: Cloncurry, Charters Towers, Winton, Barcaldine
EXPLORED: Hughenden
         expand Cloncurry
         add Mount Isa to the frontier
         move Cloncurry to the explored list

FRONTIER: Charters Towers, Winton, Barcaldine, Mount Isa
EXPLORED: Hughenden, Cloncurry
         expand Charters Towers
         add Townsville to the frontier
         move Charters Towers to the explored list
    ...
```

The frontier is initialized with the start node. At each step, the first node from the frontier is *expanded* – its neighbors added to the frontier and that node is retired to the explored set. At the last expansion

```
FRONTIER: Rockhampton, Roma
EXPLORED: Hughenden, Cloncurry, Charters Towers, Winton, Barcaldine,
Mount Isa, Townsville, Boulia, Windorah, Longreach, Blackall,
Emerald, Camooweal, Mackay, Bedourie, Birdsville, Charleville
         expand Rockhampton
         goal Brisbane reached
17 nodes explored, 1 remaining on the frontier
```

the goal is encountered, and the algorithm stops. There was one node on the frontier, still unexpanded, that might have led to a better path.

5.3.2 Uniform-Cost Search

At this point, it is useful to formalize the problem: at each step we want to choose the next node v that minimizes the cost

$$f(v) = g(v) + h(v) \tag{5.3}$$

where $g(v)$ is the cost of the path so far – the *cost to come* (from the start), and $h(v)$ is the remaining distance to the goal – the *cost to go*. However, we do not know $h(v)$ since we have not yet finished planning. For now, we settle for choosing the next node v that minimizes $f(v) = g(v)$. That is, we choose

$$v^* = \arg\min_v f(v) \tag{5.4}$$

as the next node to expand. The intuition is that in order to achieve the shortest overall path, at each step we should choose the node associated with the smallest distance so far.

This leads to another well-known graph search algorithm called uniform-cost search (UCS). In the previous search, we expanded nodes from the frontier on a first-in first-out basis, but now we will choose the frontier node that has the lowest cumulative distance from the start – the lowest cost to come. Running the UCS algorithm

```
>> [pathNodes,pathLength,searchTree] = ...
>>    g.path_UCS("Hughenden","Brisbane");
>> pathLength
pathLength =
        1659
>> pathNodes
```

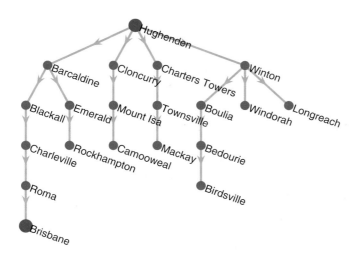

◻ Fig. 5.9   Search tree for uniform-cost search from Hughenden to Brisbane

```
pathNodes =
  1×6 string array
    "Hughenden"  "Barcaldine"  "Blackall"  "Charleville"  "Roma"
    "Brisbane"
```

we obtain a path that is 100 km shorter and shown in ◻ Fig. 5.8b. UCS has found the shortest path length whereas BFS found the path with the fewest nodes.

  As the algorithm progresses, it records, for every node, the node that it traveled from – the node's parent. This information is returned in searchTree which is a directed graph that illustrates which nodes the algorithm explored. This graph can be displayed

```
>> gPlot = searchTree.plot;
>> g.highlight_node(gPlot,["Hughenden","Brisbane"])
```

as shown in ◻ Fig. 5.9. This is the final version of the search tree, and it is built incrementally as the graph is explored.

  Once again, it is instructive to look at the step-by-step operation ▶

This method prints out many lines. The output has been cropped for inclusion in this book.

```
>> g.path_UCS("Hughenden","Brisbane",verbose=true)
FRONTIER: Hughenden(0)
EXPLORED:
    expand Hughenden
    add Cloncurry to the frontier
    add Charters Towers to the frontier
    add Winton to the frontier
    add Barcaldine to the frontier
    move Hughenden to the explored list

FRONTIER: Winton(216), Charters Towers(248), Cloncurry(401),
Barcaldine(500)
EXPLORED: Hughenden
    expand Winton
    add Boulia to the frontier
    add Windorah to the frontier
    add Longreach to the frontier
    move Winton to the explored list

FRONTIER: Charters Towers(248), Longreach(396), Cloncurry(401),
Barcaldine(500), Boulia(579), Windorah(703)
EXPLORED: Hughenden, Winton
    expand Charters Towers
    add Townsville to the frontier
    move Charters Towers to the explored list
...
```

5

In many implementations, the frontier is maintained as an ordered list, or priority queue, from smallest to largest value of $f(v)$.

The value of $f(v)$, which is equal to the cost to come $g(v)$ in UCS, is given in parentheses and at each step the node with the lowest value is chosen for expansion. ◄

Something interesting happens during expansion of the Bedourie node

```
FRONTIER: Bedourie(773), Emerald(807), Charleville(911),
Birdsville(1083), Rockhampton(1105)
EXPLORED: Hughenden, Winton, Charters Towers, Townsville, Longreach,
Cloncurry, Barcaldine, Mount Isa, Boulia, Blackall, Windorah,
Camooweal, Mackay
    expand Bedourie
 reparent Birdsville: cost 966 via Bedourie is less than cost 1083
 via Windorah, change parent from Windorah to Bedourie
    move Bedourie to the explored list
```

One of its neighbors is Birdsville, which is still in the frontier. The cost of travelling from the start to Birdsville via Windorah is 1083 km, whereas the cost of traveling via Bedourie is only

```
>> [~,costStartToBedourie] = g.path_UCS("Hughenden","Bedourie")
costStartToBedourie =
    773
>> costBedourieToBirdsville = g.cost("Bedourie","Birdsville")
costBedourieToBirdsville =
    193
```

This means that Birdsville is really only $773 + 193 = 966$ km from the start, and we have discovered a shorter route than the one already computed. The algorithm changes the parent of Birdsville in the search tree, shown in ◻ Fig. 5.9, to reflect that.

At the third-last step, our goal, Brisbane, enters the frontier

```
FRONTIER: Rockhampton(1077), Roma(1177)
EXPLORED: Hughenden, Winton, Charters Towers, Townsville, Longreach,
Cloncurry, Barcaldine, Mount Isa, Boulia, Blackall, Windorah,
Camooweal, Mackay, Bedourie, Emerald, Charleville, Birdsville
    expand Rockhampton
    add Brisbane to the frontier
    move Rockhampton to the explored list
```

```
FRONTIER: Roma(1177), Brisbane(1759)
EXPLORED: Hughenden, Winton, Charters Towers, Townsville, Longreach,
Cloncurry, Barcaldine, Mount Isa, Boulia, Blackall, Windorah,
Camooweal, Mackay, Bedourie, Emerald, Charleville, Birdsville,
Rockhampton
    expand Roma
 reparent Brisbane: cost 1659 via Roma is less than cost 1759 via
 Rockhampton, change parent from Rockhampton to Roma
    move Roma to the explored list
```

```
FRONTIER: Brisbane(1659)
EXPLORED: Hughenden, Winton, Charters Towers, Townsville, Longreach,
Cloncurry, Barcaldine, Mount Isa, Boulia, Blackall, Windorah,
Camooweal, Mackay, Bedourie, Emerald, Charleville, Birdsville,
Rockhampton, Roma
    expand Brisbane
19 nodes explored, 0 remaining on the frontier
```

but the algorithm does not stop immediately in case there is a better route to Brisbane to be discovered. In fact, at the second-last step we find a better path through Roma rather than Rockhampton. UCS has found the shortest path, but it has explored every node, and this can be expensive for very large graphs. The next algorithm we introduce can obtain the same result by exploring just a subset of the nodes.

> **Excurse 5.6: Shakey and the A* Trick**
>
> Shakey, shown in ◨ Fig. 5.1b, was a mobile robot system developed at Stanford Research Institute from 1966 to 1972. It was a testbed for computer vision, planning, navigation, and communication using natural language – artificial intelligence as it was then understood. Many novel algorithms were integrated into a physical system for the first time. The robot was a mobile base with a TV camera, rangefinder and bump sensors connected by radio links to a PDP-10 mainframe computer. Shakey can be seen in the Computer History Museum in Mountain View, CA.
>
> The A* trick was published in a 1968 paper *A formal basis for the heuristic determination of minimum cost paths* by Shakey team members Peter Hart, Nils Nilsson, and Bertram Raphael.

5.3.3 A* Search

The trick to achieving this is to rewrite (5.3) as

$$f(v) = g(v) + h^*(v) \tag{5.5}$$

where $h^*(v)$ is an *estimate* of the distance to the goal – referred to as the *heuristic* distance. A commonly used heuristic is the distance "as the crow flies" or Euclidean distance – easily computed since our graph is embedded and the position of every node is known.

The final algorithm we introduce is the very well-known A* search which incorporates this heuristic. We run it in a similar way to UCS

```
>> [pathNodes,pathLength,searchTree] = ...
>>   g.path_Astar("Hughenden","Brisbane",summary=true);
12 nodes explored, 3 remaining on the frontier
>> pathLength
pathLength =
        1659
>> pathNodes
pathNodes =
  1×6 string array
    "Hughenden"  "Barcaldine"  "Blackall"  "Charleville"  "Roma"
    "Brisbane"
```

and it returns the same optimal path as UCS but has explored only 12 nodes compared to 19 for UCS. We can visualize this by highlighting the nodes that were explored or are on the frontier. These are the nodes that are included in the returned search tree

```
>> visitedNodes = string(searchTree.Nodes.Name);
>> gPlot = g.plot(labels=true);
>> g.highlight_node(gPlot,visitedNodes,NodeColor=[0.929 0.694 0.125])
```

and shown in ◨ Fig. 5.10. The nodes Camooweal, Mount Isa, Bedourie, and Birdsville, far away from the path, were not visited.

The first two steps of A* are ▶

```
>> g.path_Astar("Hughenden","Brisbane",verbose=true)
FRONTIER: Hughenden(0)
EXPLORED:
    expand Hughenden
    add Cloncurry to the frontier
    add Charters Towers to the frontier
    add Winton to the frontier
```

This method prints out many lines. The output has been cropped for inclusion in this book.

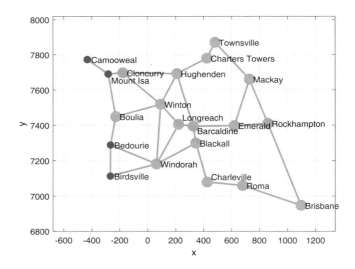

■ **Fig. 5.10** Nodes visited by A* search (either explored or in frontier) are shown in yellow

```
    add Barcaldine to the frontier
    move Hughenden to the explored list

FRONTIER: Charters Towers(1318.9011), Winton(1371.2005),
Barcaldine(1390.3419), Cloncurry(1878.2595)
EXPLORED: Hughenden
    expand Charters Towers
    add Townsville to the frontier
    move Charters Towers to the explored list
...
```

and the value in parentheses this time is the cost $g(v) + h^*(v)$ – the sum of the cost-to-come (as used by UCS) plus an *estimate* of the cost-to-go. A* explores less nodes than UCS because once the goal node has been found, we can exit the first time we see the cost of our current node exceed the cost-to-come of our goal node. The final returned path is shown in ■ Fig. 5.11a.

The performance of A* depends critically on the heuristic. A* is *guaranteed* to return the lowest-cost path only if the heuristic is admissible, that is, it *does not overestimate* the cost of reaching the goal. The Euclidean distance is an admissi-

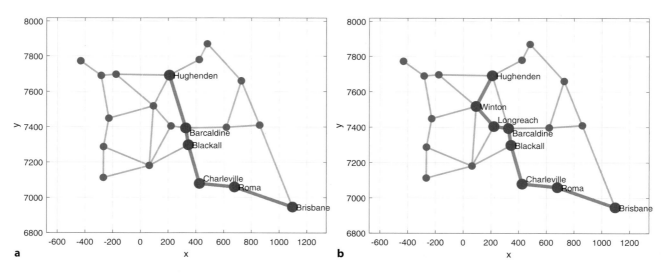

■ **Fig. 5.11** A* path from Hughenden to Brisbane for **a** minimum distance; **b** minimum time ble heuristic since it is the minimum possible cost and by definition cannot be an overestimate. Of course, $h^*(v) = 0$ is admissible but that is equivalent to UCS, and all the nodes will be visited. As the heuristic cost component increases from zero up to its maximum admissible value, the efficiency increases as fewer nodes are explored. A non-admissible heuristic may lead to a non-optimal path.

5.3.4 Minimum-Time Path Planning

For some problems, we might wish to minimize time rather than distance. Minimum time is important when there is the possibility of trading off a short slow route for a longer but faster route. We will continue the example from the last section but change the edge costs to reflect the road speeds which are either $100\,\mathrm{km\,h^{-1}}$ for a major road or $50\,\mathrm{km\,h^{-1}}$ for a minor road. We will build a new graph

```
>> g = UGraph;
>> for name = string(fieldnames(data.places))'
>>   place = data.places.(name);
>>   g.add_node(place.utm,name=place.name);
>> end
>> for route = data.routes'
>>   g.add_edge(route.start,route.end,route.distance/route.speed);
>> end
```

and the edge cost is now in units of hours. Before we run A* we need to define the heuristic cost for this case. For the minimum-distance case, the heuristic was the distance as the crow flies between the node and the goal. For the minimum-time case, the heuristic cost will be the time to travel as the crow flies from the node to the goal, but the question is at which speed? Remember that the heuristic must not *overestimate* the cost of attaining the goal. If we choose the slow speed, then on a fast road we would overestimate the time, making this an inadmissible metric. However, if we choose the fast speed, then on a slow road, we will underestimate the travel time and that *is* admissible. Therefore, our heuristic cost assumes travel at $100\,\mathrm{km\,h^{-1}}$ and is

```
>> g.measure = @(x1, x2) vecnorm(x1-x2)/100;
```

Now, we can run the A* planner

```
>> [pathNodes,time] = g.path_Astar("Hughenden","Brisbane")
pathNodes =
  1×8 string array
    "Hughenden"  "Winton"  "Longreach"  "Barcaldine"  "Blackall"
    "Charleville"  "Roma"  "Brisbane"
time =
   16.6100
```

and the length of the path is in units of hours. As we can see in ◘ Fig. 5.11b, the route is different to the minimum-distance case and has avoided the slow road between Hughenden and Barcaldine, and instead chosen the faster, but longer, route through Winton and Longreach.

5.3.5 Wrapping Up

In practice, road networks have additional complexity, particularly in urban areas. Travel time varies dynamically over the course of the day, and that requires changing the edge costs and replanning. Some roads are one-way only, and we can represent that using a directed graph, a `DGraph` instance rather than a `UGraph` instance. For two-way roads, we would have to add two edges between each pair

5

of nodes, and these could have different costs for each direction. For the one-way road, we would add just a single edge.

Planners have a number of formal properties. A planner is *complete* if it reports a solution, or the absence of a solution, in finite time. A planner is *optimal* if it reports the best path with respect to some metric. The *complexity* of a planner is concerned with the computational time and memory resources it requires.

5.4 Planning with an Occupancy Grid Map

The occupancy grid map is an array of cells, typically square, and each cell holds information about the traversability of that cell. In the simplest case of a binary occupancy grid map , a cell has two possible values: one indicates the cell is occupied and we cannot move into it – it is an obstacle; or zero if it is unoccupied. In a probabilistic occupancy grid map , the cell value represents the probability that there is an obstacle in that cell, with floating-point values in the range of $[0, 1]$.

The size of the cell depends on the application. The memory required to hold the occupancy grid map increases with the spatial area represented, and inversely with the cell size. For modern computers, this representation is feasible even for very large areas. If memory is limited, more compact map representations can be used. A 2D quadtree recursively divides the space into grid cells of different sizes. If there are large regions with the same occupancy in the map, the quadtree will use a larger cell size in that area, thus saving memory space. The `qtdecomp` function can be used to calculate this decomposition for an image-based map.

We want a planner that can return the shortest obstacle-free path through the grid, that is, the path is *optimal* and *admissible*. In this section, we make some assumptions:

- the robot operates in a grid world
- the robot occupies one grid cell
- the robot is capable of omnidirectional motion and can move to any of its eight neighboring grid cells.

The distance between two points (x_1, y_1) and (x_2, y_2) where $\Delta_x = x_2 - x_1$ and $\Delta_y = y_2 - y_1$ can be the Euclidean distance or L_2 norm $\sqrt{\Delta_x^2 + \Delta_y^2}$; or Manhattan (also known as city-block) distance or L_1 norm $|\Delta_x| + |\Delta_y|$.

5.4.1 Distance Transform

Consider a matrix of zeros with just a single nonzero element representing the goal as shown in ◘ Fig. 5.12a. The distance transform of this matrix is another matrix, of the same size, but the value of each element is its distance ◄ from the original nonzero element. Two approaches to measuring distances are commonly used for

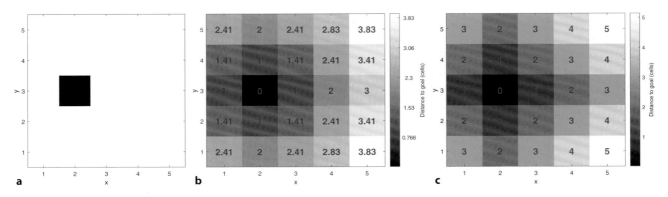

◘ **Fig. 5.12** The distance transform. **a** The goal cell is set to one, others are zero; **b** Euclidean distance (L_2) transform where cell values shown in red are the Euclidean distance from the goal cell; **c** Manhattan (L_1) distance transform

Excurse 5.7: Creating an Occupancy Grid Map

An occupancy grid map is a matrix that corresponds to a region of 2-dimensional space. Elements containing zeros are free space where the robot can move, and those with ones are obstacles where the robot cannot move. We can use many approaches to create a map. For example, we could create a matrix filled with zeros (representing all free space)

```
>> mapMatrix = false(100,100);
>> map = binaryOccupancyMap(mapMatrix);
```

and use occupancy map functions such as setOccupancy to change values in the map:

```
>> map.setOccupancy([40 20], ...
>>    true(10,60), "grid");
>> map.show
```

We can also automatically create simple occupancy maps with mapClutter and mapMaze, or use an interactive image editor.

```
>> rng(0); % obtain repeatable results
>> map = mapClutter(10,["Box", ...
>>    "Circle"],MapResolution=20);
>> figure; map.show
```

Fig. 5.13 shows the generated figure. See the online help for more detailed options to control map size, number of obstacles, etc.

The occupancy grid map in Fig. 5.5 was derived from an image file, but online building plans and street maps could also be used.

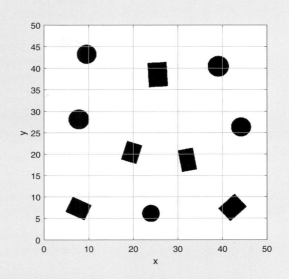

Fig. 5.13 Map generated by the mapClutter utility

There are two different ways of accessing values in an occupancy grid map. As a matrix, we use standard MATLAB® grid coordinates and index the matrix as [row,column]. A map in world coordinates (see Fig. 5.13) uses the Cartesian convention of a horizontal x-coordinate first, followed by the y-coordinate. Depending on the resolution of your map (grid cells per meter) and where the map origin is (default lower left), the conversion between grid and world coordinates can be hard. The conversion methods grid2world and world2grid of the binaryOccupancyMap can help with these conversions. See ► https://sn.pub/Sg8kV3 for more details on the different coordinate conventions for occupancy maps.

robot path planning. The RVC Toolbox default is Euclidean or L_2 distance which is shown in Fig. 5.12b – the distance to horizontal and vertical neighboring cells is one, and to diagonal neighbors is $\sqrt{2}$. The alternative is Manhattan distance, also called city-block or L_1 distance, which is shown in Fig. 5.12c. The distance to horizontal and vertical neighboring cells is one, and motion to diagonal neighbors is not permitted. The distance transform is an image processing technique that we will discuss further in ► Sect. 11.6.4.

To demonstrate the distance transform for robot navigation, we first create a simple occupancy grid map

```
>> simplegrid = binaryOccupancyMap(false(7,7));
>> simplegrid.setOccupancy([3 3],true(2,3),"grid");
>> simplegrid.setOccupancy([5 4],true(1,2),"grid");
```

which is shown in Fig. 5.14a. Next, we create a distance transform motion planner and pass it the occupancy grid map

```
>> dx = DistanceTransformPlanner(simplegrid)
dx =
DistanceTransformPlanner navigation class:
  occupancy grid: 7x7
  distance metric: euclidean
  distancemap: empty:
```

5

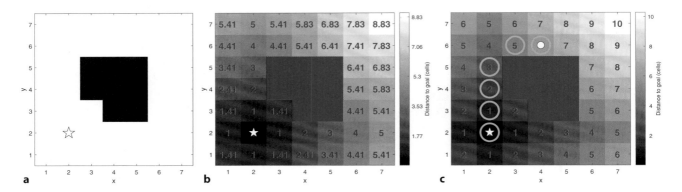

◻ Fig. 5.14 Distance transform path planning. **a** The occupancy grid map displayed as an image, white pixels are nonzero and represent obstacles; **b** Euclidean distance (L_2) transform, red pixels are obstacles; **c** Manhattan (L_1) distance transform. Numbers in red are the distance of the cell from the goal, which is denoted by a star. The path from start to goal is shown with green circles

Then we create a plan to reach a specific goal at cell $(2, 2)$

```
>> dx.plan([2 2])
```

and the distance transform can be visualized

```
>> dx.plot
```

as shown in ◻ Fig. 5.14b. The numbers in red are the distance of the cell from the goal, taking into account travel *around* obstacles. For Manhattan distance, we use a similar set of commands and result shown in ◻ Fig. 5.14c. The query method is used to calculate the actual path from a start location to the goal location we specified in plan

```
>> dx = DistanceTransformPlanner(simplegrid,metric="manhattan");
>> dx.plan([2 2]);
>> pathPoints = dx.query([4 6])
pathPoints =
      4      6
      3      6
      2      5
      2      4
      2      3
      2      2
>> dx.plot(pathPoints,labelvalues=true)
```

and the plan is overlaid on ◻ Fig. 5.14c.

The plan to reach the goal is implicit in the distance map. The optimal path to the goal, from *any* cell, is a move to the neighboring cell that has the smallest distance to the goal. The process is repeated until the robot reaches a cell with a distance value of zero which is the goal.

The optimal path from an arbitrary starting point $(6, 5)$ to the goal is obtained by *querying* the plan

```
>> pathPoints = dx.query([6 5])
pathPoints =
      6      5
      6      4
      6      3
      5      2
      4      2
      3      2
      2      2
```

There is a duality between start and goal with this type of planner. By convention, the plan is based on the goal location, and we query for a particular start location, but we could base the plan on the start position and then query for a particular goal.

and the result has one path point per row, from start to goal. ◄

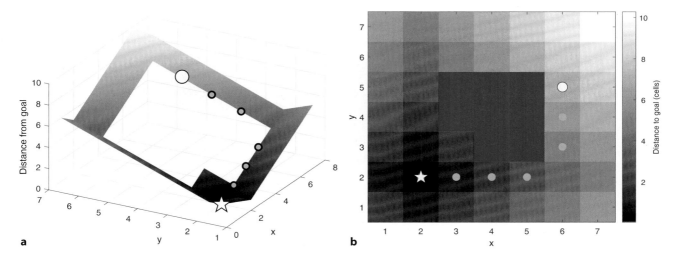

Fig. 5.15 a The distance transform as a 3D function, where height is distance from the goal. The path to the goal is simply a downhill run. Note the discontinuity in the distance transform value at the top left of the figure where two wavefronts met; **b** Distance transform with overlaid path

We can overlay this on the occupancy grid map

```
>> dx.plot(pathPoints)
```

as shown in ■ Fig. 5.15b.

If we plot distance from the goal as a 3-dimensional surface

```
>> dx.plot3d(pathPoints)
```

as shown in ■ Fig. 5.15a, then reaching the goal is simply a matter of rolling *downhill* on the distance function from *any* starting point. We have converted a fairly complex planning problem into one that can now be handled by a Braitenberg-class robot that makes local decisions based on the distance to the goal.

We can also animate the motion of the robot along the path from start to goal by

```
>> dx.query([6 5],animate=true)
```

which displays the robot's path as a series of green dots moving toward the goal.

Using the distance transform for the more realistic house navigation problem is now quite straightforward. For a goal in the kitchen

```
>> dx = DistanceTransformPlanner(floorplan);
>> dx.plan(places.kitchen)
>> dx.plot
```

the distance transform is shown in ■ Fig. 5.16. The distance map indicates the distance from any point to the goal, in units of grid cells, taking into account travel *around* obstacles. A path from bedroom three to the goal is

```
>> pathPoints = dx.query(places.br3);
>> length(pathPoints)
ans =
   338
```

which we can overlay on the occupancy grid map and distance map

```
>> dx.plot(pathPoints);
```

as shown in ■ Fig. 5.16.

This navigation algorithm has exploited its global view of the world and has, through exhaustive computation, found the shortest possible path which is 338 steps long. In contrast, the *bug2* algorithm of ▶ Sect. 5.1.2 without the global view has just bumped its way through the world and had a path length of 1300 steps

5

Fig. 5.16 The distance transform path. The background gray intensity represents the cell's distance from the goal in units of cell size as indicated by the scale on the right-hand side. Obstacles are indicated in red, unreachable cells in white

– nearly four times longer. The penalty for achieving the optimal path is upfront computational cost. This particular implementation of the distance transform is iterative. Each iteration has a cost of $O(N^2)$ and the number of iterations is at least $O(N)$, where N is the dimension of the map.

Although the plan is expensive to create, once it has been created it can be used to plan a path from *any* initial point to that goal. For large occupancy grid maps, this approach to planning will become impractical. The roadmap methods that we discuss later in this chapter provide an effective means to find paths in large maps at greatly reduced computational cost.

During the iterations of the distance transform algorithm, the distance values are propagating as a wavefront outward from the goal. The wavefront moves outward, spills through doorways into adjacent rooms and outside the house. ◄ Along the left and right edges of the distance map there are discontinuities where two wavefronts collide. The collisions delineate two different paths to the goal. On one side of the collision the optimal path is upwards toward the goal, while on the other side the optimal path is downwards toward the goal. Cells that are completely enclosed by obstacles can never be reached by the wavefront. We can again plot the distance from the goal as a 3-dimensional surface, as shown in ▣ Fig. 5.17.

The way that the distance values spread from the goal inspires its other names such as the wavefront, brushfire or grassfire algorithm.

Fig. 5.17 The distance transform of the house floorplan as a 3D function where height is the distance from the goal

◻ Fig. 5.18 Distance transform path with obstacles inflated by 6 cells. **a** Path shown with inflated obstacles; **b** path computed for inflated obstacles overlaid on original obstacle map, black regions are where no distance was computed due to obstacle inflation

The scale associated with this occupancy grid map is 45 mm per cell and we have assumed the robot occupies a single grid cell – this is a very small robot. The planner could therefore find paths through gaps that a larger real robot would fail to pass through. A common solution to this problem is to *inflate* the occupancy grid map – bigger obstacles and a small robot is equivalent to the actual obstacles and a bigger robot. For example, if the robot fits within a 600 mm circle (13 cells) then we shrink it by 6 cells in all directions, leaving just a single grid cell. To compensate, we expand or inflate all the obstacles by 6 cells in all directions – it is like applying a very thick layer of paint ▶

> Obstacle inflation is an image processing technique called morphological dilation, which is discussed in ▶ Sect. 11.6.

```
>> dx = DistanceTransformPlanner(floorMap,inflate=6);
>> dx.plan(places.kitchen);
>> pathPoints = dx.query(places.br3);
>> dx.plot(pathPoints,inflated=true);
```

and this is shown in ◻ Fig. 5.18a. The inflated obstacles are shown in red and the previous route to the kitchen has become blocked. The robot follows a different route, through the narrower corridors, to reach its goal. ▶

> A more sophisticated approach is based on an *inflation ratio* which relates robot speed to inflation radius. At high-speed, a large inflation radius is used to give plenty of clearance from obstacles.

5.4.2 D*

D* is a well-known and popular algorithm for robot path computation. Similar to the distance transform, it uses an occupancy grid map representation of the world and finds an optimal path. Under the hood, D* converts the occupancy grid map to a graph and then finds the optimal path using an A*-like algorithm. ▶ D* has several features that are useful for real-world applications. Firstly, it generalizes the occupancy grid map to a cost map that represents the cost $c \in \mathbb{R}_{>0}$ of traversing each cell in the horizontal or vertical direction. The cost of traversing the cell diagonally is $c\sqrt{2}$. For cells corresponding to obstacles, $c = \infty$ (Inf in MATLAB). If we are interested in the shortest time to reach the goal, then cost is the time to drive across the cell. If we are interested in minimizing damage to the vehicle or maximizing passenger comfort, then cost might be related to the roughness of the terrain within the cell. ▶

> D* stands for Dynamic A* and is an extension of the A* algorithm for finding minimum-cost paths through a graph that supports efficient replanning when route costs change.

Secondly, D* supports incremental replanning. This is important if, while we are moving, we discover that the world is different to our map. If we discover that a cell has a different cost, either higher or lower, then we can incrementally replan to find a better path. The incremental replanning has a lower computational cost than completely replanning as would be required using the distance transform method just discussed.

> The costs assigned to cells will also depend on the characteristics of the vehicle as well as the world, for example, a large 4-wheel drive vehicle may have a finite cost to cross a rough area, whereas for a small car this cost might be infinite.

5

◻ Fig. 5.19 The D* planner path. Obstacles are indicated by red cells and all drivable cells have a cost of 1. The background gray intensity represents the cell's distance from the goal in units of cell size as indicated by the scale on the right-hand side

To implement the D* planner using the RVC Toolbox, we use a similar pattern as before and create a D* object

```
>> ds = DStarPlanner(floorMap);
```

which converts the passed occupancy grid map into a cost map, which we can retrieve

```
>> c = ds.costmap;
```

The elements of c will be 1 or ∞ representing free and occupied cells respectively. These are the edge costs in the internal graph representation of the grid.

A plan for moving to a goal in the kitchen is generated by

```
>> ds.plan(places.kitchen)
245184 iterations
```

which creates a dense directed graph (see ▶ App. I). Every cell is a graph node and has a cost, a distance to the goal, and a link to the neighboring cell that is closest to the goal. Each cell also has a state $t \in \{NEW, OPEN, CLOSED\}$. Initially every cell is in the NEW state, the cost of the goal cell is zero and its state is OPEN. We can consider the set of all cells in the OPEN state as a wavefront propagating outward from the goal. ◀ The cost of reaching cells that are neighbors of an OPEN cell is computed and these cells in turn are set to OPEN and the original cell is removed from the open list and becomes CLOSED. In MATLAB this initial planning phase is quite slow ◀ and takes roughly half a minute and

```
>> ds.niter
ans =
        245184
```

iterations.

The path from an arbitrary starting point to the goal

```
>> ds.query(places.br3);
```

is shown in ◻ Fig. 5.19. The path is almost identical to the one given by the distance transform planner.

The real power of D* comes from being able to efficiently change the cost map during the mission. This is actually quite a common requirement in robotics since real sensors have a finite range and a robot discovers more of world as it proceeds. We inform D* about changes using the modifyCost method, for example to raise the cost of entering the kitchen via the bottom doorway

```
>> ds.modifyCost([290 335; 100 125],5);
```

The distance transform also evolves as a wavefront outward from the goal. However D* represents the frontier efficiently as a list of cells whereas the distance transform computes the frontier on a per-cell basis at every iteration – the frontier is implicitly where a cell with infinite cost (the initial value of all cells) is adjacent to a cell with finite cost.

D* is more efficient than the distance transform but it executes more slowly because it is implemented entirely in MATLAB code whereas the distance transform uses C++ code under the hood.

Fig. 5.20 Path from D* planner with the map being modified dynamically. The higher-cost region is indicated by the yellow-dashed rectangle and has changed the path compared to ◨ Fig. 5.19

we have raised the cost to 5 for a small rectangular region across the doorway. This region is indicated by the yellow-dashed rectangle in ◨ Fig. 5.20. The other drivable cells have a default cost of 1. The plan is updated by invoking the planning algorithm again

```
>> ds.plan;
```

and this time the number of iterations is only

```
>> ds.niter
ans =
        169580
```

which is 70% of that required to create the original plan. ▶ The new path for the robot

```
>> ds.query(places.br3);
```

is shown in ◨ Fig. 5.20. The cost change is relatively small, but we notice that the increased cost of driving within this region is indicated by a subtle brightening of those cells – in a cost sense these cells are now further from the goal. Compared to ◨ Fig. 5.19, the robot has taken a different route to the kitchen and avoided the bottom door. D* allows updates to the map to be made at any time while the robot is moving. After replanning the robot simply moves to the adjacent cell with the lowest cost which ensures continuity of motion even if the plan has changed.

The cost increases with the number of cells modified and the effect those changes have on the distance map. It is possible that incremental replanning takes more time than planning from scratch.

5.5 Planning with Roadmaps

5.5.1 Introduction to Roadmap Methods

In robotic path planning, the analysis of the map is referred to as the *planning phase*. The *query phase* uses the result of the planning phase to find a path from A to B. The algorithms introduced in ▶ Sect. 5.4, the distance transform and D*, require a significant amount of computation for the planning phase, but the query phase is very cheap. However, in both cases, the plan depends on the goal, so if the goal changes the expensive planning phase must be re-executed. D* does allow the path to be recomputed during query, but it does not support a changed goal.

The disparity in planning and query costs has led to the development of roadmap methods where the query can include both the start and goal positions. The planning phase performs analysis of the map that supports all starting points and goals. A good analogy is making a journey by train. We first find a local path to the nearest train station, travel through the train network, get off at the station closest to our

5

Fig. 5.21 Steps in the creation of a Voronoi roadmap. **a** Free space is indicated by white cells; **b** the skeleton of the free space is a network of adjacent white cells no more than one cell thick; **c** the skeleton (inverted) with the obstacles overlaid in red; **d** the distance transform of the obstacles, where the brightness of a point increases with its distance from the nearest obstacle

goal, and then take a local path to the goal. The train network is invariant and planning a path through the train network is straightforward using techniques like those we discussed in ▶ Sect. 5.3. Planning paths to and from the entry and exit stations respectively is also straightforward since they are likely to be short paths and we can use the occupancy-grid techniques introduced in ▶ Sect. 5.4.

The robot navigation problem then becomes one of building a network of obstacle-free paths through the environment that serve the function of the train network. In robot path planning, such a network is referred to as a *roadmap*. The roadmap need only be computed once and can then be used like the train network to get us from any start location to any goal location.

We will illustrate the principles by creating a roadmap from the occupancy grid map's free space using some image processing techniques. The essential steps in creating the roadmap are shown in ▶ Fig. 5.21. The first step is to find the free space in the map which is simply the complement of the occupied space

```
>> free = 1-floorplan;
```

and is a matrix with nonzero elements where the robot is free to move. The boundary is also an obstacle, so we mark the outermost cells as being not free

```
>> free(1,:) = 0; free(end,:) = 0;
>> free(:,1) = 0; free(:,end) = 0;
```

and this map is shown in ▶ Fig. 5.21a where free space is depicted as white.

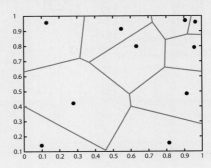

The topological skeleton of the free space is computed by a morphological image processing algorithm known as thinning ▶ applied to the free space of ◼ Fig. 5.21a

> Also known as skeletonization. We will cover this topic in ▶ Sect. 11.6.3.

```
>> skeleton = bwmorph(free,skel=Inf);
```

and the result is shown in ◼ Fig. 5.21b. We see that the obstacles have grown and the free space, the white cells, have become a thin network of connected white cells which are equidistant from the boundaries of the original obstacles.

◼ Fig. 5.21c shows the free space network overlaid on the original map. We have created a network of paths that span the free space and which could be used for obstacle-free travel around the house. These paths are the edges of a generalized Voronoi diagram. We could obtain a similar result by computing the distance transform of the obstacles, ◼ Fig. 5.21a, and this is shown in ◼ Fig. 5.21d. The value of each pixel is the distance to the nearest obstacle and the bright ridge lines correspond to the skeleton of ◼ Fig. 5.21b. Thinning or skeletonization, like the distance transform, is a computationally expensive iterative algorithm but it illustrates the principles of finding paths through free space. In the next section we will examine a cheaper alternative.

5.5.2 Probabilistic Roadmap Method (PRM)

The high computational cost of the distance transform and skeletonization methods makes them infeasible for large maps and has led to the development of probabilistic methods. These methods sparsely sample the map and the most well-known of these methods is the probabilistic roadmap method or PRM. The sampling is random, so results might be different each time you run your planner. To get reproducible results in this section's examples, we initialize the random number generator `rng` to a fixed seed.

Finding the path is a two-phase process: planning, and query. We first create a PRM planner object

```
>> rng(10) % obtain repeatable results
>> prm = mobileRobotPRM(floorMap)
prm =
  mobileRobotPRM with properties:
                  Map: [1x1 binaryOccupancyMap]
             NumNodes: 50
   ConnectionDistance: Inf
```

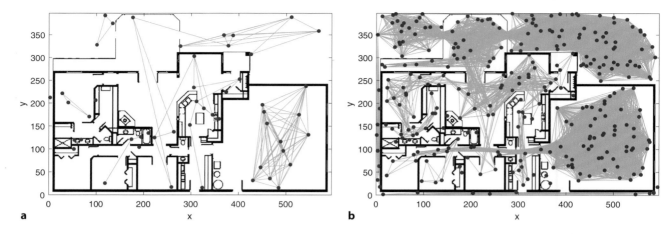

Fig. 5.22 Probabilistic roadmap (PRM) planner and the random graphs produced in the planning phase. **a** Poorly connected roadmap graph with 50 nodes; **b** well-connected roadmap graph with 300 nodes

The call to `rng(10)` ensures that the random number generator is seeded with 10, which ensures that the sequence of plans created using this planner instance will be repeatable.

The path is tested to be obstacle-free by sampling at a fixed number of points along its length. The sampling interval is picked based on the cell size of the occupancy map.

After increasing the number of `NumNodes`, the roadmap has to be recomputed. This happens automatically as a one-time operation when calling `show` or `findpath`.

and use that to generate the plan with 50 randomly chosen nodes, which is a roadmap, represented internally as an embedded graph with 50 nodes. Note that the plan – the roadmap – is independent of the start and the goal. ◄

The graph can be visualized

```
>> prm.show
```

as shown in ◨ Fig. 5.22a. The dots are the randomly chosen points and the nodes of the embedded graph. The lines are obstacle-free paths ◄ between pairs of nodes and are the edges of the graph which have a cost equal to the line length. There are two large components, but several rooms are not connected to either of those, and this means that we cannot plan a path to or from those rooms. The advantage of PRM is that relatively few points need to be tested to ascertain that the nodes and the paths between them are obstacle-free.

To improve connectivity, we can create the planner object again – the random nodes will be different and will perhaps give better connectivity. It is common practice to iterate on the planner until a path is found, noting that there is no guarantee that a valid plan will be found. We can increase the chance of success by specifying more nodes ◄

```
>> prm.NumNodes = 300;
>> prm.show;
```

which gives the result shown in ◨ Fig. 5.22b which connects all rooms in the house.

The query phase finds a path from the start point to the goal. This is simply a matter of moving to the closest node in the roadmap (the start node), following a minimum-distance A* route through the roadmap, getting off at the node closest to the goal and then traveling to the goal. An advantage of this planner is that, once the roadmap is created by the planning phase, we can change the goal and starting points very cheaply, only the query phase needs to be repeated. The path

```
>> p = prm.findpath(places.br3,places.kitchen)
p =
   50.0000    50.0000
   35.1008    73.8081
  129.8734    98.8154
  313.9926    90.5135
  308.0064   180.9964
  320.0000   190.0000
```

is a list of the node coordinates that the robot passes through – a list of waypoints or via points. The start point `places.br3` is in the first row, and the goal point `places.kitchen` is in the last row. The distance between these points is variable.

Excurse 5.9: Random Numbers

Creating good random numbers is a non-trivial problem. The MATLAB random number generator (RNG), used for `rand` and `randn`, generates a very long sequence of numbers that are an excellent approximation to a random sequence – a pseudo-random number (PRN) sequence. The generator maintains an internal state which is effectively the position within the sequence. After startup, MATLAB always generates the following random number sequence

```
>> rand
ans =
    0.8147
>> rand
ans =
    0.9058
>> rand
ans =
    0.1270
```

Many algorithms discussed in this book make use of random numbers and this means that the results can never be replicated. To solve this problem, we can *seed* the random number sequence. Before all such examples we call `rng(0)` or `rng("default")` (they are equivalent) which resets the random number generator to a known state.

```
>> rng("default")  % obtain repeatable results
>> rand
ans =
    0.8147
>> rand
ans =
    0.9058
```

and we see that the random sequence has been restarted.

We can see the path in red overlaid on the PRM roadmap in ◘ Fig. 5.23a, with a closeup of the last path segment, connecting the roadmap with the goal point, in ◘ Fig. 5.23b. For a robot to follow such a path we could use the path-following controller discussed in ▶ Sect. 4.1.1.3.

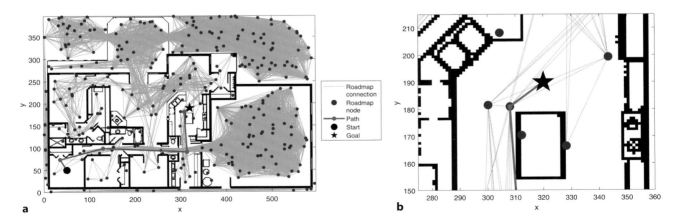

a   b

◘ **Fig. 5.23** Probabilistic roadmap (PRM) planner: **a** showing the red path taken by the robot via nodes of the roadmap; the roadmap connections are shown in light blue; **b** closeup view of goal region where the short path from a roadmap node to the goal can be seen

There are some important trade-offs in achieving this computational efficiency:

- the random sampling of the free space means that a different roadmap is created every time the planner is run, resulting in different paths and path lengths.
- the planner can fail by creating a network consisting of disconnected components as shown in ◘ Fig. 5.22a. If the start and goal points are not connected by the roadmap, that is, they are close to different graph components, the `findpath` method will report an error. The only solution is to rerun the planner and/or increase the number of nodes. We can iterate on the planner until we find a path between the start and goal.
- long narrow gaps between obstacles, such as corridors, are unlikely to be exploited since the probability of randomly choosing points that lie along such spaces is very low.
- the paths generated are not as smooth as those generated by the distance transform and D* and may involve non-optimal back-tracking or very sharp turns.

The computational advantages of PRM extend into higher-dimensional planning problems such as 3D assembly or protein folding. Of more relevance to robotics is to plan in vehicle configuration space and we revisit probabilistic sampling again in ▶ Sect. 5.6.5.

5.6 Planning Drivable Paths

The planners introduced so far can produce paths that are optimal and admissible but not necessarily *feasible*, that is, a real vehicle may not actually be able to follow the path. A path-following controller, such as described in ▶ Sect. 4.1.1.3, will guide a robot along the path but steering constraints mean that we cannot guarantee how closely the vehicle will follow the path. If we cannot precisely follow the collision-free path, then we cannot guarantee that the robot will not collide with an obstacle.

An alternative is to *design* a path from the outset that we know the vehicle can follow. The next two planners that we introduce take into account the motion model of the vehicle and relax the assumption we have made so far that the robot is capable ofomnidirectional motion.

In ▶ Chap. 4, we modeled the motion of wheeled vehicles and how their configuration evolves over time as a function of inputs such as steering and velocity. To recap, the bicycle model follows an arc of radius $R = L/\tan\psi$ where ψ is the angle of the steered wheel and L is the wheel base. This arc radius R is also commonly called turning radius.

In this section, it is useful to consider the curvature κ of the path driven by the robot which is the inverse of the arc radius R. For the bicycle model, we can write

$$\kappa = \frac{1}{R} = \frac{\tan\psi}{L} \; . \tag{5.6}$$

When $\psi = 0$, the curvature is zero, which is a straight line. For a real vehicle, $\psi \in [-\psi_m, \psi_m]$ and the highest-curvature path the vehicle can follow is

$$\kappa_m = \frac{\tan\psi_m}{L} \; . \tag{5.7}$$

The inverse of the *maximum* curvature is the *minimum* turning radius

$$R_m = \frac{1}{\kappa_m} \tag{5.8}$$

Consider the challenging problem of moving a car-like vehicle sideways, which is similar to the parallel parking problem. The start and goal configurations are

```
>> qs = [0 0 pi/2];
>> qg = [1 0 pi/2];
```

For the bicycle motion model, the nonholonomic constraint means that the robot cannot always move directly between two configurations. It may have to follow some indirect path, which we call a maneuver. The rest of this section introduces various approaches to solving this problem.

5.6.1 Dubins Path Planner

A Dubins path is the shortest curve that connects two planar configurations with a constraint on the curvature of the path and assuming forward motion only. While the restriction to forward motion might seem very limiting, we should remember that many useful vehicles such as ships and aircraft have this constraint.

The Dubins path always connects two poses as a sequence of three possible motions: a straight motion (S), a left turn at minimum turning radius (L), and a right turn at minimum turning radius (R). Any two poses in the (x, y, θ) configuration space can be connected through a Dubins path, which will come in handy for sampling-based planning approaches covered in ▶ Sect. 5.6.5.

We create an instance of a 2D Dubins planner ▶ for a minimum turning radius of one meter

```
>> dubins = dubinsConnection(MinTurningRadius=1)
dubins =
  dubinsConnection with properties:
    DisabledPathTypes: {}
     MinTurningRadius: 1
         AllPathTypes: {'LSL'  'LSR'  'RSL'  'RSR'  'RLR'  'LRL'}
```

A 3D variant of the Dubins planner is very popular for describing motions of fixed-wing UAVs. See the `uavDubinsConnection` class.

and use this to compute a Dubins path

```
>> [paths,pathLength] = dubins.connect(qs,qg)
paths =
  1×1 cell array
    {1×1 dubinsPathSegment}

pathLength =
    7.2832
```

The `paths` output is a cell array of paths; each path is an object of type `dubinsPathSegment`. Since we only passed a single start and goal pose, there is only one path returned that has a length of 7.2832 meters.

We can plot this path

```
>> p = paths{1};
>> p.show
```

which is shown in ◨ Fig. 5.24a.

The path object has additional information

```
>> p
p =
  dubinsPathSegment with properties:
          StartPose: [0 0 1.5708]
           GoalPose: [1 0 1.5708]
    MinTurningRadius: 1
       MotionLengths: [4.7124 1 1.5708]
         MotionTypes: {'L'  'S'  'L'}
              Length: 7.2832
```

which indicates that this path comprises a left turn (L), a straight motion (S), and another left turn (L). The total path length is 7.2832 and the lengths of each motion are given by `MotionLengths`. They are 75% of a circle ($3\pi/2$), a straight line of length 1, and 25% of a circle ($\pi/2$).

This path is C^1 *continuous*, so the first derivatives are continuous between the first left turn, the straight motion, and the second left turn. However, the curvature

5

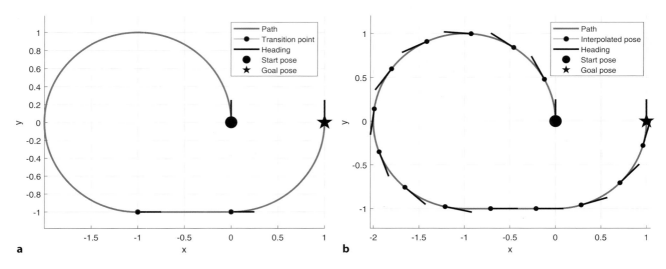

◻ Fig. 5.24 Continuous Dubins path to achieve lateral translation of vehicle. **a** Dubins path consisting of a left turn, a straight motion, and another left turn. The motion transition points are shown as black circles; **b** Dubins path interpolated at an interval of 0.5 meters. The transition points have been removed from the interpolation

There are different levels of geometric continuity between connected path segments. C^0 continuity means point continuity, so the end of one path segment is the same as the start of the next path segment. C^1 continuity means that the tangents (first derivatives with respect to arc lengths) are the same at those two points. C^2 continuity ensures that the curvature is the same at those two points as well. In this case, we have C^0 and C^1 continuity, but not C^2 continuity.

The most common MATLAB functions for generating these vectors are ":" (colon operator) for basic vector creation, `linspace` for linearly spaced vectors, and `logspace` for logarithmically spaced vectors.

is not continuous. ◄ The curvature is constant in each motion segment, but discontinuous between segments. We will discuss paths with continuous curvature in ▶ Sect. 5.6.4.

To obtain discrete (x, y, θ) samples along the path, we can interpolate at discrete intervals. To interpolate at an interval of 0.5 meters, we can use

```
>> samples = p.interpolate(0:0.5:p.Length);
>> whos samples
  Name      Size      Bytes    Class      Attributes
  samples   18x3      432      double
```

The resulting `samples` matrix has one row per sample point. Each row is a vehicle configuration (x, y, θ). All transition points are part of `samples`. To get a set of interpolated points without transition points, we can call `interpolate` without arguments

```
>> allTransPoints = p.interpolate;
>> samplesNoTrans = setdiff(samples,allTransPoints(2:end-1,:),"rows");
>> whos samplesNoTrans
  Name            Size      Bytes    Class      Attributes
  samplesNoTrans  16x3      384      double
```

to retrieve the start, goal, and transition points, and then remove the transition point rows from the original set of `samples`. The resulting points in `samplesNoTrans` are shown in ◻ Fig. 5.24b. The input vector to the `interpolate` function determines the distances at which the path is interpolated (up to the maximum path length). The sample distances can be uniform or non-uniform. ◄

5.6.2 Reeds-Shepp Path Planner

The Dubins planner assumes that the vehicle can only move forward. For vehicles that are capable of forward and backward motion we can use the Reeds-Shepp planner instead

```
>> rs = reedsSheppConnection(MinTurningRadius=1);
>> [paths,pathLength] = rs.connect(qs,qg);
>> p = paths{1}
p =
```

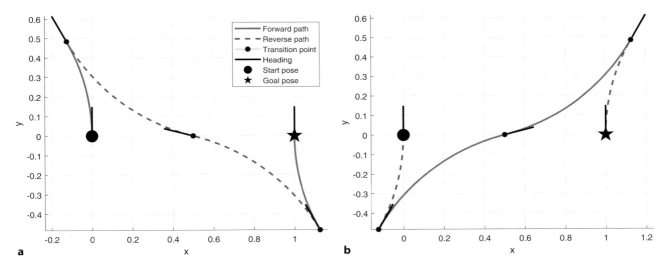

Fig. 5.25 Reeds-Shepp path to achieve lateral translation of vehicle. The start pose is shown by the black circle and the goal is shown by the black star. The vehicle headings are indicated by short black lines. **a** Reeds-Shepp path consisting of a forward left turn, a long backward right, then left motion, and a forward right turn. The motion transition points are shown as black circles; **b** When increasing the `ReverseCost` to 3, another path becomes the optimal one. This one consists of a short backward left turn, a long forward right, then left motion, and another backward left turn

```
reedsSheppPathSegment with properties:
            StartPose: [0 0 1.5708]
             GoalPose: [1 0 1.5708]
      MinTurningRadius: 1
         MotionLengths: [0.5054 0.8128 0.8128 0.5054 0]
      MotionDirections: [1 -1 -1 1 1]
           MotionTypes: {'L'  'R'  'L'  'R'  'N'}
                Length: 2.6362
```

The `p` object in this case indicates the path comprises a left turn, a right turn (but traveling backward as indicated by the sign of the corresponding `MotionDirections` property), a left turn backward, and then another right turn traveling forward. The ability to go backward significantly shortens the path length compared to the Dubins planner in ▶ Sect. 5.6.1. The Dubins path length is over 7 meters long, the Reeds-Sheep path is only 2.6362 meters.

We can interpolate this continuous path at 0.3-meter sampling ▶

The returned variables have many rows. The output has been cropped for inclusion in this book.

```
>> [samples,directions] = p.interpolate(0:0.3:p.Length)
samples =
         0         0    1.5708
   -0.0447    0.2955    1.8708
   -0.1250    0.4841    2.0762
   -0.0753    0.4036    2.1708
   ...

directions =
     1
     1
     1
    -1
   ...
```

The direction at each sampled point on the path is given by the `directions` output – these values are either 1 for forward or -1 for backward motion.

We can plot the path, with color coding for direction by

```
>> p.show
```

which is shown in ◨ Fig. 5.25a. The vehicle drives forward, makes a long reversing motion, and then drives forward to the goal configuration.

5

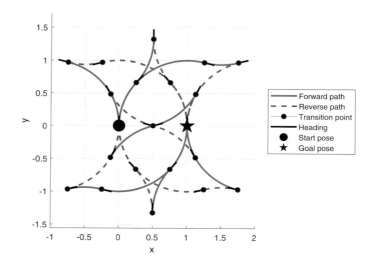

◻ **Fig. 5.26** There are 8 possible Reeds-Shepp paths between the start and goal configuration. They are all shown overlaid in one plot

When connecting two poses, the `reedsSheppConnection` class by default returns the path with the lowest cost. The cost is calculated by multiplying the length of each motion by the `ForwardCost` (for forward motion) or by the `ReverseCost` (for backward motion). For most vehicles, switching from forward to backward motion takes time, and when driving backward the top speeds are usually more restricted.

To penalize backward motion, we can set the `ReverseCost` to a higher value.

```
>> rs.ReverseCost = 3;
>> [paths,cost] = rs.connect(qs,qg);
>> p = paths{1};
>> cost
cost =
    4.6577
>> p.Length
ans =
    2.6362
```

The returned path has the same length as the previous path, but its `cost` is higher because it has segments with backward motion. The resulting path is shown in ◻ Fig. 5.25b. Instead of the long backward motion in ◻ Fig. 5.25a, the new path has a longer forward motion.

How many possible Reeds-Shepp paths are there between two configurations? It turns out that there are 44 possible paths (each with a different motion combination and a different cost). Depending on the start and goal configurations, not all 44 paths are valid. We can retrieve all valid paths by asking for all `PathSegments`.

```
>> [paths,pathLength] = rs.connect(qs,qg,PathSegments="all");
>> validPaths = find(~isnan(pathLength));
>> numel(validPaths)
ans =
    8
```

For our start and goal, there are 8 different connecting paths. They are shown in ◻ Fig. 5.26, illustrating all possible Reeds-Shepp motions to get from the start to the goal.

Both Dubins and Reeds-Shepp paths are useful to connect arbitrary poses in (x, y, θ) configuration space, which will come in handy when we talk about path planning in configuration space in ▶ Sect. 5.6.5.

5.6.3   **Lattice Planner**

The lattice planner computes a path between discrete points in the robot's 3-dimensional configuration space. Consider ◻ Fig. 5.27a that shows the robot, initially at the origin, and driving forward to the three points indicated by blue dots. ▶ Similar to the Dubins planner, each path is an arc with a constant curvature of $\kappa \in \{\kappa_m, 0, -\kappa_m\}$ and the grid pitch is such that the robot's heading direction at the end of each arc is $\theta \in \{\frac{\pi}{2}, 0, -\frac{\pi}{2}, -\pi\}$ radians. With the default grid pitch of 1, the turning radius R is 1, so $\kappa_m = 1$. We can create this by

> The pitch of the grid, the horizontal and vertical distance between nodes of the grid, is dictated by the turning radius of the vehicle.

```
>> lp = LatticePlanner;
>> lp.plan(iterations=1)
4 nodes created
>> lp.plot
```

which is shown in ◻ Fig. 5.27a.

At the end of each branch, we can add the same set of three motions suitably rotated and translated

```
>> lp.plan(iterations=2)
13 nodes created
>> lp.plot
```

and this is shown in ◻ Fig. 5.27b. The graph now contains 13 nodes and represents 9 paths each 2 segments long. Each node represents a configuration (x, y, θ), not just a position, and if we rotate the plot, we can see in ◻ Fig. 5.28 that the paths lie in the 3-dimensional configuration space.

By increasing the number of iterations

```
>> lp.plan(iterations=8)
780 nodes created
>> lp.plot
```

we can fill in more possible paths as shown in ◻ Fig. 5.27c and the paths now extend well beyond the area shown. There are only four possible values for the heading angle $\theta \in \{\frac{\pi}{2}, 0, -\frac{\pi}{2}, -\pi\}$ for each 2D grid coordinate. A left-turning arc from $\theta = \frac{\pi}{2}$ will result in a heading angle of $\theta = -\pi$.

Now that we have created the graph, we can compute a path between any two configurations using the `query` method. For the same problem we used for the

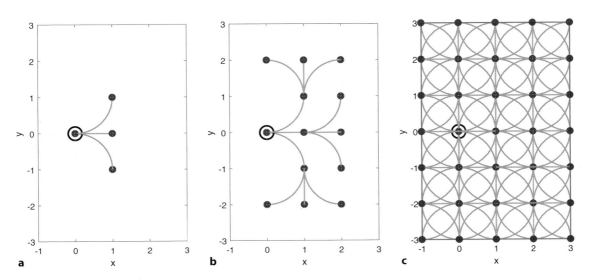

◻ **Fig. 5.27**   Lattice plan after 1, 2, and 8 iterations. The initial configuration is $(0, 0, 0)$

5

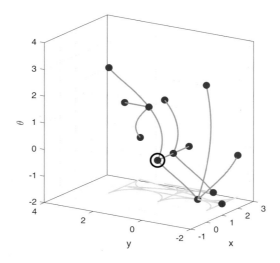

◘ **Fig. 5.28** Lattice plan after 2 iterations shown in 3-dimensional configuration space

Dubins and Reeds-Shepp planners (as in ◘ Fig. 5.24)

```
>> p = lp.query(qs,qg)
A* path cost 5

p =
   -0.0000    0.0000    1.5708
   -1.0000    1.0000    3.1416
   -2.0000         0   -1.5708
   -1.0000   -1.0000         0
   -0.0000   -1.0000         0
    1.0000    0.0000    1.5708
```

the lowest-cost path is a series of waypoints in configuration space, each row represents the configuration-space coordinates (x, y, θ) of a node in the lattice along the path from start to goal configuration. The path, overlaid on the lattice

```
>> lp.plot
```

is shown in ◘ Fig. 5.29a and is the same as the Dubins curve we saw in ◘ Fig. 5.24a.

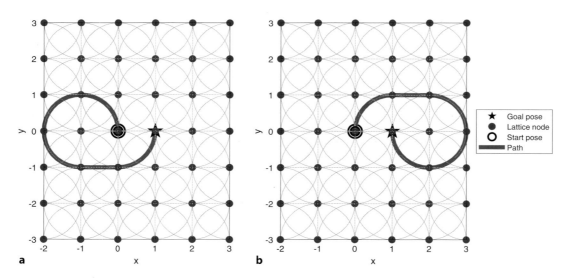

◘ **Fig. 5.29** Lattice planner path for lateral translation. The goal pose is marked with a blue star. **a** With uniform cost; **b** with increased penalty for left turns

◻ Fig. 5.30   Lattice planner path for lateral translation with obstacles

Internally, A* search is used to find a path through the lattice graph and the edge costs are the arc lengths. However, we can override those edge costs, for example we can increase the cost associated with turning left

```
>> lp.plan(cost=[1 10 1]); % S, L, R
780 nodes created
>> lp.query(qs,qg);
A* path cost 5
>> lp.plot
```

and the result is shown in ◻ Fig. 5.29b. Controlling the edge costs provides some control over the returned path compared to the Dubins planner.

Another advantage of the lattice planner over the Dubins planner is that it can be integrated with a 2D occupancy grid map. We first create an occupancy grid map object and some obstacles

```
>> og = binaryOccupancyMap(false(11,11), ...
>>    GridOriginInLocal=[-5.5 -5.5]);
>> og.setOccupancy([-2.5 -2.5],true(1,5));
>> og.setOccupancy([-2.5 -1.5],true(1,3));
>> og.setOccupancy([1.5 -0.5],true(5,2));
```

and then repeat the planning process

```
>> lp = LatticePlanner(og);
>> lp.plan
355 nodes created
```

In this case, we did not specify the number of iterations, so the planner will iterate until it can add no more nodes to the free space. We then query for a path

```
>> lp.query(qs,qg)
A* path cost 13
>> lp.plot
```

and the result is shown in ◻ Fig. 5.30.

Based on our choice of qs and qg, the start and end poses happen to align with the nodes in the lattice, but if different poses are given, we need to plan separate paths from the start pose to the closest lattice node and from the lattice node closest to the goal to the goal pose.

The lattice planner is similar to PRM in that a local planner is required to move from the start configuration to the nearest discrete configuration in the lattice, and from a different discrete lattice node to the goal. This is a very coarse lattice with just

5

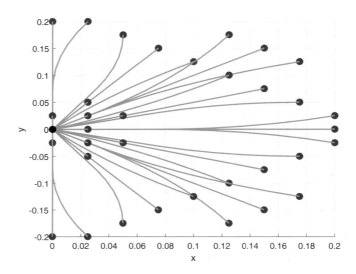

Fig. 5.31 A more sophisticated lattice generated by the package **sbpl** with 43 paths based on the kinematic model of a unicycle. Note that the paths shown here have been rotated by 90° clockwise for consistency with ◨ Fig. 5.27

three branches at every node. More sophisticated lattice planners employ a richer set of motion primitives, such as shown in ◨ Fig. 5.31. In an autonomous driving scenario, this palette of curves can be used for turning corners or lane-changing. To keep the size of the graph manageable, it is typical to generate all the forward curves from the vehicle's current configuration, for a limited number of iterations. At each iteration, curves that intersect obstacles are discarded. From this set of admissible paths, the vehicle chooses the curve that meets its current objective and takes a step along that path – the process is repeated at every control cycle.

5.6.4 Clothoids

The Dubins, Reeds-Shepp, and lattice planners all generate paths that appear smooth, and if the maximum path curvature is achievable by the vehicle, then it can follow the path. However, the curvature is discontinuous – at any point on the path $\kappa \in \{\kappa_m, 0, -\kappa_m\}$ and there are no intermediate values. If you were driving this path in a car, you would have to move the steering wheel instantaneously between the three angles corresponding to hard left turn, straight ahead, and hard right turn. Instantaneous motion is not possible, and it takes time to move the steering wheel, the resulting errors in path following can raise the possibility of a collision.

Road designers have known for a long time that path curvature must change continuously with distance. When we follow such a path with constant velocity, the acceleration will also be constant, which makes for a more enjoyable ride for humans. Humans are not great at sensing velocity, but we are sensitive to acceleration (changes in velocity). In general, roads follow clothoid curves ◀ where curvature increases linearly with distance. An example clothoid curve is shown in ◨ Fig. 5.32.

Also called Euler or Cornu spirals.

Given a start configuration, (x_0, y_0, θ_0), a clothoid curve is then defined by

$$\dot{x}(s) = \cos \theta(s) \tag{5.9}$$

$$\dot{y}(s) = \sin \theta(s) \tag{5.10}$$

$$\dot{\theta}(s) = \dot{\kappa}s + \kappa \tag{5.11}$$

with initial conditions

$$x(0) = x_0, y(0) = y_0, \theta(0) = \theta_0 \tag{5.12}$$

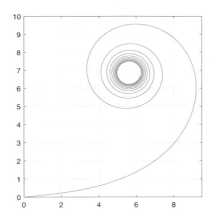

◘ Fig. 5.32 A clothoid has a linearly increasing curvature with respect to its length

The linear change in curvature is the defining characteristic of a clothoid. When $\kappa = \dot{\kappa} = 0$, the curve is a straight line. When the curvature κ is non-zero, the path is a circular segment. When $\dot{\kappa} = 0$, the curve is an arc with constant curvature (similar to the Dubins and Reeds-Shepp motion primitives).

The configuration, as a function of path length s, can then be described as a combination of Fresnel integrals

$$x(s) = x_0 + \int_0^s \cos\left(\frac{1}{2}\dot{\kappa}t^2 + \kappa t + \theta_0\right)dt \qquad (5.13)$$

$$y(s) = y_0 + \int_0^s \sin\left(\frac{1}{2}\dot{\kappa}t^2 + \kappa t + \theta_0\right)dt \qquad (5.14)$$

Given a desired start and goal configuration, we can solve (5.14) by finding the roots of a nonlinear system with the unknown curvature κ, curvature rate $\dot{\kappa}$, and path length s_f. A technique like the Newton-Raphson method can be used to find these roots. This clothoid fitting is implemented by

```
>> p = referencePathFrenet([qs;qg])
p =
  referencePathFrenet with properties:
              PathLength: 1.2743
       SegmentParameters: [0 0 1.5708 -7.2116 11.3186 0]
  DiscretizationDistance: 0.0500
               Waypoints: [2x3 double]
          MaxNumWaypoints: Inf
>> p.show;
```

and the solution is shown in ◘ Fig. 5.33. Compared to the Dubins curve of ◘ Fig. 5.24, it differs in having continuous curvature and therefore it can be fol-

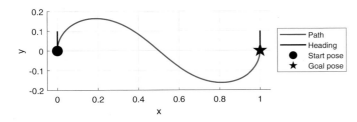

◘ Fig. 5.33 Clothoid path to achieve lateral translation between two poses. The path has C^2 continuity

5

Fig. 5.34 Curvature versus normalized distance along path for the Dubins and clothoid paths

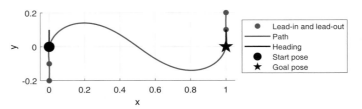

Fig. 5.35 Clothoid path consisting of 6 waypoints. We achieve an approximate heading of $\pi/2$ at qs and qg by placing additional lead-in and lead-out waypoints

lowed exactly by a wheeled vehicle. ◘ Fig. 5.34 compares the path curvature of the Dubins and clothoid approaches. We note, firstly, the curvature discontinuity in the Dubins curve. Secondly, the clothoid approach has a continuous curvature and a much shorter path length (1.2743 vs. 7.2832 meters for Dubins).

The resulting path is C^2 *continuous* between the start and the goal since the curvature is continuous. To preserve the C^2 continuity between more than 2 waypoints, we have to relax our constraints by not enforcing a desired heading angle θ. A common approach to still achieve a heading angle that is reasonably close for a particular waypoint, we create additional lead-in and lead-out waypoints along the tangent.

For example, if we want to preserve an approximate heading angle of $\pi/2$ at qs and qg, we can add 2 additional waypoints before qs and 2 additional waypoints after qg. We specify one waypoint per row, which each waypoint as $[x, y]$ coordinate

```
>> p = referencePathFrenet([0 -0.2; 0 -0.1; qs(1:2); ...
>>    qg(1:2); 1 0.1; 1 0.2]);
>> p.show;
```

and the resulting path is shown in ◘ Fig. 5.35.

Other analytic solutions to this problem can be obtained using splines such as Bézier splines or B-splines. The `bsplinepolytraj` function can generate trajectories using B-splines. Clothoid curves are commonly used in autonomous driving to describe road center lines and desired car trajectories. A real-world example of highway trajectory planning is shown in the example

```
>> openExample("autonomous_control/" + ...
>>    "HighwayTrajectoryPlanningUsingFrenetReferencePathExample");
```

▶ sn.pub/Re7Q07

◘ Fig. 5.36 Simulation of highway driving and lane changing that uses clothoid curves to describe road center lines. The possible trajectories of the blue vehicle are shown in white. They are described by longitudinal and lateral 4th and 5th order polynomials (Image reprinted with permission of The MathWorks, Inc.)

and a screenshot from that example is shown in ◘ Fig. 5.36. This example also shows how to implement obstacle avoidance with moving vehicles and dynamic replanning. An extended application that includes planning and active lane changing in different scenarios is described in ▶ Sect. 16.2.

5.6.5 Planning in Configuration Space (RRT)

The final planner that we introduce is the rapidly exploring random tree or RRT. The RRT is able to take into account the motion model of the vehicle, but unlike the lattice planner, which plans over a regular grid, RRT uses a probabilistic approach like the PRM planner.

The main steps in creating an RRT are as follows, with the notation shown in ◘ Fig. 5.37. A graph of robot configurations is maintained and each node is a configuration $q \in \mathbb{R}^2 \times \mathbf{S}^1$ which is represented by a 3-vector $q \sim (x, y, \theta)$. The first, or root, node in the graph is the start configuration of the robot. A random configuration q_{rand} is chosen, and the node with the closest configuration q_{near} is found – this configuration is near in terms of a cost function that includes distance and orientation. ▶ A control is computed that moves the robot from q_{near} toward q_{rand} over a fixed path simulation time. The configuration that it reaches is q_{new} and this is added to the graph. ▶

The distance measure must account for a difference in position and orientation but, from a strict consideration of units, it is not proper to combine units of length (meters) and angles (radians). In some cases, it might be appropriate to weight the terms to account for the different scaling of these quantities.

Besides this control-based approach, there are other RRT variants that rely on geometric connections between tree nodes (no need to simulate), for example using straight lines, Dubins paths, or Reeds-Shepp paths. See `plannerRRT` and `plannerRRTStar`.

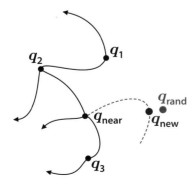

◘ Fig. 5.37 The RRT is a directed graph of configurations q_i where q_g (not shown) is the goal. Edges represent admissible (obstacle-free) and feasible (drivable) paths

We first describe the vehicle kinematics as a bicycle model that was introduced in ▶ Sect. 4.1.1

```
>> bike = mobileRobotPropagator(KinematicModel="bicycle");
>> bike.SystemParameters.KinematicModel
ans =
  struct with fields:
     WheelBase: 1
    SpeedLimit: [-1 1]
    SteerLimit: [-0.7854 0.7854]
>> bike.SystemParameters.KinematicModel.SteerLimit = [-0.5 0.5];
>> bike.SystemParameters.KinematicModel.SpeedLimit = [0 1];
```

and here we have specified a car-like vehicle with a maximum steering angle of ± 0.5 rad and a maximum speed of $1\,\text{ms}^{-1}$. The minimum speed of 0 ensures that only forward motion is allowed.

The `mobileRobotPropagator` computes the controls and simulates (propagates) the motion of the vehicle over a fixed time horizon. It has many parameters to adjust the control law, how states are integrated, and how distances are estimated.

We can then create a `plannerControlRRT` object that takes this kinematic model as an input

```
>> rrt = plannerControlRRT(bike)
rrt =
  plannerControlRRT with properties:
            StatePropagator: [1×1 mobileRobotPropagator]
    ContinueAfterGoalReached: 0
            MaxPlanningTime: Inf
             MaxNumTreeNode: 10000
            MaxNumIteration: 10000
           NumGoalExtension: 1
                   GoalBias: 0.1000
              GoalReachedFcn: @plannerControlRRT.GoalReachedDefault
```

For an obstacle-free environment, we can then create a plan for a given `start` and `goal` pose

```
>> start = [0 0 pi/2];
>> goal = [8 2 0];
>> rng(0); % obtain repeatable results
>> [p,solnInfo] = rrt.plan(start,goal);
>> showControlRRT(p,solnInfo,[],start,goal); % [] for no map
```

The resulting path `p` and the explored random tree are shown in ◩ Fig. 5.38. Without any obstacles, the planner can quickly find a path to the goal configuration.

An important part of the RRT algorithm is computing the control input that moves the robot from an existing configuration q_{near} in the graph toward q_{rand}. From ▶ Sect. 4.1 we understand the difficulty of driving a nonholonomic vehicle to a specified configuration. Rather than the complex nonlinear controller of ▶ Sect. 4.1.1.4 we will use something simpler that fits with the randomized sampling strategy used in this class of planner. The controller randomly chooses whether to drive forward or backward and randomly chooses a steering angle within the limits ◀. It then simulates motion of the vehicle model for a fixed period of time, and computes the closest distance between the path and q_{rand}. This is repeated multiple times and the control input with the best performance is chosen. The configuration on its path that was closest to q_{rand} is chosen as q_{new} and becomes a new node in the graph.

Handling obstacles with the RRT is quite straightforward. The configuration q_{rand} is discarded if it lies within an obstacle, and the point q_{new} will not be added to the graph if the path from q_{near} toward q_{rand} intersects an obstacle. The result is a set of paths, a roadmap, that is collision free and drivable by this nonholonomic vehicle. ◀

Uniformly randomly distributed between the steering angle limits.

In fact, rather than using an approximation, the paths are generated by the exact kinematic or dynamic model of the vehicle with specific control inputs and control durations. We are using a simple bicycle model here, but more complex vehicle models are possible in this framework.

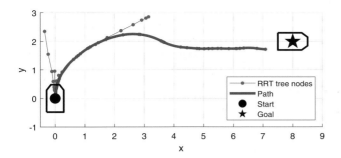

◻ Fig. 5.38 An RRT computed for an obstacle-free motion of the bicycle model. RRT nodes are shown in blue. The RRT graph is sparse here, since the control-based sampling quickly finds a set of control inputs that drive the vehicle from start to goal. The end of the path does not reach the goal configuration exactly, since we are propagating a kinematic model. We can adjust the threshold for convergence in the planner

To illustrate a more real-world problem, we can look at a parallel-parking maneuver. We start by loading a simple map and inflating it by 0.5 m to account for the vehicle size.

```
>> load parkingMap
>> inflatedMap = parkingMap.copy;
>> inflatedMap.inflate(0.5,"world");
```

We then configure the vehicle kinematic model as a bicycle model to approximate a car-like motion with a wheel base of 2 m. To estimate distances between configurations, we use the Reeds-Shepp motion primitives from ▶ Sect. 5.6.2, since this metric accounts for non-holonomic constraints and allows forward and backward motion. For the simpler example above, we used a cheaper Euclidean distance estimation metric. We also pass the inflated environment map and ensure that the RRT sampling happens only within the limits of the map.

```
>> carModel = mobileRobotPropagator( ...
>>    KinematicModel="bicycle",DistanceEstimator="reedsshepp", ...
>>    ControlPolicy="linearpursuit",Environment=inflatedMap);
>> carModel.SystemParameters.KinematicModel.WheelBase = 2;
>> carModel.StateSpace.StateBounds(1:2,:) = ...
>>    [parkingMap.XWorldLimits; parkingMap.YWorldLimits];
```

To allow for smoother motions, we allow the propagator to apply controls for up to 5 s. At a step size of 0.1 s, this corresponds to a maximum number of 50 control steps.

```
>> carModel.ControlStepSize = 0.1;
>> carModel.MaxControlSteps = 50;
```

Now we are ready to create our planner. To make rapid progress towards the goal, we allow the planner to propagate the kinematic model multiple times towards the goal after adding a new node to the tree. The planner continues propagating until no valid control is found, the planner reaches the goal, or the function has been called `NumGoalExtension` times.

```
>> rrt = plannerControlRRT(carModel);
>> rrt.NumGoalExtension = 2;
```

How does the planner know when the goal is reached? Since we are propagating a kinematic model, it is unlikely that we can achieve exact configurations. We allow the planner to finish when the final configuration is within 0.25 rad of the goal orientation and within 0.75 m of the goal configuration by defining a `GoalReachedFcn` function and assigning it to the planner.

5

```
>> rrt.GoalReachedFcn = @(planner,q,qTgt) ...
>>    abs(angdiff(q(3),qTgt(3))) < 0.25 && ...
>>    norm(q(1:2)-qTgt(1:2)) < 0.75;
```

Now, we can plan a backward parking maneuver. The result is a continuous path which will park the vehicle backward. We can then overlay the path on the occupancy grid map and RRT tree as shown in ◻ Fig. 5.39a.

```
>> start = [9.5 4 0];
>> goal = [5.5 2 0];
>> rng(0); % obtain repeatable results
>> [p,solnInfo] = rrt.plan(start,goal);
>> showControlRRT(p,solnInfo,parkingMap,start,goal);
```

We can also animate the motion along the path by interpolating the path between the states

```
>> plotPath = p.copy; plotPath.interpolate;
>> plotvehicle(plotPath.States,"box",scale=1/20, ...
>>    edgecolor=[0.5 0.5 0.5],linewidth=1)
```

Some of the intermediate poses are shown in ◻ Fig. 5.39a. Now that we have a planner set up, it is easy to vary the start and goal poses and adjust the RRT planner parameters. For example, we can easily plan a forward parallel-parking maneuver by setting the starting configuration to $[2, 4, 0]$ and the result can be seen in ◻ Fig. 5.39b. Compared to the backward-parking maneuver, the planner had to expand more tree nodes to reach the desired goal.

This example illustrates some important points about the RRT. Firstly, as for the PRM planner, there may be some distance (and orientation) between the goal configuration and the nearest node. Minimizing this requires tuning RRT parameters such as the number of nodes, path simulation time, or the `GoalReachedFcn` function. Secondly, the path is feasible but not quite optimal. In this case, the vehicle has changed direction twice before driving into the parking slot. This is due to the random choice of nodes – rerunning the planner and/or increasing the number of nodes or planning time may help. Finally, we can see that the vehicle body collides with the obstacle, and this is very apparent if you view the animation. This is due to the fact that we only inflated the occupancy map by 0.5 m. We can increase that inflation radius to the smallest circle that contains the robot or use a more sophisti-

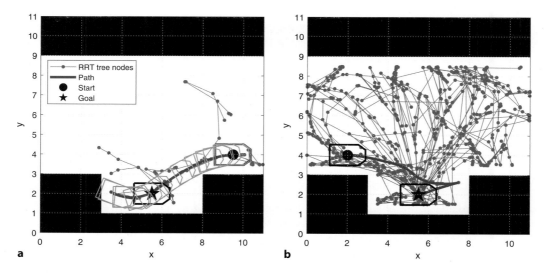

◻ **Fig. 5.39**   A simple parallel parking example based on the RRT planner with an occupancy map. **a** Backward parallel parking with a few snapshots of the intermediate vehicle configurations overlaid in gray. The car makes a long backward motion, before going forward to the final configuration; **b** forward parallel parking with `start` equal to $[2, 4, 0]$

◻ Fig. 5.40 Simulated backward driving of truck-trailer system. **a** The kinematic model of the articulated system, which is simulated through `plannerControlRRT`; **b** the final reverse trajectory across an obstacle field is shown for the truck and the trailer (Images reprinted with permission of The MathWorks, Inc.)

cated geometric collision checking algorithm, for example based on polygons (see the `polyshape` class).

This control-based RRT framework can be used with a variety of vehicle kinematic and dynamic models. One particularly complex problem that can be tackled is the reverse motion planning for truck-trailer systems

```
>> openExample("nav/" + ...
>>    "TractorTrailerPlanningUsingPlannerControlRRTExample");
```

and a screenshot from this example is shown in ◻ Fig. 5.40.

This system is inherently unstable during reverse motion, so paths returned by conventional geometric planners are unlikely to produce good results during path following. The control-based RRT can solve this problem efficiently.

► sn.pub/caw9iX

5.7 Advanced Topics

5.7.1 A* vs Dijkstra Search

A very famous approach to graph search is Dijkstra's *shortest-path algorithm* from 1956. It is framed in terms of the shortest path from a source node to all other nodes in the graph and produces a shortest-path tree. The source node could be the start or the goal. The algorithm is similar to UCS in that it visits every node and computes the distance of every node from the start. However, UCS does require a start and a goal, whereas Dijkstra's search requires only a start (or a goal). Dijkstra's algorithm maintains two sets of nodes, visited and unvisited, which are analogous to UCS's explored and frontier sets – Dijkstra's unvisited set is initialized with all nodes, whereas the UCS frontier is initialized with only the start node. The computed cost expands outwards from the starting point and is analogous to the distance transform or wavefront planner.

5.7.2 Converting Grid Maps to Graphs

We have treated grid maps and graphs as distinct representations, but we can easily convert an occupancy grid map to a graph. We place a node in every grid cell and then add edges to every neighboring cell that is not occupied. If the vehicle can drive to all neighbors, then a node can have at most eight neighbors. If it is a Manhattan-world problem, then a node can have at most four neighbors.

5

Excurse 5.10: Edsger Wybe Dijkstra

Dijkstra (1930–2002) was a pioneering Dutch computer scientist and winner of the 1972 Turing award. A theoretical physicist by training, he worked in industry and academia at the time when computing was transitioning from a craft to a discipline, with no theoretical underpinnings, and few computer languages. He was an early advocate of new programming styles to improve the quality of programs, coined the phrase "structured programming" and developed languages and compilers to support this. In the 1960s, he worked in concurrent and distributed computing on topics such as mutual exclusion, semaphores, and deadlock. He also developed an algorithm for finding shortest paths in graphs. Many of his contributions to computer science are critical to the implementation of robotic systems today.

5.7.3 Converting Between Graphs and Matrices

A graph with n nodes can be represented as an $n \times n$ distance matrix $\mathbf{D}$. For example

```
>> g = UGraph;
>> for i = 1:4 % add 4 nodes
>>    g.add_node([0 0]);
>> end
>> g.add_edge(1,2,1); % Connect nodes 1 and 2 with cost 1
>> g.add_edge(1,4,2); % Connect nodes 1 and 4 with cost 2
>> g.add_edge(2,3,3); % Connect nodes 2 and 3 with cost 3
>> D = g.adjacency
D =
      0    1    0    2
      1    0    3    0
      0    3    0    0
      2    0    0    0
```

The elements of the matrix $d_{i,j}$ are the edge cost from vertex i to vertex j or 0 if vertices i and j are not connected. For an undirected graph, as shown here, the matrix is symmetric.

5.7.4 Local and Global Planning

The global planners we have discussed are able to produce optimal paths because they have full knowledge of the state of the world, but in reality, this is not sufficient. When we drive a car from A to B, we can compute a globally optimal path through the road network, but that is critically reliant on the accuracy and currency of the map – a map is never going to reflect all the road closures and construction, let alone the cars, bicycles and pedestrians that we will encounter. When we drive, we make numerous small adjustments to accommodate those elements of the environment that are unmappable and for this we use our sensory system – most importantly, our eyes. In practice, we use both map-based planning and reactive strategies.

This insight is reflected in the common robot planning architecture shown in ◘ Fig. 5.41. The task of the local planner is to keep the vehicle as close as possible to the global optimal path, while accommodating a priori unknowable local conditions.

Local path planning is based on sensory input and is particularly challenging since sensors have limited range and therefore global information is not available. For example, in the process of avoiding one local obstacle, we might turn left to

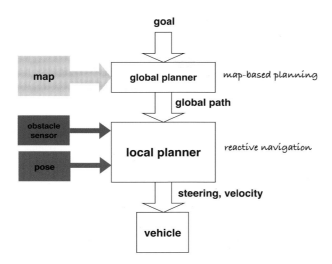

□ Fig. 5.41 Global and local path planning architecture

avoid it, and encounter another obstacle, which we would not have encountered had we turned right. This was the challenge we encountered with the *bug2* planner in ▶ Sect. 5.1.2.

5.8 Wrapping Up

Robot navigation is the problem of guiding a robot towards a goal and we have covered a spectrum of approaches. The simplest was the purely reactive Braitenberg-type vehicle. Then we added limited memory to create state machine based automata such as *bug2*, which can deal with obstacles, however, the paths that it finds can be far from optimal.

Several different map-based planning algorithms were then introduced. We started with graph-based maps and covered a number of algorithms for finding optimal paths through a road network, minimizing distance or travel time. For grid-based maps, we started with the distance transform to find an optimal path to the goal. D* finds a similar path but allows grid cells to have a continuous traversability measure rather than being considered as only free space or obstacle. D* also supports computationally cheap incremental replanning for small changes in the map. PRM reduces the computational burden by probabilistic sampling but at the expense of somewhat less optimal paths and no guarantee of finding a path even if one exists. To accommodate the motion constraints of real vehicles, we introduced the Dubins and Reeds-Shepp planners that find drivable paths comprising arcs and straight lines, and a lattice planner that also supports planning around obstacles. These paths, while technically drivable, are discontinuous in curvature and clothoids were introduced to overcome that limitation. Finally, we looked at RRT for planning a path through configuration space in the presence of obstacles and with vehicle motion constraints. All the map-based approaches require a map and knowledge of the robot's location, and these are both topics that we will cover in the next chapter.

5.8.1 Further Reading

Comprehensive coverage of planning for robots is provided by two textbooks. Choset et al. (2005) covers geometric and probabilistic approaches to planning, as well as the application to robots with dynamic and nonholonomic constraints.

LaValle (2006) covers motion planning, planning under uncertainty, sensor-based planning, reinforcement learning, nonlinear systems, trajectory planning, nonholonomic planning, and is available online for free at ► http://lavalle.pl/planning. In particular, these books provide a much more sophisticated approach to representing obstacles in configuration space and cover potential-field planning methods that we have not discussed. The powerful planning techniques discussed in these books can be applied beyond robotics to very high-order systems such as vehicles with trailers, robotic arms, or even the shape of molecules. LaValle (2011a) and LaValle (2011b) provide a concise two-part tutorial introduction. More succinct coverage of planning is given by Kelly (2013), Siegwart et al. (2011), the Robotics Handbook (Siciliano and Khatib 2016, § 7), and also by Spong et al. (2006) and Siciliano et al. (2009).

The *bug1* and *bug2* algorithms were described by Lumelsky and Stepanov (1986), and Ng and Bräunl (2007) implemented and compared eleven variations of the Bug algorithm in a number of different environments. Graph-based planning is well-covered by Russell and Norvig (2020) and the road network example in ► Sect. 5.3 was inspired by their use of a map of Romania. The distance transform is well-described by Borgefors (1986) and its early application to robotic navigation was explored by Jarvis and Byrne (1988). Efficient approaches to implementing the distance transform include the two-pass method of Hirata (1996), fast marching methods, or reframing it as a graph search problem that can be solved using Dijkstra's method; the last two approaches are compared by Alton and Mitchell (2006). The first algorithm for finding shortest paths within a graph was by Dijkstra (1959), and the later A* algorithm (Hart et al. 1968) improves the search efficiency. Any occupancy grid map can be converted to a graph which is the approach used in the D* algorithm by Stentz (1994) which allows cheap replanning when the map changes. There have been many further extensions including, but not limited to, Field D* (Ferguson and Stentz 2006) and D* lite (Koenig and Likhachev 2005). D* is used in many real-world robot systems and many implementations exist, including open source.

The ideas behind PRM started to emerge in the mid-1990s and the algorithm was first described by Kavraki et al. (1996). Geraerts and Overmars (2004) compare the efficacy of a number of subsequent variations that have been proposed to the basic PRM algorithm. Approaches to planning that incorporate the vehicle's dynamics include state-space sampling (Howard et al. 2008), and the RRT, which is described in LaValle (1998, 2001). More recently, RRT* has been proposed by Karaman et al. (2011). The approaches to connecting two poses by means of arcs and line segments were introduced by Dubins (1957) and Reeds and Shepp (1990). Owen et al. (2015) extend the Dubins algorithm to 3D which is useful for aircraft applications. Lattice planners are covered in Pivtoraiko et al. (2009). A good introduction to path curvature, clothoids, and curvature polynomials is given by Kelly (2013). For a comprehensive treatment of clothoids, see the paper from Bertolazzi and Frego (2015).

■■ **Historical and interesting**

The defining book in cybernetics was written by Wiener in 1948 and updated in 1965 (Wiener 1965). Walter published a number of popular articles (1950, 1951) and a book (1953) based on his theories and experiments with robotic tortoises, while Holland (2003) describes Walter's contributions.

The definitive reference for Braitenberg vehicles is Braitenberg's own book (1986) which is a whimsical, almost poetic, set of thought experiments. Vehicles of increasing complexity (fourteen vehicle families in all) are developed, some including nonlinearities, memory, and logic to which he attributes anthropomorphic characteristics such as love, fear, aggression, and egotism. The second part of the book outlines the factual basis of these machines in the neural structure of animals.

Early behavior-based robots included the Johns Hopkins Beast, built in the 1960s, and Genghis (Brooks 1989) built in 1989. Behavior-based robotics are described in the early paper by Brooks (1986), the book by Arkin (1999), and the Robotics Handbook (Siciliano and Khatib 2016, § 13). Matarić's Robotics Primer (Matarić 2007) and associated comprehensive web-based resources are also an excellent introduction to reactive control, behavior-based control, and robot navigation. A rich collection of archival material about early cybernetic machines, including Walter's tortoise and the Johns Hopkins Beast can be found at the Cybernetic Zoo ▶ https://cyberneticzoo.com/.

5.8.2 Resources

A very powerful set of motion planners exist in OMPL, the Open Motion Planning Library (▶ https://ompl.kavrakilab.org/) written in C++. It has a Python-based app that provides a convenient means to explore planning problems. Steve LaValle's web site ▶ http://lavalle.pl/software.html has many code resources (C++ and Python) related to motion planning. Lattice planners are included in the sbpl package from the Search-Based Planning Lab (▶ http://sbpl.net/) which has MATLAB tools for generating motion primitives and C++ code for planning over the lattice graphs. Clothoid curves and interpolations are implemented in Bertolazzi's Clothoids library (▶ https://github.com/ebertolazzi/Clothoids). It is written in C++ and has a MATLAB binding.

Python Robotics (▶ https://github.com/AtsushiSakai/PythonRobotics) is a comprehensive and growing collection of algorithms in Python for robotic path planning, as well as localization, mapping, and SLAM. You can call Python code from MATLAB to explore these algorithms.

5.8.3 Exercises

1. Braitenberg vehicles (▶ Sect. 5.1.1)
   a) Experiment with different start configurations and control gains.
   b) Modify the signs on the steering signal to make the vehicle light-phobic.
   c) Modify the `sensorfield` function so that the peak moves with time.
   d) The vehicle approaches the maxima asymptotically. Add a stopping rule so that the vehicle stops when either sensor detects a value greater than 0.95.
   e) Create a scalar field with two peaks. Can you create a starting pose where the robot gets confused?
2. Occupancy grid maps. Create some different occupancy grid maps and test them on the different planners discussed.
   a) Create an occupancy grid map that contains a maze and test it with various planners. See ▶ https://rosettacode.org/wiki/Maze_generation and `mapMaze`.
   b) Create an occupancy grid map from a downloaded floor plan.
   c) Create an occupancy grid map from a city street map, perhaps apply color segmentation (▶ Chap. 12) to segment roads from other features. Can you convert this to a cost map for D* where different roads or intersections have different costs?
   d) Experiment with obstacle inflation.
   e) At 1 m cell size, how much memory is required to represent the surface of the Earth? How much memory is required to represent just the land area of Earth? What cell size is needed in order for a map of your country to fit in 1 Gbyte of memory?
3. Bug algorithms (▶ Sect. 5.1.2)
   a) Create a map to challenge *bug2*. Try different starting points.
   b) Create an obstacle map that contains a maze. Can *bug2* solve the maze?
   c) Experiment with different start and goal locations.

5

d) Create a bug trap. Make a hollow box and start the bug inside a box with the goal outside. What happens?

e) Modify *bug2* to change the direction it turns when it hits an obstacle.

f) Implement other bug algorithms such as *bug1* and *tangent bug*. Do they perform better or worse?

4. Distance transform (▶ Sect. 5.4.1)

a) Create an obstacle map that contains a maze and solve it using distance transform.

5. D* planner (▶ Sect. 5.4.2)

a) Add a low-cost region to the living room. Can you make the robot prefer to take this route to the kitchen?

b) Block additional doorways to challenge the robot.

c) Implement D* as a mex-file to speed it up.

6. PRM planner (▶ Sect. 5.5)

a) Run the PRM planner 100 times and gather statistics on the resulting path length.

b) Vary the value of the distance threshold parameter and observe the effect.

c) Use the output of the PRM planner as input to a pure pursuit planner as discussed in ▶ Chap. 4.

7. Dubins and Reeds-Shepp planners (▶ Sects. 5.6.1 and 5.6.2)

a) Find a path to implement a 3-point turn.

b) Investigate the effect of changing the maximum curvature.

8. Lattice planner (▶ Sect. 5.6.3)

a) How many iterations are required to completely fill the region of interest shown in ◘ Fig. 5.27c?

b) How does the number of nodes and the spatial extent of the lattice increase with the number of iterations?

c) Given a car with a wheel base of 4.5 m and maximum steered-wheel angles of $\pm 30°$, what is the lattice grid size?

d) Compute and plot curvature as a function of path length for a path through the lattice such as the one shown in ◘ Fig. 5.29a.

e) Design a controller in Simulink that will take a unicycle or bicycle model with a finite steered-wheel angle rate (there is a block parameter to specify this) that will drive the vehicle along the three paths shown in ◘ Fig. 5.27a.

9. Clothoids (▶ Sect. 5.6.4)

a) Investigate Bézier, Hermite and B-splines to create smooth paths between positions or configurations.

b) Modify the clothoid planner to have an upper bound on curvature (hard).

10. RRT planner (▶ Sect. 5.6.5)

a) Find a path to implement a 3-point turn.

b) Experiment with RRT parameters such as the number of nodes, the vehicle steering angle limits, and the path integration time.

c) Additional information in the node of each graph holds the control input that was computed to reach the node. Plot the steering angle and velocity sequence required to move from start to goal pose.

d) Add a local planner to move from initial pose to the closest node, and from the final node to the goal pose.

e) Determine a path through the graph that minimizes the number of reversals of direction.

f) The collision test currently only checks that the center point of the robot does not lie in an occupied cell. Modify the collision test so that the robot is represented by a rectangular robot body and check that the entire body is obstacle-free.

Localization and Mapping

» *In order to get somewhere, we need to know where we are*

Contents

6.1   Dead Reckoning Using Odometry – 218

6.2   Landmark Maps – 226

6.3   Occupancy Grid Maps – 244

6.4   Pose-Graph SLAM – 256

6.5   Wrapping Up – 266

Fig. 6.1 Position estimation by dead reckoning. The ship's position at 3 P.M. is based on its position at 2 P.M., the estimated distance traveled since, and the average compass heading

6

chapter6.mlx

▶ sn.pub/KocMeG

Besides the GPS satellite constellation, there are many other global constellations in use today, including GLONASS (Russia), Galileo (Europe), and BeiDou (China). For purposes of this chapter, all the pros and cons of GPS apply to all of these constellations.

In our discussion of map-based navigation so far, we assumed that the robot had a means of knowing its position. In this chapter, we discuss some of the common techniques used to estimate the pose of a robot in the world – a process known as localization.

Today, GPS ◄ makes outdoor localization so easy that we often take this capability for granted. Unfortunately, GPS isn't the perfect sensor many people imagine it to be. It relies on very weak radio signals received from distant satellites, which means that GPS fails or has error where there is no *line-of-sight* radio reception, for instance indoors, underwater, underground, in urban canyons, or in deep mining pits. GPS signals can also be jammed and for some applications that is not acceptable.

GPS has only been in use since 1995 yet humankind has been navigating the planet and localizing for thousands of years. In this chapter we will introduce the *classical* navigation principles such as dead reckoning and the use of landmarks on which modern robotic navigation is founded.

Dead reckoning is the estimation of position based on estimated speed, direction, and time of travel with respect to a previous estimate. ◘ Fig. 6.1 shows how a ship's position is updated on a chart. Given the average compass heading over the previous hour and the distance traveled, the position at 3 P.M. can be found using elementary geometry from the position at 2 P.M. However, the measurements on which the update is based are subject to both systematic and random error. Modern instruments are quite precise but 500 years ago clocks, compasses, and speed measurements were primitive. The recursive nature of the process, each estimate is based on the previous one, means that errors will accumulate over time and for sea voyages of many years this approach was quite inadequate.

Excurse 6.1: Measuring Speed at Sea
A ship's log is an instrument that provides an estimate of the distance traveled. The oldest method of determining the speed of a ship at sea was the Dutchman's log – a floating object was thrown into the water at the ship's bow and the time for it to pass the stern was measured using an hourglass. Later came the chip log, a flat quarter-circle of wood with a lead weight on the circular side causing it to float upright and resist towing. It was tossed overboard and a line with knots at 50 foot intervals was payed out. A special hourglass, called a log glass, ran for 30 s, and each knot on the line over that interval corresponds to approximately 1 nmi h^{-1} or 1 knot. A nautical mile (nmi) is now defined as 1852 m. (Image modified from Text-Book of Seamanship, Commodore S. B. Luce 1891)

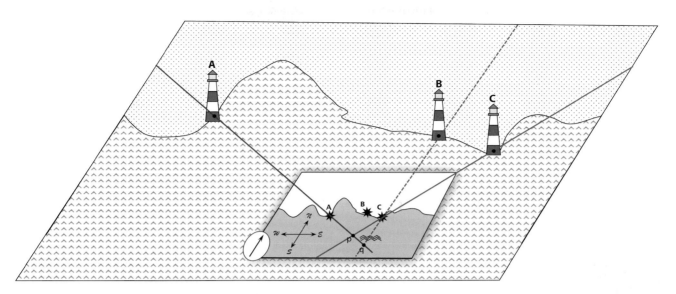

◻ Fig. 6.2 Position estimation using a map. Lines of sight from two lighthouses, **A** and **C**, and their corresponding positions on the map provide an estimate *p* of position. If we mistake lighthouse **B** for **C**, then we obtain an incorrect estimate *q*

The Phoenicians were navigating at sea more than 4000 years ago and they did not even have a compass – that was developed 2000 years later in China. The Phoenicians navigated with crude dead reckoning but wherever possible they used *additional information* to correct their position estimate – sightings of islands and headlands, primitive maps, and observations of the Sun and the Pole Star.

A landmark is a visible feature in the environment whose position is known with respect to some coordinate frame. ◻ Fig. 6.2 shows schematically a map and a number of lighthouse landmarks. We first of all use a compass to align the north axis of our map with the direction of the north pole. The direction of a single landmark constrains our position to lie along a line on the map. Sighting a second landmark places our position on another constraint line, and our position must be at their intersection – a process known as resectioning. ▸ For example, lighthouse **A** constrains us to lie along the blue line while lighthouse **C** constrains us to lie along the red line – our true position *p* is at the intersection of these two lines.

Resectioning is the estimation of position by measuring the bearing angles to known landmarks. Triangulation is the estimation of position by measuring the bearing angles to an unknown point from each of the landmarks.

However, this process is critically reliant on correctly associating the observed landmark with the corresponding feature on the map. If we mistake one lighthouse for another, for example we see **B** but think it is **C** on the map, then the red-dashed line leads to a significant error in estimated position – we would believe we were at *q* instead of *p*. This belief would lead us to overestimate our distance from the coastline. If we decided to sail toward the coast, we would run aground on rocks and be surprised since they were not where we expected them to be. This is unfortunately a very common error and countless ships have foundered because of this fundamental data association error. This is why lighthouses flash! In the eighteenth-century, technological advances enabled lighthouses to emit unique flashing patterns so that the identity of the particular lighthouse could be reliably determined and associated with a point on a navigation chart.

The earliest mariners had no maps, or lighthouses, or even compasses. They had to create maps as they navigated by incrementally adding new landmarks to their maps just beyond the boundaries of what was already known. It is perhaps not surprising that so many early explorers came to grief ▸ and that maps were tightly-kept state secrets.

Robots operating today in environments without GPS face *exactly* the same problems as the ancient navigators and, perhaps surprisingly, borrow heavily from navigational strategies that are centuries old. A robot's estimate of distance traveled

Magellan's 1519 expedition started with 237 men and 5 ships but most, including Magellan, were lost along the way. Only 18 men and 1 ship returned.

6

Excurse 6.2: Celestial Navigation

The position of celestial bodies in the sky is a predictable function of the time and the observer's latitude and longitude. This information can be tabulated and is known as ephemeris (meaning daily) and such data has been published annually in Britain since 1767 as *The Nautical Almanac* by HM Nautical Almanac Office. The elevation of a celestial body with respect to the horizon can be measured using a sextant, a handheld optical instrument.

Time and longitude are coupled, the star field one hour later is the same as the star field 15° to the east. However, the northern Pole Star, *Polaris*, or the *North Star*, is very close to the celestial pole and its elevation angle is independent of longitude and time, allowing latitude to be determined very conveniently from a single sextant measurement.

Solving the longitude problem was the greatest scientific challenge to European governments in the eighteenth century since it was a significant impediment to global navigation and maritime supremacy. The British Longitude Act of 1714 created a prize of £20,000 which spurred the development of nautical chronometers, clocks that could maintain high accuracy onboard ships. More than fifty years later John Harrison developed a suitable chronometer, a copy of which was used by Captain James Cook on his second voyage of 1772–1775. After a three year journey the error in estimated longitude was just 13 km. With accurate knowledge of time, the elevation angle of stars could be used to estimate latitude and longitude. This technological advance enabled global exploration and trade. (Image courtesy archive.org)

THE

NAUTICAL ALMANAC

AND

ASTRONOMICAL EPHEMERIS,

FOR THE YEAR 1781.

Published by ORDER of the

COMMISSIONERS OF LONGITUDE.

LONDON

PRINTED BY WILLIAM RICHARDSON, PRINTER;

AND SOLD BY

J. NOURSE, in the Strand, and Meff. MOUNT and PAGE on Tower-Hill,

Bookfellers to the faid COMMISSIONERS.

M DCC LXXIX.

[Price Three Shillings and Six Pence.]

will be imperfect and it may have no map, or perhaps an imperfect or incomplete map. Additional information from observed features in the world is critical to minimizing a robot's localization error, but the possibility of data association errors remains.

We can define the localization problem formally: x is the true, but unknown, position or pose of the robot and $\hat{x}$ is our best estimate of that. We also wish to know the *uncertainty* of the estimate that we can consider in statistical terms as the standard deviation associated with the estimate $\hat{x}$.

It is useful to describe the robot's estimated position in terms of a probability density function (PDF) over all possible positions of the robot. Some example PDFs are shown in ◘ Fig. 6.3 where the magnitude of the function $f(x, y)$ at any point (x, y) is the relative likelihood of the robot being at that position. ◄ A Gaussian function is commonly used and can be described succinctly in terms of intuitive concepts such as mean and standard deviation. The robot is most likely to be at the position of the peak (the mean) and increasingly less likely to be at positions further away from the peak. ◘ Fig. 6.3a shows a peak with a small standard deviation that indicates that the robot's position is very well known. There is an almost zero probability that the robot is at the point indicated by the vertical black line. In contrast, the peak in ◘ Fig. 6.3b has a large standard deviation that means that we are less certain about the position of the robot. There is a reasonable probability that the robot is at the point indicated by the vertical line. Using a PDF also allows for multiple hypotheses about the robot's position. For example, the PDF of ◘ Fig. 6.3c describes a robot that is quite certain that it is at one of four places. This is more useful than it seems at face value. Consider an indoor robot that has observed a door and there are four doors on the map. In the absence of any

By extension, for robot pose rather than position, the PDF $f(x, y, \theta)$ will be a 4-dimensional surface. The volume under the surface is always 1.

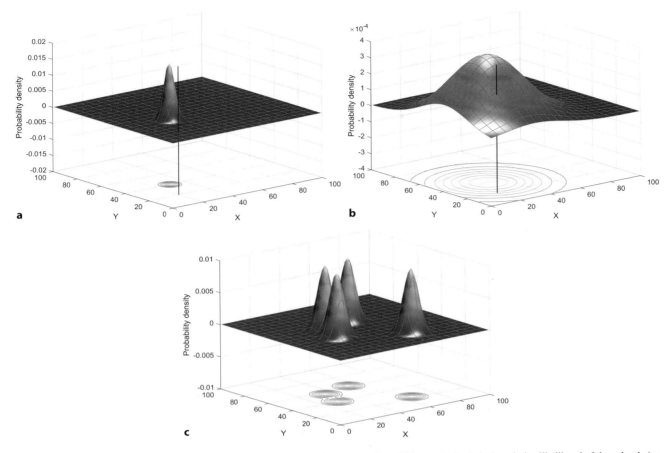

◻ Fig. 6.3 Notions of robot position and uncertainty in the xy-plane, θ is not considered. The vertical axis is the relative likelihood of the robot being at that position, sometimes referred to as belief or $bel(x, y)$. Contour lines are displayed on the lower plane. **a** The robot has low position uncertainty, $\sigma = 3$; **b** the robot has much higher position uncertainty, $\sigma = 20$; **c** the robot has multiple hypotheses for its position, each $\sigma = 4$

other information, the robot is equally likely to be in the vicinity of *any* of the four doors. We will revisit this approach in ▶ Sect. 6.2.4.

Determining the PDF based on knowledge of how the robot moves and its observations of the world is a problem in *estimation*, which we can define as:

》 *the process of inferring the value of some quantity of interest, x, by processing data that is in some way dependent on x.*

For example, a ship's navigator or a surveyor estimates position by measuring the bearing angles to known landmarks or celestial objects, and a GPS receiver estimates latitude and longitude by observing the time delay from moving satellites whose positions are known.

For our robot localization problem, the true and estimated state are vector quantities. Uncertainty is represented by a square covariance matrix whose diagonal elements are the variance of the corresponding state elements, and the off-diagonal elements are correlations that indicate how uncertainties between the state elements are connected.

❶ Probability of being at a pose

The value of a PDF at a point is *not* the probability of being at that position. Instead, consider a small region of the xy-plane, the volume under that region of the PDF is the probability of being in that region.

6

One of the earliest systems was LORAN, based on the British World War II GEE system. A network of transmitters around the world emitted synchronized radio pulses and a receiver measured the difference in arrival time between pulses from a pair of radio transmitters. Knowing the identity of two transmitters and the time difference (TD) constrains the receiver to lie along a hyperbolic curve shown on navigation charts as *TD lines*. Using a second pair of transmitters (which may include one of the first pair) gives another hyperbolic constraint curve, and the receiver must lie at the intersection of the two curves.

The Global Positioning System (GPS) was proposed in 1973 but did not become fully operational until 1995. It comprises around 30 active satellites orbiting the Earth in six planes at a distance of 20,200 km. A GPS receiver measures the time of travel of radio signals from four or more satellites whose orbital position is encoded in the GPS signal. With four known points in space and four measured time delays it is possible to compute the (x, y, z) position of the receiver and the time. If the GPS signals are received after reflecting off some surface, the distance traveled is longer and this will introduce an error in the position estimate. This is known as multi-path effect and is a common problem in large-scale industrial facilities.

Variations in the propagation speed of radio waves through the atmosphere is the main cause of error in the position estimate. However, these errors vary slowly with time and are approximately constant over large areas. This allows the error to be measured at a reference station and transmitted as an augmentation to compatible nearby receivers that can offset the error – this is known as Differential GPS (DGPS). This information can be transmitted via the internet, via coastal radio networks to ships, or by satellite networks such as WAAS, EGNOS or OmniSTAR to aircraft or other users. RTK GPS achieves much higher precision in time measurement by using phase information from the carrier signal. The original GPS system deliberately added error, euphemistically termed selective availability, to reduce its utility to military opponents but this *feature* was disabled in May 2000. Other satellite navigation systems include the European Galileo, the Chinese BeiDou, and the Russian GLONASS. The general term for these different systems is Global Navigation Satellite System (GNSS). GNSS made LORAN obsolete and much of the global network has been decommissioned. However recently, concern over the dependence on, and vulnerability of, GNSS had led to work on enhanced LORAN (eLORAN).

6.1   Dead Reckoning Using Odometry

Dead reckoning is the estimation of a robot's pose based on its estimated speed, direction, and time of travel with respect to a previous estimate.

An odometer is a sensor that measures distance traveled. For wheeled vehicles this is commonly achieved by measuring the angular rotation of the wheels using a sensor like an encoder. The direction of travel can be measured using an electronic compass, or the change in direction can be measured using a gyroscope or the difference in angular velocity of a left- and right-hand side wheel. These sensors are imperfect due to systematic errors such an incorrect wheel radius or gyroscope bias, and random errors such as slip between wheels and the ground. In robotics, the term odometry generally means measuring distance and direction of travel.

For robots without wheels, such as aerial and underwater robots, there are alternatives. Inertial navigation, introduced in ▸ Sect. 3.4, uses accelerometers and gyroscopes to estimate change in pose. Visual odometry is a computer-vision approach based on observations of the world moving past the robot, and this is covered in ▸ Sect. 14.8.3. Laser odometry is based on changes in observed point clouds measured with lidar sensors, and is covered in ▸ Sect. 6.3.2.

6.1.1   Modeling the Robot

The first step in estimating the robot's pose is to write a function, $f(\cdot)$, that describes how the robot's configuration changes from one time step to the next. A vehicle model such as (4.4) or (4.10) describes the progression of the robot's configuration as a function of its control inputs, however for real robots we rarely have direct access to these control inputs. Most robotic platforms have proprietary motion control systems that accept motion commands from the user such as desired position, and report odometry information.

Instead of using (4.4) or (4.10) directly, we will write a discrete-time model for the change of configuration based on odometry where $\boldsymbol{\delta}\langle k \rangle = (\delta_d, \delta_\theta)^\top$ is the distance traveled, δ_d, and change in heading over the preceding interval, δ_θ, and k is the time step. The initial pose is represented by an **SE**(2) matrix

$$\mathbf{T}\langle k \rangle = \begin{pmatrix} \cos\theta\langle k\rangle & -\sin\theta\langle k\rangle & x\langle k\rangle \\ \sin\theta\langle k\rangle & \cos\theta\langle k\rangle & y\langle k\rangle \\ 0 & 0 & 1 \end{pmatrix} .$$

We make a simplifying assumption that motion over the time interval is *small* so the order of applying the displacements is not significant. We choose to move forward in the robot's body frame x-direction, see ◘ Fig. 4.4, by δ_d, and then rotate by δ_θ giving the new pose

$$\mathbf{T}\langle k+1\rangle = \begin{pmatrix} \cos\theta\langle k\rangle & -\sin\theta\langle k\rangle & x\langle k\rangle \\ \sin\theta\langle k\rangle & \cos\theta\langle k\rangle & y\langle k\rangle \\ 0 & 0 & 1 \end{pmatrix} \begin{pmatrix} 1 & 0 & \delta_d \\ 0 & 1 & 0 \\ 0 & 0 & 1 \end{pmatrix} \begin{pmatrix} \cos\delta_\theta & -\sin\delta_\theta & 0 \\ \sin\delta_\theta & \cos\delta_\theta & 0 \\ 0 & 0 & 1 \end{pmatrix}$$

$$= \begin{pmatrix} \cos(\theta\langle k\rangle + \delta_\theta) & -\sin(\theta\langle k\rangle + \delta_\theta) & x\langle k\rangle + \delta_d\cos\theta\langle k\rangle \\ \sin(\theta\langle k\rangle + \delta_\theta) & \cos(\theta\langle k\rangle + \delta_\theta) & y\langle k\rangle + \delta_d\sin\theta\langle k\rangle \\ 0 & 0 & 1 \end{pmatrix}$$

which we can represent concisely as a configuration vector $\boldsymbol{q} = (x, y, \theta)$

$$\boldsymbol{q}\langle k+1\rangle = [\mathbf{T}\langle k+1\rangle]_{xy\theta} = \begin{pmatrix} x\langle k\rangle + \delta_d\cos\theta\langle k\rangle \\ y\langle k\rangle + \delta_d\sin\theta\langle k\rangle \\ \theta\langle k\rangle + \delta_\theta \end{pmatrix} \qquad (6.1)$$

which gives the new configuration in terms of the previous configuration and the odometry.

In practice, odometry is not perfect and we model the error by imagining a random number generator that corrupts the output of a perfect odometer. The measured output of the real odometer is the perfect, but unknown, odometry $(\delta_d, \delta_\theta)$ plus the output of the random number generator (v_d, v_θ). Such random errors are often referred to as noise, or more specifically as sensor noise. ▶ The random numbers are not known and cannot be measured, but we assume that we know the distribution from which they are drawn.

The robot's configuration at the next time step, including the odometry error, is

$$\boldsymbol{q}\langle k+1\rangle = \boldsymbol{f}(\boldsymbol{q}\langle k\rangle, \boldsymbol{\delta}\langle k\rangle, \boldsymbol{v}\langle k\rangle) = \begin{pmatrix} x\langle k\rangle + (\delta_d + v_d)\cos\theta\langle k\rangle \\ y\langle k\rangle + (\delta_d + v_d)\sin\theta\langle k\rangle \\ \theta\langle k\rangle + \delta_\theta + v_\theta \end{pmatrix} \qquad (6.2)$$

where k is the time step, $\boldsymbol{\delta}\langle k\rangle = (\delta_d, \delta_\theta)^\top \in \mathbb{R}^{2\times 1}$ is the odometry measurement and $\boldsymbol{v}\langle k\rangle = (v_d, v_\theta)^\top \in \mathbb{R}^{2\times 1}$ is the random measurement noise over the preceding interval. ▶

In the absence of any information to the contrary, we model the odometry noise as $\boldsymbol{v} = (v_d, v_\theta)^\top \sim N(0, \mathbf{V})$, a zero-mean multivariate Gaussian process ▶ with covariance

$$\mathbf{V} = \begin{pmatrix} \sigma_d^2 & 0 \\ 0 & \sigma_\theta^2 \end{pmatrix} .$$

We assume that the noise is additive. Another option is multiplicative noise which is appropriate if the noise magnitude is related to the signal magnitude.

The odometry noise is *inside* the model of our process (robot motion) and is referred to as process noise.

A normal distribution of angles on a circle is actually not possible since $\theta \in \mathbf{S}^1 \notin \mathbb{R}$, that is angles wrap around 2π. However, if the covariance for angular states is small, this problem is minimal. A normal-like distribution of angles on a circle is the von Mises distribution.

In reality, this is unlikely to be the case, since odometry distance errors tend to be worse when change of heading is high.

This odometry covariance matrix is diagonal, so errors in distance and heading are *independent*. ◄ Choosing **V** is not always easy but we can conduct experiments or make some reasonable engineering assumptions. In the examples which follow, we choose $\sigma_d = 2$ cm and $\sigma_\theta = 0.5°$ per sample interval which leads to a covariance matrix of

```
>> V = diag([0.02 deg2rad(0.5)].^2);
```

Objects derived from the `Vehicle` superclass provide a method `f()` that implements the appropriate configuration update equation. For the case of a robot with bicycle kinematics, with the motion model (4.4) and the configuration update (6.2), we create a `BicycleVehicle` object

```
>> robot = BicycleVehicle(Covariance=V)
robot =
BicycleVehicle object
  WheelBase=1, MaxSteeringAngle=0.5, MaxAcceleration=Inf
  Superclass: Vehicle
    MaxSpeed=1, dT=0.1, nhist=0
    Covariance=(0.0004, 7.61544e-05)
    Configuration: x=0, y=0, theta=0
```

which shows the default parameters such as the robot's wheel base, maximum speed, steering limits, and the sample interval which defaults to 0.1 s. The object provides a method to simulate motion over one time step

```
>> rng(0) % obtain repeatable results
>> odo = robot.step(1,0.3)
odo =
    0.1108    0.0469
```

We simulate the odometry noise using random numbers that have zero-mean and a covariance given by `V`. The random noise means that repeated calls to this function will return different values. To make the outputs predictable, we reset the random number generator in this example.

where we have specified a speed of 1 ms^{-1} and a steered-wheel angle of 0.3 rad. The function updates the robot's true configuration and returns a noise corrupted odometer reading. ◄ With a sample interval of 0.1 s, the robot reports that it is moving approximately 0.11 m over the interval and changing its heading by approximately 0.05 rad. The robot's true (but "unknown") configuration can be seen by

```
>> robot.q
ans =
    0.1000         0    0.0309
```

Given the reported odometry, we can estimate the configuration of the robot after one time step using (6.2). Assuming the previous configuration was [0, 0, 0], this is implemented by

```
>> robot.f([0 0 0],odo)
ans =
    0.1106    0.0052    0.0469
```

where the discrepancy with the true configuration is due to the use of a noisy odometry measurement.

For the scenarios that we want to investigate, we require the simulated robot to drive for a long time period within a defined spatial region. The `RandomDriver` class is a *driver* that steers the robot to randomly selected waypoints within a specified region. We create an instance of a driver object and connect it to the robot

```
>> robot.addDriver(RandomDriver([-10 10 -10 10],show=true))
```

where the argument specifies a working region that spans ± 10 m in the *x*- and *y*-directions. We can display an animation of the robot with its driver by

The number of history records is indicated by `nhist=` in the display of the object. The `qhist` property is a matrix that holds the vehicle state at each time step.

```
>> robot.run(100);
```

which runs the simulation for 100 time steps, or 10 s, by repeatedly calling the `step` method and saving a history of the true state of the robot within the `BicycleVehicle` object. ◄

6.1.2 Estimating Pose

The problem we face, just like the ship's navigator, is how to estimate our new pose given the previous pose and noisy odometry. The mathematical tool that we will use to solve this problem is the Kalman filter. This is a recursive algorithm that updates, at each time step, the *optimal* estimate of the unknown true configuration and the uncertainty associated with that estimate, based on the previous estimate, the control input, and noisy measurement data.

The Kalman filter is described more completely in ▶ App. H. This filter can be shown to provide the optimal estimate of the system state assuming that the noise is Gaussian and has a zero mean. The Kalman filter is formulated for linear systems, but our model of the robot's motion (6.2) is nonlinear – for this case we use the extended Kalman filter.

We will start by considering the one-dimensional example shown in ◻ Fig. 6.4a. The initial PDF has a mean value of $\hat{x}_{\langle k \rangle} = 2$ and a variance $P_{\langle k \rangle} = 0.25$. The odometry is 2 and its variance is $V = 0$ so the predicted PDF at the next time step is simply shifted to a mean value of $\hat{x}_{\langle k+1 \rangle} = 4$. In ◻ Fig. 6.4b the odometry variance is $V = 0.5$ so the predicted PDF still has a mean value of $\hat{x}_{\langle k+1 \rangle} = 4$, but the distribution is now wider, accounting for the uncertainty in odometry.

For the multi-dimensional robot localization problem, the state vector is the robot's configuration

$$\boldsymbol{x} = (x_v, \ y_v, \ \theta_v)^\top$$

Excurse 6.4: Reverend Thomas Bayes

Bayes (1702–1761) was a nonconformist Presbyterian minister who studied logic and theology at the University of Edinburgh. He lived and worked in Tunbridge-Wells in Kent and through his association with the 2nd Earl Stanhope, he became interested in mathematics and was elected to the Royal Society in 1742. After his death, his friend Richard Price edited and published his work in 1763 as *An Essay towards solving a Problem in the Doctrine of Chances*, which contains a statement of a special case of Bayes' theorem. Bayes is buried in Bunhill Fields Cemetery in London.

Bayes' theorem shows the relation between a conditional probability and its inverse: the probability of a hypothesis given observed evidence and the probability of that evidence given the hypothesis. Consider the hypothesis that the robot is at position X and it makes a sensor observation S of a known landmark. The *posterior* probability that the robot is at X given the observation S is

$$P(X|S) = \frac{P(S|X)P(X)}{P(S)}$$

where $P(X)$ is the *prior* probability that the robot is at X (not accounting for any sensory information). $P(S|X)$ is the

likelihood of the sensor observation S given that the robot is at X, and $P(S)$ is the prior probability of the observation S. The Kalman filter, and the Monte-Carlo estimator we discuss later in this chapter, are essentially two different approaches to solving this inverse problem.

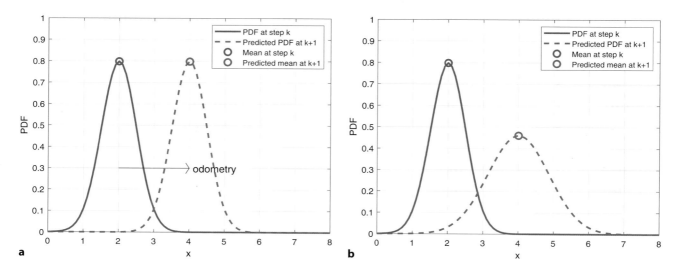

□ **Fig. 6.4** One-dimensional example of Kalman filter prediction. The initial PDF has a mean of 2 and a variance of 0.25, and the odometry value is 2. **a** Predicted PDF for the case of no odometry noise $V = 0$; **b** updated PDF for the case $V = 0.5$

The Kalman filter, □ Fig. 6.8, has two steps: *prediction* based on the model, and *update* based on sensor data. In this dead reckoning case, we use only the prediction equation.

and the prediction equations ◄

$$\hat{\boldsymbol{x}}^{+}\langle k+1 \rangle = \boldsymbol{f}(\hat{\boldsymbol{x}}\langle k \rangle, \, \hat{\boldsymbol{u}}\langle k \rangle) \tag{6.3}$$

$$\hat{\mathbf{P}}^{+}\langle k+1 \rangle = \mathbf{F}_x \hat{\mathbf{P}}\langle k \rangle \mathbf{F}_x^{\top} + \mathbf{F}_v \hat{\mathbf{V}} \mathbf{F}_v^{\top} \tag{6.4}$$

describe how the state and covariance evolve with time. The term $\hat{\boldsymbol{x}}^{+}\langle k+1 \rangle$ indicates an estimate of $\boldsymbol{x}$ at time $k + 1$ based on information up to, and including, time k. $\hat{\boldsymbol{u}}\langle k \rangle$ is the input to the process, which in this case is the measured odometry, so $\hat{\boldsymbol{u}}\langle k \rangle = \boldsymbol{\delta}\langle k \rangle$. $\hat{\mathbf{V}}$ is our *estimate* of the covariance of the odometry noise, which we do not know.

Excurse 6.5: The Kalman Filter

Rudolf Kálmán (1930–2016) was a mathematical system theorist born in Budapest. He obtained his bachelors and master's degrees in electrical engineering from MIT, and Ph.D. in 1957 from Columbia University. He worked as a Research Mathematician at the Research Institute for Advanced Study, in Baltimore, from 1958–1964 where he developed his ideas on estimation. These were met with some skepticism among his peers, and he chose a mechanical (rather than electrical) engineering journal for his paper, *A new approach to linear filtering and prediction problems*, because "When you fear stepping on hallowed ground with entrenched interests, it is best to go sideways". He received many awards including the IEEE Medal of Honor, the Kyoto Prize, and the Charles Stark Draper Prize.

Stanley F. Schmidt (1926–2015) was a research scientist who worked at NASA Ames Research Center and was an early advocate of the Kalman filter. He developed the first implementation as well as the nonlinear version now known as the extended Kalman filter. This led to its incorporation in the Apollo navigation computer for trajectory estimation.

(Extract from Kálmán's famous paper (1960) reprinted with permission of ASME)

The covariance matrix is symmetric

$$
\hat{\mathbf{P}} = \left(\begin{array}{cc|c} \sigma_x^2 & \mathrm{cov}(x,y) & \mathrm{cov}(x,\theta) \\ \mathrm{cov}(x,y) & \sigma_y^2 & \mathrm{cov}(y,\theta) \\ \hline \mathrm{cov}(x,\theta) & \mathrm{cov}(y,\theta) & \sigma_\theta^2 \end{array} \right) \tag{6.5}
$$

and represents uncertainty in the estimated robot configuration, as well as dependence between elements of the configuration. Variance ▶ is concerned with a single state and $\mathrm{cov}(u,u) = \sigma_u^2$. Covariance between the states u and v is written as $\mathrm{cov}(u,v) = \mathrm{cov}(v,u)$, and is zero if u and v are independent. The highlighted top-left corner is the covariance of robot *position* rather than configuration.

The two terms in (6.4) have a particular form, a similarity transformation, $\mathbf{F}\,\boldsymbol{\Sigma}\,\mathbf{F}^\top$ where $\boldsymbol{\Sigma}$ is a covariance matrix. For a function $\boldsymbol{y} = \boldsymbol{f}(\boldsymbol{x})$ the covariance of $\boldsymbol{x}$ and $\boldsymbol{y}$ are related by $\boldsymbol{\Sigma}_y = \mathbf{F}\,\boldsymbol{\Sigma}_x\,\mathbf{F}^\top$ where $\mathbf{F}$ is a Jacobian matrix – the vector version of a derivative which is reviewed in ▶ App. E. The Jacobian matrices are obtained by differentiating (6.2) and evaluating the result at $\boldsymbol{v} = 0$ giving ▶

> Covariance is defined as $\mathrm{cov}(u,v) = \sum_{i=1}^{N}(u_i - \bar{u})(v_i - \bar{v})$.

> The noise value $\boldsymbol{v}$ cannot be measured, so we evaluate the derivative using its mean value $\boldsymbol{v} = 0$.

$$
\mathbf{F}_x = \left.\frac{\partial \boldsymbol{f}}{\partial \boldsymbol{x}}\right|_{v=0} = \begin{pmatrix} 1 & 0 & -\delta_d \sin\theta_v \\ 0 & 1 & \delta_d \cos\theta_v \\ 0 & 0 & 1 \end{pmatrix} \tag{6.6}
$$

$$
\mathbf{F}_v = \left.\frac{\partial \boldsymbol{f}}{\partial \boldsymbol{v}}\right|_{v=0} = \begin{pmatrix} \cos\theta_v & 0 \\ \sin\theta_v & 0 \\ 0 & 1 \end{pmatrix} \tag{6.7}
$$

which are functions of the current robot state and odometry. We drop the time step notation $\langle k \rangle$ to reduce clutter. All objects of the vehicle superclass provide methods Fx and Fv to compute these Jacobians, for example

```
>> robot.Fx([0 0 0],[0.5 0.1])
ans =
    1.0000         0         0
         0    1.0000    0.5000
         0         0    1.0000
```

where the first argument is the configuration at which the Jacobian is computed and the second is the odometry, $(\delta_d, \delta_\theta)$.

To simulate the robot and the EKF using the RVC Toolbox, we must define the initial covariance. We choose a diagonal matrix

```
>> P0 = diag([0.05 0.05 deg2rad(0.5)].^2);
```

written in terms of the standard deviation of position and orientation which are respectively 5 cm and 0.5° in orientation – this implies that we have a good idea of the initial configuration. We pass this to the constructor for an EKF object

```
>> ekf = EKF(robot,V,P0);
```

along with the robot kinematic model and the odometry covariance matrix, defined earlier.

Running the filter for 400 time steps

```
>> rng(0) % obtain repeatable results
>> ekf.run(400);
```

drives the robot along a random path starting at the origin. At each time step the filter updates the state estimate using various methods provided by the robot object.

We can plot the true path taken by the robot by

```
>> clf
>> robot.plotxy
```

6

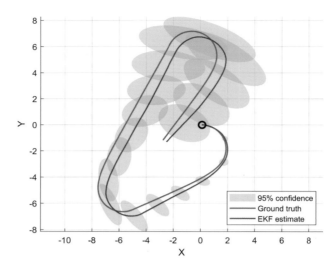

□ **Fig. 6.5** Dead reckoning using the EKF showing the true path and estimated path of the robot. 95% confidence ellipses (green) of the robot's pose are also shown. The robot starts at the origin

and the filter's estimate of the path stored within the EKF object,

```
>> hold on
>> ekf.plotxy
```

and these are shown in ■ Fig. 6.5. We see some divergence between the true and estimated robot path.

The covariance at time step 150 is

```
>> P150 = ekf.history(150).P
P150 =
    0.0868    0.0163   -0.0070
    0.0163    0.2676    0.0399
   -0.0070    0.0399    0.0115
```

The matrix is symmetric, and the diagonal elements are the estimated variance associated with the states, that is σ_x^2, σ_y^2, and σ_θ^2 respectively. The standard deviation σ_x of the PDF associated with the robot's x-coordinate is

```
>> sqrt(P150(1,1))
ans =
    0.2946
```

There is a 95% chance that the robot's x-coordinate is within the $\pm 2\sigma$ bound or ± 0.59 m in this case. We can compute uncertainty for y and θ similarly.

The off-diagonal terms are correlation coefficients and indicate that the uncertainties between the corresponding variables are related. For example, the value $p_{1,3} = p_{3,1} = -0.1253$ indicates that the uncertainties in x and θ are related – error in heading angle causes error in x-position and vice versa. Conversely, new information about θ can be used to correct θ as well as x. The uncertainty in position is described by the top-left 2×2 covariance submatrix of $\hat{\mathbf{P}}$, which can be interpreted as an ellipse defining a confidence bound on position. We can overlay such ellipses on the plot by

```
>> ekf.plotellipse(fillcolor="g",alpha=0.3)
```

The alpha name-value pair plots the ellipses with some transparency (1 is opaque, 0 is fully transparent) to distinguish them when they are overlapping.

as shown in ■ Fig. 6.5. ◄ These correspond to the default 95% confidence bound and are plotted by default every 10 time steps. The robot started at the origin and as it progresses, we see that the ellipses become larger as the estimated uncertainty increases. The ellipses only show the uncertainty in x- and y-position, but uncertainty in θ also grows.

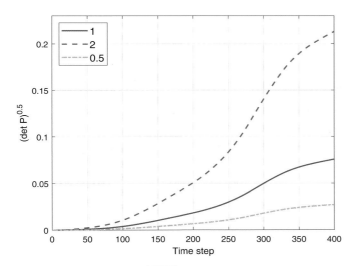

⬛ Fig. 6.6 Overall uncertainty given by $\sqrt{\det(\hat{\mathbf{P}})}$ versus time shows monotonically increasing uncertainty. Changing the magnitude of $\hat{\mathbf{V}}$ alters the rate of uncertainty growth. Curves are shown for $\hat{\mathbf{V}} = \alpha\mathbf{V}$ where $\alpha = 0.5, 1, 2$

The estimated uncertainty in position and heading ▷ is given by $\sqrt{\det(\hat{\mathbf{P}})}$, which we can plot as a function of time

```
>> clf
>> ekf.plotP;
```

as the solid curve in ⬛ Fig. 6.6. We observe that the uncertainty never decreases, and this is because in (6.4) we keep adding a positive-definite matrix ▷ (the second term) to the estimated covariance.

❗ V vs V̂

We have used the odometry covariance matrix v twice. The first usage, in the `BicycleVehicle` constructor, is the covariance $\mathbf{V}$ of the zero-mean Gaussian noise source that is added to the true odometry to *simulate* odometry error in (6.2). In a real application this noise is generated by some physical process *hidden inside* the robot and we would not know its parameters. The second usage, in the EKF

The elements of $\mathbf{P}$ have different units: m^2 and rad^2. The uncertainty is therefore a mixture of spatial and angular uncertainty with an implicit weighting. If the range of the position variables $x, y \gg \pi$ then positional uncertainty dominates.

A positive-definite matrix can be thought of as the matrix equivalent of a positive number.

Excurse 6.6: Error Ellipses
We consider the PDF of the robot's position (ignoring orientation) such as shown in ⬛ Fig. 6.3 to be a 2-dimensional Gaussian probability density function

$$p(\boldsymbol{x}) = \frac{1}{(2\pi)\det(\mathbf{P}_{xy})^{1/2}}e^{\{-\frac{1}{2}(\boldsymbol{x}-\boldsymbol{\mu}_x)^{\top}\mathbf{P}_{xy}^{-1}(\boldsymbol{x}-\boldsymbol{\mu}_x)\}}$$

where $\boldsymbol{x} = (x, y)^{\top}$ is the position of the robot, $\boldsymbol{\mu}_x = (\hat{x}, \hat{y})^{\top}$ is the estimated mean position and $\mathbf{P}_{xy} \in \mathbb{R}^{2\times2}$ is the position covariance matrix, the top left of the covariance matrix $\mathbf{P}$ as shown in (6.5). A horizontal cross-section of the PDF is a contour of constant probability which is an ellipse defined by the points $\boldsymbol{x}$ such that

$$(\boldsymbol{x} - \boldsymbol{\mu}_x)^{\top}\mathbf{P}_{xy}^{-1}(\boldsymbol{x} - \boldsymbol{\mu}_x) = s .$$

Such error ellipses are often used to represent positional uncertainty as shown in ⬛ Fig. 6.5. A large ellipse corresponds to a wider PDF peak and less certainty about position. To obtain a particular confidence contour (for example, 99%) we choose s as the inverse of the χ^2 cumulative distribution function for 2 degrees of freedom. In MATLAB®, we can use the function `chi2inv(C,2)`, where $C \in [0, 1]$ is the confidence value. Such confidence values can be passed to several EKF methods when specifying error ellipses.

A handy scalar measure of total position uncertainty is the area of the ellipse $\pi r_1 r_2$ where the radii $r_i = \sqrt{\lambda_i}$ and λ_i are the eigenvalues of $\mathbf{P}_{xy}$. Since $\det(\mathbf{P}_{xy}) = \pi\lambda_i$, the ellipse area – the scalar uncertainty – is proportional to $\sqrt{\det(\mathbf{P}_{xy})}$. See also ▷ App. C.1.4 and G.

6

constructor is $\hat{\mathbf{V}}$, which is our best *estimate* of the odometry covariance and is used in the filter's state covariance update equation (6.4).

The relative values of $\mathbf{V}$ and $\hat{\mathbf{V}}$ control the rate of uncertainty growth as shown in ◻ Fig. 6.6. If $\hat{\mathbf{V}} > \mathbf{V}$ then $\hat{\mathbf{P}}$ will be larger than it should be and the filter is pessimistic – it overestimates uncertainty and is less certain than it should be. If $\hat{\mathbf{V}} < \mathbf{V}$ then $\hat{\mathbf{P}}$ will be smaller than it should be and the filter will be *overconfident* of its estimate – the actual uncertainty is greater than the estimated uncertainty. In practice some experimentation is required to determine the appropriate value for the estimated covariance.

6.2   Landmark Maps

One common way to represent the robot's environment is to keep track of the positions and covariances of distinct landmarks, which are salient and recognizable features. In indoor environments, landmarks may be doors and hallway corners; in outdoor environments, tree trunks, or intersections. When dealing with landmarks, we typically assume that we have a sensor that can directly measure the landmark's position, or the landmark position can be determined through sensor processing. This section focuses on how to solve the localization, mapping, and SLAM problems for *landmark maps*. Another common map representation, an occupancy map, is covered in ▶ Sect. 6.3.

6.2.1   Localizing in a Landmark Map

We have seen how uncertainty in position grows without bound using dead reckoning alone. The solution, as the Phoenicians worked out 4000 years ago, is to use additional information from observations of known features, or *landmarks*, in the world.

Continuing the one-dimensional example of ◻ Fig. 6.4b, we now assume that the robot also has a sensor that can measure the robot's position with respect to some external reference. The PDF of the sensor measurement is shown as the yellow-dotted curve in ◻ Fig. 6.7a and has a mean $\hat{x} = 5$ as well as some un-

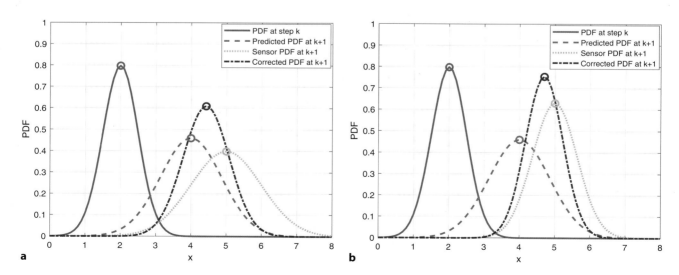

◻ **Fig. 6.7** One-dimensional example of Kalman filter prediction, the blue and red-dashed curves are from ◻ Fig. 6.4b. A sensor now provides a measurement to improve the prediction: **a** for the case of sensor variance $\hat{W} = 1$; **b** for the case of sensor variance $\hat{W} = 0.3$

certainty due to its own variance $\hat{W} = 1$. The mean of the sensor measurement is different to the mean of the prediction based on odometry, $\hat{x} = 4$. Perhaps, the odometry was under-reporting due to wheel slippage. We now have two PDFs for the robot's position, the red-dashed prediction from odometry and the yellow-dotted measurement from the sensor. The corrected PDF is the *product* of these two Gaussians, which is also a Gaussian, and this is shown as the purple-dash-dotted curve. We have combined two uncertain estimates to produce a new, optimal, estimate – there is a noticeable reduction in the uncertainty associated with the robot's position. It's notable that the corrected PDF has less uncertainty than either the prediction or the sensor PDF. This makes sense because we have two different methods that are telling us similar information (predicted and measured mean), then we should have more confidence in the combined answer. Using a better sensor, with a variance of $\hat{W} = 0.3$, as shown in ◻ Fig. 6.7b, the corrected prediction shown by the purple-dash-dotted curve has less variance and is closer to the mean of the sensor measurement. Given two estimates, one from odometry and one from the sensor, we place more trust in the estimate with lower variance.

For a robotics problem we consider that the robot is able to observe a set of landmarks using an onboard sensor. For the moment we will consider that the position of the landmarks is known, and we will use a map that contains N fixed but randomly located landmarks whose positions are known. The RVC Toolbox provides a `LandmarkMap` object

```
>> rng(0) % obtain repeatable results
>> map = LandmarkMap(20,10)
map =
LandmarkMap object
  20 landmarks
  dimension 10.0
```

that in this case contains $N = 20$ landmarks, randomly spread over a region spanning ± 10 m in the x- and y-directions, and this can be displayed by

```
>> map.plot
```

The robot is equipped with a sensor that provides observations of the landmarks *with respect to the robot* as described by

$$z = h(q, \ p_i) \tag{6.8}$$

where $q = (x_v, y_v, \theta_v)^\top$ is the robot configuration, and $p_i = (x_i, y_i)^\top$ is the *known* position of the ith landmark in the world frame.

To make this tangible, we will consider a common type of sensor that measures the range and bearing angle to a landmark in the environment, for instance a radar or lidar ▶ such as shown in ◻ Fig. 6.19b–d. The sensor is mounted onboard the robot so the observation of the ith landmark is

$$z = h(q, \ p_i) = \begin{pmatrix} \sqrt{(y_i - y_v)^2 + (x_i - x_v)^2} \\ \tan^{-1} \frac{y_i - y_v}{x_i - x_v} - \theta_v \end{pmatrix} + \begin{pmatrix} w_r \\ w_\beta \end{pmatrix} \tag{6.9}$$

where $z = (r, \beta)^\top$ and r is the range, β the bearing angle, and $w = (w_r, w_\beta)^\top$ is a zero-mean Gaussian random variable that models errors in the sensor

$$\begin{pmatrix} w_r \\ w_\beta \end{pmatrix} \sim N(0, \mathbf{W}), \quad \mathbf{W} = \begin{pmatrix} \sigma_r^2 & 0 \\ 0 & \sigma_\beta^2 \end{pmatrix}.$$

Radar (**ra**dio **d**etection **a**nd **r**anging) and lidar (**li**ght **d**etection **a**nd **r**anging) measure distance by timing the round trip time for a microwave signal or infrared pulse that reflects from an object and returns to the sensor.

It also indicates that covariance is independent of range but in reality covariance may increase with range since the strength of the return signal, lidar or radar, drops rapidly ($1/r^4$) with distance r to the target.

A subclass of Sensor.

The landmark is chosen randomly from the set of visible landmarks, those that are within the field of view and the minimum and maximum range limits. If no landmark is visible, the method returns 0 for i and [] for z.

The constant diagonal covariance matrix indicates that range and bearing errors are independent. ◄

For this example, we set the sensor uncertainty to be $\sigma_r = 0.1$ m and $\sigma_\beta = 1°$ giving a sensor covariance matrix

```
>> W = diag([0.1 deg2rad(1)].^2);
```

We model this type of sensor with a LandmarkSensor object ◄

```
>> sensor = LandmarkSensor(robot,map,covar=W, ...
>>    range=10,angle=[-pi/2 pi/2]);
```

which is connected to the robot and map objects, and we specify the sensor covariance matrix W along with the maximum range and the bearing angle limits. The reading method provides the range and bearing to a randomly selected visible ◄ landmark along with its identity, for example

```
>> [z,i] = sensor.reading
z =
    0.7754
    0.1170
i =
    14
```

The identity is an integer $i = 1, \ldots, 20$ since the map was created with 20 landmarks. We have avoided the data association problem by assuming that we know the identity of the sensed landmark. The position of landmark 11 can be looked up in the map

```
>> map.landmark(14)
ans =
    4.8626
   -2.1555
```

Using (6.9) the robot can estimate the range and bearing angle to the landmark based on its own estimated position and the known position of the landmark from the map. The difference between the observation z and the estimated observation

$$v = z_{\langle k \rangle} - h\big(\hat{x}^+_{\langle k \rangle}, p_i\big) \tag{6.10}$$

indicates an error in the robot's pose estimate $\hat{x}^+$ – the robot isn't where it *thought* it was. This difference is called the *innovation*, since it represents valuable *new* information, and is key to the operation of the Kalman filter.

The second step of the Kalman filter updates the *predicted* state and covariance computed using (6.3) and (6.4), and denoted by the superscript $^+$. The state estimate update

$$\hat{x}_{\langle k+1 \rangle} = \hat{x}^+_{\langle k+1 \rangle} + K v_{\langle k+1 \rangle} \tag{6.11}$$

uses a gain term $\mathbf{K}$ to add some *amount* of the innovation to the predicted state estimate so as to reduce the estimation error. We could choose some constant value of $\mathbf{K}$ to drive the innovation toward zero, but there is a better way

$$\mathbf{K} = \mathbf{P}^+_{\langle k+1 \rangle} \mathbf{H}_x^\top \Big(\mathbf{H}_x \mathbf{P}^+_{\langle k+1 \rangle} \mathbf{H}_x^\top + \mathbf{H}_w \hat{\mathbf{W}} \mathbf{H}_w^\top\Big)^{-1} \tag{6.12}$$

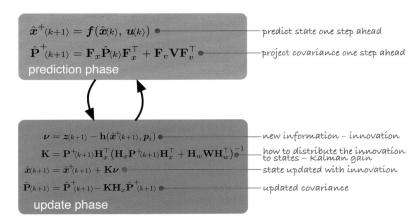

Fig. 6.8 Summary of Extended Kalman Filter (EKF) algorithm showing the prediction and update phases

where $\hat{\mathbf{W}}$ is the estimated covariance of the sensor noise and $\mathbf{H}_x$ and $\mathbf{H}_w$ are Jacobians. ▶ $\mathbf{K}$ is called the Kalman gain matrix and it *distributes* the innovation from the landmark observation, a 2-vector, to update every element of the state vector – the position and orientation of the robot. The off-diagonal elements of the covariance matrix are the error correlations between the states, and the Kalman filter uses these so that innovation in one state optimally updates the other correlated states. The size of the state estimate update depends on the relevant sensor covariance – if uncertainty is low then the state estimate has a larger update than if the uncertainty was high.

The covariance update also depends on the Kalman gain

This can also be written as

$$\mathbf{S} = \mathbf{H}_x \mathbf{P}^+{}_{\langle k+1\rangle} \mathbf{H}_x^\top + \mathbf{H}_w \hat{\mathbf{W}} \mathbf{H}_w^\top$$

$$\mathbf{K} = \mathbf{P}^+{}_{\langle k+1\rangle} \mathbf{H}_x^\top \mathbf{S}^{-1}$$

where $\mathbf{S}$ is the estimated covariance of the innovation.

$$\hat{\mathbf{P}}_{\langle k+1\rangle} = \hat{\mathbf{P}}^+{}_{\langle k+1\rangle} - \mathbf{K}\mathbf{H}_x \hat{\mathbf{P}}^+{}_{\langle k+1\rangle} \tag{6.13}$$

and it is important to note that the second term in (6.13) is *subtracted* from the estimated covariance. This provides a means for covariance to decrease, and that was not possible for the dead reckoning case of (6.4).

The Jacobians $\mathbf{H}_x$ and $\mathbf{H}_w$ are obtained by differentiating (6.9) yielding

$$\mathbf{H}_x = \left.\frac{\partial \boldsymbol{h}}{\partial \boldsymbol{x}}\right|_{\boldsymbol{w}=0} = \begin{pmatrix} -\frac{x_i - x_v}{r} & -\frac{y_i - y_v}{r} & 0 \\ \frac{y_i - y_v}{r^2} & -\frac{x_i - x_v}{r^2} & -1 \end{pmatrix} \tag{6.14}$$

which is a function of landmark and robot position, and landmark range; and

$$\mathbf{H}_w = \left.\frac{\partial \boldsymbol{h}}{\partial \boldsymbol{w}}\right|_{\boldsymbol{w}=0} = \begin{pmatrix} 1 & 0 \\ 0 & 1 \end{pmatrix}. \tag{6.15}$$

The `LandmarkSensor` object above includes methods h to implement (6.9) and Hx and Hw to compute the Jacobians. The EKF comprises two phases: prediction and update, and these are summarized in ☐ Fig. 6.8.

We now have all the pieces to build an estimator that uses odometry and observations of map features and the full implementation is

```
>> rng(0) % obtain repeatable results
>> map = LandmarkMap(20,10);
>> V = diag([0.02 deg2rad(0.5)].^2);
>> robot = BicycleVehicle(Covariance=V);
```

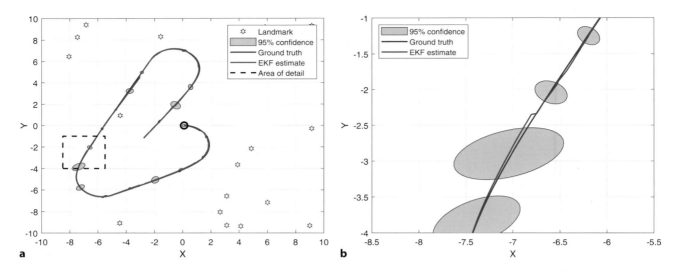

Fig. 6.9 **a** EKF localization with dead reckoning and landmarks, showing the true and estimated path of the robot. Black stars are landmarks, and the green ellipses are 95% confidence bounds. The robot starts at the origin; **b** Closeup of the robot's true and estimated path showing the effect of a landmark observation reducing the covariance ellipse

```
>> robot.addDriver(RandomDriver(map.dim,show=true));
>> sensor = LandmarkSensor(robot,map,covar=W, ...
>>   range=4,angle=[-pi/2 pi/2],animate=true);
>> P0 = diag([0.05 0.05 deg2rad(0.5)].^2);
>> ekf = EKF(robot,V,P0,sensor,W,map);
```

The second argument to the `LandmarkMap` constructor sets the map span to $\pm 10\,\text{m}$ in each dimension and this information is also used by the `RandomDriver` and `LandmarkSensor` objects. Running the simulation for 400 time steps

```
>> ekf.run(400)
```

shows an animation of the robot moving and observations being made to the landmarks. We plot the saved results

```
>> clf
>> map.plot
>> robot.plotxy("b");
>> ekf.plotxy("r");
>> ekf.plotellipse(fillcolor="g",alpha=0.3)
```

which are shown in ◘ Fig. 6.9a. We see that the error ellipses grow as the robot moves, just as we observed for the dead reckoning case of ◘ Fig. 6.5, and then shrink due to a landmark observation that corrects the robot's estimated state. The close-up view in ◘ Fig. 6.9b shows the robot's actual and estimated path, and the latter has a sharp kink at the correction (around position $[-6.75, -2.5]$). New information, in addition to odometry, has been used to correct the state in the Kalman filter update phase. The error ellipses are now much smaller, and many can hardly be seen.

◘ Fig. 6.10a shows that the overall uncertainty is no longer growing monotonically. When the robot observes a landmark, the estimated covariance is dramatically reduced. ◘ Fig. 6.10b shows the error associated with each component of pose and the pink background is the estimated 95% confidence bound (derived from the covariance matrix) and we see that the error is mostly within this envelope. The landmark observations are plotted below and we see that the confidence bounds are tight. This indicates low uncertainty while landmarks are being observed, but that they start to grow during periods with no observations.

The EKF is an extremely powerful tool that allows data from many and varied sensors to update the state, which is why the estimation problem is also referred

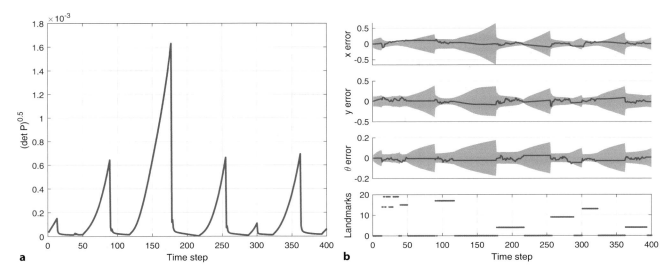

Fig. 6.10 **a** Covariance magnitude as a function of time. Overall uncertainty is given by $\sqrt{\det(\mathbf{P})}$ and shows that uncertainty does not continually increase with time; **b** Top: pose estimation error with 95% confidence bound shown in pink; bottom: scatter plot of observed landmarks and periods of no observations can be clearly seen

Excurse 6.7: Data Association

So far, we have assumed that we can determine the identity of the observed landmark, but in reality this is rarely the case. Instead, we can compare our observation to the predicted position of all currently known landmarks and decide which landmark it is most likely to be, or whether it is a new landmark. This decision needs to consider the uncertainty associated with the robot's pose, the sensor measurement, and the landmarks in the map. This is the data association problem. Errors in this step are potentially catastrophic – incorrect innovation would introduce error in the state of the robot that increases the chance of an incorrect data association on the next cycle. In practice, filters only use a landmark when there is a very high confidence in its estimated identity – a process that involves Mahalanobis distance and χ^2 confidence tests. If the situation is ambiguous, it is best not to use the landmark – a bad update can do more harm than good.

An alternative is to use a multi-hypothesis estimator, such as the particle filter that we will discuss in ▶ Sect. 6.2.4, which can model the possibility of observing landmark A or landmark B, and future observations will reinforce one hypothesis and weaken the others. The Extended Kalman filter uses a Gaussian probability model, with just one peak (or mode), which limits it to holding only one hypothesis about the robot's pose. (Picture: the wreck of the Tararua, 1881)

Excurse 6.8: Artificial Landmarks

For some applications it might be appropriate to use artificial landmarks, or fiducial markers, that can be cheaply sensed by a camera. These are not only visually distinctive in the environment, but they can also encode a unique identity. Commonly used 2-dimensional bar codes include ArUco markers, ARTags, and AprilTags (shown here). If the camera is calibrated (see ▶ Sect. 13.2), we can determine the range to the marker and its surface normal. MATLAB supports reading AprilTags with the `readAprilTag` function and its usage is discussed in ▶ Sect. 13.6.1.

to as sensor fusion. For example, heading angle from a compass, yaw rate from a gyroscope, target bearing angle from a camera, and position from GPS could all be used to update the state. For each sensor we only need to provide the observation function $h(\cdot)$, the Jacobians $\mathbf{H}_x$ and $\mathbf{H}_w$, and some estimate of the sensor covariance $\hat{\mathbf{W}}$. The function $h(\cdot)$ can be nonlinear and even noninvertible – the EKF will do the rest.

ⓘ W vs $\hat{\mathbf{W}}$

As discussed earlier for the odometry covariance matrix $\mathtt{V}$, we also use the sensor covariance $\mathtt{W}$ twice. The first usage, in the constructor for the `LandmarkSensor` object, is the covariance $\mathbf{W}$ of the zero-mean Gaussian noise that is added to the computed range and bearing to *simulate* sensor error as in (6.9). In a real application this noise is generated by some physical process *hidden inside* the sensor and we would not know its parameters. The second usage, $\hat{\mathbf{W}}$ is our best *estimate* of the sensor covariance which is used by the Kalman filter (6.12).

The relative magnitude of $\mathbf{W}$ and $\hat{\mathbf{W}}$ affects the rate of covariance growth. Large $\hat{\mathbf{W}}$ indicates that less trust should be given to the sensor measurements, so the covariance reduction at each observation will be lower.

6.2.2 **Creating a Landmark Map**

So far, we have taken the existence of the map for granted, and this is understandable given that maps today are common and available for free via the internet. Nevertheless, somebody, or something, has to create the maps we use. Our next example considers the problem of creating a map – determining the positions of the landmarks – in an environment that the robot is moving through.

As before, we have a range and bearing sensor mounted on the robot, which imperfectly measures the position of landmarks with respect to the robot. We will assume that the sensor can determine the identity of each observed landmark, and that the robot's pose is known perfectly – it has ideal localization. This is unrealistic but this scenario is an important conceptual stepping stone to the next section. ◄

A close and realistic approximation would be a high-quality RTK GPS+INS system operating in an outdoor environment with no buildings or hills to obscure satellites, or a motion capture (MoCap) system in an indoor environment.

Since the robot's pose is known perfectly, we do not need to estimate it, but we do need to estimate the coordinates of the landmarks. For this problem the state vector comprises the coordinates of the M landmarks that have been observed so far

$$\boldsymbol{x} = (x_1, \ y_1, \ x_2, \ y_2, \ \ldots \ x_M, \ y_M)^{\top} \in \mathbb{R}^{2M \times 1} \ .$$

The covariance matrix is symmetric

$$\hat{\mathbf{P}} = \begin{pmatrix} \sigma_{x_1}^2 & \mathrm{cov}(x_1, y_1) & \mathrm{cov}(x_1, x_2) & \mathrm{cov}(x_1, y_2) & \cdots \\ \mathrm{cov}(x_1, y_1) & \sigma_{y_1}^2 & \mathrm{cov}(y_1, x_2) & \mathrm{cov}(y_1, y_2) & \cdots \\ \mathrm{cov}(x_1, x_2) & \mathrm{cov}(x_1, y_2) & \sigma_{x_2}^2 & \mathrm{cov}(x_2, y_2) & \cdots \\ \mathrm{cov}(y_1, x_2) & \mathrm{cov}(y_1, y_2) & \mathrm{cov}(x_2, y_2) & \sigma_{y_2}^2 & \\ \vdots & \vdots & \vdots & \vdots & \ddots \end{pmatrix} \quad (6.16)$$

and each 2×2 block on the diagonal is the covariance of the corresponding landmark position. The off-diagonal blocks are the covariance between the landmarks.

Initially $M = 0$ and is incremented every time a previously unseen landmark is observed. The state vector has a variable length since we do not know in advance how many landmarks exist in the environment. The estimated covariance $\hat{\mathbf{P}} \in \mathbb{R}^{2M \times 2M}$ also has variable size.

The prediction equation is straightforward in this case, since the landmarks are assumed to be stationary

$$\hat{\boldsymbol{x}}^+{}_{\langle k+1\rangle} = \hat{\boldsymbol{x}}_{\langle k\rangle} \tag{6.17}$$

$$\hat{\mathbf{P}}^+{}_{\langle k+1\rangle} = \hat{\mathbf{P}}_{\langle k\rangle} \tag{6.18}$$

We introduce the function $\boldsymbol{g}(\cdot)$, which is the inverse of $\boldsymbol{h}(\cdot)$ and returns the world coordinates of the observed landmark based on the known robot pose (x_v, y_v, θ_v) and the sensor observation (this is also known as the inverse measurement function or inverse sensor model)

$$\boldsymbol{g}(\boldsymbol{x}, \boldsymbol{z}) = \begin{pmatrix} x_v + r\cos(\theta_v + \beta) \\ y_v + r\sin(\theta_v + \beta) \end{pmatrix}. \tag{6.19}$$

Since $\hat{\boldsymbol{x}}$ has a variable length we need to extend the state vector and the covariance matrix when we encounter a previously unseen landmark. The state vector is extended by the function $\boldsymbol{y}(\cdot)$

$$\boldsymbol{x}'_{\langle k\rangle} = \boldsymbol{y}(\boldsymbol{x}_{\langle k\rangle},\ \boldsymbol{z}_{\langle k\rangle},\ \boldsymbol{x}_{v\langle k\rangle}) \tag{6.20}$$

$$= \begin{pmatrix} \boldsymbol{x}_{\langle k\rangle} \\ \boldsymbol{g}(\boldsymbol{x}_{v\langle k\rangle},\ \boldsymbol{z}_{\langle k\rangle}) \end{pmatrix} \tag{6.21}$$

which appends the sensor-based estimate of the new landmark's coordinates to those already in the map. The order of feature coordinates within $\hat{\boldsymbol{x}}$ therefore depends on the order in which they are observed.

The covariance matrix also needs to be extended when a new landmark is observed, and this is achieved by

$$\hat{\mathbf{P}}'_{\langle k\rangle} = \mathbf{Y}_z \begin{pmatrix} \hat{\mathbf{P}}_{\langle k\rangle} & \mathbf{0}_{2\times M} \\ \mathbf{0}_{M\times 2} & \hat{\mathbf{W}} \end{pmatrix} \mathbf{Y}_z^\top$$

where $\mathbf{Y}_z$ is the insertion Jacobian ▶

$$\mathbf{Y}_z = \frac{\partial \boldsymbol{y}}{\partial \boldsymbol{z}} = \left(\begin{array}{cc|c} \mathbf{1}_{M\times M} & & \mathbf{0}_{M\times 2} \\ \mathbf{G}_x & \mathbf{0}_{2\times(M-3)} & \mathbf{G}_z \end{array} \right) \tag{6.22}$$

that relates the rate of change of the extended state vector to the new observation. M is the dimension of $\hat{\mathbf{P}}$ *prior* to it being extended and the Jacobians are

$$\mathbf{G}_x = \frac{\partial \boldsymbol{g}}{\partial \boldsymbol{x}} = \begin{pmatrix} 0 & 0 & 0 \\ 0 & 0 & 0 \end{pmatrix} \tag{6.23}$$

$$\mathbf{G}_z = \frac{\partial \boldsymbol{g}}{\partial \boldsymbol{z}} = \begin{pmatrix} \cos(\theta_v + \beta) & -r\sin(\theta_v + \beta) \\ \sin(\theta_v + \beta) & r\cos(\theta_v + \beta) \end{pmatrix}. \tag{6.24}$$

$\mathbf{G}_x$ is zero since $\boldsymbol{g}(\cdot)$ is independent of the map and therefore $\boldsymbol{x}$. An additional Jacobian for $\boldsymbol{h}(\cdot)$ is

$$\mathbf{H}_{p_i} = \frac{\partial \boldsymbol{h}}{\partial \boldsymbol{p}_i} = \begin{pmatrix} \frac{x_i - x_v}{r} & \frac{y_i - y_v}{r} \\ -\frac{y_i - y_v}{r^2} & \frac{x_i - x_v}{r^2} \end{pmatrix} \tag{6.25}$$

which describes how the sensor measurement changes with respect to the position of landmark i for a particular robot pose, and is implemented by the `LandmarkSensor` method `Hp`.

For the mapping case, the Jacobian $\mathbf{H}_x$ used in (6.13) describes how the landmark observation changes with respect to the full state vector. However, the observation depends only on the position of the observed landmark, so this Jacobian is mostly zeros

$$\mathbf{H}_x = \frac{\partial \boldsymbol{h}}{\partial \boldsymbol{x}} \bigg|_{w=0} = \begin{pmatrix} \mathbf{0}_{2\times 2} & \cdots & \mathbf{H}_{p_i} & \cdots & \mathbf{0}_{2\times 2} \end{pmatrix} \in \mathbb{R}^{2\times 2M} \tag{6.26}$$

The insertion Jacobian (6.22) is important but details of its formulation are difficult to find. A derivation can be found at ▶ sn.pub/XQv4jR.

where $\mathbf{H}_{p_i}$ is at the location in the state vector corresponding to landmark i. This structure represents the fact that observing a particular landmark provides information to estimate the position of that landmark, but no others.

The RVC Toolbox implementation is

```
>> rng(0) % obtain repeatable results
>> map = LandmarkMap(20,10);
>> robot = BicycleVehicle; % error free vehicle
>> robot.addDriver(RandomDriver(map.dim));
>> W = diag([0.1 deg2rad(1)].^2);
>> sensor = LandmarkSensor(robot,map,covar=W);
>> ekf = EKF(robot,[],[],sensor,W,[]);
```

and the empty matrices passed to EKF indicate that there is no estimated odometry covariance for the vehicle (the estimate is perfect), no initial vehicle state covariance (since it is initially a 0×0 matrix), and no map (since it is unknown). We run the simulation for 1000 time steps

```
>> ekf.run(1000);
```

and see an animation of the robot moving and the covariance ellipses associated with the map features evolving over time. The estimated landmark positions

```
>> map.plot;
>> ekf.plotmap("g");
>> robot.plotxy("b");
```

are shown in ◘ Fig. 6.11a as 95% confidence ellipses along with the true landmark positions and the path taken by the robot. ◘ Fig. 6.11b shows a close-up view of landmark 17 and we see that the estimate is close to the true value and well within the confidence ellipse.

The covariance matrix is shown graphically in ◘ Fig. 6.11c and has a block-diagonal structure. The off-diagonal elements are zero, which implies that the landmark estimates are uncorrelated or independent. This is to be expected since observing one landmark provides no new information about any other landmark.

Internally, the EKF object uses a table to relate the landmark's identity, returned by the LandmarkSensor object, to the position of that landmark's coordinates within the state vector. For example, landmark 17

```
>> ekf.landmarks(:,17)'  % transpose for display
ans =
     1     47
```

was seen a total of 47 times during the simulation and its estimated position can be found in the EKF state vector starting at element 1 and its estimated location is

```
>> ekf.x_est(1:2)'  % transpose for display
ans =
    -4.4534   -9.0742
```

Its estimated covariance is a submatrix within $\hat{\mathbf{P}}$

```
>> ekf.P_est(1:2,1:2)
ans =
   1.0e-03 *
    0.1831    0.0134
    0.0134    0.1587
```

6.2.3 EKF SLAM

Finally, we tackle the problem of determining our position and creating a map at the same time. This is an old problem in marine navigation and cartography – incrementally extending maps while also using the map for navigation. ◘ Fig. 6.12

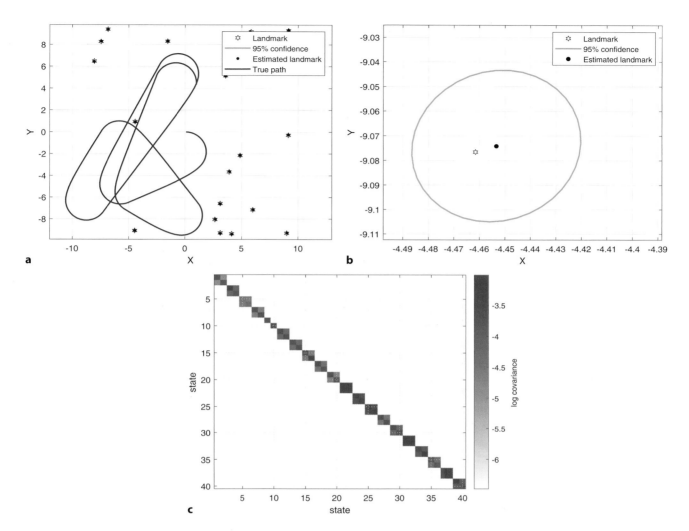

◻ Fig. 6.11 **a** EKF mapping results, the uncertainty ellipses are too small to see; **b** enlarged view of landmark 17 and its covariance ellipse; **c** the final covariance matrix has a block diagonal structure

shows what can be done without GPS from a moving ship with poor odometry and infrequent celestial position *fixes*. In robotics this problem is known as simultaneous localization and mapping (SLAM) orconcurrent mapping and localization (CML). This is a "chicken and egg" problem since we need a map to localize, and we need to localize to make the map. However, based on what we have learned in the previous sections this problem is now quite straightforward to solve.

The state vector comprises the robot configuration *and* the coordinates of the M landmarks that have been observed so far

$$\boldsymbol{x} = (x_v, \ y_v, \ \theta_v, \ x_1, \ y_1, \ x_2, \ y_2, \ \ldots \ x_M, \ y_M)^\top \in \mathbb{R}^{(2M+3)\times 1}$$

and therefore, has a variable length. The estimated covariance is a variable-sized symmetric matrix with a block structure

$$\hat{\mathbf{P}} = \begin{pmatrix} \hat{\mathbf{P}}_{vv} & \hat{\mathbf{P}}_{vm} \\ \hat{\mathbf{P}}_{vm}^\top & \hat{\mathbf{P}}_{mm} \end{pmatrix} \in \mathbb{R}^{(2M+3)\times(2M+3)}$$

where $\hat{\mathbf{P}}_{vv} \in \mathbb{R}^{3\times3}$ is the covariance of the robot pose as described by (6.5), $\hat{\mathbf{P}}_{mm} \in \mathbb{R}^{2M\times2M}$ is the covariance of the estimated landmark positions as described by (6.16), and $\hat{\mathbf{P}}_{vm} \in \mathbb{R}^{3\times2M}$ is the correlation between robot and landmark states.

6

◻ **Fig. 6.12** Map of the Eora territory coast (today, the Sydney region of eastern Australia) created by Captain James Cook in 1770. The path of the ship and the map of the coast were determined at the same time. Numbers indicate depth in fathoms (1.83 m) (National Library of Australia, MAP NK 5557 A)

The predicted robot state and covariance are given by (6.3) and (6.4) and the sensor-based update is given by (6.11) to (6.15). When a new feature is observed, the state vector is updated using the insertion Jacobian $\mathbf{Y}_z$ given by (6.22) but in this case $\mathbf{G}_x$ is not zero

$$\mathbf{G}_x = \frac{\partial \boldsymbol{g}}{\partial \boldsymbol{x}} = \begin{pmatrix} 1 & 0 & -r\sin(\theta_v + \beta) \\ 0 & 1 & r\cos(\theta_v + \beta) \end{pmatrix} \tag{6.27}$$

since the estimate of the new landmark depends on the state vector which now contains the robot's configuration.

For the SLAM case, the Jacobian $\mathbf{H}_x$ used in (6.13) describes how the landmark observation changes with respect to the state vector. The observation depends only on two elements of the state vector: the position of the robot and the position of the observed landmark

$$\mathbf{H}_x = \left.\frac{\partial \boldsymbol{h}}{\partial \boldsymbol{x}}\right|_{\boldsymbol{w}=0} = \begin{pmatrix} \mathbf{H}_{x_v} & \cdots & \mathbf{0}_{2\times2} & \cdots & \mathbf{H}_{p_i} & \cdots & \mathbf{0}_{2\times2} \end{pmatrix} \in \mathbb{R}^{2\times(2M+3)} \tag{6.28}$$

where $\mathbf{H}_{p_i}$ is at the location in the state vector corresponding to landmark i. This is similar to (6.26) but with an extra nonzero block $\mathbf{H}_{x_v}$ given by (6.14).

The Kalman gain matrix $\mathbf{K}$ *distributes* innovation from the landmark observation, a 2-vector, to update *every* element of the state vector – the configuration of the robot *and* the position of *every* landmark in the map.

The implementation is by now quite familiar

```
>> rng(0) % obtain repeatable results
>> map = LandmarkMap(20,10);
>> V = diag([0.1 deg2rad(1)].^2);
>> robot = BicycleVehicle(covar=V, q0=[3 6 deg2rad(-45)]);
>> robot.addDriver(RandomDriver(map.dim));
>> W = diag([0.1 deg2rad(1)].^2);
>> sensor = LandmarkSensor(robot,map,covar=W);
>> P0 = diag([0.05 0.05 deg2rad(0.5)].^2);
>> ekf = EKF(robot,V,P0,sensor,W,[]);
```

and once again the map object is not passed to EKF since it is unknown to the filter and P0 is the initial 3×3 covariance for the robot state. The initial configuration of the robot is $(3, 6, -45°)$ but the EKF is using the default robot configuration of $(0, 0, 0)$ specified as the x_est0 parameter to the run method.

We run the simulation for 500 time steps

```
>> ekf.run(500,x_est0=[0 0 0]);
```

and, as usual, an animation of the moving robot is shown. We also see the covariance ellipses associated with the map features evolving over time. We can plot the landmarks and the path taken by the robot

```
>> map.plot;          % plot true map
>> robot.plotxy("b"); % plot true path
```

which are shown in ◨ Fig. 6.13a. Next we will plot the EKF estimates

```
>> ekf.plotmap("g");  % plot estimated landmarks + covariances
>> ekf.plotxy("r");   % plot estimated robot path
```

which are shown in ◨ Fig. 6.13b and we see that the robot's path is quite different – it is shifted and rotated with respect to the robot path shown in ◨ Fig. 6.13a. The estimated path and landmarks are relative to the map frame. A large part of the difference is because the EKF has used a default initial robot configuration of $(0, 0, 0)$ which is different to the actual initial configuration of the robot $(3, 6, -45°)$. Sensor noise also affects the pose of the map frame: its position depends on the initial estimate of the robot's pose and the odometry error at the first

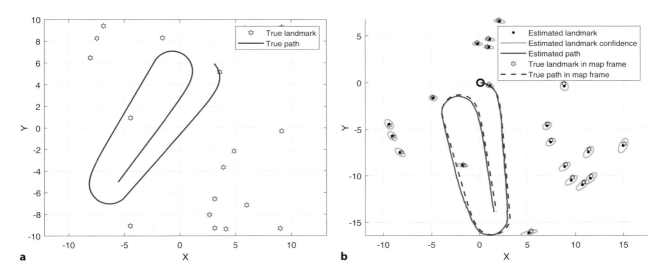

◨ **Fig. 6.13** Simultaneous localization and mapping. **a** The true robot path and landmark positions in the world reference frame; **b** the estimated robot path and landmark positions in the map frame. The true robot path and landmarks are overlaid after being transformed into the map frame. Ellipses show the 95% confidence interval

6

◘ Fig. 6.14   Simultaneous localization and mapping. **a** Covariance versus time; **b** the final covariance matrix

filter step, and the frame's orientation depends on the dead-reckoned orientation and sensor noise when the first measurement was made. Knowing the true and estimated landmark positions of at least two landmarks, we can compute an $\mathbf{SE}(2)$ transformation from the map frame to the world frame

```
>> T = ekf.transform(map)
T =
  se2
     0.7182    -0.6965     2.0471
     0.6965     0.7182    -6.4905
          0          0     1.0000
```

which we can use to transform the robot's path and the landmarks from the world frame to the map frame and these are overlaid in ◘ Fig. 6.13b. This shows very close agreement once we have brought the true and estimated values into alignment.

◘ Fig. 6.14a shows that uncertainty is decreasing over time. The final covariance matrix is shown graphically in ◘ Fig. 6.14b and we see a complex structure. Unlike the mapping case of ◘ Fig. 6.11, $\hat{\mathbf{P}}_{mm}$ is not block diagonal and the finite off-diagonal terms represent correlations *between* the landmarks in the map. The landmark uncertainties never increase, the position prediction model is that they do not move, but they can never drop below the initial position uncertainty of the robot in $\hat{\mathbf{P}}_0$. The block $\hat{\mathbf{P}}_{vm}$ is the correlation between errors in the robot configuration and the landmark positions.

A landmark's pose estimate is a function of the robot's pose, so errors in the robot position appear as errors in the landmark position – and vice versa. The correlations are used by the Kalman filter to apply the innovation from a landmark observation to improve the estimate of every other landmark in the map as well as the robot configuration. Conceptually, it is as if all the states were connected by springs and the movement of any one affects all the others.

The extended Kalman filter is a powerful tool for estimation but has a number of limitations. Firstly, the size of the matrices involved increases with the number of landmarks and can lead to memory and computational bottlenecks as well as numerical problems. Secondly, the underlying assumption of the Kalman filter is that all errors have a Gaussian distribution, and this is far from true for sensors like lidars that we will discuss later in this chapter. Thirdly, it requires good estimates of covariance of the noise sources, which in practice are challenging to estimate.

6.2.4 Sequential Monte-Carlo Localization

The estimation examples so far have assumed that the errors in sensors, such as odometry and landmark range, have a Gaussian probability density function. In practice we might find that a sensor has a one-sided distribution (like a Poisson distribution) or a multi-modal distribution with several peaks. The functions we used in the Kalman filter, such as (6.2) and (6.8), are strongly nonlinear, which means that sensor noise with a Gaussian distribution will not result in a Gaussian error distribution on the value of the function – this is discussed further in ▶ App. H.2. The probability density function associated with a robot's configuration may have multiple peaks to reflect several hypotheses that equally well explain the data from the sensors as shown in ◻ Fig. 6.3c.

The Monte-Carlo estimator that we discuss in this section makes no assumptions about the distribution of errors. It can also handle multiple hypotheses for the state of the system. The basic idea is disarmingly simple. We maintain many *different* values of the robot's configuration or state vector. When a new measurement is available, we score how well each particular value of the state explains what the sensor just observed. We keep the best fitting states and randomly sample from the prediction distribution to form a new generation of states. Collectively, these many possible states, and their scores, form a discrete approximation of the probability density function of the state we are trying to estimate. There is never any assumption about Gaussian distributions nor any need to linearize the system. While computationally expensive, it is quite feasible to use this technique with today's standard computers and it is well suited to parallel computation. If we plot these state vectors as points in the state space, we have a cloud of particles each representing a plausible robot state. Hence this type of estimator is often referred to as a particle filter.

We will apply Monte-Carlo estimation to the problem of localization using odometry and a map. Estimating only three states $x = (x, y, \theta)$ is computationally tractable to solve with straightforward MATLAB code. The estimator is initialized by creating N particles $x_i, i = 1, \ldots, N$ distributed randomly over the configuration space of the robot. All particles have the same initial weight, importance, or likelihood: $w_i = 1/N$. The steps in the main iteration of the algorithm are:

1. Apply the state update to each particle

$$x_i^+\langle k+1 \rangle = f(x_i\langle k \rangle, \, u\langle k \rangle + r\langle k \rangle)$$

   where $u\langle k \rangle$ is the input to the system or the measured odometry $u\langle k \rangle = \delta\langle k \rangle$ and $r\langle k \rangle \in \mathbb{R}^2$ is a random vector that represents uncertainty in the odometry. Often r is drawn from a Gaussian random variable with covariance $\mathbf{R}$ but any physically meaningful distribution can be used. The state update is often simplified to

$$x_i^+\langle k+1 \rangle = f(x_i\langle k \rangle, \, u\langle k \rangle) + q\langle k \rangle$$

   where $q\langle k \rangle \in \mathbb{R}^3$ represents uncertainty in the pose of the robot.

2. We make an observation z_j of landmark j that has, according to the map, co-ordinate p_j. For each particle we compute the innovation

$$v_i = h(x_i^+, \, p_j) - z_j$$

   which is the error between the predicted and actual landmark observation. A likelihood function provides a scalar measure of how well the particular particle explains this observation. In this example we choose a likelihood function

$$w_i = e^{-v_i^\top L^{-1} v_i} + w_0$$

6

Excurse 6.9: Monte-Carlo Methods

These are a class of computational algorithms that rely on repeated random sampling to compute their results. An early example of this idea is Buffon's needle problem posed in the eighteenth century by Georges-Louis Leclerc (1707–1788), Comte de Buffon: *Suppose we have a floor made of parallel strips of wood of equal width t, and a needle of length l is dropped onto the floor. What is the probability that the needle will lie across a line between the strips?* If n needles are dropped and h cross the lines, the probability can be shown to be $h/n = 2l/\pi t$ and in 1901 the Italian mathematician Mario Lazzarini performed the experiment, tossing a needle 3408 times, and obtained the estimate $\pi \approx 355/113$ (3.14159292).

Monte-Carlo methods are often used when simulating systems with a large number of coupled degrees of freedom with significant uncertainty in inputs. Monte-Carlo methods tend to be used when it is infeasible or impossible to compute an exact result with a deterministic algorithm. Their reliance on repeated computation and random or pseudo-random numbers make them well suited to calculation by a computer. The method was developed at Los Alamos as part of the Manhattan project during WW II by the mathematicians John von Neumann, Stanislaw Ulam, and Nicholas Metropolis. The name Monte-Carlo alludes to games of chance and was the code name for the secret project.

This particular type of estimator is known as a bootstrap particle filter, since the weight is computed at each step, with no dependence on previous values.

There are many resampling strategies for particle filters, both for choosing a resampling algorithm and setting a resampling frequency. Here we use the simplest strategy known variously as multinomial resampling, simple random resampling, or select with replacement, at every time step. This is also sometimes referred to as condensation. Other common strategies are residual, stratified, and systematic resampling.

to determine the weight of the particle, where $\mathbf{L}$ is a covariance-like matrix, and $w_0 > 0$ ensures that there is a finite probability of a particle being retained despite sensor error. We use a quadratic exponential function only for convenience, the function does not need to be smooth or invertible but only to adequately describe the likelihood of an observation. ◄

3. Select the particles that best explain the observation, a process known as resampling ◄ or importance sampling. A common scheme is to randomly select particles according to their weight. Given N particles $\boldsymbol{x}_i$ with corresponding weights w_i, we first normalize the weights

$$w_i' = \frac{w_i}{\sum_{i=1}^{N} w_i} \tag{6.29}$$

and construct a cumulative histogram

$$c_j = \Sigma_{i=1}^{j} w_i' \; . \tag{6.30}$$

As shown in ◼ Fig. 6.15 we then draw a uniform random number $r \in [0, 1]$ and find

$$i = \arg \min_j r < c_j$$

and select particle i for the next generation. The process is repeated N times. The end result will be a new set of N particles with multiple copies of highly weighted particles while some lower weighted particles may not be copied at all. Step 1 of the next iteration will *spread out* the identical particles through the addition of the random vector $\boldsymbol{r}\langle k \rangle$. This random spreading and resampling are critical components of the particle filter without which the filter would quickly produce a degenerate set of particles where a few have high weights and the bulk have almost zero weight which poorly represents the true distribution. This degenerate condition is also known as particle starvation.

These steps are summarized in ◼ Fig. 6.16. The implementation is broadly similar to the previous examples. We create a map

```
>> rng(0) % obtain repeatable results
>> map = LandmarkMap(20,10);
```

Fig. 6.15 Random resampling. The cumulative histogram (top) is formed from the particle weights (bottom). A random number r is used to select a particle i. The probability of selecting a highly weighted particle like 1 or 3 is clearly higher than for a low-weighted particle like 2 or 4

$$\boldsymbol{x}_i^+\langle k{+}1\rangle = \boldsymbol{f}(\boldsymbol{x}_i\langle k\rangle, \boldsymbol{u}\langle k\rangle + \boldsymbol{q}\langle k\rangle)$$

predict state of this particle one step ahead based on inputs and model

add a random value to represent uncertainty

prediction phase

$$\boldsymbol{\nu}_i = \boldsymbol{z}\langle k{+}1\rangle - \mathbf{h}(\hat{\boldsymbol{x}}_i^+\langle k{+}1\rangle, \boldsymbol{p}_j)$$

new information – innovation - from observing landmark j

$$w_i = e^{-\boldsymbol{\nu}_i^T \mathbf{L}^{-1} \boldsymbol{\nu}_i} + w_0$$

weight is a function of how likely the measurement was

$$w_i' = w_i / \sum_{i=1}^{N} w_i$$

normalize the weights 0 to 1

$$\boldsymbol{x}\langle k{+}1\rangle \leftarrow R(\boldsymbol{x}^+\langle k{+}1\rangle, \boldsymbol{w})$$

resample, weighted selection of particles to go to the next round

update phase

Fig. 6.16 The particle filter estimator showing the prediction and update phases

and a robot with noisy odometry and an initial condition

```
>> V = diag([0.1 deg2rad(1)].^2)
V =
    0.0100         0
         0    0.0003
>> robot = BicycleVehicle(covar=V);
>> robot.addDriver(RandomDriver(10,show=true));
```

and then a sensor with noisy readings

```
>> W = diag([0.02 deg2rad(0.5)].^2);
>> sensor = LandmarkSensor(robot,map,covar=W);
```

For the particle filter we need to define two covariance matrices. The first is the covariance of the random noise added to the particle states at each iteration to represent uncertainty in configuration. We choose the covariance values to be comparable with those of **W**

```
>> Q = diag([0.1 0.1 deg2rad(1)]).^2;
```

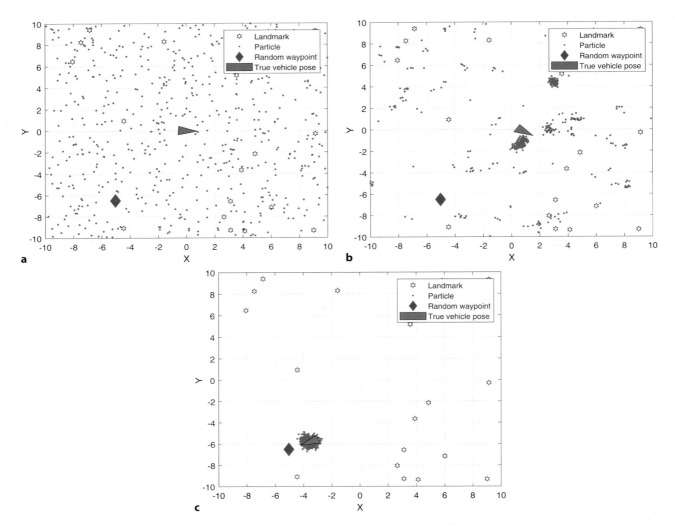

Fig. 6.17 Particle filter results showing the evolution of the particle's configuration on the xy-plane over time. **a** Initial particle distribution at step 1; **b** particle distribution after 10 time steps; **c** particle distribution after 100 time steps

and the covariance of the likelihood function applied to innovation

```
>> L = diag([0.1 0.1]);
```

Finally, we construct a `ParticleFilter` estimator

```
>> pf = ParticleFilter(robot,sensor,Q,L,1000);
```

which is configured with 1000 particles. The particles are initially uniformly distributed over the 3-dimensional configuration space. We run the simulation for 1000 time steps

```
>> pf.run(1000);
```

and watch the animation, snapshots of which are shown in ◻ Fig. 6.17. We see the particles moving as their states are updated by odometry, random perturbation, and resampling. The initially randomly distributed particles begin to aggregate around those regions of the configuration space that best *explain* the sensor observations that are made. In Darwinian fashion, these particles become more highly weighted and survive the resampling step, while the lower-weighted particles are extinguished.

The particles approximate the probability density function of the robot's configuration. The most likely configuration is the expected value or mean of all the particles. A measure of uncertainty of the estimate is the spread of the particle

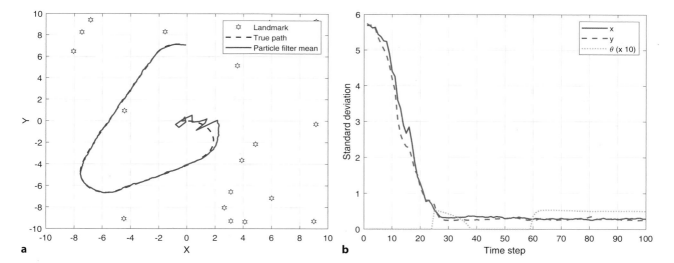

◨ Fig. 6.18 Particle filter results. **a** Robot path; **b** standard deviation of the particles versus time

cloud or its standard deviation. The `ParticleFilter` object keeps the history of the mean and standard deviation of the particle state at each time step, taking into account the particle weighting. ▸ As usual we plot the results of the simulation

```
>> map.plot;
>> robot.plotxy("b--");
```

and overlay the mean of the particle cloud

```
>> pf.plotxy("r");
```

which is shown in ◨ Fig. 6.18a. The initial part of the estimated path has quite high error since the particles have not converged on the true configuration. We can plot the standard deviation against time

```
>> clf
>> plot(1:100,abs(pf.std(1:100,:)))
```

and this is shown in ◨ Fig. 6.18b. We can see a sudden drop before time step 30 as the particles that are distant from the true solution are rapidly eliminated. As mentioned at the outset, the particles are a sampled approximation to the PDF, and we can display this as

```
>> clf
>> pf.plotpdf
```

The problem we have just solved is known in robotics as the kidnapped robot problem – a robot is placed in the world with no estimate of its initial pose and needs to localize itself as quickly as possible. To represent this large uncertainty, we uniformly distribute the particles over the 3-dimensional configuration space and their sparsity can cause the particle filter to take a long time to converge unless a very large number of particles is used. It is debatable whether this is a realistic problem. Typically, we have some approximate initial pose of the robot and the particles would be initialized to that part of the configuration space. For example, if we know the robot is in a corridor then the particles would be placed in those areas of the map that are corridors, or if we know the robot is pointing north then we would set all particles to have that orientation.

Setting the parameters of the particle filter requires a little experience and the best way to learn is to experiment. For the kidnapped robot problem we set **Q** and the number of particles high so that the particles explore the configuration space but once the filter has converged, lower values could be used. There are many

Here we have taken statistics over all particles, but other methods are possible. The simplest strategy is to just use the particle with the highest weight. Alternatively, we can estimate the kernel density at every particle – the sum of the weights of all neighbors within a fixed radius – and take the particle with the largest value.

variations of the particle filter concerned with the shape of the likelihood function and the resampling strategy.

The `stateEstimatorPF` also implements a variety of resampling algorithms, resampling policies, and state estimation methods.

Under the hood, the `ParticleFilter` uses the `stateEstimatorPF` class, which is a general-purpose particle filter used to estimate an arbitrary number of state variables. ◄ For the example in this section, the filter estimates 3 state variables, the robot state (x, y, θ), and it would be straightforward to add new state variables, such as landmark positions. The particle filter scales poorly to high-dimensional state spaces, since the number of particles required to approximate the state distribution quickly becomes intractable.

6.2.5 Rao-Blackwellized SLAM

We will briefly and informally introduce the underlying principle of Rao-Blackwellized SLAM of which FastSLAM is a popular and well known instance. The approach is motivated by the fact that the size of the covariance matrix for EKF SLAM is quadratic in the number of landmarks, and for large-scale environments becomes computationally intractable.

If we compare the covariance matrices shown in ◘ Fig. 6.11c and ◘ Fig. 6.14b we notice a stark contrast. In both cases we were creating a map of unknown landmarks but ◘ Fig. 6.11c is mostly zero with a finite block diagonal structure whereas ◘ Fig. 6.14b has no zero values at all. The difference is that for ◘ Fig. 6.11c, we assumed the robot trajectory was known exactly and that makes the landmark estimates *independent* – observing one landmark provides information about *only* that landmark. The landmarks are *uncorrelated*, hence all the zeros in the covariance matrix. If the robot trajectory is not known, the case for ◘ Fig. 6.14b, then the landmark estimates are correlated – error in one landmark position is related to errors in robot pose and other landmark positions. The Kalman filter uses the correlation information so that a measurement of any one landmark provides information to improve the estimate of all the other landmarks and the robot's pose.

Such as the particle filter that we discussed in ► Sect. 6.2.4.

In practice we don't know the true pose of the robot but imagine a multi-hypothesis estimator ◄ where every hypothesis represents a robot trajectory that we *assume* is correct. This means that the covariance matrix will be block diagonal like ◘ Fig. 6.11c – rather than a filter with a $2N \times 2N$ covariance matrix, we can have N simple filters that are each *independently* estimating the position of a single landmark and have a 2×2 covariance matrix. Independent estimation leads to a considerable saving in both memory and computation. Importantly, we are only able to do this because we *assumed* that the robot's estimated trajectory is correct.

Each hypothesis also holds an estimate of the robot's trajectory to date. Those hypotheses that best explain the landmark measurements are retained and propagated while those that don't are removed and recycled. If there are M hypotheses the overall computational burden falls from $O(N^2)$ for the EKF SLAM case to $O(M \log N)$ and in practice works well for M in the order of tens to hundreds but can work for a value as low as $M = 1$.

6.3 Occupancy Grid Maps

As we already covered in ► Sect. 5.4, an occupancy grid map, or *occupancy map*, is an array of cells, typically square, and each cell holds information about the traversability of that cell. In the simplest case of a binary occupancy grid map, a cell has two possible values: one indicates the cell is occupied and we cannot move into it – it is an obstacle; or zero if it is unoccupied. In a probabilistic occupancy grid map, the cell value represents the probability that there is an obstacle in that cell. Depending on the environment and necessary precision, grid cell sizes can vary widely, with typical values being between 0.1 m and 0.5 m.

Fig. 6.19 Robot range measuring sensors. **a** A low-cost ultrasonic rangefinder with maximum range of 6.5 m at 20 measurements per second (LV-MaxSonar-EZ1, image courtesy of SparkFun Electronics®); **b** a low-cost time-of-flight 1D-lidar with maximum range of 20 cm at 10 measurements per second (VL6180, image courtesy of SparkFun Electronics®); **c** a 2D-lidar with a maximum range of 30 m, an angular range of 270° in 0.25° intervals at 40 scans per second (UTM-30LX, image courtesy of Hokuyo Automatic Co. Ltd.); **d** a 3D-lidar with 16 beams, a 30 × 360° field of view, 100 m range and 5-20 revolutions per second (VLP-16 Puck, image courtesy of Velodyne Lidar®)

This section focuses on how to solve the localization, mapping, and SLAM problems for occupancy maps. Another common map representation, a landmark map, was covered in ▶ Sect. 6.2.

6.3.1 Application: Lidar

To implement robot localization and SLAM, we require a sensor that can make measurements of range and bearing to landmarks. Many such sensors are available and are based on a variety of physical principles, including ultrasonic ranging (■ Fig. 6.19a), computer vision, or radar. One of the most common sensors used for robot navigation is lidar (■ Fig. 6.19b–d). A lidar emits short pulses of infrared laser light and measures the time for the reflected pulses to return. Operating range can be up to 100 m with an accuracy in the order of centimeters.

Common lidar configurations are shown in ■ Fig. 6.20. A 1D-lidar sensor, as shown in ■ Fig. 6.19b, measures the distance $r \in \mathbb{R}$, along the beam, to a surface – these are also called laser distance measuring devices or laser rangefinders. A 2D-lidar, as shown in ■ Fig. 6.19c, contains a 1D-lidar that rotates around a fixed axis. It returns a planar cross-section of the world in polar coordinate form $\mathbb{R} \times \mathbf{S}^1$, and emits a fixed number of pulses per revolution. ▶ For mobile robotic applications the lidar is typically configured to scan in a plane parallel to, and slightly above, the ground plane. A 3D-lidar, as shown in ■ Fig. 6.19d, contains multiple 1D-lidars that produce a fan of beams in a vertical plane, and that plane rotates around a fixed axis. It returns spherical coordinates $\mathbb{R} \times \mathbf{S}^2$ or 3D point coordinates $\mathbb{R}^3$ which can be visualized as a point cloud, such as shown in ■ Fig. 6.21. Some lidars also

Typically every quarter, half, or one degree. Lidars may have a reduced angular range of 180 or 270°, rather than a full revolution.

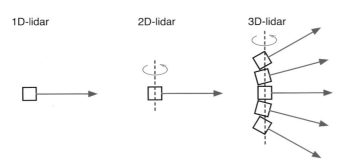

Fig. 6.20 Common lidar configurations; the red arrow represents the laser beam

6

◨ **Fig. 6.21** Point cloud created from a single scan of a Velodyne Alpha Prime 3D-lidar mounted on the vehicle shown in the foreground (this vehicle is an inserted model and not part of the sensed point cloud). The lidar has a fan of 128 beams arranged in a vertical plane which trace out concentric circles on the ground plane, and the beam angles are not uniformly spaced. The trees on the right have cast *range shadows* and only their front surfaces are seen – this is sometimes called a "2.5D" representation. Points are colored according to the intensity (remission) of the returned laser pulse (Image courtesy of Velodyne Lidar)

Solid-state lidars are an exciting emerging technology where optical beam steering replaces mechanical scanning.

measure the return signal strength or remission, which is a function of the object's distance, infrared reflectivity, and beam incidence angle.

Lidars have advantages and disadvantages compared to cameras and computer vision that we discuss in Part IV of this book. On the positive side, lidars provide metric data, that is, they return actual range to points in the world in units of meters, and they can work in the dark. However, lidars work less well than cameras outdoors since the returning laser pulse is overwhelmed by infrared light from the sun. Other disadvantages include the finite time to scan that is problematic if the sensor is moving, inability to discern fine texture or color; having moving parts; ◄ as well as being bulky, power-hungry, and expensive compared to cameras.

6.3.2 **Lidar-Based Odometry**

A common application of lidars is lidar odometry, estimating the change in robot pose using lidar scan data rather than wheel encoder data. We will illustrate this with lidar scan data from a real robot

```
>> [~,lidarData,lidarTimes,odoTruth] = g2oread("killian.g2o");
>> whos lidarData
  Name          Size          Bytes  Class        Attributes
  lidarData     1x3873     11158113  lidarScan
```

and each of the 3873 scans is a `lidarScan` object that stores range and bearing information. The range and bearing data for scan 100

```
>> p100 = lidarData(100)
p100 =
  lidarScan with properties:
       Ranges: [180×1 double]
       Angles: [180×1 double]
    Cartesian: [180×2 double]
        Count: 180
```

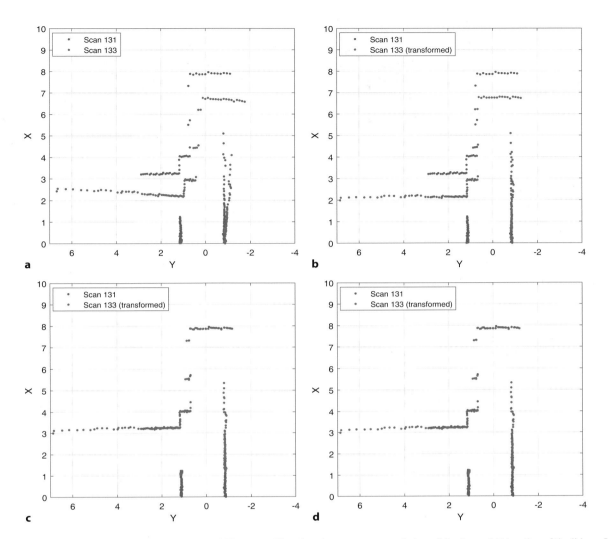

◘ Fig. 6.22 Lidar scan matching. Note that the X and Y axes are flipped to show a more natural view of the forward X location of the lidar. **a** Lidar scans 131 and 133; **b** points from location 131 and transformed points from location 133 based on `matchScans`; **c** transformed points based on `matchScans` with an initial guess; **d** transformed points based on `matchScansGrid`

is stored in the `Ranges` and `Angles` properties, each represented by a vector of 180 elements. We can plot the scan

```
>> clf
>> p100.plot
```

in Cartesian/polar form. We will use two lidar scans

```
>> p131 = lidarData(131);
>> p133 = lidarData(133);
```

and these are overlaid in ◘ Fig. 6.22a. ► The robot is moving forward (the x-axis points to the front of the lidar), so the points on the wall toward the top of the figure are closer in the second scan. The robot also seems to start a turn to the left as the angle changes between scans.

To determine the change in pose of the robot between the two scans we need to align these two sets of points, and this can be achieved with a scan matching

Note that the points close to the lidar, at coordinate $(0, 0)$ in this sensor reference frame, are much more tightly clustered, and this is a characteristic of lidars where the points are equally spaced in angle not over an area.

algorithm

```
>> pose = matchScans(p133,p131)
pose =
    0.0494    0.0158    0.0722
```

which returns a relative pose $^{131}\mathbf{T}_{133}$ as (x, y, θ) describing the motion of the robot – around 0.05 m in the x-direction and a slight rotation. We can also use $^{131}\mathbf{T}_{133}$ to transform the lidar scan from frame {133} to {131}

```
p133T = transformScan(p133, pose);
```

which can then be plotted as shown in ◘ Fig. 6.22b.

We notice that the scan matching algorithm was able to correct the rotation between the two scans, but several of the wall points are not well aligned. This scan matcher is based on the Normal Distributions Transform (NDT) algorithm, which does not rely on the points alone, but is based on their probability distributions. At each iteration, the fixed scan (scan 131 in our example) is divided into cells of constant size and a normal distribution is computed for the points in the grid cell. Starting with an initial guess of the transformation for the moving scan (scan 133 in our example), the algorithm finds the sum of the statistical likelihood of each aligned point from the moving scan based on the normal distributions of the fixed scan. To improve registration, the algorithm maximizes the probability score of the moving scan on the normal distributions of the fixed scan. This is done by iteratively optimizing angular and translational estimations with a gradient descent algorithm. The algorithm stops once the maximum iterations, or the tolerance parameters, are met.

So why did NDT not find the correct point alignment in ◘ Fig. 6.22b? It's likely that the dense wall points on the right side biased the optimization and the algorithm converged to a suboptimal, local minimum. Remember that the lidar points closest to the lidar origin are much more tightly clustered and can bias the alignment.

To break the algorithm out of the local minimum, we can provide a rough initial guess of a movement of 1 m in the x-direction (for example from odometry)

```
>> pose = matchScans(p133,p131,InitialPose=[1 0 0])
pose =
    1.1610    0.0065    0.0875
```

and the scans in ◘ Fig. 6.22c are now well-aligned. When comparing the time stamp information we can see that the scans were captured

```
>> seconds(lidarTimes(133)-lidarTimes(131))
ans =
    3.9600
```

seconds apart which indicates that the robot is moving quite slowly – a little under $0.3 \, \text{ms}^{-1}$.

In most cases, we might not have an initial pose guess from odometry available, so we can use other algorithms that are more robust than NDT. The function `matchScansGrid`

```
>> pose = matchScansGrid(p133,p131)
pose =
    1.1500         0    0.0875
```

returns a good pose estimate without any initial guess and the alignment result is shown in ◘ Fig. 6.22d. For this grid-based scan matching algorithm, lidar scans are converted into probability grids and the matching pose between the two scans is found through correlation of the grids. The search range for translation and rotation is discretized and a branch and bound strategy is used to improve the computational efficiency. This discretization is visible in the x-coordinate of the `pose`. The default `Resolution` is set to 0.05 m and the x-coordinate is a multiple of that.

Although the grid-based algorithm is more accurate and does not require an initial guess for this scan pair, it is considerably slower, in the order of 5 times for this example. ▶

```
>> timeit(@() matchScansGrid(p133,p131)) / ...
>>    timeit(@() matchScans(p133,p131))
ans =
    4.5179
```

To optimize performance of `matchScansGrid`, we can set the `TranslationSearchRange` and `RotationSearchRange` name-value pairs to limit the search space of possible translations and rotations.

In practice, the NDT-based scan matcher is known to be better at estimating rotation, whereas the grid-based scan matcher excels at estimating translation. Sophisticated lidar SLAM systems typically employ several different scan matchers that are best suited to the data that is being captured.

There are additional challenges of scan matching in real-world applications. Some lidar pulses will not return to the sensor if they fall on a surface with low reflectivity or on an oblique polished surface that specularly reflects the pulse away from the sensor – in these cases the sensor typically reports its maximum value or NaN. People moving through the environment change the shape of the world and temporarily cause a shorter range to be reported. In very large spaces all the walls may be beyond the maximum range of the sensor. Outdoors, the beams can be reflected from rain drops, absorbed by fog or smoke, or the return pulse can be overwhelmed by ambient sunlight. In long corridors, where the lidar scans at consecutive poses are just two parallel lines, there is no way to gauge distance along the corridor – we would have to rely on wheel odometry alone. Finally, the lidar, like all sensors, has measurement noise.

6.3.3 Lidar-Based Map Building

If the robot pose is sufficiently well known, through some localization process, then we can transform all the lidar scans to a global coordinate frame and build a map. Various map representations are possible but here we will outline how to build an occupancy grid as discussed in ▶ Sect. 5.4.

For a robot at a given pose, each beam in the scan is a directed ray that tells us several things. Firstly, the range measurement is the distance along the ray to the closest obstacle, and we can determine the coordinates of the cell that contains an obstacle. However, we cannot tell anything about cells further along the ray – the lidar only senses the surface of the object. Secondly, it is implicit that all the cells between the sensor and the first obstacle are obstacle-free.

We can define a probabilistic occupancy grid

```
>> cellSize = 0.1; maxRange = 40;
>> og = occupancyMap(10,10,1/cellSize, ...
>>    LocalOriginInWorld=[0 -5]);
```

with size of 10×10 meters, a cell size of 0.1 m, and a maximum lidar range of 40 m. To ensure that the y-axis runs from -5 to 5, we are also setting the `LocalOriginInWorld`. We can use the scans and relative pose from ▶ Sect. 6.3.2 to insert scans 131 and 133 into the occupancy map

```
>> og.insertRay([0 0 0],p131,maxRange)
>> og.insertRay(pose,p133,maxRange)
>> og.show; view(-90,90)
```

with the map after the first scan shown in ◘ Fig. 6.23a ▶ and the map after the second scan shown in ◘ Fig. 6.23b.

As the robot moves around the environment each cell is intersected by many rays and we use a simple voting scheme to determine whether cells are free or occupied. As shown in ◘ Fig. 6.23c, occupied cells will have a value closer to 1 and free cells have a value closer to 0. Based on the robot's pose and the beam, we

For visualization convenience and easy comparison to other lidar plots, we are flipping the x- and y-axis in the plot by calling `view(-90,90)`.

6

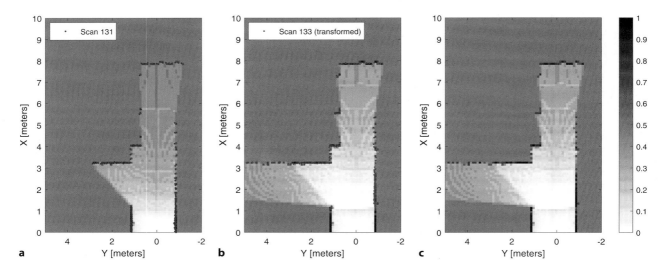

Fig. 6.23 Lidar-based map building. **a** Occupancy map after inserting scan 131 (with lidar points overlaid); **b** occupancy map after inserting the transformed scan 133; **c** occupancy grid after inserting scans 131 and 133. The colorbar shows the range of values with white (0) indicating free cells, black (1) indicating occupied cells, and all other occupancy probabilities in between. At the beginning of the mapping, each cell in the occupancy map has a value of 0.5. The grid cell size is 0.1 m

compute the equation of the ray and visit each cell in turn using the Bresenham line-drawing algorithm. All cells up to the returned range have their vote decreased (less occupied). The cell at the returned range has its vote increased (more occupied) and no further cells along the ray are visited. If the sensor receives no return laser pulse – the surface is not reflective or simply too distant – it sets the range to some maximum value, 40 m in this case, and we skip this ray.

The dataset we loaded earlier has odometry ground truth information stored in the `odoTruth` variable. Rather than adding each of the scans one-by-one, we can use the convenience `buildMap` function to build the occupancy grid in one call

```
>> omap = buildMap(num2cell(lidarData),odoTruth,1/cellSize,maxRange);
>> omap.show
```

and the result is shown in ◙ Fig. 6.24. More sophisticated approaches treat the beam as a wedge of finite angular width and employ a probabilistic model of sen-

Fig. 6.24 **a** Lidar scans rendered into an occupancy grid. An enlarged view of the area enclosed in the central square is displayed in **b**. White cells are free space, black cells are occupied, and gray cells are unknown. Grid cell size is 0.1 m

sor return versus range. The principle can be extended to creating 3-dimensional occupancy maps based on point clouds from a 3D-lidar on a moving robot as shown in ◘ Fig. 6.21.

6.3.4 Lidar-Based Localization

We have mentioned landmarks a number of times in this chapter but avoided concrete examples of what they are. They could be distinctive visual features as discussed in ▶ Sect. 12.3 or artificial markers as discussed in ▶ Sect. 13.6.1. If we consider a lidar scan such as shown in ◘ Fig. 6.22a, we see a fairly distinctive arrangement of points – a geometric signature – which we can use as a landmark. In many cases the signature will be ambiguous and of little value, for example, in the case of a long corridor where all the points are collinear, but some signatures will be highly unique and can serve as a useful landmark. Naively, we could match the current lidar scan against all others (or a map) and if the fit is good we can determine the pose of the robot within the map. However, this strategy would be expensive with a large number of scans or large maps, so in practice we need to use a better methodology.

Assume we have a map of an area, like the one we created in ◘ Fig. 6.24 and a method for matching lidar scans against map areas (like the ones we learned about in ▶ Sect. 6.3.2). The only thing missing is a mechanism for us to check different areas of the map and keep track of multiple hypotheses in parallel. This is a great fit for the the particle filter framework from ▶ Sect. 6.2.4.

Remember that the particle filter used the following high-level steps:

1. Apply the state update to each particle based on measured odometry.
2. Make an observation of landmarks and, for each particle, compute the error between the predicted and actual landmark observation. A likelihood function provides a scalar measure of how well the particular particle explains this observation.
3. Select the particles that best explain the observation, a process known as resampling.

With a few variations, we can apply this same algorithm to the lidar-based localization problem. This algorithm is widely known as Monte-Carlo Localization (MCL) or Adaptive Monte-Carlo Localization (AMCL), and we can create an object that implements it

```
>> mcl = monteCarloLocalization(UseLidarScan=true)
mcl =
  monteCarloLocalization with properties:
           InitialPose: [0 0 0]
     InitialCovariance: [3×3 double]
     GlobalLocalization: 0
         ParticleLimits: [500 5000]
            SensorModel: [1×1 likelihoodFieldSensorModel]
            MotionModel: [1×1 odometryMotionModel]
       UpdateThresholds: [0.2000 0.2000 0.2000]
     ResamplingInterval: 1
           UseLidarScan: 1
```

Step 1 is implemented by the `odometryMotionModel` object assigned to the `MotionModel` property. It approximates the kinematics of a differential-drive robot by describing a pose change as a sequence of three steps: a rotation toward the new position, a translation to the new position, and a final rotation to match the goal pose. Instead of using a motion model based on controls (see ▶ Sect. 6.1.1), this *odometry motion model* uses odometry measurements. The model uses the relative motion information, as measured by the robot's internal

6

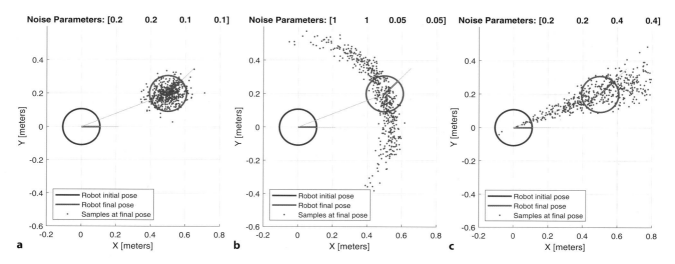

■ **Fig. 6.25** Odometry motion model. For each run, we are propagating 500 samples with different noise parameters. The noise parameters are shown in the title of each figure. **a** A typical distribution; **b** unusually large rotational noise; **c** unusually large translational noise

These standard deviations are specified in the `Noise` property, with the first two values reflecting rotational noise and the last two values reflecting translational noise.

odometry, which is usually readily available. To find the probability distribution of a motion, $p(x_{\langle k+1 \rangle} | u_{\langle k+1 \rangle}, x_{\langle k \rangle})$, we can easily sample this motion model based on a given pose change and rotational and translational standard deviations ◄

```
>> motion = mcl.MotionModel; clf;
>> motion.Noise = [0.2 0.2 0.1 0.1];
>> motion.showNoiseDistribution(OdometryPoseChange=[0.5 0.2 pi/4]);
```

and the resulting particle distribution is shown in ■ Fig. 6.25a. The red samples show possible robot poses if the measured odometry is $(0.5, 0.2, \pi/4)$ and we apply the `Noise` parameters.

We can adjust the noise parameter depending on how much we trust our odometry sensor. Some example distributions for different noise parameters are shown in ■ Fig. 6.25b and 6.25c. If this motion model is applied repeatedly, the uncertainty in the robot's pose will continue to grow and spread across an increasingly large space. This increased uncertainty is common for odometry-based dead reckoning and is also discussed in ► Sect. 6.1. To limit that spread, we need to leverage the information from the lidar scans.

Step 2 is implemented by the `likelihoodFieldSensorModel` object assigned to the `SensorModel` property. Remember that we are looking for a function that calculates how well the particular particle explains the lidar scan observation, $p(z_{\langle k+1 \rangle} | x_{\langle k+1 \rangle}, m)$, where m is the known map

```
>> sensor = mcl.SensorModel;
>> sensor.Map = omap
sensor =
  likelihoodFieldSensorModel with properties:
                         Map: [1×1 occupancyMap]
                  SensorPose: [0 0 0]
                SensorLimits: [0 12]
                    NumBeams: 60
            MeasurementNoise: 0.2000
      RandomMeasurementWeight: 0.0500
    ExpectedMeasurementWeight: 0.9500
        MaxLikelihoodDistance: 2
```

Rather than explicitly modeling the laser beams, this sensor model projects the end points of a lidar scan into the map and then assumes that the measurement likelihood is a mixture of three types of errors: measurement noise (the Gaussian standard deviation is set with `MeasurementNoise`), unexpected objects that cause measured ranges to be shorter (weighted by `RandomMeasurementWeight`), and unexplained random measurements weighted by `ExpectedMeasurementWeight`.

Step 3 handles the particle resampling. As for the normal particle filter, particles with a higher weight (likelihood) are more likely to be resampled. In addition, MCL will dynamically adjust the number of particles. For global localization problems ▶, we typically want to start out with a large number of particles, but once the filter converges, we can reduce the number of particles to a more manageable, and computationally more efficient level. To do that, MCL calculates the Kullback-Leibler divergence (KLD) to determine the required number of particles that can approximate the true posterior distribution within some error bound. To set the minimum and maximum number of particles, we can use the `ParticleLimits` property on the `monteCarloLocalization` object.

Let's look at the code for solving this global localization problem. We start with a high number of particles since the initial pose of the robot is not known. Since we are dealing with a fairly large map, we pick a larger number of particles, 20,000 in this case

Global localization is another name for the kidnapped robot problem we first encountered in ▶ Sect. 6.2.4.

```
>> rng(0) % obtain repeatable results
>> mcl.GlobalLocalization = 1;
>> mcl.ParticleLimits = [500 20000];
```

as upper limit, which is used for the initial global sampling. We can run the first iteration of the MCL algorithm ▶

The `monteCarloLocalization` object is a system object, so you can call it by using the variable name `mcl` like a function. That is equivalent to calling `mcl.step`.

```
>> [isUpdated,pose,covar] = mcl(odoTruth(100,:),lidarData(100))
isUpdated =
  logical
  1
pose =
   47.7362   51.7482    2.3342
covar =
    0.1496   -0.0607         0
   -0.0607    0.1312         0
         0         0   13.1817
>> mp = mclPlot(omap);
>> mp.plot(mcl,pose,odoTruth(100,:),lidarData(100));
```

with the resulting particle distribution shown in ◼ Fig. 6.26a. After this first iteration, the estimated `pose` is very far away from the true pose but running the algorithm for more iterations will help convergence. The current confidence in the particle filter estimate is captured in the covariance matrix `covar`. To avoid noise accumulation and enhance performance, the MCL algorithm only runs once the robot has moved a certain distance. The `isUpdated` output will indicate when an update occurred.

Running the algorithm for 50 more robot sensor readings

```
>> for i = 101:150
>>    [isUpdated,pose] = mcl(odoTruth(i,:),lidarData(i));
>>    mp.plot(mcl,pose,odoTruth(i,:),lidarData(i));
>> end
```

will converge the particles to a single estimate. The particle state after 25 more iteration is shown in ◼ Fig. 6.26b and the converged state after all 50 iterations is visualized in ◼ Fig. 6.26c.

◼ Fig. 6.27 shows the evolution of the number of particles during the 50 iterations. As we described above, the filter starts out with the upper limit of 20,000 particles and then reduces the number of particles all the way to the lower limit of 500 particles.

We just solved the global localization (kidnapped robot) problem, but in many practical applications, we have a general idea in what area of the map the robot is located. Rather than initializing the particle filter by sampling in the whole map, we can sample around an initial pose (`InitialPose` property) and with some

6

◻ Fig. 6.26 Monte-Carlo Localization. **a** The initial distribution of particles on the submap shown in ◻ Fig. 6.24b after initializing the filter; **b** after 25 steps, several particle clusters are tracking multiple hypotheses for the robot pose; the current best estimated pose is far away from the true pose, but has a good match for the lidar data; **c** after 50 steps, the filter estimate converges on a single particle cluster and the true pose

covariance (`InitialCovariance` property). To enable this local sampling, set the `GlobalLocalization` property to 0.

All the localization algorithms discussed so far assume that the world (based on a landmark or an occupancy map) is static, so each sensor reading is assumed to be independent for each time step. Most interesting environments have dynamic changes to the map, for example people walking around an office environment. The approaches we presented have some robustness to such unmodeled factors through the sensor noise terms but might fail in highly dynamic scenes. In the literature, there are two approaches to deal with dynamic environments: *outlier rejection* filters at the sensor reading level to remove measurements linked to dynamic changes and *state augmentation* expands the state estimated by the filter to track the state of these dynamic changes, for example the poses of people in the environment.

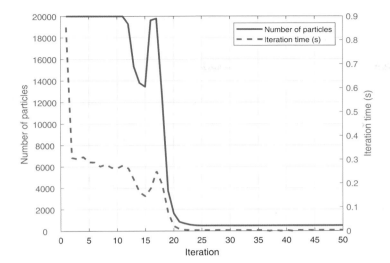

◘ Fig. 6.27 The number of particles (blue) starts high during global localization but drops to a minimum number of 500 after ∼ 20 iterations. The iteration processing time (red-dashed) is correlated with the number of particles. After convergence, each iteration only takes ∼ 3 ms

Both approaches significantly increase the computational complexity of the algorithm.

In practice, it can be difficult to tune the parameters of the motion model, sensor model, and MCL algorithm itself. It is a good idea to try to tune the parameters based on a reference data set for the desired robot and sensor system. If some ground truth localization information is available (either through external tracking or high-precision odometry), it can be used to verify which parameter settings are appropriate for the used robot.

6.3.5 Simulating Lidar Sensors

In the previous section, we talked about the importance of a reference data set that contains ground truth information that allows us to tune the performance of our localization or mapping algorithms. If that reference data set is not available, we can create simulated environments, trajectories, and sensor readings. In ▶ Sect. 6.2.1, we saw that the `LandmarkSensor` object was able to simulate the range and bearing to a landmark. But what about simulating lidar sensor readings in occupancy maps?

Let's consider the now well-known occupancy map shown in ◘ Fig. 6.24. To simulate lidar readings for any sensor pose in the map, we need to know a couple of key sensor parameters and then can use Bresenham's algorithm (see ▶ Sect. 6.3.3) to trace each ray to the nearest cell that is occupied.

The `rangeSensor` object encapsulates these capabilities

```
>> lidar = rangeSensor
lidar =
  rangeSensor with properties:
                       Range: [0 20]
             HorizontalAngle: [-3.1416 3.1416]
   HorizontalAngleResolution: 0.0244
                  RangeNoise: 0
        HorizontalAngleNoise: 0
                 NumReadings: 258
```

and we can adjust the parameters to match the physical lidar sensor that recorded the lidar readings we imported in ▶ Sect. 6.3.2. That lidar has a maximum range

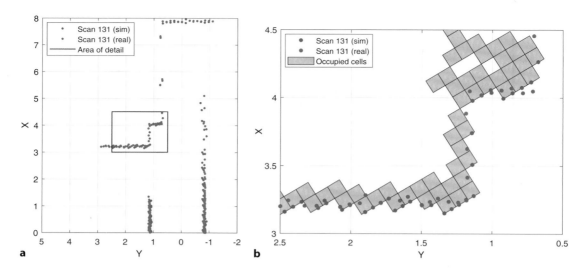

◘ Fig. 6.28 Simulated lidar scan. **a** The simulated lidar scan compared to the real lidar scan at pose 131; **b** a detailed section of the scan with occupied cells of the underlying occupancy map shown in gray. Notice that all simulated lidar points are located on the edges of occupied cells

If a beam does not intersect an occupied cell within the maximum range, a range of `NaN` will be returned for that scan angle.

of 50 meters and provides readings from -90° to 90° with an angular resolution of 1° ◄

```
>> lidar.Range = [0 50];
>> lidar.HorizontalAngle = deg2rad([-90 90]);
>> lidar.HorizontalAngleResolution = deg2rad(1);
```

We can now visualize the simulated lidar scan and compare it to the real lidar scan at the same pose

```
>> [ranges,angles] = lidar(odoTruth(131,:),omap);
>> clf
>> lidarScan(ranges,angles).plot
>> hold on
>> lidarData(131).plot
```

as shown in ◘ Fig. 6.28a. The scans from the real and simulated lidars match closely. The `rangeSensor` can be configured to simulate different types of sensor noise, including `RangeNoise` to adjust uncertainty in the range readings and `HorizontalAngleNoise` ◄ to simulate noise in the angle at which readings are taken. Taking a closer look at a section of the lidar readings in ◘ Fig. 6.28b, we can see that the precision of the simulated range readings is limited by the resolution of the underlying occupancy grid. All simulated range readings lie on the edge of occupied grid cells (shown in gray). A larger grid cell size will result in coarser resolution of simulated range readings.

Both noise quantities are specified as standard deviations of a zero-mean white noise process, with units in meters and radians, respectively.

6.4 Pose-Graph SLAM

An alternative approach to the SLAM problem is to formulate it as a pose graph as shown in ◘ Fig. 6.29, where the robot's path is considered to be a sequence of distinct poses and the task is to estimate those poses. There are two types of nodes: robot poses denoted by circles and landmark positions denoted by stars. An edge between two nodes represents a spatial constraint between them due to some observation $z_{i,j}$. The constraint between poses is typically derived from odometry – distance traveled and change in heading – or by matching successive lidar scans (as illustrated in ► Sect. 6.3.2). The constraint between a pose and a landmark is typically derived from a lidar sensor or a camera – distance and bearing angle from the robot to the landmark.

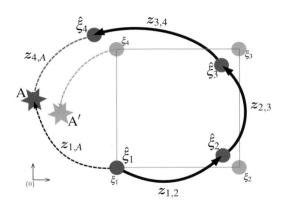

Fig. 6.29 Fundamentals of pose-graph SLAM. The robot is moving counter-clockwise around the square. The estimated poses are denoted by blue circles and the solid arrows are observed odometry as the robot moves between poses. The blue star is a landmark point and observations of that point from poses in the graph are shown as dashed lines. $z_{i,j}$ is a sensor observation associated with the edge from node i to j. We assume that $\hat{\boldsymbol{\xi}}_1 = \boldsymbol{\xi}_1$

As the robot progresses counter-clockwise around the square, it compounds an increasing number of uncertain relative poses from odometry so that the cumulative error in the estimated pose of the nodes is increasing – the problem with dead reckoning we discussed in ▶ Sect. 6.1. By the time the robot reaches the fourth position the pose error is significant. However, the robot is able to observe landmark A, which it saw previously, and this adds the constraint shown in red, which forms a loop in the pose graph – this type of event is referred to as a *loop closure*. Loop closure can also be achieved by matching lidar scans to previously encountered scans or submaps.

The function $\boldsymbol{g}(\cdot)$, for example (6.19), estimates the landmark's world coordinates given the robot pose and a sensor observation. The estimate $\boldsymbol{g}(\hat{\boldsymbol{\xi}}_1, \boldsymbol{z}_{1,A})$ will be subject to sensor error, but the estimate $\boldsymbol{g}(\hat{\boldsymbol{\xi}}_4, \boldsymbol{z}_{4,A})$ will be subject to sensor error and the accumulated odometry error in $\hat{\boldsymbol{\xi}}_4$. This discrepancy, analogous to the innovation of the Kalman filter, can be used to adjust the nodes – poses and landmarks – so as to minimize the total error. However, there is insufficient information to determine exactly where the error lies, so naively adjusting the $\hat{\boldsymbol{\xi}}_4$ node to fit the landmark observation might increase the error in another part of the graph. To minimize the error consistently over the whole graph we will formulate this as a minimization problem, which ensures that an error in one part of the graph is optimally distributed across the whole graph.

We consider first the case without landmarks, and all nodes represent poses. The state vector contains the poses of all the nodes

$$\boldsymbol{x} = \{\boldsymbol{\xi}_1, \boldsymbol{\xi}_2, \ldots, \boldsymbol{\xi}_N\} \tag{6.31}$$

and we seek an estimate $\hat{\boldsymbol{x}}$ that minimizes the error across all the edges

$$\hat{\boldsymbol{x}} = \arg\min_{\boldsymbol{x}} \sum_k F_k(\boldsymbol{x}) \tag{6.32}$$

where $F_k(\boldsymbol{x}) \in \mathbb{R}_{\geq 0}$ is a cost associated with the edge k. The term that we are minimizing is the total edge error that is analogous to the potential energy in a flexible structure which we want to *relax* into a minimum energy state.

■ Fig. 6.30 shows two nodes from the pose graph, specifically node i and node j, which represent the estimated poses $\hat{\boldsymbol{\xi}}_i$ to $\hat{\boldsymbol{\xi}}_j$, and edge k, which connects them. We have two values for the relative pose between the nodes. One is the explicit sensor measurement $\boldsymbol{z}_{i,j}$ and the other, shown dashed, is implicit and

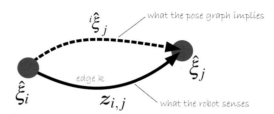

□ **Fig. 6.30** A single edge of the pose graph showing the estimated robot poses $\hat{\boldsymbol{\xi}}_i$ and $\hat{\boldsymbol{\xi}}_j$, the relative pose based on the sensor measurement $\boldsymbol{z}_{i,j}$, and the implicit relative pose shown as a dashed line

based on the current estimates from the pose graph. Any difference between the two indicates that one or both nodes need to be moved.

For simplicity, we are assuming a 2D pose graph for the rest of this section, but the calculations and workflows also scale to 3D. The first step is to express the error associated with the graph edge in terms of the sensor measurement and our best estimates of the node poses with respect to the world. For edge k the relative pose, based on the current estimates in the pose graph, is

$$^i\hat{\boldsymbol{\xi}}_j = \ominus\hat{\boldsymbol{\xi}}_i \oplus \hat{\boldsymbol{\xi}}_j$$

which we can write as an **SE**(2) matrix $^i\mathbf{T}_j$. We can also estimate the relative pose from the observation, the odometry $\boldsymbol{z}_k = \boldsymbol{z}_{i,j} = (\delta_d, \delta_\theta)$ which leads to ◄

$$\mathbf{T}_z(\boldsymbol{z}_k) = \begin{pmatrix} \cos\delta_\theta & -\sin\delta_\theta & \delta_d\cos\delta_\theta \\ \sin\delta_\theta & \cos\delta_\theta & \delta_d\sin\delta_\theta \\ 0 & 0 & 1 \end{pmatrix}.$$

This assumes that the initial pose is $[0, 0, 0]$. See ► Sect. 6.1.1 for more details on odometry modeling.

The *difference* between these two relative poses is $\mathbf{T}_e = \mathbf{T}_z^{-1}\,{}^i\mathbf{T}_j$ that can be written as

$$\boldsymbol{f}_k(\boldsymbol{x}, \boldsymbol{z}) = \left[\mathbf{T}_z^{-1}(\boldsymbol{z}_{i,j})\mathbf{T}^{-1}(\boldsymbol{x}_i)\mathbf{T}(\boldsymbol{x}_j)\right]_{xy\theta} \in \mathbb{R}^3$$

which is a configuration $(x_e, y_e, \theta_e) \in \mathbb{R}^2 \times \mathbf{S}^1$ that will be zero if the observation and the relative pose of the nodes are equal. To obtain the scalar cost required by (6.32) from the error vector $\boldsymbol{e}$ we use a quadratic expression

$$F_k(\boldsymbol{x}, \boldsymbol{z}_k) = \boldsymbol{f}_k^\top(\boldsymbol{x}, \boldsymbol{z}_k)\boldsymbol{\Omega}_k\boldsymbol{f}_k(\boldsymbol{x}, \boldsymbol{z}_k) \tag{6.33}$$

In practice this matrix is diagonal reflecting confidence in the x-, y-, and θ-directions. The *bigger* (in a matrix sense) $\boldsymbol{\Omega}$ is, the more the edge *matters* in the optimization procedure. Different sensors have different accuracy and this must be taken into account. Information from a high-quality sensor should be given more weight than information from a low-quality sensor. If no accuracy information is available from the sensor, an identity information matrix is a reasonable start.

where $\boldsymbol{\Omega}_k \in \mathbb{R}^{3\times3}$ is a positive-definite information matrix used as a weighting term. ◄ Although (6.33) is written as a function of all poses $\boldsymbol{x}$, it in fact depends only on elements i and j of $\boldsymbol{x}$ and the measurement $\boldsymbol{z}_k$. Solving (6.32) is a complex optimization problem that does not have a closed-form solution, but this kind of nonlinear least squares problem can be solved numerically if we have a good initial estimate of $\hat{\boldsymbol{x}}$. Specifically, this is a sparse nonlinear least squares problem which is discussed in ► App. F.2.4.

The edge error $\boldsymbol{f}_k(\boldsymbol{x})$ can be linearized (see ► App. E) about the current state $\boldsymbol{x}_0$ of the pose graph

$$\boldsymbol{f}'_k(\boldsymbol{\Delta}) \approx \boldsymbol{f}_{0,k} + \mathbf{J}_k\boldsymbol{\Delta}$$

where $\boldsymbol{\Delta}$ is a displacement relative to $\boldsymbol{x}_0$, $\boldsymbol{f}_{0,k} = \boldsymbol{f}_k(\boldsymbol{x}_0)$ and

$$\mathbf{J}_k = \frac{\partial\boldsymbol{f}_k(\boldsymbol{x})}{\partial\boldsymbol{x}} \in \mathbb{R}^{3\times3N}$$

is a Jacobian matrix that depends only on the pose of its two nodes $\boldsymbol{\xi}_i$ and $\boldsymbol{\xi}_j$ and is therefore mostly zeros

$$\mathbf{J}_k = \begin{pmatrix} \mathbf{0} & \cdots & \mathbf{A}_i & \cdots & \mathbf{B}_j & \cdots & \mathbf{0} \end{pmatrix},$$

$$\text{where} \quad \mathbf{A}_i = \frac{\partial\boldsymbol{f}_k(\boldsymbol{x})}{\partial\boldsymbol{\xi}_i} \in \mathbb{R}^{3\times3}, \ \mathbf{B}_j = \frac{\partial\boldsymbol{f}_k(\boldsymbol{x})}{\partial\boldsymbol{\xi}_j} \in \mathbb{R}^{3\times3}.$$

There are many ways to compute the Jacobians but here we will demonstrate the use of the Symbolic Math Toolbox™

```
>> syms xi yi ti xj yj tj xm ym tm assume real
>> xi_e = inv(tform2d(xm,ym,tm))*inv(tform2d(xi,yi,ti)) * ...
>>    tform2d(xj,yj,tj);
>> xyt = [xi_e(1,3); xi_e(2,3); ...
>>    atan2(xi_e(2,1),xi_e(1,1))]; % Extract [x;y;theta] vector
>> fk = simplify(xyt);
```

and the Jacobian that describes how the function f_k varies with respect to ξ_i is

```
>> Ai = simplify(jacobian(fk,[xi yi ti]))
Ai =
[-cos(ti+tm), -sin(ti+tm), yj*cos(ti+tm)  - yi*cos(ti+tm)
  + xi*sin(ti+tm)  - xj*sin(ti+tm)]
[ sin(ti+tm), -cos(ti+tm), xi*cos(ti+tm)  - xj*cos(ti+tm)
  + yi*sin(ti+tm)  - yj*sin(ti+tm)]
[ 0, 0, -1]
>> size(Ai)
ans =
     3    3
```

and we follow a similar procedure for $\mathbf{B}_j$. The Symbolic Math Toolbox can automatically generate code, in a variety of programming languages, to compute the Jacobians.

It is quite straightforward to solve this type of pose graph problem in MATLAB. We load a simple pose graph, similar to ◻ Fig. 6.29, from a data file ▶

```
>> pg = g2oread("pg1.g2o")
pg =
  poseGraph with properties:
              NumNodes: 4
              NumEdges: 4
    NumLoopClosureEdges: 1
     LoopClosureEdgeIDs: 4
       LandmarkNodeIDs: [1×0 double]
```

This simple text-based file format is used by the popular pose graph optimization package g²o which can be found at ▶ https://openslam-org.github.io/g2o.html.

which returns a `poseGraph` object that describes the pose graph. ▶ This is a simple pose graph with only 4 nodes, one loop closure, and no landmarks. We can show the initial guesses for the node poses, one per row, and visualize the graph

If the imported g2o file contains nodes in 3D, the function will return a `poseGraph3D` object instead.

```
>> pg.nodeEstimates
ans =
          0        0        0
    11.0000   2.0000   0.1000
     9.0000  14.0000   1.5000
     3.0000  16.0000  -2.0000
>> clf; pg.show(IDs="nodes");
```

as shown in ◻ Fig. 6.31a.

The pose graph optimization reduces the error in the network based on the measurements and information matrices associated with each edge

```
>> pgopt = optimizePoseGraph(pg,VerboseOutput="on");
Iteration 1, residual error 23511.224819.
Iteration 2, residual error 20860.194090.
...
Iteration 6, residual error 0.000000.
```

The displayed text indicates that the total error is decreasing, and the nodes are moving into a configuration that minimizes the overall error in the network. The pose graph iterations are overlaid and shown in ◻ Fig. 6.31b.

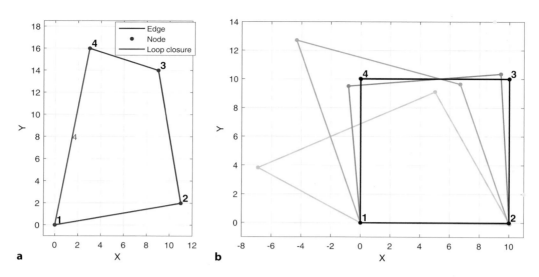

Fig. 6.31 Pose graph optimization. **a** Initial pose graph after loading `pg1.g2o`. Each node is numbered. All edges between nodes are created through odometry observations. The loop closure edge is highlighted in red; **b** shows the result of consecutive optimizations with increasing darkness; the final result is shown in black with node numbers

The simple text-based TORO file format is used by the TORO SLAM library, which can be found at ▶ https://openslam-org.github.io/toro.html.

Now let's look a much larger example based on real robot data ◄

```
>> pg = tororead("killian-small.toro")
pg =
  poseGraph with properties:
               NumNodes: 1941
               NumEdges: 3995
     NumLoopClosureEdges: 2055
      LoopClosureEdgeIDs: [127 128 130 131 133 ...]
       LandmarkNodeIDs: [1×0 double]
```

There are many nodes, and this takes a few seconds.

which we can plot ◄

```
>> pg.show(IDs="off");
```

and this is shown in ▢ Fig. 6.32a. Note the mass of edges in the center of the graph, and if you zoom in you can see these in detail. For this data set, the `NumNodes` nodes

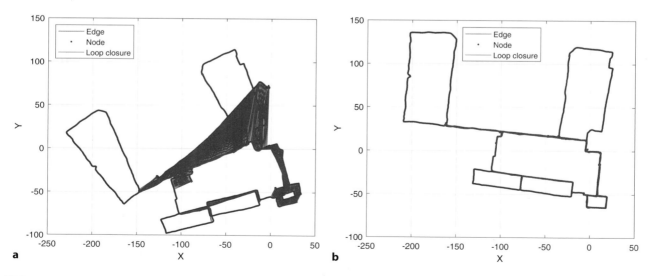

Fig. 6.32 Pose graph with 1941 nodes and 3995 edges, from the MIT Killian Court dataset. **a** Initial configuration; notice the long loop closure lines (red) that indicate large errors in node alignment; **b** final configuration after optimization

come from odometry data and the NumEdges edges are created from odometry and lidar. The lidar detects places that the robot previously visited and these edges are added as loop closures (there are a total of NumLoopClosureEdges). We optimize the pose graph by ▶

```
>> pgopt = optimizePoseGraph(pg,"g2o-levenberg-marquardt", ...
>>    VerboseOutput="on");
Iteration 1, residual error 6973602.836759
...
Iteration 16, residual error 10343.782692
```

This function prints information about each iteration. The output has been cropped for inclusion in this book.

and the final configuration is shown in ◻ Fig. 6.32b. In this call to the optimization function, we also selected a different solver (g2o-levenberg-marquardt) that can optimize this pose graph faster than the default solver. Both solvers typically give very similar results, but depending on the input pose graph, one might be faster than the other. The original pose graph had severe pose errors from accumulated odometry error that meant the two trips along the corridor were initially very poorly aligned (see the long red lines in ◻ Fig. 6.32a)

Pose graph optimization results in a graph that has optimal *relative* poses and positions, but the absolute poses and positions are not necessarily correct. ▶ We can remedy this by fixing or *anchoring* one or more nodes (poses or landmarks) and not updating them during the optimization, and this is discussed in ▶ App. F.2.4.2. For example, we can fix the pose of the first node

This is similar to the estimates in the map frame that we saw with the EKF SLAM system in ▶ Sect. 6.2.3.

```
>> pgopt = optimizePoseGraph(pg,"g2o-levenberg-marquardt", ...
>>    FirstNodePose=[-50 20 deg2rad(95)]);
>> pgopt.show(IDs="off");
```

where the first node is at $x = -50$, $y = 20$, and $\theta = 95°$. ▶ Since all other nodes in the pose graph are defined relative to the first node, we will see the whole optimized pose graph translated and rotated by this first pose.

If FirstNodePose is not specified, it is assumed to be [0,0,0].

6.4.1 Pose-Graph Landmark SLAM

The pose graph can also include landmarks as shown in ◻ Fig. 6.33. Landmarks are described by a coordinate vector $\boldsymbol{p}_j \in \mathbb{R}^2$, not a pose, and therefore differ from the nodes discussed so far. To accommodate this, we redefine the state vector to be

$$x = \{\boldsymbol{\xi}_1, \boldsymbol{\xi}_2, \ldots, \boldsymbol{\xi}_N | \boldsymbol{p}_1, \boldsymbol{p}_2, \ldots, \boldsymbol{p}_M\} \tag{6.34}$$

which includes N robot poses and M landmark positions. The robot at pose i observes landmark j at range and bearing $\boldsymbol{z}_{i,j} = (r^\#, \beta^\#)$, where $\cdot^\#$ denotes a measured value, and is converted to Cartesian form in frame {i}

$$^i\boldsymbol{p}_j^\# = \left(r^\# \cos \beta^\#, \ r^\# \sin \beta^\#\right) \in \mathbb{R}^2 \ .$$

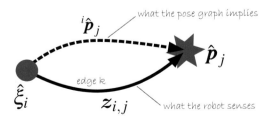

◻ **Fig. 6.33** A single edge of the pose graph showing the estimated pose $\hat{\boldsymbol{\xi}}_i$ and estimated robot landmark position $\hat{\boldsymbol{p}}_j$, the relative position based on the sensor measurement $\boldsymbol{z}_{i,j}$, and the implicit relative position shown as a dashed line

6

The estimated position of the landmark in frame {i} can also be determined from the nodes of the pose graph

$$^{i}\hat{\boldsymbol{p}}_{j} = \left(\ominus\, ^{0}\hat{\boldsymbol{\xi}}_{i} \right) \cdot \hat{\boldsymbol{p}}_{j} \in \mathbb{R}^{2}$$

and the error vector is

$$\boldsymbol{f}_{k}(\boldsymbol{x}, \boldsymbol{z}_{k}) = {}^{i}\hat{\boldsymbol{p}}_{j} - {}^{i}\boldsymbol{p}_{j}^{\#} \in \mathbb{R}^{2}\,.$$

Edge cost is given by (6.33) and $\boldsymbol{\Omega} \in \mathbb{R}^{2\times 2}$. We follow a similar approach as earlier, but the Jacobian matrix is now

$$\mathbf{J}_{k} = \frac{\partial \boldsymbol{f}_{k}(\boldsymbol{x})}{\partial \boldsymbol{x}} \in \mathbb{R}^{2\times(3N+2M)}$$

which again is mostly zero

$$\mathbf{J}_{k} = \begin{pmatrix} \mathbf{0} & \cdots & \mathbf{A}_{i} & \cdots & \mathbf{B}_{j} & \cdots & \mathbf{0} \end{pmatrix}$$

but the two nonzero blocks now have different widths

$$\mathbf{A}_{i} = \frac{\partial \boldsymbol{f}_{k}(\boldsymbol{x})}{\partial \boldsymbol{\xi}_{i}} \in \mathbb{R}^{2\times 3}, \quad \mathbf{B}_{j} = \frac{\partial \boldsymbol{f}_{k}(\boldsymbol{x})}{\partial \boldsymbol{p}_{j}} \in \mathbb{R}^{2\times 2}$$

and the solution can be achieved as before, see ▶ App. F.2.4 for more details.

The `poseGraph` object we used earlier is also able to handle landmark nodes. One popular artificial landmark is based on fiducial markers called AprilTags that can be easily sensed by a camera (see ▶ Sect. 13.6.1). When the camera is calibrated, it is straightforward to estimate the pose of the marker relative to the camera. Based on the estimated pose of these markers, we can add them as landmarks to the pose graph. A full example of this workflow is given in

```
>> openExample("nav/LandmarkSLAMUsingAprilTagMarkersExample")
```

and a screenshot is shown in ◻ Fig. 6.34.

▶ sn.pub/Vqs8Co

a

b

◻ **Fig. 6.34** Pose graph SLAM with odometry and landmarks. **a** AprilTag landmarks are mounted on different surfaces and detected in the camera image (red dots mark detected corners); **b** the optimized pose graph is overlaid on a blueprint of the corridor. As expected, the optimized landmark positions (green diamonds) are clearly placed close to walls in the blueprint (Images reprinted with permission of The MathWorks, Inc.)

6.4.2 Pose-Graph Lidar SLAM

In practice, a pose-graph SLAM system comprises two asynchronous subsystems as shown in ◨ Fig. 6.35: a *front end* and a *back end*, connected by a pose graph. The front end adds new nodes as the robot travels ▶ as well as edges that define constraints between nodes. For example, when moving from one place to another, wheel odometry gives an estimate of distance and change in orientation which is a constraint. In addition, the robot's exteroceptive sensors may observe the relative position of a landmark and this also adds a constraint. Every measurement adds a constraint – an edge in the graph. There is no limit to the number of edges entering or leaving a node. The back end runs periodically to optimize the pose graph and adjusts the poses of the nodes so that the constraints are satisfied as well as possible, that is, that the sensor observations are best explained. Since the graph is only ever extended in the region around the robot, it is possible to optimize just a local subset of the pose graph and less frequently optimize the entire graph. If nodes are found to be equivalent after optimization, they can be merged.

Let's see how this system would work in practice assuming that we have information from a lidar sensor. We can load some lidar scans that were prerecorded in an indoor environment

> Typically a new node is added by the front end every meter or so of travel, or after a sharp turn.

```
>> load("indoorSLAMData.mat","scans");
```

which creates a `scans` variable in the workspace.

We can use the techniques from ▶ Sect. 6.3.2 to align lidar scans and determine the relative robot motion. One popular technique for large-scale loop closure detection with lidar scans is to build smaller submaps of the environment based on several successive scans, which is sufficiently accurate for short periods of time. Loop closures are then detected by matching new scans against this set of submaps to recognize places we previously visited. To optimize performance, only submaps that are considered close to the current pose estimate are used for matching. Once loop closures are detected and added to the pose graph, we can optimize the pose graph.

This algorithm is implemented in the `lidarSLAM` object

```
>> maxLidarRange = 9;  % meters
>> mapResolution = 20; % cells per meter, cell size = 5 cm
>> slam = lidarSLAM(mapResolution,maxLidarRange)
slam =
  lidarSLAM with properties:
                    PoseGraph: [1×1 poseGraph]
                MapResolution: 20
                MaxLidarRange: 9
                OptimizationFcn: @optimizePoseGraph
          LoopClosureThreshold: 100
      LoopClosureSearchRadius: 8
```

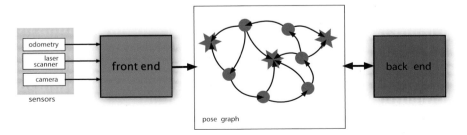

◨ **Fig. 6.35** Pose graph SLAM system. The front end creates nodes as the robot travels, and creates edges based on sensor data. The back end adjusts the node positions to minimize total error

6

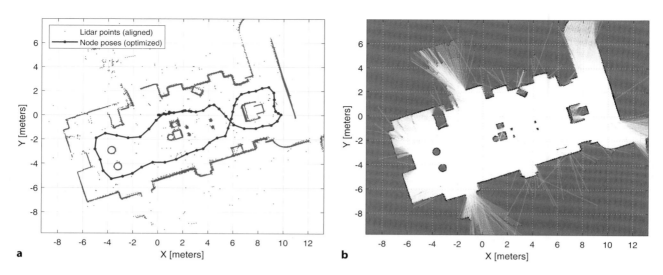

◻ Fig. 6.36 Pose graph SLAM with lidar scans. **a** The optimized poses and raw lidar scans that are aligned based on the poses; **b** the generated occupancy map

```
   LoopClosureMaxAttempts: 1
  LoopClosureAutoRollback: 1
     OptimizationInterval: 1
        MovementThreshold: [0 0]
   ScanRegistrationMethod: 'BranchAndBound'
    TranslationSearchRange: [4.5000 4.5000]
       RotationSearchRange: 1.5708
```

which we initialize with a resolution of 20 cells per meter and a modest lidar sensor range of 9 m. The object has many parameters, but some of the most critical ones are related to loop closure detection. As mentioned before, new lidar scans are matched against smaller submaps and the `LoopClosureThreshold` determines the minimum match score from the scan matching algorithm before a loop closure is detected. This threshold is set empirically based on the environment. Using a higher threshold helps reject false positives in loop closure detection, although setting the threshold too high might prevent valid loop closures from being detected. Only submaps that are considered close to the current pose estimate are used for matching and the `LoopClosureSearchRadius` (in meters) determines in which area submaps are matched. Using a higher radius allows the algorithm to search a wider range of the submaps around the current pose estimate for loop closures. For this set of lidar scans, the parameter settings

```
>> slam.LoopClosureThreshold = 200;
>> slam.LoopClosureSearchRadius = 8;
```

This can take up to a minute to process all the data

work well. We can now add lidar scans by calling `addScan` ◀

```
>> for i = 1:70
>>    slam.addScan(scans{i});
>> end
```

which adds new nodes to the pose graph based on scan-to-scan matching, finds loop closures based on matches to submaps, and regularly optimizes the pose graph. Once this process completes, we can visualize the optimized trajectory and the aligned lidar scans

```
>> slam.show;
```

as shown in ◻ Fig. 6.36a.

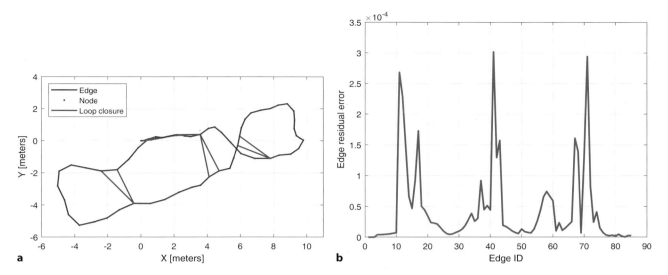

☐ **Fig. 6.37** The optimized pose graph. **a** The pose graph showing several loop closures; **b** the residual errors for all the edges; note that the residuals are small, indicating a good global fit

As we discussed in ▶ Sect. 6.3.3, the optimized scans and poses can be used to generate an `occupancyMap`

```
>> [scansAtPoses,optimizedPoses] = slam.scansAndPoses;
>> map = buildMap(scansAtPoses,optimizedPoses, ...
>>    mapResolution,maxLidarRange);
>> map.show;
```

which represents the environment as a probabilistic occupancy grid and is shown in ☐ Fig. 6.36b. The occupancy map removes a lot of the lidar scan outliers and shows good global alignment.

Let's take a closer look at the optimized pose graph. As we discussed in the context of (6.33), we can find a scalar residual error for each edge that will be zero if the observation (measured odometry) and the relative pose of the two nodes are equal. We can view the pose graph and residual errors

```
>> pg = slam.PoseGraph;
>> pg.show;
>> figure; plot(pg.edgeResidualErrors)
```

in ☐ Fig. 6.37a and b respectively. We can see that the edge residuals are exceptionally low, which indicates a good overall optimization. ▶

An incorrectly detected loop closure will distort the entire graph and raise $\sum_k F_k(\hat{x})$. Let's illustrate this by intentionally inserting an incorrect loop closure into the pose graph and assessing its effect on the final result. We can add a new loop closure between nodes 59 and 44 ▶

```
>> pgwrong = pg.copy;
>> [~,id] = pgwrong.addRelativePose([0 0 0],[1 0 0 1 0 1],59,44)
id =
    86
```

by telling the pose graph that they are coincident nodes (relative pose is (0,0,0)) and that we have high confidence in this fact (information matrix ▶ has large diagonal elements). The arguments to `addRelativePose` are the relative pose between the nodes, the compact information matrix, the start node, and the goal node. This adds a new loop closure edge with ID 86. We can then optimize this pose graph to see the effect on the pose graph and the residual errors.

```
>> pgwrongopt = optimizePoseGraph(pgwrong);
>> pgwrongopt.show;
>> figure; plot(pgwrongopt.edgeResidualErrors)
```

Remember that when calculating the scalar residual, the pose error (vector) is multiplied by the information matrix, which could be different for each edge. That means that we should interpret this residual plot with caution and realize that the pose errors might be weighted differently for different parts of the pose graph.

We can plot all the node and edge IDs by calling `pg.show(IDs="all")`.

The information matrix is the inverse of the covariance matrix and is specified as a compact 6-element vector that represents the upper triangle of the square information matrix. Here, an input of `[1 0 0 1 0 1]` corresponds to the information matrix

$$\begin{pmatrix} 1 & 0 & 0 \\ 0 & 1 & 0 \\ 0 & 0 & 1 \end{pmatrix}$$

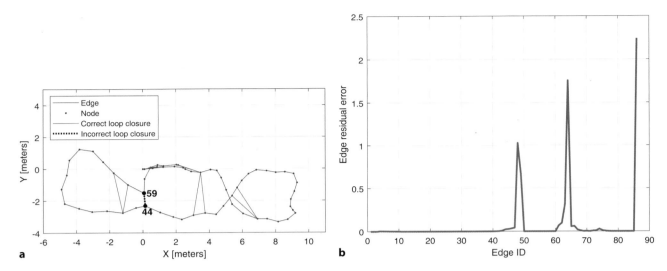

Fig. 6.38 Pose graph with erroneous loop closure. **a** The optimized pose graph with correct and erroneous loop closures; **b** the residual errors are significantly larger than in ▢ Fig. 6.37b (note the scale difference on the y axis), indicating the presence of wrong loop closures

The resulting (distorted) pose graph is shown in ▢ Fig. 6.38a. The corresponding residual errors in ▢ Fig. 6.38b are much larger than the original residuals in ▢ Fig. 6.37b.

One strategy to resolve this problem is to progressively remove the highest cost edges until the graph is able to *relax* into a low energy state. Based on ▢ Fig. 6.38b, this would work in this case, since edge 86 has the highest residual error. In practice, we rarely know how many loop closure edges might be erroneous and a naive approach of iteratively removing the highest cost edges might remove correct loop closures as well. A more sophisticated approach is implemented in the trimLoopClosures function, which is based on the graduated non-convexity (GNC) method. Running this on the wrong pose graph with an empirically chosen threshold

```
>> trimParams.TruncationThreshold = 2;
>> trimParams.MaxIterations = 10;
>> [pgFixed,trimInfo] = trimLoopClosures(pgwrongopt, ...
>>    trimParams,poseGraphSolverOptions);
>> trimInfo
trimInfo =
  struct with fields:
            Iterations: 3
    LoopClosuresToRemove: 86
```

removes the erroneous loop closure (edge ID 86) while preserving all correct ones. This function can deal with a large number of erroneous loop closures and is a powerful tool to prevent pose graph corruption.

Throughout this whole section, we have used many parameters and thresholds to ensure that our lidar SLAM solution is accurate. Experimenting with these parameters can be tedious. Visual inspection of intermediate results is critical and manual modifications of the pose graph might be needed. To assist with this task, we can use the interactive SLAM Map Builder app (slamMapBuilder) and a screenshot is shown in ▢ Fig. 6.39.

6.5 Wrapping Up

In this chapter we learned about two very different ways of estimating a robot's position: by dead reckoning, and by observing landmarks or lidar features whose true position is known from a map. Dead reckoning is based on the integration

Fig. 6.39 The SLAM Map Builder app operating on the same data that we used in this section. The app is currently processing scan 35. On the right side, we can see the incremental scan match (from scan 34 to scan 35) and the loop closure that was just identified (from scan 35 to scan 18)

of odometry information, the distance traveled, and the change in heading angle. Over time, errors accumulate leading to increased uncertainty about the pose of the robot.

We modeled the error in odometry by adding noise to the sensor outputs. The noise values are drawn from some distribution that describes the errors of that particular sensor. For our simulations, we used zero-mean Gaussian noise with a specified covariance, but only because we had no other information about the specific sensor. The most realistic noise model available should be used. We then introduced the Kalman filter that provides an optimal estimate, in the least-squares sense, of the true configuration of the robot based on noisy measurements. The Kalman filter is however only optimal for the case of zero–mean Gaussian noise and a linear model. The model that describes how the robot's configuration evolves with time can be nonlinear in which case we approximate it with a linear model that included some partial derivatives expressed as Jacobian matrices – an approach known as extended Kalman filtering.

The Kalman filter also estimates uncertainty associated with the pose estimate and we see that the magnitude can never decrease and typically grows without bound. Only additional sources of information can reduce this growth and we looked at how observations of landmarks, with known positions, can be used. Once again, we use the Kalman filter, but in this case, we use both the prediction and the update phases of the filter. We see that in this case the uncertainty can be decreased by a landmark observation, and that over the longer term, the uncertainty does not grow. We then applied the Kalman filter to the problem of estimating the positions of the landmarks given that we knew the precise position of the robot. In this case, the state vector of the filter comprised the coordinates of the landmarks.

Next, we brought all this together and estimated the robot's configuration, the position of the landmarks and their uncertainties – simultaneous localization and mapping (SLAM). The state vector in this case contained the configuration of the robot and the coordinates of the landmarks.

An important problem when using landmarks is data association – determining which landmark has been observed by the sensor so that its position can be looked up in a map of known or estimated landmark positions. If the wrong landmark is looked up, then an error will be introduced in the robot's position that may lead to future data associations errors and catastrophic divergence between the estimated and true state.

Secondly, we looked at odometry, mapping, localization, and SLAM approaches that operate on data from a lidar sensor, which is a very common sensor for mobile robots. Rather than relying on discrete landmark maps, the lidar-based algorithms use occupancy grid maps that subdivide the world into grid cells with obstacle probabilities.

We looked at Monte-Carlo estimation and introduced the particle filter. This technique is computationally intensive but makes no assumptions about the distribution of errors from the sensor or the linearity of the robot motion model and supports multiple hypotheses. Particle filters can be considered as providing an approximate solution to the true system model, whereas a Kalman filter provides an exact solution to an approximate system model. We used the particle filter to localize a robot in an occupancy grid map, an algorithm called Monte-Carlo Localization. Rao-Blackwellized SLAM combines Monte-Carlo and Kalman filter techniques.

We also introduced an alternative approach to filter-based techniques. Pose graph SLAM involves solving a large but sparse nonlinear least squares problem, turning the problem from one of (Kalman) filtering to one of optimization.

6.5.1 Further Reading

■ ■ **Localization and SLAM**

The tutorials by Bailey and Durrant-Whyte (2006) and Durrant-Whyte and Bailey (2006) are a good introduction to this topic, while the textbook *Probabilistic Robotics* (Thrun et al. 2005) is a readable and comprehensive coverage of all the material touched on in this chapter. The book by Barfoot (2017) covers all the concepts of this chapter and more, albeit with different notation. The book by Siegwart et al. (2011) also has a good treatment of robot localization.

Particle filters are described by Thrun et al. (2005), Stachniss and Burgard (2014), and the tutorial introduction by Rekleitis (2004). There are many variations such as fixed or adaptive number of particles and when and how to resample – Li et al. (2015) provide a comprehensive review of resampling strategies. Determining the most likely pose was demonstrated by taking the weighted mean of the particles but many more approaches have been used. The kernel density approach takes the particle with the highest weight of neighboring particles within a fixed-size surrounding hypersphere. The Adaptive Monte Carlo Localization (AMCL) was originally introduced by Dellaert et al. (1999) and covered in more detail in Thrun et al. (2005).

Pose graph optimization, also known as GraphSLAM, has a long history starting with Lu and Milios (1997). More recently there has been significant progress with many publications (Grisetti et al. 2010) and open-source tools including g^2o (Kümmerle et al. 2011), $\sqrt{SAM}$ (Dellaert and Kaess 2006), iSAM (Kaess et al. 2007), and factor graphs leading to GTSAM ► https://gtsam.org/tutorials/intro.html. Agarwal et al. (2014) provides a gentle introduction to pose graph SLAM and discusses the connection to land-based geodetic survey which is centuries old. Factor graphs have recently become very popular, since they abstract the concept of poses in the pose graph to arbitrary *factors* that represent probabilistic constraints on the nodes of the factor graph. This enables a more natural representation of data coming from multiple sensors with different sensing modalities. Factor graphs for robotics applications were popularized in Dellaert (2012) and, in more depth, in the book by Dellaert and Kaess (2017). The Graduated Non-Convexity (GNC) method that we used in ▶ Sect. 6.4.2 for removing wrong loop closures was introduced by Yang et al. (2020).

Rao-Blackwellized SLAM implementations include FastSLAM (Montemerlo et al. 2003; Montemerlo and Thrun 2007) as well as gmapping (▶ https://openslam-org.github.io/gmapping.html).

Popular implementations include Hector SLAM (▶ https://wiki.ros.org/hector_slam) and Cartographer (▶ https://github.com/cartographer-project/cartographer), both of which, along with gmapping, are available as part of the ROS ecosystem.

There are many online resources related to SLAM. A collection of open-source SLAM implementations is available from OpenSLAM at ▶ http://www.openslam.org. An implementation of smoothing and mapping using factor graphs is available at ▶ https://gtsam.org and has C++, Python, and MATLAB bindings. MATLAB implementations of SLAM algorithms can be found at ▶ https://www.iri.upc.edu/people/jsola/JoanSola/eng/toolbox.html and ▶ https://www-personal.acfr.usyd.edu.au/tbailey. Great teaching resources available online include ▶ https://www.dis.uniroma1.it/~grisetti and Paul Newman's free ebook *C4B Mobile Robots and Estimation Resources* at ▶ https://www.free-ebooks.net/ebook/C4B-Mobile-Robotics.

V-SLAM or Visual SLAM is able to estimate 3-dimensional camera motion from images using natural images features (covered in ▶ Chap. 12) without the need for artificial landmarks. VI-SLAM is a V-SLAM system that also incorporates inertial sensor information from gyroscopes and accelerometers. VIO-SLAM is a VI-SLAM system that also incorporates odometry information. One of the first V-SLAM system was PTAM (parallel tracking and mapping) system by Klein and Murray (2007). PTAM has two parallel computational threads. One is the map builder that performs the front- and back-end tasks, adding landmarks to the pose graph based on estimated camera (robot) pose and performing graph optimization. The other thread is the localizer that matches observed landmarks to the estimated map to estimate the camera pose. ORB-SLAM is a powerful and mature family of V-SLAM implementations and the current version, ORB-SLAM3 (Campos 2021), supports V- and VI-SLAM with monocular, stereo, and RGB-D cameras with regular and fisheye lenses. PRO-SLAM (▶ https://gitlab.com/srrg-software/srrg_proslam) is a simple stereo V-SLAM system designed for teaching.

■ ■ Lidar scan matching and map making

In this chapter's sections, for example ▶ Sect. 6.3.2, we used the Normal Distribution Transform (NDT) for lidar scan matching and SLAM, originally proposed for 2D lidars by Biber and Straßer (2003), extended to 3D by Magnusson et al. (2007). Several implementations are available at ▶ https://pointclouds.org. Other scan matchers are based on the Iterative Closest Point (ICP) algorithm which is covered in more detail in ▶ Sect. 14.7. Many versions and variants of ICP exist. A comparison of ICP and NDT for a field robotic application is described by Magnusson et al. (2009). A fast and popular approach to lidar scan matching is that of Censi (2008) with additional resources at ▶ https://censi.science/research/robot-perception/plicp and this forms the basis of the canonical scan matcher in ROS. The grid-based scan matching algorithm covered in ▶ Sect. 6.3.2 and (for loop closure detection) in ▶ Sect. 6.4.2 was introduced by Hess et al. (2016). This algorithm is also an integral part of the lidar processing in the Cartographer library (▶ https://github.com/cartographer-project/cartographer).

Other approaches to scan matching rely on point cloud features, such as FPFH introduced by Rusu et al. (2009) and eigenvalue-based features introduced by

Weinmann et al. (2014). These methods can be faster and more robust than ICP and NDT for environments with distinct features and salient points.

When attempting to match a local geometric signature in a large point cloud (2D or 3D) to determine loop closure, we often wish to limit our search to a local spatial region. An efficient way to achieve this is to organize the data using a kd-tree. FLANN (Muja and Lowe 2009) is a fast approximation with MATLAB bindings and available at ▶ https://github.com/flann-lib/flann.

For creating a map from robotic lidar scan data in ▶ Sect. 6.3.1 we used the beam model or likelihood field as described in Thrun et al. (2005).

■ ■ **Kalman filtering**

There are many published and online resources for Kalman filtering. Kálmán's original paper, Kálmán (1960), over 50 years old, is quite readable. The first use for aerospace navigation in the 1960s is described by McGee and Schmidt (1985). The book by Zarchan and Musoff (2005) is a very clear and readable introduction to Kalman filtering. I have always found the classic book, recently republished, Jazwinski (2007) to be very readable. Bar-Shalom et al. (2001) provide comprehensive coverage of estimation theory and also the use of GPS. Groves (2013) also covers Kalman filtering. Welch and Bishop's online resources at ▶ https://www.cs.unc.edu/~welch/kalman have pointers to papers, courses, software, and links to other relevant web sites.

A significant limitation of the EKF is its first-order linearization, particularly for processes with strong nonlinearity. Alternatives include the iterated EKF described by Jazwinski (2007) or the Unscented Kalman Filter (UKF) (Julier and Uhlmann 2004) which uses discrete sample points (sigma points) to approximate the PDF. Some of these topics are covered in the Handbook (Siciliano and Khatib 2016, § 5 and § 35). The information filter is an equivalent filter that maintains an inverse covariance matrix which has some useful properties and is discussed in Thrun et al. (2005) as the sparse extended information filter.

The insertion Jacobian (6.22) is important but details of its formulation are difficult to find, a derivation can be found at ▶ https://sn.pub/XQv4jR.

■ ■ **Data association**

SLAM techniques are critically dependent on accurate data association between observations and mapped landmarks, and a review of data association techniques is given by Neira and Tardós (2001). FastSLAM (Montemerlo and Thrun 2007) is capable of estimating data association as well as landmark position. Fiducials, such as AprilTags and ArUco markers, discussed in ▶ Sect. 13.6.1, can be used as artificial landmarks that eliminate the data association problem since the identity is encoded in the fiducial. Mobile robots can uniquely identify places based on their visual appearance using tools such as OpenFABMAP (Glover et al. 2012) or DBoW2 (Gálvez-López and Tardós 2012).

Data association for Kalman filtering is covered in the Robotics Handbook (Siciliano and Khatib 2016). Data association in the tracking context is covered in considerable detail in the, now very old, book by Bar-Shalom and Fortmann (1988).

■ ■ **Sensors**

The book by Kelly (2013) has a good coverage of sensors particularly lidar range finders. For aerial and underwater robots, odometry cannot be determined from wheel motion and an alternative, also suitable for wheeled robots, is visual odometry (VO). This is introduced in the tutorials by Fraundorfer and Scaramuzza (2012) and Scaramuzza and Fraundorfer (2011) and will be covered in ▶ Chap. 14. The Robotics Handbook (Siciliano and Khatib 2016) has good coverage of a wide range of robotic sensors. The principles of GPS and other radio-based localization systems are covered in some detail in the book by Groves (2013), and a number of links to GPS technical data are provided from this book's web site. The SLAM problem can be formulated in terms of bearing-only or range-only measurements. A camera is effectively a bearing-only sensor, giving the direction to a feature in the world. A V-SLAM system is one that performs SLAM using bearing-only visual information, just a camera, and an introduction to the topic is given by Neira et al. (2008) and the associated special issue. Interestingly, the robotic V-SLAM problem is the same as the bundle adjustment problem known to the computer vision community and which will be discussed in ▶ Chap. 14.

The book by Borenstein et al. (1996), although very dated, has an excellent discussion of robotic sensors in general and odometry in particular. It is out of print but can be found online. The book by Everett (1995) covers odometry, range and bearing sensors, as well as radio, ultrasonic, and optical localization systems. Unfortunately, the discussion of range and bearing sensors is now quite dated since this technology has evolved rapidly over the last decade.

■ ■ **General interest**

Bray (2014) gives a very readable account of the history of techniques to determine our location on the planet. If you ever wondered how to navigate by the stars or use a sextant, Blewitt (2011) is a slim book that provides an easy-to-understand introduction.

The book *Longitude* (Sobel 1996) is a very readable account of the longitude problem and John Harrison's quest to build a marine chronometer.

6.5.2 **Exercises**

1. What is the value of the Longitude Prize in today's currency?
2. Implement a driver object (▶ Sect. 6.1.1) that drives the robot inside a circle with specified center and radius.
3. Derive an equation for heading change in terms of the rotational rate of the left and right wheels for the car-like and differential-steer robot models.
4. Dead Reckoning using Odometry (▶ Sect. 6.1)
   a) Experiment with different values of $\mathbf{P}_0$, $\mathbf{V}$ and $\hat{\mathbf{V}}$.
   b) ◘ Fig. 6.5 compares the actual and estimated position. Plot the actual and estimated heading angle.
   c) Compare the variance associated with heading to the variance associated with position. How do these change with increasing levels of range and bearing angle variance in the sensor?
   d) Derive the Jacobians in (6.6) and (6.7) for the case of a differential-steer robot.
5. Localizing in a Landmark Map (▶ Sect. 6.2.1)
   a) Vary the characteristics of the sensor (covariance, sample rate, range limits, and bearing angle limits) and investigate the effect on performance.
   b) Vary $\mathbf{W}$ and $\hat{\mathbf{W}}$ and investigate what happens to estimation error and final covariance.
   c) Modify the `LandmarkSensor` to create a bearing-only sensor, that is, as a sensor that returns angle but not range. The implementation includes all the Jacobians. Investigate performance.
   d) Modify the sensor model to return occasional errors (specify the error rate) such as incorrect range or beacon identity. What happens?
   e) Modify the EKF to perform data association instead of using the landmark identity returned by the sensor.
   f) ◘ Fig. 6.9 compares the actual and estimated position. Plot the actual and estimated heading angle.

g) Compare the variance associated with heading to the variance associated with position. How do these change with increasing levels of range and bearing angle variance in the sensor?

6. Creating a Landmark Map (▶ Sect. 6.2.2)
    a) Vary the characteristics of the sensor (covariance, sample rate, range limits, and bearing angle limits) and investigate the effect on performance.
    b) Use the bearing-only sensor from above and investigate performance relative to using a range and bearing sensor.
    c) Modify the EKF to perform data association instead of using identity returned by the sensor.

7. Simultaneous localization and mapping (▶ Sect. 6.2.3)
    a) Vary the characteristics of the sensor (covariance, sample rate, range limits, and bearing angle limits) and investigate the effect on performance.
    b) Use the bearing-only sensor from above and investigate performance relative to using a range and bearing sensor.
    c) Modify the EKF to perform data association instead of using the landmark identity returned by the sensor.
    d) ◾ Fig. 6.13 compares the actual and estimated position. Plot the actual and estimated heading angle.
    e) Compare the variance associated with heading to the variance associated with position. How do these change with increasing levels of range and bearing angle variance in the sensor?

8. Modify the pose graph optimizer and test using the simple graph `pg1.g2o`
    a) Anchor one node at a particular pose.
    b) Add one or more landmarks. You will need to derive the relevant Jacobians first and add the landmark positions, constraints, and information matrix to the data file.

9. Create a simulator for Buffon's needle problem, and estimate π for 10, 100, 1000 and 10,000 needle throws. How does convergence change with needle length?

10. Particle filter (▶ Sect. 6.2.4)
    a) Run the filter numerous times. Does it always converge?
    b) Vary the parameters $\mathbf{Q}$, $\mathbf{L}$, w_0 and N and understand their effect on convergence speed and final standard deviation.
    c) Investigate variations to the kidnapped robot problem. Place the initial particles around the initial pose. Place the particles uniformly over the xy-plane but set their orientation to its actual value.
    d) Use a different type of likelihood function, perhaps inverse distance, and compare performance.

11. Experiment with ArUCo markers or AprilTags. Print some tags and extract them from images using the functions described in ▶ Sect. 13.6.1.

12. Implement a lidar odometer and test it over the entire path saved in `killian.g2o`. Compare your odometer with the relative pose changes in the file.

13. In order to measure distance using lidar, what timing accuracy is required to achieve 1 cm depth resolution?

14. Reformulate the localization, mapping, and SLAM problems using a bearing-only landmark sensor.

15. Implement a localization or SLAM system using an external simulator such as CoppeliaSim. Obtain range measurements from the simulated robot, do lidar odometry and landmark recognition, and send motion commands to the robot. You can communicate with CoppeliaSim using its MATLAB API. Alternatively, use the Gazebo simulator and communicate using the ROS protocol.

Robot Manipulators

Contents

Chapter 7   Robot Arm Kinematics – 275

Chapter 8   Manipulator Velocity – 329

Chapter 9   Dynamics and Control – 355

Robot manipulator arms are a common and familiar type of robot. They are most commonly found in factories doing jobs such as assembly, welding and handling tasks. Other applications include picking items from shelves for e-commerce, packing fruit into trays, performing surgical procedures under human supervision, or even grappling cargo at the international space station. The first of these robots started work over 60 years ago and they have been enormously successful in practice – many millions of them are working in the world today, and they have assembled, packed, or handled many products that we buy. Unlike the mobile robots we discussed in the previous part, robot manipulators do not move through the world. They have a static base and therefore operate within a limited workspace.

The chapters in this part cover the fundamentals of serial-link manipulators, a subset of all robot manipulator arms. ► Chap. 7 introduces the topic and is concerned with kinematics – the geometric relationship between the robot's joint angles or positions, and the pose of its end effector. We discuss the creation of smooth paths that the robot can follow and present an example of a robot drawing a letter on a plane and a 4-legged walking robot. ► Chap. 8 introduces the relationship between the rate of change of joint coordinates and the end-effector velocity which is described by the manipulator Jacobian matrix. We also cover alternative methods of generating paths in 3D space and introduce the relationship between forces on the end effector, and torques or forces at the joints. ► Chap. 9 discusses independent joint control and some performance limiting factors such as weight and variable inertia. This leads to a discussion of the full nonlinear dynamics of serial-link manipulators – including inertia, gyroscopic forces, friction, and gravity – and more sophisticated model-based and task-space control approaches.

Robot Arm Kinematics

> **»** *Take to kinematics. It will repay you.*
> *It is more fecund than geometry; it adds a fourth dimension to space.*
> – Chebyshev to Sylvester 1873

Contents

7.1   Forward Kinematics – 277

7.2   Inverse Kinematics – 297

7.3   Trajectories – 308

7.4   Applications – 315

7.5   Advanced Topics – 320

7.6   Wrapping Up – 325

© The Author(s), under exclusive license to Springer Nature Switzerland AG 2023
P. Corke, W. Jachimczyk, R. Pillat, *Robotics, Vision and Control*, Springer Tracts in Advanced Robotics 147,
https://doi.org/10.1007/978-3-031-07262-8_7

7

⬛ **Fig. 7.1**   Robot end effectors. **a** A vacuum gripper holds a sheet of glass; **b** a human-like robotic hand (© Shadow Robot Company 2008)

chapter7.mlx

▶ sn.pub/AfMT1t

Degrees of freedom were covered in ▶ Sect. 2.4.9.

A robot manipulator arm comprises mechanical links and computer-controllable joints. The base of the robot is generally fixed and motion of the joints changes the pose of the tool or end effector in space in order to perform useful work. End effectors range in complexity from simple 2-finger or parallel-jaw grippers to complex human-like hands with multiple actuated finger joints and an opposable thumb – some examples are shown in ⬛ Fig. 7.1.

There are many different types of robot manipulators and ⬛ Fig. 7.2 shows some of that diversity. Most common is the six degree of freedom (DoF) ◀ robot manipulator arm comprising a series of rigid links and six actuated rotary joints. The SCARA (Selective Compliance Assembly Robot Arm) is rigid in the vertical direction and compliant in the horizontal plane which is advantageous for planar tasks such as electronic circuit board assembly. A gantry robot has two DoF of motion along overhead rails which gives it a very large working volume. In these robots the joints are connected in series – they are serial-link manipulators – and each joint has to support and move all the joints between itself and the end effector. In contrast, a parallel-link manipulator has all the motors at the base and connected directly to the end effector by mechanical linkages. Low moving mass, combined with the inherent stiffness of the parallel link structure, makes this class of robot capable of very high acceleration.

Kinematics is derived from the Greek word for motion.

This chapter is concerned with *kinematics* – the branch of mechanics that studies the motion of a body, or a system of bodies, without considering mass or force. ◀ The pose of an end effector is a function of the state of each of the robot's joints and this leads to two important kinematic problems. Forward kinematics, covered in ▶ Sect. 7.1, is about determining the pose of the end effector from the joint configuration. Inverse kinematics, covered in ▶ Sect. 7.2, is about determining the joint configuration given the end-effector pose. ▶ Sect. 7.3 describes methods for generating smooth paths for the end effector. ▶ Sect. 7.4 works two complex applications: writing on a planar surface and a four-legged walking robot whose legs are simple robotic arms. ▶ Sect. 7.5 concludes with a discussion of some advanced topics.

◼ Fig. 7.2 **a** A 6 DoF general purpose industrial manipulator (source: ABB); **b** SCARA robot which has 4 DoF, typically used for electronic assembly (Image of Adept Cobra s600 SCARA robot courtesy of Adept Technology, Inc.); **c** a gantry robot; the arm moves along an overhead rail (Image courtesy of Güdel Group AG Switzerland, Mario Rothenbühler, ► https://www.gudel.com/); **d** a parallel-link manipulator, the end effector is driven by 6 parallel links (source: ABB)

7.1 Forward Kinematics

The robots shown in ◼ Fig. 7.2a–c are serial-link manipulators and comprise a series of rigid links connected by actuated joints. Each joint controls the relative pose of the two links it connects, and the pose of the end effector is a function of all the joints. A series of links and joints is a kinematic chain or a rigid-body chain. Most industrial robots like ◼ Fig. 7.2a have six joints but increasingly we are seeing robots with seven joints that enable greater dexterity and a larger usable workspace – we will discuss that topic in the next chapter. The YuMi® robot shown in ◼ Fig. 7.3a has two kinematic chains, each with seven joints. The Atlas™ robot in ◼ Fig. 7.3b has four chains: two of the chains are arms, and two of the chains are legs, with the end effectors being hands and feet respectively. Branched robots are covered more extensively in ► Sect. 7.1.3.

In robotics the most common type of joint is revolute, where one link rotates relative to another, about an axis fixed to both. Less common is a prismatic joint, also known as a sliding or telescopic joint, where one link slides relative to another along an axis fixed to both. The robots in ◼ Fig. 7.2b, c have one and three prismatic joints respectively. Many other types of joints are possible such as screw joints (1 DoF), cylindrical joints (2 DoF, one revolute and one prismatic) or spherical joints (3 DoF) but they are rare in robotics and will not be discussed here. ►

We are interested in understanding the relationship between the joints, which we control, and the pose of the end effector which does the work. This is the forward kinematics, a mapping from joint coordinates, or robot configuration, to

In mechanism theory, all these joints are collectively known as lower pairs and involve contact between two surfaces. Higher pairs involve contact between a point and a surface or a line and a surface.

◻ Fig. 7.3 **a** ABB YuMi® has two kinematic chains and 14 joints (source: ABB); **b** Boston Dynamics Atlas™ humanoid robot has four kinematic chains and 28 joints (source: Boston Dynamics)

end-effector pose. We start in ► Sect. 7.1.1 by describing the problem in terms of pose graphs, in ► Sect. 7.1.2 we introduce the notion of rigid-body chains, and in ► Sect. 7.1.3 we introduce chains with multiple branches or rigid-body trees. ► Sect. 7.1.4 introduces the URDF format for exchange of robot models and ► Sect. 7.1.5 introduces the classical Denavit-Hartenberg notation. In each section we treat the simpler 2D case first, then the 3D case.

7.1.1 Forward Kinematics from a Pose Graph

See ► Sect. 2.1.3 for an introduction to pose graphs.

In this section we apply our knowledge of pose graphs ◄ to solve the forward kinematics problem for robots that move in 2D and 3D.

7.1.1.1 2-Dimensional (Planar) Robotic Arms

We will start simply with a single kinematic chain operating in the plane. Consider the very simple example shown in ◻ Fig. 7.4a which comprises one revolute joint and one link and is somewhat like the hand of a clock. The pose graph depicts the composition of two relative poses. In the pose graphs discussed in ► Sect. 2.1.3 edges represented constant, relative poses. For the kinematic chain here, one edge (shown in red) is a variable pose, since the pose is a function of a joint variable. The edges or arrows represent, in order, a rotation by the joint angle q and then a fixed translation in the x-direction by a distance a_1. The pose of the end effector is simply

$$^0\boldsymbol{\xi}_E = \boldsymbol{\xi}^r(q) \oplus \boldsymbol{\xi}^{t_x}(a_1)$$

where the relative pose $\boldsymbol{\xi}$ is written as a function with a parameter which is either a variable, for example q, or a constant, for example a_1. The superscript indicates the specific function: $\boldsymbol{\xi}^r$ is a relative pose that is a planar rotation and $\boldsymbol{\xi}^{t_x}$ is a relative pose that is a planar translation in the x-direction.

Using the RVC Toolbox we represent this by an elementary transformation sequence (ETS) ◄ which we create by composing two elementary transformations which, for the 2-dimensional case, are created by methods of the ETS2 class

So called because the rigid-body transformations are the simplest, or most elemental, transformations possible: a rotation, a translation in the x-direction, or a translation in the y-direction.

```
>> a1 = 1;
>> e = ETS2.Rz("q1")*ETS2.Tx(a1)
e =
Rz(q1)Tx(1)
```

◘ Fig. 7.4 Some simple planar robotic arms and their associated pose graphs. The end effector is shown as a black sphere. The generalized joint variables are q_i. In the pose graph, edges can be variable due to joint variables (red) or constant (black). Frame {0} is the fixed world-reference frame. **a** One link and one revolute joint; **b** two links and two revolute joints; **c** two links with one revolute and one prismatic joint

and we overload $\star$ to denote composition since $\oplus$ is not a MATLAB® operator. The result is a 2D elementary transformation sequence encapsulated in an `ETS2` array of ►

```
>> size(e)
ans =
     1     2
```

An object array in MATLAB allows you to store multiple objects as vectors and matrices, just like numeric values. In the case of `ETS2`, it will always be a row vector.

that contains two rigid-body transforms: one variable and one constant. The display of the object array is customized, so it is easy to see which transforms are contained within. The argument to `Rz` is a string which indicates that its parameter is a joint variable. The argument to `Tx` is a constant numeric robot dimension, in this case equal to one, and the translation is defined relative to the x axis of the previous frame.

For a particular value of the joint variable, say $q = \frac{\pi}{6}$ radians, the end-effector pose is

```
>> e.fkine(pi/6)
ans =
    0.8660   -0.5000    0.8660
    0.5000    0.8660    0.5000
         0         0    1.0000
```

which is an $\mathbf{SE}(2)$ matrix computed by composing the two relative poses in the ETS. The argument to `fkine` is substituted for the joint variable in the expression and is equivalent to

```
>> se2(pi/6,"theta")*se2(eye(2),[a1 0])
ans =
  se2
    0.8660   -0.5000    0.8660
    0.5000    0.8660    0.5000
         0         0    1.0000
```

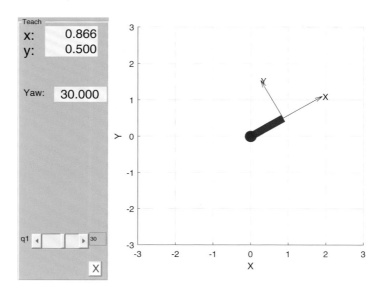

Fig. 7.5 RVC Toolbox depiction of the 1-joint planar robot from ▪ Fig. 7.4a using the `teach` method. The left-hand panel displays the position and orientation (yaw angle in degrees) of the end effector and contains the joint angle slider (in degrees). The right-hand panel shows the robot and its end-effector coordinate frame

which is the forward kinematics for our very simple robot. The arguments to `se2` are a 2D rotation matrix and an (x, y) translation vector, so the first `se2` object represents the rotation around the z axis and the second object represents the translation along the x axis.

An easy and intuitive way to understand how this simple robot behaves is by

```
>> e.teach;
```

which generates an interactive graphical representation of the robot arm as shown in ▪ Fig. 7.5. As we adjust the joint angle using the slider, the arm's shape changes, and the end-effector position and orientation are updated. This is not a particularly useful robot arm since its end effector can only reach points that lie on a circle.

Consider now the robot shown in ▪ Fig. 7.4b which has two revolute joints q_1 and q_2. From the pose graph, the pose of the end effector is

$$^{0}\boldsymbol{\xi}_E = \boldsymbol{\xi}^r(q_1) \oplus \boldsymbol{\xi}^{t_x}(a_1) \oplus \boldsymbol{\xi}^r(q_2) \oplus \boldsymbol{\xi}^{t_x}(a_2)$$

and using the RVC Toolbox we write

```
>> a1 = 1; a2 = 1;
>> e = ETS2.Rz("q1")*ETS2.Tx(a1)*ETS2.Rz("q2")*ETS2.Tx(a2)
e =
Rz(q1)Tx(1)Rz(q2)Tx(1)
```

When the object is displayed, we see that the joint variables `q1` and `q2` are associated with their transform. We can evaluate this expression for specific joint angles, in this case given in degrees, ◄ by passing in a vector

To easily convert between degrees and radians use the functions `deg2rad` and `rad2deg`.

```
>> T = e.fkine(deg2rad([30 40]));
>> printtform2d(se2(T),unit="deg")
t = (1.21, 1.44), 70 deg
```

and the result is the end-effector pose when $q_1 = 30°$ and $q_2 = 40°$. This is equivalent to

```
>> T = se2(deg2rad(30),"theta")*se2(eye(2),[a1 0]) * ...
>>    se2(deg2rad(40),"theta")*se2(eye(2),[a2 0]);
>> printtform2d(T,unit="deg")
 T: t = (1.21, 1.44), 70 deg
```

The `ETS2` object has many methods. We can find the number of joints

```
>> e.njoints
ans =
    2
```

The joint structure of a robot is often given in a shorthand notation comprising the letters R (for revolute) or P (for prismatic) to indicate the number and types of its joints. For this kinematic chain

```
>> e.structure
ans =
    'RR'
```

indicates a revolute-revolute sequence of joints. We could display the robot interactively as in the previous example, or non-interactively by

```
>> e.plot(deg2rad([30 40]));
```

The `ETS2` object array acts like a normal vector. We can access specific elements

```
>> etx = e(2)
etx =
Tx(1)
```

and this single object, an `ETS2` object, has a number of methods

```
>> etx.param
ans =
    1
>> etx.T
ans =
        1        0        1
        0        1        0
        0        0        1
```

which return respectively, the transformation parameter passed to the constructor, `a1` in this case, and the corresponding $\mathbf{SE}(2)$ matrix which in this case represents a translation of 1 in the x-direction.

This simple planar robot arm has some interesting characteristics. Firstly, most end-effector *positions* can be reached with two different sets of joint angles. Secondly, the robot can position the end effector at any point within its reach but cannot achieve an arbitrary end-effector orientation as well.

We describe the configuration of a robot manipulator with N joints by a vector of generalized coordinates where $q_j \in \mathbf{S}^1$ is an angle for a revolute joint or $q_j \in \mathbb{R}$ is a length ▶ for a prismatic joint. We refer to q as the *joint configuration* or *joint coordinates* and for an all-revolute robot as the *joint angles*

> This is a real interval since the joint length will have a minimum and maximum possible value.

❶ This is sometimes, confusingly, referred to as the pose of the manipulator, but in this book we use the term pose to mean ξ, the position and orientation of a body in space.

Recalling our discussion of configuration and task space from ▶ Sect. 2.4.9 this robot has 2 degrees of freedom and its configuration space is $C \in \mathbf{S}^1 \times \mathbf{S}^1$ and $q \in C$. This is sufficient to reach points in the task space $\mathcal{T} \subset \mathbb{R}^2$ since $\dim C = \dim \mathcal{T}$. However if our task space includes orientation $\mathcal{T} \subset \mathbb{R}^2 \times \mathbf{S}^1$ then it is underactuated since $\dim C < \dim \mathcal{T}$ and the robot can access only a subset of the task space, for example, positions but not orientation.

We will finish up with the robot shown in ◻ Fig. 7.4c which includes a prismatic joint and is commonly called a polar-coordinate robot arm. The end-effector pose is

$$^0\xi_E = \xi^r(q_1) \oplus \xi^{t_x}(q_2)$$

7

For a prismatic joint we need to specify the range of motion `qlim` so that the `teach` method can determine the dimensions of the plot surface and the scale of the slider. Unless otherwise specified, revolute joints are assumed to have a range $[-\pi, \pi)$.

and using the RVC Toolbox this is ◄

```
>> e = ETS2.Rz("q1")*ETS2.Tx("q2",qlim=[1 2])
e =
Rz(q1)Tx(q2)
>> e.structure
ans =
    'RP'
```

and like the previous objects it has `fkine`, `teach`, and `plot` methods.

We could easily add more joints and use the now familiar RVC Toolbox functionality to explore the capability of robots we create. A robot with three or more revolute joints is able to access all points in the task space $\mathcal{T} \subset \mathbb{R}^2 \times \mathbf{S}^1$, that is, achieve any pose in the plane (limited by reach).

> A serial-link manipulator with N joints (numbered from 1 to N) has $N + 1$ links (numbered from 0 to N). Joint j connects link $j - 1$ to link j and moves them relative to each other. Link ℓ connects joint ℓ to joint $\ell + 1$. Link 0 is the base of the robot, typically fixed, and link N, is the last link of the robot and carries the end effector or tool.

7.1.1.2 3-Dimensional Robotic Arms

Most real-world robot manipulators have a task space $\mathcal{T} \subset \mathbb{R}^3 \times \mathbf{S}^3$ which allows arbitrary position and orientation of the end effector within its 3D working envelope or workspace. This requires a robot with a configuration space $\dim \mathcal{C} \geq \dim \mathcal{T}$ which can be achieved by a robot with six or more joints. Using the 3D version of the code example above, we can define a robot that is a simplified human arm

```
>> a1 = 1; a2 = 1;
>> e = ETS3.Rz("q1")*ETS3.Ry("q2")* ...
>>    ETS3.Tz(a1)*ETS3.Ry("q3")*ETS3.Tz(a2)* ...
>>    ETS3.Rz("q4")*ETS3.Ry("q5")*ETS3.Rz("q6");
```

which has a 2 DoF spherical shoulder joint (line 2); an upper-arm, elbow joint, and lower-arm (line 3); and a 3 DoF spherical wrist joint which is a ZYZ Euler angle sequence (line 4). `e` is an `ETS3` object, the 3-dimensional version of `ETS2`, which

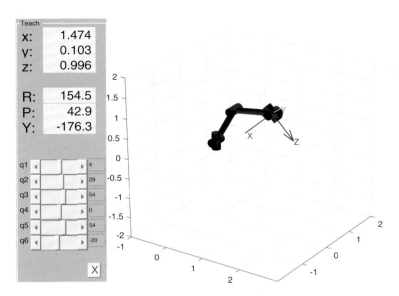

◘ Fig. 7.6 The 3-dimensional manipulator from ▶ Sect. 7.1.1.2 visualized using the `teach` method. The left-hand panel displays the position and orientation (roll, pitch, and yaw angles in degrees) of the end effector and contains the sliders for all 6 joint angles (in degrees). The right-hand panel shows the robot and its end-effector coordinate frame

supports the same methods as `ETS2` used in the previous section

```
>> e.njoints
ans =
     6
>> e.structure
ans =
    'RRRRRR'
```

and the end-effector pose for all zero joint angles is

```
>> e.fkine(zeros(1,6))
ans =
     1      0      0      0
     0      1      0      0
     0      0      1      2
     0      0      0      1
```

which is an **SE**(3) matrix computed by composing the eight relative poses in the ETS. To understand how the joint angles affect the end effector pose we can show a graphical representation

```
>> e.teach;
```

with an interactive panel to adjust all the 6 joint angles. An example configuration is shown in ◘ Fig. 7.6.

7.1.2 Forward Kinematics as a Chain of Robot Links

The bulk of a robot's physical structure is due to its links and these dictate its shape and appearance, as well as having important properties such as mass, inertia, and collision potential. ◘ Fig. 7.7 shows the relationship between joints and link frames. The frame for link ℓ is indicated as $\{\ell\}$ and its pose is a function of the joint coordinates $\{q_j, j = 1, \ldots, \ell\}$. We say that link ℓ is the parent of link $\ell + 1$ and the child of link $\ell - 1$. The fixed base is considered to be link 0 and has no parent, while the end effector has no child. The pose of the end effector, the forward kinematics

7

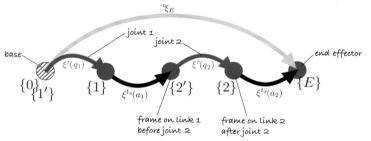

◻ Fig. 7.7 Pose graph showing link coordinate frames. Each relative pose in this graph is a function of a joint variable. Frame {0} is the base which is fixed and frame {E} holds the end effector

is

$$
{}^{0}\boldsymbol{\xi}_N = {}^{0}\boldsymbol{\xi}_1(q_1) \oplus {}^{1}\boldsymbol{\xi}_2(q_2) \oplus \cdots \oplus {}^{N-1}\boldsymbol{\xi}_N(q_N) \tag{7.1}
$$

where ${}^{j}\boldsymbol{\xi}_{j+1}(q_{j+1})$ is the pose of link frame {j + 1} with respect to link frame {j}. We can write this in functional form as

$$
{}^{0}\boldsymbol{\xi}_N = \mathcal{K}(\boldsymbol{q}) \tag{7.2}
$$

where $\mathcal{K}(\cdot)$ is specific to the robot and incorporates the joint and link parameters.

7.1.2.1 Robots as Rigid Body Trees

Up until now, we focused on describing robot configurations through elementary transform sequences in 2D and 3D. For the rest of this chapter, we will leverage dedicated objects and functions that encapsulate the concepts of links, joints, and robots.

There are three objects that are essential to describing robots:

- A `rigidBodyTree` represents the robot structure. This can either be a single chain (as we have seen so far) or a branched or tree robot (as we will introduce in ► Sect. 7.1.3)

- A rigid body tree is made up of `rigidBody` objects. These are the links of the robot and have properties such as mass, inertia, and collision geometries. The terms *rigid body* and *link* are synonymous. For brevity, we will use *link* when referring to `rigidBody` objects for the remainder of the book.

— Each link is connected to its parent with a `rigidBodyJoint`. This joint defines how that link moves relative to its parent. Common joint types are revolute, prismatic, and fixed. Joints describe the transformation between links.

In the rest of this section, we focus on how to use these objects to describe robot structures. The `rigidBodyTree` is used as an input by many kinematic and dynamic algorithms that we will encounter throughout the next chapters. ▶

The structure of the rigid body tree is further described and visually illustrated in ▶ https://sn.pub/2bUGZE.

7.1.2.2 2-Dimensional (Planar) Case

We consider again the robot of ◘ Fig. 7.4b and rigidly attach a coordinate frame to each link as shown in ◘ Fig. 7.8. The pose of the link 1 frame shown in red depends only on q_1, while the pose of the link 2 frame shown in blue depends on q_1 and q_2, as well as the length of link 1.

◘ Fig. 7.9 shows a more detailed pose graph for this robot with the joints and links highlighted. Frame $\{\ell'\}$ is the parent side of the joint, that is the joint is a transformation from $\{\ell'\}$ on the parent link to frame $\{\ell\}$ on the child. When $q_\ell = 0$ these two frames are identical.

◘ Fig. 7.8 Coordinate frames are shown attached to the links of the robot from ◘ Fig. 7.4b. Link 1 is shown in red, and link 2 is in blue

▶ sn.pub/ui9HYd

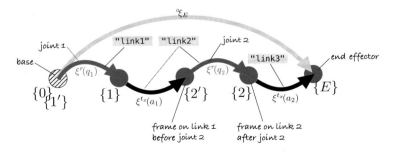

◘ Fig. 7.9 Pose graph for the robot from ◘ Fig. 7.4b highlighting the joints and link frames. Gray text boxes connect elements of the pose graph to the named `rigidBody` objects

In the code, we commonly use `linkN` as name for links and `jointN` as name for joints, but arbitrary strings will work, as long as there is a unique set of link names and a unique set of joint names in the robot.

We can describe the robot by first creating the relevant link and joint objects, each with a unique name ◄

```
>> link1 = rigidBody("link1");
>> link1.Joint = rigidBodyJoint("joint1","revolute");
>> link2 = rigidBody("link2");
>> link2.Joint = rigidBodyJoint("joint2","revolute");
>> link2.Joint.setFixedTransform(se3([a1 0 0],"trvec"));
>> link3 = rigidBody("link3");
>> link3.Joint.setFixedTransform(se3([a2 0 0],"trvec"));
```

which creates instances of `rigidBody` objects. Each link object contains a single `rigidBodyJoint` object to describe the transformation from the parent's link frame to its own frame. The method `setFixedTransform` sets an object property, `JointToParentTransform`, which describes where the parent side of the joint is, with respect to the parent link frame. In the example above, we have also created a link frame for the end effector (`link3`), which has a fixed joint with a constant transformation with respect to its parent. ◄

When creating a `rigidBody` object, the `Joint` property contains a fixed joint by default.

We can now create a robot object and add these links to it. For each link, we need to define its parent. The first link is attached to the robot's fixed base.

```
>> myRobot = rigidBodyTree(DataFormat="row");
>> myRobot.addBody(link1,myRobot.BaseName);
>> myRobot.addBody(link2,link1.Name);
>> myRobot.addBody(link3,link2.Name);
```

Alternatively, we can also tell the object to represent q as a column vector or a struct. The choice depends on your preference and how your other code is structured.

The `DataFormat` argument defines that our joint configuration q is represented as a row vector. ◄ For example, this simple robot has 2 revolute joints and its home configuration q_{home} is a row vector with 2 elements

```
>> myRobot.homeConfiguration
ans =
     0     0
```

The 0 values in this vector are the default home positions for the two revolute joints, which can be changed by setting the `HomePosition` property of the joints.

The result of this method is a wide table which has been cropped for inclusion in this book.

We can now display the rich structure that we created ◄

```
>> myRobot.showdetails
--------------------
Robot: (3 bodies)
```

Idx	Body Name	Joint Name	Joint Type	Parent Name(Idx)	...
					...
1	link1	joint1	revolute	base(0)	...
2	link2	joint2	revolute	link1(1)	...
3	link3	link3_jnt	fixed	link2(2)	...

```
--------------------
```

The `Idx` values are assigned automatically and will change depending on the structure of the rigid body chain or tree.

We see the number of bodies, with each body listed in the table with its unique `Idx` value ◄ and body name. Each body has a joint and lists its parent and child bodies. `link3` here is an end effector – it has no children and has a fixed transformation with respect to its parent (since it is connected to a fixed joint). ◄

A variant of this robot with 3 links and 2 revolute joints can also be created by calling `twoJointRigidBodyTree`.

The `rigidBodyTree` object has many methods. ◄ Calling `getTransform` calculates the forward kinematics and returns a 4×4 homogeneous transformation matrix

Calling `help rigidBodyTree` on the command-line will display some basic documentation along with a complete property and method list.

```
>> T = myRobot.getTransform(deg2rad([30 40]),"link3");
>> printtform(T,unit="deg")
T: t = (1.21, 1.44, 0), RPY/zyx = (70, 0, 0) deg
```

representing the pose of the `link3` end-effector frame relative to the base frame. Since we are working with a 2D manipulator in the XY plane, the Z translation will

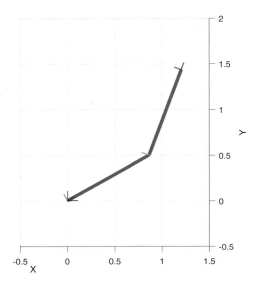

◻ Fig. 7.10 A 2-link robot in the XY-plane at configuration $q = (30°, 40°)$. The line segments (blue) connect the link frames rather than depicting the physical links of the robot

always be 0. We can plot the robot at this configuration as shown in ◻ Fig. 7.10

```
>> myRobot.show(deg2rad([30 40]));
>> view(0,90)   % show only XY (2D) plane
```

but we can also plot a sequence of joint configurations as an animation. The sequence is provided as an array with the rows being consecutive configurations

```
>> q = [linspace(0,pi,100)' linspace(0,-2*pi,100)'];
>> whos q
  Name    Size      Bytes   Class      Attributes
   q      100x2      1600   double
>> r = rateControl(10);
>> for i = 1:size(q,1)
>>    myRobot.show(q(i,:),FastUpdate=true,PreservePlot=false);
>>    r.waitfor;
>> end
```

The `rateControl` object ensures that the animation runs at 10 Hz, or frames per second. We use the `FastUpdate` parameter to ensure that we are only updating the robot, not redrawing it at every configuration.

We can retrieve the links from the robot object

```
>> link2 = myRobot.Bodies{2}
link2 =
  rigidBody with properties:
            Name: 'link2'
           Joint: [1×1 rigidBodyJoint]
            Mass: 1
    CenterOfMass: [0 0 0]
         Inertia: [1 1 1 0 0 0]
          Parent: [1×1 rigidBody]
        Children: {[1×1 rigidBody]}
         Visuals: {}
      Collisions: {}
```

which returns link 2, or look them up by name

```
>> link2 = myRobot.getBody("link2");
```

which maps a link name to a link object.

7

The `Parent` and `Children` properties are only populated once links are added to the `rigidBodyTree`

The link objects also have many properties and methods. We can obtain a handle to its parent link ◄

```
>> parentLink = link2.Parent;
```

as well as to its children

```
>> childLinks = link2.Children;
```

which is a cell array of link objects. In this case there is only one child, but for branched robots (see ► Sect. 7.1.3) there can be more.

We can retrieve the joint type for `link2`

```
>> link2.Joint.Type
ans =
    'revolute'
```

which indicates a revolute joint. Joints can have upper and lower movement limits, but in this case the joint is using the default full motion range of $[-\pi, \pi]$.

```
>> link2.Joint.PositionLimits
ans =
   -3.1416    3.1416
```

For forward kinematics an important method is

```
>> myRobot.getTransform(deg2rad([0 30]),"link2","link1")
ans =
    0.8660   -0.5000        0    1.0000
    0.5000    0.8660        0         0
         0         0   1.0000         0
         0         0        0    1.0000
```

which is the relative pose of the `link2` frame with respect to its parent's frame, or $^1\boldsymbol{\xi}_2$. Forward kinematics is simply the product of the link transforms from a source link to a target link, given the robot's joint configuration. We have seen before that if the target link is not specified, `getTransform` will return the forward kinematics with respect to the robot's fixed base.

7.1.2.3 3-Dimensional Case

For the 3D case we can create a robot model following a similar pattern. However, we can take some shortcuts.

Firstly, we can create a `rigidBodyTree` robot directly from an ETS expression, in this case the example from ► Sect. 7.1.1.2.

```
>> a1 = 1; a2 = 1;
>> robot6 = ets2rbt(ETS3.Rz("q1")*ETS3.Ry("q2")* ...
>>     ETS3.Tz(a1)*ETS3.Ry("q3")*ETS3.Tz(a2)* ...
>>     ETS3.Rz("q4")*ETS3.Ry("q5")*ETS3.Rz("q6"));
>> robot6.homeConfiguration
ans =
     0     0     0     0     0     0
```

The `ets2rbt` function has automatically partitioned the ETS expression and created link and joint objects. We can now apply familiar methods like `getTransform` and `show`.

When writing code in the Editor or Live Editor, MATLAB suggests and completes the names of functions, parameters, and options. To show these suggestions, or tab completions, on the command-line, press the "Tab" key. This can be a tremendous time-saver, as it reduces the need to check the help or documentation.

The second shortcut is to use one of the many models available in MATLAB. In the next function call, `<TAB>` indicates that you should press the "Tab" key on your keyboard. This will display all preconfigured robots (over 40 manipulators and mobile robots) in a small popup window. ◄

```
>> loadrobot("<TAB>
abbIrb120
abbIrb120T
abbIrb1600
```

```
abbYuMi
amrPioneer3AT
amrPioneer3DX
amrPioneerLX
atlas
...
```

Each model is a `rigidBodyTree` object. For example, to instantiate a model of the Franka Emika Panda robot ▶

The result of this method is a wide table which has been cropped for inclusion in this book.

```
>> panda = loadrobot("frankaEmikaPanda",DataFormat="row");
>> panda.showdetails
--------------------
Robot: (11 bodies)
  Idx      Body Name         Joint Name        Joint Type   ...
  ---      ---------         ----------        ----------   ...
    1     panda_link1       panda_joint1         revolute   ...
    2     panda_link2       panda_joint2         revolute   ...
    3     panda_link3       panda_joint3         revolute   ...
    4     panda_link4       panda_joint4         revolute   ...
    5     panda_link5       panda_joint5         revolute   ...
    6     panda_link6       panda_joint6         revolute   ...
    7     panda_link7       panda_joint7         revolute   ...
    8     panda_link8       panda_joint8            fixed   ...
    9     panda_hand        panda_hand_joint        fixed   ...
   10     panda_leftfinger  panda_finger_joint1  prismatic  ...
   11     panda_rightfinger panda_finger_joint2  prismatic  ...
--------------------
```

and the format of the displayed table should be familiar by now. ▶

We can compute the forward kinematics from the robot's base to the Panda's hand link for a given configuration `qr`

The `loadrobot` function also has a second struct output that contains information about the robot version, the source of its meshes and parameters, and other data.

```
>> qr = [0 -0.3 0 -2.2 0 2 0.7854 0 0];
>> T = panda.getTransform(qr,"panda_hand");
>> printtform(T,unit="deg")
T: t = (0.474, -4.18e-12, 0.516),
RPY/zyx = (-0.000106, -5.73, -180) deg
```

or plot the robot for this same configuration

```
>> panda.show(qr);
```

which creates the 3D rendered plot shown in ◘ Fig. 7.11a. Panda and the other robots included in MATLAB already have visual and collision meshes assigned,

◘ Fig. 7.11 3D plot of the Panda robot. **a** The visual meshes for the links are provided by the manufacturer. The link frames are also shown as smaller lines for the link's *x*- (red), *y*- (green), and *z*-axis (blue); **b** Clicking on a link mesh displays additional information (yellow)

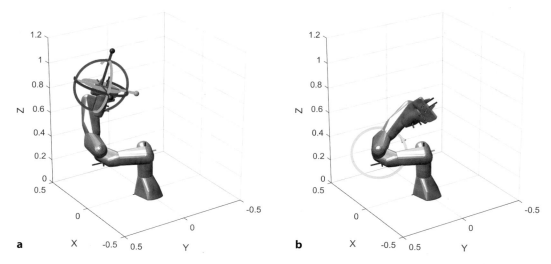

7

□ **Fig. 7.12**   The `interactiveRigidBodyTree` visualization. **a** The interactive marker can be attached to any body in the tree and is used to adjust the body's position and orientation. All joint positions will be computed by inverse kinematics and are subject to joint limits, so some desired body poses cannot be achieved exactly; **b** Joint angles can also be adjusted individually (see yellow ring) to achieve finer control

The prismatic joints (q_8 and q_9) are the fingers of the Panda's gripper. Changing their position has no effect on the pose of the `panda_hand` link.

This visualization opens in a new window and is independent of the `show` method.

which make them very easy to use. In the standard 3D plot shown in □ Fig. 7.11a, we can click on individual link meshes to see more information about the name of the body, its index, and the associated joint. An example for the `panda_hand` link is shown in □ Fig. 7.11b.

The configuration `qr` for the Panda robot has 9 elements, since the robot has 9 non-fixed joints (7 revolute and 2 prismatic ◄). How can we find other possible configurations for this robot? In ► Sect. 7.2, we will discuss how to use inverse kinematics to calculate these configurations, but it is also possible to find them interactively.

You can use a dedicated interactive visualization tool ◄ to explore a manipulator's workspace and find suitable configurations

```
>> intPanda = interactiveRigidBodyTree(panda,Configuration=qr);
```

by moving and rotating the interactive marker attached to the end effector. The robot's configuration will automatically adjust to achieve this pose. We can then retrieve the current robot configuration, store the current configuration in the object to define a trajectory, and retrieve that trajectory

```
>> intPanda.Configuration'  % transpose for display
ans =
         0   -0.3000    0   -2.2000    0    2.0000   0.7854   0    0
>> intPanda.addConfiguration;      % store configuration
>> intPanda.StoredConfigurations;  % retrieve stored trajectory
```

Examples of different configurations of the interactive marker can be seen in □ Fig. 7.12a and □ Fig. 7.12b.

7.1.2.4 Tools and Bases

It is useful to extend the forward kinematic expression of (7.1) by adding two extra transforms

$$^{W}\boldsymbol{\xi}_E = \underbrace{^{W}\boldsymbol{\xi}_0}_{\boldsymbol{\xi}_B} \oplus {}^{0}\boldsymbol{\xi}_1 \oplus {}^{1}\boldsymbol{\xi}_2 \oplus \cdots \oplus {}^{N-1}\boldsymbol{\xi}_N \oplus \underbrace{{}^{N}\boldsymbol{\xi}_E}_{\boldsymbol{\xi}_T} \tag{7.3}$$

which is shown in □ Fig. 7.13. We have used W to denote the world frame since here 0 designates link 0, the base link.

Fig. 7.13 The kinematic chain from ◘ Fig. 7.7 extended to include a base and tool transform

Conventionally, the start of the kinematic chain is the base of the robot, but the base transformation ξ_B allows us to place the base of the robot at an arbitrary pose within the world coordinate frame. In a manufacturing cell, the robot's base is fixed and defined relative to the cell. If the arm is mounted on a moving mobile robot, the base transformation is time varying.

There is no standard for kinematic models that dictates where the tool frame {N} is physically located on the robot. For a URDF robot model it is is likely to be the tool-mounting flange on the physical end of the robot as shown in ◘ Fig. 7.14b. For a Denavit-Hartenberg model (▶ Sect. 7.1.5) it is frequently the center of the spherical wrist mechanism which is physically inside the robot. The tool transformation ξ_T describes the pose of the tool tip – the bit that does the work and sometimes called the tool center point – with respect to frame {N}. A strong convention is that the robot's tool points in the z-direction as shown in ◘ Fig. 2.20. In practice, ξ_T might vary from application to application, for instance a gripper, a screwdriver, or a welding torch, or it might vary within an application if it involves tool changing. It might also consist of several components, perhaps a relative pose to a tool holder and then a relative pose specific to the selected tool.

7.1.3 Branched Robots

The simple robots discussed so far have a single end effector, but increasingly we see robots with multiple end effectors and some examples are shown in ◘ Fig. 7.3. The robot in ◘ Fig. 7.3a has two arms in a human-like configuration.

Let's use the humanoid robot in ◘ Fig. 7.3b to illustrate key points of branched robots. The robot has a torso with four manipulator arms (two of which we call

Fig. 7.14 Panda robot **a** with gripper; **b** without gripper and showing the DIN ISO 9409-1-A50 mounting flange to which tools can be attached (Image by Dorian Tsai)

◻ **Fig. 7.15**   Visualization of the Boston Dynamics Atlas™ humanoid robot

legs). We can load it into MATLAB and display it, as shown in ◻ Fig. 7.15.

```
>> atlas = loadrobot("atlas",DataFormat="row");
>> atlas.show;
```

We see that multiple links including `"l_clav"` (attachment point for left arm, similar to the human clavicle), `"r_clav"` (attachment point for right arm), and `"head"` share `"utorso"` as their parent. Conversely, the link named `"utorso"` has 5 children

```
>> childLinks = atlas.getBody("utorso").Children';
>> size(childLinks)
ans =
     5     1
>> cellfun(@(c) string(c.Name),childLinks)
ans =
  5×1 string array
    "l_clav"
    "l_situational_awareness_camera_link"
    "head"
    "r_clav"
    "r_situational_awareness_camera_link"
```

We can extract the subtree of all links attached to a body. For example, to extract the subtree for the right arm, we can call

```
>> rightArm = atlas.subtree("r_clav");
>> rightArm.show;
```

which will show the arm visualization in ◻ Fig. 7.16, which is a link chain with `"utorso"` as its base link.

This subtree of the right arm contains one end effector, a link named `"r_hand_force_torque"`, which has a fixed joint and represents the tool mounting point described in ▶ Sect. 7.1.2.4. The Atlas robot has 3 additional end effectors: `"l_hand_force_torque"` (end of left arm), `"l_foot"` (left foot), and `"r_foot"` (right foot).

As with the previous examples, we can interactively move this robot and find suitable configurations.

```
>> intAtlas = interactiveRigidBodyTree(atlas);
```

◘ Fig. 7.16 Subtree for the right arm extracted from the Atlas robot. The black frame is the base link at `"utorso"`. Connected to that is the root of our subtree, the frame for `"l_clav"` (blue). The fixed frame (pink) on the end of the arm is the tool end-effector frame `"r_hand_force_torque"`

For forward kinematics, we now need to specify which end effector we are interested in. To retrieve the pose of the left foot ►

The target end effector can be specified by its name as a string.

```
>> atlas.getTransform(atlas.homeConfiguration,"l_foot");
```

and the relative pose of `"r_hand_force_torque"` with respect to `"r_clav"` as shown in ◘ Fig. 7.16 is simply

```
>> T = atlas.getTransform(atlas.homeConfiguration, ...
>>    "r_hand_force_torque","r_clav");
>> printtform(T,unit="deg")
T: t = (0, -0.84, -0.261), RPY/zyx = (180, 0, 0) deg
```

7.1.4 Unified Robot Description Format (URDF)

So far we have used code to create rigid-body trees, but that is very specific to the programming language and software tools being used. To enable exchange of models we use URDF which is a portable XML-based file format. ► The format is widely used, and models of many robots can be found online. We can parse a URDF file that is part of MATLAB ►

Another popular format for describing scenes with robots is SDF (Simulation Description Format). This can also be imported to extract a robot description.

```
>> sawyer = importrobot("sawyer.urdf");
```

which returns a rigid body tree initialized with parameters extracted from the URDF file. Inertial parameters, if present in the file, will be included in the link objects. ►

URDF models often represent the gripper fingers as links with their own joint variable – essentially creating a branched robot. In practice, gripper fingers are treated as part of an independent subsystem and not part of the kinematic chain. At most, we would represent the gripper by a fixed distance to the middle of the fingertips – the tool center point – and then open or close the fingers about that point.

MATLAB provides many URDF files that can be loaded directly and correspond to the `loadrobot` robots introduced in ► Sect. 7.1.2.3.

If a relative path is given it is first looked for relative to the current folder, and if not found, it checks folders on the MATLAB path.

Such as mass (`Mass`), center of mass (`CenterOfMass`), and inertia tensor (`Inertia`).

7

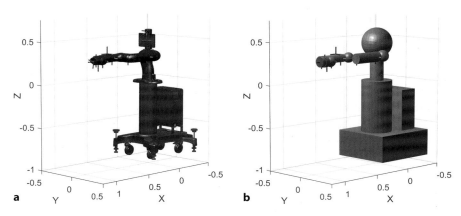

◘ **Fig. 7.17** Visualizing the different meshes of a Rethink Robotics® Sawyer robot. **a** The visual meshes are very detailed (with a large number of vertices) and have color information; **b** the collision meshes (green) are much simpler and convex to speed up collision checks during motion planning

The mesh files are typically COLLADA™ (.dae) or STL format (.stl). COLLADA files also contain color and surface texture data which leads to more realistic rendering than for STL files which are geometry only.

The geometry data allows for realistic 3D rendering of the robot as shown in ◘ Fig. 7.17. Each link is described by a mesh file ◄ which includes the 3D shape and optionally the surface color and texture. If the robot model includes this geometry and appearance data, then

```
>> sawyer.show;
```

will display a fully rendered view of the robot, like that shown in ◘ Fig. 7.17a. If available in the URDF file, collision meshes are also imported. Collision meshes are simple geometric approximations of the link shape that can be tested for intersection with other objects during simulation, and this is discussed in ► Sect. 7.5.4. We can display just the collision meshes for this robot with

```
>> sawyer.show(Visuals="off",Collisions="on");
```

and the result can be seen in ◘ Fig. 7.17b.

The inertial parameters for one of the bodies can be listed

```
>> sawyer.getBody("head")
ans =
  rigidBody with properties:
            Name: 'head'
           Joint: [1×1 rigidBodyJoint]
            Mass: 1.5795
    CenterOfMass: [0.0053 -2.6549e-05 0.1021]
         Inertia: [0.0283 0.0248 0.0050 4.7027e-06 -8.0863e-04
                   -4.2438e-06]
          Parent: [1×1 rigidBody]
        Children: {[1×1 rigidBody]  [1×1 rigidBody]}
         Visuals: {'Mesh:head.stl'}
      Collisions: {'Sphere Radius 0.18'}
```

The six unique elements of the symmetric inertia tensor relative to the link frame: $I_{xx}, I_{yy}, I_{zz}, I_{yz}, I_{xz}, I_{xy}$.

where `Mass` is the link's mass, `CenterOfMass` is the link's center of gravity with respect to the link frame, `Inertia` lists the unique elements of the symmetric link inertia tensor. ◄ Robot dynamics is the topic of ► Chap. 9.

⊕ Inertia tensor reference frame

The inertia tensor is defined with respect to the link frame. However, this is not a standard and will vary across software packages. URDF files describe inertia with respect to the inertia frame whose origin is at the center of mass. The software used in the Python version of this book stores the inertia tensor with respect to the center of mass of the link which is convenient for many dynamics algorithms.

7.1.5   **Denavit-Hartenberg Parameters**

From the 1960s, well before the advent of URDF, the way to share a robot model was as a simple table of numbers like that shown in ◘ Tab. 7.1. There is one row per link-joint pair to describe the kinematics of the robot. If known, the inertial parameters of the link can be added as additional columns.

This compact notation is known as Denavit-Hartenberg (DH) notation. Each row in the table defines the spatial relationship between two consecutive link frames as shown in ◘ Fig. 7.18. A manipulator with N joints numbered from 1 to N has $N + 1$ links, numbered from 0 to N. Joint j connects link $j - 1$ to link j. It follows that link ℓ connects joint ℓ to joint $\ell + 1$. Link 0 is the base of the robot, typically fixed and link N, the last link of the robot, carries the end effector or tool.

Each link is described by four parameters. The relationship between two link coordinate frames would ordinarily entail six parameters, three each for translation and rotation. Denavit-Hartenberg notation uses only four parameters but there are also two constraints: axis x_j intersects z_{j-1} and axis x_j is perpendicular to z_{j-1}. One consequence of these constraints is that sometimes the link coordinate frames are not actually located on the physical links of the robot. Another consequence is that the robot must be placed into a particular configuration – the zero-angle configuration – in order to assign the link frames. The Denavit-Hartenberg parameters are summarized in ◘ Tab. 7.2.

◘ **Table 7.1**   Standard Denavit-Hartenberg parameters for the PUMA 560 robot

θ_j	d_j	a_j	α_j	σ_j
q_1	0.6718	0	90°	0
q_2	0	0.4318	0°	0
q_3	0.1505	0.0203	-90°	0
q_4	0.4318	0	90°	0
q_5	0	0	-90°	0
q_6	0	0	0°	0

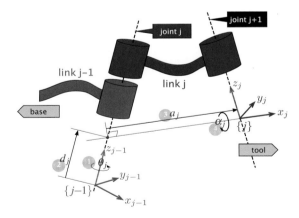

◘ **Fig. 7.18**   Definition of standard Denavit-Hartenberg link parameters. The colors red and blue denote all things associated with links $j - 1$ and j respectively. The numbers in circles represent the order in which the elementary transforms are applied. x_j is parallel to $z_{j-1} \times z_j$ and if those two axes are parallel then d_j can be arbitrarily chosen

7

> **Table 7.2** Denavit-Hartenberg parameters: their physical meaning, symbol, and formal definition

Joint angle	θ_j	The angle between the x_{j-1} and x_j axes about the z_{j-1} axis	Revolute joint variable or constant
Link offset	d_j	The distance from the origin of frame {j − 1} to the x_j axis along the z_{j-1} axis	Prismatic joint variable or constant
Link length	a_j	The distance between the z_{j-1} and z_j axes along the x_j axis; for intersecting axes, a_j is parallel to $z_{j-1} \times z_j$	Constant
Link twist	α_j	The angle from the z_{j-1} axis to the z_j axis about the x_j axis	Constant
Joint type	σ_j	$\sigma = R$ for a revolute joint, $\sigma = P$ for a prismatic joint. By convention $R = 0$ and $P = 1$	Constant

The coordinate frame {j} is attached to the far (distal) end of link j. The z-axis of frame {j} is aligned with the axis of joint $j + 1$.

The transformation from link coordinate frame {j − 1} to frame {j} is defined in terms of elementary transformations as

$$^{j-1}\xi_j = \xi^{r_z}(\theta_j) \oplus \xi^{t_z}(d_j) \oplus \xi^{t_x}(a_j) \oplus \xi^{r_x}(\alpha_j) \tag{7.4}$$

which can be expanded as an **SE**(3) matrix

$$^{j-1}\mathbf{A}_j = \begin{pmatrix} \cos\theta_j & -\sin\theta_j \cos\alpha_j & \sin\theta_j \sin\alpha_j & a_j \cos\theta_j \\ \sin\theta_j & \cos\theta_j \cos\alpha_j & -\cos\theta_j \sin\alpha_j & a_j \sin\theta_j \\ 0 & \sin\alpha_j & \cos\alpha_j & d_j \\ 0 & 0 & 0 & 1 \end{pmatrix} \tag{7.5}$$

The parameters α_j and a_j are always constant. For a revolute joint, θ_j is the joint variable and d_j is constant, while for a prismatic joint, d_j is variable, and θ_j is constant. The generalized joint coordinates are

$$q_j = \begin{cases} \theta_j & \text{if } \sigma_j = R \\ d_j & \text{if } \sigma_j = P \end{cases}$$

A revolute robot joint and link can be created by

```
>> a = 1; alpha = 0; d = 0; theta = 0;
>> link = rigidBody("link1");
>> link.Joint = rigidBodyJoint("joint1","revolute");
>> link.Joint.setFixedTransform([a alpha d theta],"dh");
```

The `"dh"` parameter in the call to `setFixedTransform` ensures that standard Denavit-Hartenberg (DH) conventions are used. ◄

A slightly different notation, *modified* Denavit-Hartenberg notation, is discussed in ▶ Sect. 7.5.2 and can be used by replacing `"dh"` with `"mdh"`.

In the Denavit-Hartenberg representation link 0 is the base of the robot and commonly for the first link $d_1 = 0$, but we could set $d_1 > 0$ to represent the height of the first joint above the world coordinate frame. For the final link, link N, the parameters d_N, a_N and α_N provide a limited means to describe the tool-tip pose with respect to the {N} frame. It is common to add a more general tool transformation as described in ▶ Sect. 7.1.2.4.

Determining the Denavit-Hartenberg parameters for a particular robot is challenging, but Denavit-Hartenberg kinematic descriptions of many robots can be found in manufacturer datasheets and in the literature. They are very compact compared to a URDF model, and a table of Denavit-Hartenberg parameters is sufficient to compute the forward and inverse kinematics. They can also be used to compute Jacobians, which will be covered in ► Chap. 8, and when combined with inertial parameters can be used to compute the rigid-body dynamics which will be covered in ► Chap. 9.

7.2 Inverse Kinematics

A problem of real practical interest is the inverse of that just discussed: given the desired pose of the end effector ξ_E what are the required joint coordinates? For example, if we know the Cartesian pose of an object, what should the robot's joint coordinates be to reach it? This is the inverse kinematics problem which is written in functional form as

$$q = \mathcal{K}^{-1}(\xi_E) \tag{7.6}$$

and, in general, is not unique, that is, a particular end-effector pose can be achieved by more than one joint configuration.

Fig. 7.19 The Inverse Kinematics Designer app operating on the NASA Valkyrie robot. The app can be used to find inverse kinematics solutions for specific body poses or general constraints. On the bottom, we see a table of stored configurations that can be used as waypoints for trajectory generation

Two approaches can be used to determine the inverse kinematics. Firstly, a closed-form or analytical solution can be found using geometric or algebraic techniques. However, this becomes increasingly challenging as the number of robot joints increases and for some serial-link manipulators no closed-form solution exists. Secondly, an iterative numerical solution can be used. In ▶ Sect. 7.2.1 we again use the simple 2-dimensional case to illustrate the principles and then in ▶ Sect. 7.2.2 extend these to robot arms that move in 3 dimensions.

Throughout this section, we will see that determining an inverse kinematics solution depends on many constraints and parameters. Visual inspection of inverse kinematics results and the ability to store trajectories of joint configurations is critical. To assist with this task, we can use the interactive Inverse Kinematics Designer app (`inverseKinematicsDesigner`) and a screenshot is shown in ◻ Fig. 7.19.

7.2.1 2-Dimensional (Planar) Robotic Arms

We will solve the inverse kinematics for the 2-joint robot of ◻ Fig. 7.4b in two ways: algebraic closed-form and numerical.

7.2.1.1 Closed-Form Solution

We start by computing the forward kinematics symbolically and will use the Symbolic Math Toolbox™ to help us with the algebra. We define some symbolic constants for the robot's lengths and then create an ETS for the robot of ◻ Fig. 7.4b in the now familiar way

```
>> syms a1 a2 real
>> e = ETS2.Rz("q1")*ETS2.Tx(a1)*ETS2.Rz("q2")*ETS2.Tx(a2);
```

Next, we define some symbolic variables to represent the joint angles

```
>> syms q1 q2 real
```

and then compute the forward kinematics as an **SE**(2) matrix

```
>> TE = e.fkine([q1 q2])
TE =
[cos(q1 + q2), -sin(q1 + q2), a2*cos(q1 + q2) + a1*cos(q1)]
[sin(q1 + q2),  cos(q1 + q2), a2*sin(q1 + q2) + a1*sin(q1)]
[           0,             0,                            1]
>> transl = TE(1:2,3)';
```

which is an algebraic representation of the robot's forward kinematics – the end-effector pose as a function of the joint variables.

Finally, we define two more symbolic variables to represent the desired end-effector position (x, y)

```
>> syms x y real
```

and equate them with the results of the forward kinematics ▶

With the Symbolic Math Toolbox™ the == operator denotes equality, as opposed to = which denotes assignment.

```
>> e1 = x == transl(1)
e1 =
x == a2*cos(q1 + q2) + a1*cos(q1)
>> e2 = y == transl(2)
e2 =
y == a2*sin(q1 + q2) + a1*sin(q1)
```

which gives two scalar equations that we can solve simultaneously

```
>> [s1,s2] = solve([e1 e2],[q1 q2]);
```

where the arguments are respectively the set of equations and the set of unknowns to solve for. The outputs are the solutions for q_1 and q_2 respectively. We observed in ▶ Sect. 7.1.1.1 that two (or more) different sets of joint angles give the same end-effector position, and this means that the inverse kinematics does not have a unique solution. Here MATLAB has returned

```
>> length(s2)
ans =
     2
```

indicating two solutions. One solution for q_2 for a given link length $a_1 = 1$ and $a_2 = 1$ is

```
>> subs(s2(2),[a1 a2],[1 1])
ans =
2*atan((-(x^2 + y^2)*(x^2 + y^2 - 4))^(1/2)/(x^2 + y^2))
```

and would be used in conjunction with the corresponding element of the solution vector for q_1, which is s1(2).

From ▶ Sect. 7.1.1.1, we recall that $q_1 = 30°$ and $q_2 = 40°$ resulted in an end-effector position of $x = 1.21$ and $y = 1.44$. Let's see if one of our algebraic solutions matches the original q_1 and q_2 angles.

```
>> xfk = eval(subs(transl(1), [a1 a2 q1 q2], ...
>>    [1 1 deg2rad(30) deg2rad(40)]))
xfk =
    1.2080
>> yfk = eval(subs(transl(2), [a1 a2 q1 q2], ...
>>    [1 1 deg2rad(30) deg2rad(40)]))
yfk =
    1.4397
```

7

Excurse 7.5: Spherical Wrist

A spherical wrist is a key component of many modern robot manipulators. The wrist has three axes of rotation that are orthogonal and intersect at a common point. This is a gimbal-like mechanism, and as discussed in ▶ Sect. 2.3.1.2 will have a singularity.

The robot end-effector pose, its position and an orientation, is defined at the center of the wrist for mathematical convenience. Since the wrist axes intersect at a common point, wrist motion causes zero translational motion, therefore the end-effector position is a function only of the first three joints. This is a critical simplification that makes it possible to find closed-form inverse kinematic solutions for 6-axis industrial robots. An arbitrary end-effector orientation is achieved independently of position by means of the three wrist joints.

Substituting these values and the known link lengths a_1 and a_2 allows us to numerically evaluate the required q_1 and q_2 angles to reach that configuration.

```
>> q1r = eval(subs(s1(2),[a1 a2 x y],[1 1 xfk yfk]));
>> q1 = rad2deg(q1r)
q1 =
   30.0000
>> q2r = eval(subs(s2(2),[a1 a2 x y],[1 1 xfk yfk]));
>> q2 = rad2deg(q2r)
q2 =
   40.0000
```

As expected, we get the result of $q = (30°, 40°)$. If we perform the same substitution for s1(1) and s2(1), the result would be $q = (70°, -40°)$. Depending on the physical joint limits of the robot, this might be a second possible configuration.

As mentioned earlier, the complexity of the algebraic solution increases dramatically with the number of joints, and this is hard to completely automate.

7.2.1.2 Numerical Solution

We can think of the inverse kinematics problem as that of adjusting the joint coordinates until the forward kinematics matches the desired pose. More formally this is an optimization problem – to minimize the error between the forward kinematic solution and the desired end-effector pose $\boldsymbol{\xi}_E^*$

$$\boldsymbol{q}^* = \arg \min_{\boldsymbol{q}} \| \mathcal{K}(\boldsymbol{q}) \ominus \boldsymbol{\xi}_E^* \|$$

Once again, we use the robot from ◘ Fig. 7.4b

```
>> e = ETS2.Rz("q1")*ETS2.Tx(1)*ETS2.Rz("q2")*ETS2.Tx(1);
```

and define an error function based on the end-effector position error, not its orientation

$$E(q) = \| [\mathcal{K}(\boldsymbol{q})]_t - (x^*, \ y^*)^\top \|$$

We can solve this using the multi-variable minimization function `fminsearch`

```
>> pstar = [0.6 0.7];  % Desired position
>> q = fminsearch(@(q) norm(se2(e.fkine(q)).trvec-pstar),[0 0])
q =
   -0.2295    2.1833
```

where the first argument is the error function, expressed here as a MATLAB anonymous function, that incorporates the desired end-effector position; and the second argument is the initial guess at the joint coordinates. The computed joint angles indeed give the desired end-effector position

```
>> printtform2d(e.fkine(q),unit="deg")
t = (0.6, 0.7), 112 deg
```

As already discussed, there are two solutions for q, but the solution that is found using this approach depends on the initial guess of q passed to the minimizer.

7.2.2 3-Dimensional Robotic Arms

7.2.2.1 Closed-Form Solution

Closed-form solutions have been developed for most common types of 6-axis industrial robots, typically those defined by Denavit-Hartenberg parameters. A necessary condition for a closed-form inverse kinematics (also called analytical inverse kinematics) solution of a 6-axis robot is a spherical wrist mechanism. ▶ Some robot models have closed-form inverse kinematics and we will illustrate this for the ABB IRB 1600 industrial robot. It is shown in its home configuration in ◲ Fig. 7.20 ▶

Even robots with more than 6 DoF might have an analytical solution if we can extract a subtree that fulfills the same properties, for example an arm of a humanoid robot.

The result of this method is a wide table which has been cropped for inclusion in this book.

```
>> abb = loadrobot("abbIrb1600",DataFormat="row");
>> abb.show;
>> abb.showdetails
-------------------
Robot: (7 bodies)
  Idx   Body Name    Joint Name    Joint Type    Parent Name(Idx)   ...
  ---   ---------    ----------    ----------    ----------------   ...
    1     link_1       joint_1      revolute        base_link(0)    ...
    2     link_2       joint_2      revolute          link_1(1)     ...
    3     link_3       joint_3      revolute          link_2(2)     ...
    4     link_4       joint_4      revolute          link_3(3)     ...
    5     link_5       joint_5      revolute          link_4(4)     ...
```

◲ **Fig. 7.20** The ABB IRB 1600 industrial robot in its home configuration. It has a spherical wrist, which makes it a candidate for analytical inverse kinematics

```
        6       link_6          joint_6       revolute        link_5(5)  ...
        7       tool0      joint_6-tool0         fixed         link_6(6)  ...
--------------------
```

Let's take a closer look at all 6 DoF kinematic structures of the robot

```
>> aIK = analyticalInverseKinematics(abb);
>> aIK.showdetails
--------------------
Robot: (7 bodies)
Index    Base Name    EE Body Name    Type              Actions
-----    ---------    ------------    ----              -------
    1    base_link          link_6    RRRSSS  Use this kinematic group
    2    base_link           tool0    RRRSSS  Use this kinematic group
```

that are analytically solvable. These 6 DoF structures, called kinematic groups, usually start at the base link and end with an end-effector link, but they can exist anywhere in the robot's tree structure. A kinematic group is a special type of kinematic chain.

We can see that there are 2 possible kinematic groups that can be used to calculate the inverse kinematics solution. Both start at `base_link`, but one has `link_6` as an end effector and the other has `tool0`. Why are there 2 kinematic groups for a 6 DoF robot? Both groups solve for the same joint angles but use slightly different end-effector coordinate frames. The `tool0` link is attached to `link_6` by a fixed joint with only a rotational offset

```
>> abb.getBody("tool0").Joint.Type
ans =
    'fixed'
>> T = abb.getBody("tool0").Joint.JointToParentTransform;
>> printtform(T)
T: t = (0, 0, 0), RPY/zyx = (0, 1.57, 0) rad
```

so the serial chain to the base still satisfies the spherical wrist constraint.

The `Type` column shows the joint structure of the kinematic group, with `R` for revolute joint, `P` for prismatic joint, and `S` for a spherical joint that consists of three revolute joints with intersecting axes. On the MATLAB command-line, the `Use this kinematic group` strings are links that select which kinematic group the object should use.

By default, the object picks the kinematic group from the base to the last body in the tree (`tool0`). We will use that kinematic group for the rest of this section. Before we can find a closed-form solution, we need to generate an inverse kinematics function specific to this robot ◄

The MATLAB file for this function will be placed in the current folder. Make sure that MATLAB has write access and that the current folder is not a package folder (starting with a +).

```
>> abbIKFcn = aIK.generateIKFunction("ikIRB1600")
abbIKFcn =
  function_handle with value:
    @ikIRB1600
```

which returns a function handle that we can call like any other function. For a given pose of body `tool0`, we can now find the closed-form solutions for the joint angles

```
>> tgtPose = trvec2tform([0.93 0 0.5])*eul2tform([0 pi/2 0]);
>> qsol = abbIKFcn(tgtPose)
qsol =
    0.0000    0.5695    0.1306    3.1416    0.7000    3.1416
    0.0000    0.5695    0.1306         0   -0.7000    0.0000
```

which returns two possible configurations (one in each row). The first one

```
>> qn = qsol(1,:);
>> abb.show(qn);
```

is shown in ◙ Fig. 7.21a. The second configuration looks quite different from the first, but we can verify that both solutions

```
>> T1 = abb.getTransform(qsol(1,:),"tool0");
>> printtform(T1)
T1: t = (0.93, 0, 0.5), RPY/zyx = (0, 1.57, 0) rad
>> T2 = abb.getTransform(qsol(2,:),"tool0");
>> printtform(T2)
T2: t = (0.93, 0, 0.5), RPY/zyx = (0, 1.57, 0) rad
```

result in the *same* end-effector pose (see `tgtPose`). In the second configuration, the spherical wrist is simply rotated differently.

These two configurations are both physically achievable within the joint limits defined by the `abb` robot. In general, there are many more configurations that give the same end-effector pose. We can retrieve all of them by passing another `false` argument to the function, which indicates that joint limits should be ignored ▶

The function returns a 11 × 6 matrix. It has been cropped for inclusion in this book.

```
>> qsol = abbIKFcn(tgtPose,false)
qsol =
     0.0000    2.5344    3.0110    0.0000    0.7378    0.0000
    -3.1416    4.3471   -0.8952    3.1416    0.3103    0.0000
     0.0000    0.5695    0.1306    3.1416    0.7000    3.1416
     ...
```

This returns 11 different solutions. Some of them are shown in ◙ Fig. 7.21. Since we ignored joint limits, most of the configurations are not physically achievable by this particular robot, but the closed-form solution allows us to retrieve all possible configurations and assess the impact of different joint limits on the solution space.

◙ **Fig. 7.21** A subset of all possible analytical inverse kinematics solutions for the ABB IRB 1600 robot. The target translation for the end effector is at [0.93, 0, 0.5]. **a** Joint configuration that can be achieved within physical joint limits; **b**, **c**, **d** Configurations that are only valid when joint limits are disabled

It is also possible that no solution can be achieved. For example

```
>> abbIKFcn(trvec2tform([3,0,0]))
ans =
  0×6 empty double matrix
```

has failed because the arm is simply not long enough to achieve the requested pose. A pose may also be unachievable due to motion limits of the robot's joints.

At a singularity, the alignment of axes reduces the effective degrees of freedom (the gimbal lock problem again). The IRB 1600 has a wrist singularity when q_5 is equal to zero and the axes of joints 4 and 6 become aligned. In this case the best that the closed-form solution can do is to constrain $q_4 + q_6$ but their individual values are arbitrary. For example, consider the configuration

```
>> q = [0 pi/4 0 0.1 0 0.4];
```

The inverse kinematic solution is

```
>> qsol = abbIKFcn(abb.getTransform(q,"tool0"))
qsol =
   0.0000    0.7854    0.0000   -0.5354        0    1.0354
```

which has quite different values for q_4 and q_6 but the sum $q_4 + q_6 = 0.5$ is the same for both cases.

Closed-form solutions might also be possible for robots that have more than 6 DoF, if we can identify a kinematic group within the rigid body tree that fulfills the spherical wrist requirement. For example, we can solve the inverse kinematics for the Willow Garage PR2 robot. It has 45 DoF, but each of its arms contains analytically solvable kinematic groups. ◄

```
>> pr2 = loadrobot("willowgaragePR2",DataFormat="row");
>> aIK2 = analyticalInverseKinematics(pr2);
>> aIK2.showdetails
--------------------
Robot: (94 bodies)
```

Index	Base Name	EE Body Name	Type	...
				...
1	l_shoulder_pan_link	l_wrist_roll_link	RSSSSS	...
2	r_shoulder_pan_link	r_wrist_roll_link	RSSSSS	...

```
...
```

A complete example of how to compute and visualize different PR2 configurations is shown in the example

```
>> openExample("robotics/SolveAnalyticalIKForLargeDOFRobotExample")
```

and a screenshot from that example is shown in ◻ Fig. 7.22.

7.2.2.2 Numerical Solution

For the case of robots which do not have six joints and a spherical wrist or for which no analytical solution is available, we need to use an iterative numerical solution. Continuing with the example of the ABB IRB 1600 robot from the previous section,

```
>> abbHome = abb.homeConfiguration;
>> T = abb.getTransform(qn,"tool0");
>> printtform(T)
T: t = (0.93, 0, 0.5), RPY/zyx = (0, 1.57, 0) rad
>> abbIK = inverseKinematics(RigidBodyTree=abb);
```

we use the `inverseKinematics` object to compute the inverse kinematics solution numerically, where `inverseKinematics` is a MATLAB system object, so you can

The result of this method is a tall and wide table which has been cropped for inclusion in this book.

► sn.pub/AduZj0

◻ Fig. 7.22 An example of how to compute a closed-form solution for the left arm of the Willow Garage PR2 robot (Image reprinted with permission of The MathWorks, Inc.)

call it by using the variable name `abbIK` like a function. ▶

Calling `abbIK()` like a function is equivalent to calling `abbIK.step`.

```
>> [qsol,solinfo] = abbIK("tool0",T,ones(1,6),abbHome)
qsol =
   -0.0000    0.5695    0.1306    0.0000   -0.7000    0.0000

solinfo =
  struct with fields:
           Iterations: 13
    NumRandomRestarts: 0
        PoseErrorNorm: 6.6805e-09
             ExitFlag: 1
               Status: 'success'
```

which indicates success after just 13 iterations and with a very low solution error or residual (see the field `PoseErrorNorm`). The third input to the `abbIK` function is the vector of weights for orientation and position. This tells the optimizer what aspect of the final pose to prioritize when calculating the cost. A vector of 6 ones gives equal weight to all orientations and positions, and ▶ Sect. 7.2.3 shows an example where the weights need to be adjusted. The fourth input to the function is an initial guess for the joint configuration – we have used the home configuration, which is a vector of all zeros.

The joint configuration however is different from the original value

```
>> qn
qn =
    0.0000    0.5695    0.1306    3.1416    0.7000    3.1416
```

but does result in the correct end-effector pose

```
>> T2 = abb.getTransform(qsol,"tool0");
>> printtform(T2)
T2: t = (0.93, 0, 0.5), RPY/zyx = (0, 1.57, 0) rad
```

A limitation of this general numeric approach is that it does not provide explicit control over which solution is found as the analytical approach did – the only control is implicit via the initial estimate of joint coordinates (we used the home configuration with all zeros). If we specify different initial joint coordinates

```
>> qsol = abbIK("tool0",T,ones(1,6),[0 0 0 pi 0 0])
qsol =
    0.0000    0.5695    0.1306    3.1416    0.7000   -3.1416
```

we have nudged the minimizer to converge on the other valid configuration. Note that q_6 is a freely rotating joint, so the joint angle $-\pi$ (as in `qsol`) is the same as the joint angle π (as in `qn`).

By default, the `inverseKinematics` object will obey the joint limits defined for the robot, but that can be changed in the `SolverParameters` property, along with other parameters that affect the optimization.

As expected, the general numerical approach of `inverseKinematics` is considerably slower than the analytical approach of `analyticalInverseKinematics`. For our example, the numerical approach is an order of magnitude ($\approx 11.96\times$) slower than the analytical one. ◄

The specific times shown here will be different from machine to machine, but their ratio will generally be similar regardless of your hardware.

```
>> tNumericalIK = timeit(@() abbIK("tool0",T,ones(1,6),abbHome))
tNumericalIK =
    0.0083
>> tAnalyticalIK = timeit(@() abbIKFcn(T))
tAnalyticalIK =
    6.9415e-04
>> tNumericalIK / tAnalyticalIK
   11.9571
```

The analytical approach is essentially a constant time operation for a given robot while the numerical approach is iterative and its runtime depends strongly on the initial guess.

However, the numerical solver has the great advantage of being able to work with manipulators at singularities and for manipulators with less than, or more than, six joints. The principles behind numerical inverse kinematics are discussed in ► Sect. 8.5.

7.2.3 Underactuated Manipulator

An underactuated manipulator is one that has fewer than six joints, and SCARA robots such as shown in ◘ Fig. 7.2b are a common example. They have an RRPR joint structure which is optimized for planar-assembly tasks which require control of end-effector position in 3D and orientation in the xy-plane. The task space is $\mathcal{T} \subset \mathbb{R}^3 \times \mathbf{S}^1$ and the configuration space is $C \subset (\mathbf{S}^1)^2 \times \mathbb{R} \times (\mathbf{S}^1)$. Since $\dim C < 6$ the robot is limited in the end-effector poses that it can achieve. We will load a model of the Omron eCobra 600 SCARA robot

```
>> cobra = loadrvcrobot("cobra");
```

and then define a desired end-effector pose

```
>> TE = se3(deg2rad([170 0 30]),"eul","XYZ",[0.4 -0.3 0.2]);
```

where the end-effector approach vector is pointing downward but not parallel to the vertical axis – it is $10°$ off vertical. This pose is *over-constrained* for the 4-joint SCARA robot, the robot physically cannot meet the orientation requirement for an approach vector that is not vertical and the inverse kinematic solution

```
>> cobraHome = cobra.homeConfiguration;
>> cobraIK = inverseKinematics(RigidBodyTree=cobra);
>> weights = ones(1,6);
>> rng(0); % obtain repeatable results
>> [qsol,solinfo] = cobraIK("link4",TE.tform,weights,cobraHome)
qsol =
   -1.1760    1.1760    0.1870    0.5236

solinfo =
  struct with fields:
          Iterations: 1500
    NumRandomRestarts: 86
        PoseErrorNorm: 0.1745
```

```
        ExitFlag: 2
         Status: 'best available'
```

has failed to converge, ▶ but gives us the "best available" configuration. We need to relax the requirement and ignore error in rotation about the x- and y-axes, and we achieve that by specifying a different weight vector

```
>> weights = [0 0 1 1 1 1];
>> [qsol,solinfo] = cobraIK("link4",TE.tform,weights,cobraHome)
qsol =
   -0.1110   -1.1760    0.1870   -0.7634

solinfo =
  struct with fields:
           Iterations: 25
    NumRandomRestarts: 0
         PoseErrorNorm: 2.0509e-08
             ExitFlag: 1
               Status: 'success'
```

The `PoseErrorNorm` is still quite high. This measure of the residual as a scalar is imperfect, since it combines errors in position and orientation.

The elements of the weight vector correspond respectively to the three orientations and three translations: $r_x, r_y, r_z, t_x, t_y, t_z$, in the base coordinate frame. In this example we specified that error in rotation about the x- and y-axes is to be ignored (the zero elements). The resulting joint angles correspond to an achievable end-effector pose

```
>> T4 = cobra.getTransform(qsol,"link4");
>> printtform(T4,unit="deg",mode="xyz")
T4: t = (0.4, -0.3, 0.2), RPY/xyz = (180, 0, 30) deg
```

which has the desired translation and y and z rotation angles, but the x angle is incorrect, as we *allowed* it to be. They are what the robot mechanism actually permits. We can also compare the desired and achievable poses graphically

```
>> plotTransforms(TE,FrameColor="red"); hold on
>> plotTransforms(se3(T4),FrameColor="blue")
```

as shown in ❑ Fig. 7.23a.

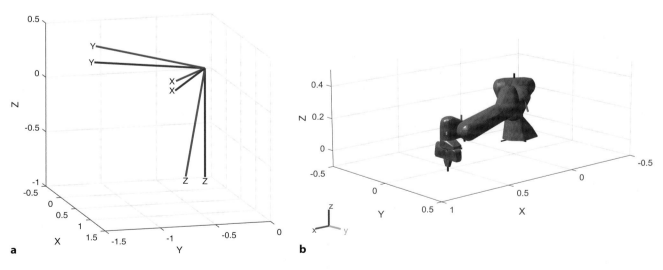

❑ **Fig. 7.23** Underactuated and redundant manipulators. **a** The desired (red) and achievable (blue) end-effector poses for the Omron eCobra 600 robot. The 4-joint SCARA robot is underactuated and cannot achieve the desired pose; **b** the redundant Panda robot achieves a desired target pose

7.2.4   **Overactuated (Redundant) Manipulator**

An overactuated or redundant manipulator is a robot with more than six joints. As mentioned previously, six joints are theoretically sufficient to achieve any desired pose in a Cartesian task space $\mathcal{T} \subset \mathbb{R}^3 \times \mathbf{S}^3$. In practice, issues such as joint limits, self-collision, and singularities mean that not all poses within the robot's reachable space can be achieved. Adding additional joints is one way to overcome this problem but results in an infinite number of joint-coordinate solutions. To find a single solution we need to introduce constraints – a common one is the minimum-norm constraint which returns a solution where the norm of the joint-coordinate vector has the smallest magnitude.

We will illustrate this with the Panda robot shown in ◘ Fig. 7.11 which has 7 joints. We load the model ◄

This model can be defined using modified Denavit-Hartenberg notation which is covered in ▶ Sect. 7.5.2

```
>> panda = loadrobot("frankaEmikaPanda",DataFormat="row");
```

then define the desired end-effector pose

```
>> TE = se3(trvec2tform([0.7 0.2 0.1]))* ...
>>    se3(oa2tform([0 1 0],[0 0 -1]));
```

which has its approach vector pointing downward. The numerical inverse kinematics solution is

```
>> pandaHome = panda.homeConfiguration;
>> pandaIK = inverseKinematics(RigidBodyTree=panda);
>> rng(0); % obtain repeatable results
>> qsol = pandaIK("panda_hand",TE.tform,ones(1,6),pandaHome)
qsol =
    1.0166   1.5964   -1.2999   -1.2547   1.5099   1.8361   -2.5897   0   0
```

which is the joint configuration vector with the smallest norm that results in the desired end-effector pose, as we can verify

```
>> handT = panda.getTransform(qsol,"panda_hand");
>> printtform(handT,mode="axang",unit="deg")
handT: t = (0.7, 0.2, 0.1), R = (180deg | 1.49e-09, -1, -6.45e-09)
>> panda.show(qsol);
```

The resulting robot configuration is shown in ◘ Fig. 7.23b.

7.3   **Trajectories**

A very common requirement in robotics is to move the end effector smoothly from one pose to another. Building on what we learned in ▶ Sect. 3.3 we will discuss two approaches to generating such trajectories: in configuration space and in task space. These are known respectively as joint-space and Cartesian motion.

7.3.1   **Joint-Space Motion**

Consider the end effector moving between two poses

```
>> TE1 = se3(3,"rotx",[0.6 -0.5 0.1]);
>> TE2 = se3(2,"rotx",[0.4 0.5 0.1]);
```

which describe two points in the xy-plane with different end-effector orientations. The joint configurations for these poses are ◄

We could have used numerical inverse kinematics here as well.

```
>> sol1 = abbIKFcn(TE1.tform);
>> sol2 = abbIKFcn(TE2.tform);
>> waypts = [sol1(1,:)' sol2(1,:)'];
```

Fig. 7.24 Joint-space trajectory based on a trapezoidal velocity profile. A subset of the 101 configurations is shown here. The time step between the configurations is constant, so it's easy to see the acceleration at the beginning of the trajectory and the deceleration at the end

There are two possible configurations for each pose, so we will only use the first solutions as waypoints. We require the motion to occur over a time period of 2 seconds in 20 ms time steps, and we create an array of time values

```
>> t = 0:0.02:2;
```

A joint-space trajectory is formed by smoothly interpolating between the two joint configurations. We can either interpolate the joint angles with a 5th-order polynomial

```
>> [q,qd,qdd] = quinticpolytraj(waypts,[0 2],t);
```

or apply a trapezoidal velocity profile

```
>> [q,qd,qdd] = trapveltraj(waypts,numel(t),EndTime=2);
```

which each result in a 6×101 array `q` with one column per time step and one row per joint. A subset of robot configurations for the output from `trapveltraj` is shown in ◨ Fig. 7.24. For this example, the `waypts` input only contains 2 waypoints, the start and the end, but the trajectory functions can handle an arbitrary number of intermediate waypoints. We can obtain the joint velocity and acceleration vectors, as a function of time, through optional output arguments `qd` and `qdd`.

The trajectory is best viewed as an animation

```
>> r = rateControl(50);
>> for i = 1:size(q,2)
>>   abb.show(q(:,i)',FastUpdate=true,PreservePlot=false);
>>   r.waitfor;
>> end
```

but we can also plot the joint angles versus time

```
>> xplot(t,q')
```

as shown in ◨ Fig. 7.25a. The joint-space trajectory is smooth, which was one of our criteria, but we do not know how the robot's end effector will move in Cartesian space. We can easily determine this by applying forward kinematics to the joint coordinate trajectory

```
>> for i = 1:size(q,2)
>>   trajT(i) = se3(abb.getTransform(q(:,i)',"tool0"));
```

7

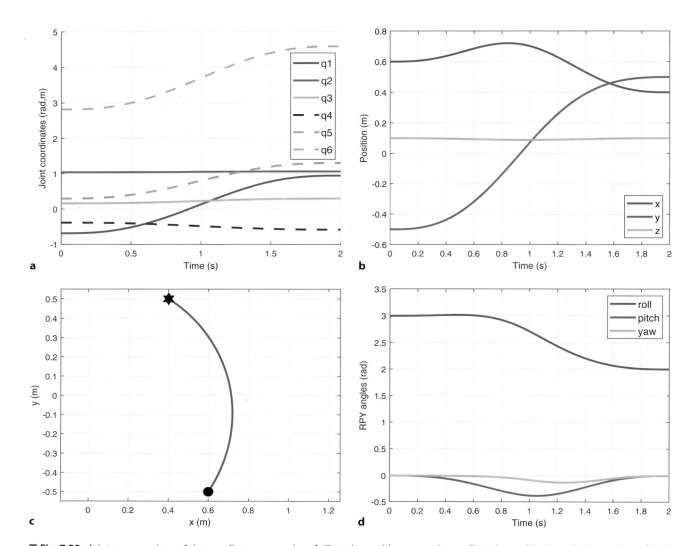

a   **b**   **c**   **d**

◘ **Fig. 7.25** Joint-space motion. **a** Joint coordinates versus time; **b** Cartesian position versus time; **c** Cartesian position locus in the *xy*-plane; **d** roll-pitch-yaw angles versus time

```
>> end
>> size(trajT)
ans =
     1   101
```

which results in a vector of 101 `se3` instances. The translational part of this trajectory

```
>> p = trajT.trvec;
>> size(p)
ans =
   101     3
```

is an array with one row per time step and one column per coordinate. This is plotted against time

```
>> plot(t,p);
>> legend("x","y","z")
```

in ◘ Fig. 7.25b. The path of the end effector in the *xy*-plane is shown in ◘ Fig. 7.25c and it is clear that the path is not a straight line. This is to be expected since we only specified the Cartesian coordinates of the end-points, but interpolated in configuration space. As the robot rotates about its waist joint during the motion the end effector will naturally follow a circular arc. In practice this could lead to collisions between the robot and nearby objects, even if they do not lie on the direct path between the start and end poses. The orientation of the end effector, in XYZ roll-pitch-yaw angle form, can also be plotted against time

```
>> plot(t,trajT.eul("XYZ"))
>> legend("roll","pitch","yaw")
```

as shown in ■ Fig. 7.25d. Note that the roll angle ▶ varies from 3 to 2 radians as we specified. The yaw and pitch angles have met their boundary conditions but have deviated along the path.

Besides the 5th-order polynomial and trapezoidal velocity profile trajectories shown in this section, we can also use other interpolants: third-order polynomial (`cubicpolytraj`), B-splines (`bsplinepolytraj`), and minimum jerk polynomial (`minjerkpolytraj`). Minimum jerk trajectories are especially interesting for practical robot applications. Jerk is the third derivative of position, and it has been shown that minimum jerk trajectories ▶ can be tracked efficiently by controllers and minimize wear and tear on the robot hardware.

Rotation about the z-axis for a robot end effector from ▶ Sect. 2.3.1.2.

It turns out that humans (subconsciously) favor minimum jerk trajectories when moving their arms.

7.3.2 Cartesian Motion

For many applications we require straight-line motion in Cartesian space which is known as Cartesian motion. This is implemented by

```
>> T = transformtraj(TE1,TE2,[0 2],t);
```

where the arguments are the initial and final pose and the time vector. It returns the trajectory as an array of `se3` objects. The default interpolation is linear, so the task space velocity is constant through the trajectory, which is not a desirable trajectory for a manipulator, and would cause discontinuous motion at the start and end of the trajectory.

We can use the `TimeScaling` to define the trajectory time using an intermediate parameterization, s, such that `transformtraj` is defined using $s(t)$ as time. In the earlier example, time scaling was uniform, so $s(t) = t$. The result is a linear motion between each pose. Instead, we can use time scaling defined by a different interpolant. For example, we can generate a minimum-jerk trajectory with `s = minjerkpolytraj(t)` or a trapezoidal velocity profile with `s = trapveltraj(t)`. $s(t)$ is defined on the normalized interval $[0, 1]$.

```
>> [s,sd,sdd] = minjerkpolytraj([0 1],[0 2],numel(t));
>> Ts = transformtraj(TE1,TE2,[0 2],t, ...
>>    TimeScaling=[s;sd;sdd]);
```

As for the previous joint-space example, we extract and plot the translation

```
>> plot(t,Ts.trvec);
>> legend("x","y","z")
```

and orientation components

```
>> plot(t,Ts.eul("XYZ"));
>> legend("roll","pitch","yaw")
```

which are shown in ■ Fig. 7.26b and ■ Fig. 7.26d.

■ Fig. 7.26c shows that the task-space path is now a straight line. The corresponding joint-space trajectory is obtained by applying the analytical inverse kinematics repeatedly. It is noteworthy that for trajectories, we want to ensure that there are no sudden jumps in joint angles. To accomplish that we can pass additional arguments to `abbIKFcn` to sort multiple solutions based on the distance to the previous trajectory configuration. ▶ This is handled internally by the `ikineTraj` function

```
>> qj = ikineTraj(abbIKFcn,Ts);
```

The `abbIKFcn` generated function has 2 optional arguments to specify a reference configuration and return the inverse kinematics solutions sorted by distance to the reference configuration. The distance is computed as Euclidean norm of the difference in joint angles (for revolute joints) and joint positions (for prismatic joints).

7

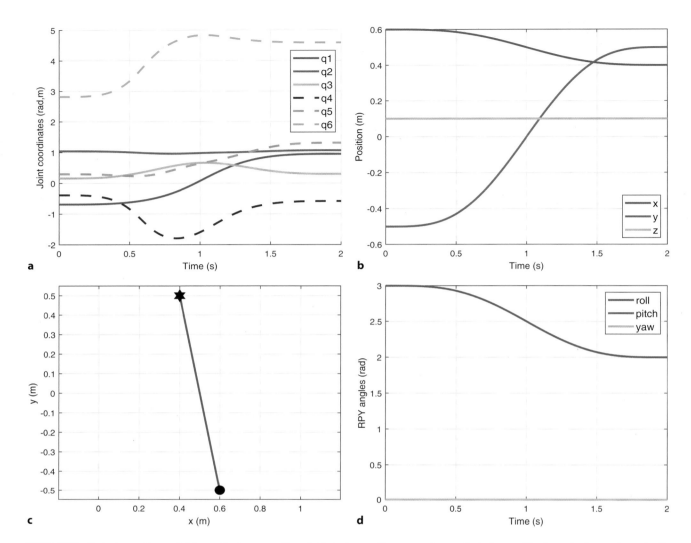

a

b

c

d

⬤ Fig. 7.26 Cartesian motion. **a** Joint coordinates versus time; **b** Cartesian position versus time; **c** Cartesian position locus in the *xy*-plane; **d** roll-pitch-yaw angles versus time

When solving for a trajectory using numerical inverse kinematics, the solution for one point is used to initialize the solution for the next point in the trajectory.

and is shown in ⬤ Fig. 7.26a. ◄ While broadly similar to ⬤ Fig. 7.25a the minor differences are what make the difference between a curved and straight line in the task space.

Additional trajectory generators for joint-space and Cartesian motion are described in the example

```
>> openExample("robotics/" + ...
>>    "ChooseATrajectoryForAManipulatorApplicationExample");
```

► sn.pub/TA0c8B

7.3.3 Kinematics in Simulink

We can also implement the example shown in ► Sect. 7.3.1 using Simulink®

```
>> sl_jointspace
```

and the block diagram model is shown in ⬤ Fig. 7.27. The inputs of the `Trapezoidal Velocity Profile Trajectory` block are the time and the initial and final values for the joint coordinates as columns. The smoothly varying

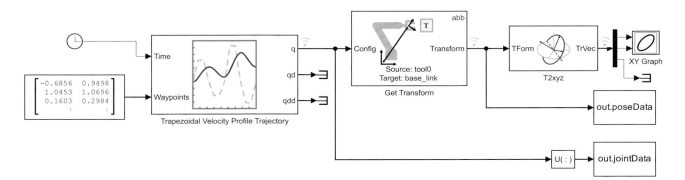

$$\begin{bmatrix} -0.6856 & 0.9498 \\ 1.0453 & 1.0696 \\ 0.1603 & 0.2984 \\ \vdots & \vdots \end{bmatrix}$$

Fig. 7.27 Simulink model `sl_jointspace` for joint-space motion

`sl_jointspace`

► sn.pub/er8VOO

joint angles are wired to a `Get Transform` block to compute the forward kinematics. This block has a parameter to specify the robot object to be used, in this case, set to `abb`. The Cartesian position of the end-effector pose is extracted using the `T2xyz` block which is analogous to the `tform2trvec` function. The `XY Graph` block plots `y` against `x`.

When running this model, the robot will follow the same trajectory as shown in ◘ Fig. 7.24. The `robotmaniplib` Simulink library also contains a block for inverse kinematics, which is useful for Cartesian trajectories.

7.3.4 Motion Through a Singularity

We briefly touched on the topic of singularities in ► Sect. 7.2.2.1 and we will revisit them again in the next chapter. In this next example we deliberately choose a trajectory that moves through a robot wrist singularity. We repeat the previous example but choose different start and end poses relative to a known singular configuration `qsing`

```
>> qsing = [0 pi/4 0 0.1 0 0.4];
>> TG = se3(abb.getTransform(qsing,"tool0"));
>> TE1 = se3(trvec2tform([0 -0.3 0]))*TG;
>> TE2 = se3(trvec2tform([0 0.3 0]))*TG;
```

which results in motion in the y-direction. The Cartesian path is

```
>> Ts = transformtraj(TE1,TE2,[0 2],t, ...
>>    TimeScaling=[s;sd;sdd]);
```

which we convert to joint coordinates

```
>> qj = ikineTraj(abbIKFcn,Ts);
```

and plot against time

```
>> xplot(t,qj,unwrap=true)
```

which is shown in ◘ Fig. 7.28a. We have used the `unwrap` option to remove the distracting jumps when angles pass through $\pm\pi$.

At time $t \approx 1$ s we observe that the wrist joint angles q_4 and q_6 change very rapidly. ► At this time q_5 is close to zero which means that the q_4 and q_6 rotational axes of the wrist are almost aligned – another gimbal lock situation or singularity. The robot has lost one degree of freedom and is now effectively a 5-axis robot.

As discussed in ► Sect. 7.2.2.1 this axis alignment means that we can only solve for the sum $q_4 + q_6$ meaning that there are an infinite number of solutions for

q_6 has increased rapidly, while q_4 has decreased rapidly. This counter-rotational motion of the two joints means that the gripper does not actually rotate but the two motors are working hard.

7

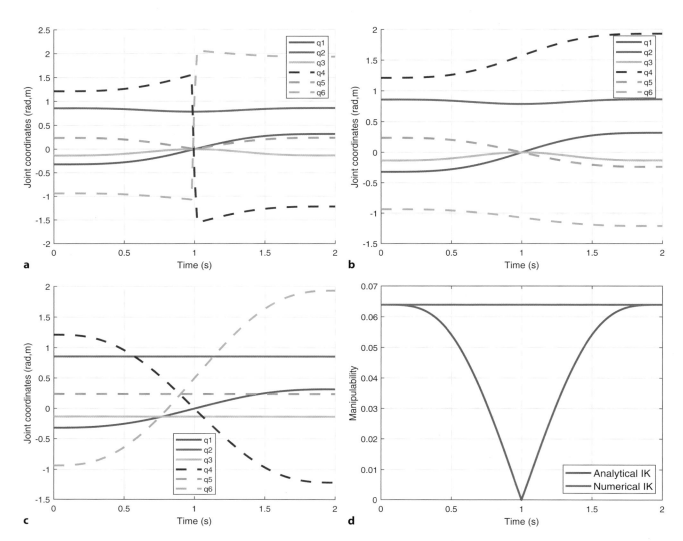

■ **Fig. 7.28** Joint angles for a path through a wrist singularity. **a** Cartesian trajectory computed using analytical inverse kinematics (`ikineTraj`); **b** Cartesian trajectory computed using numerical inverse kinematics (`ikineTrajNum`); **c** joint-space motion; **d** manipulability for the two Cartesian trajectories

q_4 and q_6 that have this sum. From ■ Fig. 7.28b we observe that numerical inverse kinematics

```
>> qj_num = ikineTrajNum(abb,Ts);
```

handles the singularity with far less unnecessary joint motion. This is a consequence of the minimum-norm solution which has returned q_4 and q_6 which have the correct sum but the smallest magnitude. The `ikineTrajNum` function also uses the previous solution as an initial guess for the next trajectory point, which biases the solver.

The joint-space motion between the two poses, ■ Fig. 7.28c, is immune to this problem since it is does not involve inverse kinematics. However, it will not maintain the orientation of the tool for the whole path – only at the two end points.

The dexterity of a manipulator is concerned with how *easily* its end effector can translate along, or rotate about, any axis. A common scalar measure of dexterity is manipulability which can be computed for each point along the trajectory

```
>> m = manipulability(abb,qj);
```

and is plotted in ◼ Fig. 7.28d. At around $t = 1$ s the manipulability for the path computed using the analytical inverse kinematics (blue line) was almost zero, indicating a significant loss of dexterity. A consequence of this is the very rapid wrist joint motion, seen in ◼ Fig. 7.28a, which is required to follow the end-effector trajectory. We can see that for the path computed using numerical inverse kinematics (red line), the robot was able to keep manipulability high throughout the trajectory since it implicitly minimizes the velocity of all the joints. Manipulability and the numerical inverse kinematics function are based on the manipulator's Jacobian matrix which is the topic of the next chapter, ▶ Chap. 8.

7.4 Applications

7.4.1 Writing on a Surface

Our goal is to create a trajectory that will allow a robot to draw a letter. The RVC Toolbox comes with a preprocessed version of the Hershey font ▶

Developed by Dr. Allen V. Hershey at the Naval Weapons Laboratory in 1967, data from ▶ http://paulbourke. net/dataformats/hershey.

```
>> load hershey
```

as a cell array of character descriptors. For an upper-case 'B'

```
>> B = hershey{'B'}
B =
  struct with fields:
    stroke: [2×23 double]
     width: 0.8400
       top: 0.8400
    bottom: 0
```

the structure describes the dimensions of the character, vertically from 0 to 0.84 and horizontally from 0 to 0.84. ▶

The path to be drawn is ▶

This is a variable-width font and all characters fit within a unit-grid.

The result is a tall and wide matrix which has been cropped for inclusion in this book.

```
>> B.stroke
ans =
  Columns 1 through 9
    0.1600    0.1600        NaN    0.1600    ...
    0.8400         0        NaN    0.8400    ...
    ...
```

where the rows are the x- and y-coordinates respectively, and a column of NaNs indicates the end of a segment – the pen is lifted and placed down again at the beginning of the next segment. We perform some processing

```
>> p = [0.5*B.stroke; zeros(1,size(B.stroke,2))];
>> k = find(isnan(p(1,:)));
>> p(:,k) = p(:,k-1); p(3,k) = 0.2;
```

to scale the path p by 0.5 so that the character is around 40 cm tall, append a row of zeros (add z-coordinates to this 2-dimensional path), find the columns that contain NaNs and replace them with the preceding column but with the z-coordinate set to 0.2 in order to lift the pen off the surface.

Next, we convert this to a continuous trajectory

```
>> dt = 0.02;  % sample interval
>> traj = mstraj(p(:,2:end)',[0.5 0.5 0.5],[],p(:,1)',dt,0.2);
```

where the second argument is the maximum speed in the x-, y-, and z-directions, the fourth argument is the initial coordinate followed by the sample interval and the acceleration time. The number of steps in the interpolated path is

```
>> whos traj
  Name     Size     Bytes   Class      Attributes
  traj     558x3    13392   double
```

Fig. 7.29 Drawing the letter 'B'. **a** The end-effector path (blue) with waypoints shown as red crosses; **b** The robot tracing the letter shape

and will take

```
>> size(traj,1)*dt
ans =
   11.1600
```

seconds to execute at the 20 ms sample interval. The trajectory can be plotted

```
>> plot3(traj(:,1),traj(:,2),traj(:,3))
```

as shown in ◘ Fig. 7.29a.

We now have a sequence of 3-dimensional points but for inverse kinematics we require a sequence of end-effector poses. We will create a coordinate frame at every point and assume that the robot is writing on a horizontal surface, so these frames must have their approach vector pointing downward, that is, $a = [0, 0, -1]$. For this application the gripper can be arbitrarily ◄ oriented and we will choose alignment with the y-axis, that is, $o = [0, 1, 0]$. The character will be placed at $(0.5,0,0)$ in the workspace and we assume that the pencil has a length of 0.075 m. All this is achieved by

The orientation of the pen around its longitudinal axis is not important in this application.

```
>> Tp = se3(trvec2tform([0.5 0 0.075]))*se3(eye(3),traj)* ...
>>    se3(oa2tform([0 1 0],[0 0 -1]));
```

Now we can apply inverse kinematics

```
>> qj = ikineTraj(abbIKFcn,Tp);
```

to determine the joint coordinates and then animate it

```
>> r = rateControl(1/dt);
>> for i = 1:size(qj,1)
>>    abb.show(qj(i,:),FastUpdate=true,PreservePlot=false);
>>    r.waitfor;
>> end
```

The ABB robot is drawing the letter 'B' and lifting its pen in between strokes! The approach is quite general, and we could easily change the size of the letter, write whole words and sentences, write on an arbitrary plane or use a robot with quite different kinematics. ◄

We have not considered the force that the robot-held pen exerts on the paper, we cover force control in ► Chap. 9. In a real implementation of this example, it would be prudent to use a spring to push the pen against the paper with sufficient force to allow it to write.

7.4.2   A 4-Legged Walking Robot

◘ Fig. 7.3b shows a sophisticated bipedal walking robot. In this example we will tackle something simpler, a four-legged (quadruped) walking robot with a gait that ensures it is always statically stable.

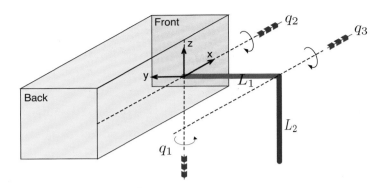

◻ **Fig. 7.30** The coordinate frame and axis rotations for the simple leg. The leg is shown in its zero angle pose

Kinematically, a robot leg is the same as a robot arm. For this application a three-joint serial-link manipulator is sufficient since the foot has point contact with the ground and orientation is not important. As always, we start by defining our coordinate frames. This is shown in ◻ Fig. 7.30 along with the robot leg in its zero-angle configuration. We have chosen a coordinate convention with the x-axis forward and the z-axis upward, constraining the y-axis to point to the left-hand side. The dimensions of the leg are

```
>> L1 = 0.5;
>> L2 = 0.5;
```

The robot has a 2 degree-of-freedom spherical hip. The first joint creates forward and backward motion, which is rotation about the z-axis. The second joint is hip up and down motion, which is rotation about the x-axis. The knee is translated by L1 in the $-y$-direction, and the third joint is knee motion, toward and away from the body, which is rotation about the x-axis. Finally, the foot is translated by L2 in the $-z$-direction. In ETS format this is

```
>> e = ETS3.Rz("q1")*ETS3.Rx("q2")*ETS3.Ty(-L1)* ...
>>    ETS3.Rx("q3")*ETS3.Tz(-L2);
>> leg = ets2rbt(e);
```

where leg is a rigidBodyTree. A quick sanity check shows that for zero joint angles the foot is at

```
>> se3(leg.getTransform([0 0 0],"link5")).trvec
ans =
         0   -0.5000   -0.5000
```

as we designed it. We could also visualize the zero-angle pose by

```
>> leg.show;
```

7.4.2.1 Motion of One Leg

Next, we define the path that the end effector of the leg, its foot, will follow. The first consideration is that the end effectors of all feet move backwards at the same speed in the ground plane – propelling the robot's body forward without its feet slipping. Each leg has a limited range of movement so it cannot move backward for very long. At some point we must reset the leg – lift the foot, move it forward and place it on the ground again. The second consideration comes from static stability – the robot must have always at least three feet on the ground so each leg must take its turn to reset. This requires that any leg is in contact with the ground for 75% of the cycle and is resetting for 25% of the cycle. A consequence of this is that the leg has to move much faster during reset since it has a longer path and less time to do it in. This pattern of coordinated leg motion is known as a gait – in this case a crawl gait or a wave gait.

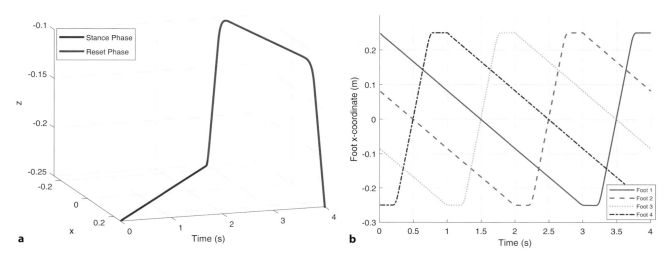

7

■ **Fig. 7.31** **a** Trajectory taken by a single foot. Recall from ■ Fig. 7.30 that the z-axis is upward; **b** The x-direction motion of each leg (offset vertically) to show the gait. The leg reset is the period of high x-direction velocity

During the stance phase, the foot is 25 cm below the hip and in contact with the ground. It moves from 25 cm forward of the hip to 25 cm behind, pushing the robot's body forward. During the reset phase the foot is raised to 10 cm below the hip, swung forward, and placed on the ground. The complete cycle – the start of the stance phase, the end of stance, top of the leg lift, top of the leg return and the start of stance – is defined by the via points

```
>> xf = 0.25; xb = -xf;   y = -0.25; zu = -0.1; zd = -0.25;
>> via = [xf y zd
>>        xb y zd
>>        xb y zu
>>        xf y zu
>>        xf y zd];
```

where `xf` and `xb` are the forward and backward limits of leg motion in the x-direction (in units of m), `y` is the distance of the foot from the body in the y-direction, and `zu` and `zd` are respectively the height of the foot motion in the z-direction for foot up and foot down. Next, we sample the multi-segment path at 100 Hz

```
>> x = mstraj(via,[],[3 0.25 0.5 0.25],[],0.01,0.1);
```

and we have specified a vector of desired segment times rather than maximum joint velocities to ensure that the reset takes exactly one quarter of the cycle. The final three arguments are the initial leg configuration, the sample interval, and the acceleration time. This trajectory is shown in ■ Fig. 7.31a and has a total time of 4 s and therefore comprises 400 points.

We apply inverse kinematics to determine the joint angle trajectories required for the foot to follow the computed Cartesian trajectory. This robot is underactuated, so we use numerical inverse kinematics and set the mask so as to solve only for end-effector translation

```
>> qcycle = ikineTrajNum(leg,se3(eye(3),x),"link5", ...
>>    weights=[0 0 0 1 1 1]);
```

We can view the motion of the leg in animation

```
>> r = rateControl(50);
>> for i = 1:size(qcycle,1)
>>    leg.show(qcycle(i,:),FastUpdate=true,PreservePlot=false);
>>    r.waitfor;
>> end
```

to verify that it does what we expect: slow motion along the ground, then a rapid lift, forward motion, and foot placement.

7.4.2.2 Motion of Four Legs

Our robot has width and length

```
>> W = 0.5; L = 1;
```

We create multiple instances of the leg by copying the leg object we created earlier, and providing different base transformations so as to attach the legs to different points on the body

```
>> Tflip = se3(pi,"rotz");
>> legs = [ ...
>>    rbtTform(leg,se3(eye(3),[L/2 W/2 0])*Tflip), ...
>>    rbtTform(leg,se3(eye(3),[-L/2 W/2 0])*Tflip), ...
>>    rbtTform(leg,se3(eye(3),[L/2 -W/2 0])), ...
>>    rbtTform(leg,se3(eye(3),[-L/2 -W/2 0]))];
```

The result is a vector of `rigidBodyTree` objects. Note that legs 1 and 2, on the left-hand side of the body have been rotated about the z-axis so that they point away from the body.

As mentioned earlier, each leg must take its turn to reset. Since the trajectory is a cycle, we achieve this by having each leg run the trajectory with a phase shift equal to one quarter of the total cycle time. Since the total cycle has 400 points, each leg's trajectory is offset by 100, and we use modulo arithmetic to index into the cyclic gait for each leg. The result is the gait pattern shown in ◘ Fig. 7.31b. The core of the walking program is

```
>> r = rateControl(50);
>> for i = 1:500
>>    legs(1).show(gait(qcycle,i,0,false)); hold on;
>>    legs(2).show(gait(qcycle,i,100,false));
>>    legs(3).show(gait(qcycle,i,200,true));
>>    legs(4).show(gait(qcycle,i,300,true)); hold off;
>>    r.waitfor;
>> end
```

where the function

```
>> function q = gait(cycle, k, phi, flip)
>>    k = mod(k+phi-1,size(cycle,1))+1;
>>    q = cycle(k,:);
>>    if flip
>>       q(1) = -q(1);    % for left-side legs
>>    end
```

returns the $(k+phi)^{th}$ element of q with modulo arithmetic that considers q as a cycle. The argument `flip` reverses the sign of the joint 1 motion for legs on the

◘ **Fig. 7.32** The walking robot

left-hand side of the robot. A snapshot from the simulation is shown in ◘ Fig. 7.32. The entire implementation, with some additional refinement and explanation, is in the file `examples/walking.m`.

7.5   Advanced Topics

7.5.1   Creating the Kinematic Model for a Robot

The classical method of determining Denavit-Hartenberg parameters is to systematically assign a coordinate frame to each link. The link frames for the PUMA 560 robot using the standard Denavit-Hartenberg formalism are shown in ◘ Fig. 7.33. There are constraints on placing each frame since joint rotation must be about the z-axis and the link displacement must be in the x-direction. This in turn imposes constraints on the placement of the coordinate frames for the base and the end effector, and ultimately dictates the zero-angle pose. Determining the Denavit-Hartenberg parameters and link coordinate frames for a completely new mechanism is therefore more difficult than it should be – even for an experienced roboticist. Here we will introduce an alternative approach.

We will use the classic PUMA 560 as an exemplar of the class of all-revolute six-axis robot manipulators with $C = (S^1)^6$. ◘ Fig. 7.34 shows a side view of the robot in the desired zero-joint configuration shape. Its key dimensions are shown and have the values

```
>> L1 = 0.672; L2 = -0.2337; L3 = 0.4318;
>> L4 = 0.0203; L5 = 0.0837; L6 = 0.4318;
```

We will create a kinematic model of this robot by stepping along the kinematic chain from base to tip and recording the elementary transformations as we go. Starting with the world frame 0 we move up a distance of L_1, perform the waist rotation about the z-axis by q_1, move to the left ($-y$-direction) by L_2, perform the shoulder rotation about the y-axis by q_2, move up (z-direction) by L_3, move forward (x-direction) by L_4, move sideways by L_5, perform the elbow rotation

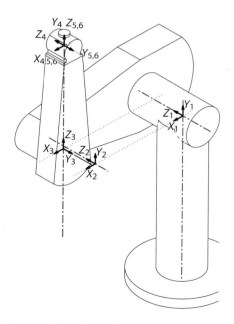

◘ **Fig. 7.33**   PUMA 560 robot coordinate frames. Standard Denavit-Hartenberg link coordinate frames for the PUMA in the zero-angle pose (Corke 1996b)

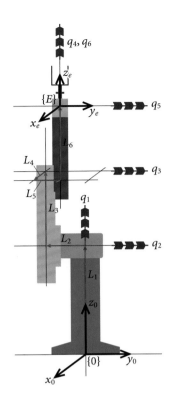

◻ Fig. 7.34 PUMA 560 robot in the zero-joint-angle configuration showing dimensions and joint axes (indicated by blue triple arrows) (after Corke 2007)

about the y-axis by q_3, move up by L_6, then rotate by q_4 about the z-axis, q_5 about the y-axis and q_6 about the z-axis. The result of our journey is

```
>> e = ETS3.Tz(L1)*ETS3.Rz("q1")*ETS3.Ty(L2)*ETS3.Ry("q2") ...
>>     *ETS3.Tz(L3)*ETS3.Tx(L4)*ETS3.Ty(L5)*ETS3.Ry("q3") ...
>>     *ETS3.Tz(L6)*ETS3.Rz("q4")*ETS3.Ry("q5")*ETS3.Rz("q6");
```

and we turn this into a fully-fledged robot object

```
>> robot = ets2rbt(e);
```

which we can use for kinematics and plotting. This approach avoids the complexities associated with the Denavit-Hartenberg convention, particularly in choosing the robot configuration and link frames.

7.5.2 Modified Denavit-Hartenberg Parameters

The Denavit-Hartenberg notation introduced in ▶ Sect. 7.1.5 dates to the 1960s and is covered in many robotics textbooks. The modified Denavit-Hartenberg notation was introduced later ▶ and differs by placing the link coordinate frames at the near (proximal), rather than the far (distal), end of each link as shown in ◻ Fig. 7.35. This modified notation is in some ways clearer and tidier and is now quite commonly used.

It first appeared in Craig (1986).

> ❶ **Standard and modified Denavit-Hartenberg confusion**
>
> Having two Denavit-Hartenberg notations is confusing, particularly for those who are new to robot kinematics. Algorithms for kinematics, Jacobians and dynamics depend on the kinematic conventions used. Knowing the "Denavit-Hartenberg parameters" for a robot is useless unless you know whether they are standard or modified parameters.

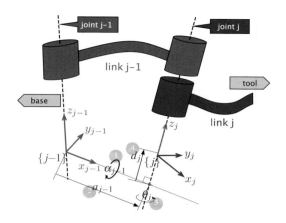

□ **Fig. 7.35** Definition of modified Denavit and Hartenberg link parameters associated with links $j-1$ (red) and j (blue). The numbers in circles represent the order in which the elementary transformations are applied

In modified Denavit-Hartenberg notation the link transformation matrix is

$$
{}^{j-1}\boldsymbol{\xi}_j = \boldsymbol{\xi}^{t_x}(a_{j-1}) \oplus \boldsymbol{\xi}^{r_x}(\alpha_{j-1}) \oplus \boldsymbol{\xi}^{r_z}(\theta_j) \oplus \boldsymbol{\xi}^{t_z}(d_j) \tag{7.7}
$$

which has the same terms as (7.4) but in a different order – remember rotations are not commutative. In both conventions a_j is the link length but in the standard form it is the displacement between the origins of frame $\{j-1\}$ and frame $\{j\}$ while in the modified form it is from frame $\{j\}$ to frame $\{j+1\}$.

MATLAB can handle either form, it only needs to be specified, and this is achieved by choosing the appropriate parameter when setting the transform of a `rigidBodyJoint`

```
>> a = 1; alpha = 0; d = 0; theta = 0;
>> link = rigidBody("link1");
>> link.Joint = rigidBodyJoint("joint1","revolute");
>> link.Joint.setFixedTransform([a alpha d theta],"mdh");
```

Everything else from here on, creating the robot object, kinematic and dynamic functions works as previously described.

The two forms can be interchanged by considering the link transformation as a string of elementary rotations and translations as in (7.4) or (7.7). Consider the transformation chain for standard Denavit-Hartenberg notation

$$
\underbrace{\boldsymbol{\xi}^{r_z}(\theta_1) \oplus \boldsymbol{\xi}^{t_z}(d_1) \oplus \boldsymbol{\xi}^{t_x}(a_1) \oplus \boldsymbol{\xi}^{r_x}(\alpha_1)}_{\mathrm{DH}_1 = {}^0\boldsymbol{\xi}_1} \oplus \underbrace{\boldsymbol{\xi}^{r_z}(\theta_2) \oplus \boldsymbol{\xi}^{t_z}(d_2) \oplus \boldsymbol{\xi}^{t_x}(a_2) \oplus \boldsymbol{\xi}^{r_x}(\alpha_2)}_{\mathrm{DH}_2 = {}^1\boldsymbol{\xi}_2} \cdots
$$

Excurse 7.6: What sort of DH parameters are these?

If you intend to build a robot model from a table of kinematic parameters provided in an online forum or research paper you need to know which convention is being used. Too often, this important fact is omitted and you need to do a little sleuthing.

An important clue lies in the column headings. If they all have the same subscript, i.e. θ_j, d_j, a_j and α_j then this is standard Denavit-Hartenberg notation. If half the subscripts are different, i.e. θ_j, d_j, a_{j-1} and α_{j-1} then you are dealing with modified Denavit-Hartenberg notation.

You can help the field when you share code or publish papers, by clearly stating which kinematic convention you use.

which we can regroup as

$$\underbrace{\boldsymbol{\xi}^{r_z}(\theta_1) \oplus \boldsymbol{\xi}^{t_z}(d_1)}_{\text{MDH}_1 = {}^0\boldsymbol{\xi}_1} \oplus \underbrace{\boldsymbol{\xi}^{t_x}(a_1) \oplus \boldsymbol{\xi}^{r_x}(\alpha_1) \oplus \boldsymbol{\xi}^{r_z}(\theta_2) \oplus \boldsymbol{\xi}^{t_z}(d_2)}_{\text{MDH}_2 = {}^1\boldsymbol{\xi}_2} \oplus \underbrace{\boldsymbol{\xi}^{t_x}(a_2) \oplus \boldsymbol{\xi}^{r_x}(\alpha_2)}_{\text{MDH}_3 = {}^2\boldsymbol{\xi}_3} \cdots$$

where the terms marked as MDH_j have the form of (7.7).

7.5.3 Products of Exponentials

We introduced 2D and 3D twists in ▶ Chap. 2 and recall that a twist is a rigid-body transformation defined by a screw-axis direction, a point on the screw axis, and a screw pitch. A 2D twist is a vector $\boldsymbol{S} \in \mathbb{R}^3$ and a 3D twist is a vector $\boldsymbol{S} \in \mathbb{R}^6$.

For the case of the single-joint robot of ◻ Fig. 7.4a we can place a screw along the joint axis that rotates the end-effector frame about the screw axis

$$\mathbf{T}_E(q) = e^{q[\hat{S}]}\mathbf{T}_E(0)$$

where $\hat{S}$ is a unit twist representing the screw axis and $\mathbf{T}_E(0)$ is the pose of the end effector when $q = 0$.

For the 2-joint robot of ◻ Fig. 7.4b we would write

$$\mathbf{T}_E(\boldsymbol{q}) = e^{\mathring{q}_1[\hat{S}_1]}\left(e^{q_2[\hat{S}_2]}\mathbf{T}_E(0)\right)$$

where $\hat{S}_1$ and $\hat{S}_2$ are unit twists representing the screws aligned with the joint axes at the zero-joint configuration $q_1 = q_2 = 0$, and $\mathbf{T}_E(0)$ is the end-effector pose at that configuration. The term in parentheses is similar to the single-joint robot above, and the first twist rotates that joint and link about $\hat{S}_1$. Using the RVC Toolbox we define the link lengths and compute $\mathbf{T}_E(0)$

```
>> a1 = 1; a2 = 1;
>> TE0 = se2(eye(2),[a1+a2 0]);
```

define the two 2-dimensional twists, in $\mathbf{SE}(2)$, for this example

```
>> S0 = Twist2d.UnitRevolute([0 0]);
>> S1 = Twist2d.UnitRevolute([a1 0]);
```

and apply them to $\boldsymbol{T}_E(0)$

```
 >> TE = S0.exp(deg2rad(30))*S1.exp(deg2rad(40))*TE0
TE =
  se2
    0.3420   -0.9397    1.2080
    0.9397    0.3420    1.4397
         0         0    1.0000
```

For a general robot we can write the forward kinematics in product of exponential (PoE) form as

$$\mathbf{T}_E = \left(e^{q_1[\hat{S}_1]} \cdots e^{q_N[\hat{S}_N]}\right)\mathbf{T}_E(0)$$

The tool and base transformations are effectively included in $\mathbf{T}_E(0)$, but an explicit base transformation could be added if the screw axes are defined with respect to the robot's base rather than the world coordinate frame, or use the adjoint matrix to transform the screw axes from base to world coordinates.

where ${}^0\mathbf{T}_E(0)$ is the end-effector pose when the joint coordinates are all zero and $\hat{S}_j$ is the unit twist for joint j expressed in the world frame at that configuration. ▶ This can also be written as

$$\mathbf{T}_E = \mathbf{T}_E(0)\left(e^{q_1\left[{}^E\hat{S}_1\right]} \cdots e^{q_N\left[{}^E\hat{S}_N\right]}\right)$$

and $e^{q_j[{}^{E}\boldsymbol{S}_j]}$ is the twist for joint j expressed in the end-effector frame which is related to the twists above by ${}^{E}\hat{\boldsymbol{S}}_j = \mathrm{Ad}({}^{E}\boldsymbol{\xi}_0)\hat{\boldsymbol{S}}_j$.

A serial-link manipulator can be succinctly described by a table listing the 6 screw parameters for each joint as well as the zero-joint-coordinate end-effector pose.

7.5.4 Collision Checking

For many robotic applications we require the ability to check for potential collisions between the robot and objects that are fixed or moving in the world. This is typically achieved by modeling every object with one or more collision shapes as shown in ◻ Fig. 7.36. These shapes can be simple geometric primitives such as rectangular prisms, spheres or cylinders whose intersections can be computed cheaply, or more general meshes. ◄

As defined by an STL or COLLADA file.

To illustrate this, we will instantiate a model of the Panda robot and visualize its collision models

```
>> panda = loadrobot("frankaEmikaPanda",DataFormat="row");
>> qp = [0 -0.3 0 -2.2 0 2 0.7854 0, 0];
>> panda.show(qp,Visuals="off",Collisions="on",Frames="off");
```

as green meshes in ◻ Fig. 7.36a. ◄

In comparison to ◻ Fig. 7.17, the collision meshes for the Panda robot are quite detailed, which can negatively affect the collision checking performance.

We can now create a box in the environment with 1 m sides and centered at $(1.1, -0.5, 0)$ in the workspace

```
>> box = collisionBox(1,1,1);
>> box.Pose = se3([1.1 -0.5 0],"trvec");
```

We test for a collision between the robot in a specified configuration and the box by

```
>> panda.checkCollision(qp,{box},IgnoreSelfCollision="on")
ans =
  logical
  0
```

which in this case is false – none of the robot's links will intersect the box. In this call, we are also ignoring self-collisions between the different links of the robot,

a   **b**

◻ **Fig. 7.36** Collision checking with a Panda robot. **a** Its collision meshes shown in green and a box-shaped, orange obstacle; **b** When not in collision, we can retrieve the minimum separation distances and witness points. The witness points are shown as circles for some of the robot's links since we want to focus on environment collisions. If we set the center of the box to $(1.0, -0.5, 0)$

```
>> box.Pose = se3([1.0 -0.5 0],"trvec");
>> hold on; box.show
>> panda.checkCollision(qp,{box},IgnoreSelfCollision="on")
ans =
  logical
    1
```

then there is a collision and shown in ◖ Fig. 7.36a.

Even if objects are not in collision with the robot, we can still get important information about their distance from the robot. For example, we might want to reduce the robot's velocity when we get within a safety distance to an obstacle.

Consider the following scenario where we import a mesh from an STL file

```
>> vehicleMesh = stlread("groundvehicle.stl");
>> vehicle = collisionMesh(vehicleMesh.Points/100);
>> vehicle.Pose = se3([0.5 0 0],"trvec");
>> [isColl,sepDist,witPts] = panda.checkCollision(qp,{vehicle}, ...
>>    IgnoreSelfCollision="on")
isColl =
  logical
    0
sepDist =
    0.3448
    0.3733
    0.5390
    ...
witPts =
    0.0552    0.4000
   -0.0036   -0.0036
    0.1410    0.1410
    ...
```

and calculate the separating distances `sepDist` between each robot link (including the base) and the vehicle. The exact points on the robot and the vehicle that have the minimum separation distance are returned as witness points in `witPts`, with a 3×2 matrix for each robot link. This information is visualized in ◖ Fig. 7.36b.

Besides box-shaped obstacles, we can also define collision objects as spheres (`collisionSphere`), cylinders (`collisionCylinder`), or arbitrary convex meshes (`collisionMesh`). The `checkCollision` method takes a cell array of collision objects, so we can check against a single object or against multiple objects at once.

For most motion planning applications, collision checking is the most expensive step in the planning process. There are multiple ways that this cost is usually reduced. Firstly, if we just want to know if a collision occurred or not, we don't have to check against all combinations of robot links and environment objects but exit early when the first collision is detected. By default, the function returns on the first collision, but this behavior can be controlled in `checkCollision` by passing the `Exhaustive` argument. Secondly, collision checkers typically only deal with *convex* meshes. ▶ The `collisionMesh` class will automatically create a convex hull ▶ if the input vertices describe a concave mesh. Thirdly, the collision meshes for robots are typically more conservative and simpler than the high-resolution meshes used for visualization. As an example of this simplification, see ◖ Fig. 7.17.

A mesh is convex if it contains the line segments connecting each pair of its vertices.

The smallest convex mesh that contains all input vertices.

7.6 Wrapping Up

In this chapter we have learned how to determine the forward and inverse kinematics of a serial-link manipulator arm. Forward kinematics involves compounding the relative poses due to each joint and link, giving the pose of the robot's end effector relative to its base. We have considered this from the perspective of pose graphs and rigid body trees in both 2D and 3D, and we have discussed other ways of representing models such as URDF files, Denavit-Hartenberg parameters, and the product of exponentials using twists.

Inverse kinematics is the problem of determining the joint coordinates given the end-effector pose. This inverse is not unique, and the robot may have several joint configurations that result in the same end-effector pose. For simple robots, or those with six joints and a spherical wrist we can compute the inverse kinematics using an analytical solution which provides explicit control over which solution is found.

For robots which do not have six joints and a spherical wrist we can use an iterative numerical approach to solve the inverse kinematic problem. We showed how this could be applied to an underactuated 4-joint SCARA robot and a redundant 7-link robot. We also touched briefly on the topic of singularities which are due to the alignment of joint axes.

We also learned about creating paths to move the end effector smoothly between poses. Joint-space paths are simple to compute but in general do not result in straight-line paths in Cartesian space which may be problematic for some applications. Straight-line paths in Cartesian space can be generated but singularities in the workspace may lead to very high joint rates.

7.6.1 **Further Reading**

Serial-link manipulator kinematics are covered in all the standard robotics textbooks such as the Robotics Handbook (2016), Siciliano et al. (2009), Spong et al. (2006), and Paul (1981). Craig's text (2005) is also an excellent introduction to robot kinematics and uses the modified Denavit-Hartenberg notation, and the examples in the third edition are based on an older version of the RVC Toolbox. Lynch and Park (2017) and Murray et al. (1994) cover the product of exponential approach. An emerging alternative to Denavit-Hartenberg notation is URDF (Unified Robot Description Format) which is described at ▶ https://wiki.ros.org/urdf.

Siciliano et al. (2009) provide a very clear description of the process of assigning Denavit-Hartenberg parameters to an arbitrary robot. An alternative approach based on symbolic factorization of elementary transformation sequences was described in detail by Corke (2007). The definitive values for the parameters of the PUMA 560 robot are described in the paper by Corke and Armstrong-Hélouvry (1994).

Robotic walking is a huge field and the example given here is very simplistic. Machines have been demonstrated with complex gaits such as running and galloping that rely on dynamic rather than static balance. A good introduction to legged robots is given in the Robotics Handbook (Siciliano and Khatib 2016, § 17). Robotic hands, grasping, and manipulation are large topics which we have not covered – there is a good introduction in the Robotics Handbook (Siciliano and Khatib 2016, § 37, 38).

Parallel-link manipulators have not been covered in this book. They have advantages such as increased actuation force and stiffness (since the actuators form a truss-like structure) and are capable of very high-speed motion. For this class of mechanism, the inverse kinematics is usually closed-form and it is the forward kinematics that requires numerical solution. Useful starting points for this class of robots are the handbook (2016, § 18), a brief section in Siciliano et al. (2009) and Merlet (2006).

Closed-form inverse kinematic solutions can be derived algebraically by writing down a number of kinematic relationships and solving for the joint angles, as described in Paul (1981). The `analyticalInverseKinematics` class used in this chapter is based on the work by Pieper (1968) for 6 DoF kinematic groups. Raghavan and Roth (1990) explore how to find analytical solutions for additional manipulator configurations. Other software packages to automatically generate the forward and inverse kinematics for a given robot have been developed and these include SYMORO (Khalil and Creusot 1997) and OpenRAVE (▶ http://openrave. org/), which contains the IKFast kinematic solver.

▪ ▪ Historical

The original work by Denavit and Hartenberg was their 1955 paper (1955) and their textbook (Hartenberg and Denavit 1964). The book has an introduction to the field of kinematics and its history but is currently out of print, although a version can be found online. The first full description of the kinematics of a six-link arm with a spherical wrist was by Paul and Zhang (1986). Lipkin (2005) provides a discussion about the various forms of Denavit-Hartenberg notation beyond the two variants covered in this chapter. Spong et al. (2006) clarify that although the Denavit-Hartenberg notation only has 4 parameters, the constraints on axis alignment and intersection ensure that these 4 parameters are sufficient to specify any homogeneous transformation. In other words, for any given manipulator, we can choose coordinate frames in such a way that the manipulator kinematics can be described by Denavit-Hartenberg notation.

7.6.2 Exercises

1. Forward kinematics for planar robot from ◼ Fig. 7.4.
    a) For the 2-joint robot use the `teach` method to determine the two sets of joint angles that will position the end effector at (0.5, 0.5).
    b) Experiment with the three different models in ◼ Fig. 7.4 using the `fkine` and `teach` methods.
    c) Vary the models: adjust the link lengths, create links with a translation in the y-direction, or create links with a translation in the x- and y-direction.
2. Experiment with the `interactiveRigidBodyTree` object for the Panda robot.
3. Inverse kinematics for the 2-link robot (▶ Sect. 7.2.1).
    a) What happens to the solution when a point is out of reach?
    b) Most end-effector positions can be reached by two different sets of joint angles. What points can be reached by only one set?
    c) Find a solution given x and θ where y is unconstrained.
4. Compare the solutions generated by `analyticalInverseKinematics` and `inverseKinematics` for the ABB IRB 1600 robot at different poses. Is there any difference in accuracy? How much slower is `inverseKinematics`?
5. For a ABB IRB 1600 robot investigate the errors in end-effector pose due to manufacturing errors.
    a) Make link 2 longer by 0.5 mm. For 100 random joint configurations what is the mean and maximum error in the components of end-effector pose?
    b) Introduce an error of $0.1°$ in the joint 2 angle and repeat the analysis above.
6. Use `loadrobot` to investigate the redundant robot models `kinovaGen3` and `abbYuMi`. Manually control them using the `interactiveRigidBodyTree` object, compute forward kinematics and numerical inverse kinematics.
7. Experiment with inverse kinematics for the case where the joint coordinates have limits (modeling mechanical end stops). Joint limits are set with the `PositionLimits` property of the `rigidBodyJoint` class.
8. Drawing a 'B' (▶ Sect. 7.4.1)
    a) Change the size of the letter.
    b) Write a word or sentence.
    c) Write on a vertical plane.

    d) Write on an inclined plane.

    e) Change the robot from an ABB IRB 1600 to a Panda.

    f) Write on a sphere. Hint: Write on a tangent plane, then project points onto the sphere's surface.

    g) This writing task does not require 6DoF since the rotation of the pen about its axis is not important. Remove the final link from the ABB IRB 1600 robot model and repeat the exercise.

9. Walking robot (▶ Sect. 7.4.2)

    a) Shorten the reset trajectory by reducing the leg lift during reset.

    b) Increase the stride of the legs.

    c) Determine the position of the center of the robot as a function of time.

    d) Figure out how to steer the robot by changing the stride length on one side of the body.

    e) Change the gait so the robot moves sideways like a crab.

    f) Add another pair of legs. Change the gait to reset two legs or three legs at a time.

    g) Currently in the simulation the legs move but the body does not move forward. Modify the simulation so the body moves.

10. A robot hand comprises a number of fingers, each of which is a small serial-link manipulator. Create a model of a hand with 2, 3 or 5 fingers and animate the finger motion.

11. Create a simulation with two robot arms next to each other, whose end effectors are holding a basketball at diametrically opposite points in the horizontal plane. Write code to move the robots so as to rotate the ball about the vertical axis.

Manipulator Velocity

Contents

8.1   Manipulator Jacobian – 330

8.2   Application: Resolved-Rate Motion Control – 335

8.3   Jacobian Condition and Manipulability – 338

8.4   Force Relationships – 346

8.5   Numerical Inverse Kinematics – 348

8.6   Advanced Topics – 350

8.7   Wrapping Up – 351

P. Corke, W. Jachimczyk, R. Pillat, *Robotics, Vision and Control*, Springer Tracts in Advanced Robotics 147,
https://doi.org/10.1007/978-3-031-07262-8_8

8

The velocity of a robot manipulator's end effector – the rate of change of its pose – is known as spatial velocity and was introduced in ▶ Sect. 3.1. This task-space velocity has a rotational and translational component and for 3-dimensional motion is represented by a 6-vector. That velocity is a consequence of the rate of change of the joint coordinates, the joint-space velocity, or simply joint velocity.

In this chapter, we introduce the manipulator Jacobian matrix, which describes the relationship between the joint velocity and the spatial velocity of the end effector. ▶ Sect. 8.1 uses a simple planar robot to introduce the manipulator Jacobian matrix and then extends the principle to more general robots. ▶ Sect. 8.2 uses the inverse Jacobian to generate straight-line Cartesian paths without requiring inverse kinematics. ▶ Sect. 8.3 discusses the numerical properties of the Jacobian matrix which provide insight into the dexterity of the manipulator – the directions in which it can move easily and those in which it cannot. It also introduces the concepts of over- and underactuated robots. ▶ Sect. 8.4 demonstrates how the Jacobian transpose is used to transform forces and moments on the end effector to forces and moments at the joints. ▶ Sect. 8.5 revisits the problem of inverse kinematics with a more detailed discussion of the numerical solution that was introduced in the previous chapter, and its dependence on the Jacobian matrix is fully described. We finish in ▶ Sect. 8.6 with some advanced topics.

8.1   Manipulator Jacobian

In the last chapter we discussed the relationship between joint coordinates and end-effector pose – the manipulator kinematics. Now we investigate the relationship between the rate of change of these quantities – between joint velocity and the end-effector spatial velocity. This is the differential kinematics of the manipulator.

8.1.1   Jacobian in the World Coordinate Frame

We illustrate the basics with our now familiar 2-dimensional example, see ◘ Fig. 8.1, this time defined with symbolic rather than numeric parameters

```
>> syms a1 a2 real
>> e = ETS2.Rz("q1")*ETS2.Tx(a1)*ETS2.Rz("q2")*ETS2.Tx(a2);
```

and define two symbolic parameters to represent the joint angles

```
>> syms q1 q2 real
```

The forward kinematics are

```
>> TE = e.fkine([q1 q2]);
```

◘ **Fig. 8.1** Two-link robot showing the end-effector position P, represented by the coordinate vector $p = (x, y)$, and the task-space velocity vector $v = \mathrm{d}p/\mathrm{d}t$. This is the same as the robot in ◘ Fig. 7.4b

> **Excurse 8.1: Jacobian Matrix**
>
> A Jacobian is the multidimensional form of the derivative – the derivative of a vector-valued function of a vector with respect to a vector. If $y = f(x)$ and $x \in \mathbb{R}^n$ and $y \in \mathbb{R}^m$ then the Jacobian is the $m \times n$ matrix
>
> $$J = \frac{\partial f}{\partial x} = \begin{pmatrix} \frac{\partial y_1}{\partial x_1} & \cdots & \frac{\partial y_1}{\partial x_n} \\ \vdots & \ddots & \vdots \\ \frac{\partial y_m}{\partial x_1} & \cdots & \frac{\partial y_m}{\partial x_n} \end{pmatrix}$$
>
> The Jacobian is named after Carl Jacobi, and more details are given in ▶ App. E.

and the position of the end effector $p = (x, y) \in \mathbb{R}^2$ as a column vector is

```
>> p = TE(1:2,3)
p =
a2*cos(q1 + q2) + a1*cos(q1)
a2*sin(q1 + q2) + a1*sin(q1)
```

and now we compute the derivative of p with respect to the joint configuration q. Since p and q are both vectors, the derivative

$$\frac{\partial p}{\partial q} = \mathbf{J}(q) \tag{8.1}$$

will be a matrix – a Jacobian (see ▶ App. E.4) matrix

```
>> J = jacobian(p,[q1 q2])
J =
[- a2*sin(q1 + q2) - a1*sin(q1), -a2*sin(q1 + q2)]
[  a2*cos(q1 + q2) + a1*cos(q1),  a2*cos(q1 + q2)]
```

which is typically denoted by the symbol $\mathbf{J}$ and in this case is 2×2.

To determine the relationship between *joint* velocity and *end-effector* velocity we rearrange (8.1) as

$$\partial p = \mathbf{J}(q) \, \partial q$$

and dividing through by dt we obtain

$$\dot{p} = \mathbf{J}(q) \, \dot{q}$$

where the Jacobian matrix maps velocity from the joint configuration space to the end-effector velocity in task space and is itself a function of the joint configuration.

To obtain the general form, we write the forward kinematics as a function, repeating (7.2)

$${}^0\xi = \mathcal{K}(q)$$

where $q \in \mathbb{R}^N$ and $N = \dim C$ is the dimension of the configuration space – the number of non-fixed robot joints. Taking the derivative, we write the spatial velocity of the end effector

$${}^0v = {}^0\mathbf{J}(q) \, \dot{q} \in \mathbb{R}^M \tag{8.2}$$

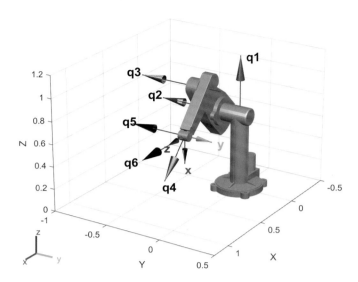

☐ **Fig. 8.2** PUMA 560 robot in its nominal pose `qn`. The end-effector z-axis points in the world x-direction, and the x-axis points downward

where $M = \dim \mathcal{T}$ is the dimension of the task space. For the common case of operation in 3D space where $\boldsymbol{\xi} \in \mathbf{S}^3 \times \mathbb{R}^3$ then, as discussed in ▶ Sect. 3.1.2, $^0\boldsymbol{v} = (\omega_x, \omega_y, \omega_z, v_x, v_y, v_z) \in \mathbb{R}^6$ and comprises rotational and translational velocity components. The matrix $^0\mathbf{J}(\boldsymbol{q}) \in \mathbb{R}^{6 \times N}$ is the manipulator Jacobian or the geometric Jacobian. This relationship is sometimes referred to as the instantaneous forward kinematics.

The Jacobian matrix can be computed by the `geometricJacobian` method of any robot object. For the PUMA robot in the configuration shown in ☐ Fig. 8.2 the Jacobian is

```
>> [puma,conf] = loadrvcrobot("puma");
>> J = puma.geometricJacobian(conf.qn,"link6")
J =
         0    0.0000    0.0000    0.7071    0.0000    1.0000
   -0.0000   -1.0000   -1.0000   -0.0000   -1.0000   -0.0000
    1.0000    0.0000    0.0000   -0.7071    0.0000   -0.0000
    0.1501    0.0144    0.3197    0.0000         0         0
    0.5963    0.0000    0.0000   -0.0000         0         0
         0    0.5963    0.2910    0.0000         0         0
```

and is a 6×6 matrix. Each row corresponds to a task-space degree of freedom in the order ω_x, ω_y, ω_z, v_x, v_y, and v_z. For example, the first row describes ω_x, the rotational motion about the world x-axis.

❗ Some references and software tools (including the text and code for other editions of this book) place the translational components first, that is, the rows correspond to v_x, v_y, v_z, ω_x, ω_y, and ω_z.

Each column corresponds to a joint-space degree of freedom – it is the end-effector spatial velocity created by unit velocity of the corresponding joint. The total end-effector velocity is the linear sum of the columns weighted by the joint velocities. In this configuration, motion of joint 1, the first column, causes translation in the world y- and x-directions and rotation about the z-axis. Motion of joint 3 cause translation in the world x- and z-directions and negative rotation about the y-axis.

The 3×3 block in the bottom right of the Jacobian relates wrist velocity $(\dot{q}_4, \dot{q}_5, \dot{q}_6)$ to end-effector translational velocity. These values are zero due to the spherical wrist mechanism.

◐ Fig. 8.2 shows the joint axes in space, and you can imagine how the end effector moves in space as each joint is slightly rotated. You could also use the `interactiveRigidBodyTree` object

```
>> irbt = interactiveRigidBodyTree(puma,Configuration=conf.qn);
```

to adjust individual joint angles and observe the corresponding change in end-effector pose.

8.1.2 Jacobian in the End Effector Coordinate Frame

The Jacobian computed by the method `geometricJacobian` maps joint velocity to the end-effector spatial velocity expressed in the world coordinate frame. We can use the velocity transformation (3.5) from the world frame to the end-effector frame, which is a function of the end-effector pose, to obtain the spatial velocity in the end-effector coordinate frame

$$
{}^{E}v = {}^{E}\mathbf{J}_0({}^{E}\boldsymbol{\xi}_0)\,{}^{0}\mathbf{J}(q)\,\dot{q} = \begin{pmatrix} {}^{E}\mathbf{R}_0 & \mathbf{0}_{3\times3} \\ \mathbf{0}_{3\times3} & {}^{E}\mathbf{R}_0 \end{pmatrix} {}^{0}\mathbf{J}(q)\,\dot{q} = {}^{E}\mathbf{J}(q)\,\dot{q} \tag{8.3}
$$

which results in a new Jacobian for end-effector velocity.

For the PUMA robot at the pose used above, we can compute this as follows ▶

We can find ${}^{E}\boldsymbol{\xi}_0$ by `getTransform(q,0,E)` which gives the pose of the base frame `"base"` (0) relative to the end-effector frame `"link6"` (E).

```
>> T = se3(puma.getTransform(conf.qn,"base","link6"));
>> Je = velxform(T)*J
Je =
   -1.0000   -0.0000   -0.0000    0.7071   -0.0000         0
   -0.0000   -1.0000   -1.0000   -0.0000   -1.0000   -0.0000
   -0.0000    0.0000    0.0000    0.7071    0.0000    1.0000
   -0.0000   -0.5963   -0.2910   -0.0000         0         0
    0.5963    0.0000    0.0000   -0.0000         0         0
    0.1500    0.0144    0.3197    0.0000         0         0
```

8.1.3 Analytical Jacobian

In (8.2) and (8.3) the spatial velocity is expressed in terms of angular and translational velocity vectors; although angular velocity is not a very intuitive concept.

For some applications it can be more useful to consider the rotational velocity in terms of rates of change of roll-pitch-yaw angles, Euler angles, or exponential co-ordinates. Analytical Jacobians are those where the rotational velocity is expressed in a representation other than angular velocity.

Consider the case of XYZ roll-pitch-yaw angles $\boldsymbol{\Gamma} = (\gamma, \beta, \alpha) \in (\mathbf{S}^1)^3$ for which the rotation matrix is

$$\mathbf{R} = \mathbf{R}_x(\gamma)\,\mathbf{R}_y(\beta)\,\mathbf{R}_z(\alpha)$$

$$= \begin{pmatrix} c\beta c\alpha & -c\beta s\alpha & s\beta \\ c\gamma s\alpha + c\alpha s\beta s\gamma & -s\beta s\gamma s\alpha + c\gamma c\alpha & -c\beta s\gamma \\ s\gamma s\alpha - c\gamma c\alpha s\beta & c\gamma s\beta s\alpha + c\alpha s\gamma & c\beta c\gamma \end{pmatrix} \in \mathbf{SO}(3)$$

where we use the shorthand $c\theta$ and $s\theta$ to mean $\cos\theta$ and $\sin\theta$ respectively. Using the chain rule and a bit of effort we can write the derivative $\dot{\mathbf{R}}$, and recalling (3.1)

$$\dot{\mathbf{R}} = [\boldsymbol{\omega}]_\times \mathbf{R}$$

we can solve for the elements of $\boldsymbol{\omega}$ in terms of roll-pitch-yaw angles and their rates to obtain

$$\begin{pmatrix} \omega_x \\ \omega_y \\ \omega_z \end{pmatrix} = \begin{pmatrix} s\beta\dot{\alpha} + \dot{\gamma} \\ -c\beta s\gamma\dot{\alpha} + c\gamma\dot{\beta} \\ c\beta c\gamma\dot{\alpha} + s\gamma\dot{\beta} \end{pmatrix}$$

which can be factored as

$$\boldsymbol{\omega} = \begin{pmatrix} s\beta & 0 & 1 \\ -c\beta s\gamma & c\gamma & 0 \\ c\beta c\gamma & s\gamma & 0 \end{pmatrix} \begin{pmatrix} \dot{\alpha} \\ \dot{\beta} \\ \dot{\gamma} \end{pmatrix}$$

and written concisely as

$$\boldsymbol{\omega} = \mathbf{A}(\boldsymbol{\Gamma})\dot{\boldsymbol{\Gamma}}$$

The matrix $\mathbf{A}$ is itself a Jacobian that maps XYZ roll-pitch-yaw angle velocity to angular velocity. It can be computed by

```
>> A = rpy2jac(0.3,0.2,0.1,"xyz")
A =
    0.1987          0     1.0000
   -0.2896     0.9553          0
    0.9363     0.2955          0
```

where the arguments are the yaw, pitch, and roll angles. Provided that $\mathbf{A}$ is not singular, the analytical Jacobian is

$$\mathbf{J}_a(\boldsymbol{q}) = \begin{pmatrix} \mathbf{1}_{3\times 3} & \mathbf{0}_{3\times 3} \\ \mathbf{0}_{3\times 3} & \mathbf{A}^{-1}(\boldsymbol{\Gamma}) \end{pmatrix} \mathbf{J}(\boldsymbol{q})\,.$$

$\mathbf{A}$ is singular when $\cos\beta = 0$ (pitch angle $\beta = \pm\frac{\pi}{2}$) and this condition is referred to as a representational singularity. The analytical Jacobian is computed by

```
>> Ja = [ones(3) zeros(3); zeros(3) inv(A)]*J;
```

and the associated spatial velocity vector is with respect to the world frame, but the rotational velocity is expressed as the rate of change of XYZ roll-pitch-yaw angles. A similar approach can be taken for Euler angles using the corresponding function `eul2jac`.

Another useful analytical Jacobian relates angular velocity to the rate of change of exponential coordinates $s = \hat{v}\theta \in \mathbb{R}^3$ by

$$\omega = \mathbf{A}(s)\dot{s}$$

where

$$\mathbf{A}(s) = \mathbf{1}_{3\times3} - \frac{1 - \cos\theta}{\theta}[\hat{v}]_\times + \frac{\theta - \sin\theta}{\theta}[\hat{v}]_\times^2$$

and $\hat{v}$ and θ can be determined from the end-effector rotation matrix via the matrix logarithm. This is singular when $\theta = 0$, which corresponds to a null rotation.

8.2 Application: Resolved-Rate Motion Control

We have discussed how the Jacobian matrix maps joint velocity to end-effector task-space velocity, but the inverse problem has strong practical use – what joint velocities are needed to achieve a particular end-effector task-space velocity v? Provided that the Jacobian matrix is square and nonsingular we can invert (8.2) and write

$$\dot{q} = \mathbf{J}^{-1}(q)v \tag{8.4}$$

which is the essence of resolved-rate motion control – a simple and elegant algorithm that generates constant velocity end-effector motion without requiring inverse kinematics. It uses the inverse Jacobian matrix to map or *resolve* the desired task-space velocity to joint velocity. In this example we will assume that the Jacobian is square (6×6) and nonsingular, but we will relax these constraints later.

The motion control scheme is typically implemented in discrete-time form as

$$\dot{q}^*_{\langle k\rangle} = \mathbf{J}^{-1}(q_{\langle k\rangle})v^*$$
$$q^*_{\langle k+1\rangle} \leftarrow q^*_{\langle k\rangle} + \delta_t\dot{q}^*_{\langle k\rangle} \tag{8.5}$$

where δ_t is the sample interval and $\langle\cdot\rangle$ indicates the time step. The first equation computes the required joint velocity as a function of the current joint configuration and the desired end-effector spatial velocity v^*. The second computes the desired joint coordinates at the next time step, using forward-rectangular integration.

An example of the algorithm is implemented by the Simulink® model

```
>> sl_rrmc
```

shown in ◘ Fig. 8.3. The desired task-space velocity is 0.05 m s⁻¹ in the world y-direction. The inputs to the Get Jacobian block are the current manipulator joint

sl_rrmc

▶ sn.pub/FEv7VI

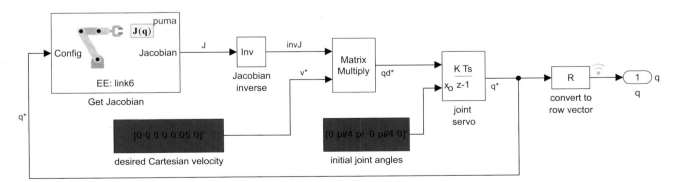

◘ **Fig. 8.3** Simulink model sl_rrmc for resolved-rate motion control for constant end-effector velocity. The Jacobian inverse block does a standard matrix inversion, since the Jacobian for the PUMA robot is square. See ▶ Sect. 8.3.4 for how to deal with non-square Jacobians

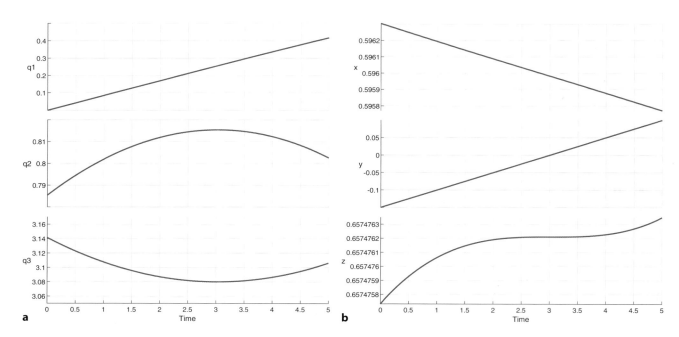

◻ Fig. 8.4   Robot motion using resolved-rate motion control of ◻ Fig. 8.3: **a** joint-space motion; **b** task-space end-effector motion. Note the small, but unwanted motion in the x- and z-directions

This models an ideal robot whose actual joint velocity precisely matches the demanded joint velocity. A real robot exhibits tracking error and this is discussed in ▶ Sect. 9.1.7.

coordinates and the output is a 6×6 Jacobian matrix. This is inverted and multiplied by the desired task-space velocity to form the required joint velocity, and the robot is modeled by an integrator. ◀

Running the simulation

```
>> r = sim("sl_rrmc");
```

will calculate a manipulator motion with its end effector moving at constant velocity in task space. Simulation results are returned in the simulation object `r` from which we extract time and joint coordinates

```
>> t = r.tout;
>> q = r.yout;
```

The motions of the first three joints

```
>> figure;
>> stackedplot(t,q(:,1:3),DisplayLabels=["q1","q2","q3"], ...
>>    GridVisible=true,XLabel="Time")
```

are shown in ◻ Fig. 8.4a and they are not linear with time – reflecting the changing kinematic configuration of the arm.

We apply forward kinematics to determine the end-effector position

```
>> for i = 1:size(q,1)
>>    Tfk(i) = se3(puma.getTransform(q(i,:),"link6"));
>> end
```

which we plot as a function of time

```
>> stackedplot(t,Tfk.trvec,DisplayLabels=["x","y","z"], ...
>>    GridVisible=true,XLabel="Time");
```

and this is shown in ◻ Fig. 8.4b. The task-space motion is $0.05 \, \mathrm{m \, s^{-1}}$ in the y-direction as specified.

The approach just described, based purely on integration, suffers from an accumulation of error which we observe as small but unwanted motions in the x- and z-directions in ◻ Fig. 8.4b. The Jacobian is the gradient of a very nonlinear function – the forward kinematics – which is a function of configuration. The root

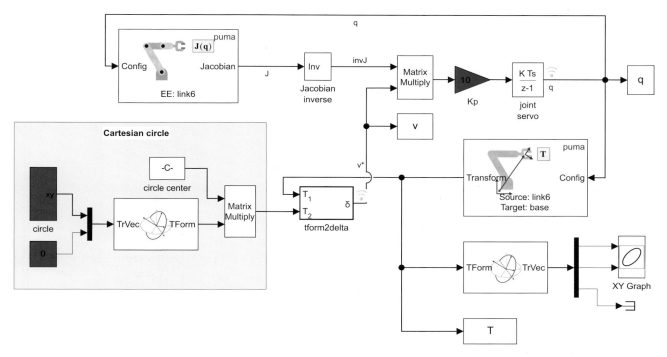

Fig. 8.5 Simulink model `sl_rrmc2` for closed-loop resolved-rate motion control with circular end-effector motion

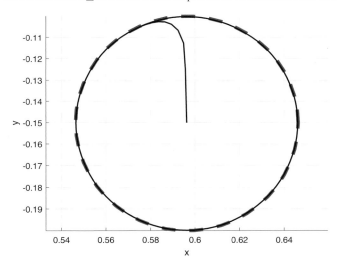

Fig. 8.6 End-effector trajectory using closed-loop resolved-rate motion control of ▪ Fig. 8.5 with circular end-effector motion. The desired circular motion is shown as a red-dashed circle

cause of the error is that, with a finite timestep, as the robot moves, the previously computed Jacobian diverges from the true gradient.

To follow a more complex path we can change the algorithm to a *closed-loop* form based on the difference between the actual pose and the time-varying desired pose

$$\dot{\boldsymbol{q}}^*\langle k \rangle \leftarrow K_p \, \mathbf{J}^{-1}(\boldsymbol{q}\langle k \rangle) \, \Delta(\mathcal{K}(\boldsymbol{q}\langle k \rangle), \boldsymbol{\xi}^*\langle k \rangle) \tag{8.6}$$

where K_p is a proportional gain, $\Delta(\cdot) \in \mathbb{R}^6$ is the spatial displacement from ▶ Sect. 3.1.5 and the desired pose $\boldsymbol{\xi}^*\langle k \rangle$ is a function of time.

A block-diagram model to demonstrate this for a circular path is

```
>> sl_rrmc2
```

sl_rrmc2

▶ sn.pub/j5i5GO

shown in ◘ Fig. 8.5, where the tool of the robot traces out a circle of radius 50 mm in the horizontal plane. The desired pose is a frame whose origin follows a circular path as a function of time. The *difference*, using the $\Delta(\cdot)$ operator, between the desired pose and the current pose from forward kinematics is computed by the `tform2delta` block. The result is a spatial displacement, a rotation and a translation described by a 6-vector which is scaled by a proportional gain, to become the demanded spatial velocity. This is transformed by the inverse manipulator Jacobian matrix to become the demanded joint velocity that will drive the end effector toward the time-varying pose. To run this

```
>> sim("sl_rrmc2");
```

and the results are shown in ◘ Fig. 8.6. This closed-loop system will also prevent the unbounded growth of integration error as shown in ◘ Fig. 8.4b.

8.3 Jacobian Condition and Manipulability

We have assumed, so far, that the Jacobian is square and non-singular, but in practice this is not always the case. The Jacobian is a $\dim \mathcal{T} \times \dim C$ matrix so for a robot operating in the task space $\mathcal{T} \in \mathbf{S}^3 \times \mathbb{R}^3$ where $\dim \mathcal{T} = 6$, a square Jacobian requires a robot with 6 joints.

8.3.1 Jacobian Singularities

If the Jacobian is square, a robot configuration q where $\det(\mathbf{J}(q)) = 0$ is described as singular or degenerate. Singularities occur when the robot is at maximum reach, or when the axes of one or more joints become aligned. This results in the loss of degrees of freedom of motion, and this is the gimbal lock problem again.

For example, the PUMA robot at its *ready* configuration has a Jacobian matrix

```
>> J = puma.geometricJacobian(conf.qr,"link6")
J =
         0         0         0         0         0         0
         0   -1.0000   -1.0000         0   -1.0000         0
    1.0000    0.0000    0.0000    1.0000    0.0000    1.0000
    0.1500   -0.8636   -0.4318         0         0         0
    0.0203    0.0000    0.0000         0         0         0
         0    0.0203    0.0203         0         0         0
```

which is singular

```
>> det(J)
ans =
     0
```

and digging a little deeper we see that the Jacobian rank has become

```
>> rank(J)
ans =
     5
```

When $q_5 = 0$ the axes of the first and last wrist joints (joints 4 and 6) are aligned. The PUMA 560 robot has several mechanical offsets that prevent other joint axes from becoming aligned.

compared to a maximum of six for a 6×6 matrix. The rank deficiency of $6 - 5 = 1$ means that one column is equal to some linear combination of the other columns. Looking at the Jacobian, we see that columns 4 and 6 are identical. This means that motion of joint 4 or joint 6 results in the same task-space velocity, leaving motion in one task-space direction inaccessible. ◄

If the robot is close to, but not actually at, a singularity we still encounter problems. Consider a configuration close to the singularity at `qr` where q_5 has a small but nonzero value of 5°

```
>> qns = conf.qr;
>> qns(5) = deg2rad(5)
qns =
         0    1.5708   -1.5708         0    0.0873         0
```

and the Jacobian is now

```
>> J = puma.geometricJacobian(qns,"link6");
```

If we use resolved-rate motion control to generate a modest end-effector velocity of $0.1 \, \mathrm{m \, s^{-1}}$ in the z-direction

```
>> qd = inv(J)*[0 0 0 0 0 0.1]';
>> qd'  % transpose for display
ans =
   -0.0000   -4.9261    9.8522    0.0000   -4.9261   -0.0000
```

the result is very high-speed motion of the shoulder and elbow – the elbow joint would have a velocity of $9.85 \, \mathrm{rad \, s^{-1}}$ or nearly 2 revolutions per second. ▶ The reason is that although the robot is no longer at a singularity, the determinant of the Jacobian is still very small

We observed this effect in ▢ Fig. 7.28.

```
>> det(J)
ans =
   1.5509e-05
```

The closer we get to the singularity, the higher the joint rates will be – at the singularity those rates will go to infinity. Alternatively, we can say that the condition number of the Jacobian is very high

```
>> cond(J)
ans =
   235.2498
```

and the Jacobian is *poorly conditioned*.

For some motions, such as rotation in this case, the poor condition of the Jacobian is not problematic. If we wished to rotate the tool about the y-axis

```
>> qd = inv(J)*[0 0.2 0 0 0 0]';
>> qd'  % transpose for display
ans =
   -0.0000        0   -0.0000    0.0000   -0.2000   -0.0000
```

the required joint rates are modest and achievable. This particular joint configuration is therefore good for certain end-effector motions, but poor for others.

8.3.2 Velocity Ellipsoid and Manipulability

It would be useful to have some measure of how well the robot is able to achieve a particular task-space velocity. We start by considering the set of generalized joint velocities $\dot{q} \in \mathbb{R}^N$ with a unit norm $\|\dot{q}\| = 1$, that is

$$\dot{q}^\top \dot{q} = 1$$

Geometrically, if we consider $\dot{q}$ to be a point in the N-dimensional joint-velocity space, this equation describes all points which lie on the surface of a hypersphere. Substituting (8.4) we can write

$$v^\top \left(\mathbf{J}(q) \mathbf{J}^\top(q) \right)^{-1} v = 1 \tag{8.7}$$

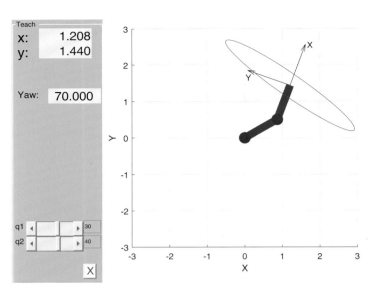

◘ Fig. 8.7   Two-link robot with overlaid velocity ellipse (red)

which is the equation of points v on the surface of a hyper-ellipsoid within the dim $\mathcal{T}$-dimensional task-velocity space. Each point represents a possible velocity of the end effector and ideally these velocities are large in all possible task-space directions – an isotropic end-effector velocity space. That implies that the ellipsoid is close to spherical, and its radii are of the same order of magnitude. However, if one or more radii are very small this indicates that the end effector cannot achieve velocity in the directions corresponding to those small radii.

For the planar robot arm of ◘ Fig. 8.1

```
>> planar2 = ETS2.Rz("q1")*ETS2.Tx(1)*ETS2.Rz("q2")*ETS2.Tx(1);
```

we can interactively explore how the shape of the velocity ellipse changes with configuration using the `teach` method

```
>> planar2.teach(deg2rad([30 40]),"vellipse");
```

and the initial view is shown in ◘ Fig. 8.7.

For a robot operating in 3-dimensional task space, (8.7) describes a 6-dimensional ellipsoid which we cannot visualize. However, we can extract that part of the Jacobian relating to *translational* velocity in the world frame. For the PUMA robot at its nominal configuration the Jacobian is

```
>> J = puma.geometricJacobian(conf.qn,"link6");
```

and the bottom three rows, which correspond to translational velocity in task space, are

```
>> Jt = J(4:6,:);
```

from which we can compute and plot the *translational* velocity ellipsoid

```
>> E = inv(Jt*Jt')
E =
   12.6638   -3.1866   -2.9220
   -3.1866    3.6142    0.7353
   -2.9220    0.7353    2.9457
>> plotellipsoid(E);
```

See ► App. C.1.4 for a refresher on ellipses and ellipsoids. Plotting an ellipsoid involves inverting the passed matrix, which we have computed using the matrix inverse. We can save ourselves two unnecessary matrix inversions by `plotellipsoid(Jt*Jt', inverted=true)`.

which is shown in ◘ Fig. 8.8a. ◄ Ideally this would be a sphere, indicating an ability to generate equal velocity in any direction – isotropic velocity capability. In this case, the lenticular shape indicates that the end effector cannot achieve equal velocity in all directions. The ellipsoid radii, in the directions of the principal axes, are given by

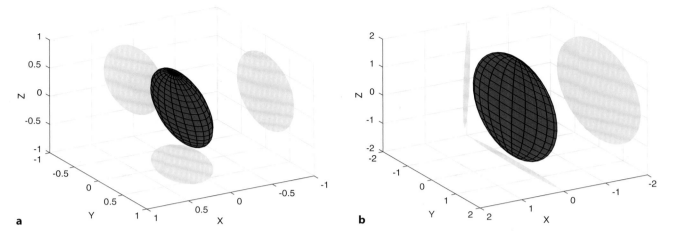

◻ **Fig. 8.8** End-effector velocity ellipsoids for the PUMA 560 robot shown in red. **a** Translational velocity ellipsoid for the nominal pose (m s^{-1}); **b** rotational velocity ellipsoid for a near singular pose (rad s^{-1}), the ellipsoid is an elliptical plate. The gray ellipses are shadows and show the shape of the ellipse as seen by looking along the x-, y-, and z-directions

```
>> [v,e] = eig(E);
>> radii = 1./sqrt(diag(e)')
radii =
    0.6902    0.6125    0.2630
```

and we see that one radius is smaller than the other two.

The *rotational* velocity ellipsoid at the near singular configuration

```
>> J = puma.geometricJacobian(qns,"link6");
>> Jr = J(1:3,:);
>> E = inv(Jr*Jr');
>> plotellipsoid(E);
```

is shown in ◻ Fig. 8.8b. It is an almost elliptical plate with a very small thickness. ▶ This is much easier to see if you The radii are change the viewpoint interactively.

```
>> [v,e] = eig(E);
>> radii = 1./sqrt(diag(e)')
radii =
    1.7321    1.7306    0.0712
```

and the small radius corresponds to the direction

```
>> v(:,3)'  % transpose for display
ans =
    0.9996    0.0000    0.0291
```

which is close to parallel with the x-axis. This indicates the robot's inability to rotate about the x-axis which is the degree of freedom that has been lost. Both joints 4 and 6 provide rotation about the world z-axis, joint 5 provides rotation about the world y-axis, but no joint generates rotation about the world x-axis. At the singularity, the ellipsoid will have zero thickness.

We can use a shorthand for displaying the velocity ellipsoid, which for the example above is

```
>> vellipse(puma,qns,mode="rot");
```

The size and shape of the ellipsoid describes how *well-conditioned* the manipulator is for making certain motions and manipulability is a succinct scalar measure of that. The `manipulability` function computes Yoshikawa's manipulability measure

$$m(q) = \sqrt{\det\left(\mathbf{J}(q)\mathbf{J}^\top(q)\right)} \in \mathbb{R} \tag{8.8}$$

The determinant is the product of the eigenvalues, and the ellipsoid radii are the square roots of the eigenvalues of $\mathbf{J}(\boldsymbol{q})\mathbf{J}^\top(\boldsymbol{q})$. ▶ App. C.1.4 is a refresher on these topics.

We need to specify for which link we are computing the manipulability. For the puma robot in this example, "link6" is the end-effector link.

Passing [] as the third argument tells the function to automatically determine the end-effector link (typically the last body in the rigid body tree). For the puma robot in this example, "link6" is the end-effector link. We can compute the manipulability for any link in the robot.

which is proportional to the volume of the ellipsoid. ◀ At the near singular configuration ◀

```
>> m = manipulability(puma,qns)
m =
    1.5509e-05
```

In these examples, manipulability is the volume of the 6D task-space velocity ellipsoid, but it is frequently useful to look at translational and rotational manipulability separately. We can compute the translational, rotational or manipulability values using the axes parameter with the options "trans", "rot", or "all". For example ◀

```
>> manipulability(puma,qns,[],axes="all")
Manipulability: translation 0.00017794, rotation 0.213487
```

returns translational and rotational manipulability. Manipulability is higher at the nominal pose

```
>> m = manipulability(puma,conf.qn)
m =
     0.0786
```

In practice we find that the seemingly large workspace of a robot is greatly reduced by joint limits, self-collision, singularities, and regions of low manipulability. The manipulability measure discussed here is based only on the kinematics parameters of the robot. The fact that it is easier to move a small wrist joint than the larger waist joint suggests that mass and inertia should be considered, and such manipulability measures are discussed in ▶ Sect. 9.2.6.

8.3.3 Dealing with Jacobian Singularity

For the case of a square Jacobian where $\det\big(\mathbf{J}(\boldsymbol{q})\big) = 0$ we cannot solve (8.2) directly for $\dot{\boldsymbol{q}}$. One strategy to deal with singularity is to replace the inverse with the damped inverse

$$\dot{\boldsymbol{q}} = \big(\mathbf{J}(\boldsymbol{q}) + \lambda\mathbf{1}\big)^{-1}\boldsymbol{v}$$

where λ is a small constant added to the diagonal which places a *floor* under the determinant. This will introduce some error in $\dot{\boldsymbol{q}}$ which, integrated over time, could lead to a significant discrepancy in tool position, but a closed-loop resolved-rate motion scheme like (8.6) would minimize that.

An alternative is to use the pseudo-, generalized-, or Moore-Penrose-pseudoinverse of the Jacobian

$$\dot{\boldsymbol{q}} = \mathbf{J}(\boldsymbol{q})^+\boldsymbol{v} \tag{8.9}$$

A matrix expression like $\boldsymbol{b} = \mathbf{A}\boldsymbol{x}$ is a system of scalar equations which we can solve for the unknowns $\boldsymbol{x}$. At singularity, some of the equations become equivalent, so there are fewer unique equations than there are unknowns – we have an under-determined system which may have an infinite number of solutions or no solution at all.

which provides a solution that minimizes $\|\mathbf{J}(\boldsymbol{q})\dot{\boldsymbol{q}} - \boldsymbol{v}\|$, ◀ which is the error between the desired task-space velocity $\boldsymbol{v}$ and the actual velocity $\mathbf{J}(\boldsymbol{q})\dot{\boldsymbol{q}}$ in a least-squares sense. It is readily computed using the function pinv.

Yet another approach is to delete from the Jacobian all those columns that are linearly dependent on other columns. This is effectively the same as locking the joints corresponding to the deleted columns and we now have an underactuated system which we discuss in the next section.

8.3.4 Dealing with a Non-Square Jacobian

The Jacobian is a $\dim\mathcal{T} \times \dim\mathcal{C}$ matrix and, so far we have assumed that the Jacobian is square. For the non-square cases it is helpful to consider the velocity

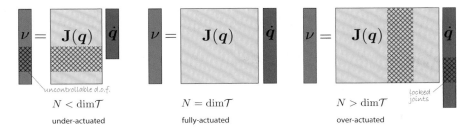

☐ Fig. 8.9 Schematic of Jacobian, $\boldsymbol{v}$ and $\dot{\boldsymbol{q}}$ for different numbers of robot joints N. The *hatched* areas represent matrix regions that could be deleted in order to create a square subsystem capable of solution

relationship

$$\boldsymbol{v} = \mathbf{J}(\boldsymbol{q})\,\dot{\boldsymbol{q}} \tag{8.10}$$

in the diagrammatic form shown in ☐ Fig. 8.9 for the common 3D case where $\dim\mathcal{T} = 6$. The Jacobian will be a $6{\times}N$ matrix, the joint velocity an N-vector, and $\boldsymbol{v}$ a 6-vector.

The case of $N < 6$ is referred to as an underactuated robot and there are too few joints to control the number of degrees of freedom in task space. Equation (8.10) is an over-determined system of equations ▶ so we can only approximate a solution, and the pseudoinverse will yield the solution which minimizes the error in a least-squares sense. Alternatively, we could *square up* the Jacobian by deleting some rows of $\boldsymbol{v}$ and $\mathbf{J}$ – accepting that some task-space degrees of freedom are not controllable given the insufficient number of joints.

The case of $N > 6$ is referred to as overactuated or redundant and there are more joints than we actually need to control the number of degrees of freedom in task space. Equation (8.10) is an under-determined system of equations ▶ so we typically choose the solution with the smallest norm. Alternatively, we can *square up* the Jacobian by deleting some columns of $\mathbf{J}$ and rows of $\dot{\boldsymbol{q}}$ – effectively locking the corresponding joints.

An over-determined set of equations has more equations than unknowns and for which there is potentially no solution.

An under-determined set of equations has more unknowns than equations and for which there is an infinite number of solutions or no solution.

8.3.4.1 Jacobian for Underactuated Robot

An underactuated robot has $N < 6$, and a Jacobian that is taller than it is wide. For example, a 2-joint manipulator at a nominal pose

```
>> qn = [1 1];
```

with $\mathcal{T} = \mathbb{R}^2 \times \mathbf{S}^1$ has the Jacobian

```
>> J = planar2.jacob0(qn)
J =
    1.0000    1.0000
   -1.7508   -0.9093
    0.1242   -0.4161
```

Each row corresponds to a 2D task-space degree of freedom in the order ω_z, v_x, and v_y. If the Jacobian has full-column rank ▶ then we can find a solution using the pseudoinverse which has the property

$$\mathbf{J}^+ \mathbf{J} = \mathbf{1}$$

just as the inverse does, and is defined as

$$\mathbf{J}^+ = \left(\mathbf{J}^\top \mathbf{J}\right)^{-1}\mathbf{J}^\top$$

and is implemented by the function `pinv`. ▶ For example, if the desired end-effector velocity is $0.1\,\text{ms}^{-1}$ in the x-direction and $0.2\,\text{ms}^{-1}$ in the y-direction then the spatial velocity is

```
>> xd_desired = [0 0.1 0.2];
```

A matrix with full-column rank is one where the columns are all linearly independent, and the rank of the matrix equals the number of columns.

This is the left generalized- or pseudoinverse, since it is applied to the left of $\mathbf{J}$.

8

and the pseudoinverse solution for joint velocity is

```
>> qd = pinv(J)*xd_desired'
qd =
    0.0831
   -0.1926
```

which results in an end-effector velocity of

```
>> J*qd
ans =
   -0.1095
    0.0297
    0.0905
```

which has *approximated* the desired velocity in the x- and y-directions and has an undesired z-axis rotational velocity. Given that the robot is underactuated, there is no exact solution and this one is best in the least-squares sense – it is a compromise. The error in this case is

```
>> norm(xd_desired - J*qd)
ans =
    0.4319
```

We have to confront the reality that we have only two inputs and we want to use them to control v_x and v_y. We rewrite (8.2) in partitioned form as

$$\begin{pmatrix} \omega_z \\ v_x \\ v_y \end{pmatrix} = \begin{pmatrix} \mathbf{J}_\omega \\ \mathbf{J}_{xy} \end{pmatrix} \begin{pmatrix} \dot{q}_1 \\ \dot{q}_2 \end{pmatrix}$$

and taking the bottom partition, the last two rows, we write

$$\begin{pmatrix} v_x \\ v_y \end{pmatrix} = \mathbf{J}_{xy} \begin{pmatrix} \dot{q}_1 \\ \dot{q}_2 \end{pmatrix}$$

where $\mathbf{J}_{xy}$ is a 2×2 matrix which we can invert if $\det(\mathbf{J}_{xy}) \neq 0$

$$\begin{pmatrix} \dot{q}_1 \\ \dot{q}_2 \end{pmatrix} = \mathbf{J}_{xy}^{-1} \begin{pmatrix} v_x \\ v_y \end{pmatrix}$$

The joint velocity is now

```
>> Jxy = J(2:3,:);
>> qd = inv(Jxy)*xd_desired(2:3)'
qd =
    0.1667
   -0.4309
```

which results in end-effector velocity

```
>> xd = J*qd
xd =
   -0.2642
    0.1000
    0.2000
```

that has exactly the desired x- and y-direction velocity. The z-axis rotation is unavoidable since we have explicitly used the two degrees of freedom to control x- and y-translation, not z-rotation. The solution error

```
>> norm(xd_desired-xd)
ans =
    0.6626
```

is $\approx 50\%$ higher than for the pseudoinverse approach, but now we have chosen the task-space constraints we wish to satisfy rather than have a compromise made for us. This same approach can be used for a robot operating in a higher-dimensional task space.

8.3.4.2 Jacobian for Overactuated Robot

An overactuated or redundant robot has $N > 6$, and a Jacobian that is wider than it is tall. In this case we can again apply the pseudoinverse but now the solution has an extra term

$$\dot{q} = \mathbf{J}(q)^{+}v + \mathbf{P}(q)\dot{q}_0 \tag{8.11}$$

The first term is, of the infinite number of solutions possible, the minimum-norm solution – the one with the smallest $\|\dot{q}\|$. The second term allows an infinite number of solutions since an arbitrary joint-space velocity $\dot{q}_0$ is added. The matrix

$$\mathbf{P}(q) = \mathbf{1}_{N \times N} - \mathbf{J}(q)^{+}\mathbf{J}^{\top}(q) \in \mathbb{R}^{N \times N} \tag{8.12}$$

projects any joint velocity $\dot{q}_0$ into the null space of $\mathbf{J}$ so that it will cause zero end-effector motion.

We will demonstrate this for the 7-axis Panda robot from ▶ Sect. 7.2.4 in the home configuration. We load the robot and remove the two gripper links (with prismatic joints) to reduce the DoF to 7.

```
>> panda = loadrobot("frankaEmikaPanda",DataFormat="row");
>> panda.removeBody("panda_leftfinger");
>> panda.removeBody("panda_rightfinger");
```

We can use inverse kinematics to calculate the robot's configuration for a target pose

```
>> ik = inverseKinematics(RigidBodyTree=panda);
>> TE = se3(eye(3),[0.5 0.2 -0.2])*se3(pi,"roty");
>> qh = panda.homeConfiguration;
>> solq = ik("panda_hand",TE.tform,ones(1,6),qh);
```

and its Jacobian

```
>> J = panda.geometricJacobian(solq,"panda_hand");
>> size(J)
ans =
     6     7
```

is a 6×7 matrix. Now consider that we want the end effector to have the spatial velocity

```
>> xd_desired = [0 0 0 0.1 0.2 0.3];
```

then using the first term of (8.12) we can compute the required joint rates

```
>> qd = pinv(J)*xd_desired';
>> qd'  % transpose for display
ans =
    0.3292   0.6347  -0.2654  -0.4960  -0.3903  -0.3264   0.3715
```

We see that all joints have nonzero velocity and contribute to the desired end-effector motion ▶ as specified

```
>> (J*qd)'  % transpose for display
ans =
    0.0000    0.0000   -0.0000    0.1000    0.2000    0.3000
```

The Jacobian in this case has seven columns and a rank of six

```
>> rank(J)
ans =
     6
```

If we use the pseudoinverse for resolved-rate motion control and the robot end effector follows a repetitive path, the joint coordinates will not generally follow a repetitive path. They are under-constrained and will drift over time and potentially hit joint limits. We can use null-space control to provide additional constraints to prevent this.

See ▶ App. B.2.2.

and therefore has a null space ◄ with just one basis vector, given as a column vector

```
>> N = null(J);
>> size(N)
ans =
     7     1
>> N'  % transpose for display
ans =
    0.4118  -0.3311  -0.4559  -0.0170   0.5608  -0.4421   0.0530
```

Any joint velocity that is a linear combination of its null-space basis vectors will result in zero end-effector motion. For this robot there is only one basis vector and we can easily show that this *null-space joint motion* causes zero end-effector motion

```
>> norm(J*N)
ans =
    2.1346e-16
```

If the basis vectors of the null space are arranged as columns in the matrix $\mathbf{N}(q) \in \mathbb{R}^{N \times R}$, where $R = N - \dim \mathcal{T}$, then an alternative expression for the projection matrix is

$$\mathbf{P}(q) = \mathbf{N}(q)\mathbf{N}(q)^{+}$$

Null-space motion can be used for redundant robots to avoid collisions between the links and obstacles (including other links), or to keep joint coordinates away from their mechanical limit stops. Consider that in addition to the desired task-space velocity, we wish to simultaneously increase q_5 in order to move the arm away from some obstacle. We set a desired joint velocity

```
>> qd_0 = [0 0 0 0 1 0 0];
```

and project it into the null space

```
>> qd = N * pinv(N)*qd_0'';
>> qd'  % transpose for display
ans =
    0.2309  -0.1857  -0.2556  -0.0096   0.3144  -0.2479   0.0297
```

A scaling has been introduced but this joint velocity, or a scaled version of this, will increase q_5 without changing the end-effector pose. Other joints move as well – they provide the required compensating motion in order that the end-effector pose is not disturbed as shown by

```
>> norm(J*qd)
ans =
    1.7326e-16
```

A highly redundant snake robot like that shown in ◘ Fig. 8.10 would have a null space with $20 - 6 = 14$ dimensions. This could be used to provide fine control over the shape of the arm independently of the end-effector pose. This is a critical capability when operating in confined spaces.

8.4 Force Relationships

In ▶ Sect. 3.2.2 we introduced wrenches $\boldsymbol{w} = (m_x, m_y, m_z, f_x, f_y, f_z) \in \mathbb{R}^6$ which are a vector of moments and forces. A wrench at the end effector is related to the joint torques or forces by the manipulator Jacobian matrix.

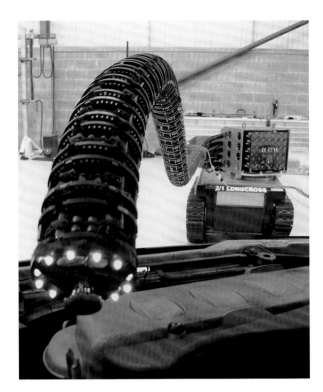

◘ Fig. 8.10 20-DoF snake-arm robot: 2.5 m reach, 90 mm diameter, and payload capacity of 25 kg (Image courtesy of OC Robotics)

8.4.1 Transforming Wrenches to Joint Space

A wrench $^0\boldsymbol{w}_E$ applied at the end effector, and defined in the world coordinate frame, is transformed to joint space by the transpose of the manipulator Jacobian

$$\boldsymbol{Q} = {^0}\mathbf{J}^\top(\boldsymbol{q})\,{^0}\boldsymbol{w}_E \tag{8.13}$$

to the generalized joint force vector $\boldsymbol{Q}$ whose elements are joint torque or force for revolute or prismatic joints respectively.

If the wrench is defined in the end-effector coordinate frame then we use instead

$$\boldsymbol{Q} = {^E}\mathbf{J}^\top(\boldsymbol{q})\,{^E}\boldsymbol{w}_E \tag{8.14}$$

For the PUMA 560 robot in its nominal pose, see ◘ Fig. 8.2, a force of 20 N in the world y-direction results in joint torques of

```
>> tau = puma.geometricJacobian(conf.qn,"link6")'*[0 0 0 0 20 0]';
>> tau'  % transpose for display
ans =
   11.9261    0.0000    0.0000   -0.0000    0    0
```

The force is pushing the end effector *sideways* and generating a torque on the waist joint of 11.93 Nm due to a lever arm effect – this will cause the robot's waist joint to rotate. A force of 20 N applied in the world x-direction results in joint torques of

```
>> tau = puma.geometricJacobian(conf.qn, "link6")'*[0 0 0 20 0 0]';
>> tau'  % transpose for display
ans =
    3.0010    0.2871    6.3937    0.0000    0    0
```

which is pulling the end effector away from the base which results in torques being applied to the first three joints.

The mapping between a wrench applied to the end effector and generalized joint force involves the transpose of the Jacobian, and this can never be singular. This contrasts with the velocity relationship which involves the Jacobian inverse that can be singular. We exploit this property of the Jacobian transpose in ▶ Sect. 8.5 to solve the inverse-kinematic problem numerically.

8.4.2 Force Ellipsoids

In ▶ Sect. 8.3.2 we introduced the velocity ellipse and ellipsoid which describe the directions in which the end effector is best able to move. We can perform a similar analysis for the forces and torques at the end effector – the end-effector wrench. We start with a set of generalized joint forces with a unit norm $\| \boldsymbol{Q} \| = 1$

$$\boldsymbol{Q}^\top \boldsymbol{Q} = 1$$

and substituting (8.13) we can write

$$\boldsymbol{w}^\top \big(\mathbf{J}(q) \mathbf{J}^\top(q) \big) \boldsymbol{w} = 1$$

which is the equation of points $\boldsymbol{w}$ on the surface of a 6-dimensional hyper-ellipsoid in the end-effector wrench space. For the planar robot arm of ◘ Fig. 8.1 we can interactively explore how the shape of the force ellipse changes with configuration using the teach method

```
>> planar2.teach(deg2rad([30 40]),"fellipse")
```

If this ellipse is close to circular, that is, its radii are of the same order of magnitude then the end effector can achieve an arbitrary wrench – an isotropic end-effector wrench space. However, if one or more radii are very small this indicates that the end effector cannot exert a force along, or a moment about, the axes corresponding to those small radii.

The force and velocity ellipsoids provide complementary information about how well suited the configuration of the arm is to a particular task. We know from personal experience that to throw an object quickly we have our arm outstretched and orthogonal to the throwing direction, whereas to lift something heavy we hold our arms close into our body.

8.5 Numerical Inverse Kinematics

In ▶ Sect. 7.2.2.2 we introduced the numerical approach to inverse kinematics for robots that move in three dimensions. This section reveals how that is implemented and shows the important role that the Jacobian plays in the solution.

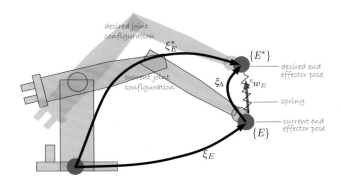

Fig. 8.11 Schematic of the numerical inverse kinematic approach, showing the current and desired manipulator configuration. A pose graph is overlaid showing the current end-effector pose $\boldsymbol{\xi}_E$, the desired end-effector pose $\boldsymbol{\xi}_E^*$, and the required change in pose $\boldsymbol{\xi}_\Delta$

Imagine a *special* spring between the end effector at the two poses which is pulling (and twisting) the robot's end effector toward the desired pose with a wrench proportional to the spatial displacement (see **Fig. 8.11**)

$$^E\boldsymbol{w} = \gamma \, ^E\boldsymbol{\Delta} \tag{8.15}$$

which is *resolved* to generalized joint forces

$$\boldsymbol{Q} = \, ^E\mathbf{J}^\top(\boldsymbol{q}) \, ^E\boldsymbol{w}$$

using the Jacobian transpose (8.14). We assume that this virtual robot has no joint motors, only viscous dampers, so the joint velocity will be proportional to those joint forces

$$\dot{\boldsymbol{q}} = \frac{\boldsymbol{Q}}{B}$$

where B is the joint damping coefficient (assuming all dampers are the same). Putting all this together, and combining constants $\alpha = \gamma/B$, we can write

$$\dot{\boldsymbol{q}} = \alpha \mathbf{J}^\top(\boldsymbol{q}) \Delta\big(\mathcal{K}(\boldsymbol{q}),\ \boldsymbol{\xi}_E^*\big)$$

which is the joint velocity that will drive the forward kinematic solution toward the desired end-effector pose. This can be solved iteratively by

$$\boldsymbol{\delta}_{q\langle k\rangle} = \alpha \mathbf{J}^\top(\boldsymbol{q}\langle k\rangle) \, \Delta\big(\mathcal{K}(\boldsymbol{q}),\ \boldsymbol{\xi}_E^*\big)$$
$$\boldsymbol{q}\langle k+1\rangle \leftarrow \boldsymbol{q}\langle k\rangle + \boldsymbol{\delta}_{q\langle k\rangle} \tag{8.16}$$

until the norm of the update $\|\boldsymbol{\delta}_{q\langle k\rangle}\|$ is sufficiently small and where $\alpha > 0$ is a well-chosen constant. Since the solution is based on the Jacobian transpose rather than the inverse, the algorithm works when the Jacobian is non-square or singular. In practice however, this algorithm is slow to converge and very sensitive to the choice of α.

More practically we can formulate this as a least-squares problem in the world coordinate frame and minimize the scalar cost

$$e = \, ^E\boldsymbol{\Delta}^\top \mathbf{M} \, ^E\boldsymbol{\Delta}$$

where $\mathbf{M} = \mathrm{diag}(\boldsymbol{m}) \in \mathbb{R}^{6\times 6}$ and $\boldsymbol{m}$ is the weight vector introduced in ▶ Sect. 7.2.3. The update becomes

$$\boldsymbol{\delta}_{q\langle k\rangle} = \big(\mathbf{J}^\top(\boldsymbol{q}\langle k\rangle) \, \mathbf{M} \, \mathbf{J}(\boldsymbol{q}\langle k\rangle)\big)^{-1} \mathbf{J}^\top(\boldsymbol{q}\langle k\rangle) \, \mathbf{M} \, \Delta\big(\mathcal{K}(\boldsymbol{q}\langle k\rangle),\ \boldsymbol{\xi}_E^*\big)$$

which is much faster to converge but can behave poorly near singularities. We remedy this by introducing a damping constant λ

$$\delta_{q\langle k \rangle} = \left(\mathbf{J}^\top(\boldsymbol{q}\langle k \rangle)\,\mathbf{M}\,\mathbf{J}(\boldsymbol{q}\langle k \rangle) + \lambda \mathbf{1}_{N \times N}\right)^{-1}\mathbf{J}^\top(\boldsymbol{q}\langle k \rangle)\,\mathbf{M}\,\Delta\big(\mathcal{K}(\boldsymbol{q}\langle k \rangle),\,\boldsymbol{\xi}_E^*\big)$$

which ensures that the term being inverted can never be singular.

An effective way to choose λ is to test whether or not an iteration reduces the error, that is if $\|\boldsymbol{\delta}_{q\langle k \rangle}\| < \|\boldsymbol{\delta}_{q\langle k-1 \rangle}\|$. If the error is reduced, we can decrease λ in order to speed convergence. If the error has increased, we revert to our previous estimate of $\boldsymbol{q}\langle k \rangle$ and increase λ. This adaptive damping factor scheme is the basis of the well-known Levenberg-Marquardt optimization algorithm.

This algorithm is implemented by the `inverseKinematics` object when selecting `SolverAlgorithm="LevenbergMarquardt"` and works well in practice. As with all optimization algorithms, it requires a reasonable initial estimate of $\boldsymbol{q}$.

An important takeaway is that to compute the inverse kinematics, all we need is an iterative solver as discussed above, the forward kinematics, and the Jacobian. These are straightforward to compute for a serial-link manipulator with an arbitrary number of joints.

8.6 Advanced Topics

8.6.1 Computing the Manipulator Jacobian Using Twists

In ▶ Sect. 7.5.3, we computed the forward kinematics as a product of exponentials based on the screws representing the joint axes in a zero joint-angle configuration. It is easy to differentiate the product of exponentials with respect to motion about each screw axis which leads to the Jacobian matrix

$$^0\bar{\mathbf{J}} = \left(\hat{\boldsymbol{S}}_1 \quad \mathrm{Ad}\big(\mathrm{e}^{[\hat{S}_1]q_1}\big)\hat{\boldsymbol{S}}_2 \quad \cdots \quad \mathrm{Ad}\big(\mathrm{e}^{[\hat{S}_1]q_1} \cdots \mathrm{e}^{[\hat{S}_{N-1}]q_{N-1}}\big)\hat{\boldsymbol{S}}_N\right)$$

for velocity in the world coordinate frame. The Jacobian is very elegantly expressed and can be easily built up column by column. Velocity in the end-effector coordinate frame is related to joint velocity by the Jacobian matrix

$$^E\bar{\mathbf{J}} = \mathrm{Ad}\big(^E\xi_0\big)\,^0\bar{\mathbf{J}}$$

where $\mathrm{Ad}(\cdot)$ is the adjoint matrix introduced in ▶ Sect. 3.1.3.

❗ The Jacobian $\bar{\mathbf{J}}$, distinguished by a bar, gives the velocity of the end effector as a velocity twist, not a spatial velocity. The difference is described in ▶ Sect. 3.1.3.

To obtain the Jacobian that gives spatial velocity as described in ▶ Sect. 8.1 we must apply a velocity transformation

$$^0\mathbf{J} = \begin{pmatrix}\mathbf{1}_{3\times 3} & -\left[^0\boldsymbol{t}_E\right]_\times \\ \mathbf{0}_{3\times 3} & \mathbf{1}_{3\times 3}\end{pmatrix}\,^0\bar{\mathbf{J}}$$

8.6.2 Manipulability, Scaling, and Units

The elements of the Jacobian matrix have values that depend on the choice of physical units used to describe link lengths and joint coordinates. Therefore, so too does the Jacobian determinant and the manipulability measure. A smaller robot will have lower manipulability than a larger robot, even if it has the same joint structure.

Yoshikawa's manipulability measure (8.8) is the volume of the velocity ellipsoid but not does measure velocity isotropy. It would give the same value for a large flat ellipsoid or a smaller more spherical ellipsoid. Many other measures have been proposed that better describe isotropy, for example, the ratio of maximum to minimum singular values.

For the redundant robot case we used the pseudoinverse to find a solution that minimizes the joint velocity norm. However, for robots with a mixture of revolute and prismatic joints the joint velocity norm is problematic since it involves quantities with different units. If the units are such that the translational and rotational velocities have very different magnitudes, then one set of velocities will dominate the other. A rotational joint velocity of $1 \, \text{rad s}^{-1}$ is comparable in magnitude to a translational joint velocity of $1 \, \text{m s}^{-1}$. However, for a robot one-tenth the size the translational joint velocity would scale to $0.1 \, \text{m s}^{-1}$ while rotational velocities would remain the same and dominate.

8.7 Wrapping Up

Jacobians are an important concept in robotics, relating velocity in one space to velocity in another. We previously encountered Jacobians for estimation in ▶ Chap. 6 and will use them later for computer vision and control.

In this chapter we have learned about the manipulator Jacobian which describes the relationship between the rate of change of joint coordinates and the spatial velocity of the end effector expressed in either the world frame or the end-effector frame. We showed how the inverse Jacobian can be used to resolve desired Cartesian velocity into joint velocity as an alternative means of generating task-space paths for under- and overactuated robots. For overactuated robots we showed how null-space motions can be used to move the robot's joints without affecting the end-effector pose. The numerical properties of the Jacobian tell us about manipulability, that is how well the manipulator is able to move in different task-space directions, which we visualized as the velocity ellipsoid. At a singularity, indicated by linear dependence between columns of the Jacobian, the robot is unable to move in certain directions. We also created Jacobians to map angular velocity to roll-pitch-yaw or Euler angle rates, and these were used to form the analytical Jacobian matrix.

The force ellipsoid describes the forces that the end effector can exert in different directions. The Jacobian transpose is used to map wrenches applied at the end effector to joint torques, and also to map wrenches between coordinate frames. It is also the basis of numerical inverse kinematics for arbitrary robots and singular poses.

8.7.1 Further Reading

The manipulator Jacobian is covered by almost all standard robotics texts such as the Robotics Handbook (Siciliano and Khatib 2016), Lynch and Park (2017), Siciliano et al. (2009), Spong et al. (2006), Craig (2005), and Paul (1981). An excellent discussion of manipulability and velocity ellipsoids is provided by Siciliano et al. (2009), and the most common manipulability measure is that proposed by Yoshikawa (1984). Patel and Sobh (2015) provide a comprehensive survey of robot performance measures including manipulability and velocity isotropy measures. Computing the manipulator Jacobian based on Denavit-Hartenberg parameters was first described by Paul and Shimano (1978).

The resolved-rate motion control scheme was proposed by Whitney (1969). Extensions such as pseudoinverse Jacobian-based control are reviewed by Klein and Huang (1983) and damped least-squares methods are reviewed by Deo and Walker (1995).

8.7.2 Exercises

1. For the simple 2-link example (▶ Sect. 8.1.1), compute the determinant symbolically and determine when it is equal to zero. What does this mean physically?
2. Add a tool to the PUMA 560 robot that is 10 cm long in the end-effector z-direction. What happens to the 3×3 block of zeros in the Jacobian?
3. For the PUMA 560 robot, can you devise a configuration in which three joint axes are parallel?
4. Derive the analytical Jacobian for Euler angles.
5. Resolved-rate motion control (▶ Sect. 8.2)
   a) Experiment with different Cartesian rotational and translational velocity demands, and combinations.
   b) Extend the Simulink model of ◘ Fig. 8.3 to also record the determinant of the Jacobian matrix.
   c) In ◘ Fig. 8.4, the robot's motion is simulated for 5 s. Extend the simulation time to 10 s and explain what happens.
   d) Set the initial pose and direction of motion to mimic that of ▶ Sect. 7.3.4. What happens when the robot reaches the singularity?
   e) Replace the Jacobian inverse block in ◘ Fig. 8.3 with the MATLAB® function `pinv`.
   f) Replace the Jacobian inverse block in ◘ Fig. 8.3 with a damped least squares function, and investigate the effect of different values of the damping factor.
   g) Replace the Jacobian inverse block in ◘ Fig. 8.3 with a block based on the MATLAB function `lscov`.
6. Use `interactiveRigidBodyTree` for the PUMA 560 robot at its singular configuration. Adjust joints 4 and 6 and see how they result in the same end-effector motion.
7. Velocity and force ellipsoids for the two link manipulator (▶ Sects. 8.3.2 and 8.4.2). Determine, perhaps using the interactive `teach` method:
   a) What configuration gives the best manipulability?
   b) What configuration is best for throwing a ball in the positive x-direction?
   c) What configuration is best for carrying a heavy weight if gravity applies a force in the $-y$-direction?
   d) Plot the velocity ellipse (x- and y-velocity) for the two-link manipulator at a grid of end-effector positions in its workspace. Each ellipsoid should be centered on the end-effector position.
8. Velocity and force ellipsoids for the PUMA manipulator
   a) For the PUMA 560 manipulator find a configuration where manipulability is greater than at `qn`.
   b) As above but find a configuration that maximizes manipulability.
   c) Repeat for a redundant robot such as the Panda.
9. Load a mobile manipulator with `loadrvcrobot("xypanda")`. It describes a 9-joint robot (PPRRRRRRR) comprising an xy-base (PP) carrying a Panda arm (RRRRRRR).
   a) Compute the null space and show that combinations of its basis vectors cause zero end-effector motion. Can you physically interpret these vectors?

b) Compute a task-space end-effector path and use numerical inverse kinematics to solve for the joint coordinates. Analyze how the motion is split between the base and the robot arm.

c) With the end effector at a constant pose, explore null-space control. Set a velocity for the mobile base and see how the arm configuration accommodates that.

d) Develop a null-space controller that keeps the last robot arm joints in the middle of their working range by using the first two joints to position the base of the arm. Modify this so as to maximize the manipulability of the XYPanda robot.

e) Consider now that the Panda robot arm is mounted on a nonholonomic robot, create a controller that generates appropriate steering and velocity inputs to the mobile robot (challenging).

f) For an arbitrary pose and end-point spatial velocity we will move six joints and lock two joints. Write an algorithm to determine which two joints should be locked.

10. Prove, or demonstrate numerically, that the two equations for null-space projection $1 - \mathbf{J}(\boldsymbol{q})^+\mathbf{J}^\top(\boldsymbol{q})$ and $\mathbf{N}(\boldsymbol{q})\mathbf{N}(\boldsymbol{q})^+$ are equivalent.

11. Load a hyper-redundant robot with `loadrvcrobot("hyper3d",20)`. It is a 20-joint robot that moves in 3-dimensional space.

a) Explore the capabilities of this robot.

b) Compute a task-space end-effector trajectory that traces a circle on the ground and use numerical inverse kinematics to solve for the joint coordinates.

c) Add a null-space control strategy that keeps all joint angles close to zero while it is moving.

d) Define an end-effector target pose on the ground that the robot must reach after passing through a hole in a wall. Can you determine the joint configuration that allows this? Can you do this for holes in two walls?

12. Write code to compute the Jacobian of a `rigidBodyTree` object using twists as described in ▶ Sect. 8.6.1 and compare results.

13. Consider a vertical plane in the workspace of a PUMA 560 robot. Divide the plane into 2-cm grid cells and for each cell determine if it is reachable by the robot, and if it is then determine the manipulability for the first three joints of the robot arm and place that value in the corresponding grid cell. Display a heat map of the robot's manipulability in the plane. Move the plane with respect to the robot and see how the heatmap changes.

14. a) Create a version of the PUMA 560 model with all the linear dimensions in inches instead of meters. For the same joint configuration as the metric version, how does the manipulability differ?

b) Create a version of the PUMA 560 model with all linear dimensions reduced by a factor of ten. For the same joint configuration, how does the manipulability differ between the two sized robots?

c) As above but make the robot ten times bigger.

Dynamics and Control

Contents

9.1　Independent Joint Control – 356

9.2　Rigid-Body Equations of Motion – 369

9.3　Forward Dynamics – 378

9.4　Rigid-Body Dynamics Compensation – 380

9.5　Task-Space Dynamics and Control – 382

9.6　Application – 387

9.7　Wrapping Up – 389

P. Corke, W. Jachimczyk, R. Pillat, *Robotics, Vision and Control*, Springer Tracts in Advanced Robotics 147,
https://doi.org/10.1007/978-3-031-07262-8_9

9

This chapter introduces the topics of dynamics and control for a serial-link manipulator arm. We start with control of the individual joints of a robot arm, and this is closely related to the problem of controlling the speed of wheels and propellers in mobile robots. ► Sect. 9.1 describes the key elements of a robot joint control system that enables a single joint to follow a desired trajectory, and introduces the challenges found in practice such as friction, gravity load, and varying inertia.

The motion of the end effector is the composition of the motion of each link, and the links are ultimately moved by forces and torques exerted on them by the joint actuators. Each link in the serial-link manipulator is supported by a reaction force and torque from the preceding link and is subject to its own weight as well as forces and torques from the links that it supports. ► Sect. 9.2 introduces the *rigid-body* equations of motion, a set of coupled nonlinear differential equations, that describe the joint torques necessary to achieve a particular manipulator state. These equations can be factored into terms describing inertia, gravity load, and gyroscopic coupling, which provide insight into how the motion of one joint exerts a disturbance force on other joints, and how those terms vary with configuration and payload. ► Sect. 9.3 introduces the forward dynamics which describe how the manipulator moves, that is, how its configuration evolves over time in response to the applied forces and torques that are due to the joint actuators, end-effector contact, and gravity.

► Sect. 9.4 introduces control systems that explicitly account for the rigid-body dynamics which leads to improved position control. Many applications require force and position control, and ► Sect. 9.5 is an introduction to task- or operational-space dynamics where the robot's dynamics are abstracted to a point mass, and the dynamics and control are expressed in task space. ► Sect. 9.6 provides an application example of a series-elastic actuator for human-safe robots.

9.1 Independent Joint Control

A robot drive train comprises an actuator or motor, and a transmission to connect it to the robot link. We start by considering a common approach to robot joint control where each joint or axis is an independent control system that attempts to accurately follow its trajectory. However, as we shall see, this is complicated by various *disturbance* torques due to gravity and friction that act on the joint, as well as coupling torques due to the motion of other joints. A very common control structure is the nested control loop. The outer loop is responsible for maintaining position and determines the velocity of the joint that will minimize position error. The inner loop is responsible for maintaining the joint velocity as demanded by the outer loop.

9.1.1 Actuators

Current control is implemented by an electronic constant current source, a variable voltage source with feedback of actual motor current. A variable voltage source is most commonly implemented by a pulse-width modulated (PWM) switching circuit. The alternative is voltage control, but this requires that the electrical dynamics of the motor due to its resistance and inductance, as well as back EMF (the voltage produced by a spinning motor that opposes the applied voltage), be taken into account when designing the control system.

Most robots today are driven by rotary electric motors. Large industrial robots typically use brushless servo motors, while small laboratory or hobby robots use brushed DC motors or stepper motors. Pneumatic actuation, using cylinders or bladders, is possible though uncommon but has the advantage of being able to generate motion while being naturally compliant. For applications that require very high forces it is common to use electro-hydraulic actuation – hydraulic actuation with electrically operated hydraulic valves. Applications include manipulators for very large payloads as used in mining, forestry or construction, or highly dynamic humanoids such as the Atlas™ robot shown in ☐ Fig. 7.3b.

The key elements of an electric drive system are shown in ☐ Fig. 9.1. Electric motors can be either current or voltage controlled. ◄ Here we assume current

Excurse 9.1: Servo Mechanism

A servo mechanism, or servo, is an automatic device that uses feedback of error between the desired and actual position of a mechanism to drive the device to the desired position. Perhaps the first such device was the ship steering engine (image courtesy WikiCommons) designed by John McFarlane Gray for Isambard Kingdom Brunel's Great Eastern in 1866. The word servo is derived from the Latin root *servus* meaning slave and the first usage was by J. J. L. Farcot in 1868 – "Le Servomoteur" – to describe the hydraulic and steam engines used for steering ships.

Error in position is measured by a sensor, then amplified to drive a motor that generates a force to move the device to reduce the error. Servo system development was spurred by WW II with the development of electrical servo systems for fire-control applications that used electric motors and electro-mechanical *amplidyne* power amplifiers. Later servo amplifiers used vacuum tubes and more recently solid state power amplifiers (electronic speed controllers). Today servo mechanisms are ubiquitous and are used to position the read/write heads in optical and magnetic disk drives, the lenses in autofocus cameras, remote control toys, satellite-tracking antennas, automatic machine tools, and robot joints.

"Servo" is properly a noun or adjective but has become a verb "to servo". In the context of vision-based control we use the compound verb "visual servoing".

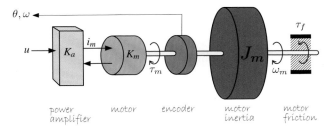

power amplifier   motor   encoder   motor inertia   motor friction

Fig. 9.1 Key components of a joint actuator. A demand voltage u controls the current i_m flowing into the motor which generates a torque τ_m that accelerates the rotational inertia J_m and is opposed by friction τ_f. The encoder measures rotational speed ω and angle θ

control where a motor driver or amplifier provides current

$$i_m = K_a u$$

that is proportional to the applied control voltage u, and where K_a is the transconductance of the amplifier with units of A V^{-1}. The torque generated by the motor is proportional to the current

$$\tau_m = K_m i_m \tag{9.1}$$

where K_m is the motor torque constant ▶ with units of N m A^{-1}. This torque accelerates the rotating part of the motor – the armature or rotor – which has rotational inertia J_m and a rotational velocity of ω_m. Motion is opposed by the frictional torque τ_f which has many components.

Which is a function of the strength of the permanent magnets and the number of turns of wire in the motor's coils.

9.1.2 Friction

Any rotating machinery, motor, or gearbox will be affected by friction – a force or torque that *opposes* motion. The net torque from the motor is

$$\tau' = \tau_m - \tau_f$$

where τ_f is the friction torque which is a function of the rotational velocity

$$\tau_f = B\omega_m + \tau_C \tag{9.2}$$

where the slope $B > 0$ is the viscous friction coefficient, and the offset is Coulomb friction which is generally modeled by

$$\tau_C = \begin{cases} \tau_C^+, & \omega_m > 0 \\ 0, & \omega_m = 0 \\ \tau_C^-, & \omega_m < 0 \end{cases} \tag{9.3}$$

In general, the friction coefficients depend on the direction of rotation and this asymmetry is more pronounced for Coulomb than for viscous friction.

The total friction torque as a function of rotational velocity is shown in ◘ Fig. 9.2. At very low speeds, highlighted in gray, Stribeck friction (pronounced streebeck) is dominant. The applied torque must exceed the stiction torque before rotation can occur – a process known as *breaking stiction*. Once the joint is moving, the Stribeck friction force rapidly decreases and viscous and Coulomb friction dominate. ◄

There are several sources of friction *experienced* by the motor. The first component is due to the motor itself: its bearings and, for a brushed motor, the brushes rubbing on the commutator. The friction parameters are often provided in the motor manufacturer's data sheet. Other sources of friction are the gearbox and the bearings that support the link. When modeling the robot dynamics in this chapter, we will ignore friction, since it introduces unwanted non-linearities.

> Stribeck friction is difficult to parameterize and in robotics is generally ignored. Even viscous and Coulomb friction data for robot joints is rare, despite the significant dynamic effect of friction.

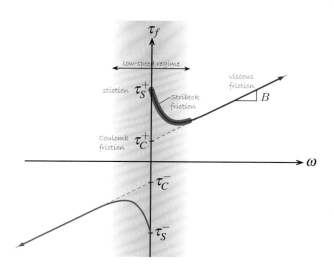

◘ **Fig. 9.2** Typical friction versus speed characteristic. The dashed lines depict a simple piecewise-linear friction model characterized by slope (viscous friction) and intercept (Coulomb friction). The low-speed regime is shaded and shown in exaggerated fashion

9.1.3 Link Mass

A motor in a robot arm does not exist in isolation, it is connected to a link as shown
schematically in ◘ Fig. 9.3. The link has two significant effects on the motor – it
adds extra inertia, and it adds a torque due to the weight of the arm. Both vary with
the configuration of the joint.

◘ Fig. 9.4 shows a simple robot with two revolute joints that operates in the
vertical plane. Joint 1 rotates the red link with respect to the fixed base frame. We
assume the mass of the red link is concentrated at its center of mass (CoM), so the
extra inertia of the link will be $m_1 r_1^2$. The motor will also experience the inertia of
the blue link, and this will depend on the value of q_2 ▶ – the inertia of the arm is
greatest when the arm is straight, and decreases as the arm is folded.

Gravity acts on the center of mass ▶ of the red link to create a torque on the
joint 1 motor which will be proportional to $\cos q_1$. Gravity acting on the center of
mass of the blue link also creates a torque on the joint 1 motor, and this is more
pronounced since it is acting at a greater distance from the motor – the *lever arm
effect* is greater.

These effects are clear from even a cursory examination of ◘ Fig. 9.4 but the
reality is even more complex. Jumping ahead to material we will cover in the next
section, we can express the torque acting on each of the joints as a function of the

How the inertia changes based on
different robot configurations is
explored in more detail in
▶ Sect. 9.2.2.

Also referred to as the center of
gravity.

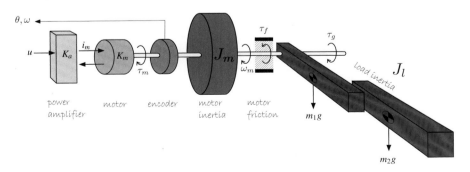

◘ **Fig. 9.3** Robot joint actuator with attached links. The center of mass of each link is indicated by ⊕

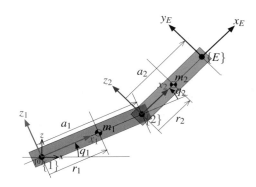

◨ **Fig. 9.4**   A two-link arm, similar to ◨ Fig. 7.8, that moves in the vertical plane, and gravity acts in the downward direction. Link frames and relevant dimensions are shown. The center of mass (CoM) of each link is indicated by ◐ and is a distance of r_i in the x-direction of the frame {i}

position, velocity, and acceleration of the joints and we can summarize as

$$\tau_1 = m_{11}(q_2)\ddot{q}_1 + \underbrace{m_{12}(q_2)\ddot{q}_2 + c_1(q_2)\dot{q}_1\dot{q}_2 + c_2(q_2)\dot{q}_2^2 + g(q_1, q_2)}_{\text{disturbance}}$$

$$m_{11}(q_2) = m_1 r_1^2 + m_2(a_1^2 + r_2^2 + 2a_1 r_2 \cos q_2),$$
$$m_{12}(q_2) = m_2 r_2(r_2 + a_1 \cos q_2),$$
$$c_1(q_2) = 2a_1 m_2 r_2 \sin q_2,$$
$$c_2(q_2) = a_1 m_2 r_2 \sin q_2,$$
$$g(q_1, q_2) = (m_1 r_1 + a_1 m_2)\text{g} \cos q_1 + m_2 r_2 \cos(q_1 + q_2)\text{g}$$

$$(9.4)$$

where g is the gravitational acceleration which acts downward.

This is Newton's second law where the inertia is dependent on q_2, plus a number of disturbance terms.

The gravity torque $g(\cdot)$ is dependent on q_1 and q_2. What is perhaps most surprising is that the torque applied to joint 1 depends on the velocity and the acceleration of joint 2 and this will be discussed further in ▶ Sect. 9.2.

In summary, the effect of joint motion in a series of mechanical links is nontrivial. Each joint experiences a torque related to the position, velocity, and acceleration of *all* the other joints and for a robot with many joints this becomes quite complex.

9.1.4   **Gearbox**

Electric motors are compact and efficient and can rotate at very high speed, but produce very low torque. Therefore, it is common to use a reduction gearbox to trade off speed for increased torque. For a prismatic joint, the gearbox might also convert rotary motion to linear. The disadvantage of a gearbox is increased cost, weight, friction, backlash, mechanical noise and, for harmonic gears, torque ripple. Very high-performance robots, such as those used in high-speed electronic assembly, use expensive high-torque motors with a direct drive or a very low gear ratio achieved using cables or thin metal bands rather than gears.

◨ Fig. 9.5 shows the complete drivetrain of a typical robot joint. For a $G{:}1$ reduction drive, the torque at the link is G times the torque at the motor. The quantities measured at the link, reference frame l, are related to the motor referenced quantities, reference frame m, as shown in ◨ Tab. 9.1. The inertia of the load is reduced by a factor of G^2 and the disturbance torque by a factor of G. ◄

If you turned the motor shaft by hand, you would *feel* the inertia of the load through the gearbox but it would be reduced by G^2.

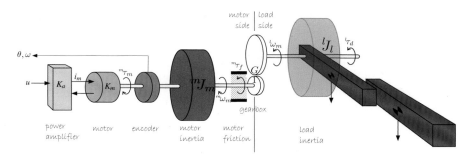

■ **Fig. 9.5** Schematic of complete robot joint including gearbox. The reference frame for all quantities is explicitly indicated by an m or l pre-superscript for motor or load side respectively. The effective inertia of the links is shown as ^lJ_l and the disturbance torque due to the link motion is $^l\tau_d$

■ **Table 9.1** Relationship between load (^{l}·) and motor (^{m}·) referenced quantities for reduction gear ratio G

$^lJ = G^2\,{}^mJ$	$^l\tau = G\,{}^m\tau$	$^l\omega = {}^m\omega/G$
$^lB = G^2\,{}^mB$	$^l\tau_C = G\,{}^m\tau_C$	$^l\dot{\omega} = {}^m\dot{\omega}/G$

There are two components of inertia *felt* by the motor. The first is due to the rotating part of the motor itself, its rotor, which is denoted by ^mJ_m. It is a constant intrinsic characteristic of the motor, and the value is provided in the motor manufacturer's data sheet. The second component is the variable load inertia ^lJ_l which is the inertia of the driven link and all the other links that are attached to it. For joint j, this is element m_{jj} of the configuration dependent inertia matrix which will be introduced in ▶ Sect. 9.2.2.

9.1.5 Modeling the Robot Joint

The complete motor drive comprises the motor to generate torque, the gearbox to amplify the torque and reduce the effects of the load, and an encoder to provide feedback of position and velocity. A schematic of a typical integrated motor unit is shown in ■ Fig. 9.6.

Collecting the various equations above, we can write the torque balance on the motor shaft, referenced to the motor, as

$$K_m K_a u - B'\omega - \tau_C'(\omega) - \frac{\tau_d(\boldsymbol{q})}{G} = J'\dot{\omega} \qquad (9.5)$$

where B', τ_C' and J' are the effective total viscous friction, Coulomb friction, and inertia due to the motor, gearbox, bearings, and the load

$$B' = B_m + \frac{B_l}{G^2}, \quad \tau_C' = \tau_{C,m} + \frac{\tau_{C,l}'}{G}, \quad J' = J_m + \frac{J_l}{G^2}\,. \qquad (9.6)$$

In order to analyze the dynamics of (9.5) we must first linearize it, and this can be done simply by setting all additive constants to zero ▶

$$J'\dot{\omega} + B'\omega = K_m K_a u$$

and then applying the Laplace transformation

$$sJ'\Omega(s) + B'\Omega(s) = K_m K_a U(s)$$

Rather than ignoring Coulomb friction, which is generally quite significant, it can be replaced by additional equivalent viscous friction $B' = B_m + \frac{B_l}{G^2} + \frac{\tau_C'}{\omega}$ at the operating point.

9

encoder

commutator

magnet

ironless winding

brush

reduction gearhead

output shaft

Fig. 9.6 Schematic of an integrated motor-encoder-gearbox assembly (courtesy of maxon precision motors, inc.)

where $\Omega(s)$ and $U(s)$ are the Laplace transform of the time domain signals $\omega(t)$ and $u(t)$ respectively. This can be rearranged as a linear transfer function

$$\frac{\Omega(s)}{U(s)} = \frac{K_m K_a}{J's + B'} \tag{9.7}$$

relating motor speed to control input. It has a single pole, the mechanical pole, at $s = -B'/J'$.

We will use data for joint 2 – the shoulder – of the PUMA 560 robot since its parameters are well known and are listed in ◘ Tab. 9.2. We have no data about link friction, so we will assume $B' = B_m$. The link inertia m_{22} experienced by the joint 2 motor as a function of configuration is shown in ◘ Fig. 9.16c and we see that it varies significantly – from 1.34 to 2.88 kg m². Using the mean value of the extreme inertia values, which is 2.11 kg m², the effective inertia is

$$\begin{aligned}
J' &= J_m + \frac{1}{G^2} m_{22} \\
&= 200 \times 10^{-6} + \frac{2.11}{(107.815)^2} \\
&= 200 \times 10^{-6} + 182 \times 10^{-6} = 382 \times 10^{-6} \text{kg m}^2
\end{aligned}$$

and we see that the inertia of the link referred to the motor side of the gearbox is comparable to the inertia of the motor itself.

9.1.6   Velocity Control Loop

A common approach to controlling the position output of a motor is the nested control loop. The outer loop is responsible for maintaining position and determines the velocity of the joint that will minimize position error. The inner loop – the velocity loop – is responsible for maintaining the velocity of the joint as demanded by the outer loop. Motor speed control is important for all types of robots, not just manipulators, for example, controlling the wheel speeds for a mobile robot, the

◻ Table 9.2 Motor and drive parameters for PUMA 560 joint 2 (shoulder) with respect to the motor side of the gearbox (Corke 1996b)

Motor torque constant	K_m	0.228	$N\,m\,A^{-1}$
Motor inertia	J_m	200×10^{-6}	$kg\,m^2$
Motor & joint viscous friction	B_m	817×10^{-6}	$N\,m\,s\,rad^{-1}$
Motor & joint Coulomb friction	τ_C^+	0.126	$N\,m$
	τ_C^-	-0.0709	$N\,m$
Gear ratio	G	107.815	
Maximum torque	τmax	0.900	$N\,m$
Maximum speed	$\dot{q}$max	165	$rad\,s^{-1}$

rotor speeds of a quadrotor as discussed in ▸ Sect. 4.2, or the propeller speeds of an underwater robot. We will start with the inner velocity loop and work our way outward.

The Simulink® model is shown in ◻ Fig. 9.7. The input to the motor power amplifier is based on the error between the demanded and actual velocity. ▸ A delay of 1 ms is included to model the computational time of the velocity loop control algorithm.

To test this velocity controller, we use the test harness

```
>> sl_vloop_test
```

shown in ◻ Fig. 9.8 which provides a trapezoidal velocity demand.

We first consider the case of proportional control where $K_i = 0$ and

$$u^* = K_v(\dot{q}^* - \dot{q}) \ . \tag{9.8}$$

Simulating the velocity controller and its test harness

```
>> sim("sl_vloop_test");
```

allows us to experiment, and we find that a gain of $K_v = 0.6$ gives satisfactory performance as shown in ◻ Fig. 9.9. We have chosen a gain that results in no overshoot at the discontinuity, but less gain leads to increased velocity error while more gain leads to oscillation – control engineering is all about tradeoffs.

The motor velocity is typically computed by taking the difference in motor position at each sample time, and the position is measured by a shaft encoder. This can be problematic at very low speeds where the encoder tick rate is lower than the sample rate. For slow-speed motion, a better strategy is to measure the time between encoder ticks.

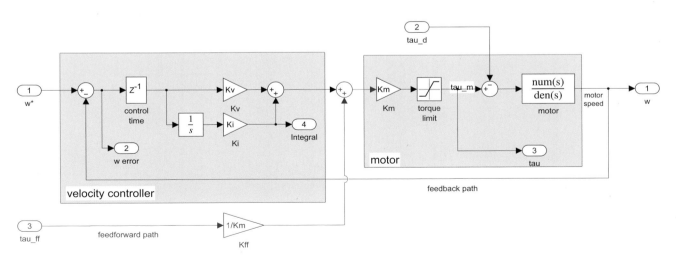

◻ Fig. 9.7 The `roblocks/Joint vloop` block subsystem modeling a robot joint velocity control loop. The `torque limit` Saturation block models the finite maximum torque that the motor can deliver

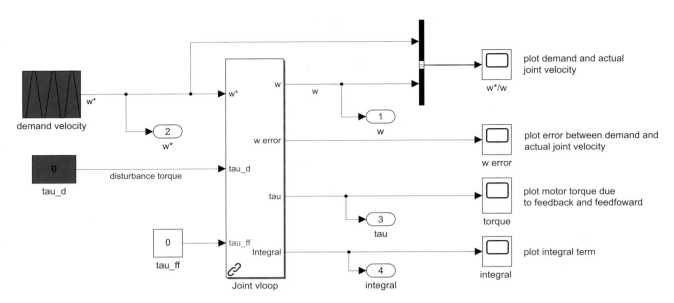

Fig. 9.8 Simulink model `sl_vloop_test` of the test harness for the velocity-control loop shown in ☐ Fig. 9.7

sl_vloop_test

▶ sn.pub/QpnzLO

We also observe a very slight steady-state error – the actual velocity is always less than the demand. From a classical control system perspective, this velocity loop is a Type 0 system since it contains no integrator. A characteristic of such systems is that they exhibit a finite error for a constant input. Intuitively, for the motor to move at constant speed, it must generate a finite torque to overcome friction, and since motor torque is proportional to velocity error, there must be a finite velocity error.

Now, we will investigate the effect of inertia variation on the closed-loop response. Using (9.6) and the data from ☐ Fig. 9.16c, we find that the minimum and maximum joint inertia at the motor are 315×10^{-6} and 448×10^{-6} kg m^2 respectively. ☐ Fig. 9.10 shows the velocity tracking error using the control gains chosen above for various values of link inertia. We can see that the tracking error decays more slowly for larger inertia and is showing signs of instability for the case of zero link inertia. For a case where the inertia variation is more extreme, the gain should be chosen to achieve satisfactory closed-loop performance at both extremes.

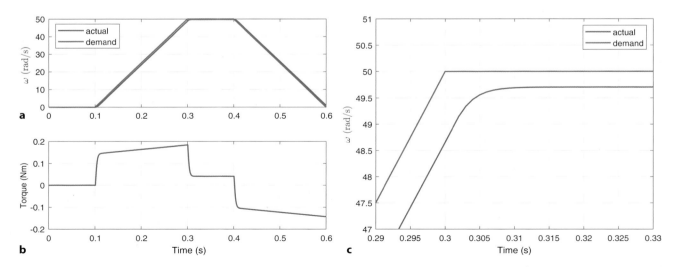

Fig. 9.9 Velocity-loop response for a trapezoidal demand. **a** Velocity tracking; **b** motor torque; **c** close-up view of velocity tracking

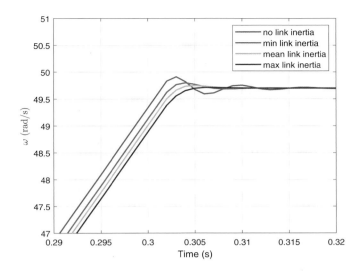

◘ Fig. 9.10 Velocity-loop response with a trapezoidal demand for varying inertia m_{22}

Excurse 9.3: Motor Limits

Electric motors are limited in both torque and speed. The maximum torque is defined by the maximum current the power amplifier can provide. A motor also has a maximum rated current beyond which the motor can be damaged by overheating or demagnetization of its permanent magnets, which irreversibly reduces its torque constant K_m. As speed increases, so does friction, and the maximum speed is $\omega_{max} = \tau_{max}/B$.

The product of motor torque and speed is the mechanical output power, and this also has an upper bound. Motors can tolerate some overloading, peak power, and peak torque, for short periods of time but the sustained rating is significantly lower than the peak.

◘ Fig. 9.15a shows that the gravity torque on this joint varies from approximately −40 to 40 N m. We now add a disturbance torque equal to just half that maximum amount, 20 N m applied on the load side of the gearbox. We do this by setting a nonzero value in the `tau_d` disturbance torque block of ◘ Fig. 9.8 and rerunning the simulation. The results shown in ◘ Fig. 9.11 indicate that the control performance has been badly degraded – the tracking error has increased to almost 2 rad s^{-1}. This has the same root cause as the very small error we saw in ◘ Fig. 9.9 – a Type 0 system exhibits a finite error for a constant input or a constant disturbance.

There are three common approaches to counter this error. The first, and simplest, is to increase the gain. This will reduce the tracking error but will also push the system toward instability and increase the overshoot.

The second approach, commonly used in industrial motor drives, is to add integral action – adding an integrator changes the system to Type 1, which has zero error for a constant input or constant disturbance. We change (9.8) to a proportional-integral (PI) controller

$$u^* = K_v(\dot{q}^* - \dot{q}) + K_i \int_0^t (\dot{q}^* - \dot{q})dt, \quad K_i > 0 .$$

In the block diagram of ◘ Fig. 9.7 this is achieved by setting `Ki` to a value greater than zero. With some experimentation, we find the gains $K_v = 1$ and $K_i =$

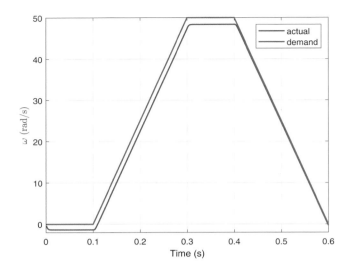

Fig. 9.11 Velocity-loop response to a trapezoidal demand with a gravity disturbance of 20 N m, mean link inertia, and proportional (P) control

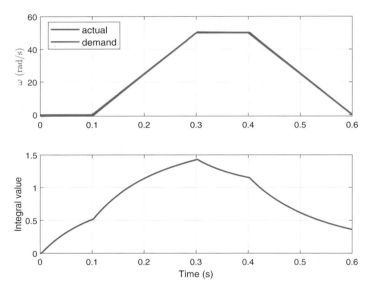

Fig. 9.12 Velocity loop response to a trapezoidal demand with a gravity disturbance of 20 N m and proportional-integral (PI) control

10 work well and the performance is shown in ▢ Fig. 9.12. The integrator state evolves over time to cancel out the disturbance term and we can see the error decaying to zero. In practice, the disturbance varies over time, and the integrator's ability to track it depends on the value of the integral gain K_i. Other disturbances affect the joint, for instance, Coulomb friction and torques due to velocity and acceleration coupling. The controller needs to be well tuned so that these have minimal effect on the tracking performance.

As always in engineering, there are tradeoffs. The integral term can lead to increased overshoot, so increasing K_i usually requires some compensating reduction of K_v. If the joint actuator is pushed to its performance limit, for instance, the torque limit is reached, then the tracking error will grow with time since the motor acceleration will be lower than required. The integral of this increasing error will grow and can lead to a condition known as integral windup. When the joint finally reaches its destination, the accumulated integral keeps driving the motor until the integral decays – leading to overshoot. Various strategies are employed to combat

> **Excurse 9.4: Back EMF**
>
> A spinning motor acts like a generator and produces a voltage V_b called the back EMF which opposes the current flowing into the motor. Back EMF is proportional to motor speed $V_b = K_m \omega$, where K_m is the motor torque constant whose units can also be interpreted as $V\,s\,rad^{-1}$. When this voltage equals the maximum possible voltage from the power amplifier, then no more current can flow into the motor and torque falls to zero – this sets an upper bound on motor speed. As speed increases, the reduced current flow reduces the torque available, and this looks like an additional source of damping.

this, such as limiting the maximum value of the integrator, or only allowing integral action when the motor is close to its setpoint. The approaches just mentioned are collectively referred to as disturbance rejection since they are concerned with reducing the effect of an unknown disturbance.

In the robotics context, the gravity disturbance is actually known. In ▶ Sect. 9.1.3, we showed that the torque due to gravity acting on the joints is a function of the joint angles, and, in a robot, these angles are always known. If we know this torque, and the motor torque constant, we can *add* a compensating control to the output of the PI controller. ▶ Predicting the disturbance and canceling it out – a strategy known as torque feedforward control – is an alternative to disturbance rejection.

Even if the gravity load is known imprecisely, this trick will reduce the magnitude of the disturbance.

We can experiment with this control approach by setting the `tau_ff` feedforward torque block of ◘ Fig. 9.8 to the same, or approximately the same, value as the disturbance.

9.1.7 Position Control Loop

The outer loop is responsible for maintaining position and we use a proportional controller ▶ based on the error between demanded $q^*(t)$ and actual $q^\#$ position to compute the desired speed of the motor

$$\dot{q}^* = K_p\big(q^*(t) - q^\#\big) \tag{9.9}$$

which is the input to the velocity loop. In many cases, particularly when following a trajectory, we also know the desired velocity $\dot{q}^*(t)$ and we can use this as a feedforward signal and the position control becomes

$$\dot{q}^* = K_p\big(q^*(t) - q^\#\big) + K_{ff}\,\dot{q}^*(t) \tag{9.10}$$

where K_{ff} would have a value ranging from 0 (no feedforward) to G (maximum feedforward), where G is the gearbox ratio.

The position loop model is shown in ◘ Fig. 9.13 and the position and velocity demand, $q^*(t)$ and $\dot{q}^*(t)$, are smooth functions of time computed by a trajectory generator ▶ that moves from 0 to 1 rad in 1 s. Joint position is obtained by integrating joint velocity, obtained from the motor velocity loop, and scaled by the inverse gearbox ratio. The error between the motor and desired position provides the velocity demand for the inner loop.

We run the test harness shown in ◘ Fig. 9.13b by

Another common approach is to use a proportional-integral-derivative (PID) controller for position but it can be shown that the D gain of this controller is related to the P gain of the inner velocity loop.

We use a trapezoidal trajectory as discussed in ▶ Sect. 3.3.1.

```
>> sl_ploop_test
>> sim("sl_ploop_test");
```

and its performance is tuned by adjusting the four gains: K_p, K_v, K_i, K_{ff} in order to achieve good tracking performance along the trajectory. For $K_p = 40$, $K_v = 0.6$,

9

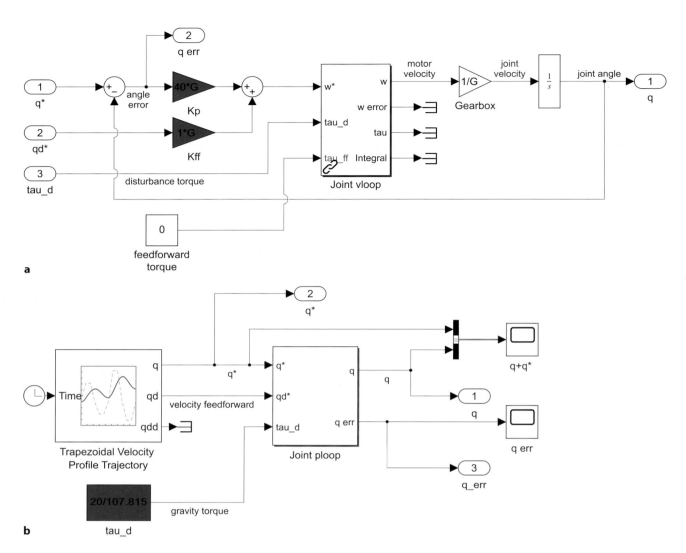

Fig. 9.13 Simulink models for position control. **a** The position loop `Joint ploop`, which is a proportional (P) controller around the inner velocity loop of ◻ Fig. 9.7; **b** test harness for following a trapezoidal trajectory `sl_ploop_test`

$K_i = 0$ and $K_{ff} = 0$, the tracking and error responses are shown in ◻ Fig. 9.14a. We see that the final error is zero but there is some tracking error along the path where the motor position lags behind the demand. The error between the demand and actual curves is due to the cumulative velocity error of the inner loop which has units of angle.

The position loop, like the velocity loop, is based on classical negative feedback. Having zero position error while tracking a ramp would mean zero demanded velocity to the inner loop which is actually contradictory. This model contains an integrator, after the velocity loop, which makes it a Type 1 system that will exhibit a constant error to a *ramp* input. There are two common approaches to reducing this tracking error. Firstly, we can add an integrator to the position loop – making it a proportional-integral (PI) controller – which adds an extra integrator and creates a Type 2 system. However, this presents yet another parameter to tune. Secondly, we can introduce feedforward control – by setting $K_{ff} = G$, where G is the gearbox ratio, we add the desired velocity to the output of the proportional controller, which is the input to the velocity loop. Conveniently, the trapezoidal trajectory function computes velocity as a function of time as well as position. The time response with velocity feedforward is shown in ◻ Fig. 9.14b and we see that the tracking error is greatly reduced.

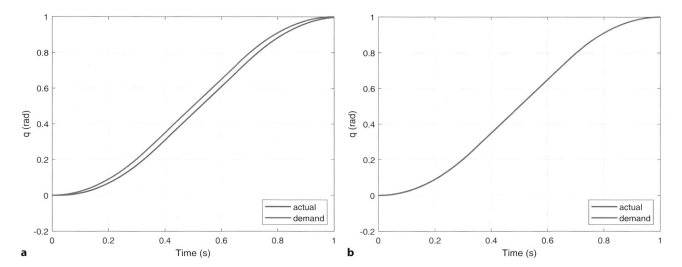

Fig. 9.14 Position loop following a trapezoidal trajectory. **a** Proportional (P) control only; **b** proportional (P) control plus velocity demand feedforward

9.1.8 Summary

A common structure for robot joint control is the nested control loop. The inner loop uses a proportional or proportional-integral control law to generate a torque so that the actual velocity closely follows the velocity demand. The outer loop uses a proportional control law to generate the velocity demand so that the actual position closely follows the position demand. Disturbance torques due to gravity and other dynamic coupling effects impact the performance of the velocity loop as do variation in the parameters of the plant being controlled, and this in turn leads to errors in position tracking. Gearing reduces the magnitude of disturbance torques by $1/G$ and the variation in inertia and friction by $1/G^2$ but at the expense of cost, weight, increased friction, and mechanical noise.

The velocity-loop performance can be improved by adding an integral control term, or by feedforward of the disturbance torque which is largely predictable. The position-loop performance can also be improved by feedforward of the desired joint velocity. In practice, control systems use both feedforward and feedback control. Feedforward is used to inject signals that we can compute, in this case the joint velocity, and in the earlier case the gravity torque. Feedback control compensates for all remaining sources of error including variation in inertia due to manipulator configuration and payload, changes in friction with time and temperature, and all the disturbance torques due to velocity and acceleration coupling. In general, the use of feedforward allows the feedback gain to be reduced since a large part of the control signal now comes from the feedforward.

9.2 Rigid-Body Equations of Motion

Consider the motor which actuates joint $j \in \{1, \ldots, N\}$ of a serial-link manipulator. From ◘ Fig. 7.18, we recall that joint j connects link $j - 1$ to link j. The motor exerts a torque that causes the outward link (link j) to rotationally accelerate but it also exerts a reaction torque on the inward link (link $j - 1$). The outward links j to N exert a weight force due to gravity, and rotating links also exert gyroscopic forces on each other. The inertia that the motor *experiences* is a function of the configuration of the outward links. When modeling the robot dynamics in this chapter, we will ignore friction, since it introduces unwanted non-linearities.

The situation at the individual link is quite complex but, for the *series* of links, the result can be written elegantly and concisely as a set of coupled differential equations in matrix form

$$Q = \mathbf{M}(q)\ddot{q} + \mathbf{C}(q,\dot{q})\dot{q} + g(q) + \mathbf{J}^\top(q)w \tag{9.11}$$

where $q, \dot{q}, \ddot{q} \in \mathbb{R}^N$ are respectively the vector of generalized joint coordinates, velocities, and accelerations, $\mathbf{M} \in \mathbb{R}^{N \times N}$ is the joint-space inertia matrix, $\mathbf{C} \in \mathbb{R}^{N \times N}$ is the Coriolis and centripetal coupling matrix, $g \in \mathbb{R}^N$ is the gravity loading, $w \in \mathbb{R}^6$ is a wrench applied at the end effector, $\mathbf{J}(q) \in \mathbb{R}^{6 \times N}$ is the manipulator Jacobian, and $Q \in \mathbb{R}^N$ is the vector of generalized actuator forces. This equation describes the manipulator rigid-body dynamics and is known as the inverse dynamics – given the configuration, velocity, and acceleration, it computes the required joint forces or torques.

These equations can be derived using any classical dynamics method such as Newton's second law and Euler's equation of motion, as discussed in ▶ Sect. 3.2.1, or a Lagrangian energy-based approach. A very efficient way of computing (9.11) is the recursive Newton-Euler algorithm which starts at the base, and working outward using the velocity and acceleration of each joint, computes the velocity and acceleration of each link. Then, working from the tool back to the base, it computes the forces and moments acting on each link and thus the joint torques.

The recursive Newton-Euler algorithm has $O(N)$ complexity and can be written in functional form as

$$Q = \mathcal{D}^{-1}(q,\dot{q},\ddot{q}) \; . \tag{9.12}$$

In MATLAB®, this is implemented by the `inverseDynamics` method of `rigidBodyTree` objects. ◀ Consider the PUMA 560 robot

Not all robot arm models returned by `loadrobot` or `loadrvcrobot` have dynamic parameters. The PUMA 560 robot is used for the examples in this chapter since it has a full dynamic model that includes inertia.

```
>> [puma,conf] = loadrvcrobot("puma");
```

at the nominal configuration, and with zero joint velocity and acceleration. To achieve this state, the required generalized joint forces, or joint torques in this case, must be

```
>> qz = zeros(1,6);
>> Q = puma.inverseDynamics(conf.qn,qz,qz)
Q =
    0.0000   31.6399    6.0351    0.0000    0.0283         0
```

Since the robot is not moving (we specified $\dot{q} = \ddot{q} = 0$), these torques must be those required to *hold the robot up* against gravity.

Let's confirm our understanding by computing the torques required in the absence of gravity. For convenience, we create a separate robot object `pumaZeroG` that is the same robot, but with no gravity

```
>> pumaZeroG = puma.copy;
>> pumaZeroG.Gravity = [0 0 0];
```

by copying the robot and changing the object's `Gravity` property to all zeros. Now, we can easily compute the torques in the absence of gravity

`quinticpolytraj` returns the configurations as column vectors, so we need to transpose them before passing to `inverseDynamics`.

```
>> Q = pumaZeroG.inverseDynamics(conf.qn,qz,qz)
Q =
     0    0    0    0    0    0
```

With a little more effort, we can also calculate the inverse dynamics on a trajectory ◀

```
>> t = linspace(0,1,10);
>> [q,qd,qdd] = quinticpolytraj([qz;conf.qr]',[0 1],t);
>> for i = 1:10
>>    Q(i,:) = puma.inverseDynamics(q(:,i)',qd(:,i)',qdd(:,i)');
>> end
```

which has returned

```
>> size(Q)
ans =
    10    6
```

a 10×6 matrix with each row representing the generalized force required for the corresponding column of q. The joint torques corresponding to the fifth time step are

```
>> Q(5,:)
ans =
    -5.6047    37.5492    3.3898    0.0000    0.0102         0
```

Consider now a case where the robot is moving. It is *instantaneously* at the nominal configuration but joint 1 is moving at 1 rad s^{-1} and the acceleration of all joints is zero. Then, in the absence of gravity, some of the required joint torques

```
>> pumaZeroG.inverseDynamics(conf.qn,[1 0 0 0 0 0],qz)
ans =
    0.0000    0.6280    -0.3607    -0.0003    -0.0000         0
```

are nonzero. Nonzero torque needs to be exerted on joints 2, 3, and 4 to oppose the gyroscopic torques that joint 1 motion is exerting on those joints.

The elements of $\mathbf{M}$, $\mathbf{C}$, $\boldsymbol{f}$ and $\boldsymbol{g}$ are complex functions of the links' kinematic and inertial parameters. For example, the second link

```
>> puma.Bodies{2}
ans =
  rigidBody with properties:

             Name: 'link2'
            Joint: [1×1 rigidBodyJoint]
             Mass: 17.4000
     CenterOfMass: [-0.3638 0.0060 0.2275]
          Inertia: [1.0312 3.7275 2.8425 -0.0238 1.4401 0.0380]
           Parent: [1×1 rigidBody]
         Children: {[1×1 rigidBody]}
          Visuals: {'Mesh:puma_link2.stl'}
       Collisions: {}
```

has ten independent inertial parameters: the link mass $m_i \in \mathbb{R}$ (Mass property); the center of mass (CoM) $\boldsymbol{r}_i \in \mathbb{R}^3$ with respect to the link coordinate frame (CenterOfMass property); and six second moments which represent the inertia tensor about the link frame {j} (Inertia property).

The remainder of this section examines the various matrix components of (9.11) in order of decreasing significance for most robotics applications.

❗ Although friction plays an important role in real-world motors (see ▶ Sect. 9.1.2), the rigidBodyTree representation does not include any specific friction modeling. Some references and software tools (including the text and code for other editions of this book) include a friction component in the rigid-body equations of motion, which will affect dynamics calculations.

❗ The Inertia property of the rigidBody object represents the inertia tensor relative to the link frame {j}. In contrast, some references and software tools (including the text and code for other editions of this book) specify the inertia tensor relative to the center of mass (CoM) of the link, but with axes aligned with the link frame. We can convert between the two representations with a similarity transform for the rotation and tensor generalization of the parallel axis theorem to account for the translation.

9.2.1 **Gravity Term**

$$Q = \mathbf{M}(q)\ddot{q} + \mathbf{C}(q,\dot{q})\dot{q} + \underline{g(q)} + \mathbf{J}^\top(q)w$$

We start our detailed discussion with the gravity term because it is generally the dominant term in (9.11) and is present even when the robot is stationary. Some robots use counterbalance weights ◀ or even springs to reduce the gravity torque that needs to be provided by the motors – this allows the motors to be smaller and thus lower in cost.

Counterbalancing will however increase the inertia associated with a joint since it adds additional mass at the end of a lever arm and will also increase the total mass of the robot.

In the previous section, we used the `inverseDynamics` method to compute the gravity load by setting the joint velocity and acceleration to zero. A more convenient approach is to use the `gravityTorque` method

```
>> g = puma.gravityTorque(conf.qn)
g =
    0.0000   31.6399    6.0351    0.0000    0.0283         0
```

The `rigidBodyTree` object contains a default gravitational acceleration vector which is initialized to the nominal value for Earth ◀

```
>> puma.Gravity
ans =
         0         0   -9.8100
```

This is the Newtonian gravitational acceleration, as discussed in ▶ Sect. 3.4.2.1, which is vertically downward. We can change that value by adjusting the `Gravity` property.

We could change gravity to the lunar value

```
>> puma.Gravity = puma.Gravity/6;
```

resulting in reduced joint torques

```
>> g = puma.gravityTorque(conf.qn)
g =
    0.0000    5.2733    1.0059    0.0000    0.0047         0
```

Before proceeding, we bring our robot back to Earth.

```
>> puma.Gravity = [0 0 -9.81];
```

The torque exerted on a joint, due to gravity acting on the robot, depends very strongly on the robot's configuration. Intuitively, the torque on the shoulder joint (joint 2) is much greater when the arm is stretched out horizontally

```
>> g = puma.gravityTorque(conf.qs)
g =
   -0.0000   46.0069    8.7722    0.0000    0.0283         0
```

than when the arm is pointing straight up

```
>> g = puma.gravityTorque(conf.qr)
g =
    0.0000   -0.7752    0.2489         0         0         0
```

The gravity torque on the elbow (joint 3) is also very high in the first configuration since it has to support the lower arm and the wrist. We can investigate how the gravity load on joints 2 and 3 varies with joint configuration by

```
>> [Q2,Q3] = meshgrid(-pi:0.1:pi,-pi:0.1:pi);
>> for i = 1:size(Q2,2)
>>   for j = 1:size(Q3,2)
>>     g = puma.gravityTorque([0 Q2(i,j) Q3(i,j) 0 0 0]);
>>     g2(i,j) = g(2);   % Shoulder gravity load
>>     g3(i,j) = g(3);   % Elbow gravity load
>>   end
>> end
>> surfl(Q2,Q3,g2); figure; surfl(Q2,Q3,g3);
```

and the results are shown in ◘ Fig. 9.15. The gravity torque on joint 2 varies approximately between ±40 N m and for joint 3 varies between ±10 N m. This type of analysis is very important in robot design to determine the required torque capacity for the motors.

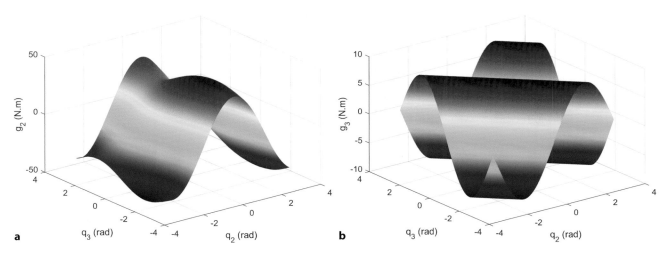

a

b

◻ **Fig. 9.15** Gravity load variation with manipulator configuration. **a** Shoulder gravity load, $g_2(q_2, q_3)$; **b** elbow gravity load $g_3(q_2, q_3)$

9.2.2 Inertia (Mass) Matrix

$$\boldsymbol{Q} = \underline{\mathbf{M}(\boldsymbol{q})\ddot{\boldsymbol{q}}} + \mathbf{C}(\boldsymbol{q}, \dot{\boldsymbol{q}})\dot{\boldsymbol{q}} + \boldsymbol{g}(\boldsymbol{q}) + \mathbf{J}^\top(\boldsymbol{q})\boldsymbol{w}$$

The joint-space inertia matrix, sometimes called the mass matrix, is symmetric and positive definite. It is a function of the manipulator configuration

```
>> M = puma.massMatrix(conf.qn)
M =
    2.8753   -0.4044    0.1006   -0.0025    0.0000   -0.0000
   -0.4044    2.0889    0.3509    0.0000    0.0024    0.0000
    0.1006    0.3509    0.3610    0.0000    0.0015    0.0000
   -0.0025    0.0000    0.0000    0.0017    0.0000    0.0000
    0.0000    0.0024    0.0015    0.0000    0.0006    0.0000
   -0.0000    0.0000    0.0000    0.0000    0.0000    0.0000
```

and the diagonal elements m_{jj} describe the inertia *experienced* by joint j, that is, $\boldsymbol{Q}_j = m_{jj}\ddot{\boldsymbol{q}}_j$. Note that the first two diagonal elements, corresponding to the robot's waist and shoulder joints, are large since motion of these joints involves rotation of the heavy upper- and lower-arm links. The off-diagonal terms $m_{ij} = m_{ji}, i \neq j$ are

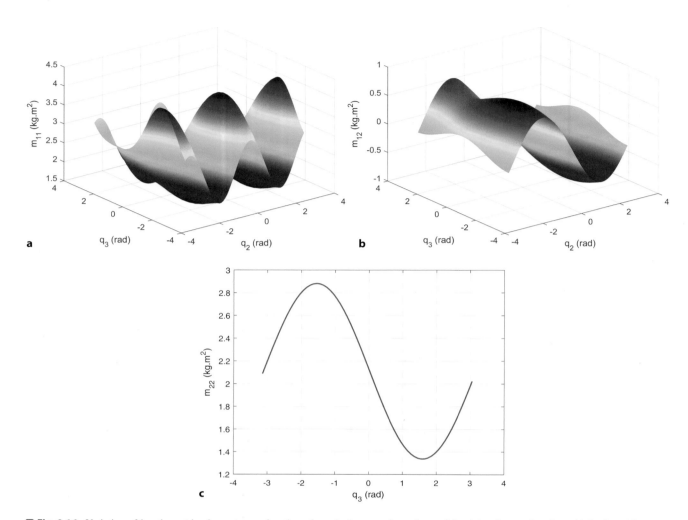

Fig. 9.16 Variation of inertia matrix elements as a function of manipulator configuration. **a** Joint 1 inertia as a function of joint 2 and 3 angles $m_{11}(q_2, q_3)$; **b** product of inertia $m_{12}(q_2, q_3)$; **c** joint 2 inertia as a function of joint 3 angle $m_{22}(q_3)$

the products of inertia which couple the acceleration of joint j to the generalized force on joint i.

We can investigate some of the elements of the inertia matrix and how they vary with robot configuration quite simply

```
>> [Q2,Q3] = meshgrid(-pi:0.1:pi,-pi:0.1:pi);
>> for i = 1:size(Q2,2)
>>   for j = 1:size(Q3,2)
>>     M = puma.massMatrix([0 Q2(i,j) Q3(i,j) 0 0 0]);
>>     m11(i,j) = M(1,1);
>>     m12(i,j) = M(1,2);
>>     m22(i,j) = M(2,2);
>>   end
>> end
>> surfl(Q2,Q3,m11); figure; surfl(Q2,Q3,m12);
```

The results are shown in ◘ Fig. 9.16, and we see significant variation in the value of m_{11} which changes by a factor of

```
>> max(m11(:))/min(m11(:))
ans =
    2.7392
```

This is important for robot design since, for a fixed maximum motor torque, inertia sets the upper bound on acceleration which in turn affects motion time and path-following accuracy.

The off-diagonal term m_{12} represents coupling between the angular acceleration of joint 2 and the torque on joint 1. That is, if joint 2 accelerates, then a torque will be exerted on joint 1 and vice versa.

9.2.3 Coriolis and Centripetal Matrix

$$Q = \mathbf{M}(q)\ddot{q} + \underline{\mathbf{C}(q,\dot{q})\dot{q}} + g(q) + \mathbf{J}(q)^{\top}w$$

The matrix $\mathbf{C}$ is a function of joint coordinates and joint velocity, and element c_{ij} couples the velocity of joint j to a generalized force acting on joint i. The coupling is due to gyroscopic effects: the centripetal torques are proportional to $\dot{q}_j^2$, while the Coriolis torques are proportional to $\dot{q}_i\dot{q}_j$.

For example, at the nominal configuration with the elbow joint moving at 1 rad s^{-1}

```
>> qd = [0 0 1 0 0 0];
```

the Coriolis matrix is

```
>> C = coriolisMatrix(puma,conf.qn,qd)
C =
     0.3607    -0.0957    -0.0957     0.0005    -0.0004     0.0000
    -0.0000     0.3858     0.3858          0    -0.0000          0
          0     0.0000    -0.0000     0.0000    -0.0009          0
    -0.0000     0.0000     0.0000     0.0000     0.0000    -0.0000
     0.0000     0.0009     0.0009          0    -0.0000    -0.0000
     0.0000    -0.0000     0.0000     0.0000    -0.0000          0
```

Element c_{23} represents significant coupling from joint 3 velocity to torque on joint 2 – rotational velocity of the elbow exerting a torque on the shoulder. The elements of this matrix represent a coupling from velocity to joint force and have the same units as viscous friction.

Although some insights can be gleaned from the Coriolis matrix, most applications need to know how the velocity of joints induces torques in all other joints. This relationship is encapsulated in the whole term, $\mathbf{C}(q,\dot{q})\dot{q}$, which is usually referred to as the velocity product. It turns out that this term is much faster to compute than the individual elements of the Coriolis matrix and is preferred for practical applications. The velocity product for the same elbow joint movement, qd, is

```
>> puma.velocityProduct(conf.qn,qd)
ans =
    -0.0957     0.3858          0     0.0000     0.0009     0.0000
```

and highlights the same strong coupling between joints 3 and 2 that we discussed earlier.

9.2.4 Effect of Payload

Any real robot has a specified maximum payload which is dictated by two dynamic effects. The first is that a mass at the end of the robot increases the inertia *experienced* by all the joint motors, and this reduces acceleration and dynamic performance. The second is that the mass generates a weight force which all the joints need to support. In the worst case, the increased gravity torque component might exceed the rating of one or more motors. However, even if the rating is not exceeded, there is less torque available for acceleration which again reduces dynamic performance.

To quantify these effects, we will take a snapshot of the robot's gravity torque and mass matrix

```
>> g = puma.gravityTorque(conf.qn);
>> M = puma.massMatrix(conf.qn);
```

and then add a 2.5 kg point mass to the PUMA 560 which is its rated maximum payload. To do this, we modify the mass and center-of-mass of the last link ◄

```
>> puma.getBody("link6").Mass = 2.5;
>> puma.getBody("link6").CenterOfMass = [0 0 0.1];
```

The center of mass of the payload cannot be at the center of the wrist coordinate frame, that is inside the wrist mechanism, so we offset it by 0.1 m in the z-direction of the wrist frame.

The inertia at the nominal configuration is now

```
>> M_loaded = puma.massMatrix(conf.qn);
```

and the ratio with respect to the unloaded case is ◄

```
>> M(abs(M)<1e-6) = nan;
>> M_loaded./M
ans =
     1.4194      0.9872      2.1490     42.3985         NaN         NaN
     0.9872      1.5516      2.8481         NaN     63.4529         NaN
     2.1490      2.8481      2.6460         NaN     49.5794         NaN
    42.3985         NaN         NaN      1.0000         NaN      1.0000
        NaN     63.4529     49.5794         NaN      1.0000         NaN
        NaN         NaN         NaN      1.0000         NaN      1.0000
```

We see that the diagonal elements have increased, for instance the joint 2 inertia has increased by 55% which reduces its maximum acceleration by nearly 50%. Reduced acceleration impairs the robot's ability to accurately follow high-speed paths. The inertia of joint 6 is unaffected since this added mass lies on the axis of this joint's rotation. Some of the off-diagonal terms have increased significantly, particularly in row and column 5, which indicates that motion of that wrist joint swinging the offset mass creates large reaction forces that are *felt* by all the other joints.

The gravity load has also increased by some significant factors

```
>> g(abs(g)<1e-6) = nan;
>> puma.gravityTorque(conf.qn)./g
ans =
        NaN      1.5222      2.5416         NaN     86.8056         NaN
```

at the elbow and wrist. We set the payload of the robot back to zero before proceeding

```
>> puma.getBody("link6").Mass = 0;
```

9.2.5 Base Wrench

A moving robot exerts a wrench on its base – its weight as well as reaction forces and torques as the arm moves around.

This wrench needs to be applied to the base to keep it in equilibrium. If we were to calculate this wrench, it would show that a vertical force of 230 N opposes the downward weight force of the robot

```
>> sum(cellfun(@(b) b.Mass, puma.Bodies))*puma.Gravity(3)
ans =
 -229.1616
```

Margin notes:

The last (wrist link) typically has negligible mass, so changing the mass to emulate payload is reasonable. As an alternative, we could also add a new `rigidBody` representing the payload as child of `link6` to the `puma`.

Some elements of **M** are very small which would result in anomalous ratios, so we set those small values to `nan` to make this explicit.

If the center of mass of the robot is not over the origin of the base coordinate frame, we would also see non-zero moments. We can calculate the center of mass of the whole robot for a given configuration qn with

```
>> puma.centerOfMass(conf.qn)
ans =
    0.1356   -0.2106    0.7622
```

which would induce moments about the *x*- and *y*-axes. If the robot is moving quickly, additional moments are induced that are dependent on $\dot{q}$.

The base wrench is important in situations where the robot does not have a rigid base such as on a satellite in space, on a boat, an underwater vehicle, or even on a ground vehicle with soft suspension.

9.2.6 Dynamic Manipulability

In ▶ Sect. 8.3.2, we discussed a kinematic measure of manipulability that describes, how well-configured the robot is to achieve velocity in any task-space direction. The force ellipsoid of ▶ Sect. 8.4.2 describes how well the manipulator is able to accelerate in different task-space directions but is based on the kinematic, not dynamic, parameters of the robot arm. Following a similar approach, we consider the set of generalized joint forces with unit norm

$$Q^\top Q = 1$$

From (9.11), ignoring gravity and assuming $\dot{q} = 0$, we can write

$$Q = \mathbf{M}(q)\,\ddot{q}$$

Differentiating (8.2) and assuming $\dot{q} = 0$, we write

$$\dot{v} = \mathbf{J}(q)\,\ddot{q}$$

where $\dot{v}$ is spatial acceleration – a vector comprising angular and translational acceleration. Combining these, we write ▶

If the matrix is not square, we use the pseudoinverse.

$$\dot{v}^\top \left(\mathbf{J}(q)\,\mathbf{M}^{-1}(q)\,\mathbf{M}^{-\top}(q)\,\mathbf{J}^\top(q)\right)^{-1}\dot{v} = 1$$

which is the equation of points $\dot{v}$ on the surface of a hyperellipsoid in the task acceleration space.

If we consider just the translational acceleration, we take the translational part of the Jacobian

```
>> J = puma.geometricJacobian(conf.qn,"link6");
>> Jt = J(4:6,:);   % last 3 rows
```

from which we can compute and plot the translational-acceleration ellipsoid

```
>> M = puma.massMatrix(conf.qn);
>> E = Jt*inv(M)*inv(M)'*Jt';
>> plotellipsoid(E,inverted=true)
```

which is shown in ◼ Fig. 9.17. The major axis of this ellipsoid is the direction in which the manipulator has maximum acceleration at this configuration. The radii of the ellipsoid are the square roots of the eigenvalues

```
>> radii = sqrt(eig(E))'  % transpose for display
radii =
    1.8078    0.2630    0.3500
```

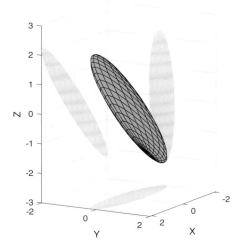

■ **Fig. 9.17** Translational acceleration ellipsoid for PUMA 560 robot in its nominal configuration. The gray ellipses are shadows and show the shape of the ellipse as seen by looking along the x-, y-, and z-directions

9

and the direction of maximum acceleration is given by the corresponding eigenvector. The ratio of the minimum to maximum radii

```
>> min(radii)/max(radii)
ans =
    0.1455
```

The 6-dimensional ellipsoid has dimensions with different units: ms^{-2} and rad s^{-2}. This makes comparison of all 6 radii problematic.

is a measure of the nonuniformity of end-effector acceleration. ◄ It would be unity for isotropic acceleration capability. In this case, acceleration capability is good in the x- and z-directions, but poor in the y-direction.

Another common scalar manipulability measure takes the inertia into account, and considers the ratio of the minimum and maximum eigenvalues of the generalized inertia matrix

$$\mathbf{M}_x = \mathbf{J}^{-\top}(\boldsymbol{q})\,\mathbf{M}(\boldsymbol{q})\,\mathbf{J}(\boldsymbol{q})^{-1}$$

which is a measure $m \in [0, 1]$ where 1 indicates uniformity of acceleration in all directions. For this example, the translational dynamic manipulability ◄

For the dynamic manipulability for translation and rotation, we can call this function with `axes="all"`.

```
>> m = manipulability(puma,conf.qn,"link6", ...
>>    method="asada",axes="trans")
m =
    0.1951
```

9.3 Forward Dynamics

To determine the motion of the manipulator in response to the forces and torques applied to its joints, we require the forward or integral dynamics. Rearranging the equations of motion (9.11), we obtain the joint acceleration

$$\ddot{\boldsymbol{q}} = \mathbf{M}^{-1}(\boldsymbol{q})\Big(\boldsymbol{Q} - \mathbf{C}(\boldsymbol{q},\dot{\boldsymbol{q}})\dot{\boldsymbol{q}} - \boldsymbol{g}(\boldsymbol{q}) - \mathbf{J}^\top(\boldsymbol{q})\boldsymbol{w}\Big) \qquad (9.13)$$

and $\mathbf{M}(\boldsymbol{q})$ is positive-definite and therefore always invertible. Let's use the UR5e robot from Universal Robots for demonstrating different dynamics use cases. Equa-

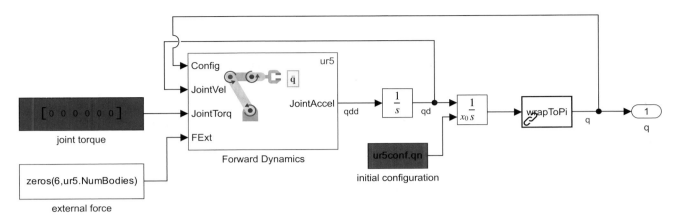

Fig. 9.18 Simulink model `sl_zerotorque` for the UR5e manipulator with zero applied joint torques. This model integrates the accelerations from the `Forward Dynamics` block to compute the joint velocities and joint angles. The `FExt` input allows us to specify a $6 \times n$ matrix of external forces, with each row representing a wrench of forces and torques that acts on one of the n bodies of the robot. For this example, we are not applying any external forces. To create the `FExt` matrix, use the `externalForce` function

tion (9.13) is computed by the `forwardDynamics` method of the `rigidBodyTree` class

sl_zerotorque

▶ sn.pub/FpvjE3

```
>> ur5 = loadrobot("universalUR5e",DataFormat="row", ...
>>    Gravity=[0 0 -9.81]);
>> ur5conf.qn = [0 -1.07 1.38 -0.3 0 -2.3];
>> qdd = ur5.forwardDynamics(ur5conf.qn);
```

given the joint coordinates, joint velocities, and applied joint torques. In this case, the joint velocities and applied joint torques are 0, so the function computes the joint accelerations induced by gravity for pose `qn`.

A simple demonstration of forward dynamics is the Simulink model

```
>> sl_zerotorque
```

which is shown in ◘ Fig. 9.18. The torques applied to the robot are zero and the initial joint angles are set as `ur5conf.qn` to the second integrator block. The simulation is run by

```
>> r = sim("sl_zerotorque");
```

which integrates (9.13) over time. The joint angles as a function of time are returned in the object `r`

```
>> t = r.find("tout");
>> q = r.find("yout");
```

We can show the robot's motion in animation

```
>> rc = rateControl(10);
>> for i = 1:size(q,1)
>>    ur5.show(q(i,:),FastUpdate=true,PreservePlot=false);
>>    rc.waitfor;
>> end
```

Since there are no torques to counter gravity and hold it upright, we see that the robot collapses and oscillates under its own weight. The robot's upper arm falls, and swings back and forth as does the lower arm, while the waist joint rotates because of Coriolis coupling. Some snapshots from this animation are shown in ◘ Fig. 9.19b, c.

To see this in more detail, we can plot the first three joint angles, returned by the simulation as a function of time

```
>> plot(t,q(:,1:3))
```

which is shown in ◘ Fig. 9.19a.

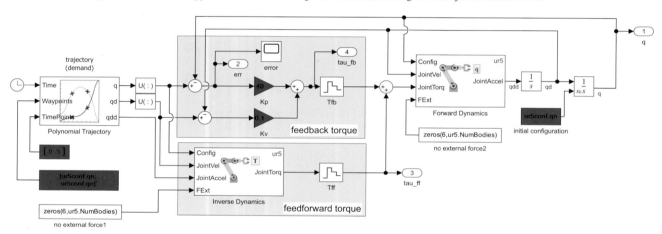

Fig. 9.19 UR5e robot collapsing under gravity when simulating ☐ Fig. 9.18. **a** Joint angle trajectory for q_1, q_2, and q_3; **b** towards the beginning of the simulation the upper arm falls; **c** both upper and lower arms swing back and forth, causing the waist joint to rotate as well

Fig. 9.20 Simulink model `sl_feedforward` for a UR5e robot with torque feedforward control. The blocks with the staircase icons are zero-order holds. We can easily adjust the start and end configuration of the robot

sl_feedforward

► sn.pub/PMTNH7

This example is simplistic and typically the joint torques would be computed by some control law as a function of the actual and desired robot joint coordinates and rates. This is the topic of the next section.

9.4 Rigid-Body Dynamics Compensation

In ► Sect. 9.1, we discussed some of the challenges for independent joint control and introduced the concept of feedforward to compensate for the gravity disturbance torque. Inertia variation and other dynamic coupling forces were not explicitly dealt with and were left for the feedback controller to handle as a disturbance. However, inertia and coupling torques can be computed according to (9.11) given knowledge of joint angles, joint velocities, joint accelerations, and the inertial parameters of the links. We can incorporate these torques into the control law using one of two *model-based* approaches: feedforward control and computed-torque control. The structural differences are contrasted in ☐ Figs. 9.20 and 9.21.

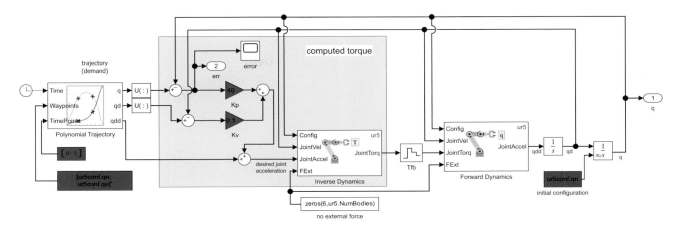

Fig. 9.21 Simulink model `sl_computed_torque` for a UR5e robot with computed-torque control

9.4.1 Feedforward Control

sl_computed_torque

▶ sn.pub/iCJDEB

The torque feedforward controller shown in ◨ Fig. 9.20 is

$$Q^* = \underbrace{\hat{\mathcal{D}}^{-1}(q^*, \dot{q}^*, \ddot{q}^*)}_{\text{feedforward}} + \underbrace{\{K_v(\dot{q}^* - \dot{q}) + K_p(q^* - q)\}}_{\text{feedback}}$$

$$= \hat{M}(q^*)\ddot{q}^* + \hat{C}(q^*, \dot{q}^*)\dot{q}^* + \hat{g}(q^*) + \{K_v(\dot{q}^* - \dot{q}) + K_p(q^* - q)\}$$

(9.14)

The feedforward term is the estimated joint force required to achieve the desired manipulator motion state of position q^*, velocity $\dot{q}^*$, and acceleration $\ddot{q}^*$. The feedback term generates joint forces related to tracking error due to errors such as uncertainty in the inertial parameters, unmodeled forces, or external disturbances. K_p and K_v are respectively the position and velocity gain (or damping) matrices and are typically diagonal. $\hat{M}$, $\hat{C}$, and $\hat{g}$ are respectively estimates of the mass matrix, Coriolis coupling matrix, and gravity. ▶

To test this controller in Simulink, we can define an end point for our trajectory and load the model

```
>> ur5conf.qe = [0.9765 -0.8013 0.9933 -0.2655 0.8590 -2.2322];
>> sl_feedforward
```

which is shown in ◨ Fig. 9.20.

In the model, the feedforward torque is computed by the `Inverse Dynamics` block and added to the feedback torque computed from position and velocity errors. The demanded joint angles, velocity, and acceleration are generated using quintic polynomials by the `Polynomial Trajectory` block, which has the initial and final joint angles, and trajectory time as inputs. Since the robot configuration changes relatively slowly, the feedforward torque can be evaluated at a greater interval, T_{ff}, than the error feedback loops, T_{fb}. In this example, we use a zero-order hold block sampling the feedforward torque at a rate of 100 Hz, whereas the feedback torque is sampled at 500 Hz.

We run the simulation by pushing the Simulink `Run` button or

```
>> r = sim("sl_feedforward");
```

and the simulation results are saved in the variable `r`.

The feedforward term linearizes the nonlinear dynamics about the operating point $(q^*, \dot{q}^*, \ddot{q}^*)$. If the linearization is ideal – that is $\hat{M} = M$, $\hat{C} = C$, and $\hat{g} = g$

This assumes that we have accurate knowledge of the robot's inertial parameters: link mass, link center of mass, and link inertia tensor.

– then the dynamics of the error $e = q^* - q$ can be obtained by combining (9.11) and (9.14)

$$\mathbf{M}(q^*)\ddot{e} + \mathbf{K}_v\dot{e} + \mathbf{K}_p e = 0 \qquad (9.15)$$

For well chosen $\mathbf{K}_p$ and $\mathbf{K}_v$, the error will decay to zero but the joint errors are coupled due to the nondiagonal matrix $\mathbf{M}$, and their dynamics are dependent on the manipulator configuration.

9.4.2   Computed-Torque Control

The computed-torque controller is shown in ◘ Fig. 9.21. It belongs to a class of controllers known as inverse dynamic control in which a nonlinear system is cascaded with its inverse so that the overall system has unity gain. The computed torque control is

$$
\begin{aligned}
Q &= \hat{\mathcal{D}}^{-1}\Big(q^*, \dot{q}^*, \ddot{q}^* + \mathbf{K}_v(\dot{q}^* - \dot{q}) + \mathbf{K}_p(q^* - q)\Big) \\
&= \hat{\mathbf{M}}(q)\{\ddot{q}^* + \mathbf{K}_v(\dot{q}^* - \dot{q}) + \mathbf{K}_p(q^* - q)\} + \hat{\mathbf{C}}(q^*, \dot{q}^*)\dot{q}^* + \hat{g}(q^*)
\end{aligned}
$$
$$(9.16)$$

where $\hat{\mathcal{D}}^{-1}(\cdot)$ is the estimated joint force required to achieve the desired manipulator motion state. In practice, the inverse is not perfect and there will be tracking errors in position and velocity. These contribute to a correctional acceleration via the position and velocity gain (or damping) matrices $\mathbf{K}_p$ and $\mathbf{K}_v$ respectively, and these are typically diagonal. $\hat{\mathbf{M}}$, $\hat{\mathbf{C}}$, and $\hat{g}$ are respectively estimates of the mass, Coriolis, and gravity. ◄

This assumes that we have accurate knowledge of the robot's inertial parameters: link mass, link center of mass, and link inertia tensor.

The desired joint angles, velocities, and accelerations are generated using quintic polynomials by the `Polynomial Trajectory` block, which has the initial and final joint angles, and trajectory time as inputs. In this case, the inverse dynamics must be evaluated at each servo interval, although the coefficient matrices $\hat{\mathbf{M}}$, $\hat{\mathbf{C}}$, and $\hat{g}$ could be evaluated at a lower rate since the robot configuration changes relatively slowly.

To test this controller in Simulink, we load the model

```
>> sl_computed_torque
```

We run the simulation

```
>> r = sim("sl_computed_torque");
```

and the simulation results are saved in the variable `r`.

If the linearization is ideal – that is $\hat{\mathbf{M}} = \mathbf{M}$, $\hat{\mathbf{C}} = \mathbf{C}$, and $\hat{g} = g$ – then the dynamics of the error $e = q^* - q$ are obtained by combining (9.11) and (9.16)

$$\ddot{e} + \mathbf{K}_v\dot{e} + \mathbf{K}_p e = 0 \qquad (9.17)$$

Unlike (9.15), the joint errors here are uncoupled and their dynamics are independent of manipulator configuration. In the case of model error, there will be some coupling between axes, and the right-hand side of (9.17) will be a nonzero forcing function.

9.5   Task-Space Dynamics and Control

Our focus so far has been on accurate control of the robot's joints which is important for industrial applications where speed and precision in positioning are critical.

Fig. 9.22 We consider the end effector as a mass depicted by the black sphere, and we are interested in how it moves in response to applied forces and torques. w is the total wrench applied by the environment and the joint actuators, and $\dot{x}$ and $\ddot{x}$ are the resulting task-space velocity and acceleration (robot image courtesy of FrankaEmika)

Today, however, we expect robots to safely perform everyday tasks like opening doors or drawers, cleaning a surface, or exchanging objects with people. These are tasks that humans perform with ease, but they involve a complex interplay of forces. When we open a door, our hand exerts forces on the door to accelerate it, but the door exerts reaction forces that guide our hand along a circular arc. Exerting forces and responding to reaction forces is critical to many real-world applications.

To consider this class of robot problems, it is useful to take an end effector centric view, as shown in ▢ Fig. 9.22. We have abstracted away the robot and consider the end effector as a point mass in 3D space. The environment can exert a wrench on the mass – that could be from a human it is interacting with, or it could be a reaction wrench from some contact task. The robot's actuators also exert a wrench on the mass. In response to the net wrench, the mass moves with some velocity and acceleration.

To understand these dynamics, we start with the relationship between joint-space and task-space velocity given in (8.2)

$$v = \mathbf{J}(q)\dot{q} \in \mathbb{R}^M \tag{9.18}$$

where $\mathbf{J}(q) \in \mathbb{R}^{M \times N}$ is the manipulator Jacobian, N is the number of joints, and $M = \dim \mathcal{T}$ is the dimension of the task space. The temporal derivative is

$$\dot{v} = \mathbf{J}(q)\ddot{q} + \dot{\mathbf{J}}(q)\dot{q} \tag{9.19}$$

where $\dot{v}$ is spatial acceleration which is a vector comprising angular and translational acceleration, and $\ddot{q}$ is the joint acceleration.

$\dot{\mathbf{J}}(q)$ is the time derivative of the Jacobian, or the rate of change of the elements of the Jacobian matrix. ▶ We can expand this as

$$\dot{\mathbf{J}}(q) = \frac{d\mathbf{J}(q)}{dt} = \frac{\partial \mathbf{J}(q)}{\partial q}\frac{dq}{dt} \in \mathbb{R}^{M \times N}$$

If $\mathbf{J}(q)$ is square we can rearrange (9.19) as

$$\ddot{q} = \mathbf{J}^{-1}(q)\left(\dot{v} - \dot{\mathbf{J}}(q)\dot{q}\right). \tag{9.20}$$

Substituting (9.20) and the inverse of (9.18) into (9.11), we can rewrite the rigid-body dynamics in task space as

$$\mathbf{M}_x(q)\ddot{x} + \mathbf{C}_x(q,\dot{q})\dot{x} + g_x(q) = w \tag{9.21}$$

This could be derived directly by symbolic differentiation of the elements of the Jacobian matrix expressed as functions of joint coordinates.

9

The rotational component of spatial velocity is angular velocity $\boldsymbol{\omega}$ as introduced in ► Sect. 3.1.1, whereas in this case it is $\dot{\boldsymbol{\varGamma}}$.

This assumes that we have accurate knowledge of the robot's inertial parameters: link mass and link center of mass. Accurate dynamic models for robots are rare and only a small subset of RVC Toolbox robot models have dynamic parameters. The most accurate dynamic model is the PUMA 560 which can be loaded with `loadrvcrobot("puma")`.

The wrench here is different to that described in ► Sects. 3.2.2 and 8.4.1. The gravity wrench computed by (9.22) is a function of an analytical Jacobian which in turn depends on the representation chosen for $\boldsymbol{x}$. Therefore, to transform the wrench to joint space, we must use the analytical Jacobian rather than the geometric Jacobian of (8.13).

where $\boldsymbol{x} \in \mathbb{R}^M$ is the task-space pose as a vector and $\boldsymbol{w} \in \mathbb{R}^M$ is the total wrench applied to the end effector. This formulation is also called the operational-space dynamics. The dynamics of the arm are abstracted into the mass depicted by the black sphere in ◻ Fig. 9.22. However, because the mass is connected to the robot arm, it *inherits* the dynamics of the robot. For example, if we let go of the sphere it would fall, but not straight down.

The terms $\mathbf{M}_x$, $\mathbf{C}_x$, $\boldsymbol{g}_x$ are respectively the task-space inertia matrix, Coriolis and centripetal coupling, and gravity load which, for a non-redundant robot, are related to the corresponding terms in (9.11) by

$$\mathbf{M}_x(\boldsymbol{q}) = \mathbf{J}_a^{-\top}(\boldsymbol{q})\,\mathbf{M}(\boldsymbol{q})\,\mathbf{J}_a^{-1}(\boldsymbol{q}) \in \mathbb{R}^{M \times M}$$

$$\mathbf{C}_x(\boldsymbol{q}, \dot{\boldsymbol{q}}) = \left(\mathbf{J}_a^{-\top}(\boldsymbol{q})\mathbf{C}(\boldsymbol{q}, \dot{\boldsymbol{q}}) - \mathbf{M}_x(\boldsymbol{q})\dot{\mathbf{J}}_a(\boldsymbol{q})\right)\mathbf{J}_a^{-1}(\boldsymbol{q}) \in \mathbb{R}^{M \times M} \quad (9.22)$$

$$\boldsymbol{g}_x(\boldsymbol{q}) = \mathbf{J}_a^{-\top}(\boldsymbol{q})\,\boldsymbol{g}(\boldsymbol{q}) \in \mathbb{R}^M$$

where $\mathbf{J}_a(\boldsymbol{q})$ is the analytical Jacobian introduced in ► Sect. 8.1.3. A common vector representation of pose in task space is $\boldsymbol{x} = (\boldsymbol{\varGamma}, \boldsymbol{p}) \in \mathbb{R}^6$ where $\boldsymbol{\varGamma} \in \mathbb{R}^3$ is its orientation represented by Euler angles, roll-pitch-yaw angles, or exponential coordinates and $\boldsymbol{p} \in \mathbb{R}^3$ is the position of the end effector. The derivatives are simply $\dot{\boldsymbol{x}} = (\dot{\boldsymbol{\varGamma}}, \dot{\boldsymbol{p}})$ and $\ddot{\boldsymbol{x}} = (\ddot{\boldsymbol{\varGamma}}, \ddot{\boldsymbol{p}})$. Using this representation means that $\dot{\boldsymbol{x}}$ is *not* spatial velocity $\boldsymbol{v}$ as used in ► Chap. 3 and 8, ◄ and therefore we must use the appropriate analytical Jacobian $\mathbf{J}_a \in \mathbb{R}^{M \times N}$.

For a redundant robot where $M \neq N$, we use instead

$$\mathbf{M}_x(\boldsymbol{q}) = \left(\mathbf{J}_a(\boldsymbol{q})\,\mathbf{M}^{-1}(\boldsymbol{q})\,\mathbf{J}_a^\top(\boldsymbol{q})\right)^{-1} \in \mathbb{R}^{M \times M}$$

$$\mathbf{C}_x(\boldsymbol{q}, \dot{\boldsymbol{q}}) = \left(\bar{\mathbf{J}}_a^\top(\boldsymbol{q})\mathbf{C}(\boldsymbol{q}, \dot{\boldsymbol{q}}) - \mathbf{M}_x(\boldsymbol{q})\dot{\mathbf{J}}_a(\boldsymbol{q})\right)\bar{\mathbf{J}}_a(\boldsymbol{q}) \in \mathbb{R}^{M \times M} \quad (9.23)$$

$$\boldsymbol{g}_x(\boldsymbol{q}) = \bar{\mathbf{J}}_a^\top(\boldsymbol{q})\,\boldsymbol{g}(\boldsymbol{q}) \in \mathbb{R}^M$$

where $\bar{\mathbf{J}}_a(\boldsymbol{q}) = \mathbf{M}^{-1}(\boldsymbol{q})\mathbf{J}_a^\top(\boldsymbol{q})\mathbf{M}_x(\boldsymbol{q}) \in \mathbb{R}^{N \times M}$ is the dynamically consistent inverse of $\mathbf{J}_a(\boldsymbol{q})$ – a generalized inverse that minimizes the manipulator's instantaneous kinetic energy.

The wrench applied to the mass in ◻ Fig. 9.22 has two components

$$\boldsymbol{w} = \boldsymbol{w}_e + \boldsymbol{w}_c \quad (9.24)$$

where $\boldsymbol{w}_e$ is the wrench applied by the environment, for example, a person pushing on the end effector or a reaction force arising from some contact task. $\boldsymbol{w}_c$ is the control, the wrench that we apply to the end effector by controlling the torques applied at the joints.

We start our discussion of task-space control by applying the simple controller

$$\boldsymbol{w}_c = \hat{\mathbf{g}}_x(\boldsymbol{q}) \quad (9.25)$$

where $\hat{\mathbf{g}}_x$ is our best estimate of the operational-space gravity wrench. ◄ Substituting this into (9.24) and (9.21), the dynamics of the end effector become

$$\mathbf{M}_x(\boldsymbol{q})\ddot{\boldsymbol{x}} + \mathbf{C}_x(\boldsymbol{q}, \dot{\boldsymbol{q}})\dot{\boldsymbol{x}} = \boldsymbol{w}_e$$

which has no gravity term – the end effector is weightless. With zero applied wrench, that is $\boldsymbol{w}_e = 0$, and if the robot is not moving, $\dot{\boldsymbol{q}} = 0$, the end effector will not accelerate. If we pushed on the end effector, that is, we applied a nonzero $\boldsymbol{w}_e$, it would accelerate, and we would feel inertia. We would also feel some effect like damping due to the $\dot{\boldsymbol{x}}$ term.

To implement this controller, the control wrench from (9.25) is mapped to the generalized joint forces using the analytical Jacobian ◄

$$\boldsymbol{Q}_c = \mathbf{J}_a^\top(\boldsymbol{q})\,\boldsymbol{w}_c \quad (9.26)$$

to become the torque or force demand for the joint actuators.

Damping, or viscous friction, is a force proportional to velocity that opposes motion. Superficially, the term $\mathbf{C}_x(\boldsymbol{q},\dot{\boldsymbol{q}})$ looks like a viscous friction coefficient, but it is a complex non-linear function of joint configuration and the products of joint-space velocities.

As we did with gravity, we can effectively remove this velocity coupling, as well as friction, using the control

$$\boldsymbol{w}_c = \hat{\mathbf{C}}_x(\boldsymbol{q},\dot{\boldsymbol{q}})\dot{\boldsymbol{x}} + \hat{\mathbf{g}}_x(\boldsymbol{q})$$

and the dynamics of the end effector becomes simply

$$\mathbf{M}_x(\boldsymbol{q})\ddot{\boldsymbol{x}} = \boldsymbol{w}_e$$

which looks simple, just Newton's second law. However, in this multi-dimensional case, the inertia matrix is not diagonal so we will still experience cross coupling, but this time acceleration cross coupling.

To follow a trajectory $\boldsymbol{x}^*(t)$ in task space we choose the control

$$\boldsymbol{w}_c = \hat{\mathbf{M}}_x(\boldsymbol{q})\Big(\ddot{\boldsymbol{x}}^* + \mathbf{K}_p(\boldsymbol{x}^* - \boldsymbol{x}) + \mathbf{K}_v(\dot{\boldsymbol{x}}^* - \dot{\boldsymbol{x}})\Big) + \hat{\mathbf{C}}_x(\boldsymbol{q},\dot{\boldsymbol{q}})\dot{\boldsymbol{x}} + \hat{\boldsymbol{g}}_x(\boldsymbol{q})$$

(9.27)

which is driven by the error between desired and actual task space pose and its derivatives. If the estimates of the task-space inertial terms are exact, then the dynamics of the task-space error $\boldsymbol{e} = \boldsymbol{x}^* - \boldsymbol{x}$ are

$$\ddot{\boldsymbol{e}} + \mathbf{K}_v\dot{\boldsymbol{e}} + \mathbf{K}_p\boldsymbol{e} = -\boldsymbol{w}_e$$

and will converge to zero for well-chosen values of $\mathbf{K}_p > 0$ and $\mathbf{K}_v > 0$ if the external wrench $\boldsymbol{w}_e = 0$. The non-linear rigid-body dynamics have been linearized, and this control approach is called *feedback linearization*. ▶ This controller maps a displacement $\boldsymbol{x}^* - \boldsymbol{x}$ to a force $\boldsymbol{w}_c$ and therefore acts as a mechanical impedance – this type of robot control is called *impedance control*.

This is analogous to the computed torque control discussed in ▶ Sect. 9.4.2.

We can go one step further and actually specify the desired inertia $\mathbf{M}^*$ of the closed-loop dynamics using the control

$$\boldsymbol{w}_c = \hat{\mathbf{M}}_x(\boldsymbol{q})\mathbf{M}_x^{*-1}\Big(\mathbf{M}_x^*\ddot{\boldsymbol{x}}^* + \mathbf{K}_p(\boldsymbol{x}^* - \boldsymbol{x}) + \mathbf{K}_v(\dot{\boldsymbol{x}}^* - \dot{\boldsymbol{x}})\Big)$$
$$+ \hat{\mathbf{C}}_x(\boldsymbol{q},\dot{\boldsymbol{q}})\dot{\boldsymbol{x}} + \hat{\boldsymbol{g}}_x(\boldsymbol{q})$$

(9.28)

and, assuming exact linearization, the closed-loop error dynamics become

$$\mathbf{M}_x^*\ddot{\boldsymbol{e}} + \mathbf{K}_v\dot{\boldsymbol{e}} + \mathbf{K}_p\boldsymbol{e} = -\underline{\mathbf{M}_x^*\mathbf{M}_x^{-1}(\boldsymbol{q})}\boldsymbol{w}_e \,.$$

This technique is sometimes referred to as *inertia shaping*. We can now specify the dynamics of the end effector in task space in terms of apparent inertia, damping, and stiffness. However, the external wrench is coupled by the underlined term which is not guaranteed to be diagonal, and this will lead to undesirable motion in response to the wrench.

To remedy this, we need to measure the external wrench and several approaches are commonly used. Joint motor current can be measured and is related to joint torque by (9.1), but in practice, friction and other effects mask the signal we are interested in. More sophisticated robots have torque sensors built into their joint actuators, for example as shown in ◻ Fig. 9.23. Joint torques are related to the end-effector wrench by the Jacobian transpose according to (8.13). An alternative

9

■ **Fig. 9.23** An integrated sensor-actuator-controller unit that can measure and control joint position and torque (Image courtesy Tamim Asfour and Karlsruhe Institute of Technology)

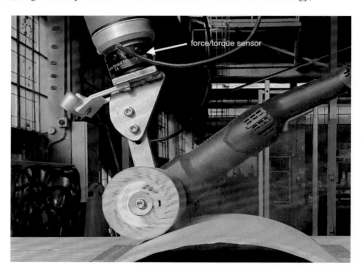

■ **Fig. 9.24** A robot performing a finishing operation with a wrist force/torque sensor (Image courtesy Amelia Luu)

approach is to use a force/torque sensing wrist such as shown in ■ Fig. 9.24. If we choose a control that also includes the measured external wrench $\boldsymbol{w}_e^{\#}$

$$
\boldsymbol{w}_c = \hat{\mathbf{M}}_x(\boldsymbol{q})\mathbf{M}_x^{*-1}\Big(\mathbf{K}_p(\boldsymbol{x}^* - \boldsymbol{x}) - \mathbf{K}_v(\dot{\boldsymbol{x}}^* - \dot{\boldsymbol{x}})\Big) + \hat{\mathbf{C}}_x(\boldsymbol{q},\dot{\boldsymbol{q}})\dot{\boldsymbol{x}} + \hat{\boldsymbol{g}}_x(\boldsymbol{q})
$$
$$
+ (\mathbf{1} - \hat{\mathbf{M}}_x(\boldsymbol{q})\mathbf{M}_x^{*-1})\boldsymbol{w}_e^{\#}
$$

(9.29)

then the closed-loop error dynamics become

$$
\mathbf{M}_x^*\ddot{\boldsymbol{e}} + \mathbf{K}_v\dot{\boldsymbol{e}} + \mathbf{K}_p\boldsymbol{e} = -\boldsymbol{w}_e .
$$

The robot now acts like a 6-dimensional spring and if we pushed on the end effector that is, we applied a nonzero $\boldsymbol{w}_e$, we would experience the dynamics of a physical mass-spring-damper system. The translational and torsional damping and stiffness are specified by $\mathbf{K}_v$ and $\mathbf{K}_p$ respectively. ◄

$\mathbf{K}_v$ and $\mathbf{K}_p$ are typically diagonal matrices so that the robot behaves like a physically realizable spring-mass-damper system. The diagonal elements of these matrices have units of N s m^{-1} or N s m rad^{-1}, and N m^{-1} or N m rad^{-1} respectively.

So far, we have assumed that the robot's joints can be commanded to exert a specified force or torque as computed by (9.26) and the particular control law. While some robots are capable of this, the majority of robots today are position controlled which reflects the heritage of robot arms for performing accurate position-controlled tasks in factories.

High-quality joint-torque control requires sophisticated joint actuators, such as shown in ◘ Fig. 9.23, that can measure torque in order to compensate for those nonidealities.

A standard position-controlled robot is very unforgiving when it comes into contact with the environment and interaction forces will rise very quickly, potentially damaging the robot or its environment. To perform compliant tasks with such a robot, we need to measure the environmental wrench $\boldsymbol{w}_e^\#$ and use that to modify the end-effector position. Using a force/torque sensor, such as shown in ◘ Fig. 9.24, we can choose a control

$$\Delta \boldsymbol{x} = \mathbf{K}_s^{-1}(\boldsymbol{w}_e^* - \boldsymbol{w}_e^\#) - \mathbf{K}_v\dot{\boldsymbol{x}}$$

to determine the change in the robot's position to achieve the desired wrench. The desired robot stiffness matrix is $\mathbf{K}_s$, and the damping is $\mathbf{K}_v > 0$. This controller maps a wrench $\boldsymbol{w}_e^\#$ to a displacement $\Delta \boldsymbol{x}$ and therefore acts as a mechanical admittance – this type of robot control is called admittance or compliance control. In practice, it requires a very high sample rate and low latency between the force/torque measurements and the corrective position commands being sent to the joint actuators.

9.6 Application

9.6.1 Series-Elastic Actuator (SEA)

For high-speed robots, the elasticity of the links and the joints becomes a significant dynamic effect which will affect the path-following accuracy. Joint elasticity is typically caused by elements of the transmission such as: a harmonic gearbox which is inherently elastic, torsional elasticity of a motor shaft, spiral couplings, or longitudinal elasticity of a toothed belt or cable drive.

There are advantages in having some flexibility between the motor and the load. Imagine a robot performing a task that involves the gripper picking an object off a table whose height is uncertain. ▸ A simple strategy to achieve this is to move down until the gripper touches the table, close the gripper, and then lift up. However, at the instant of contact a large and discontinuous force will be exerted on the robot which has the potential to damage the object or the robot. This is particularly problematic for robots with large inertia that are moving quickly – the kinetic energy must be instantaneously dissipated. An elastic element – a spring – between the motor and the joint would help here. At the moment of contact, the spring would start to compress, and the kinetic energy is transferred to potential energy in the spring – the robot control system has time to react and stop or reverse the motors. We have changed the problem from a damaging hard impact to a soft impact. In addition to shock absorption, the deformation of the spring provides a means of determining the force that the robot is exerting. This capability is particularly useful for robots that interact closely with people since it makes the robot less dangerous in case of collision, and a spring is simple technology that cannot fail.

Unfortunately, position control is now more challenging because there is an elastic element between the motor and the load.

Consider the 1-dimensional case shown in ◘ Fig. 9.25 where the motor is represented by a mass m_1 to which a controllable force u is applied. ▸ It is connected

Or the robot is not very accurate.

In a real robot, this is most commonly a rotary system with a torsional spring.

■ **Fig. 9.25** Schematic of a series-elastic actuator. The two masses represent the motor and the load, and they are connected by an elastic element or spring. We use a control force applied to m_1 to indirectly control the position of m_2. f_s is the spring force, and f_e is the environmental force

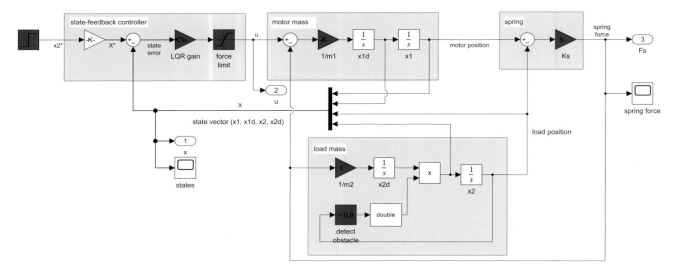

■ **Fig. 9.26** Simulink model `sl_sea` of a series-elastic actuator colliding with an obstacle

`sl_sea`

▶ sn.pub/LNZJkR

via a linear elastic element or spring to the load mass m_2. If we apply a positive force to m_1, it will move to the right and compress the spring, and this will exert a positive force on m_2 which will also move to the right. Controlling the position of m_2 is not trivial since this system has no friction and is marginally stable. It can be stabilized by feedback of position and velocity of the motor *and* of the load – all of which are potentially measurable.

In robotics, such a system built into a robot joint, is known as a series-elastic actuator or SEA.

A Simulink model of an SEA system can be loaded by

```
>> sl_sea
```

and is shown in ■ Fig. 9.26. The model attempts to drive the load m_2 to a position $x_2^* = 1$. A state-feedback LQR controller has been designed, which computes the control force based on motor position x_1, motor velocity $\dot{x}_1$, load position x_2, and load velocity $\dot{x}_2$, which form the state vector x in the Simulink model. We can simulate this model by

```
>> r = sim("sl_sea");
```

and results are shown in ■ Fig. 9.27. In the first case, there is no obstacle, and the controller achieves its goal with minimal overshoot, but note the complex force profile applied to m_1. In the second case, the load mass is stopped at $x_2 = 0.8$ and the elastic force changes to accommodate this.

Knowing x_1 and x_2 and the spring stiffness K_s allows us to compute the environmental force f_e applied to m_2. The actuator is also a sensor, and some of these principles are employed by the torque-controlled actuator shown in ■ Fig. 9.23.

A more complex problem to analyze is elasticity of the links. Most robots have very stiff links to avoid this problem, but for robots operating with extreme acceleration it may need to be taken into account.

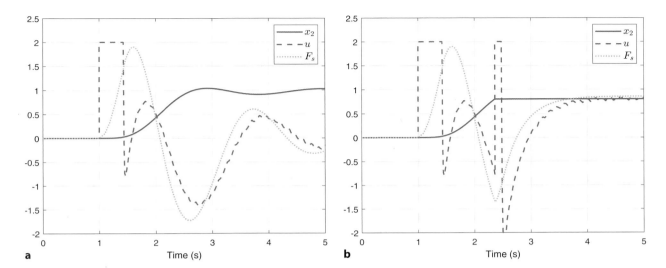

Fig. 9.27 Response of the series-elastic actuator to a unit-step demand at $t = 1$ s, showing load position x_2 (m), motor force u (N), and spring force F_s (N). **a** Moving to $x_2^* = 1$ with no collision; **b** moving to $x_2^* = 1$ with an obstacle at $x_2 = 0.8$ which is reached at $t \approx 2.3$

9.7 Wrapping Up

This chapter has introduced several different approaches to robot manipulator control. We started with the simplest case of independent joint control, explored the effect of disturbance torques and variation in inertia, and showed how feedforward of disturbances such as gravity could provide significant improvement in performance. We then learned how to model the forces and torques acting on the individual links of a serial-link manipulator. The equations of motion, or inverse dynamics, allow us to determine the joint forces required to achieve particular joint velocity and acceleration. The equations have terms corresponding to inertia, gravity, velocity coupling, and externally applied forces. We looked at the significance of these terms and how they vary with manipulator configuration and payload. The equations of motion provide insight into important issues such as how the velocity or acceleration of one joint exerts a disturbance force on other joints which is important for control design. We then discussed the forward dynamics which describe how the configuration evolves with time in response to forces and torques applied at the joints by the actuators and by external forces such as gravity. We extended the feedforward notion to full model-based control using torque feedforward, and computed-torque controllers. We then reformulated the rigid-body dynamics in task space which allowed us to design a pose controller where we can also specify the dynamic characteristics of the end effector as a spring-mass-damper system. Finally, we discussed series-elastic actuators where a compliant element between the robot motor and the link enables force control and people-safe operation.

9.7.1 Further Reading

The engineering design of motor control systems is covered in mechatronics textbooks such as Bolton (2015). The dynamics of serial-link manipulators is well covered by all the standard robotics textbooks such as Paul (1981), Spong et al. (2006), Siciliano et al. (2009), and the Robotics Handbook (Siciliano and Khatib 2016). The efficient recursive Newton-Euler method we use today is the culmination of much research in the early 1980s and described in Hollerbach (1982). The equations of motion can be derived via a number of techniques, including

9

Lagrangian (energy-based), Newton-Euler, d'Alembert (Fu et al. 1987, Lee et al. 1983), or Kane's method (Kane and Levinson 1983). The computational cost of Lagrangian methods (Uicker 1965; Kahn 1969) is enormous, $O(N^4)$, which made it infeasible for real-time use on computers of that era and many simplifications and approximation had to be made. Orin et al. (1979) proposed an alternative approach based on the Newton-Euler (NE) equations of rigid-body motion applied to each link. Armstrong (1979) then showed how recursion could be applied resulting in $O(N)$ complexity. Luh et al. (1980) provided a recursive formulation of the Newton-Euler equations with linear and angular velocities referred to link coordinate frames which resulted in a thousand-fold improvement in execution time making it practical to implement in real-time. Hollerbach (1980) showed how recursion could be applied to the Lagrangian form and reduced the computation to within a factor of 3 of the recursive NE form, and Silver (1982) showed the equivalence of the recursive Lagrangian and Newton-Euler forms, and that the difference in efficiency was due to the representation of angular velocity.

The recursive Newton-Euler algorithm is typically computed using vector representation of translational and angular velocities and accelerations, but this is cumbersome. Featherstone's spatial vector notation (Featherstone 1987, 2010a, 2010b) is elegant and compact.

The forward dynamics, ▸ Sect. 9.3, is computationally more expensive. An $O(N^3)$ method was proposed by Walker and Orin (1982). Featherstone's (1987) articulated-body method, which is used in the MATLAB implementation of the `forwardDynamics` function, has $O(N)$ complexity but for $N < 9$ is more expensive than Walker and Orin's method.

Critical to any consideration of robot dynamics is knowledge of the inertial parameters, ten per link, as well as the motor's parameters. Corke and Armstrong-Hélouvry (1994, 1995) published a meta-study of PUMA parameters and provided a consensus estimate of inertial and motor parameters for the PUMA 560 robot. Some of this data was obtained by painstaking disassembly of the robot and determining the mass and dimensions of the components. Inertia of components can be estimated from mass and dimensions by assuming mass distribution, or it can be measured using a bifilar pendulum as discussed in Armstrong et al. (1986).

Alternatively, the parameters can be estimated by measuring the joint torques or the base reaction force and moment as the robot moves. A number of early works in this area include Mayeda et al. (1990), Izaguirre and Paul (1985), Khalil and Dombre (2002), and a more recent summary is Siciliano and Khatib (2016, § 6). Key to successful identification is that the robot moves in a way that is sufficiently exciting (Gautier and Khalil 1992; Armstrong 1989). Friction is an important dynamic characteristic and is well-described in Armstrong's (1988) thesis. The survey by Armstrong-Hélouvry et al. (1994) is a very readable and thorough treatment of friction modeling and control. Motor parameters can be obtained directly from the manufacturer's data sheet or determined experimentally, without having to remove the motor from the robot, as described by Corke (1996a). The parameters used in the RVC Toolbox PUMA model are the best estimates from Corke and Armstrong-Hélouvry (1995) and Corke (1996a).

The discussion on control has been quite brief and has strongly emphasized the advantages of feedforward control. Robot joint control techniques are well covered by Spong et al. (2006), Craig (2005), Siciliano et al. (2009), and summarized in Siciliano and Khatib (2016, § 8). Siciliano et al. (2009) have a good discussion of actuators and sensors as does the, now quite old, book by Klafter et al. (1989). The control of flexible joint robots is discussed in Spong et al. (2006). Adaptive control can be used to accommodate the time-varying inertial parameters and there is a large literature on this topic but some good early references include the book by Craig (1987) and key papers include Craig et al. (1987), Spong (1989), Middleton and Goodwin (1988), and Ortega and Spong (1989). An operational-space control structure was proposed in Khatib (1987). There has been considerable recent interest in series-elastic as well as variable stiffness actuators (VSA) whose position and stiffness can be independently controlled much like our own muscles – a good collection of articles on this technology can be found in the special issue by Vanderborght et al. (2008).

Dynamic manipulability is discussed in Spong et al. (2006) and Siciliano et al. (2009). The Asada measure used earlier is described in Asada (1983).

▪▪ Historical and general

Newton's second law is described in his 1687 master work *Principia Naturalis* (mathematical principles of natural philosophy), written in Latin, but an English translation is available online at ▶ https://www.archive.org/details/newtonspmathema00newtrich. His writing on other subjects, including transcripts of his notebooks, can be found online at ▶ https://www.newtonproject.ox.ac.uk/.

9.7.2 Exercises

1. Independent joint control (▶ Sect. 9.1)
   a) Investigate different values of `Kv` and `Ki` as well as demand signal shape and amplitude.
   b) Perform a root-locus analysis of `sl_vloop_test` to determine the maximum permissible gain for the proportional case. Repeat this for the PI case.
   c) Consider that the motor is controlled by a voltage source instead of a current source, and that the motor's impedance is 1 mH and 1.6 Ω. Modify `sl_vloop_test` accordingly. Extend the model to include the effect of back EMF.
   d) Increase the required speed of motion so that the motor torque becomes saturated. With integral action, you will observe a phenomenon known as integral windup – examine what happens to the state of the integrator during the motion. Various strategies are employed to combat this, such as limiting the maximum value of the integrator, or only allowing integral action when the motor is close to its setpoint. Experiment with some of these.
   e) Create a Simulink model of the PUMA robot with each joint controlled by `sl_ploop_test`. Parameters for the different motors in the PUMA are described in Corke and Armstrong-Hélouvry (1995).
2. The motor torque constant has units of $N\,m\,A^{-1}$ and is equal to the back EMF constant which has units of $V\,s\,rad^{-1}$. Show that these units are equivalent.
3. Simple two-link robot arm of ◻ Fig. 9.4
   a) Plot the gravity load as a function of both joint angles. Assume $m_1 = 0.45\,kg$, $m_2 = 0.35\,kg$, $r_1 = 8\,cm$, and $r_2 = 8\,cm$.
   b) Plot the inertia for joint 1 as a function of q_2. To compute link inertia, assume that we can model the link as a point mass located at the center of mass.
4. Run the code that generates ◻ Fig. 9.15 to compute gravity loading on joints 2 and 3 as a function of configuration. Add a payload and repeat.
5. Run the code that generates ◻ Fig. 9.16 to show how the inertia of joints 1 and 2 vary with payload.
6. Generate the curve of ◻ Fig. 9.16c. Add a payload and compare the results. By what factor does this inertia vary over the joint angle range?
7. Why is the manipulator inertia matrix symmetric?
8. The robot exerts a wrench on the base as it moves (▶ Sect. 9.2.5). Consider that the robot is sitting on a frictionless horizontal table (say on a large air puck). Create a simulation model that includes the robot arm dynamics and the sliding dynamics on the table. Show that moving the arm causes the robot to translate and spin. Can you devise an arm motion that moves the robot base from one position to another and stops?

9. Overlay the dynamic manipulability ellipsoid on the display of the robot. Compare this with the force ellipsoid from ► Sect. 8.4.2.

10. Model-based control (► Sect. 9.4)

    a) Compute and display the joint tracking error for the torque feedforward and computed torque cases. Experiment with different motions, control parameters, and sample rate T_{fb}.

    b) Reduce the rate at which the feedforward torque is computed and observe its effect on tracking error.

11. Series-elastic actuator (► Sect. 9.6.1)

    a) Experiment with different values of stiffness for the elastic element and control parameters. Try to reduce the settling time.

    b) Modify the simulation so that the robot arm moves to touch an object at unknown distance and applies a force of 5 N to it.

    c) Plot the frequency response function $X_2(s)/X_1(s)$ for different values of K_s, m_1, and m_2.

    d) Simulate the effect of a collision between the load and an obstacle by adding a step to the spring force.

Computer Vision

Contents

Chapter 10   Light and Color – 395

Chapter 11   Images and Image Processing – 435

Chapter 12   Image Feature Extraction – 493

Chapter 13   Image Formation – 543

Chapter 14   Using Multiple Images – 593

As discussed in ▶ Chap. 1, almost all animal species have eyes and our own experience is that eyes are very effective, perhaps invaluable, sensors for recognition, navigation, obstacle avoidance and manipulation. Vision for robots – robotic vision – uses cameras to mimic the function of an eye and computer vision algorithms to mimic the function of the brain.

In ▶ Chap. 10 we start by discussing light, and in particular color because it is such an important characteristic of the world that we perceive visually. Although we learn about color at kindergarten it is a complex topic that is often not well understood. ▶ Chap. 11 introduces digital images and the fundamentals of image processing algorithms, and this provides the foundation for the feature extraction algorithms discussed in ▶ Chap. 12. Feature extraction is a problem in data reduction, in extracting the essence of the scene from the massive amount of pixel data. For example, how do we determine the coordinate of the round red object in the scene, which can be described with perhaps just 4 bytes, given the millions of bytes that comprise an image. To solve this we must address many important subproblems – what is red? how do we distinguish red pixels from nonred pixels? how do we describe the shape of the red pixels? what if there is more than one red object? and so on. ▶ Chap. 12 additionally introduces semantic segmentation and object classification using powerful deep learning techniques.

▶ Chap. 13 discusses how a camera forms an image of the world on a sensor and converts it to a digital image. A number of different types of cameras are introduced, and the limitations of using just a single camera to view the world are discussed. Once again biology shows the way – multiple eyes are common and have great utility. This leads us to consider using multiple views of the world, from a single moving camera or multiple cameras observing the scene from different viewpoints. This is discussed in ▶ Chap. 14 and is particularly important for understanding the 3-dimensional structure of the world. All of this sets the scene for describing how vision can be used for closed-loop control of arm-type and mobile robots which is the subject of the next and final part of the book.

Light and Color

» *In nature, light creates the color.*
In the picture, color creates the light.
– Hans Hoffman

Contents

10.1   Spectral Representation of Light – 396

10.2   Color – 401

10.3   Advanced Topics – 420

10.4   Application: Color Images – 427

10.5   Wrapping Up – 430

10

In ancient times it was believed that the eye radiated a cone of visual flux which mixed with visible objects in the world to create a sensation in the observer – like the sense of touch, but at a distance – which we call the extromission theory of vision. Today, the intromission theory considers that light from an illuminant falls on the scene, some of which is reflected into the eye of the observer to create a perception about that scene. The light that reaches the eye, or the camera, is a function of the illumination impinging on the scene and a material property of the scene known as reflectivity.

This chapter is about light itself and our perception of light in terms of brightness and color. ► Sect. 10.1 describes light in terms of electromagnetic radiation, and mixtures of light as continuous spectra. ► Sect. 10.2 provides a brief introduction to colorimetry, the science of color perception, human trichromatic color perception and how colors can be represented in various color spaces. ► Sect. 10.3 covers a number of advanced topics such as color constancy, gamma correction and white balancing. ► Sect. 10.4 considers two applications: distinguishing different colored objects in an image, and the removal of shadows in an image.

10.1 Spectral Representation of Light

Around 1670, Sir Isaac Newton discovered that white light was a mixture of different colors. We now know that each of these colors is a single frequency or wavelength of electromagnetic radiation. We perceive the wavelengths between 400 and 700 nm as different colors as shown in ◘ Fig. 10.1.

In general, the light that we observe is a mixture of many wavelengths and can be represented as a function $E(\lambda)$ that describes intensity as a function of wavelength λ. Monochromatic light, such as emitted by an LED or laser, comprises a single wavelength, in which case E is an impulse or Dirac function.

The most common source of light is incandescence, the emission of light from a hot body such as the Sun or the filament of an old-fashioned incandescent light

◘ **Fig. 10.1** The spectrum of visible colors as a function of wavelength in nanometers. The visible range depends on viewing conditions and the individual but is generally accepted as being 400–700 nm. Wavelengths greater than 700 nm are termed infrared and those below 400 nm are ultraviolet

Excurse 10.1: Spectrum of Light
During the plague years of 1665–1666 Isaac Newton developed his theory of light and color. He demonstrated that a prism could decompose white light into a spectrum of colors, and that a lens and a second prism could recompose the multi-colored spectrum into white light. Importantly, he showed that the color of the light did not change when it was reflected from different objects, from which he concluded that color is an intrinsic property of light and not the object. The sketch is from Newton's notebook.

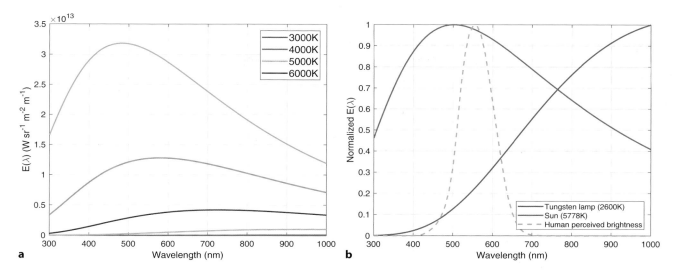

■ **Fig. 10.2** Blackbody spectra. **a** Blackbody emission spectra for temperatures from 3000–6000 K. **b** Blackbody emissions for the Sun (5778 K), a tungsten lamp (2600 K) and the perceived brightness to the human eye – all normalized to unity for readability

bulb. In physics, this is modeled as a blackbody radiator or Planckian source. The emitted power as a function of wavelength λ is given by Planck's radiation formula

$$E(\lambda) = \frac{2hc^2}{\lambda^5 \left(e^{hc/k\lambda T} - 1\right)} \text{ W sr}^{-1}\text{ m}^{-2}\text{ m}^{-1} \qquad (10.1)$$

where T is the absolute temperature (K) of the source, h is Planck's constant, k is Boltzmann's constant, and c the speed of light. ► This is the power emitted per steradian ► per unit area per unit wavelength.

$c = 2.998 \times 10^8 \text{ ms}^{-1}$
$h = 6.626 \times 10^{-34} \text{ Js}$
$k = 1.381 \times 10^{-23} \text{ JK}^{-1}$

Solid angle is measured in steradians. A full sphere is 4π sr.

We can plot the emission spectra for a blackbody at different temperatures. First we define a range of wavelengths

```
>> lambda = [300:10:1000]*1e-9;
```

in this case from 300 to 1000 nm, and then compute the blackbody spectra

```
>> for T=3000:1000:6000
>>    plot(lambda,blackbody(lambda,T)), hold on
>> end
>> hold off
```

as shown in ■ Fig. 10.2a. We can see that as temperature increases the maximum amount of power increases, and the wavelength at which the peak occurs decreases. The total amount of power radiated (per unit area) is the area under the blackbody curve and is given by the Stefan-Boltzman law

$$P(\lambda) = \frac{2\pi^5 k^4}{15c^2 h^3} T^4 \text{ W m}^{-2}$$

and the wavelength corresponding to the peak of the blackbody curve is given by Wien's displacement

$$\lambda_{\max} = \frac{2.8978 \times 10^{-3}}{T} \text{ m }.$$

The inverse relationship between peak wavelength and temperature is familiar to us in terms of what we observe when heating an object. As shown in ■ Tab. 10.1, an object starts to glow faintly red at around 800 K, and moves through orange and yellow toward white as temperature increases.

Excurse 10.2: Infrared Radiation

Infrared radiation was discovered in 1800 by William Herschel (1738–1822) the German-born British astronomer. He was Court Astronomer to George III; built a series of large telescopes; with his sister Caroline performed the first sky survey discovering double stars, nebulae and the planet Uranus; and studied the spectra of stars. Using a prism and thermometers to measure the amount of heat in the various colors of sunlight he observed that temperature increased from blue to red, and increased even more beyond red where there was no visible light. (Image from Herschel 1800)

Excurse 10.3: The Incandescent Lamp

Sir Humphry Davy demonstrated the first electrical incandescent lamp using a platinum filament in 1802. Sir Joseph Swan demonstrated his first light bulbs in 1850 using carbonized paper filaments. However, it was not until advances in vacuum pumps in 1865 that such lamps could achieve a useful lifetime. Swan patented a carbonized cotton filament in 1878 and a carbonized cellulose filament in 1881. His lamps came into use after 1880 and the Savoy Theater in London was completely lit by electricity in 1881. In the USA Thomas Edison did not start research into incandescent lamps until 1878, but he patented a long-lasting carbonized bamboo filament the next year and was able to mass produce them. The Swan and Edison companies merged in 1883.

The light bulb subsequently became the dominant source of light on the planet, but is now being phased out due to its poor energy efficiency. (Image by Douglas Brackett, Inv., Edisonian.com)

□ Table 10.1 Color of a heated object versus temperature

Incipient red heat	770–820 K
Dark red heat	920–1020 K
Bright red heat	1120–1220 K
Yellowish red heat	1320–1420 K
Incipient white heat	1520–1620 K
White heat	1720–1820 K

The filament of a tungsten lamp has a temperature of 2600 K and glows *white hot*. The Sun has a surface temperature of 5778 K. The spectra of these sources

```
>> lamp = blackbody(lambda,2600);
>> sun = blackbody(lambda,5778);
>> plot(lambda, [lamp/max(lamp) sun/max(sun)])
```

are compared in ◘ Fig. 10.2b. The tungsten lamp curve is much lower in magnitude, but has been scaled up (by 56) for readability. The peak of the Sun's emission is around 500 nm and it emits a significant amount of power in the visible part of the spectrum. The peak for the tungsten lamp is at a much longer wavelength and perversely most of its power falls in the infrared band which we perceive as heat not light.

10.1.1 Absorption

The Sun's spectrum at ground level on the Earth has been measured and tabulated

```
>> sunGround = loadspectrum(lambda,"solar");
>> plot(lambda,sunGround)
```

and is shown in ◘ Fig. 10.3a. It differs markedly from that of a blackbody since some wavelengths have been absorbed more than others by the atmosphere. Our eye's peak sensitivity has evolved to be closely aligned to the peak of the spectrum of atmospherically filtered sunlight.

Transmittance T is the inverse of absorptance, and is the fraction of light passed as a function of wavelength and distance traveled. It is described by Beer's law

$$T = 10^{-Ad} \tag{10.2}$$

where A is the absorption coefficient in units of m^{-1} which is a function of wavelength, and d is the optical path length. The absorption spectrum $A(\lambda)$ for water is loaded from tabulated data

```
>> [A,lambda] = loadspectrum([400:10:700]*1e-9,"water");
```

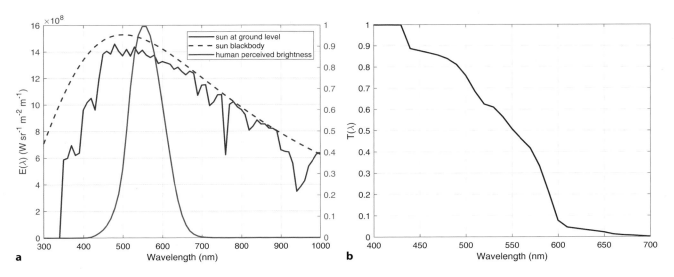

◘ **Fig. 10.3** **a** Modified solar spectrum at ground level (blue). The dips in the solar spectrum correspond to various water absorption bands. CO_2 absorbs radiation in the infrared region, and ozone O_3 absorbs strongly in the ultraviolet region. **b** Transmission through 5 m of water. The longer wavelengths, reds, have been strongly attenuated and the transmission through 5 m of water is

```
>> d = 5;
>> T = 10.^(-A*d);
>> plot(lambda,T)
```

which is plotted in ■ Fig. 10.3b. We see that the red light is strongly attenuated, which makes the scene appear more blue. Differential absorption of wavelengths is a significant concern when imaging underwater, and we revisit this topic in ► Sect. 10.3.4.

10.1.2 Reflectance

Surfaces reflect incoming light. The reflection might be specular (as from a mirror-like surface, see ► Exc. 13.11), or Lambertian (diffuse reflection from a matte surface, see ► Exc. 10.15). The fraction of light that is reflected $R(\lambda) \in [0, 1]$ is the reflectivity, reflectance or albedo of the surface, and is a function of wavelength. White paper for example has a reflectance of around 70 %. The reflectance spectra of many materials have been measured and tabulated. ◄ For example, consider the reflectivity of a red house brick over a wide spectral range

From ► http://speclib.jpl.nasa.gov/ weathered red brick (0412UUUBRK).

```
>> [R,lambda] = loadspectrum([100:10:10000]*1e-9,"redbrick");
>> plot(lambda,R)
```

which is plotted in ■ Fig. 10.4 and shows that it reflects red light more than blue.

10.1.3 Luminance

The light reflected from a surface, its luminance, has a spectrum given by

$$L(\lambda) = E(\lambda)R(\lambda) \, \mathrm{W \, m^{-2}} \tag{10.3}$$

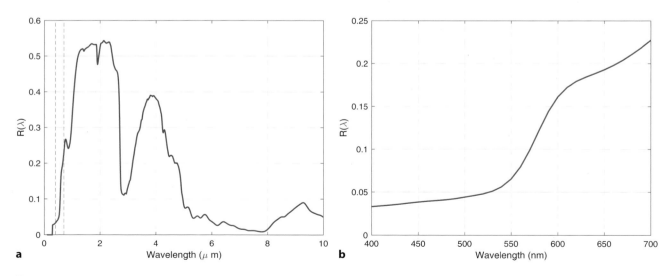

a   **b**

■ **Fig. 10.4** Reflectance of a weathered red house brick (data from ASTER, Baldridge et al. 2009). **a** Full range measured from 300 nm visible to 10,000 nm (infrared); **b** closeup view of the visible region

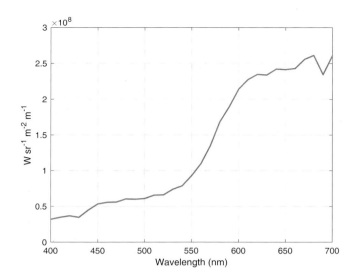

◘ Fig. 10.5 Luminance of the weathered red house brick under illumination from the Sun at ground level, based on data from ◘ Figs. 10.3a and 10.4b

where E is the incident illumination and R is the reflectance. The illuminance of the Sun in the visible region is

```
>> lambda = [400:700]*1e-9;
>> E = loadspectrum(lambda,"solar");
```

at ground level. The reflectivity of the brick is

```
>> R = loadspectrum(lambda,"redbrick");
```

and the light reflected from the brick is

```
>> L = E.*R;
>> plot(lambda,L)
```

which is shown in ◘ Fig. 10.5. It is this spectrum that is interpreted by our eyes as the color red.

10.2 Color

» *Color is the general name for all sensations arising from the activity of the retina of the eye and its attached nervous mechanisms, this activity being, in nearly every case in the normal individual, a specific response to radiant energy of certain wavelengths and intensities.*
– T.L. Troland, Report of Optical Society of America
Committee on Colorimetry 1920–1921

We have described the spectra of light in terms of power as a function of wavelength, but our own perception of light is in terms of subjective quantities such as brightness and color. Light that is visible to humans lies in the range of wavelengths from 400 nm (violet) to 700 nm (red) with the colors blue, green, yellow and orange in between, as shown in ◘ Fig. 10.1.

The brightness we associate with a particular wavelength is given by the luminosity function with units of lumens per watt. For our daylight (photopic) vision the luminosity as a function of wavelength has been experimentally determined, tabulated and forms the basis of the 1931 CIE standard that represents the average human observer. ▶ The photopic luminosity function is

```
>> human = luminos(lambda);
>> plot(lambda,human)
```

This is the photopic response for a light-adapted eye using the cone photoreceptor cells. The dark adapted, or scotopic response, using the eye's monochromatic rod photoreceptor cells is different, and peaks at around 510 nm.

10

and is shown in ◻ Fig. 10.7a. Consider two light sources emitting the same power (in watts) but one has a wavelength of 550 nm (green) and the other has a wavelength of 450 nm (blue). The perceived brightness of these two lights is quite different, in fact the blue light appears only

```
>> luminos(450e-9)/luminos(550e-9)
ans =
    0.0382
```

3.8 % as bright as the green one. The silicon sensors used in digital cameras have strong sensitivity in the red and infrared part of the spectrum. ◄

The LED on an infrared remote control can be seen as a bright light in many digital cameras – try this with your mobile phone camera and TV remote. Some security cameras provide infrared scene illumination for covert night time monitoring. Note that some cameras are fitted with infrared filters to prevent the sensor from becoming saturated by ambient infrared radiation.

10.2.1 The Human Eye

Our eyes contain two types of light-sensitive photoreceptors shown in ◻ Fig. 10.6: rod cells and cone cells. Rod cells are much more sensitive than cone cells, respond to intensity only, and are used at night. In normal daylight conditions our cone cells are active and these are color sensitive.

◻ **Fig. 10.6**   A colored scanning electron micrograph of rod cells (white) and cone cells (yellow) in the human eye. The cells diameters are in the range 0.5–4 μm. The cells contain different types of light-sensitive opsin proteins. Surprisingly, the rods and cones are not on the surface of the retina, they are behind that surface which is a network of nerves and blood vessels

Humans, like most primates, are trichromats and have three types of cones that respond to different parts of the spectrum. They are referred to as long (L), medium (M) and short (S) cones according to the wavelength of their peak response, or more commonly as red, green and blue cones. Most other mammals are dichromats, they have just two types of cone cells. Many birds and fish are tetrachromats and have a type of cone cell that is sensitive to ultra violet light. Dragonflies have up to thirty different types of cone cells.

Human rod and cone cells have been extensively studied and the spectral response of cone cells can be loaded

```
>> cones = loadspectrum(lambda,"cones");
>> plot(lambda,cones)
```

where `cones` has three columns corresponding to the L, M and S cone responses and each row corresponds to the wavelength in `lambda`. The spectral responses of the cones $L(\lambda)$, $M(\lambda)$ and $S(\lambda)$ are shown in ◧ Fig. 10.7b. ▶

The different spectral characteristics are due to the different photopsins in the cone cell.

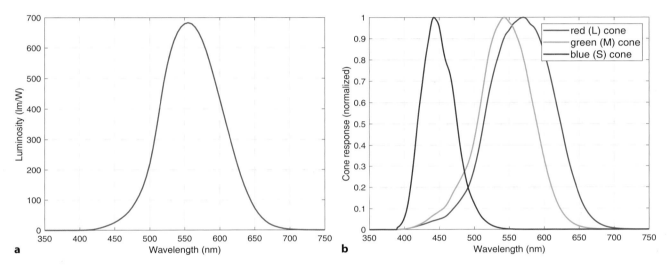

◧ **Fig. 10.7** **a** Luminosity curve for the standard human observer. The peak response is $683\,\mathrm{lm W^{-1}}$ at 555 nm (green). **b** Spectral response of human cones (normalized)

Excurse 10.5: Opsins

Opsins are the photoreceptor molecules used in the visual systems of all animals. They belong to the class of G protein-coupled receptors (GPCRs) and comprise seven helices that pass through the cell's membrane. They change shape in response to particular molecules outside the cell and initiate a cascade of chemical signaling events inside the cell that results in a change in cell function. Opsins contain a chromophore, a light-sensitive molecule called retinal derived from vitamin A, that stretches across the opsin. When retinal absorbs a photon its shape changes, deforming the opsin and activating the cell's signalling pathway. The basis of all vision is a fortuitous genetic mutation 700 million years ago that made a chemical sensing receptor light sensitive. There are many opsin variants across the animal kingdom – our rod cells contain rhodopsin and our cone cells contain photopsins. The American biochemist George Wald (1906–1997) received the 1967 Nobel Prize in Medicine for his discovery of retinal and characterizing the spectral absorbance of photopsins. (Image by Dpryan from Wikipedia; the chromophore is indicated by the arrow)

Excurse 10.6: Trichromatic Theory

The trichromatic theory of color vision suggests that our eyes have three discrete types of receptors that when stimulated produce the sensations of red, green and blue, and that all color sensations are "psychological mixes" of these fundamental colors. It was first proposed by the English scientist Thomas Young (1773–1829) in 1802 but made little impact. It was later championed by Hermann von Helmholtz and James Clerk Maxwell. The figure shows how beams of red, green and blue light mix. Helmholtz (1821–1894) was a prolific German physician and physicist. He invented the opthalmascope for examining the retina in 1851, and in 1856 he published the *Handbuch der physiologischen Optik* (Handbook of Physiological Optics) which contained theories and experimental data relating to depth perception, color vision, and motion perception. Maxwell (1831–1879) was a Scottish scientist best known for his electromagnetic equations, but who also extensively studied color perception, color blindness, and color theory. His 1860 paper *On the Theory of Colour Vision* won a Rumford medal, and in 1861 he demonstrated color photography in a Royal Institution lecture.

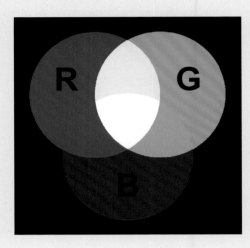

Excurse 10.7: Opponent Color Theory

The opponent color theory holds that colors are perceived with respect to two axes: red–green and blue–yellow. One clue comes from color after-images – staring at a red square and then a white surface gives rise to a green after-image. Another clue comes from language – we combine color words to describe mixtures, for example reddish-blue, but we never describe a reddish-green or a blueish-yellow. The theory was first mooted by the German writer Johann Wolfgang von Goethe (1749–1832) in his 1810 *Theory of Colours* but later had a strong advocate in Karl Ewald Hering (1834–1918), a German physiologist who also studied binocular perception and eye movements. He advocated opponent color theory over trichromatic theory and had acrimonious debates with Helmholtz on the topic.

In fact, both theories hold. Our eyes have three types of color sensing cells but the early processing in the retinal ganglion layer appears to convert these signals into an opponent color representation.

The retina of the human eye has a central or foveal region which is only 0.6 mm in diameter, has a 5 degree field of view and contains most of the 6 million cone cells: 65 % sense red, 33 % sense green and only 2 % sense blue. We unconsciously scan our high-resolution fovea over the world to build a large-scale mental image of our surroundings. In addition, there are 120 million rod cells distributed over the retina. Rod cells are stimulated by light over wide range of intensities and are responsible for perceiving the size, shape, and brightness of visual images.

10.2.2   Camera Sensor

The photosites generally correspond directly with pixels in the resulting image.

The sensor in a digital camera is analogous to the retina, but instead of rod and cone cells there is a regular array of light-sensitive photosites on a silicon chip. ◄ Each photosite is of the order 1–10 µm square and outputs a signal proportional to the intensity of the light falling over its area.

Fig. 10.8 Bayer filtering. The gray blocks represent the array of light-sensitive silicon photosites over which is an array of red, green and blue filters. Invented by Bryce E. Bayer of Eastman Kodak, U.S. Patent 3,971,065

Most common color cameras have an array of tiny color filters over the photosites as shown in ◻ Fig. 10.8. These filters pass either red, green or blue light to the photosites and the spectral response of these filters is chosen to be similar to that of human cones $M(\lambda)$ shown in ◻ Fig. 10.7b. A very common arrangement of color filters is the Bayer pattern – a regular 2×2 pattern comprising two green filters, one red and one blue.

Each photosite is the analog of a cone cell, sensitive to just one color. A consequence of this arrangement is that we cannot make independent measurements ▶ of red, green and blue at every pixel, but it can be estimated. For example, the amount of red at a blue sensitive pixel is obtained by interpolation from its red filtered neighbors. Digital camera *raw image* files contain the actual outputs of the Bayer-filtered photosites.

Larger filter patterns are possible and 3×3 or 4×4 arrays of filters are known as assorted pixel arrays. Using more than 3 different color filters leads to a multi-spectral camera with better color resolution, a range of neutral density (gray) filters leads to a high-dynamic-range camera, or these various filters can be mixed to give a camera with better dynamic range and color resolution.

More expensive "3 CCD" (charge coupled device) cameras make independent measurements at each pixel since the light is split by a set of prisms, filtered and presented to one CCD image sensor for each primary color.

10.2.3 Measuring Color

The path taken by the light entering the eye is shown in ◻ Fig. 10.9a. The spectrum of the luminance $L(\lambda)$ is a function of the light source and the reflectance of the object as given by (10.3). The response from each of the three cones is

$$\rho = \int_\lambda L(\lambda) M_r(\lambda) d\lambda$$

$$\gamma = \int_\lambda L(\lambda) M_g(\lambda) d\lambda$$

$$\beta = \int_\lambda L(\lambda) M_b(\lambda) d\lambda \tag{10.4}$$

where $M_r(\lambda)$, $M_g(\lambda)$ and $M_b(\lambda)$ are the spectral response of the red, green and blue cones respectively as shown in ◻ Fig. 10.7b. The response is a 3-vector (ρ, γ, β) which is known as a tristimulus value.

For the case of the red brick, the integrals correspond to the areas of the solid color regions in ◻ Fig. 10.9b. We can compute tristimulus values by approximating the integrals of (10.4) as a summation with $d\lambda = 1$ nm.

□ **Fig. 10.9** The tristimulus response of the human eye. **a** Path of light from illuminant to the eye. **b** Within the eye three filters are applied and the total output of these filters, the areas shown in solid color, are the tristimulus values

```
>> lambda = [400:700]*1e-9;
>> E = loadspectrum(lambda,"solar");
>> R = loadspectrum(lambda,"redbrick");
>> L = E.*R;  % light reflected from the brick
>> cones = loadspectrum(lambda,"cones");
>> sum((L*ones(1,3)).*cones*1e-9)
ans =
   16.3571   10.0665    2.8225
```

The dominant response is from the L cone, which is unsurprising since we know that the brick is red.

An arbitrary continuous spectrum is an infinite-dimensional vector and cannot be uniquely represented by just three parameters. Spectrally different objects or color stimuli that have the same tristimulus values are called metamers . As long as the tristimulus values are the same, we perceive them as having the same color despite them potentially having different spectra. This weakness in our ability to

---

Excurse 10.8: Lightmeters, Illuminance, and Luminance

A photographic lightmeter measures luminous flux which has units of $lm\,m^{-2}$ or lux (lx). The luminous intensity I of a point light source (emitting light in all directions) is the luminous flux per unit solid angle measured in $lm\,sr^{-1}$ or candelas (cd). The illuminance E falling normally onto a surface is

$$E = \frac{I}{d^2}\,\text{lx}$$

where d is the distance between source and the surface. Outdoor illuminance on a bright sunny day is approximately 10,000 lx. Office lighting levels are typically around 1000 lx and moonlight is 0.1 lx.

The luminance or *brightness* of a surface is

$$L_s = E_i \cos\theta \,\text{nt}$$

which has units of $cd\,m^{-2}$ or nit (nt), and where E_i is the incident illuminance at an angle θ to the surface normal.

Excurse 10.9: Color Blindness

Color blindness, or color deficiency, is the inability to perceive differences between some of the colors that others can distinguish. Protanopia, deuteranopia, and tritanopia refer to the absence of the L, M and S cones respectively. More common conditions are protanomaly, deuteranomaly and tritanomaly where the cone pigments are mutated and the peak response frequency changed. It is most commonly a genetic condition since the red and green photopsins are coded in the X chromosome. The most common form (occurring in 6 % of males including one of the authors) is deuteranomaly where the M-cone's response is shifted toward the red end of the spectrum resulting in reduced sensitivity to greens and poor discrimination of hues in the red, orange, yellow and green region of the spectrum.

The English scientist John Dalton (1766–1844) confused scarlet with green, and pink with blue. He hypothesized that the vitreous humor in his eyes was tinted blue and instructed that his eyes be examined after his death. This revealed that the humors were perfectly clear but DNA recently extracted from his preserved eye showed that he was a deuteranope. Color blindness is sometimes referred to as Daltonism.

◻ **Table 10.2** The CIE 1976 primaries (Commission Internationale de L'Éclairage 1987) are spectral colors corresponding to the emission lines in a mercury vapor lamp

	red	green	blue
λ (nm)	700.0	546.1	435.8

distinguish some spectra has not stopped our species from thriving in diverse environments. More important is the corollary – many, but not all, color stimuli can be generated by a mixture of just three monochromatic stimuli. These are the three primary colors we learned about as children. ▶ There is no unique set of primaries – any three will do so long as none of them can be matched by a combination of the others. The CIE has defined a set of monochromatic primaries, and their wavelengths are given in ◻ Tab. 10.2.

Primary colors are not a fundamental property of light – they are a fundamental property of the observer. There are three primary colors only because we, as trichromats, have three types of cones. Birds would have four primary colors and dogs would have two.

10.2.4 Reproducing Colors

A computer or television display is able to produce a variable amount of each of three primaries at every pixel. For a liquid crystal display (LCD) the colors are obtained by color filtering and attenuating white light emitted by the backlight, and an OLED display comprises a stack of red, green, and blue LEDs at each pixel. The important problem is to determine how much of each primary is required to match a given color stimulus.

We start by considering a monochromatic stimulus of wavelength λ_S which is defined as

$$L(\lambda) = \begin{cases} L_\lambda & \text{if } \lambda = \lambda_S \\ 0 & \text{otherwise} \end{cases}.$$

10

Excurse 10.10: What are Primary Colors?

The notion of primary colors is very old, but their number (anything from two to six) and their color is the subject of much debate. Much of the confusion is due to there being additive primaries (red, green, and blue) that are used when mixing lights, and subtractive primaries (cyan, magenta, yellow) used when mixing paints or inks. Whether or not black and white are primary colors was also debated.

The response of the cones to this stimulus is given by (10.4) but because $L(\lambda)$ is an impulse we can drop the integral to obtain the tristimulus values

$$
\begin{aligned}
\rho &= L_\lambda M_r(\lambda_S) \\
\gamma &= L_\lambda M_g(\lambda_S) \\
\beta &= L_\lambda M_b(\lambda_S) \ .
\end{aligned}
\tag{10.5}
$$

The units are chosen such that equal quantities of the primaries appear to be gray.

Consider next three monochromatic primary light sources denoted **R**, **G** and **B** with wavelengths λ_r, λ_g and λ_b and intensities R, G and B respectively. ◀ The tristimulus values from these light sources are

$$
\begin{aligned}
\rho &= RM_r(\lambda_r) + GM_r(\lambda_g) + BM_r(\lambda_b) \\
\gamma &= RM_g(\lambda_r) + GM_g(\lambda_g) + BM_g(\lambda_b) \\
\beta &= RM_b(\lambda_r) + GM_b(\lambda_g) + BM_b(\lambda_b) \ .
\end{aligned}
\tag{10.6}
$$

For the perceived color of these three light sources combined to match that of the monochromatic stimulus, the two tristimuli must be equal. We equate (10.5) and (10.6) and write compactly in matrix form as

$$
L_\lambda \begin{pmatrix} M_r(\lambda_S) \\ M_g(\lambda_S) \\ M_b(\lambda_S) \end{pmatrix} = \begin{pmatrix} M_r(\lambda_r) & M_r(\lambda_g) & M_r(\lambda_b) \\ M_g(\lambda_r) & M_g(\lambda_g) & M_g(\lambda_b) \\ M_{br}(\lambda_r) & M_b(\lambda_g) & M_b(\lambda_b) \end{pmatrix} \begin{pmatrix} R \\ G \\ B \end{pmatrix}
$$

and then solve for the required amounts of primary colors

$$
\begin{pmatrix} R \\ G \\ B \end{pmatrix} = L_\lambda \begin{pmatrix} M_r(\lambda_r) & M_r(\lambda_g) & M_r(\lambda_b) \\ M_g(\lambda_r) & M_g(\lambda_g) & M_g(\lambda_b) \\ M_{br}(\lambda_r) & M_b(\lambda_g) & M_b(\lambda_b) \end{pmatrix}^{-1} \begin{pmatrix} M_r(\lambda_S) \\ M_g(\lambda_S) \\ M_b(\lambda_S) \end{pmatrix} \ .
\tag{10.7}
$$

This visual stimulus has a spectrum comprising three impulses (one per primary), yet has the same visual appearance as the original continuous spectrum – this is the basis of trichromatic matching. The 3×3 matrix is constant, but depends upon the spectral response of the cones to the chosen primaries $(\lambda_r, \lambda_g, \lambda_b)$.

The right-hand side of (10.7) is simply a function of λ_S which we can write in an even more compact form

$$
\begin{pmatrix} R \\ G \\ B \end{pmatrix} = \begin{pmatrix} \bar{r}(\lambda_S) \\ \bar{g}(\lambda_S) \\ \bar{b}(\lambda_S) \end{pmatrix}
\tag{10.8}
$$

where $\bar{r}(\lambda), \bar{g}(\lambda), \bar{b}(\lambda)$ are known as color matching functions. These functions have been empirically determined from human test subjects and tabulated for the

Excurse 10.11: Color Matching Experiments

Color matching experiments are performed using a light source comprising three adjustable lamps that correspond to the primary colors and whose intensity can be individually adjusted. The lights are mixed and diffused and compared to some test color. In color matching notation the primaries, the lamps, are denoted by **R**, **G** and **B**, and their intensities are R, G and B respectively. The three lamp intensities are adjusted by a human subject until they appear to match the test color. This is denoted

$$\mathbf{C} \equiv R\mathbf{R} + G\mathbf{G} + B\mathbf{B}$$

which is read as the visual stimulus **C** (the test color) is matched by, or looks the same as, a mixture of the three primaries with brightness R, G and B. The notation $R\mathbf{R}$ can be considered as the lamp **R** at intensity R.

Experiments show that color matching obeys the algebraic rules of additivity and linearity which are known as Grassmann's laws. For example two light stimuli $\mathbf{C}_1$ and $\mathbf{C}_2$

$$\mathbf{C}_1 \equiv R_1\mathbf{R} + G_1\mathbf{G} + B_1\mathbf{B}$$
$$\mathbf{C}_2 \equiv R_2\mathbf{R} + G_2\mathbf{G} + B_2\mathbf{B}$$

when mixed will match

$$\mathbf{C}_1 + \mathbf{C}_2 \equiv (R_1 + R_2)\mathbf{R} + (G_1 + G_2)\mathbf{G} + (B_1 + B_2)\mathbf{B} \ .$$

Fig. 10.10 The 1931 color matching functions for the standard observer, based on the CIE 1976 standard primaries

standard CIE primaries listed in Tab. 10.2. They can be loaded as follows

```
>> lambda = [400:700]*1e-9;
>> cmf = cmfrgb(lambda);
>> plot(lambda,cmf)
```

and are shown graphically in Fig. 10.10. Each curve indicates how much of the corresponding primary is required to match the monochromatic light of wavelength λ.

For example, to create the sensation of light at 500 nm (green) we would need

```
>> green = cmfrgb(500e-9)
green =
   -0.0714    0.0854    0.0478
```

Surprisingly, this requires a significant *negative* amount of the red primary and this is problematic since a light source cannot have a negative luminance.

We reconcile this by adding some white light ($R = G = B = w$, see ▶ Sect. 10.2.8) so that the tristimulus values are all positive. For instance

```
>> white = -min(green)*[1 1 1]
white =
    0.0714    0.0714    0.0714
>> feasible_green = green + white
feasible_green =
         0    0.1567    0.1191
```

If we looked at this color side-by-side with the desired 500 nm green we would say that the generated color had the correct hue but was not as *saturated*. Saturation refers to the purity of the color. Spectral colors are *fully saturated* but become less saturated (more pastel) as increasing amounts of white is added. In this case we have mixed in a stimulus of light-gray light (7 % gray).

This leads to a very important point about color reproduction – it is *not* possible to reproduce every possible color using just three primaries. This makes intuitive sense since a color is properly represented as an infinite-dimensional spectral function and a 3-vector can only approximate it. To understand this more fully we need to consider chromaticity spaces.

The RVC Toolbox function cmfrgb can also compute the CIE tristimulus values for an arbitrary spectrum. The luminance spectrum of the redbrick illuminated by sunlight at ground level was computed in ▶ Sect. 10.1.3, and its tristimulus values are

```
>> rgbBrick = cmfrgb(lambda,L)
rgbBrick =
    0.0155    0.0066    0.0031
```

These are the respective amounts of the three CIE primaries that are perceived – by the average human – as having the same color as the original brick under those lighting conditions.

10.2.5 **Chromaticity Coordinates**

The tristimulus values describe color as well as brightness. For example, if we double R, G and B we would perceive approximately the same color but describe it as brighter.

Relative tristimulus values are obtained by normalizing the tristimulus values

$$r = \frac{R}{R + B + G}, \quad g = \frac{G}{R + B + G}, \quad b = \frac{B}{R + B + G} \tag{10.9}$$

which results in chromaticity coordinates r, g and b that are invariant to overall brightness. By definition $r + g + b = 1$ so one coordinate is redundant and typically only r and g are considered. Since the effect of intensity has been eliminated, the 2-dimensional quantity (r, g) represents color, or more precisely, *chromaticity*.

We can plot the locus of spectral colors, the colors of the rainbow, on the chromaticity diagram using a variant of the color matching functions

```
>> [r,g] = lambda2rg([400:700]*1e-9);
>> plot(r,g)
>> rg_addticks
```

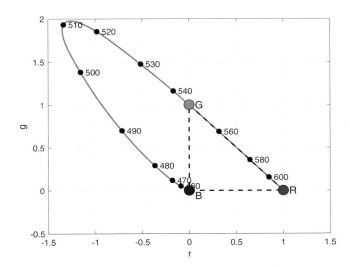

■ **Fig. 10.11** The spectral locus on the r–g chromaticity plane. Monochromatic stimuli lie on the locus and the wavelengths (in nm) are marked. The straight line joining the extremities is the purple boundary and is the locus of saturated purples. All possible colors lie on, or within, this locus. The CIE standard primary colors are marked and the dashed line indicates the gamut of colors that can be represented by these primaries

which results in the horseshoe-shaped curve shown in ■ Fig. 10.11. The `lambda2rg` function computes the color matching function shown in ■ Fig. 10.10 for the specified wavelength and then converts the tristimulus values to chromaticity coordinates using (10.9).

The CIE primaries listed in ■ Tab. 10.2 can be plotted as well

```
>> hold on
>> primaries = lambda2rg(cie_primaries());
>> plot(primaries(:,1),primaries(:,2),"o")
```

and are shown as circles in ■ Fig. 10.11. The complete plot shown in ■ Fig. 10.11, with wavelength annotations, is produced by

```
>> plotSpectralLocus
```

Grassmann's center of gravity law states that a mixture of two colors lies along a line between those two colors on the chromaticity plane. A mixture of N colors lies within a region bounded by those colors. Considered with respect

---

Excurse 10.12: Colorimetric Standards

Colorimetry is a complex topic and standards are very important. Two organizations, CIE and ITU, play a leading role in this area.

The Commission Internationale de L'Éclairage (CIE) or International Commission on Illumination was founded in 1913 and is an independent nonprofit organization that is devoted to worldwide cooperation and the exchange of information on all matters relating to the science and art of light and lighting, color and vision, and image technology. The CIE's eighth session was held at Cambridge, UK, in 1931 and established international agreement on colorimetric specifications and formalized the XYZ color space. The CIE is recognized by ISO as an international standardization body. See ▶ http://www.cie.co.at for more information and CIE datasets.

The International Telecommunication Union (ITU) is an agency of the United Nations and was established to standardize and regulate international radio and telecommunications. It was founded as the International Telegraph Union in Paris on 17 May 1865. The International Radio Consultative Committee or CCIR (Comité Consultatif International des Radiocommunications) became, in 1992, the Radiocommunication Bureau of ITU or ITU-R. It publishes standards and recommendations relevant to colorimetry in its BT series (broadcasting service television). See ▶ http://www.itu.int for more detail.

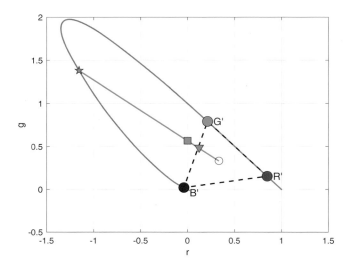

◼ Fig. 10.12 Chromaticity diagram showing the color gamut for nonstandard primaries at 600, 555 and 450 nm. 500 nm green (star), equal-energy white (circle), a feasible green (square) and a displayable green (triangle). The locus of different saturated greens in shown as a green line

We could increase the gamut by choosing different primaries, perhaps using a different green primary would make the gamut larger, but there is the practical constraint of finding a light source (LED or phosphor) that can efficiently produce that color.

to ◼ Fig. 10.11 this has significant implications. Firstly, since all color stimuli are combinations of spectral stimuli, all real color stimuli must lie on or inside the spectral locus. Secondly, any colors we create from mixing the primaries can only lie *within* the triangle bounded by the primaries – the color gamut. It is clear from ◼ Fig. 10.11 that the CIE primaries define only a small subset of all possible colors – within the dashed triangle. Many real colors *cannot* be created using these primaries, in particular the colors of the rainbow which lie on the spectral locus from 460–545 nm. In fact no matter where the primaries are located, not all possible colors can be produced. ◂ In geometric terms, there are no three points within the gamut that form a triangle that includes the entire gamut. Thirdly, we observe that much of the locus requires a negative amount of the red primary and cannot be represented.

We spend much of our life looking at screens, and this theoretical inability to display all real colors is surprising and perplexing. To illustrate how this is resolved, we revisit the problem from ▶ Sect. 10.2.4 which is concerned with displaying 500 nm green – the green we see in a rainbow. ◼ Fig. 10.12 shows the chromaticity of spectral green

```
>> green_cc = lambda2rg(500e-9)
green_cc =
   -1.1558    1.3823
>> plot(green_cc(:,1),green_cc(:,2),"kp")
```

as a star-shaped marker. White is by definition $R = G = B = 1$ and its chromaticity

```
>> white_cc = tristim2cc([1 1 1])
white_cc =
    0.3333    0.3333
>> plot(white_cc(:,1),white_cc(:,2),"o")
```

is plotted as a circle. According to Grassmann's law the mixture of our desired green and white must lie along the indicated green line. The chromaticity of the feasible green computed earlier is indicated by a square, but is outside the *displayable* gamut of the nonstandard primaries used in this example. The least saturated displayable green lies at the intersection of the green line and the gamut boundary and is indicated by the triangular marker.

Earlier we said that there are no three points within the gamut that form a triangle that includes the entire gamut. The CIE therefore proposed, in 1931, a system

◘ Fig. 10.13 **a** The color matching functions for the standard observer, based on the imaginary primaries **X**, **Y** (intensity) and **Z** are tabulated by the CIE. **b** Colors on the *xy*-chromaticity plane

of *imaginary nonphysical primaries* known as **X**, **Y** and **Z** that totally enclose the spectral locus of ◘ Fig. 10.11. **X** and **Z** have zero luminance – the luminance is contributed entirely by **Y** ▶. All real colors can thus be matched by positive amounts of these three primaries. ▶ The corresponding tristimulus values are denoted (x, y, z).

The *XYZ* color matching functions defined by the CIE

```
>> cmf = cmfxyz(lambda);
>> plot(lambda,cmf)
```

are shown graphically in ◘ Fig. 10.13a. This shows the amount of each CIE XYZ primary that is required to match a spectral color and we note that points on these curves always have positive coordinates. The corresponding chromaticity coordinates are

$$x = \frac{X}{X + Y + Z}, \quad y = \frac{Y}{X + Y + Z}, \quad z = \frac{Z}{X + Y + Z} \quad (10.10)$$

and once again $x + y + z = 1$ so only two parameters are required – by convention y is plotted against x in a chromaticity diagram. The spectral locus can be plotted in a similar way as before

```
>> [x,y] = lambda2xy(lambda);
>> plot(x,y)
```

A more sophisticated plot, showing the colors within the spectral locus, can be created

```
>> plotChromaticity
```

and is shown in ◘ Fig. 10.13b. ▶ These coordinates are a *standard* way to represent color for graphics, printing and other purposes. For example the chromaticity coordinates of peak green (550 nm) is

```
>> lambda2xy(550e-9)
ans =
    0.3016    0.6923
```

and the chromaticity coordinates of a standard tungsten illuminant at 2600 K is

```
>> lamp = blackbody(lambda,2600);
>> lambda2xy(lambda,lamp)
ans =
    0.4677    0.4127
```

Luminance here has different meaning to that defined in ▶ Sect. 10.1.3 and can be considered synonymous to brightness.

The units are chosen such that equal quantities of the primaries are required to match the equal-energy white stimulus.

The colors depicted in figures such as ◘ Figs. 10.1 and 10.13b can only approximate the true color due to the gamut limitation of the technology you use to view this book: the inks used to print the page or your computer's display. No display technology has a gamut large enough to present an accurate representation of the chromaticity at every point.

10.2.6 **Color Names**

For example, on Linux, this file is named: `/usr/share/X11/rgb.txt`.

Chromaticity coordinates provide a quantitative way to describe and compare colors, however humans refer to colors by name. Many computer operating systems contain a database or file ◄ that maps human understood names of colors to their corresponding (*R*, *G*, *B*) tristimulus values. The RVC Toolbox provides a copy of such a file and an interface function `colorname`. For example, we can query a color name that includes a particular substring

```
>> colorname("?burnt")
ans =
  1×2 string array
    "burntsienna"    "burntumber"
```

The RGB tristimulus values of burnt Sienna are

```
>> colorname("burntsienna")
ans =
    0.5412    0.2118    0.0588
```

with the values normalized to the interval [0,1]. We could also request *xy*-chromaticity coordinates

```
>> bs = colorname("burntsienna","xy")
bs =
    0.5568    0.3783
```

With reference to ◘ Fig. 10.13, we see that this point is in the red-brown part of the colorspace and not too far from the color of chocolate

```
>> colorname("chocolate","xy")
ans =
    0.5318    0.3988
```

We can also solve the inverse problem. Given the tristimulus values

```
>> colorname([0.2 0.3 0.4])
ans =
    "darkslateblue"
```

we obtain the name of the closest, in Euclidean terms, color.

Excurse 10.13: Colors
Colors are important to human beings and there are over 4000 color-related words in the English language (Steinvall 2002). All languages have words for black and white, and red is the next most likely color word to appear followed by yellow, green, blue and so on. The ancient Greeks only had words for black, white, red and yellowish-green. We also associate colors with emotions, for example red is angry and blue is sad, but this varies across cultures. In Asia, orange is generally a positive color whereas in the west it is the color of road hazards and bulldozers. Technologies, including chemistry, have made a huge number of colors available to us in the last 700 years, yet with this choice comes confusion about color naming – people may not necessarily agree on the linguistic tag to assign to a particular color. (Word cloud created using the `wordcloud` function from the Text Analytics Toolbox™ and based on color frequency data from Steinvall 2002)

10.2.7 Other Color and Chromaticity Spaces

A color space is a 3-dimensional space that contains all possible tristimulus values – all colors and all levels of brightness. There are an infinite number of choices of Cartesian coordinate frame with which to define colors and we have already discussed two of these: RGB and XYZ. However, we could also use polar, spherical or hybrid coordinate systems.

The 2-dimensional rg- or xy-chromaticity spaces do not account for brightness – we normalized it out in (10.9) and (10.10). Brightness, frequently referred to as luminance in this context, is denoted by Y and the definition from ITU Recommendation 709

$$Y^{709} = 0.2126R + 0.7152G + 0.0722B \qquad (10.11)$$

is a weighted sum of the RGB-tristimulus values, and reflects the eye's high sensitivity to green and low sensitivity to blue. Chromaticity plus luminance leads to 3-dimensional color spaces such as rgY or xyY.

Humans seem to more naturally consider chromaticity in terms of two characteristics: hue and saturation. Hue is the dominant color, the closest spectral color, and saturation refers to the purity, or absence of mixed white. Stimuli on the spectral locus are completely saturated while those closer to its centroid are less saturated. The concepts of hue and saturation are illustrated in geometric terms in ◘ Fig. 10.14.

The color spaces that we have discussed lack easy interpretation in terms of hue and saturation, so alternative color spaces have been proposed. The two most commonly known are HSV and CIE L*C*h. In color-space notation H is hue, S is saturation which is also known as C or chroma. The intensity dimension is named either V for value or L for lightness but they are computed quite differently. ▶

There are several functions, including `rgb2hsv`, that can be used to convert between different color spaces. For example the hue, saturation and intensity for

L^* is a nonlinear function of relative luminance and approximates the nonlinear response of the human eye. Value is given by $V = \frac{1}{2}(\min(R, G, B) + \max(R, G, B))$.

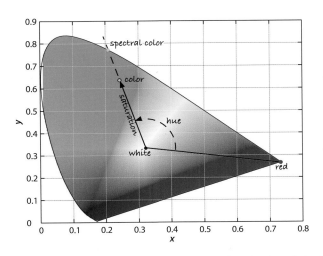

◘ **Fig. 10.14** Hue and saturation. A line is extended from the white point through the chromaticity in question to the spectral locus. The angle of this line is hue, and saturation is the length of the vector normalized with respect to distance to the locus

This function assumes that *RGB* values are linear, that is, not gamma encoded, see ► Sect. 10.3.6. The particular numerical values chosen here are invariant under gamma encoding.

each of pure red, green and blue RGB tristimulus values ◄ are

```
>> rgb2hsv([1 0 0])
ans =
    0    1    1
>> rgb2hsv([0 1 0])
ans =
    0.3333    1.0000    1.0000
>> rgb2hsv([0 0 1])
ans =
    0.6667    1.0000    1.0000
```

In each case the saturation is 1, the colors are pure, and the intensity is 1. As shown in ◘ Fig. 10.14 hue is represented as an angle in the range $[0, 360)°$ with red at $0°$ increasing through the spectral colors associated with decreasing wavelength (orange, yellow, green, blue, violet). If we reduce the amount of the green primary

```
>> rgb2hsv([0 0.5 0])
ans =
    0.3333    1.0000    0.5000
```

For very dark colors numerical problems lead to imprecise hue and saturation coordinates.

we see that intensity drops but hue and saturation are unchanged. ◄ For a medium gray

```
>> rgb2hsv([0.4 0.4 0.4])
ans =
    0    0    0.4000
```

the saturation is zero, it is only a mixture of white, and the hue has no meaning since there is no color. If we add the green to the gray

```
>> rgb2hsv([0 0.5 0] + [0.4 0.4 0.4])
ans =
    0.3333    0.5556    0.9000
```

we have the green hue and a medium saturation value.

The `rgb2hsv` function can also be applied to a color image

```
>> flowers = im2double(imread("flowers4.png"));
>> whos flowers
    Name          Size         Bytes    Class      Attributes
    flowers     426x640x3     6543360   double
```

which is shown in ◘ Fig. 10.15a and comprises several different colored flowers and background greenery. The image `flowers` has 3 dimensions and the third is the color plane that selects the red, green or blue pixels.

To convert the image to hue, saturation and value is simply

```
>> hsv = rgb2hsv(flowers);
>> whos hsv
    Name       Size         Bytes    Class      Attributes
    hsv      426x640x3     6543360   double
```

and the result is another 3-dimensional matrix but this time the color planes represent hue, saturation and value. We can display these planes

```
>> imshow(hsv(:,:,1))
>> imshow(hsv(:,:,2))
>> imshow(hsv(:,:,3))
```

as images which are shown in ◘ Fig. 10.15b, c and d respectively. In the hue image, dark represents red and bright white represents violet. The red flowers appear as both a very small hue angle (dark) and a very large angle close to $360°$. The yellow flowers and the green background can be seen as distinct hue values. The saturation image shows that the red and yellow flowers are highly saturated, while the green leaves and stems are less saturated. The white flowers have very low saturation since, by definition, the color white contains a lot of white.

◘ Fig. 10.15 Flower scene. **a** Original color image; **b** hue image; **c** saturation image. Note that the white flowers have low saturation (they appear dark); **d** intensity or monochrome image; **e** a* image (green to red); **f** b* image (blue to yellow)

A limitation of many color spaces is that the *perceived* color difference between two points is not directly related to their Euclidean distance. In some parts of the chromaticity space two distant points might appear quite similar, whereas in another region two close points might appear quite different. This has led to the development of perceptually uniform color spaces such as the CIE L*u*v*(CIELUV) and L*a*b*(CIELAB) spaces. ▶

To convert this image to L*a*b* color space follows the same pattern

```
>> Lab = rgb2lab(flowers);
>> whos Lab
  Name       Size        Bytes  Class    Attributes
  Lab        426x640x3   6543360  double
```

which again results in an image with 3 dimensions. The chromaticity ▶ is encoded in the a* and b* planes.

```
>> imshow(Lab(:,:,2),[])
>> imshow(Lab(:,:,3),[])
```

and these are shown in ◘ Fig. 10.15e and f respectively. L*a*b* is an opponent color space where a* spans colors from green (black) to red (white) while b* spans blue (black) to yellow (white), with white at the origin where a* = b* = 0.

L*u*v* is a 1976 update to the 1960 CIE Luv uniform color space (UCS). L*a*b* is a 1976 CIE standardization of Hunter's Lab color space defined in 1948.

Relative to a white illuminant, which this function assumes as CIE D_{65} with Y = 1. a*b* are not invariant to overall luminance.

10.2.8 Transforming Between Different Primaries

The CIE standards were defined in 1931 which was well before the introduction of color television in the 1950s. The CIE primaries in ◘ Tab. 10.2 are based on the emission lines of a mercury lamp which are highly repeatable and suitable for laboratory use. Early television receivers used cathode ray tube (CRT) monitors where the primary colors were generated by phosphors that emit light when bombarded by electrons. The phosphors used, and their colors have varied over the years in pursuit of brighter displays. An international agreement, ITU recommendation 709, defines the primaries for high definition television (HDTV) and these are listed in ◘ Tab. 10.3.

This raises the problem of converting tristimulus values from one set of primaries to another. Using the notation we introduced earlier we define two sets of

> **Table 10.3** *xyz*-chromaticity of standard primaries and whites. The CIE primaries of
> **Tab. 10.2** and the more recent ITU recommendation 709 primaries defined for HDTV. D_{65} is
> the white of a blackbody radiator at 6500 K, and E is equal-energy white

	R_{CIE}	G_{CIE}	B_{CIE}	R_{709}	G_{709}	B_{709}	D_{65}	E
x	0.7347	0.2738	0.1666	0.640	0.300	0.150	0.3127	0.3333
y	0.2653	0.7174	0.0089	0.330	0.600	0.060	0.3290	0.3333
z	0.0000	0.0088	0.8245	0.030	0.100	0.790	0.3582	0.3333

primaries: $\mathbf{P}_1, \mathbf{P}_2, \mathbf{P}_3$ with tristimulus values (S_1, S_2, S_3), and $\mathbf{P}'_1, \mathbf{P}'_2, \mathbf{P}'_3$ with tristimulus values (S'_1, S'_2, S'_3). We can always express one set of primaries as a linear combination ◄ of the other

The coefficients can be negative so the new primaries do not have to lie within the gamut of the old primaries.

$$\begin{pmatrix} \mathbf{P}_1 \\ \mathbf{P}_2 \\ \mathbf{P}_3 \end{pmatrix} = \begin{pmatrix} a_{11} & a_{21} & a_{31} \\ a_{12} & a_{22} & a_{32} \\ a_{13} & a_{23} & a_{33} \end{pmatrix} \begin{pmatrix} \mathbf{P}'_1 \\ \mathbf{P}'_2 \\ \mathbf{P}'_3 \end{pmatrix} \tag{10.12}$$

and since the two tristimuli match then

$$\begin{pmatrix} S'_1 & S'_2 & S'_3 \end{pmatrix} \begin{pmatrix} \mathbf{P}'_1 \\ \mathbf{P}'_2 \\ \mathbf{P}'_3 \end{pmatrix} = \begin{pmatrix} S_1 & S_2 & S_3 \end{pmatrix} \begin{pmatrix} \mathbf{P}_1 \\ \mathbf{P}_2 \\ \mathbf{P}_3 \end{pmatrix} \tag{10.13}$$

Substituting (10.12), equating tristimulus values and then transposing we obtain

$$\begin{pmatrix} S'_1 \\ S'_2 \\ S'_3 \end{pmatrix} = \begin{pmatrix} a_{11} & a_{21} & a_{31} \\ a_{12} & a_{22} & a_{32} \\ a_{13} & a_{23} & a_{33} \end{pmatrix}^{\top} \begin{pmatrix} S_1 \\ S_2 \\ S_3 \end{pmatrix} = \mathbf{C} \begin{pmatrix} S_1 \\ S_2 \\ S_3 \end{pmatrix} \tag{10.14}$$

which is simply a linear transformation of tristimulus values.

Consider the concrete problem of transforming from CIE primaries to XYZ tristimulus values. We know from **Tab. 10.3** the CIE primaries in terms of XYZ primaries

```
>> C = [0.7347 0.2653 0; ...
>>      0.2738 0.7174 0.0088; ...
>>      0.1666 0.0089 0.8245]'
C =
    0.7347    0.2738    0.1666
    0.2653    0.7174    0.0089
         0    0.0088    0.8245
```

which is exactly the first three columns of **Tab. 10.3**. The transform is therefore

$$\begin{pmatrix} X \\ Y \\ Z \end{pmatrix} = \mathbf{C} \begin{pmatrix} R \\ G \\ B \end{pmatrix}$$

Recall from ► Sect. 10.2.5 that luminance is contributed entirely by the $\mathbf{Y}$ primary. It is common to apply the constraint that unity R, G, B values result in unity luminance Y and a white with a specified chromaticity. We will choose D_{65} white whose chromaticity is given in **Tab. 10.3** and which we will denote $(x^{\mathrm{w}}, y^{\mathrm{w}}, z^{\mathrm{w}})$. We can now write

$$\frac{1}{y_{\mathrm{w}}} \begin{pmatrix} x_{\mathrm{w}} \\ y_{\mathrm{w}} \\ z_{\mathrm{w}} \end{pmatrix} = \mathbf{C} \begin{pmatrix} J_R & 0 & 0 \\ 0 & J_G & 0 \\ 0 & 0 & J_B \end{pmatrix} \begin{pmatrix} 1 \\ 1 \\ 1 \end{pmatrix}$$

where the left-hand side has $Y = 1$ and we have introduced a diagonal matrix $\mathbf{J}$ which scales the luminance of the primaries. We can solve for the elements of $\mathbf{J}$

$$\begin{pmatrix} J_R \\ J_G \\ J_B \end{pmatrix} = \mathbf{C}^{-1} \begin{pmatrix} x_{\mathrm{w}} \\ y_{\mathrm{w}} \\ z_{\mathrm{w}} \end{pmatrix} \frac{1}{y_{\mathrm{w}}}$$

Substituting real values we obtain

```
>> J = inv(C)*[0.3127 0.3290 0.3582]'*(1/0.3290)
J =
    0.5609
    1.1703
    1.3080
>> C*diag(J)
ans =
    0.4121    0.3204    0.2179
    0.1488    0.8395    0.0116
         0    0.0103    1.0785
```

The middle row of this matrix leads to the luminance relationship

$$Y = 0.1488R + 0.8395G + 0.0116B$$

which is similar to (10.11). The small variation is due to the different primaries used – CIE in this case versus Rec. 709 for (10.11).

The *RGB* tristimulus values of the redbrick were computed earlier and we can determine its *XYZ* tristimulus values

```
>> xyzBrick = C*diag(J)*rgbBrick'
xyzBrick =
    0.0092
    0.0079
    0.0034
```

which we convert to chromaticity coordinates by (10.10)

```
>> chromBrick = tristim2cc(xyzBrick')
chromBrick =
    0.4483    0.3859
```

Referring to ◨ Fig. 10.13b we see that this *xy*-chromaticity lies in the red region and is named

```
>> colorname(chromBrick,"xy")
ans =
    "sandybrown"
```

which is plausible for a "weathered red brick".

10.2.9 What Is White?

In the previous section we touched on the subject of white. White is both the absence of color and also the sum of all colors. One definition of white is *standard daylight* which is taken as the mid-day Sun in Western/Northern Europe which has been tabulated by the CIE as illuminant D_{65}. It can be closely approximated by a blackbody radiator at 6500 K

```
>> d65 = blackbody(lambda,6500);
>> lambda2xy(lambda,d65)
ans =
    0.3136    0.3243
```

which we see is close to the D_{65} chromaticity given in ◨ Tab. 10.3.

Another definition is based on white light being an equal mixture of all spectral colors. This is represented by a uniform spectrum

```
>> ee = ones(size(lambda));
```

which is also known as the equal-energy stimulus and has chromaticity

```
>> lambda2xy(lambda,ee)
ans =
    0.3334    0.3340
```

which is close to the defined value of (1/3, 1/3).

10.3 Advanced Topics

Color is a large and complex subject, and in this section we will briefly introduce a few important remaining topics. Color temperature is a common way to describe the spectrum of an illuminant, and the effect of illumination color on the apparent color of an object is the color constancy problem which is very real for a robot using color cues in an environment with natural lighting. White balancing is one way to overcome this. Another source of color change, in media such as water, is the absorption of certain wavelengths. Most cameras actually implement a nonlinear relationship, called gamma correction, between actual scene luminance and the output tristimulus values. Finally, we look at a more realistic model of surface reflection which has both specular and diffuse components, each with different spectral characteristics.

10.3.1 Color Temperature

Photographers often refer to the color temperature of a light source – the temperature of a black body whose spectrum according to (10.1) is most similar to that of the light source. The color temperature of a number of common lighting conditions are listed in ◘ Tab. 10.4. We describe low-color-temperature illumination as warm – it appears reddish orange to us. High-color-temperature is more harsh – it appears as brilliant white, perhaps with a tinge of blue.

◘ **Table 10.4** Color temperatures of some common light sources

Light source	Color temperature (K)
Candle light	1900
Dawn/dusk sky	2000
40 W tungsten lamp	2600
100 W tungsten lamp	2850
Tungsten halogen lamp	3200
Direct sunlight	5800
Overcast sky	6000–7000
Standard daylight (sun + blue sky)	6500
Hazy sky	8000
Clear blue sky	10,000–30,000

Excurse 10.14: Scene Luminance

Scene luminance is the product of illuminance and reflectance but reflectance is key to scene understanding since it can be used as a proxy for the type of material. Illuminance can vary in intensity and color across the scene and this complicates image understanding. Unfortunately separating luminance into illuminance and reflectance is an ill-posed problem yet humans are able to do this very well as the shown image illustrates – the squares labeled A and B have the same gray level.

The American inventor and founder of Polaroid Corporation Edwin Land (1909–1991) proposed the retinex theory (retinex=retina + cortex) to explain how the human visual system factorizes reflectance from luminance. Land realized that both the eyes and the brain are involved in perceiving colors. Humans can identify the same color in different light levels or conditions. The brain can override raw wavelength information to make us realize what are the actual colors in the observed scene. (Checker shadow illusion courtesy of Edward H. Adelson, ▶ http://persci.mit.edu/gallery)

10.3.2 Color Constancy

Studies show that human perception of what is white is adaptive and has a remarkable ability to *tune out* the effect of scene illumination so that white objects always appear to be white. ▶ For example at night, under a yellowish tungsten lamp, the pages of a book still appear white to us, but a photograph of that scene viewed later under different lighting conditions will look yellow. All of this poses real problems for a robot that is using color to understand the scene, because the observed chromaticity varies with lighting. Outdoors, a robot has to contend with an illumination spectrum that depends on the time of day and cloud cover, as well as colored reflections from buildings and trees. This affects the luminance and apparent color of the object. To illustrate this problem we revisit the red brick

We adapt our perception of color so that the integral, or average, over the entire scene is gray. This works well over a color temperature range 5000–6500 K.

```
>> lambda = [400:10:700]*1e-9;
>> R = loadspectrum(lambda,"redbrick");
```

under two different illumination conditions, the Sun at ground level

```
>> sun = loadspectrum(lambda,"solar");
```

and a tungsten lamp

```
>> lamp = blackbody(lambda,2600);
```

and compute the *xy*-chromaticity for each case

```
>> xy_sun = lambda2xy(lambda,sun.*R)
xy_sun =
    0.4760    0.3784
>> xy_lamp = lambda2xy(lambda,lamp.*R)
xy_lamp =
    0.5724    0.3877
```

and we can see that the chromaticity, or apparent color, has changed significantly. These values are plotted on the chromaticity diagram in ◻ Fig. 10.16.

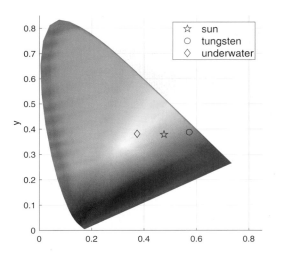

□ **Fig. 10.16** Chromaticity of the red-brick under different illumination conditions

10.3.3 White Balancing

Photographers need to be aware of the illumination color temperature. An incandescent lamp appears more yellow than daylight, so a photographer would place a blue filter on the camera to attenuate the red part of the spectrum to compensate. We can achieve a similar function by choosing the matrix **J**

$$
\begin{pmatrix} R' \\ G' \\ B' \end{pmatrix} = \begin{pmatrix} J_R & 0 & 0 \\ 0 & J_G & 0 \\ 0 & 0 & J_B \end{pmatrix} \begin{pmatrix} R \\ G \\ B \end{pmatrix}
$$

Typically $J_G = 1$ and J_R and J_B are adjusted.

to adjust the gains of the color channels. ◄ For example, increasing J_B would compensate for the lack of blue under tungsten illumination. This is the process of white balancing – ensuring the appropriate chromaticity of objects that we know are white (or gray).

Some cameras allow the user to set the color temperature of the illumination through a menu, typically with options for tungsten, fluorescent, daylight and flash, which select different preset values of **J**. In manual white balancing the camera is pointed at a gray or white object and a button is pressed. The camera adjusts its channel gains **J** so that equal tristimulus values are produced $R' = G' = B'$, which as we recall results in the desired white chromaticity. For colors other than white, this correction introduces some color error but nevertheless has a satisfactory appearance to the eye. Automatic white balancing is commonly used and involves heuristics to estimate the color temperature of the light source, but it can be fooled by scenes with a predominance of a particular color.

The most practical solution is to use the tristimulus values of three objects with known chromaticity in the scene. This allows the matrix C in (10.14) to be estimated directly, mapping the tristimulus values from the sensor to XYZ coordinates which are an absolute lighting-independent representation of surface reflectance. From this the chromaticity of the illumination can also be estimated. This approach has been used for the panoramic camera on Mars rovers where the calibration target shown in □ Fig. 10.17 was imaged periodically to update the white balance under changing Martian illumination.

Fig. 10.17 The calibration target used for the Spirit and Opportunity Mars Rovers' PanCam. Regions of known reflectance and chromaticity (red, yellow, green, blue and shades of gray) are used to set the white balance of the camera. The central stalk has a very low reflectance and also serves as a sundial. In the best traditions of sundials it bears a motto (Image courtesy NASA/JPL/Cornell University/Jim Bell)

10.3.4 Color Change Due to Absorption

A final, and extreme, example of problems with color occurs underwater. Consider a robot trying to find a docking station identified by colored targets. As discussed earlier in ▶ Sect. 10.1.1, water acts as a filter that absorbs more red light than blue light. For an object underwater, this filtering affects both the illumination falling on the object and the reflected light, the luminance, on its way to the camera. Consider again the red brick

```
>> [R,lambda] = loadspectrum([400:5:700]*1e-9,"redbrick");
```

which is now 1 m underwater and with a camera a further 1 m from the brick. The illumination on the water's surface is that of sunlight at ground level

```
>> sun = loadspectrum(lambda,"solar");
```

The absorption spectrum of water is

```
>> A = loadspectrum(lambda,"water");
```

and the total optical path length through the water is

```
>> d = 2;
```

The transmission T is given by Beer's law (10.2).

```
>> T = 10.^(-d*A);
```

and the resulting luminance of the brick is

```
>> L = sun.*R.*T;
```

Excurse 10.15: Lambertian Reflection
A non-mirror-like or matte surface is a diffuse reflector and
the amount of light reflected at a particular angle from the
surface normal is proportional to the cosine of the reflection
angle θ_r. This is known as Lambertian reflection after the
Swiss mathematician and physicist Johann Heinrich Lambert
(1728–1777). A consequence is that the object has the same
apparent brightness at all viewing angles. See also specular
reflection in ▶ Exc. 13.11.

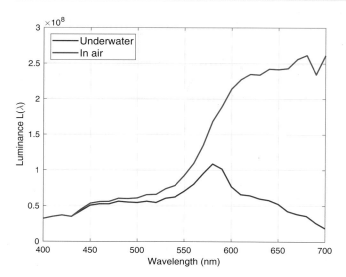

■ **Fig. 10.18** Spectrum of the red brick luminance when viewed underwater. The spectrum without
the water absorption is shown in red

which is shown in ■ Fig. 10.18. We see that the longer wavelengths, the reds, have
been strongly attenuated. The apparent color of the brick is

```
>> xy_water = lambda2xy(lambda,L)
xy_water =
   0.3733   0.3811
```

which is also plotted in the chromaticity diagram of ■ Fig. 10.16. The brick appears much more blue than it did before. In reality, underwater imaging is more
complex than this due to the scattering of light by tiny suspended particles which
reflect ambient light into the camera that has not been reflected from the target.

10.3.5 Dichromatic Reflection

The simple reflection model introduced in ▶ Sect. 10.1.3 is suitable for objects with
matte surfaces (e.g. paper, unfinished wood) but if the surface is somewhat shiny
the light reflected from the object will have two components – the dichromatic
reflection model – as shown in ■ Fig. 10.19a. One component is the illuminant
specularly reflected from the surface without spectral change – the interface or
Fresnel reflection. The other is light that interacts with the surface: penetrating,

Fig. 10.19 Dichromatic reflection. **a** Some incoming light undergoes specular reflection from the surface, while light that penetrates the surface is scattered, filtered and re-emitted in all directions according to the Lambertian reflection model. **b** Specular surface reflection can be seen clearly in the nonred highlight areas on the two tomatoes, these are reflections of the ceiling lights (Image courtesy of Distributed Robot Garden project, MIT)

scattering, undergoing selective spectral absorbance and being re-emitted in all directions as modeled by Lambertian reflection. The relative amounts of these two components depend on the material and the geometry of the light source, observer and surface normal.

A good example of this can be seen in ◨ Fig. 10.19b. Both tomatoes appear red which is due to the scattering lightpath where the light has interacted with the surface of the fruit. However, each fruit has an area of specular reflection that appears to be white, the color of the light source, not the surface of the fruit.

The real world is more complex due to inter-reflections. For example, green light reflected from the leaves will fall on the red fruit and be scattered. Some of that light will be reflected off the green leaves again, and so on – nearby objects influence each other's color in complex ways. To achieve photorealistic results in computer graphics all these effects need to be modeled based on detailed knowledge of surface reflection properties and the geometry of all surfaces. In robotics we rarely have this information, so we need to develop algorithms that are robust to these effects.

10.3.6 Gamma

CRT monitors were once ubiquitous and the luminance produced at the face of the display was nonlinearly related to the control voltage V according to

$$L = V^\gamma \tag{10.15}$$

where $\gamma \approx 2.2$. To correct for this, early video cameras applied the inverse nonlinearity $V = L^{1/\gamma}$ to their output signal which resulted in a system that was linear from end to end. ▶ Both transformations are commonly referred to as gamma correction though more properly the camera-end operation is gamma encoding and the display-end operation is gamma decoding. ▶

LCD displays have a stronger nonlinearity than CRTs, but correction tables are applied within the display to make it follow the *standard* $\gamma = 2.2$ behavior of the obsolete CRT. ▶

Some cameras have an option to choose gamma as either 1 or 0.45 ($= 1/2.2$).

Gamma encoding and decoding are often referred to as gamma compression and gamma decompression respectively, since the encoding operation compresses the range of the signal, while decoding decompresses it.

Macintosh computers are an exception and prior to macOS 10.6 used $\gamma = 1.8$ which made colors appear brighter and more vivid.

◘ Fig. 10.20   The linear intensity wedge

To show the effect of display gamma we create a simple test pattern

```
>> wedge = [0:0.1:1];
>> imshow(wedge,InitialMagnification=10000)
```

that is shown in ◘ Fig. 10.20 and is like a photographer's *grayscale step wedge*. If we display this on our computer screen it will appear differently to the one printed in the book. We will most likely observe a large change in brightness between the second and third block – the effect of the gamma decoding nonlinearity (10.15) in the display of your computer.

If we apply gamma encoding

```
>> imshow(wedge.^(1/2.2),InitialMagnification=10000)
```

we observe that the intensity changes appear to be more linear and closer to the one printed in the book.

❶ Color Spaces and Gamma

The chromaticity coordinates of (10.9) and (10.10) are computed as ratios of tristimulus values which are linearly related to luminance in the scene. The nonlinearity applied to the camera output must be corrected, gamma decoded, *before* any colometric operations. The function `imadjust` performs this operation.

The JPEG file header (JFIF file format) has a tag `Color Space` which is set to either `sRGB` or `Uncalibrated` if the gamma or color model is not known. See ► Sect. 12.1.1.

Today most digital cameras ◄ encode images in sRGB format (IEC 61966-2-1 standard) which uses the ITU Rec. 709 primaries and a gamma encoding function of

$$E' = \begin{cases} 12.92L, & L \leq 0.0031308 \\ 1.055L^{1/2.4} - 0.055, & L > 0.0031308 \end{cases}$$

which comprise a linear function for small values and a power law for larger values. The overall gamma is approximately 1/2.2.

❶ Video Color Spaces: YUV and YC_BC_R

For most colorspaces such as *HSV* or *xyY* the chromaticity coordinates are invariant to changes in intensity. Many digital video devices provide output in YUV or YC_BC_R format. It has a luminance component Y and two chromaticity components which are color difference signals such that $U, C_B \propto B' - Y'$ and $V, C_R \propto R' - Y'$ where R', B' are gamma-*encoded* tristimulus values, and Y' is gamma-*encoded* intensity. The gamma nonlinearity means that UV or C_BC_R will not be a constant as overall lighting level changes.

The tristimulus values from the camera must be first converted to linear tristimulus values, by applying the appropriate gamma decoding, and then computing chromaticity.

10.4   Application: Color Images

10.4.1   Comparing Color Spaces

In this section we bring together many of the concepts and tools introduced in this chapter. We will compare the chromaticity coordinates of the colored squares (top three rows) of the Color Checker chart shown in ◘ Fig. 10.21 using the *xy*- and $L^*a^*b^*$-color spaces. We compute chromaticity from first principles using the spectral reflectance information for each square

```
>> lambda = [400:5:700]*1e-9;
>> macbeth = loadspectrum(lambda,"macbeth");
```

which has 24 columns, one per square of the test chart. We load the relative power spectrum of the D_{65} standard white illuminant

```
>> d65 = loadspectrum(lambda,"D65")*3e9;
```

and scale it to a brightness comparable to sunlight as shown in ◘ Fig. 10.3a. Then for each nongray square

```
>> XYZ = []; Lab = [];
>> for i=1:18
>>    L = macbeth(:,i).*d65;
>>    tristim = max(cmfrgb(lambda,L),0);
>>    RGB = imadjust(tristim,[],[],0.45);
>>    XYZ(i,:) = rgb2xyz(RGB);
>>    Lab(i,:) = rgb2lab(RGB);
>> end
```

we compute the luminance spectrum (line 3). We then use the CIE color matching functions to determine the eye's tristimulus values response and impose the gamut limits (line 4). Finally, we apply a gamma encoding (line 5) since the `rgb2xyz` and `rgb2lab` functions expect gamma encoded RGB data. This is converted to the XYZ color space (line 6), and the $L^*a^*b^*$ color space (line 7). Next, we convert XYZ to xy by dividing X and Y each by $X + Y + Z$, and extract the a^*b^* columns

```
>> xy = XYZ(:,1:2)./(sum(XYZ,2)*[1 1]);
>> ab = Lab(:,2:3);
```

◘ **Fig. 10.21** The GretagMacbeth ColorChecker® is an array of 24 printed color squares (numbered left to right, top to bottom), which includes different grays and colors as well as spectral simulations of skin, sky, foliage etc. Spectral data for the squares is provided with the RVC Toolbox

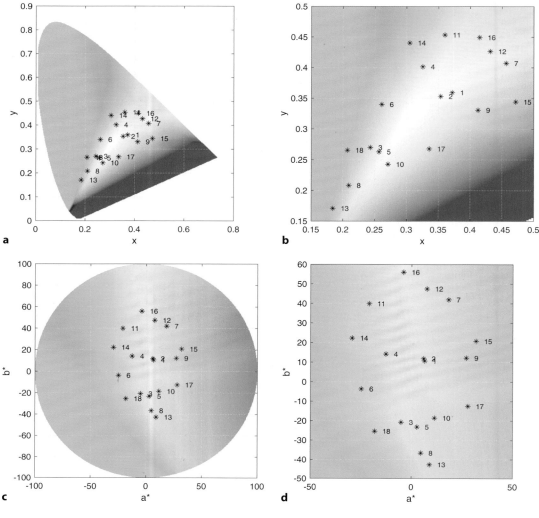

■ **Fig. 10.22** Color Checker chromaticities. **a** *xy*-space; **b** *xy*-space zoomed; **c** *a*b**-space; **d** *a*b**-space zoomed

giving two matrices, each 18 × 2, with one row per colored square. Finally, we plot these points on their respective color planes

```
>> showcolorspace(xy,"xy")
>> showcolorspace(ab,"Lab")
```

and the results are displayed in ■ Fig. 10.22. We see, for example, that square 15 is closer to 9 and further from 7 in the a*b* plane. The L*a*b* color space was designed so that the Euclidean distance between points is approximately proportional to the color difference perceived by humans. If we are using algorithms to distinguish objects by color, then L*a*b* would be preferred over RGB or XYZ.

10.4.2 Shadow Removal

For a robot vision system that operates outdoors, shadows are a significant problem as we can see in ■ Fig. 10.23a. Shadows cause surfaces of the same type to appear quite different and this is problematic for a robot trying to use vision to understand the scene and plan where to drive. Even more problematic is that this effect is not constant; it varies with the time of day and cloud condition. The image in

◘ Fig. 10.23 Shadows create confounding effects in images. **a** View of a park with strong shadows; **b** the shadow invariant image in which the lighting variation has been almost entirely removed (Corke et al. 2013)

◘ Fig. 10.23b has had the effects of shadowing removed, and we can now see very clearly the different types of terrain – grass and gravel.

The key to removing shadows comes from the observation that the bright parts of the scene are illuminated directly by the sun while the darker shadowed regions are illuminated by the sky. Both the sun and the sky can be modeled as blackbody radiators with color temperatures as listed in ◘ Tab. 10.4. Shadows therefore have two defining characteristics: they are dark and they have a slight blue tint.

We model the camera using (10.4) but assume that the spectral response of the camera's color sensors are Dirac functions $M(\lambda) = \delta(\lambda - \lambda_x)$ which allows us to eliminate the integrals

$$R = E(\lambda_R)R(\lambda_R)M_R(\lambda_R)$$
$$G = E(\lambda_G)R(\lambda_G)M_G(\lambda_G)$$
$$B = E(\lambda_B)R(\lambda_B)M_B(\lambda_B)$$

For each pixel we compute chromaticity coordinates $r = R/G$ and $b = B/G$ which are invariant to change in illumination magnitude.

$$r = \frac{E(\lambda_R)R(\lambda_R)M_R(\lambda_R)}{E(\lambda_G)R(\lambda_G)M_R(\lambda_G)} = \frac{\frac{2hc^2}{\lambda^5\left(e^{hc/k\lambda_R T}-1\right)}R(\lambda_R)M_R(\lambda_R)}{\frac{2hc^2}{\lambda^5\left(e^{hc/k\lambda_G T}-1\right)}R(\lambda_G)M_R(\lambda_G)}$$

To simplify further, we apply the Wien approximation, eliminating the -1 term, which is a reasonable approximation for color temperatures in the range under consideration, and now we can write

$$r \approx \frac{e^{hc/k\lambda_G T}R(\lambda_R)M_R(\lambda_R)}{e^{hc/k\lambda_R T}R(\lambda_G)M_G(\lambda_G)} = e^{hc(1/\lambda_G-1/\lambda_R)/kT}\frac{M_R(\lambda_R)}{M_G(\lambda_G)}\frac{R(\lambda_R)}{R(\lambda_G)}$$

which is a function of color temperature T and various constants: physical constants c, h and k; sensor response wavelength λ_x and magnitude $M_x(\lambda_x)$, and material properties $R(\lambda_x)$. Taking the logarithm, we obtain the very simple form

$$\log r = c_1 - \frac{c_2}{T} \tag{10.16}$$

and repeating the process for blue chromaticity we can write

$$\log b = c'_1 - \frac{c'_2}{T} \tag{10.17}$$

Every color pixel $(R, G, B) \in \mathbb{R}^3$ can be mapped to a point $(\log r, \log b) \in \mathbb{R}^2$ and as the color temperature changes the points will all move along lines with a slope of c_2'/c_2. Therefore a projection onto the orthogonal direction, a line with slope c_2/c_2', results in a 1-dimensional quantity

$$ s = -c_2 \log r + c_2' \log b $$

that is invariant to the color temperature of the illuminant. We can compute this for every pixel in an image

```
>> im = imread("parks.jpg");
>> im = rgb2lin(im);
>> gs = shadowRemoval(im,0.7,"noexp");
>> imshow(gs,[]) % [] - scale display based on range of pixel values
```

and the result is shown in ◻ Fig. 10.23b. The pixels have a grayscale value that is a complex function of material reflectance and camera sensor properties. The arguments to the function are the color image, the slope of the line in radians and a flag to return the logarithm s rather than its exponent.

To achieve this result we have made some approximations and a number of rather strong assumptions: the camera has a linear response from scene luminance to *RGB* tristimulus values, the color channels of the camera have non-overlapping spectral response, and the scene is illuminated by blackbody light sources. The first assumption means that we need to use a camera with $\gamma = 1$ or apply gamma decoding to the image before we proceed. The second is far from true, especially for the red and green channels of a color camera, yet the method works well in practice. The biggest effect is that the points move along a line with a slope different to c_2'/c_2 but we can estimate the slope empirically by looking at a set of shadowed and nonshadowed pixels corresponding to the same material in the scene

```
>> theta = esttheta(im);
```

which will prompt you to select a region and returns an angle which can be passed to the `shadowRemoval` function. The final assumption means that the technique will not work for nonincandescent light sources, or where the scene is partly illuminated by reflections from colored surfaces. More details are provided in the MATLAB® function source code.

10.5   Wrapping Up

We have learned that the light we see is electromagnetic radiation with a mixture of wavelengths, a continuous spectrum, which is modified by reflectance and absorption. The spectrum elicits a response from the eye which we interpret as color – for humans the response are tristimulus values, a 3-vector that represents the outputs of the three different types of cones in our eye. A digital color camera is functionally equivalent. The tristimulus values can be considered as a 1-dimensional brightness coordinate and a 2-dimensional chromaticity coordinate which allows colors to be plotted on a plane. The spectral colors form a locus on this plane and all real colors lie within this locus. Any three primary colors form a triangle on this plane which is the gamut of those primaries. Any color within the triangle can be matched by an appropriate mixture of those primaries. No set of primaries can define a gamut that contains all colors. An alternative set of imaginary primaries, the CIE XYZ system, does contain all real colors and is the standard way to describe colors. Tristimulus values can be transformed using linear transformations to account for different sets of primaries. Nonlinear transformations can be used to describe tristimulus values in terms of human-centric qualities such as hue and saturation. We also discussed the definition of white, color temperature, color constancy, the problem of white

> **Excurse 10.16: Beyond Human Vision**
>
> Consumer cameras are functionally equivalent to the human eye and are sensitive to the visible spectrum and return tristimulus values. More sophisticated cameras do not have these limitations.
>
> **■ ■ Infrared cameras**
>
> are sensitive to infrared radiation and a number of infrared bands are defined by CIE:
> - IR-A (700–1400 nm), near infrared (NIR)
> - IR-B (1400–3000 nm), short-wavelength infrared (SWIR)
> - IR-C (3000 nm–1000 μm) which has subbands:
>   - medium-wavelength (MWIR, 3000–8000 nm)
>   - long-wavelength (LWIR, 8000–15,000 nm), Cameras in this band are also called thermal or thermographic.
>
> **■ ■ Ultraviolet cameras**
>
> are sensitive to ultraviolet region (NUV, 200–380 nm) and are used in industrial applications such as detecting corona discharge from high-voltage electrical systems.
>
> **■ ■ Hyperspectral cameras**
>
> have more than three classes of photoreceptors. They sample the incoming spectrum, typically from infrared to ultraviolet, at tens or even hundreds of wavelengths. Hyperspectral cameras are used for applications including aerial survey classification of land-use and identification of the mineral composition of rocks.

balancing, the nonlinear response of display devices and how this affects the common representation of images and video.

We learned that the colors and brightness we perceive is a function of the light source and the surface properties of the object. While humans are quite able to "factor out" illumination change this remains a significant challenge for robotic vision systems. We finished up by showing how to remove shadows in an outdoor color image.

10.5.1 Further Reading

At face value color is a simple concept that we learn in kindergarten, but as we delve in we find it is a fascinating and complex topic with a massive amount of literature. In this chapter we have only begun to scratch the surface of photometry and colorimetry. Photometry is the part of the science of radiometry concerned with measurement of visible light. It is challenging for engineers and computer scientists since it makes use of uncommon units such as lumen, steradian, nit, candela and lux. One source of complexity is that words like intensity and brightness are synonyms in everyday speech but have very specific meanings in photometry. Colorimetry is the science of color perception, and is also a large and complex area since human perception of color depends on the individual observer, ambient illumination and even the field of view. Colorimetry is however critically important in the design of cameras, computer displays, video equipment and printers. Comprehensive online information about computer vision is available through CVonline at ▶ https://homepages.inf.ed.ac.uk/rbf/CVonline, and the material in this chapter is covered by the section *Image Physics*.

The computer vision textbooks by Gonzalez and Woods (2018) and Forsyth and Ponce (2012) each have a discussion on color and color spaces. The latter also has a discussion on the effects of shading and inter-reflections. The book by Gevers et al. (2012) is a solid introduction to color vision theory and covers the dichromatic reflection model in detail. It also covers computer vision algorithms that deal with the challenges of color constancy. The Retinex theory is described in Land and McCann (1971). Other resources related to color constancy can be found at ▶ http://colorconstancy.com.

Readable and comprehensive books on color science include Koenderink (2010), Roy S. Berns (2019), Hunt (1987) and from a television or engineering perspective Benson (1986). A more conversational approach is given by Hunter and Harold (1987), which also covers other aspects of appearance such as gloss and luster. The CIE standard (Commission Internationale de L'Éclairage 1987) is definitive but hard reading. The work of the CIE is ongoing and its standards are periodically updated at ▶ http://www.cie.co.at. The color matching functions were first tabulated in 1931 and revised in 1964.

Charles Poynton has for a long time maintained excellent online tutorials about color spaces and gamma at ▶ http://www.poynton.com. His book (Poynton 2012) is an excellent and readable introduction to these topics while also discussing digital video systems in great depth.

▪▪ General Interest

Crone (1999) covers the history of theories of human vision and color. How the human visual system works, from the eye to perception, is described in two very readable books Stone (2012) and Gregory (1997). Land and Nilsson (2002) describe the design principles behind animal eyes and how characteristics such as acuity, field of view and low light capability are optimized for different species.

10.5.2 Data Sources

The RVC Toolbox contains a number of data files describing various spectra which are summarized in ◘ Tab. 10.5. Each file has as its first column the wavelength in meters. The files have different wavelength ranges and intervals but the helper function `loadspectrum` interpolates the data to the user specified wavelengths.

Several internet sites contain spectral data in tabular format and this is linked from the book's web site. This includes reflectivity data for over 2000 materials provided by NASA's online ASTER spectral library 2.0 (Baldridge et al. 2009) at ▶ https://speclib.jpl.nasa.gov and the Spectral Database from the University of Eastern Finland Color Research Laboratory at ▶ https://uef.fi/en/spectral. Data on cone responses and CIE color matching functions is available from the Colour & Vision Research Laboratory at University College London at ▶ http://www.cvrl.org. CIE data is also available online at ▶ https://cie.co.at.

◘ **Table 10.5** Various spectra provided with the RVC Toolbox. Relative luminosity values lie in the interval [0,1], and relative spectral power distribution (SPD) are normalized to a value of 1.0 at 550 nm. These files can be loaded using the `loadspectrum` function

Filename	Units	Description
cones	Rel. luminosity	Spectral response of human cones
bb2	Rel. luminosity	Spectral response of Sony ICX 204AK sensor used in Point Grey BumbleBee2 camera
photopic	Rel. luminosity	CIE 1924 photopic response
scotopic	Rel. luminosity	CIE 1951 scoptic response
redbrick	Reflectivity	Reflectivity spectrum of a weathered red brick
macbeth	Reflectivity	Reflectivity of the Gretag-Macbeth Color Checker array (24 squares), see ◘ Fig. 10.21
solar	W m^{-2} m^{-1}	Solar spectrum at ground level
water	m^{-1}	Light absorption spectrum of water
D65	Rel. SPD	CIE Standard D_{65} illuminant

10.5.3 Exercises

1. You are a blackbody radiator! Plot your own blackbody emission spectrum. What is your peak emission frequency? What is the name of that region of the electromagnetic spectrum? What sort of sensor would you use to detect this?
2. Consider a sensor that measures the amount of radiated power P_1 and P_2 at wavelengths λ_1 and λ_2 respectively. Write an equation to give the temperature T of the blackbody in terms of these quantities.
3. Using the Stefan-Boltzman law compute the power emitted per square meter of the Sun's surface. Compute the total power output of the Sun.
4. Use numerical integration to compute the power emitted in the visible band 400–700 nm per square meter of the Sun's surface.
5. Why is the peak luminosity defined as $683\,\mathrm{lmW}^{-1}$?
6. Given typical outdoor illuminance as per ▶ Sect. 10.2.3 determine the luminous intensity of the Sun.
7. Sunlight at ground level. Of the incoming radiant power determine, in percentage terms, the fraction of infrared, visible and ultraviolet light.
8. Use numerical integration to compute the power emitted in the visible band 400–700 nm per square meter for a tungsten lamp at 2600 K. What fraction is this of the total power emitted?
9. Plot and compare the human photopic and scotopic spectral response. Compare the response curves of human cones and the RGB channels of a color camera. Use `cones.dat` and `bb2.dat`.
10. Can you create a metamer for the red brick?
11. Prove Grassmann's center of gravity law mentioned in ▶ Sect. 10.2.4.
12. On the xy-chromaticity plane plot the locus of a blackbody radiator with temperatures in the range 1000–10,000 K.
13. Plot the XYZ primaries on the rg-plane.
14. For ◘ Fig. 10.12 determine the chromaticity of the feasible green.
15. Determine the tristimulus values for the red brick using the Rec. 709 primaries.
16. Take a picture of a white object using incandescent illumination. Determine the average RGB tristimulus values and compute the xy-chromaticity. How far off white is it? Determine the color balance matrix J to correct the chromaticity. What is the chromaticity of the illumination?
17. What is the name of the color of the red brick, from ▶ Sect. 10.2.3 when viewed underwater.
18. Imagine a target like ◘ Fig. 10.17 that has three colored patches of known chromaticity. From their observed chromaticity determine the transform from observed tristimulus values to Rec. 709 primaries. What is the chromaticity of the illumination?
19. Consider an underwater application where a target d meters below the surface is observed through m meters of water, and the water surface is illuminated by sunlight. From the observed chromaticity can you determine the true chromaticity of the target? How sensitive is this estimate to incorrect estimates of m and d? If you knew the true chromaticity of the target could you determine its distance?
20. Is it possible that two different colors look the same under a particular lighting condition? Create an example of colors and lighting that would cause this.
21. Use one of your own pictures and the approach of ▶ Sect. 10.4.1. Can you distinguish different objects in the picture?
22. Show analytically or numerically that scaling tristimulus values has no effect on the chromaticity. What happens if the chromaticity is computed on gamma encoded tristimulus values?
23. Create an interactive tool with sliders for R, G and B that vary the color of a displayed patch. Now modify this for sliders X, Y and Z or x, y and Y.

24. Take a color image and determine how it would appear through 1, 5 and 10 m of water.
25. Determine the names of the colors in the Gretag-Macbeth color checker chart.
26. Plot the color-matching function components shown in ◖ Fig. 10.10 as a 3D curve. Rotate it to see the locus as shown in ◖ Fig. 10.11.
27. What temperature does a blackbody need to be at so that its peak emission matches the peak sensitivity of the human blue cone? Is there any metal still solid at that temperature?
28. Research the retinex algorithm, implement and experiment with it.

10

Images and Image Processing

Contents

11.1   Obtaining an Image – 436

11.2   Image Histograms – 447

11.3   Monadic Operations – 448

11.4   Dyadic Operations – 451

11.5   Spatial Operations – 455

11.6   Mathematical Morphology – 474

11.7   Shape Changing – 482

11.8   Wrapping Up – 488

© The Author(s), under exclusive license to Springer Nature Switzerland AG 2023
P. Corke, W. Jachimczyk, R. Pillat, *Robotics, Vision and Control*, Springer Tracts in Advanced Robotics 147,
https://doi.org/10.1007/978-3-031-07262-8_11

chapter11.mlx

► sn.pub/H0K7ox

Image processing is a computational process that transforms one or more input images into an output image. Image processing is frequently used to enhance an image for *human* viewing or interpretation, for example to improve contrast. Alternatively, and of more interest to robotics, it is the foundation for the process of feature extraction which will be discussed in much more detail in the next chapter.

An image is a rectangular array of picture elements (pixels) so we will use a MATLAB® matrix to represent an image in the workspace. This allows us to use MATLAB's powerful and efficient armory of matrix operators and functions.

We start in ► Sect. 11.1 by describing how to load images from sources such as files (images and videos), cameras, and the internet. Next, in ► Sect. 11.2, we introduce image histograms which provide information about the distribution of pixel values. We then discuss various classes of image processing algorithms. These algorithms operate pixel-wise on a single image, a pair of images, or on local groups of pixels within an image, and we refer to these as monadic, dyadic, and spatial operations respectively. Monadic and dyadic operations are covered in ► Sects. 11.3 and 11.4. Spatial operators are described in ► Sect. 11.5 and include operations such as smoothing, edge detection, and template matching. A closely related technique is shape-specific filtering, or mathematical morphology, and this is described in ► Sect. 11.6. Finally, in ► Sect. 11.7 we discuss shape changing operations such as cropping, shrinking, expanding, as well as more complex operations such as scaling, rotation and generalized image warping.

Robots will always gather imperfect images of the world due to noise, shadows, reflections and uneven illumination. In this chapter we discuss some fundamental tools and "tricks of the trade" that can be applied to real-world images.

11.1   Obtaining an Image

Today, digital images are ubiquitous since cameras are built into our digital devices and images cost almost nothing to create and share. We each have ever growing personal collections, as well as access to massive online collections of digital images such as Flickr, Instagram and Google Images. We also have access to live image streams from other people's cameras – there are tens of thousands of webcams around the world broadcasting images to the internet, as well images of Earth from space, the Moon and Mars.

11.1.1   Images from Files

Note that most functions, including `imread`, find images and other data files that are either in the current folder or on MALTLAB path, therefore it is often not necessary to specify the full path.

Lossless means that the compressed image, when uncompressed, will be exactly the same as the original image.

The example images are kept within the `images` folder of the RVC Toolbox distribution which is automatically searched by the `imread` function.

Gamma encoding and decoding is discussed in ► Sect. 10.3.6.

We start with images stored in files since it is very likely that you already have lots of images stored on your computer. In this chapter we will work with some images provided with the RVC Toolbox, but you can easily substitute your own images or use images from the Image Processing Toolbox™ or Computer Vision Toolbox™. We import an image using the `imread` function ◄

```
>> street = imread("street.png");
```

The image was read from a file called `street.png` which is in portable network graphics (PNG) format – a lossless compression format ◄ widely used on the internet. ◄ The `imread` function returns a matrix

```
>> whos street
  Name          Size                   Bytes  Class     Attributes
  street       851x1280              1089280  uint8
```

that belongs to the class `uint8` – the elements of the matrix are unsigned 8-bit integers in the interval [0, 255]. The elements are referred to as pixel values or gray values and are the gamma-encoded ◄ luminance of that point in the original scene.

Excurse 11.1: Image File Formats

A very large number of image file formats have been developed and are comprehensively cataloged at ► https://en.wikipedia.org/wiki/Image_file_formats. The most popular is JPEG which is used for digital cameras and webcams. TIFF is common in many computer systems and often used for scanners. PNG and GIF are widely used on the web. The internal format of these files are complex but a large amount of good quality open-source software exists in a variety of languages to read and write such files. MATLAB can read most of these image file formats.

A much simpler set of formats, widely used on Linux systems, are PBM, PGM, and PPM (generically PNM) which represent images without compression, and optionally as readable ASCII text. A host of open-source tools such as ImageMagick provide format conversions and image manipulation under Linux, MacOS X and Windows. (Word cloud created using the `wordcloud` function from the Text Analytics Toolbox™)

For this 8-bit image the pixel values vary from 0 (darkest) to 255 (brightest). The image is shown in ■ Fig. 11.1a. The matrix has 851 rows and 1280 columns. We normally describe the dimensions of an image in terms of its width × height, so this would be a 1280 × 851 pixel image.

🛈 **Pixel Coordinates**

We write the coordinates of a pixel as (u, v) which are the horizontal and vertical coordinates respectively. In MATLAB this is the matrix element (v, u) – note the reversal of coordinates. Note also that the top-left pixel is $(1, 1)$ in MATLAB and not $(0, 0)$. Further explanation of pixel and spatial coordinates in the MATLAB *Image Coordinate System* can be found at ► https://sn.pub/B62lt8.

For example, the pixel at image coordinate $(300, 200)$ is

```
>> street(200,300)
ans =
  uint8
  42
```

which is quite dark – the pixel corresponds to a point in the closest doorway.

There are some subtleties when working with `uint8` values in MATLAB which we can demonstrate by defining two `uint8` values

```
>> a = uint8(100)
a =
  uint8
  100
>> b = uint8(200)
b =
  uint8
  200
```

Arithmetic on `uint8` values obeys the rules of `uint8` arithmetic

```
>> a+b
ans =
  uint8
  255
```

```
>> a-b
ans =
  uint8
    0
```

and values are clipped to the interval 0 to 255. For division

```
>> a/b
ans =
  uint8
    1
```

the result has been rounded up to an integer value. For some image processing operations that we will consider later, it is useful to consider the pixel values as floating-point numbers for which more familiar arithmetic rules apply. In this case each pixel is an 8-byte MATLAB double precision number in the range [0, 1]. To obtain a double precision image, we can apply the function `im2double` to the integer image. It automatically scales the data into the range [0, 1] regardless of the input data type.

```
>> streetd = im2double(street);
>> class(streetd)
ans =
  'double'
```

Since this particular image has no color, it is called a grayscale or monochromatic image.

Two image display tools that we will use often in this part of the book are `imtool` and `imshow`.

```
>> imtool(street)
```

The `imtool` function displays the matrix as an image and allows interactive inspection of pixel values as shown in ◘ Fig. 11.1. Hovering the cursor over the image pixels will display the pixel coordinate and its gray value – integer or floating point – at the bottom left corner of the window. The image can be zoomed using controls in the toolstrip. Additional features of `imtool` let you crop the image, measure distances in pixels, inspect a pixel region and obtain information such as image size, data type, and data range. Pixel region inspection is very handy when you are debugging your own image processing algorithms. It lets you easily look at a rectangular region of pixel values while simultaneously visualizing their colors. `imtool` has many options and these are described in the online documentation. The `imshow` function, on the other hand, has a much simpler interface and is meant to be used from the command line. It also integrates well with other MATLAB graphics functions such as `plot`, `impixelinfo`, and many more.

◘ **Fig. 11.1** The `imtool` image browsing window. The bottom left, shows the coordinate and value of a pixel below the mouse pointer. **a** Grayscale image; **b** color image

We can just as easily load and display a color image

```
>> flowers = imread("flowers8.png");
>> imtool(flowers)
>> whos flowers
  Name          Size              Bytes  Class    Attributes
  flowers       426x640x3        817920  uint8
```

which is a $426 \times 640 \times 3$ matrix of `uint8` values as shown in ◘ Fig. 11.1b. We can think of this as a 426×640 matrix of RGB tristimulus values, each of which is a 3-vector. For example, the pixel at $(318, 276)$

```
>> pix = flowers(276,318,:)
  1×1×3 uint8 array
pix(:,:,1) =
   57
pix(:,:,2) =
   91
pix(:,:,3) =
  198
```

has a tristimulus value $(57, 91, 198)$ but has been displayed by MATLAB in an unusual and noncompact manner. This is because the pixel value is

```
>> size(pix)
ans =
     1  1  3
```

a $1 \times 1 \times 3$ matrix. The first two dimensions are called singleton dimensions and we can *squeeze* them out

```
>> squeeze(pix)' % transpose for display
ans =
  1x3 uint8 row vector
   57    91   198
```

which results in a more familiar 3-vector. This pixel corresponds to one of the small blue flowers and has a large blue component. We can display the image and examine it interactively using `imtool`. Hovering the cursor over a pixel will display its tristimulus value as shown in ◘ Fig. 11.1.

The tristimulus values are of type `uint8` in the range $[0, 255]$ but the image can be converted to double precision values in the range $[0, 1]$ using the `im2double` function, just as for a grayscale image.

The image is a matrix with three dimensions and the third dimension as shown in ◘ Fig. 11.2 is known as the color plane index. For example,

```
>> imtool(flowers(:,:,1))
```

will display the red color plane as a grayscale image that shows the red stimulus at each pixel. The index 2 or 3 would select the green or blue plane respectively. ▶

The red, green, blue color plane order is almost ubiquitous today, but a popular open source library, OpenCV, uses a blue, green, red ordering.

The `imread` function also accepts a URL allowing it to load an image from the web. The function can read most common image file formats including JPEG, TIFF, GIF, PNG, and several more formats. You can obtain a complete list of supported file formats from the `imformats` function.

Many image file formats also contain rich metadata – data about the data in the file. The JPEG files generated by most digital cameras are particularly comprehensive and the metadata can be retrieved using the `imfinfo` function.

```
>> md = imfinfo("roof.jpg")
md =
  struct with fields:
          Filename: '...\images\roof.jpg'
       FileModDate: '10-Mar-2018 20:53:05'
          FileSize: 810777
            Format: 'jpg'
             Width: 2009
```

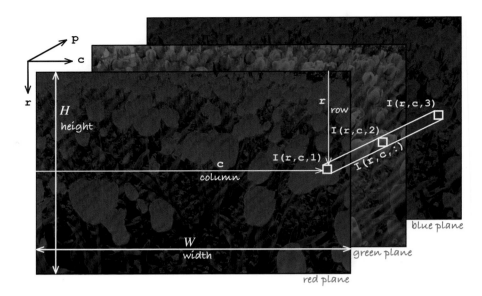

◧ **Fig. 11.2** Color image shown as a 3-dimensional structure with dimensions: row, column, and color plane

```
          Height: 1668
        BitDepth: 24
       ColorType: 'truecolor'
 NumberOfSamples: 3
   CodingMethod: 'Huffman'
  CodingProcess: 'Sequential'
            Make: 'Panasonic'
           Model: 'DMC-FZ30'
     Orientation: 1
     XResolution: 72
     YResolution: 72
  ResolutionUnit: 'Inch'
        DateTime: '2009:05:03 17:41:39'
   DigitalCamera: [1×1 struct]
```

Excurse 11.2: JPEG

The JPEG standard defines a *lossy* compression to reduce the size of the file. Unlike normal file compression (eg. zip, rar, etc.) and decompression, the decompressed image isn't the same as the original image and this allows much greater levels of compression. JPEG compression exploits limitations of the human eye, discarding information that won't be noticed such as very small color changes (which are perceived less accurately than small changes in brightness) and fine texture. It is very important to remember that JPEG is intended for compressing images that will be *viewed by humans*. The loss of color detail and fine texture may be problematic for computer algorithms that analyze images.

JPEG was designed to work well for natural scenes, but it does not do so well on lettering and line drawings with high spatial-frequency content. The degree of loss can be varied by adjusting the "quality factor" which allows a tradeoff between image quality and file size. JPEG can be used for grayscale or color images.

What is commonly referred to as a JPEG file, often with an extension of `.jpg` or `.jpeg`, is more correctly a JPEG EXIF (Exchangeable Image File Format) file. EXIF is the format of the file that holds a JPEG-compressed image as well as metadata such as focus, exposure time, aperture, flash and so on. This metadata can be accessed using the `imfinfo` function, or by command-line utilities such as `exiftool` (▶ https://exiftool.org). See the Independent JPEG group web site ▶ http://www.ijg.org for more details.

and the `DigitalCamera` substructure has additional details about the camera settings for the particular image

```
>> md.DigitalCamera
ans =
  struct with fields:
              ExposureTime: 0.0025
                   FNumber: 7.1000
           ExposureProgram: 'Normal program'
           ISOSpeedRatings: 80
               ExifVersion: [48 50 50 48]
          DateTimeOriginal: '2009:05:03 17:41:39'
         DateTimeDigitized: '2009:05:03 17:41:39'
    CompressedBitsPerPixel: 4
         ExposureBiasValue: 0
         MaxApertureValue: 3
               MeteringMode: 'Pattern'
                       ...
```

More details and options for `imread`, `imformats`, and `imfinfo` are described in the documentation.

11.1.2 Images from an Attached Camera

Most laptop computers today have a builtin camera for video conferencing. For computers without a builtin camera an external camera can be easily attached via a USB connection. The means of accessing a camera is operating system specific. ▶ A list of all attached cameras and their resolution can be obtained by

```
>> webcamlist % list webcams available on a local computer
```

We open a particular camera

```
>> cam = webcam("name") % name is a placeholder for a local webcam
```

which returns an instance of a `webcam` object. If name is not provided the first camera found is used. Once the object is constructed, you can inspect and set a number of properties of your webcam object, such as `Sharpness`, `Contrast`, `Brightness`, and others.

The dimensions of the image returned by the camera are given by the `Resolution` property

```
>> cam.Resolution
```

and an image is obtained using the `snapshot` method

```
>> im = cam.snapshot;
```

which waits until the next frame becomes available. You can also preview a live stream by using the `preview` method

```
>> cam.preview
```

MATLAB provides a simple interface via the `webcam` function to webcams for MacOS, Linux and Windows but more general support requires the Image Acquisition Toolbox™.

11.1.3 Images from a Video File

Image sequences can be read into memory from separate image files using the `imread` function introduced in ▶ Sect. 11.1.1. More commonly, image sequences are stored in a video file and it may not be practical or possible to keep the whole sequence in memory.

11

Excurse 11.3: Photons to Pixel Values

A lot goes on inside a camera. Photosites are the light-sensitive elements of the sensor chip, and are arranged in a dense array. Each photosite is typically square with a side length in the range $1-10\mu\mathrm{m}$. Over a fixed time interval, the number of photons falling on a photosite follows a Poisson distribution where the mean *and* the variance are proportional to the luminance – this variance appears as *shot noise* on the pixel value. A fraction of these photons are converted to electrons – this is the quantum efficiency of the sensor – and they accumulate in a charge well at the photosite. The number of photons captured is proportional to surface area, but not all of a photosite is light sensitive due to the presence of transistors and other devices. The fraction of the photosite's area that is sensitive is called the fill factor and for CMOS sensors can be less than 50%, but this can be improved by fabricating microlenses above each photosite.

The charge well also accumulates thermally generated electrons, the dark current, which is proportional to temperature and is a source of noise – extreme low-light cameras are cooled to reduce the noise. Another source of noise is pixel nonuniformity due to adjacent pixels having a different gain or offset – uniform illumination therefore leads to pixels with different values which appears as additive noise. The charge well has a maximum capacity and with excessive illumination surplus electrons can overflow into adjacent charge wells leading to artifacts such as flaring and streaking in the image.

At the end of the exposure interval the accumulated charge (thermal- and photo-electrons) is read. For low-cost CMOS sensors the charge wells are connected sequentially via a switching network to one or more on-chip analog to digital converters. This results in a rolling shutter and for high speed relative motion this leads to tearing or jello effect as shown in the propeller image below. More expensive CMOS and CCD sensors have a global shutter – they make a temporary snapshot copy of the charge in a buffer which is then digitized sequentially.

The exposure on the sensor is

$$H = qL\frac{T}{N^2} \text{ lx s}$$

where L is scene luminance (in nit), T is exposure time, N is the f-number (inverse aperture diameter) and $q \approx 0.7$ is a function of focal length, lens quality and vignetting. Exposure time T has an upper bound equal to the inverse frame rate. To avoid motion blur, a short exposure time is needed, but this leads to darker and noisier images.

The integer pixel value is

$$x = kH$$

where k is a gain related to the ISO setting[a] of the camera. To obtain an sRGB (see ▶ Sect. 10.3.6) image with an average value of 118[b] the required exposure is

$$H = \frac{10}{S_{\mathrm{SOS}}} \text{ lx s}$$

where S_{SOS} is the ISO rating – standard output sensitivity (SOS) – of the digital camera. Higher ISO increases image brightness by greater amplification of the measured charge, but the various noise sources are also amplified leading to increased image noise which is manifested as graininess.

In photography, the camera settings that control image brightness can be combined into a single exposure value (EV)

$$EV = \log_2 \frac{N^2}{T}$$

and all combinations of f-number and shutter speed that have the same EV value yield the same exposure. This allows a tradeoff between aperture (depth of field) and exposure time (motion blur). For most low-end cameras the aperture is fixed and the camera controls exposure using T instead of relying on an expensive, and slow, mechanical aperture. A difference of 1 EV is a factor of two change in exposure, which photographers refer to as a *stop*. Increasing EV – "stopping down" – results in a darker image. Most DSLR cameras allow you to manually adjust EV relative to what the camera's light meter has determined.

[a] – Which is backward compatible with historical scales (ASA, DIN) devised to reflect the sensitivity of chemical films for cameras – a higher number reflected a more sensitive or "faster" film.

[b] – 18% saturation, middle gray, of 8-bit pixels with gamma of 2.2.

Excurse 11.4: Dynamic Range

The dynamic range of a sensor is the ratio of its largest value to its smallest value. For images, it is useful to express the $\log_2$ of this ratio which makes it equivalent to the photographic concepts of stops or exposure value. Each photosite contains a charge well in which photon-generated electrons are captured during the exposure period (see ▶ Exc. 11.3). The charge well has a finite capacity beyond which the photosite saturates and this defines the maximum value. The minimum number of electrons is not zero, it is a finite number of thermally generated electrons.

An 8-bit image has a dynamic range of around 8 stops, a high-end 10-bit camera has a range of 10 stops, and photographic film is perhaps in the range 10–12 stops but is quite nonlinear.

At a particular state of adaptation, the human eye has a range of 10 stops, but the total adaptation range is an impressive 20 stops. This is achieved by using the iris and slower (tens of minutes) chemical adaptation of the sensitivity of rod cells. Dark adaptation to low luminance is slow, whereas adaptation from dark to bright is faster but sometimes painful.

Excurse 11.5: Video File Formats

Just as for image files there are a large number of different file formats for videos. The most common formats are MPEG4 and AVI. It is important to distinguish between the format of the file (the *container*), and the type of compression (the *codec*). An mp4 file is an MPEG-4 container and uses the H.264 codec. Video codecs compress the individual frames as well as exploiting redundant information between consecutive frames.

MPEG and, AVI and many other video file formats files can be read using the `VideoReader` class. (Word cloud created using the `wordcloud` function from the Text Analytics Toolbox)

MATLAB supports reading frames from a video file stored in any of the popular formats such as AVI, MPEG and MPEG4. For example, we can open a video file

```
>> vid = VideoReader("traffic_sequence.mpg");
vid =
  VideoReader with properties:
    General Properties:
            Name: 'traffic_sequence.mpg'
            Path: '.../images'
        Duration: -1
     CurrentTime: 0
       NumFrames: <Calculating...> learn more
```

```
Video Properties:
                 Width: 720
                Height: 576
             FrameRate: 30
          BitsPerPixel: 24
           VideoFormat: 'RGB24'
```

which returns a `VideoReader` object. There is a delay while the number of video frames is calculated but you can always access it using the `NumFrames` property of the object

```
>> vid.NumFrames
ans =
   351
```

and it shows that the video has 351 frames captured at 30 frames per second. The properties `NumFrames` and `FrameRate` provide the total number of frames and the number of frames per second. The size of each frame within the video is

```
>> [vid.Width,vid.Height]
ans =
   720   576
```

and the next frame is read from the video file by

```
>> im = vid.readFrame();
>> whos im
  Name        Size                 Bytes  Class     Attributes
   im        576x720x3            1244160  uint8
```

which is a 720×576 color image. With these few functions we can write a very simple video player

```
>> videoPlayer = vision.VideoPlayer;
>> cont = vid.hasFrame;
>> while cont
>>    im = vid.readFrame;
>>    videoPlayer.step(im);
>>    cont = vid.hasFrame && videoPlayer.isOpen;
>>    pause(0.05);
>> end
```

The loop terminates when the last frame is read or when the video player window is closed. The `CurrentTime` property of the reader allows you to rewind the video or move to a specific time in the video prior to invoking `readFrame` method. You can also use the `read` method to read a specific frame of the video using a frame number.

11.1.4 Images from the Web

Webcams support a variety of options that can be embedded in the URL and there is no standard for these.

If the .jp domain for this webcam seems puzzling, note that this webcam is in Sweden but can be remotely controlled from Japan (▶ https://uk.jokkmokk.jp/).

The term "web camera" has come to mean any USB-connected local camera, but here we use it in the traditional sense as an *internet*-connected camera that runs a web server that can deliver images on request. There are tens of thousands of these web cameras around the world that are pointed at scenes from the mundane to the spectacular. Given the URL of a webcam ◀ we can display an image from a camera anywhere in the world and place it in a matrix in our MATLAB workspace.

For example, using the `imread` function, we can connect to a camera in the town of Porjus, in Lappland, a province in northernmost Sweden. ◀

```
>> img = imread("http://uk.jokkmokk.jp/photo/nr4/latest.jpg");
```

◻ Fig. 11.3 An image from the Swedish town of Porjus

Excurse 11.6: Aspect Ratio

The aspect ratio of an image is the ratio of its width to its height. It varies widely across different imaging and display technologies. For 35 mm film it is 3 : 2 (1.5) which matches a 4 × 6″ (1.5) print. Other print sizes have different aspect ratios: 5 × 7″ (1.4), and 8 × 10″ (1.25) which require cropping the vertical edges of the image in order to fit.

TV and early computer monitors used 4 : 3 (1.33), for example the ubiquitous 640 × 480 VGA format. HDTV has settled on 16 : 9 (1.78). Modern digital SLR cameras typically use 1.81 which is close to the ratio for HDTV. In movie theaters very-wide images are preferred with aspect ratios of 1.85 or even 2.39. CinemaScope was developed by 20th Century Fox from the work of Henri Chrétien in the 1920s. An anamorphic lens on the camera compresses a wide image into a standard aspect ratio in the camera, and the process is reversed at the projector.

The image size in this case is

```
>> size(img)
ans =
   480   720   3
```

and the image is displayed by

```
>> imshow(img)
```

which returns a color image such as the one shown in ◻ Fig. 11.3. Webcams are configured by their owner to take pictures periodically, anything from once per second to once per minute. Repeated access will return the same image until the camera takes its next picture.

11.1.5 Images from Code

When debugging an algorithm, it can be very helpful to start with a perfect and simple image before moving on to more challenging real-world images. You could

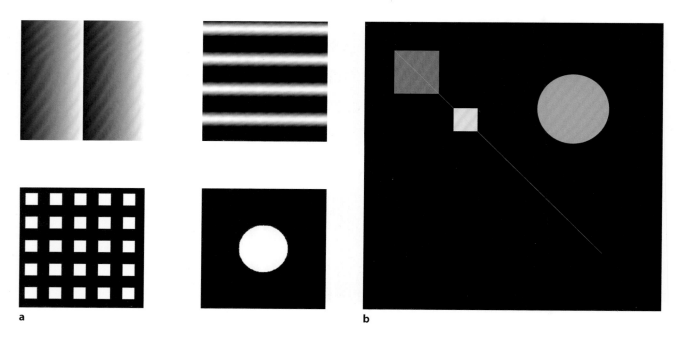

Fig. 11.4 Images from code. **a** Some RVC Toolbox generated test patterns; **b** Simple image created from graphical primitives

11

draw such an image with your favorite drawing package and import it, or construct it using functions shown in this section. The RVC Toolbox function `testpattern` generates simple images with a variety of patterns including lines, grids of dots or squares, intensity ramps and intensity sinusoids. For example

```
>> im = testpattern("rampx",256,2);
>> im = testpattern("siny",256,2);
>> im = testpattern("squares",256,50,25);
>> im = testpattern("dots",256,256,100);
```

are shown in ◻ Fig. 11.4a. The second argument is the size of the created image, in this case they are all 256 × 256 pixels, and the remaining arguments are specific to the type of pattern requested. See the online documentation for details.

We can also construct an image from simple graphical primitives. ◄ First we create a blank `canvas` containing all black pixels (pixel value of zero)

An image/matrix can be edited using the command `openvar("canvas")` which brings up a spreadsheet-like interface.

```
>> canvas = zeros(1000,1000);
```

and then we use the `insertShape` function to add two squares into the canvas. Unlike MATLAB graphics plotting functions, the `insertShape` function modifies the data in the canvas matrix, thus creating a new image.

```
>> canvas = insertShape(canvas,"FilledRectangle",[100 100 150 150], ...
>>    Opacity=1,Color=[0.5 0.5 0.5]);
>> canvas = insertShape(canvas,"FilledRectangle",[300 300 80 80], ...
>>    Opacity=1,Color=[0.9 0.9 0.9]);
```

The first square is 150 × 150 and has pixel values of 0.5 (medium gray). Its upper-left corner is located at [100, 100]. The second square is smaller (80 × 80) but brighter with pixel values of 0.9. The returned canvas is 1000 × 1000 × 3 to accommodate RGB color values. We explicitly override the default opacity of 0.6 so that the color values are not blended with the background. We can also create a circle

```
>> canvas = insertShape(canvas,"FilledCircle",[700 300 120], ...
>>    Opacity=1,Color=[0.7 0.7 0.7]);
```

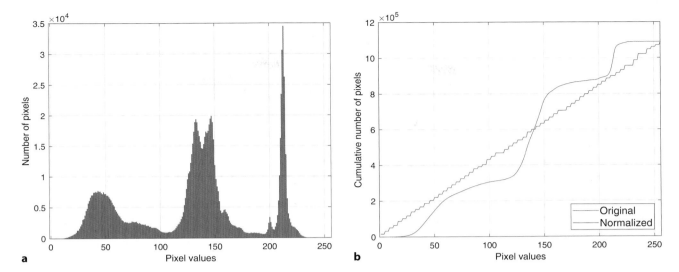

Fig. 11.5 Church scene from ☐ Fig. 11.7a. **a** Histogram, **b** cumulative histogram before and after normalization

centered at [700,300] and of radius 120 pixels with a gray value of 0.7. Finally, we draw a line segment onto our canvas

```
>> canvas = insertShape(canvas,"Line",[100 100 800 800], ...
>>    Opacity=1,Color=[0.8 0.8 0.8]);
```

which extends from (100, 100) to (800, 800) and its pixels are set to gray value of 0.8. The result

```
>> imshow(canvas)
```

is shown in ☐ Fig. 11.4b. We can clearly see that the shapes have different brightness, and we note that the line and the circle show the effects of quantization which results in a *steppy* or jagged shape. ▶

In computer graphics it is common to apply anti-aliasing where edge pixels and edge-adjacent pixels are set to fractional gray values which give the impression of a smoother edge.

11.2 Image Histograms

The distribution of pixel values provides useful information about the quality of the image and the composition of the scene. We obtain the distribution by computing the histogram of the image which indicates the number of times each pixel value occurs. For example, the histogram of the image, shown in ☐ Fig. 11.7a, is computed and displayed by

```
>> church = rgb2gray(imread("church.png"));
>> imhist(church)
```

and the result is shown in ☐ Fig. 11.5a. We see that the pixel values (horizontal axis) span the range from 5 to 238 which is close to the full range of possible values. If the image was underexposed the histogram area would be shifted to the left. If the image was overexposed the histogram would be shifted to the right and many pixels would have the maximum value. A cumulative histogram is shown in ☐ Fig. 11.5b and its use will be discussed in the next section. Histograms can also be computed for color images, in which case the result is three histograms – one for each color channel.

In this case, distribution of pixel values is far from uniform and we see that there are three significant peaks. However, if we look more closely, we see lots of very minor peaks. The concept of a peak depends on the scale at which we consider the data. We can obtain the histogram as a pair of vectors

```
>> [counts,x] = imhist(church);
```

where the elements of `counts` are the number of times gray values occur with the value of the corresponding element of `x`. The function `findpeaks` will automatically find the position of the peaks

```
>> [~,p] = findpeaks(counts,x); % obtain only peak indices
>> size(p)
ans =
    30    1
```

and in this case it found 30 peaks most of which are quite minor. Peaks that are *significant* are not only greater than their immediate neighbors, they are greater than all other values *nearby* – the problem now is to specify what we mean by nearby. For example, the peaks that are separated from all other peaks by at least 60 histogram bins

```
>> [~,p] = findpeaks(counts,x,MinPeakDistance=60);
>> p' % transpose for display
ans =
    43    147    213
```

which are the three significant peaks that we observe by eye. The critical part of finding the peaks is choosing the appropriate criteria for separating them. Peak finding is a topic that we will encounter again later and is also discussed in ▶ App. J.

The peaks in the histogram correspond to populations of pixels in the image. The left-most peak corresponds to the dark pixels which generally belong to the ground and the roof. The middle peak generally corresponds to the sky pixels, and the right-most peak generally corresponds to the white walls. However, each of the scene elements has a distribution of gray values and for most real scenes we cannot simply map gray level to a scene element. For example, some sky pixels are brighter than some wall pixels, and a very small number of ground pixels are brighter than some sky and wall pixels.

11.3 Monadic Operations

Monadic image-processing operations are shown schematically in ◻ Fig. 11.6. The result **Y** is an image of the same size $W \times H$ as the input image **X**, and each output pixel is a function of the corresponding input pixel

$$y_{u,v} = f(x_{u,v}), \quad \forall (u,v) \in \mathbf{X} .$$

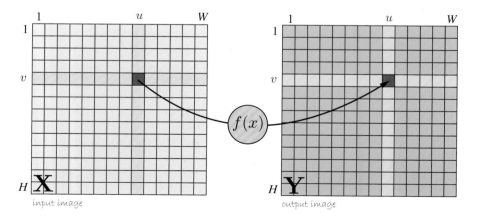

◻ **Fig. 11.6** Monadic image processing operations. Each output pixel is a function of the corresponding input pixel (shown in red)

One useful class of monadic functions changes the type of the pixel data. For example, to change from `uint8` (integer pixels in the range [0, 255]) to double precision values in the range [0, 1] we use the function `im2double`

```
>> imd = im2double(church);
```

and vice versa

```
>> im = im2uint8(imd);
```

A color image has 3-dimensions which we can also consider as a 2-dimensional image where each pixel value is a 3-vector. A monadic operation can convert a color image to a grayscale image where each output pixel value is a scalar representing the luminance of the corresponding input pixel

```
>> flowers = rgb2gray(imread("flowers8.png"));
```

To convert the image into RGB representation we can run

```
>> flowersRGB = repmat(flowers,[1 1 3]);
```

which returns a 3-dimensional RGB image where each color plane is equal to `flowers` – when displayed it still appears as a monochrome image. We can create a color image where the red plane is equal to the input image by

```
>> flowersRGB = zeros(size(flowersRGB),like=flowersRGB);
>> flowersRGB(:,:,1) = flowers;
```

which is a red tinted version of the original image. The `like` option in the call to `zeros` function causes the output array to inherit the data type of `flowersRGB` array.

A very common monadic operation is thresholding. ▶ This is a logical monadic operation which separates the pixels into two classes according to their intensity

```
>> bright = (church >= 180);
>> imshow(bright)
```

and the resulting image is shown in ◘ Fig. 11.7b where all pixels that lie in the interval [180, 255] are shown as white. Such images, where the pixels have only two values are known as binary images. Looking at the image histogram in ◘ Fig. 11.5a we see that the gray value of 180 lies midway between the second and third peak which is a good approximation to the optimal strategy for separating pixels belonging to these two populations.

The variable `bright` is of type logical where the pixels have values of only logical true (1) or false (0). MATLAB automatically converts these to numerical one and zero respectively when used in arithmetic operations and the `imshow` function does likewise.

Many monadic operations are concerned with altering the distribution of gray levels within the image. Sometimes an image does not span the full range of available gray levels, for example the image is under- or over-exposed. We can apply a linear mapping to the gray-scale values

```
>> im = imadjust(church);
```

which ensures that pixel values span the full range ▶ which is either [0, 1] or [0, 255] depending on the class of the image. The `imadjust` function saturates 1% of the data at low and high intensities of the image to eliminate potential outliers. This increases the odds of improving the contrast even for noisy images.

A more sophisticated version is histogram normalization or histogram equalization

```
>> im = histeq(church);
```

which is based on the cumulative distribution

```
>> counts = imhist(church);
>> plot(cumsum(counts))
```

MATLAB provides the Color Thresholder app, a very convenient tool for thresholding images. It can be invoked using the `colorThresholder` function.

The histogram of such an image will have gaps. If M is the maximum possible pixel value, and $N < M$ is the maximum value in the image then the stretched image will have at most N unique pixel values, meaning that $M - N$ values cannot occur.

◘ Fig. 11.7 Some monadic image operations: **a** original image, **b** thresholded, **c** histogram normalized, **d** gamma corrected, **e** after brightness increase, **f** posterized. Inset in each figure is a graph showing the mapping from image gray level on the horizontal axis to the output value on the vertical axis

as shown in ◘ Fig. 11.5b. Mapping the original image via the normalized cumulative distribution ensures that the cumulative distribution of the resulting image is linear – all gray values occur an equal number of times. The result is shown in ◘ Fig. 11.7c and now details of wall and sky texture which had a very small gray-level variation have been accentuated.

> ⊘ **Image enhancement does not add information to the image**
> Operations such as `imadjust` and `histeq` can *enhance* the image from the perspective of a human observer, but it is important to remember that no new information has been added to the image. Subsequent image processing steps will not be improved.

As discussed in ▶ Sect. 10.3.6 the output of a camera is generally gamma encoded so that the pixel value is a nonlinear function L^γ of the luminance sensed at the photosite. Such images can be gamma decoded by a nonlinear monadic operation

```
>> im = imadjust(church,[],[],1/0.45);
```

that raises each pixel to the specified power as shown in ◘ Fig. 11.7d, or

```
>> im = lin2rgb(church);
```

to decode images with the sRGB standard gamma encoding. ▶

Another simple nonlinear monadic operation is posterization or banding. This pop-art effect is achieved by reducing the number of gray levels

```
>> imshow(church/64,[])
```

as shown in ◘ Fig. 11.7f. Since integer division is used, the resulting image has pixels with values in the range [0, 3] and therefore just four different shades of gray. Finally, since an image is represented by a matrix, any MATLAB element-wise matrix function or operator is a monadic operator, for example unary negation, scalar multiplication or addition, or functions such `abs` or `sqrt`. It is worth mentioning that many monadic operations can be sped up by use of a lookup table (LUT). A LUT is a predetermined array of numbers that provides a shortcut for a specific computation. For example, instead of using a potentially expensive computation, all results for uint8 pixel values can be looked up and substituted from a LUT.

When displayed, the resulting image appears to have unnaturally high contrast. The gamma correction has now been applied twice: once by the `imadjust` function and once in the display device.

11.4 Dyadic Operations

Dyadic operations are shown schematically in ◘ Fig. 11.8. Two input images result in a single output image, and all three images are of the same size. Each output pixel is a function of the corresponding pixels in the two input images

$$y_{u,v} = f(x_{1:u,v}, x_{2:u,v}), \quad \forall (u, v) \in \mathbf{X}_1, \mathbf{X}_2 .$$

Examples of common dyadic operations include binary arithmetic operators such as addition, subtraction, element-wise multiplication, or dyadic matrix functions such as `max`, `min`, `atan2`.

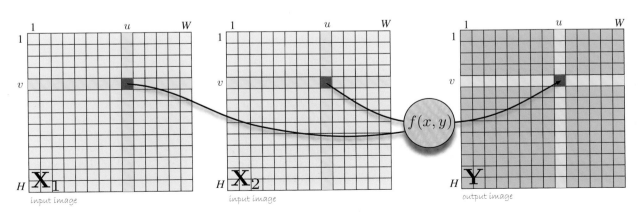

◘ **Fig. 11.8** Dyadic image processing operations. Each output pixel is a function of the two corresponding input pixels (shown in red)

> ❗ Subtracting one `uint8` image from another results in another `uint8` image even though the result is potentially negative. MATLAB quite properly clamps values to the interval $[0, 255]$ so subtracting a larger number from a smaller number will result in zero not a negative value. With addition a result greater than 255 will be set to 255. To remedy this, the images should be first converted to signed integers using the function `cast` or to floating-point values using the `im2double` function.

We will illustrate dyadic operations with two examples.

11.4.1 Application: Chroma Keying

The first example is chroma-keying – a technique commonly used in television to superimpose the image of a person over some background, for example a weather presenter superimposed over a weather map. The subject is filmed against a blue or green background which makes it quite easy, using just the pixel values, to distinguish between background and the subject. We load an image of a subject taken in front of a green screen

```
>> subject = im2double(imread("greenscreen.jpg"));
>> imshow(subject)
```

and this is shown in ◨ Fig. 11.9a. We convert the image to L*a*b* color space which makes it simpler to separate the green background in the a* channel. The negative values in the a* channel correspond to green colors while the positive values correspond to red colors. The easiest way to choose a suitable colorspace and thresholds to segment an image based on color is to use the Color Thresholder (`colorThresholder`) app. The app provides a choice of color spaces in which you can perform thresholding. It then lets you interactively set the thresholds using sliders movable over color histograms corresponding to each color channel.

```
>> imlab = rgb2lab(subject);
>> astar = imlab(:,:,2);
```

In this case, the a* channel alone is sufficient to distinguish the background pixels. A histogram of values

```
>> histogram(astar(:))
```

shown in ◨ Fig. 11.9b indicates a large population of pixels around -32 which is the background and another population which belongs to the subject. We can safely say that the subject corresponds to any pixel for which a* > -15 and create a *mask* image

```
>> mask = astar > -15;
>> imshow(mask)
```

where a pixel is true (equal to one and displayed as white) if it is part of the subject as shown in ◨ Fig. 11.9c.

The image of the subject without the background is

```
>> imshow(mask.*subject);
```

Note that `mask` is 2-dimensional and `subject` is 3-dimensional. Despite that mismatch, MATLAB automatically replicates the mask in the third dimension to complete the multiplication. Next, we load the desired background image

```
>> bg = im2double(imread("desertRoad.png"));
```

and scale and crop it to be the same size as our original image

```
>> bg = imresize(bg,size(subject,[1 2]));
```

and display it with a *cutout* for the subject

```
>> imshow(bg.*(1-mask))
```

◻ Fig. 11.9 Chroma-keying. **a** The subject against a green background; **b** a histogram a* channel values; **c** the computed mask image where true is white; **d** the subject masked into a background scene (green screen image courtesy of Fiona Corke)

Finally, we add the subject with no background, to the background with no subject to obtain the subject superimposed over the background

```
>> imshow(subject.*mask + bg.*(1-mask))
```

which is shown in ◻ Fig. 11.9d. The technique will of course fail if the subject contains any colors that match the color of the background. ▸

Distinguishing foreground objects from the background is an important problem in robot vision but the terms foreground and background are ill-defined and application specific. In robotics we rarely have the luxury of a special background as we did for the chroma-key example. We could instead take a picture of the scene without a foreground object present and consider this to be the background, but that requires that we have special knowledge about when the foreground object is not present. It also assumes that the background does not vary over time. Variation is a significant problem in real-world scenes where ambient illumination and shadows change over quite short time intervals, and the scene may be structurally modified over very long time intervals.

In the early days of television a blue screen was used. Today a green background is more popular because of problems that occur with blue eyes and blue denim clothing.

11.4.2 Application: Motion detection

In this example we process an image sequence and *estimate* the background, even though there are a number of objects moving in the scene. We will use a recursive algorithm that updates the estimated background image $\hat{\mathbf{B}}$ at each time step, based on the previous estimate and the current image

$$\hat{\mathbf{B}}_{\langle k+1 \rangle} \leftarrow \hat{\mathbf{B}}_{\langle k \rangle} + c(\mathbf{X}_{\langle k \rangle} - \hat{\mathbf{B}}_{\langle k \rangle})$$

where k is the time step and $c(\cdot)$ is a monadic image saturation function

$$c(x) = \begin{cases} \sigma, & x > \sigma \\ x, & -\sigma \leq x \leq \sigma \\ -\sigma, & x < -\sigma \end{cases}$$

To demonstrate this, we open a video showing traffic moving through an intersection

```
>> vid = VideoReader("traffic_sequence.mpg");
```

and initialize the background to the first image in the sequence

```
>> bg = im2double(im2gray(vid.readFrame()));
```

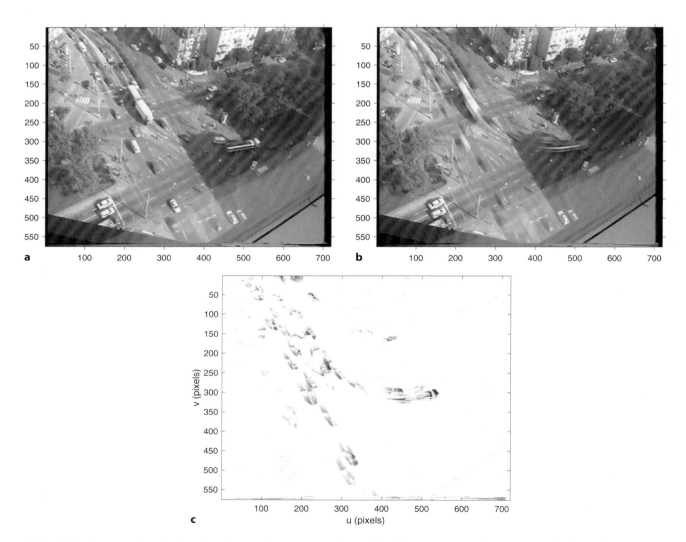

□ **Fig. 11.10** Example of motion detection for the traffic sequence at frame 200. **a** The current image; **b** the estimated background image; **c** the difference between the current and estimated background images where white is zero, red and blue are negative and positive values respectively and magnitude is indicated by color intensity which was converted to grayscale double format. The main loop is

```
>> sigma = 0.02;
>> while vid.hasFrame()
>>    im = im2double(im2gray(vid.readFrame()));
>>    d = im-bg;
>>    d = max(min(d,sigma),-sigma); % apply c(x)
>>    bg = bg+d;
>>    imshow(bg);
>> end
```

One frame from this sequence is shown in ◻ Fig. 11.10a. The estimated background image shown in ◻ Fig. 11.10b reveals the static elements of the scene and the moving vehicles have become a faint blur. Subtracting the scene from the estimated background creates an image where pixels are bright where they are different to the background as shown in ◻ Fig. 11.10c. Applying a threshold to the absolute value of this difference image shows the area of the image where there is motion. Of course, where the cars are stationary for long enough, they will become part of the background.

11.5 Spatial Operations

Spatial operations are shown schematically in ◻ Fig. 11.11. Each pixel in the output image is a function of all pixels in a *region* surrounding the corresponding pixel in the input image

$$y_{u,v} = f(\mathcal{W}_{u,v}), \quad \forall (u,v) \in \mathbf{X}$$

where $\mathcal{W}_{u,v}$ is known as the window, typically a $w \times w$ square region with odd side length $w = 2h + 1$ where $h \in \mathbb{N}$ is the half-width. In ◻ Fig. 11.11 the window includes all pixels in the red shaded region. Spatial operations are powerful because of the variety of possible functions $f(\cdot)$, linear or nonlinear, that can be applied. The remainder of this section discusses linear spatial operators such as smoothing and edge detection, and some nonlinear functions such as order-statistic filtering and template matching. The following section covers a large and important class of nonlinear spatial operators known as mathematical morphology.

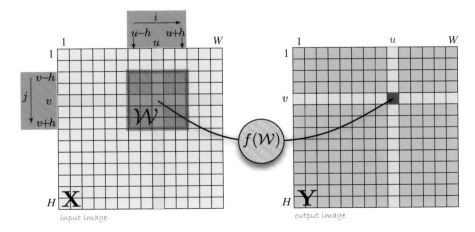

◻ **Fig. 11.11** Spatial image processing operations. The red shaded region shows the window $\mathcal{W}$ that is the set of pixels used to compute the output pixel (shown in red)

11.5.1   Linear Spatial Filtering

A very important linear spatial operator is correlation

$$y_{u,v} = \sum_{(i,j)\in\mathcal{W}} x_{u+i,v+j}\, k_{i,j}, \quad \forall(u,v)\in\mathbf{X} \tag{11.1}$$

where $\mathbf{K}\in\mathbb{R}^{w\times w}$ is the kernel and the elements are referred to as the filter coefficients. For every output pixel the corresponding window of pixels from the input image $\mathcal{W}$ is multiplied element-wise with the kernel $\mathbf{K}$. The center of the window and kernel is coordinate $(0,0)$ and $i,j = -h,\dots,h$. This can be considered as the weighted sum of pixels within the window where the weights are defined by the kernel $\mathbf{K}$. Correlation is often written in operator form as

$$\mathbf{Y} = \mathbf{K}\otimes\mathbf{X}$$

A closely related operation is convolution

$$y_{u,v} = \sum_{(i,j)\in\mathcal{W}} x_{u-i,v-j}\, k_{i,j}, \quad \forall(u,v)\in\mathbf{X} \tag{11.2}$$

where $\mathbf{K}\in\mathbb{R}^{w\times w}$ is the convolution kernel. Note that the sign of the i and j indices has changed in the first term. Convolution is often written in operator form as

$$\mathbf{Y} = \mathbf{K}*\mathbf{X}$$

As we will see, convolution is the workhorse of image processing and the kernel $\mathbf{K}$ can be chosen to perform functions such as smoothing, gradient calculation or edge detection.

Convolution is computationally expensive – an $N\times N$ input image with a $w\times w$ kernel requires $w^2 N^2$ multiplications and additions. Convolution is performed using the function `imfilter`, which is specifically designed for images. ◄

By default, the `imfilter` function performs correlation. Filtering is defined by convolution and the default choice of correlation in `imfilter` was left for historical reasons.

```
>> Y = imfilter(X,K,"conv");
```

If `X` has multiple color planes then so will the output image – each output color plane is the convolution of the corresponding input plane with the kernel `K`.

Excurse 11.7: Correlation or Convolution?

These two terms are often used loosely and they have similar, albeit distinct, definitions. Convolution is the spatial domain equivalent of frequency domain multiplication and the kernel is the impulse response of a frequency domain filter. Convolution also has many useful mathematical properties outlined in ▶ Exc. 11.8.

The difference in indexing between (11.1) and (11.2) is equivalent to reflecting the kernel – flipping it horizontally and vertically about its center point. If you prefer, instead of kernel reflection, you can also think of it as 180-degree rotation of the kernel. Many kernels are symmetric in which case correlation and convolution yield the same result. However edge detection is always based on nonsymmetric kernels so we must take care to apply convolution. We will only use correlation for template matching in ▶ Sect. 11.5.2.

> **Excurse 11.8: Properties of Convolution**
>
> Convolution obeys the familiar rules of algebra. It is commutative
>
> $$\mathbf{A} * \mathbf{B} = \mathbf{B} * \mathbf{A}$$
>
> associative
>
> $$\mathbf{A} * \mathbf{B} * \mathbf{C} = (\mathbf{A} * \mathbf{B}) * \mathbf{C} = \mathbf{A} * (\mathbf{B} * \mathbf{C})$$
>
> distributive (superposition applies)
>
> $$\mathbf{A} * (\mathbf{B} + \mathbf{C}) = \mathbf{A} * \mathbf{B} + \mathbf{A} * \mathbf{C}$$
>
> linear
>
> $$\mathbf{A} * (\alpha\mathbf{B}) = \alpha(\mathbf{A} * \mathbf{B})$$
>
> and shift invariant, which means that the result of the operation is the same everywhere in the image. Shift invariance is the spatial equivalent of time invariance in 1D signal processing.

11.5.1.1 Image Smoothing

Consider a convolution kernel which is a square 21×21 matrix containing equal elements

```
>> K = ones(21,21) / 21^2;
```

and of unit volume, that is, its values sum to one. This kernel is commonly known as a box filter. The result of convolving an image with this kernel is an image where each output pixel is the mean of the pixels in a corresponding 21×21 neighborhood in the input image. As you might expect this averaging

```
>> mona = rgb2gray(imread("monalisa.png"));
>> imshow(imfilter(mona,K,"conv","replicate"));
```

leads to smoothing, blurring or *defocus* ▶ which we see in ◘ Fig. 11.12b. We used the `replicate` padding option of `imfilter` to avoid a black band of pixels on the image border. Boundary effects are discussed in more detail in ▶ Sect. 11.5.1.2. A more suitable kernel for smoothing is the 2-dimensional Gaussian function

Defocus involves a kernel which is a 2-dimensional Airy pattern or sinc function. The Gaussian function is similar in shape, but is always positive whereas the Airy pattern has low amplitude negative going rings.

$$\mathbf{G}(u, v) = \frac{1}{2\pi\sigma^2} e^{-\frac{u^2+v^2}{2\sigma^2}} \tag{11.3}$$

which is symmetric about the origin and the volume under the curve is unity. For the Gaussian we will use a kernel size of 31×31. The spread of the Gaussian is controlled by the standard deviation parameter σ. Applying this kernel to the image

```
>> K = fspecial("gaussian",31,5);
>> imshow(imfilter(mona,K,"replicate"));
```

produces the result shown in ◘ Fig. 11.12c ▶. Here we have specified the standard deviation of the Gaussian to be 5 pixels. Smoothing can be achieved conveniently and quickly by using the highly optimized function `imgaussfilt`

Note that the edge kernels returned by the `fspecial` function are arranged to work with the default operation of `imfilter` which is correlation.

```
>> imshow(imgaussfilt(mona,5))
```

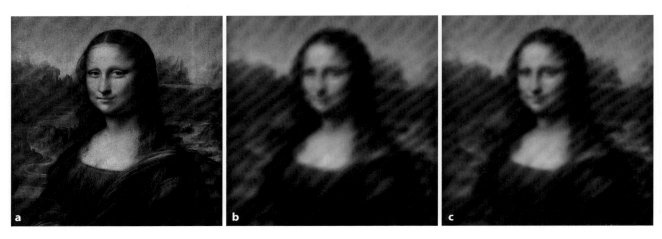

□ Fig. 11.12   Smoothing. **a** Original image; **b** smoothed with a 21×21 averaging kernel; **c** smoothed with a 31×31 Gaussian $G(\sigma = 5)$ kernel

11

Excurse 11.9: How Wide is My Gaussian?

When choosing a Gaussian kernel we need to consider the standard deviation, usually defined by the task, and the dimensions of the kernel $\mathcal{W} \in \mathbb{R}^{w \times w}$ that contains the discrete Gaussian function. Computation time is proportional to w^2 so ideally we want the window to be no bigger than it needs to be. The Gaussian decreases monotonically in all directions but never reaches zero. Therefore, we choose the half-width h of the window such that value of the Gaussian is less than some threshold outside the $w \times w$ convolution window. By default, the `imgaussfilt` function chooses $h = \text{ceil}(2\sigma)$.

Excurse 11.10: Properties of the Gaussian

The Gaussian function $G(\cdot)$ has some special properties. The convolution of two Gaussians is another Gaussian

$$\mathbf{G}(\sigma_1) * \mathbf{G}(\sigma_2) = \mathbf{G}\left(\sqrt{\sigma_1^2 + \sigma_2^2} \right)$$

For the case where $\sigma_1 = \sigma_2 = \sigma$ then

$$\mathbf{G}(\sigma) * \mathbf{G}(\sigma) = \mathbf{G}\left(\sqrt{2}\sigma \right)$$

A Gaussian also has the same shape in the spatial and frequency domains. Additionally, the 2-dimensional Gaussian can be written as the product of two 1-dimensional Gaussians which we can show by

$$\mathbf{G}(u, v) = \left(\frac{1}{\sigma\sqrt{2\pi}} e^{-\frac{u^2}{2\sigma^2}} \right) \left(\frac{1}{\sigma\sqrt{2\pi}} e^{-\frac{v^2}{2\sigma^2}} \right)$$

This implies that convolution with a 2-dimensional Gaussian can be computed by convolving each row with a 1-dimensional Gaussian, and then each column. A filter, such as the Gaussian, that can be decomposed into row and column kernels is called a separable filter. The total number of operations is reduced to $2wN^2$, better by a factor of w. Any filter kernel that has rank of 1 is a separable filter.

Blurring is a counter-intuitive image processing operation since we typically go to a lot of effort to obtain a clear and crisp image. To deliberately *ruin it* seems, at face value, somewhat reckless. However, as we will see later, Gaussian smoothing turns out to be extremely useful.

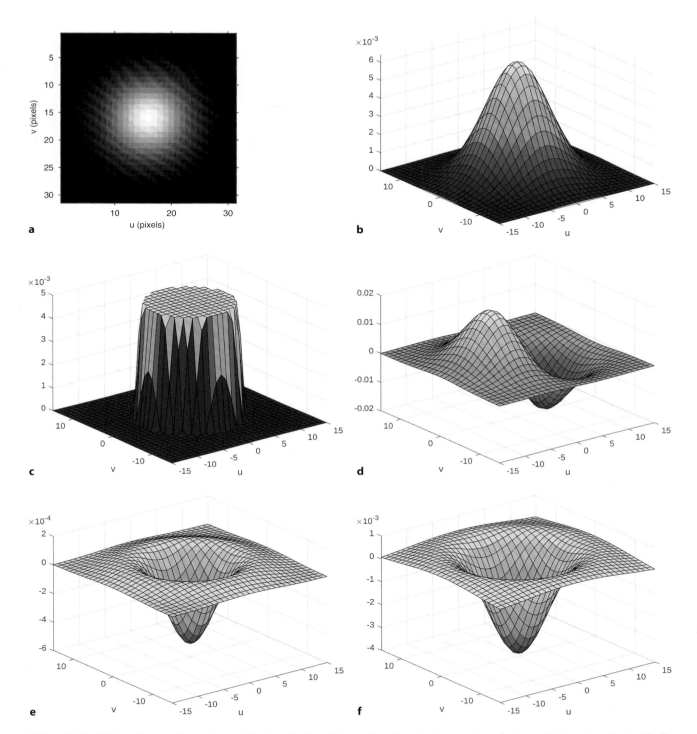

Fig. 11.13 Gallery of commonly used convolution kernels ($h = 15$, $\sigma = 5$). **a** Gaussian shown as intensity image; **b** Gaussian; **c** Top hat (r=8); **d** Derivative of Gaussian; **e** Laplacian of Gaussian (LoG); **f** Difference of Gaussian (DoG)

The kernel is itself a matrix and therefore we can display it as an image

```
>> imshow(K,[],InitialMagnification=800);
```

which is shown in ◘ Fig. 11.13a. Note that we added a second argument `[]` to "stretch the contrast" of K for display, otherwise it would show as black since its

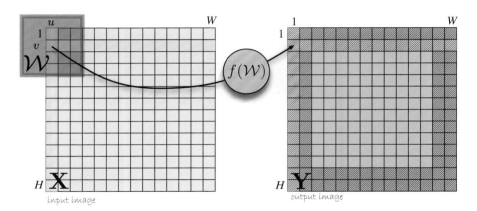

□ **Fig. 11.14**   The output pixel at (u, v) is not defined when the window $\mathcal{W}$ extends beyond the border of the input image. The hatched pixels in the output image are all those for which the output value is not defined

values are small. We clearly see the large value at the center of the kernel and that it falls off smoothly in all directions. We can also display the kernel as a surface

```
>> surfl(-15:15,-15:15,K);
```

as shown in □ Fig. 11.13b. A crude approximation to the Gaussian is the top hat kernel which is cylinder with vertical sides rather than a smooth and gentle fall off in amplitude. The function `fspecial` creates a kernel which can be considered a unit height cylinder of specified radius

```
>> K = fspecial("disk",8);
```

which in this case is 8 pixels as shown in □ Fig. 11.13c.

11.5.1.2   Boundary Processing

A difficulty with all spatial operations occurs when the window is close to the edge of the input image as shown in □ Fig. 11.14. In this case the output pixel is a function of a window that contains pixels *beyond the border* of the input image – these pixels have no defined value. There are several common remedies to this problem. Firstly, we can assume the pixels beyond the image have a particular value. A common choice is zero and this is the default behavior implemented by the function `imfilter`. For example, we can observe the effect of zero padding by reproducing □ Fig. 11.12b without using the `replicate` option, which produces dark borders of the smoothed image due to the influence of the zeros. In most cases, the `replicate` option of `imfiter` is the best and recommended choice since it reduces the border artifacts. With this option, pixels outside of image bounds are set to nearest image border value. The `imfilter` function has a handful of padding options and they are well described in the documentation.

Another option is to consider that the result is invalid when the window crosses the boundary of the image. Invalid output pixels are shown hatched out in □ Fig. 11.14. The result is an output image of size $(W - 2h) \times (H - 2h)$ which is slightly smaller than the input image. This result can be obtained by passing the `"valid"` option to the `conv2` function.

11.5.1.3   Edge Detection

Frequently we are interested in finding the edges of objects in a scene. Consider the image

```
>> castle = im2double(imread("castle.png"));
```

Fig. 11.15 Edge intensity profile. **a** Original image with overlaid line at $v = 360$; **b** graylevel profile along horizontal line $v = 360$; **c** closeup view of the spike at $u \approx 580$; **d** derivative of **c** (Image from the ICDAR 2005 OCR dataset; Lucas 2005)

shown in ◧ Fig. 11.15a. It is informative to look at the pixel values along a 1-dimensional profile through the image. A horizontal profile of the image at $v = 360$ is

```
>> p = castle(360,:);
```

which is a vector that we can plot

```
>> plot(p);
```

against the horizontal coordinate u in ◧ Fig. 11.15b. The clearly visible tall spikes correspond to the white letters and other markings on the sign. Looking at one of the spikes more closely, ◧ Fig. 11.15c, we see the intensity profile across the vertical stem of the letter T. The background intensity ≈ 0.3 and the bright intensity ≈ 0.9 but will depend on lighting levels. However, the very rapid increase over the space of just a few pixels is distinctive and a more reliable indication of an edge than any decision based on the actual gray levels.

The discrete first-order derivative along this cross-section is

$$\frac{dp}{du}\bigg|_u = p_u - p_{u-1}$$

which can be computed using the MATLAB function `diff`

```
>> plot(diff(p))
```

and is shown in ■ Fig. 11.15d. The signal is nominally zero with clear nonzero responses at the edges of an object, in this case the edges of the stem of the letter T.

The derivative at point u can also be written as a *symmetrical* first-order difference

$$\frac{dp}{du}\bigg|_u = \frac{1}{2}(-p_{u-1} + p_{u+1})$$

where the terms have been ordered to match the left-to-right increase in the pixel's u-coordinate. This is equivalent to a dot product or *correlation* with the 1-dimensional kernel of odd length

$$\mathbf{K}_{\text{corr}} = \left(\begin{array}{ccc} -\frac{1}{2} & 0 & \frac{1}{2} \end{array}\right)$$

or *convolution* with the kernel

$$\mathbf{K} = \left(\begin{array}{ccc} \frac{1}{2} & 0 & -\frac{1}{2} \end{array}\right)$$

due to the difference in the indexing between (11.1) and (11.2). Convolving the image with this kernel

```
>> K = [0.5 0 -0.5];
>> imshow(imfilter(castle,K,"conv"),[]);
```

produces a result very similar to that shown in ■ Fig. 11.16b in which vertical edges, high horizontal gradients, are clearly seen.

> ❗ Since this kernel has signed values, the result of the convolution will also be signed, that is, the gradient at a pixel can be positive or negative as shown in ■ Fig. 11.16a, b. The `imshow` function with the second input of `[]`, displays the minimum, most negative, value as black and the maximum, most positive, value as white. Therefore, zero appears as middle gray.

Many convolution kernels have been proposed for computing vertical gradient. A popular choice is the Sobel kernel

```
>> Dv = fspecial("sobel")
Dv =
     1     2     1
     0     0     0
    -1    -2    -1
```

and we see that each column is a scaled version of the 1-dimensional kernel `K` defined above. The overall result is a weighted sum of the vertical gradient for the current column, and the columns to the left and right. Convolving our image with this kernel

```
>> imshow(imfilter(castle,Dv,"conv"),[])
```

generates the vertical gradient image shown in ■ Fig. 11.16a which highlights horizontal edges. Horizontal gradient is computed using the transpose of the kernel

```
>> imshow(imfilter(castle,Dv',"conv"),[])
```

Filters can be designed to respond to edges at any arbitrary angle. The Sobel kernel itself can be considered as an image and rotated using `imrotate`. To obtain angular precision generally requires a larger kernel.

and highlights vertical edges ◄ as shown in ■ Fig. 11.16b. The notation used for gradients varies considerably across the literature. Most commonly the vertical and horizontal gradient are denoted respectively as $\partial\mathbf{X}/\partial v$, $\partial\mathbf{X}/\partial u$; $\nabla_v\mathbf{X}$, $\nabla_u\mathbf{X}$ or $\mathbf{X}_v$, $\mathbf{X}_u$. In operator form this is written

$$\mathbf{X}_v = \mathbf{D}_v * \mathbf{X}$$

$$\mathbf{X}_u = \mathbf{D}_v^\top * \mathbf{X}$$

where $\mathbf{D}_v$ is a vertical gradient kernel such as Sobel.

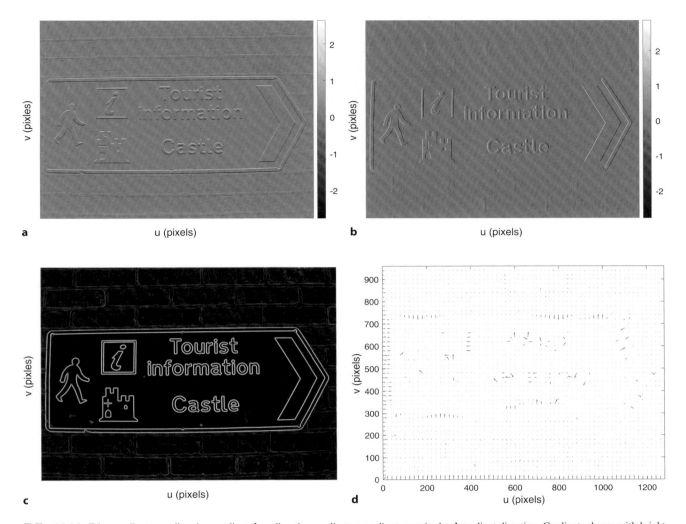

◻ Fig. 11.16 Edge gradient. **a** v-direction gradient; **b** u-direction gradient; **c** gradient magnitude; **d** gradient direction. Gradients shown with bright values as positive, dark values as negative

Taking the derivative of a signal accentuates high-frequency noise, and all images have noise as discussed in ▶ Sect. 11.1.2. At the pixel level noise is a stationary random process – the values are not correlated between pixels. However, the features that we are interested in such as edges have correlated changes in pixel value over a larger spatial scale as shown in ◻ Fig. 11.15c. We can reduce the effect of noise by smoothing the image before taking the derivative

$$\mathbf{X}_v = \mathbf{D}_v * (\mathbf{G}(\sigma) * \mathbf{X})$$

Instead of convolving the image with the Gaussian and *then* the derivative, we exploit the associative property of convolution to write

$$\mathbf{X}_v = \mathbf{D}_v * (\mathbf{G}(\sigma) * I) = (\mathbf{D}_v * \mathbf{G}(\sigma)) * \mathbf{X}$$

We convolve the image with the *derivative of the Gaussian* which can be obtained numerically by

```
>> Xv = imfilter(Dv,fspecial("gaussian",hsize,sigma), ...
>>              "conv","full");
```

or analytically by taking the derivative, in the v-direction, of the Gaussian (11.3) yielding

$$\mathbf{G}_v(u, v) = -\frac{v}{2\pi\sigma^4}e^{-\frac{u^2+v^2}{2\sigma^2}} \; . \tag{11.4}$$

The derivative of Gaussian is shown in ■ Fig. 11.13d. The standard deviation σ controls the *scale* of the edges that are detected. For large σ, which implies increased smoothing, edges due to fine texture will be attenuated, leaving only the edges of large features. This ability to find edges at different spatial scale is important, and underpins the concept of scale space that we will discuss in ▶ Sect. 12.3.2. Another interpretation of this operator is as a spatial *bandpass filter* – a cascade of a low-pass filter (smoothing) with a high-pass filter (differentiation).

Computing the horizontal and vertical components of gradient at each pixel

```
>> derOfGKernel = imfilter(Dv,fspecial("gaussian",5,2), ...
>>    "conv","full");
>> Xv = imfilter(castle,derOfGKernel,"conv");
>> Xu = imfilter(castle,derOfGKernel',"conv");
```

allows us to compute the magnitude of the gradient at each pixel

```
>> m = sqrt(Xu.^2 + Xv.^2);
```

This *edge-strength* image shown in ■ Fig. 11.16c reveals the edges very distinctly. The direction of the gradient at each pixel is

```
>> th = atan2(Xv,Xu);
```

and is best viewed as a sparse quiver plot

```
>> quiver(1:20:size(th,2), 1:20:size(th,1), ...
>>    Xu(1:20:end,1:20:end), Xv(1:20:end,1:20:end))
```

as shown in ■ Fig. 11.16d. The edge direction plot is much noisier than the magnitude plot. Where the edge gradient is strong, on the border of the sign or the edges of letters, the direction is normal to the edge, but the fine-scale brick texture appears as almost random edge direction.

◨ **Fig. 11.17** Closeup of gradient magnitude around the letter T shown as a 3-dimensional surface

A well-known and very effective edge detector is the Canny edge operator. It uses the edge magnitude and direction that we have just computed and performs two additional steps. The first is nonlocal maxima suppression. Consider the gradient magnitude image of ◨ Fig. 11.16c as a 3-dimensional surface where height is proportional to brightness as shown in ◨ Fig. 11.17. We see a series of hills and ridges and we wish to find the pixels that lie along the *ridge lines*. By examining pixel values in a local neighborhood *normal* to the edge direction, that is in the direction of the edge gradient, we can find the maximum value and set all other pixels to zero. The result is a set of nonzero pixels corresponding to peaks and ridge lines. The second step is hysteresis thresholding. For each nonzero pixel that exceeds the upper threshold, a chain is created of adjacent pixels that exceed the lower threshold. Any other pixels are set to zero.

To apply the Canny operator to our example image is straightforward

```
>> edges = edge(castle,"canny");
```

and returns a binary image with edges marked by ones where edges were found and zeros elsewhere as shown in ◨ Fig. 11.18a. We observe that the edges are much thinner than those for the derivative of Gaussian operator which is shown in

a   **b**

■ **Fig. 11.18** Comparison of two edge operators: **a** Canny edge detector with default parameters; **b** Thresholded magnitude of derivative of Gaussian kernel ($\sigma = 2$). The Gaussian derivative magnitude operator requires less computation than Canny but generates thicker edges. For both cases, results are shown inverted, white is zero

■ Fig. 11.18b. In this example, $\sigma = 2$ for the derivative of Gaussian operation. The hysteresis threshold parameters can be set with optional arguments.

So far we have considered an edge as a point of high gradient, and nonlocal-maxima suppression has been used to *search* for the maximum value in local neighborhoods. An alternative means to find the point of maximum gradient is to compute the second derivative and determine where it is equal to zero. The Laplacian operator

$$\nabla^2 \mathbf{X} = \frac{\partial^2 \mathbf{X}}{\partial u^2} + \frac{\partial^2 \mathbf{X}}{\partial v^2} = \mathbf{X}_{uu} + \mathbf{X}_{vv} \tag{11.5}$$

is the sum of the second spatial derivative in the horizontal and vertical directions. This is an isotropic filter that responds equally to edges in any direction.

For a discrete image this can be computed by convolution with the Laplacian kernel ◄

The shape parameter to `fspecial` function for the Laplacian varies the kernel element weights and can be adjusted empirically depending on your needs. The default is 0.2, but here we use 0 to reproduce the classical filter weights. If you were to think of the shape as a paraboloid, it lowers or heightens its extremum.

```
>> shape = 0;
>> L = fspecial("laplacian",shape)
L =
     0     1     0
     1    -4     1
     0     1     0
```

The second derivative is even more sensitive to noise than the first derivative and is again commonly used in conjunction with a Gaussian smoothed image

$$\nabla^2 \mathbf{X} = \mathbf{L} * (\mathbf{G}(\sigma) * \mathbf{X}) = \underbrace{(\mathbf{L} * \mathbf{G}(\sigma))}_{\text{LoG}} * \mathbf{X} \tag{11.6}$$

Excurse 11.13: Difference of Gaussians

The Laplacian of Gaussian (LoG) can be approximated by the difference of two Gaussian functions

$$\mathbf{DoG}(u, v; \sigma_1, \sigma_2) = \mathbf{G}(\sigma_1) - \mathbf{G}(\sigma_2) = \frac{1}{2\pi\sigma_1^2\sigma_2^2}\left(\sigma_2^2 e^{-\frac{u^2+v^2}{2\sigma_1^2}} - \sigma_1^2 e^{-\frac{u^2+v^2}{2\sigma_2^2}}\right)$$

where $\sigma_1 > \sigma_2$ and commonly $\sigma_1 = 1.6\sigma_2$. This kernel is shown in ◘ Fig. 11.13f.

This approximation is useful in scale-space sequences which will be discussed in ▶ Sect. 12.3.2. Consider an image sequence $\mathbf{X}_{\langle k \rangle}$ where $\mathbf{X}_{\langle k+1 \rangle} = \mathbf{G}(\sigma) \otimes \mathbf{X}_{\langle k \rangle}$, that is, the images are increasingly smoothed. The difference between any two images in the sequence is therefore equivalent to $\mathbf{DoG}(\sqrt{2}\sigma, \sigma)$ applied to the original image.

which we combine into the Laplacian of Gaussian (LoG) kernel, and $\mathbf{L}$ is the Laplacian kernel given above. This can be written analytically as

$$\mathbf{LoG}(u, v) = \frac{\partial^2 \mathbf{G}}{\partial u^2} + \frac{\partial^2 \mathbf{G}}{\partial v^2} \tag{11.7}$$

$$= \frac{1}{\pi\sigma^4}\left(\frac{u^2 + v^2}{2\sigma^2} - 1\right)e^{-\frac{u^2+v^2}{2\sigma^2}} \tag{11.8}$$

which is known as the Marr-Hildreth operator or the *Mexican hat* kernel and is shown in ◘ Fig. 11.13e.

We apply this kernel to our image by

```
>> lap = imfilter(castle,fspecial("log",[],2));
```

and the result is shown in ◘ Fig. 11.19a, b. The maximum gradient occurs where the second derivative is zero but a significant edge is a zero crossing from a strong positive value to a strong negative value. Consider the closeup view of the Laplacian of the letter T shown in ◘ Fig. 11.19b. We generate a horizontal cross-section of the stem of the letter T at $v = 360$

```
>> p = lap(360,570:600);
>> plot(570:600,p,"-o");
```

which is shown in ◘ Fig. 11.19c. We see that the zero values of the second derivative lies *between* the pixels. For LoG, the edge function uses a zero crossing detector. It selects pixels adjacent to the zero crossing points

```
>> bw = edge(castle,"log");
```

and this is shown in ◘ Fig. 11.19d. We see that there is only one edge even though there are two neighboring zero crossings for every edge. Referring again to ◘ Fig. 11.19c we observe a weak zero crossing in the interval $u \in [573, 574]$ and a much more definitive zero crossing in the interval $u \in [578, 579]$. The edge function automatically sets the threshold to eliminate weaker zero crossings and thus avoids doubling of the edges.

❗ A fundamental limitation of all edge detection approaches is that intensity edges do not necessarily delineate the boundaries of objects. The object may have poor contrast with the background which results in weak boundary edges. Conversely, the object may have a stripe on it which is not its edge. Shadows frequently have very sharp edges but are not real objects. Object texture will result in a strong output from an edge detector at points not just on its boundary, as for example with the bricks in ◘ Fig. 11.15b.

◘ Fig. 11.19 Laplacian of Gaussian. **a** Laplacian of Gaussian; **b** closeup of **a** around the letter T where bright and dark shades indicate positive and negative values respectively; **c** a horizontal cross-section of the LoG through the stem of the T; **d** closeup of the LoG edge detector output at the letter T

11.5.2 Template Matching

In our discussion so far we have used kernels that represent mathematical functions such as the Gaussian and its derivative, or its Laplacian. We have also considered the convolution kernel as a matrix, as an image and as a 3-dimensional surface as shown in ◘ Fig. 11.13. In this section we will consider that the kernel *is an image*, or a part of an image, which we refer to as a template. In template matching we wish to find which parts of the input image are most similar to the template.

Template matching is shown schematically in ◘ Fig. 11.20. Each pixel in the output image is given by

$$y_{u,v} = s(\mathcal{W}_{u,v}, \mathbf{T}), \quad \forall (u,v) \in \mathbf{X}$$

where $\mathbf{T}$ is the $w \times w$ template, the pattern of pixels we are looking for, with odd side length $w = 2h + 1$, and $\mathcal{W}$ is the $w \times w$ window centered at (u, v) in the input

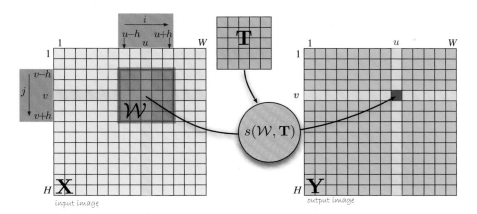

Fig. 11.20 Template matching. The similarity $s(\mathcal{W}, \mathbf{T})$ between the red shape region $\mathcal{W}$ and the template $\mathbf{T}$ is computed as the output pixel (shown in red)

Table 11.1 Similarity measures for two equal sized image regions $\mathbf{A}$ and $\mathbf{B}$. The Z-prefix indicates that the measure accounts for the zero-offset or the difference in mean of the two images (Banks and Corke 2001). $\overline{\mathbf{A}}$ and $\overline{\mathbf{B}}$ are the mean of image regions $\mathbf{A}$ and $\mathbf{B}$ respectively

Sum of absolute differences

SAD	$s = \sum_{(u,v) \in \mathbf{A}, \mathbf{B}} |a_{u,v} - b_{u,v}|$
ZSAD	$s = \sum_{(u,v) \in \mathbf{A}, \mathbf{B}} |(a_{u,v} - \overline{\mathbf{A}}) - (b_{u,v} - \overline{\mathbf{B}})|$

Sum of squared differences

SSD	$s = \sum_{(u,v) \in \mathbf{A}, \mathbf{B}} (a_{u,v} - b_{u,v})^2$
ZSSD	$s = \sum_{(u,v) \in \mathbf{A}, \mathbf{B}} \left((a_{u,v} - \overline{\mathbf{A}}) - (b_{u,v} - \overline{\mathbf{B}})\right)^2$

Cross correlation

NCC	$s = \dfrac{\sum_{(u,v) \in \mathbf{A}, \mathbf{B}} a_{u,v} b_{u,v}}{\sqrt{\sum_{(u,v) \in \mathbf{A}\mathbf{X}} a_{u,v}^2 \cdot \sum_{(u,v) \in \mathbf{B}} b_{u,v}^2}}$
ZNCC	$s = \dfrac{\sum_{(u,v) \in \mathbf{A}, \mathbf{B}} (a_{u,v} - \overline{\mathbf{A}})(b_{u,v} - \overline{\mathbf{B}})}{\sqrt{\sum_{(u,v) \in \mathbf{A}} (a_{u,v} - \overline{\mathbf{A}})^2 \cdot \sum_{(u,v) \in \mathbf{B}} (b_{u,v} - \overline{\mathbf{B}})^2}}$

image. The function $s(\mathbf{A}, \mathbf{B})$ is a scalar measure that describes the *similarity* of two equally-sized images $\mathbf{A}$ and $\mathbf{B}$.

Several common similarity measures ▶ are given in ◻ Tab. 11.1. The most intuitive are computed simply by computing the pixel-wise difference $\mathbf{A} - \mathbf{B}$ and taking the sum of the absolute differences (SAD) or the sum of the squared differences (SSD). These metrics are zero if the images are identical and increase with dissimilarity. It is not easy to say what value of the measure constitutes a poor match, but a ranking of similarity measures can be used to determine the *best* match.

More complex measures such as normalized cross-correlation yield a score in the interval $[-1, +1]$ with $+1$ for identical regions. In practice a value greater than 0.8 is considered to be a good match. Normalized cross correlation is computationally more expensive – requiring multiplication, division and square root operations. Note that it is possible for the result to be undefined if the denominator is zero, which occurs if either $\mathbf{A}$ or $\mathbf{B}$ are all zero for NCC, or uniform (all pixels have the same value) for ZNCC.

If $\mathbf{B} \equiv \mathbf{A}$ then it is easily shown that $\mathrm{SAD} = \mathrm{SSD} = 0$ and $\mathrm{NCC} = 1$ indicating a perfect match. To illustrate we will use the Mona Lisa's eye as a 51×51 template

These measures can be augmented with a Gaussian weighting to de-emphasize the differences that occur at the edges of the two windows.

```
>> mona = im2double(rgb2gray(imread("monalisa.png")));
>> A = mona(170:220,245:295);
```

11

and evaluate the three common measures that we will define as anonymous functions

```
>> sad = @(I1,I2)(sum(abs(I1(:)-I2(:))));
>> ssd = @(I1,I2)(sum((I1(:)-I2(:)).^2));
>> ncc = @(I1,I2)(dot(I1(:),I2(:))/ ...
>>    sqrt(sum(I1(:).^2)*sum((I2(:)).^2)));
>> B = A;
>> sad(A,B)
ans =
      0
>> ssd(A,B)
ans =
      0
>> ncc(A,B)
ans =
      1.0000
```

Now consider the case where the two images are of the same scene, but one image is darker than the other – the illumination or the camera exposure has changed. In this case $\mathbf{B} = \alpha\mathbf{A}$ and now

```
>> B = 0.9*A;
>> sad(A,B)
ans =
    111.1110
>> ssd(A,B)
ans =
      5.6462
```

these measures indicate a degree of dissimilarity. However, the normalized cross-correlation

```
>> ncc(A,B)
ans =
      1.0000
```

is invariant to the change in intensity.

Next, consider that the pixel values have an offset ◀ so that $\mathbf{B} = \mathbf{A} + \beta$ and we find that

This could be due to an incorrect black level setting. A camera's black level is the value of a pixel corresponding to no light and is often > 0.

```
>> B = A+0.1;
>> sad(A,B)
ans =
    260.1000
>> ssd(A,B)
ans =
     26.0100
```

```
>> ncc(A,B)
ans =
    0.9974
```

all measures now indicate a degree of dissimilarity. The problematic offset can be dealt with by first subtracting from each of **A** and **B** their mean value

```
>> submean = @(I)(I-mean(I(:)));
>> zsad = @(I1,I2)(sad(submean(I1),submean(I2)));
>> zssd = @(I1,I2)(ssd(submean(I1),submean(I2)));
>> zncc = @(I1,I2)(ncc(submean(I1),submean(I2)));
>> zsad(A,B)
ans =
    2.4605e-13
>> zssd(A,B)
ans =
    2.4884e-29
>> zncc(A,B)
ans =
    1.0000
```

and these measures now all indicate a perfect match. The z-prefix denotes variants of the similarity measures described above that are invariant to intensity offset. Only the ZNCC measure

```
>> B = 0.9*A + 0.1;
>> zncc(A,B)
ans =
    1.0000
```

is invariant to both gain and offset variation. All these methods will provide imperfect results if the images have a change in relative rotation or scale. Finding the location of a template with a small amount of rotation in the image, approximately $\pm\,10°$, often returns acceptable results. When the rotation is greater or when a significant scale change is present, a different and more complex approach is required.

Consider the problem from the well-known children's book "Where's Wally" or "Where's Waldo" – the fun is trying to find Wally's face in a crowd

```
>> crowd = imread("wheres-wally.png");
>> imshow(crowd)
```

Fortunately, we know roughly what he looks like and the template

```
>> wally = imread("wally.png");
>> imshow(wally);
```

was extracted from a different image and scaled so that the head is approximately the same width as other heads in the crowd scene (around 21 pixel wide).

The similarity of our template `wally` to every possible window location is computed by

```
>> S = normxcorr2(wally,crowd);
```

using the matching measure ZNCC, which we defined earlier. The result

```
>> imshow(S,[]), colorbar
```

is shown in ◗ Fig. 11.21 and the pixel gray level indicates the ZNCC similarity. We can see a number of spots of high similarity (white) which are candidate positions for Wally. The peak values, with respect to a local 3×3 window, are

```
>> peakFinder = vision.LocalMaximaFinder(MaximumNumLocalMaxima=5, ...
>>    NeighborhoodSize=[3 3],Threshold=min(S(:)));
>> uvLoc = peakFinder.step(S);
>> vals = S(sub2ind(size(S),uvLoc(:,2),uvLoc(:,1)))'
vals =
    0.5258    0.5230    0.5222    0.5032    0.5023
```

■ **Fig. 11.21** Similarity image S with top five Wally candidates marked and showing their similarity scores. Note the border of indeterminate values where the template window falls off the edge of the input image

in descending order. The largest value 0.5258 is the similarity of the strongest match found. These matches occur at the coordinates (u, v) given by the uvLoc and we can highlight these points on the scene

```
>> labels = "Wally " + string(1:5) + ": " + string(vals);
>> markedSim = insertObjectAnnotation(S,"circle",[uvLoc, ...
>>    10*ones(5,1)],labels,FontSize=21);
>> imshow(markedSim)
```

using yellow circles that are numbered sequentially and annotated with five highest scores. The best match at $(271, 389)$ is in fact the correct answer – we found Wally! ◄ It is interesting to look at the other highly ranked candidates. Numbers two and three at the bottom of the image are people also wearing baseball caps who look quite similar.

There are some important points to note from this example. The images have quite low resolution and the template is only 21×25 – it is a very crude likeness to Wally. The match is not a strong one – only 0.5258 compared to the maximum possible value of 1.0 and there are several contributing factors. The matching measure is not invariant to scale, that is, as the relative scale (zoom) changes the similarity score falls quite quickly. In practice perhaps a 10–20% change in scale between **T** and $\mathcal{W}$ can be tolerated. For this example the template was only approximately scaled. Secondly, not all Wallys are the same. Wally in the template is facing forward but the Wally we found in the image is looking to our left. Another problem is that the square template typically includes pixels from the background as well as the object of interest. As the object moves the background pixels may change, leading to a lower similarity score. This is known as the mixed pixel problem. Ideally, the template should bound the object of interest as tightly as possible. In practice another problem arises due to perspective distortion. A square pattern of pixels in the center of the image will appear keystone shaped at the edge of the image and thus will match less well with the square template.

A common problem with template matching is that false matches can occur. In the example above the second candidate had a similarity score only 0.5% lower than the first, the fifth candidate was only 5% lower. In practice, several rules are

Note that the normxcorr2 function pads the input image so that the template center slides to the very edge of the input image. The resulting output is therefore larger than the input image.

applied before a match is accepted: the similarity must exceed some threshold and the first candidate must exceed the second candidate by some factor to ensure there is no ambiguity – this is called a ratio test.

Another approach is to bring more information to bear on the problem such as known motion of the camera or object. For example, if we were tracking Wally from frame to frame in an image sequence then we would pick the best Wally closest to the previous location he was found. Alternatively, we could create a motion model, typically a constant velocity model which assumes he moves approximately the same distance and direction from frame to frame. In this way we could predict his future position and pick the Wally closest to that predicted position, or only search in the vicinity of the predicted position to reduce computation. We would also have to deal with practical difficulties such as Wally stopping, changing direction, or being temporarily obscured.

11.5.3 Nonlinear Operations

Another class of spatial operations is based on nonlinear functions of pixels within the window. For example

```
>> out = nlfilter(mona,[7 7],@(x)var(x(:)));
```

computes the variance of the pixels in *every* 7×7 window. The arguments specify the window size and an anonymous function that uses the function `var`. The `var` function is called with a 49×1 vector argument comprising the pixels in the window arranged as a column vector and the function's return value becomes the corresponding output pixel value. This operation acts as an edge detector since it has a low value for homogeneous regions irrespective of their brightness. It is however computationally expensive because the `var` function is called over 470,000 times. Any MATLAB function, builtin or your own M-file, that accepts a matrix input and returns a scalar can be used in this way.

Order filters sort the pixels within the window by value and return the specified element from the sorted list. The maximum value over a 5×5 window about each pixel is the first ordered pixel in the window

```
>> mx = ordfilt2(mona,1,ones(5,5));
```

where the arguments are the order and the domain specifying ones for pixels to process and zeros for pixels to omit. The median over a 5×5 window is the twelfth in order

```
>> med = ordfilt2(mona,12,ones(5,5));
```

and is useful as a filter to remove impulse-type noise and for the Mona Lisa image this significantly reduces the fine surface cracking. A more powerful demonstration is to add significant impulse noise to a copy of the Mona Lisa image image

```
>> spotty = imnoise(mona,"salt & pepper");
>> imshow(spotty)
```

and this is shown in ◘ Fig. 11.22a. This type of noise is often referred to as impulse noise or salt and pepper noise. We apply a 3×3 median filter using the function `medfilt2` which is a more efficient implementation of the median filter.

```
>> imshow(medfilt2(spotty,[3 3]))
```

and the result shown in ◘ Fig. 11.22b is considerably improved. A similar effect could have been obtained by smoothing but that would tend to blur the image – median filtering preserves edges in the scene and does not introduce a blur.

The third argument to `ordfilt2` specifies the domain which allows for some very powerful operations. For example

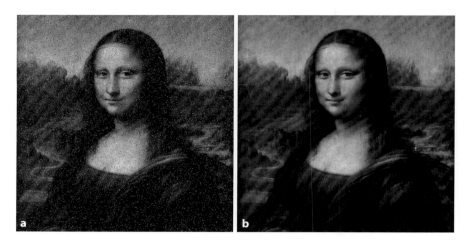

�‌ Fig. 11.22   Median filter cleanup of impulse noise. **a** Noise corrupted image; **b** median filtered result

```
>> M = ones(3,3);
>> M(2,2) = 0
M =
     1     1     1
     1     0     1
     1     1     1
>> mxn = ordfilt2(mona,1,M);
```

specifies the first in rank (maximum) over a *subset* of pixels from the window corresponding to the nonzero elements of M. In this case M specifies the eight neighboring pixels but not the center pixel. The result mxn is the maximum of the eight neighbors of each corresponding pixel in the input image. We can use this

```
>> imshow(mona > mxn)
```

to display all those points where the pixel value is greater than its local neighbors which performs nonlocal maxima suppression. These correspond to local maxima, or peaks if the image is considered as a surface. This mask matrix is very similar to a structuring element which we will encounter in the next section.

11.6 Mathematical Morphology

Mathematical morphology is a class of nonlinear spatial operators shown schematically in ◌ Fig. 11.23. Each pixel in the output matrix is a function of a *subset* of pixels in a region surrounding the corresponding pixel in the input image

$$y_{u,v} = f(\mathcal{W}_{u,v}, \mathbf{S}), \qquad \forall (u,v) \in \mathbf{X} \tag{11.9}$$

where $\mathbf{S}$ is the structuring element, an arbitrary small binary image. For implementation purposes this is embedded in a rectangular window with odd side lengths. The structuring element is similar to the convolution kernel discussed previously except that now it controls *which pixels* in the neighborhood are processed by the function $f(\cdot)$ – it specifies a subset of pixels within the window. The selected pixels are those for which the corresponding values of the structuring element are nonzero – these are shown in red in ◌ Fig. 11.23. Mathematical morphology, as its name implies, is concerned with the form or *shape* of objects in the image.

The easiest way to explain the concept is with a simple example, in this case a synthetic binary image loaded by

```
>> load eg_morph1.mat
>> imshow(im,InitialMagnification=800)
```

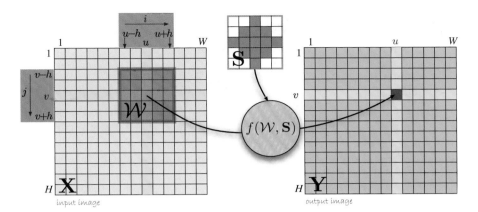

Fig. 11.23 Morphological image processing operations. The operation is defined only for the elements of $\mathcal{W}$ selected by the corresponding elements of the structuring element **S** shown in red

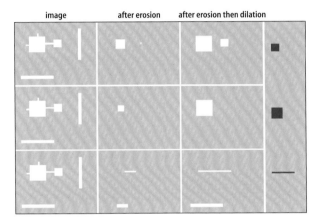

Fig. 11.24 Mathematical morphology example. Pixels are either 0 (gray) or 1 (white). Each column corresponds to processing using the structuring element, shown at the end in red. The first column is the original image, the second column is after erosion by the structuring element, and the third column is after the second column is dilated by the structuring element

which is shown, repeated, down the first column of ■ Fig. 11.24. The structuring element is shown in red at the end of each row. If we consider the top-most row, the structuring element is a square of side length of 5

```
>> S = strel("square",5);
```

and is applied to the original image using the minimum operation

```
>> mn = imerode(im,S);
```

and the result is shown in the second column. For each pixel in the input image we take the *minimum* of all pixels in the 5×5 window. If *any* of those pixels are zero the resulting pixel will be zero. We can see this in animation by

```
>> morphdemo(im,S,"min")
```

The result is dramatic – two objects have disappeared entirely, and the two squares have become separated and smaller. The two objects that disappeared were not *consistent* with the shape of the structuring element. This is where the connection to morphology or shape comes in – only shapes that could *contain* the structuring element will be present in the output image.

The structuring element could define any shape: a circle, an annulus, a 5-pointed star, a line segment 20 pixels long at 30° to the horizontal, or the silhouette of a duck. Mathematical morphology allows very powerful shape-based filters to be created. The second row shows the results for a larger 7×7 structuring element which has resulted in the complete elimination of the small square and the further reduction of the large square. The third row shows the results for a structuring element which is a horizontal line segment 14 pixel wide, and the only remaining shapes are long horizontal lines.

The operation we just performed is known as *erosion*, since large objects are eroded and become smaller – in this case the 5×5 structuring element has caused two pixels ◄ to be *shaved off* all the way around the perimeter of each shape. The small square, originally 5×5, is now only 1×1. If we repeated the operation the small square would disappear entirely, and the large square would be reduced even further.

The half width of the structuring element.

The inverse operation is dilation which makes objects larger. In ◘ Fig. 11.24 we apply dilation to the second column results

```
>> mx = imdilate(mn,S);
```

and the results are shown in the third column. For each pixel in the input image, we take the *maximum* of all pixels in the 5×5 window. If *any* of those neighbors is one the resulting pixel will be one. In this case we see that the two squares have returned to their original size, but the large square has lost its protrusions.

Morphological operations are often written in operator form. Erosion is

$$\mathbf{Y} = \mathbf{X} \ominus \mathbf{S}$$

where in (11.9) $f(\cdot) = \min(\cdot)$, and dilation is

$$\mathbf{Y} = \mathbf{X} \oplus \mathbf{S}$$

where in (11.9) $f(\cdot) = \max(\cdot)$. These operations are also known as Minkowski subtraction and addition respectively.

Erosion and dilation are related by

$$\mathbf{X} \oplus \mathbf{S} = \overline{\overline{\mathbf{X}} \ominus \mathbf{S}'}$$

where the bar denotes the logical complement of the pixel values, and the prime denotes reflection about the center pixel. Essentially this states that eroding the white pixels is the same as dilating the dark pixels and vice versa. For morphological operations

$$(\mathbf{X} \oplus \mathbf{S}_1) \oplus \mathbf{S}_2 = \mathbf{X} \oplus (\mathbf{S}_1 \oplus \mathbf{S}_2)$$
$$(\mathbf{X} \ominus \mathbf{S}_1) \ominus \mathbf{S}_2 = \mathbf{X} \ominus (\mathbf{S}_1 \oplus \mathbf{S}_2)$$

For example a 3×3 square structuring element applied twice is equivalent to 5×5 square structuring element. The former involves $2 \times (3 \times 3 \times N^2) = 18N^2$ operations whereas the later involves $5 \times 5 \times N^2 = 25N^2$ operations.

which means that successive erosion or dilation with a structuring element is equivalent to the application of a single larger structuring element, but the former is computationally cheaper. ◄

The sequence of operations, erosion then dilation, is known as opening since it opens gaps. In operator form it is written as

$$\mathbf{X} \circ \mathbf{S} = (\mathbf{X} \ominus \mathbf{S}) \oplus \mathbf{S}$$

Not only has the opening selected particular shapes but it has also *cleaned up* the image: the squares have been separated and the protrusions on the large square have been removed since they are not consistent with the shape of the structuring element.

In ◘ Fig. 11.25 we perform the operations in the opposite order, dilation then erosion. In the first row no shapes have been lost, they grew then shrank, and the large square still has its protrusions. The hole has been filled since it is not consistent with the shape of the structuring element. In the second row, the larger

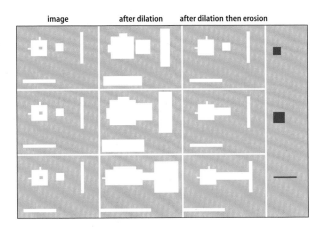

◻ Fig. 11.25 Mathematical morphology example. Pixels are either 0 (gray) or 1 (white). Each row corresponds to processing using the structuring element, shown at the end in red. The first column is the original image, the second column is after dilation by the structuring element, and the third column is after the second column is eroded by the structuring element

structuring element has caused the two squares to join together. This sequence of operations is referred to as closing since it closes gaps and is written in operator form as

$$\mathbf{X} \bullet \mathbf{S} = (\mathbf{X} \oplus \mathbf{S}) \ominus \mathbf{S}$$

Note that in the bottom row the two line segments have remained attached to the edge, this is due to the default behavior in handling edge pixels.

Opening and closing ▶ are implemented by the functions `imopen` and `imclose` respectively. Unlike erosion and dilation, repeated application of opening or closing is futile since those operations are idempotent

$$(\mathbf{X} \circ \mathbf{S}) \circ \mathbf{S} = \mathbf{X} \circ \mathbf{S}$$
$$(\mathbf{X} \bullet \mathbf{S}) \bullet \mathbf{S} = \mathbf{X} \bullet \mathbf{S}$$

These operations can also be applied to grayscale images to emphasize particular shaped objects in the scene prior to an operation like thresholding. Dilation is the maximum of the elements selected by the sliding structuring element, and erosion is the minimum.

These names make sense when considering what happens to white objects against a black background. For black objects the operations perform the inverse function.

11.6.1 Noise Removal

A common use of morphological opening is to remove noise from an image. The image

```
>> objects = imread("segmentation.png");
```

Excurse 11.15: Dealing with Edge Pixels

The problem of a convolution window near the edge of an input image was discussed in ▶ Sect. 11.5.1.2. Similar problems exist for morphological spatial operations, and the functions such as `nlfilter` or `ordfilt2` provide padding options that make sense for the chosen operation, in a similar way as for `imfilter`. `ordfilt2`, for example, provides zero padding and symmetric padding.

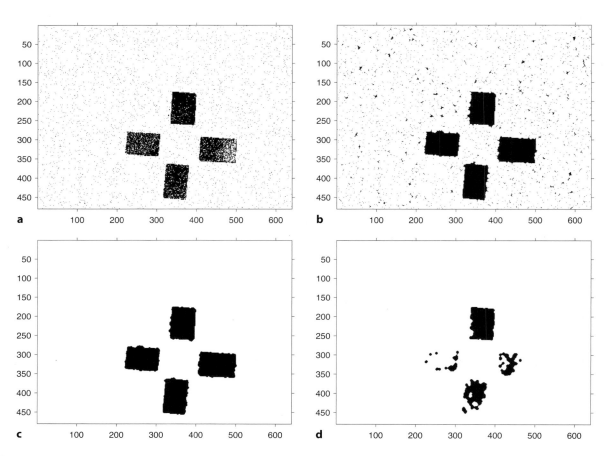

◧ Fig. 11.26   Morphological cleanup. **a** Original image, **b** original after closing, **c** closing then opening, **d** opening then closing. Structuring element is a circle of radius 4. Color map is inverted, set pixels are shown as black

Image segmentation and binarization is discussed in ▶ Sect. 13.1.1.

shown in ◧ Fig. 11.26a is a noisy binary image from the output of a rather poor object segmentation operation. ◄ We wish to remove the dark pixels that do not belong to the objects and we wish to fill in the holes in the four dark rectangular objects.

We choose a symmetric *circular* structuring element of radius 4

```
>> S = strel("disk",4);
>> S.Neighborhood
ans =
  7×7 logical array
   0   0   1   1   1   0   0
   0   1   1   1   1   1   0
   1   1   1   1   1   1   1
   1   1   1   1   1   1   1
   1   1   1   1   1   1   1
   0   1   1   1   1   1   0
   0   0   1   1   1   0   0
```

and apply a closing operation to fill the holes in the objects

```
>> closed = imclose(objects,S);
```

and the result is shown in ◧ Fig. 11.26b. The holes have been filled, but the noise pixels have grown to be small circles and some have agglomerated. We eliminate these by an opening operation

```
>> clean = imopen(closed,S);
```

and the result shown in ◐ Fig. 11.26c is a considerably cleaned up image. If we apply the operations in the inverse order, opening then closing

```
>> opened = imopen(objects,S);
>> closed = imclose(opened,S);
```

the results shown in ◐ Fig. 11.26d are much poorer. Although the opening has removed the isolated noise pixels it has removed large chunks of the targets which cannot be restored.

11.6.2 Boundary Detection

The top-hat transform uses morphological operations to detect the edges of objects. Continuing the example from above, and using the image `clean` shown in ◐ Fig. 11.26c we compute its erosion using a circular structuring element

```
>> eroded = imerode(clean,strel("disk",1));
```

The objects in this image are slightly smaller since the structuring element has caused one pixel to be *shaved off* the outside of each object. Subtracting the eroded image from the original

```
>> imshow(clean-eroded)
```

results in a layer of pixels around the edge of each object as shown in ◐ Fig. 11.27.

11.6.3 Hit or Miss Transform

The hit or miss transform uses a composite of two structuring elements, or a variation on the morphological structuring element, called an interval. To illustrate use of the hit or miss transform, we will use an interval whose values are 1, 0, or -1 where zero corresponds to *don't care* as shown in ◐ Fig. 11.28a. The 1 elements must match foreground pixels of the processed image and the -1 pixels must match the background pixels of the processed image in order for the result to be a one. If there is any mismatch of a one or minus one then the result will be zero. ◐ Fig. 11.28b shows an example binary input image and the result of applying the hit or miss transform is shown in ◐ Fig. 11.28c. The syntax with interval input is as follows

```
out = bwhitmiss(img,interval);
```

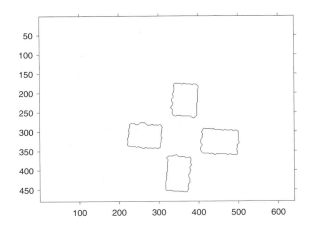

◐ **Fig. 11.27** Boundary detection by morphological processing. Results are shown inverted, white is zero

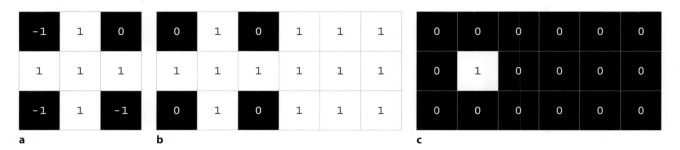

a   **b**   **c**

☐ **Fig. 11.28** Hit or miss transform. **a** The interval; **b** an example input binary image; **c** result of processing the input image using the interval

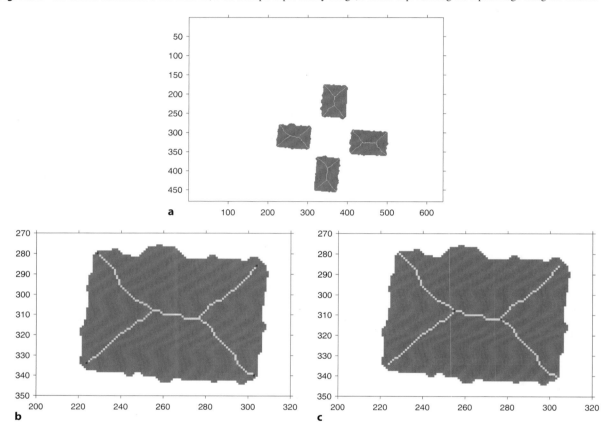

a

b   **c**

☐ **Fig. 11.29** Hit or miss transform operations. **a** Skeletonization; **b** endpoint detection; **c** branch point detection. The images are shown inverted with the original binary image superimposed in gray. The end- and triplepoints are shown as red pixels

The hit or miss transform can be used iteratively with a sequence of intervals to perform complex operations such as skeletonization and linear feature detection. The skeleton of the objects is computed by

```
>> skeleton = bwmorph(clean,"skel",Inf);
```

and is shown in ☐ Fig. 11.29a. The lines are a single pixel wide and are the edges of a generalized Voronoi diagram – they delineate sets of pixels according to the shape boundary they are closest to. We can then find the endpoints of the skeleton

```
>> ends = bwmorph(skeleton,"endpoints");
```

and also the triplepoints

```
>> joints = bwmorph(skeleton,"branchpoints");
```

which are points at which the lines join. These are shown in ☐ Fig. 11.29b, c respectively.

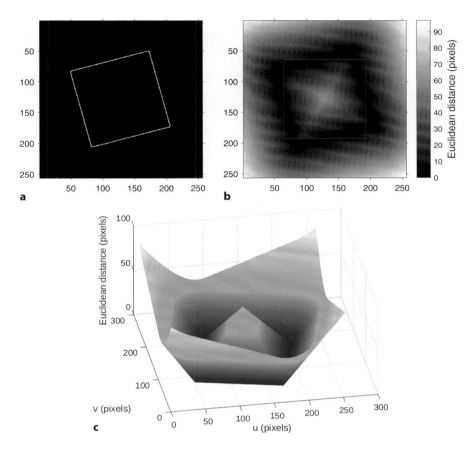

Fig. 11.30 Distance transform. **a** Input binary image; **b** distance transformed input image with overlaid model square; **c** distance transform as a surface

11.6.4 Distance Transform

We discussed the distance transform in ▶ Sect. 5.4.1 for robot path planning. Given an occupancy grid it computed the distance of every free cell from the goal location. The distance transform we discuss here operates on a binary image and the output value, corresponding to every zero pixel in the input image, is the Euclidean distance to the nearest nonzero pixel.

Consider the problem of fitting a model to a shape in an image. We create the outline of a rotated square

```
>> im = zeros(256);
>> im = insertShape(im,"FilledRectangle",[64 64 128 128], ...
>>   Color="w",Opacity=1);
>> im = imrotate(im(:,:,1),15,"crop");
>> edges = edge(im,"canny");
```

which is shown in ◨ Fig. 11.30a and then compute the distance transform

```
>> dx = bwdist(edges);
```

which is shown in ◨ Fig. 11.30b.

An initial estimate of the square is shown with a red line. The value of the distance transform at any point on this red square indicates how far away it is from the nearest point on the original square. If we summed the distance transform for every point on the red square, or even just the vertices, we obtain a total distance measure which will only be zero when our model square overlays the original square. The total distance is a cost function which we can minimize using an optimization

Excurse 11.16: Distance Transform

The **distance transform** of a binary image has a value at each pixel equal to the distance from that pixel to the nearest nonzero pixel in the input image. The distance metric is typically either Euclidean (L_2 norm) or Manhattan distance (L_1 norm). It is zero for pixels that are nonzero in the input image. This transform is closely related to the **signed distance function** whose value at any point is the distance of that point to the nearest boundary of a shape, and is positive inside the shape and negative outside the shape. The figure shows the signed distance function for a unit circle, and has a value of zero, indicated by the red plane, at the object boundary. If we consider a shape to be defined by its signed distance transform then its zero contour defines the shape boundary.

routine that adjusts the position, orientation and size of the square. Considering the distance transform as a 3-dimensional surface in ■ Fig. 11.30c, our problem is analogous to dropping an extensible square hoop into the valley of the distance transform. Note that the distance transform only needs to be computed once, and during model fitting the cost function is simply a lookup of the computed distance. This is an example of chamfer matching and a full example, with optimization that uses Global Optimization Toolbox, is given in `examples/chamfer_match.m`.

11.7 Shape Changing

The final class of image processing operations that we will discuss are those that change the shape or size of an image.

11.7.1 Cropping

The simplest shape change of all is selecting a rectangular region from an image which is the familiar *cropping* operation. Consider the image

```
>> mona = imread("monalisa.png");
```

shown in ■ Fig. 11.31a from which we interactively specify a region of interest or ROI

```
>> [eyes,roi] = imcrop(mona);
>> imshow(eyes)
```

by clicking and dragging a selection box over the image. Double-click inside the ROI to finalize your selection. In this case we selected the eyes, and the ROI of the selected region can be optionally returned and it was

```
>> roi
roi =
    239.5100   170.5100   134.9800    42.9800
```

where the first two values are the (u, v) coordinates for the top-left corner and the next two values are the width and height of the ROI respectively. ◄ The function can be used noninteractively by specifying an ROI

```
>> smile = imcrop(mona,[265 264 77 22]);
```

which in this case selects the Mona Lisa's smile shown in ■ Fig. 11.31b.

The returned `roi` values are non-integer and correspond to the mouse pointer locations within the square image pixels.

a   b

■ **Fig. 11.31** Example of region of interest or image cropping. **a** Original image; **b** selected region of interest

11.7.2   Image Resizing

Often, we wish to reduce the dimensions of an image, perhaps because the large number of pixels results in long processing time or requires too much memory. We demonstrate this with a high-resolution image

```
>> roof = rgb2gray(imread("roof.jpg"));
>> whos roof
  Name         Size              Bytes  Class      Attributes
  roof       1668x2009         3351012  uint8
```

which is shown in ■ Fig. 11.32a. The simplest means to reduce image size is sub-sampling or decimation which selects every m-th pixel in the u- and v-direction, where $m \in \mathbb{N}$ is the subsampling factor. For example with $m = 2$ an $N \times N$ image becomes an $N/2 \times N/2$ image which has one quarter the number of pixels of the original image.

For this example, we will reduce the image size by a factor of seven in each direction

```
>> smaller = roof(1:7:end,1:7:end);
```

using standard MATLAB indexing syntax to select every seventh row and column. The result is shown is shown in ■ Fig. 11.32b and we observe some pronounced curved lines on the roof which were not in the original image. These are artifacts of the sampling process. Subsampling reduces the spatial sampling rate of the image which can lead to spatial aliasing of high-frequency components due to texture or sharp edges. To ensure that the Shannon-Nyquist sampling theorem is satisfied an anti-aliasing low-pass spatial filter must be applied to reduce the spatial bandwidth of the image before it is subsampled. ▶ This is another use for image blurring and the Gaussian kernel is a suitable low-pass filter for this purpose. The combined operation of smoothing and subsampling is implemented by

```
>> smaller = imresize(roof,1/7);
```

and the results for $m = 7$ are shown in ■ Fig. 11.32c. We note that the curved line artifacts are no longer present.

The inverse operation is image upsampling

```
>> bigger = imresize(smaller,7);
```

Any realizable low-pass filter has a finite response above its *cutoff* frequency. In practice the cutoff frequency is selected to be far enough below the theoretical cutoff that the filter's response at the Nyquist frequency is *sufficiently* small. As a rule of thumb for subsampling by m, a Gaussian with $\sigma = m/2$ is used.

□ Fig. 11.32 Image scaling example. **a** Original image; **b** subsampled with $m = 7$, note the axis scaling; **c** subsampled with $m = 7$ after smoothing; **d** image **c** restored to original size by pixel replication

which is shown in □ Fig. 11.32d and appears a little *blurry*. The decimation stage removed 98% of the pixels and restoring the image to its original size has not added any new information.

11.7.3 **Image Pyramids**

An important concept in computer vision, and one that we return to in the next chapter is scale space. The function impyramid can be used to construct a pyramidal decomposition of the input image

```
>> p0 = impyramid(rgb2gray(mona),"reduce");
>> p1 = impyramid(p0,"reduce");
>> p2 = impyramid(p1,"reduce");
>> p3 = impyramid(p2,"reduce");
```

Four levels of the pyramid are pasted into a composite image which is displayed in □ Fig. 11.33.

An image pyramid is the basis of many so-called coarse-to-fine strategies. Consider the problem of looking for a pattern of pixel values that represent some object of interest. The smallest image can be searched very quickly for the object since

◻ Fig. 11.33 Image pyramid, a succession of images each approximately half (by side length) the resolution of the one to the left

it comprises only a small number of pixels. The search is then refined using the next larger image, but we now know which area of that larger image to search. The process is repeated until the object is located in the highest resolution image.

11.7.4 Image Warping

Image warping is a transformation of the pixel *coordinates* rather than the pixel values. Warping can be used to scale an image up or down in size, rotate an image or apply quite arbitrary shape changes. The coordinates of a pixel in the original view (u, v) are expressed as functions

$$u = f_u(u', v'), \quad v = f_v(u', v') \tag{11.10}$$

of the coordinates (u', v') in the new view.

Consider the simple example, shown in ◻ Fig. 11.34a, where the image is reduced in size by a factor of 4 in both dimensions and offset so that its origin, its top-left corner, is shifted to the coordinate $(100, 200)$. We can express this concisely as

$$u' = \frac{u}{4} + 100, \quad v' = \frac{v}{4} + 200 \tag{11.11}$$

which we rearrange into the form of (11.10) as

$$u = 4(u' - 100), \quad v = 4(v' - 200) \tag{11.12}$$

First, we read the image and establish a pair of coordinate matrices ▶ that span the domain of the input image, the set of all possible (u, v)

```
>> mona = im2double(rgb2gray(imread("monalisa.png")));
>> [Ui,Vi] = meshgrid(1:size(mona,2),1:size(mona,1));
```

The coordinate matrices are such that U(u,v) = u and V(u,v) = v and are a common construct in MATLAB, see the documentation for meshgrid.

and another pair that span the domain of the output image, which we choose arbitrarily to be 400×400, the set of all possible (u', v')

```
>> [Up,Vp] = meshgrid(1:400,1:400);
```

Now, for *every* pixel in the output image the corresponding coordinate in the input image is given by (11.12)

```
>> U = 4*(Up-100); V = 4*(Vp-200);
```

Fig. 11.34 Warped images. **a** Scaled and shifted; **b** rotated by 30° about its center. Pixels displayed as red were set to a value of NaN by `interp2` – they were not interpolated from any image pixels

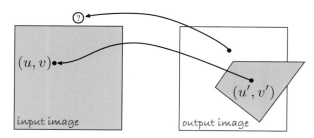

Fig. 11.35 Coordinate notation for image warping. The pixel (u', v') in the output image is sourced from the pixel at (u, v) in the input image as indicated by the arrow. The warped image is not necessarily polygonal, nor entirely contained within the output image

We can now warp the input image using the MATLAB function `interp2`

```
>> little_mona = interp2(Ui,Vi,mona,U,V);
```

and the result is shown in ◘ Fig. 11.34a. Note that `interp2` requires a floating-point image.

Some subtle things happen under the hood. Firstly, while (u', v') are integer coordinates the input image coordinates (u, v) will not necessarily be integers. The pixel values must be interpolated ◄ from neighboring pixels in the input image. Secondly, not all pixels in the output image have corresponding pixels in the input image as illustrated in ◘ Fig. 11.35. Fortunately for us `interp2` handles all these issues and pixels that do not exist in the input image are set to NaN in the output image which we have displayed as red. In case of mappings that are extremely distorted it may be that many adjacent output pixels map to the same input pixel and this leads to pixelation or *blockyness* in the output image.

Now let's try something a bit more ambitious and rotate the image by 30° into an output image of the same size as the input image

```
>> [Up,Vp] = meshgrid(1:size(mona,2),1:size(mona,1));
```

We want to rotate the image about its center but since the origin of the input image is the top-left corner we must first change the origin to the center, then rotate and then move the origin back to the top-left corner. ◄ The warp equation is therefore

$$\begin{pmatrix} u' \\ v' \end{pmatrix} = \underbrace{\begin{pmatrix} \cos \pi/6 & -\sin \pi/6 \\ \sin \pi/6 & \cos \pi/6 \end{pmatrix}}_{\mathbf{R}(\pi/6)} \begin{pmatrix} u - u_c \\ v - v_c \end{pmatrix} + \begin{pmatrix} u_c \\ v_c \end{pmatrix} \tag{11.13}$$

Different interpolation modes can be selected by a trailing argument to `interp2` but the default option is bilinear interpolation. A pixel at coordinate $(u + \delta_u, v + \delta_v)$ where $u, v \in \mathbb{N}$ and $\delta_u, \delta_v \in [0, 1)$ is a linear combination of the pixels (u, v), $(u + 1, v)$, $(u, v + 1)$ and $(u + 1, v + 1)$. The interpolation function acts as a weak anti-aliasing filter, but for very large reductions in scale the image should be smoothed first using a Gaussian kernel.

This is the application of a twist as discussed in ► Chap. 2.

where (u_c, v_c) is the coordinate of the image center and $\mathbf{R}(\frac{\pi}{6})$ is a rotation matrix. This can be rearranged into the *inverse form* and implemented as

```
>> R = rotm2d(pi/6);
>> uc = 256; vc = 256;
>> U = R(1,1)*(Up-uc)+R(2,1)*(Vp-vc)+uc;
>> V = R(1,2)*(Up-uc)+R(2,2)*(Vp-vc)+vc;
>> twisted_mona = interp2(Ui,Vi,mona,U,V);
```

and the result is shown in ◘ Fig. 11.34b. Note the direction of rotation – our definition of the x- and y-axes (parallel to the u- and v-axes respectively) is such that the z-axis is defined as being into the page making a clockwise rotation a positive angle. Also note that the corners of the original image have been lost, they fall outside the bounds of the output image.

The function `imresize` uses image warping to change image scale, and the function `imrotate` uses warping to perform rotation. The example above could be achieved far more easily by

```
>> twisted_mona = imrotate(mona,-rad2deg(pi/6),"crop");
```

More general image warping can also be performed using the `imwarp` function.

Finally, we will describe, in detail, the image undistortion procedure that we will later use in ▶ Sect. 13.2.4. The distorted image from the camera is the input image and will be warped to remove the distortion. We are in luck since the distortion model (13.13) is already in the inverse form. Recall that

$$u' = u + \delta_u$$
$$v' = v + \delta_v$$

where the distorted coordinates are denoted with a prime and δ_u and δ_v are functions of (u, v).

First, we load the distorted image and build the coordinate matrices for the distorted and undistorted images

```
>> distorted = imread(fullfile(toolboxdir("vision"),"visiondata", ...
>>    "calibration","mono","image01.jpg"));
>> distorted = im2double(rgb2gray(distorted));
>> [Ui,Vi] = imeshgrid(distorted);
>> Up = Ui; Vp = Vi;
```

and then load the results of the camera calibration.

```
>> d = load("cameraIntrinsics");
```

For readability we unpack the required parameters into variables with short names

```
>> k = [d.intrinsics.RadialDistortion 0];
>> p = d.intrinsics.TangentialDistortion;
>> u0 = d.intrinsics.PrincipalPoint(1);
>> v0 = d.intrinsics.PrincipalPoint(2);
>> fpix_u = d.intrinsics.FocalLength(1);
>> fpix_v = d.intrinsics.FocalLength(2);
```

for radial and tangential distortion vectors, principal point, and focal length in pixels. During the calibration of our camera, we used two lens distortion coefficients but up to three can be obtained. Therefore, for illustration purposes, we padded `k` with an additional zero. Next, we convert pixel coordinates to normalized image coordinates ▶

```
>> u = (Up-u0)/fpix_u;
>> v = (Vp-v0)/fpix_v;
```

In units of meters with respect to the camera's principal point.

The radial distance of the pixels from the principal point is then

```
>> r = sqrt(u.^2+v.^2);
```

◘ Fig. 11.36 Warping to undistort an image. **a** Original distorted image; **b** corrected image. Note that the side edge of the white board has become a straight line (Images reprinted with permission of The MathWorks, Inc.)

and the pixel coordinate errors due to distortion are

```
>> delta_u = u.*(k(1)*r.^2 + k(2)*r.^4 + k(3)*r.^6) + ...
>>    2*p(1)*u.*v + p(2)*(r.^2 + 2*u.^2);
>> delta_v = v.*(k(1)*r.^2 + k(2)*r.^4 + k(3)*r.^6) + ...
>>    p(1)*(r.^2 + 2*v.^2) + 2*p(2)*u.*v;
```

The distorted pixel coordinates in metric units are

```
>> ud = u+delta_u; vd = v+delta_v;
```

which we convert back to pixel coordinates

```
>> U = ud*fpix_u + u0;
>> V = vd*fpix_v + v0;
```

and finally apply the warp

```
>> undistorted = interp2(Ui,Vi,distorted,U,V);
>> imshowpair(distorted,undistorted,"montage")
```

The results are shown in ◘ Fig. 11.36. The change is quite visible, and is most noticeable along the left-side edge of the white board. ◄

To illustrate core concepts, this section demonstrates how to undistort an image by showing all the steps. A far more efficient and compact way of undistorting an image is provided in the undistortImage function.

11.8 Wrapping Up

In this chapter we learned how to acquire images from a variety of sources such as image files, video files, video cameras and the internet, and load them into the MATLAB workspace. Once there we can treat them as matrices, the principal MATLAB datatype, and conveniently manipulate them. The elements of the image matrices can be integer, floating-point or logical values. Next, we discussed many processing operations. Operations on a single image include unary arithmetic operations, type conversion, various color transformations and gray-level stretching; nonlinear operations such as histogram normalization and gamma encoding or decoding; and logical operations such as thresholding. We also discussed operations on pairs of images such as green screening, background estimation and moving object detection.

The largest and most diverse class of operations are spatial operators. We discussed convolution which can be used to smooth an image and to detect edges. Linear operations are defined by a kernel matrix which can be chosen to perform functions such as image smoothing (to reduce the effect of image noise or as a low-pass anti-aliasing filter prior to decimation) or for edge detection. Nonlinear spatial operations were used for template matching, computing rank statistics (including the median filter which eliminates impulse noise) and mathematical morphology which filters an image based on shape and can be used to cleanup binary images. A variant form, the hit or miss transform, can be used iteratively to perform functions such as skeletonization.

Finally, we discussed shape changing operations such as regions of interest, scale changing and the problems that can arise due to aliasing, and generalized image warping which can be used for scaling, translation, rotation or undistorting an image. All these image processing techniques are the foundations of feature extraction algorithms that we discuss in the next chapter.

11.8.1 Further Reading

Image processing is a large field and this chapter introduced many of the most useful techniques from a robotics perspective. More comprehensive coverage of the topics introduced here and others such as grayscale morphology, image restoration, wavelet and frequency domain methods, and image compression can be found in Gonzalez, Woods and Eddins (2020), Szeliski (2022), Soille (2003), Nixon and Aguado (2012) and Forsyth and Ponce (2011). Online information about computer vision is available through CVonline at ▶ http://homepages.inf.ed.ac.uk/rbf/ CVonline, and the material in this chapter is covered under the section *Image Transformations and Filters*.

Edge detection is a subset of image processing but one with huge literature of its own. Forsyth and Ponce (2011) have a comprehensive introduction to edge detection and a useful discussion on the limitations of edge detection. Nixon and Aguado (2012) also cover phase congruency approaches to edge detection and compare various edge detectors. The Sobel kernel for edge detection was described in an unpublished 1968 publication from the Stanford AI lab by Irwin Sobel and Jerome Feldman: *A 3×3 Isotropic Gradient Operator for Image Processing*. The Canny edge detector was originally described in Canny (1983, 1987).

Mathematical morphology is another very large topic and we have only scraped the surface and important techniques such as grayscale morphology and watersheds have not been covered at all. The general image processing books mentioned above have useful discussion on this topic. Most of the specialist books in this field are now quite old but Shih (2009) is a good introduction and the book by Dougherty and Lotufo (2003) has a more hands on tutorial approach. Another good choice is Soille (2003) which includes an introduction and many practical techniques.

The approach to computer vision covered in this book is often referred to as bottom-up processing. This chapter has been about *low-level* vision techniques which are operations on pixels. The next chapter is about *high-level* vision techniques where sets of pixels are grouped and then described so as to represent objects in the scene.

11.8.2 Sources of Image Data

All the images used in this part of the book are provided in the *images* folder of the RVC Toolbox.

There are thousands of online webcams as well as a number of sites that aggregate them and provide lists categorized by location, for example Opentopia and EarthCam. Most of these sites do not connect you directly to the web camera so the URL of the camera must be dug out of the HTML page source. Some of the content on these list pages can be rather dubious – so beware.

11.8.3 Exercises

1. Become familiar with `imshow` and `imtool` for grayscale and color images. Explore pixel values in the image as well as the zoom inspect pixel values, measure distance, crop image buttons. Use `imcrop` to extract the Mona Lisa's smile.

2. Look at the histogram of grayscale images that are under, well and over exposed. For a color image look at the histograms of the RGB color channels for scenes with different dominant colors. Combine real-time image capture with computation and display of the histogram.

3. Create two copies of a grayscale image into workspace variables A and B. Write code to time how long it takes to compute the difference of A and B using the MATLAB shorthand A-B or using two nested `for` loops. Use the functions `tic` and `toc` to perform the timing.

4. Grab some frames from the camera on your computer or from a video file and display them.

5. Write a loop that grabs a frame from your camera and displays it. Add some effects to the image before display such as "negative image", thresholding, posterization, false color, edge filtering etc.

6. Given a scene with luminance of 800 nit and a camera with ISO of 1000, $q = 0.7$ and f-number of 2.2 what exposure time is needed so that the average gray level of the 8-bit image is 150?

7. Motion detection
   a) Modify the Traffic example in ▶ Sect. 11.1.3 and highlight the moving vehicles.
   b) Write a loop that performs background estimation using frames from your camera. What happens as you move objects in the scene, or let them sit there for a while? Explore the effect of changing the parameter σ.
   c) Combine concepts from motion detection and chroma-keying to put pixels from the camera where there is motion into the desert scene.

8. Convolution
   a) Compare the results of smoothing using a 21×21 uniform kernel and a Gaussian kernel. Can you observe the smoother output of the latter?
   b) Why do we choose a smoothing kernel that sums to one?
   c) Compare the performance of the simple horizontal gradient kernel $\mathbf{K} = \begin{pmatrix} 0.5 & 0 & -0.5 \end{pmatrix}$ with the Sobel kernel.
   d) Investigate filtering with the Gaussian kernel for different values of σ and kernel size.
   e) Create a 31×31 kernel to detect lines at 60 deg.
   f) Derive analytically the derivative of the Gaussian in the x-direction (11.4).
   g) Derive analytically the Laplacian of Gaussian (11.8).
   h) Show the difference between difference of Gaussian and derivative of Gaussian.

9. Show analytically the effect of an intensity scale error on the SSD and NCC similarity measures.

10. Template matching using the Mona Lisa image; convert it first to grayscale.
    a) Use `imcrop` to select one of Mona Lisa's eyes as a template. The template should have odd dimensions.
    b) Use `normxcorr2` to compute the similarity image. What is the best match and where does it occur? What is the similarity to the other eye? Where does the second-best match occur and what is its similarity score?
    c) Scale the intensity of the Mona Lisa image and investigate the effect on the peak similarity.
    d) Add an offset to the intensity of the Mona Lisa image and investigate the effect on the peak similarity.

e) Repeat steps (c) and (d) for different similarity measures such as SAD, SSD, rank and census.

f) Scale the template size by different factors (use `imresize`) in the range 0.5 to 2.0 in steps of 0.05 and investigate the effect on the peak similarity. Plot peak similarity vs scale.

g) Repeat (f) for rotation of the template in the range -0.2 to 0.2 rad in steps of 0.05.

11. Perform the sub-sampling example from ▶ Sect. 11.7.2 and examine aliasing artifacts around sharp edges and the regular texture of the roof tiles.

12. Write a function to create ◼ Fig. 11.33 from the output of `impyramid`.

13. Warp the image to polar coordinates (r, θ) with respect to the center of the image, where the horizontal axis is r and the vertical axis is θ.

14. Create a warp function that mimics your favorite funhouse mirror.

Image Feature Extraction

Contents

12.1   Region Features – 496

12.2   Line Features – 521

12.3   Point Features – 525

12.4   Applications – 534

12.5   Wrapping Up – 539

© The Author(s), under exclusive license to Springer Nature Switzerland AG 2023
P. Corke, W. Jachimczyk, R. Pillat, *Robotics, Vision and Control*, Springer Tracts in Advanced Robotics 147,
https://doi.org/10.1007/978-3-031-07262-8_12

`chapter12.mlx`

▶ sn.pub/D6tGzn

12

In the last chapter we discussed the acquisition and processing of images. We learned that images are simply large arrays of pixel values but for robotic applications, images have too much *data* and not enough *information*. We need to be able to answer pithy questions such as what is the pose of the object? what type of object is it? how fast is it moving? how fast am I moving? and so on. The answers to such questions are *measurements* obtained from the image and which we call image features. Features are the *gist* of the scene and the raw material that we need for robot control. There are many manually crafted feature types as well as features derived automatically by end-to-end learning systems based on deep learning. In this chapter, we will mainly focus on manually crafted features and also show two examples of using deep learning techniques.

The image processing operations from the last chapter operated on one or more input images and returned another image. In contrast, *feature extraction* operates on a single image and returns one or more *image features*. Features are typically scalars (for example area or aspect ratio) or short vectors (for example the coordinate of an object or the parameters of a line). Image feature extraction is a necessary first step in using image data to control a robot. It is an *information concentration* step that reduces the data rate from 10^6–10^8 bytes/sec at the output of a camera to something of the order of tens of features per frame that can be used as input to a robot's control system.

In this chapter we discuss features and how to extract them from images. Drawing on image processing techniques from the last chapter we will discuss three classes of features: regions, lines and interest points. ▶ Sect. 12.1 discusses region features which are contiguous groups of pixels that are homogeneous with respect to some pixel property. For example, the set of pixels that represent a red object against a non-red background. ▶ Sect. 12.2 discusses line features which describe straight lines in the world. Straight lines are distinct and very common in human-made environments – for example the edges of doorways, buildings or roads. The final class of features, discussed in ▶ Sect. 12.3, are point features. These are distinctive *points* in an image that can be reliably detected in different views of the same scene and are critical to the content of ▶ Chap. 14. We also briefly introduce deep-learning-based approaches to the problems of finding semantically homogeneous pixel regions and detecting objects.

▶ Sect. 12.4 presents two applications: optical character recognition, and image retrieval. The latter is the problem of finding which image, from an existing set of images, is most similar to some new image. This can be used by a robot to determine whether it has visited a particular place, or seen the same object, before.

❶ It is important to always keep in mind that image features are a summary of the information present in the pixels that comprise the image, and that the mapping from the world to pixels involves significant information loss – the perspective projection discussed in ▶ Chap. 13. The information lost by summarization is countered by assumptions based on our knowledge of the scene, but our system will only ever be as good as the validity of our assumptions. For example, we might use image features to describe the position and shape of a group of red pixels that correspond to a red object. However, the feature size, typically the number of pixels, does not say anything about the size of the red object in the world – we need extra information such as the distance between the camera and the object, and the camera's focal length. We also need to assume that the object is not partially occluded – that would make the observed size less than the true size. Further, we need to assume that the illumination is such that the chromaticity of the light reflected from the object is considered to be red. We might also find features in an image that do not correspond to a physical object – decorative markings, shadows, or reflections from bright light sources.

◘ Fig. 12.1 Examples of pixel classification. The left-hand column is the input image and the right-hand column is the classification. The classification is application specific and the pixels have been classified into C categories represented as different colors. The objects of interest are **a** the individual letters on the sign; **b** the red tomatoes. **c** a segmentation of the grains using lazy snapping, a graph-based segmentation technique; **d** semantic segmentation using Deeplab v3+ deep learning network showing complex classification results

12.1   Region Features

Image segmentation is the process of partitioning an image into *application mean-ingful* regions as illustrated in ◘ Fig. 12.1. The aim is to segment or separate those pixels that represent objects of interest from all other pixels in the scene. This is one of the oldest approaches to scene understanding and while conceptually straight-forward it is very challenging in practice. A key requirement is *robustness* which is how gracefully the method degrades as the underlying assumptions are violated, for example changing scene illumination or viewpoint. Segmentation of an image into regions comprises three sub-problems:

Classification – which is a decision process applied to each pixel that assigns the pixel to one of C classes $c = 0, \ldots, C-1$. Commonly we use $C = 2$ which is known as binary classification or binarization and some examples are shown in ◘ Fig. 12.1a–c. The pixels have been classified as object ($c = 1$) or not-object ($c = 0$) which are displayed as white or black pixels respectively. ◘ Fig. 12.1d are multi-level classifications where $C > 2$ and the pixel's class is reflected in its displayed color.

The underlying *assumption* in the examples of ◘ Fig. 12.1 is that regions are *ho-mogeneous* with respect to some characteristic such as brightness, color, texture or semantic meaning – the classification is *always application specific*. In practice we accept that this stage is imperfect and that pixels may be misclassified – subsequent processing steps will have to deal with this.

Representation – where adjacent pixels of the same class are *connected* to form m spatial sets $S_1 \ldots S_m$. For example in ◘ Fig. 12.1b there are two sets, each rep-resenting an individual tomato – a tomato *instance*. The sets can be represented by assigning a set label to each pixel, by lists of the coordinates of all pixels in each set, or lists of pixels that define the perimeter of each set.

Description – where the sets S_i are *described* in terms of compact scalar or vector-valued *features* such as size, position, and shape. In some applications, there is an additional step – semantic segmentation – where we assign an applica-tion meaningful label to the pixel classes or sets, for example, "cat", "dog", "coffee cup", etc.

These topics are further explained in the next three subsections.

12.1.1   Pixel Classification

Pixels, whether grayscale or color, are assigned to a class that is represented by an integer $c = 0, \ldots, C-1$ where C is the number of classes. In many of the examples we will use binary classification with just two classes corresponding to object and not-object, or foreground and background.

12.1.1.1   Grayscale Image Classification

A common approach to binary classification of grayscale pixels is the monadic operator introduced in the last chapter

$$c_{u,v} = \begin{cases} 0, & \text{if} \quad x_{u,v} \le t \\ 1, & \text{if} \quad x_{u,v} > t \end{cases} \quad \forall (u,v) \in \mathbf{X}$$

where the decision is based simply on the value of the pixel $x_{u,v}$. This approach is called thresholding and t is the *threshold*.

Thresholding is very simple to implement. Consider the image

```
>> castle = im2double(imread("castle.png"));
```

Fig. 12.2 Binary classification. **a** Original image (Image sourced from the ICDAR collection; Lucas 2005); **b** histogram of grayscale pixel values with Otsu threshold in red; **c** binary classification with threshold of 0.7; **d** binary classification with Otsu threshold

which is shown in ▢ Fig. 12.2a. The thresholded image

```
>> imshow(castle > 0.7)
```

is shown in ▢ Fig. 12.2c. The pixels have been quite accurately classified as corresponding to white paint or not. This classification is based on the seemingly reasonable *assumption* that the white paint objects are brighter than everything else in the image.

In the early days of computer vision, when computer power was limited, this approach was widely used – it was easier to contrive a world of white objects and dark backgrounds than to implement more sophisticated classification. Many industrial vision inspection systems still use this simple approach since it allows the use of modest embedded computers – it works very well if the objects are on a conveyor belt of a suitable contrasting color or in silhouette at an inspection station. In an unconstrained robot environment we generally have to work harder to achieve useful classification. An important question, and a hard one, is where did the threshold value of 0.7 come from? The most common approach is trial and error! We can explore the possibilities most easily by using the Image Segmenter app. It offers a variety of approaches to segment a grayscale image

```
>> imageSegmenter(castle)
```

It displays the image and offers a toolstrip with several button choices. To experience the effect of manual thresholding, select the Threshold button from the toolbar. Once you enter the thresholding mode, choose Manual Threshold method from the toolstrip. This will give you access to a threshold slider that can be adjusted until a satisfactory result is obtained. To see the binary image, you can press the Show Binary button. However, on a day with different lighting condition the intensity profile of the image would change.

```
>> imageSegmenter(castle*0.8)
```

A more principled approach than trial and error is to analyze the histogram of the image

```
>> imhist(castle);
```

which is shown in ◘ Fig. 12.2b. The histogram has two clearly defined peaks, a bimodal distribution, which correspond to two *populations* of pixels. The smaller peak around 0.9 corresponds to the pixels that are bright and it has quite a small range of variation in value. The wider and taller peak around 0.3 corresponds to pixels in the darker background of the sign and the bricks, and has a much larger variation in brightness.

To separate the two classes of pixels we choose the decision boundary, the threshold, to lie in the valley between the peaks. The choice of $t = 0.7$ works well. Since the valley is very wide, we actually have quite a range of choice for the threshold, for example $t = 0.75$ would also work well.

The optimal threshold can be computed using Otsu's method

```
>> t = graythresh(castle)
t =
    0.6039
```

which separates an image into two classes of pixels, see ◘ Fig. 12.2d, in a way that minimizes the variance of values within each class and maximizes the variance of values between the classes – assuming that the histogram has just two peaks. Sadly, as we shall see, the real world is rarely this facilitating.

Consider a different image of the same scene which has a flash reflection

```
>> castle = im2double(imread("castle2.png"));
```

and is shown in ◘ Fig. 12.3a. The histogram shown in ◘ Fig. 12.3b is similar – it is still bimodal – but we see that the peaks are wider and the valley is less deep. The pixel gray-level populations are now overlapping and unfortunately for us no single threshold can separate them. Otsu's method computes a threshold of

```
>> t = graythresh(castle)
t =
    0.5961
```

and the result of applying this threshold is shown in ◘ Fig. 12.3c. The pixel classification is poor and the highlight overlaps several of the characters. The result of using a higher threshold of 0.75 is shown in ◘ Fig. 12.3d – the highlight is reduced, but not completely, and some other characters are starting to break up.

❶ Thresholding approach is often unreliable

Segmentation approaches based on thresholding are notoriously brittle – a slight change in illumination of the scene means that the thresholds we chose would no longer be appropriate. In most real scenes there is no simple mapping from pixel values to particular objects – we cannot, for example, choose a threshold that would select a motorbike or a duck. Distinguishing an object from the background is a difficult computer vision problem. Most recent advances in deep learning using semantic segmentation do a much better job but are also considerably more complex. See semantic segmentation example in ▶ Sect. 12.1.1.3 for more details.

◧ Fig. 12.3 Binary segmentation example. **a** Grayscale image with intensity highlight; **b** histogram; **c** thresholded with Otsu's threshold at 0.596; **d** thresholded at 0.75

One alternative is to choose a local rather than a global threshold. Bradley's method is widely used in optical character recognition systems and computes a local threshold

$$t[u, v] = \mu(\mathcal{W})$$

where $\mathcal{W}$ is a region about the point (u, v) and $\mu(\cdot)$ is the mean of $\mathcal{W}$. The size of the window $\mathcal{W}$ is a critical parameter and should be of a similar size to the objects we are looking for. For this example, using the measuring tool of `imtool`, we found out that the characters are approximately 50–80 pixels tall. We chose an odd-sized window of 65 pixels ▶

```
>> t = adaptthresh(castle,NeighborhoodSize=65);
>> imshow(t)
```

The resulting local threshold t is shown in ◧ Fig. 12.4a. We apply the threshold pixel-wise to the original image

```
>> imshow(imbinarize(castle,t))
```

resulting in the classification shown in ◧ Fig. 12.4b. All the pixels belonging to the letters have been correctly classified but compared to ◧ Fig. 12.3c there are many false positives – nonobject pixels classified as objects. Later in this section, we will discuss techniques to eliminate these false positives. Note that the classification

Many functions that request neighborhood size specification as input, such as `adaptthresh`, require the size to be odd. This makes the neighborhood's center unambiguous.

Fig. 12.4 Bradley's thresholding method. **a** The local threshold displayed as an image; **b** the binary segmentation

process is no longer a function of just the input pixel, it is now a more complicated function of the pixel and its neighbors. While we no longer need to choose t we now need to choose the window size, and again this is usually a trial and error process that can be made to work well only for a particular type of scene.

The results shown in ◘ Fig. 12.3c and d are disappointing at first glance, but we see that every character object is correctly classified at some, but not all, thresholds. In fact, each object is correctly segmented for some *range* of thresholds and what we would like is the union of regions classified over the range of all thresholds. The maximally stable extremal region or MSER algorithm does exactly this. It is implemented by the `detectMSERFeatures` function.

```
>> [regions,CC] = detectMSERFeatures(castle, ...
>>    RegionAreaRange=[100 20000]);
```

and for this image

```
>> regions.Count
ans =
    402
```

402 stable regions were found. ◄

The other return value is a connected components ◄ structure that holds a compact representation of the regions that were found, including a pixel index list for each region. To visualize the regions, we first convert the connected components structure into a label matrix using the `labelmatrix` function. ◄ It can then be turned into an image that gives a unique color to each labeled region using the `label2rgb` function

```
>> L = labelmatrix(CC);
>> imshow(label2rgb(L));
```

and the result is shown in ◘ Fig. 12.5. Each nonzero pixel in the label matrix corresponds to a stable set and the value is the *label* assigned to that stable set which is displayed as a unique color. All the character objects were correctly classified.

12.1.1.2 Color Image Classification

Color is a powerful cue for segmentation but roboticists tend to shy away from using it because of the problems with color constancy discussed in ▶ Sect. 10.3.2. In this section, we consider two examples that use color images. The first is a simple scene with four yellow markers that were used for an indoor drone landing experiment

```
>> imTargets = imread("yellowtargets.png");
```

shown in ◘ Fig. 12.6a. The second is from the MIT Robotic Garden project

```
>> imGarden = imread("tomato_124.jpg");
```

Although no explicit threshold has been given, the `detectMSERFeatures` function has a number of parameters and in this case their default values have given a satisfactory result. See the online documentation for details of the parameters and detailed description of returned structures.

In a binary image, connected components are regions of logical true pixels which touch. Connectivity and connected component analysis are discussed in more detail in ▶ Sect. 12.1.2.

A label matrix assigns unique integer values to objects or connected components in an image.

12

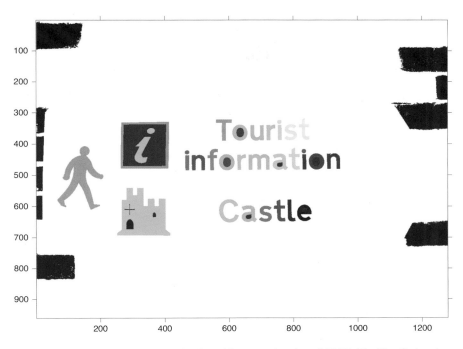

Fig. 12.5 Segmentation using maximally stable extremal regions (MSER). The identified regions are uniquely color coded using `label2rgb` function

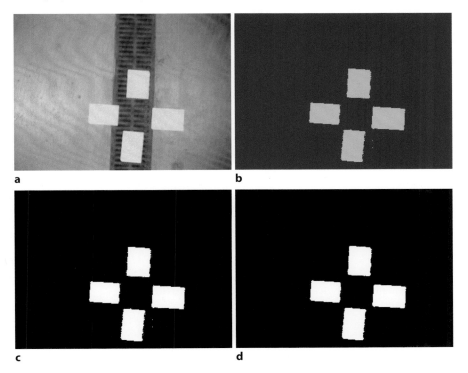

Fig. 12.6 Target image example. **a** Original image; **b** pixel classification shown in false color; **c** all pixels of class $c = 2$; **d** after morphological opening with a disk of radius 2

and is shown in ■ Fig. 12.7a. Our objective is to determine the centroids of the yellow targets and the red tomatoes respectively. The initial stages of processing are the same for each image but we will illustrate the process in detail for the image of the yellow targets as shown in ■ Fig. 12.6.

K-means clustering is computationally expensive and therefore an efficient implementation of it is necessary for use in real-time applications.

Using the `rgb2lab` function, we map each color pixel onto the two-dimensional a*b*-chromaticity plane. Then, using the `imsegkmeans` function, we will separate the 2-D chromaticity space into distinct color clusters. This function uses the *k*-means algorithm for segmentation. ◄ A limitation of the *k*-means algorithm is that we must specify the number of clusters to find. We will use our knowledge that this particular scene has essentially two differently colored elements: yellow targets and gray floor, metal drain cover and shadows. The pixels are clustered into *two* chromaticity classes by

```
>> im_lab = rgb2lab(imTargets);
>> im_ab = single(im_lab(:,:,2:3)); % imsegkmeans requires singles
>> [L,cab] = imsegkmeans(im_ab,2);
```

We used a*b*-chromaticity since Euclidean distance in this space, used by *k*-means to determine the clusters, matches the human perception of difference between colors. The function returns the a*b*-chromaticity of the cluster centers

```
>> cab
cab =
  2×2 single matrix
    0.5251   -4.1324
   -1.4775   57.4939
```

as one row per cluster. We can confirm that cluster 2 is the closest to yellow since

```
>> colorname(cab(2,:),"ab")
ans =
    "gold4"
```

The function `imsegkmeans` also returns the label matrix which we can display as an image

```
>> imshow(label2rgb(L))
```

as shown in ◻ Fig. 12.6b. The pixels in the label matrix, *L*, have values $c = \{1, 2\}$ indicating to which class the corresponding input pixels have been assigned. ◄ The yellow target pixels have been assigned to class $c = 2$ which is displayed as cyan.

The pixels belonging to class 2 can be selected

```
>> cls2 = (L==2);
```

The `imsegkmeans` uses labels that start with 1, while labels produced by the `bwlabel` start with 0, where 0 implies background since the `bwlabel` operates on binary images and objects are assumed to be the logical value of `true`.

which is a *logical image* that can be displayed

```
>> imshow(cls2)
```

as shown in ◻ Fig. 12.6c. All pixels of class 2 are displayed as white and correspond to the yellow targets in the original image. This binary image shows a good classification but there are a few minor imperfections: some rough edges, and some stray pixels.

A morphological opening operation as discussed in ► Sect. 11.6 will eliminate these. We apply a symmetric structuring element of radius 2

```
>> targetsBin = imopen(cls2,strel("disk",2));
>> imshow(targetsBin)
```

and the result is shown in ◻ Fig. 12.6d. It shows a clean binary segmentation of the pixels into the two classes: target and not-target.

For the garden image we follow a very similar procedure. We classify the pixels into three clusters ($C = 3$) based on our knowledge that the scene contains: red tomatoes, green leaves, and dark background

```
>> im_lab = rgb2lab(imGarden);
>> im_ab = single(im_lab(:,:,2:3));
>> [L,cab] = imsegkmeans(im_ab,3);
>> cab
cab =
  3×2 single matrix
```

Fig. 12.7 Garden image example. **a** Original image (courtesy of Distributed Robot Garden project, MIT); **b** pixel classification ($C = 3$) shown in false color; **c** all pixels of class $c = 3$; **d** after morphological closing with a disk of radius 15

```
-16.2562    28.4769
 -1.2220     3.1538
 44.0557    26.6873
```

The pixel classes are shown in false color in ◘ Fig. 12.7b. Pixels corresponding to the tomato have been assigned to class $c = 3$ and are displayed as yellow. The name of the color closest to cluster 3 is

```
>> colorname(cab(3,:),"ab")
ans =
    "brown4"
```

The red pixels can be selected

```
>> cls3 = (L==3);
>> imshow(cls3)
```

and the resulting logical image is shown in ◘ Fig. 12.7c.

This segmentation is far from perfect. Both tomatoes have holes due to specular reflection as discussed in ▶ Sect. 10.3.5. A few pixels at the bottom left have been erroneously classified as a tomato. We can improve the result by applying a morphological closing operation with a large disk kernel which is consistent with the shape of the tomato

```
>> tomatoesBin = imclose(cls3,strel("disk",15));
>> imshow(tomatoesBin)
```

and the result is shown in ◘ Fig. 12.7d. The closing operation has somewhat restored the shape of the fruit, but with the unwanted consequence that the group of misclassified pixels in the bottom-left corner have been enlarged. Nevertheless, this image contains an acceptable classification of pixels into two classes: tomato and not-tomato.

The garden image illustrates two common real-world imaging artifacts: specular reflection and occlusion. The surface of the tomato is sufficiently shiny and

Excurse 12.1: k-**Means Clustering**

k-means clustering is an iterative algorithm for grouping n-dimensional points into k spatial clusters. Each cluster is defined by a center point which is an n-vector $\boldsymbol{c}_i$, $i = 1, \ldots, k$. At each iteration, all points are assigned to the *closest* cluster center, and then each cluster center is updated to be the mean of all the points assigned to the cluster. The mean is computed on the per-dimension basis.

The algorithm is implemented by the `kmeans` function in the Statistics and Machine Learning Toolbox™. The distance metric used is Euclidean distance. By default, `kmeans` randomly selects k of the provided points as initial cluster centers.

To demonstrate, we choose 500 random 2-dimensional points

```
>> % set random seed for repeatable results
>> rng(0)
>> a = rand(500,2);
```

where `a` is a 500×2 matrix with one point per row. We will cluster this data into three sets

```
>> [cls,center] = kmeans(a,3);
```

where `cls` is a matrix whose elements specify the class of the corresponding row of `a`. `center` is a 3×2 matrix whose rows specify the center of each 2-dimensional cluster.

We plot the points in each cluster with different colors

```
>> hold on
```

```
>> for i=1:3
>>   h = plot(a(cls==i,1),a(cls==i,2),".");
>>   % Maintain previous plot color
>>   % while showing cluster center
>>   h.Parent.ColorOrderIndex = ...
>>     h.Parent.ColorOrderIndex-1;
>>   plot(center(i,1),center(i,2),".", ...
>>     MarkerSize=32)
>> end
```

and it is clear that the points have been sensibly partitioned. The centroids center have been superimposed as large dots.

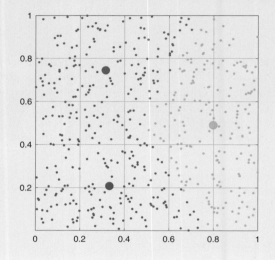

Observe that they have the same chromaticity as the black background, class 2 pixels, which are situated close to the white point on the a^*b^*-plane.

oriented in such a way that the camera sees a reflection of the room light – these pixels are white rather than red. ◀ The top tomato is also partly obscured by leaves and branches. Depending on how the application works, this may or may not be a problem. Since the tomato cannot be reached from the direction the picture was taken, because of the occluding material, it might in fact be appropriate to not classify this as a tomato.

These examples have achieved an acceptable classification of the image pixels into object and not-object. The resulting groups of white pixels are commonly

Excurse 12.2: Specular Highlights

Specular highlights in images are reflections of bright light sources and can complicate segmentation as shown in ◘ Fig. 12.3.

As discussed in ▶ Sect. 10.3.5 the light reflected by most real objects has two components: the specular surface reflectance which does not change the spectrum of the light; and the diffuse body reflectance which filters the reflected light.

There are several ways to reduce the problem of specular highlights. Firstly, move or remove the problematic light source, or move the camera. Secondly, use a diffuse light source near the camera, for instance a ring illuminator that fits around the lens of the camera. Thirdly, attenuate the specular reflection using a polarizing filter since light that is specularly reflected from a dielectric surface will be polarized.

known as blobs. It is interesting to note that we have not specified any threshold or any definition of the object color, but we did have to specify the number of classes and determine which of those classes corresponded to the objects of interest. ▶ We have also had to choose the sequence of image processing steps and the parameters for each of those steps, for example, the radius of the structuring element. Pixel classification is a difficult problem but we can get quite good results by exploiting knowledge of the problem, having a good collection of image processing tricks, and experience.

This is a relatively easy problem. The color of the object of interest is known (we could use the `colorname` function to find it) so we could compute the distance between each cluster center and the color of interest by name and choose the cluster that is closest.

12.1.1.3 Semantic Segmentation

For many applications, deep learning techniques have overtaken previous approaches to segmenting complex scenes. Deep learning involves the use of convolutional neural networks loosely inspired by the biological structure of a brain. Artificial neural networks contain many elemental computational units organized in layers. The word *deep* refers to the fact that these networks include many layers that in turn have millions of tunable parameters. These complex/deep neural networks were made practical by revolutionary improvements in performance of Graphical Processing Units (GPUs). For example, for a popular network called Resnet-50, the number of layers is 50 and the number of parameters is approximately, 25 million. A deep network is trained using large amounts of labeled data, that is images where the outlines of objects have been drawn by humans and the object type determined by humans. ▶ The training process uses a feedback loop and an optimization process that adjusts millions of weights in the neural network so that the predicted output for a set of training images best matches the human labeled data. Training may take from hours to days on GPU hardware, and the result is a neural network model, literally a set of millions of parameters that tune the network structure.

This is the basis of a whole new industry in data labeling. MATLAB provides several tools for labeling including `imageLabeler`, `videoLabeler`, and `groundTruthLabeler` which combines labeling of video and point cloud data in a single app.

Using the trained neural network to map an input image to a set of outputs is referred to as *inference* and is relatively fast, typically a small fraction of a second, and does not require a GPU although use of the GPU improves performance. Neural networks have the ability to generalize, that is, to recognize an object even if it has never encountered that exact image before. They are not perfect, but can give very impressive results, most often superior to classical approaches especially for complex natural scenes.

The details of neural network architectures and training networks is beyond the scope of this book, but it is instructive to take a pretrained network and apply it to an image. Semantic segmentation classifies each pixel according to the *type* of object it belongs to, for example a person, car, road etc. We will use the Deep Learning Toolbox™ and the specific network we will use is Deeplab v3+ based on ResNet-18, a fully convolutional network that has been pretrained using the CamVid dataset. ▶ Although CamVid contains 32 semantic classes including car, pedestrian and road, the pretrained network that we will use reduces that number to 11, for example, by combining passenger cars with SUVs, etc. You can learn about how this network was trained by following the example ▶ https://sn.pub/MajWCj.

To learn about the CamVid dataset, see ▶ https://sn.pub/1GFbeC.

The first step is to load the image we wish to segment

```
>> I = imread("streetScene.png");
>> imshow(I)
```

which is shown in ◻ Fig. 12.8a. This is a complex scene and all the approaches we have discussed so far would be unable to segment this scene in a meaningful way.

Next, we download the pretrained Deeplab v3+ network,

```
>> pretrainedURL = "https://ssd.mathworks.com/supportfiles" + ...
>>    "/vision/data/deeplabv3plusResnet18CamVid.zip";
>> pretrainedFolder = fullfile(tempdir,"pretrainedNetwork");
>> pretrainedNetworkZip = fullfile(pretrainedFolder, ...
>>    "deeplabv3plusResnet18CamVid.zip");
>> mkdir(pretrainedFolder);
```

◘ Fig. 12.8 Semantic segmentation using Deeplab v3+ based on ResNet-18. **a** the original scene; **b** pixel classification (Image courtesy of Gabriel Brostow; ► http://www0.cs.ucl.ac.uk/staff/G.Brostow/)

```
>> disp("Downloading pretrained network (58 MB)...")
>> websave(pretrainedNetworkZip,pretrainedURL);
>> unzip(pretrainedNetworkZip, pretrainedFolder);
```

we load it into workspace and invoke the inference on the input image

```
>> pretrainedNetwork = fullfile(pretrainedFolder, ...
>>    "deeplabv3plusResnet18CamVid.mat");
>> data = load(pretrainedNetwork);
>> C = semanticseg(I,data.net);
```

then visualize the results

```
>> classes = ["Sky","Building","Pole","Road","Pavement", ...
>>    "Tree","SignSymbol","Fence","Car","Pedestrian","Bicyclist"];
>> labels = 1:size(classes,2);
>> cmap = im2single(squeeze(label2rgb(labels,"colorcube")));
>> B = labeloverlay(I,C,Colormap=cmap,Transparency=0.4);
>> imshow(B)
>> pixelLabelColorbar(cmap,classes);
```

The outputs are shown in ◘ Fig. 12.8b and are quite impressive. For example, to find all the road pixels is now easy

```
>> road = (C == "Road");
```

since `"Road"` is the label assigned to the road pixels by this pretrained network.

12.1.2 Representation: Distinguishing Multiple Objects

In the previous section we took grayscale or color images and processed them to produce binary images that contain one or more regions of white pixels. When we look at the binary image in ◘ Fig. 12.9a we see five white objects but the image is just an array of black and white pixels – there is no notion of objects. What we mean when we talk about objects is a set of pixels, of the same class, that are *adjacent* to, or *connected* to, each other. In this section we consider the problem of representation or allocating pixels to spatial sets $S_1 \ldots S_m$ which correspond to the human notion of objects, and where m is the number of objects in the scene.

12.1.2.1 Creating Binary Blobs

Consider the binary image

```
>> im = imread("multiblobs.png");
```

which is shown

```
>> imshow(im)
```

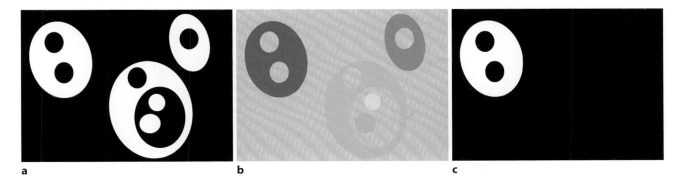

Fig. 12.9 Image labeling example. **a** Binary image; **b** labeled image; **c** all pixels with the label 1

in ▪ Fig. 12.9a. We quickly identify a number of white and black blobs in this scene but what defines a blob? It is a set of pixels of the same class that are *connected* to each other. More formally we could say a blob is a spatially contiguous region of pixels of the same class. Blobs are also known as regions or connected components.

Connected component or connectivity analysis can be applied to this binary image using the `bwlabel` function ▶

```
>> [label,m] = bwlabel(im);
```

The number of sets, or components, in this image is

```
>> m
m =
    5
```

comprising five white blobs. ▶ These blobs are labeled from 1 to 5. The returned label matrix has the same size as the original image and each element contains the label $1, \ldots, m$ of the set to which the corresponding input pixel belongs. The label matrix can be displayed as an image ▶ in false color

```
>> imshow(label2rgb(label))
```

as shown in ▪ Fig. 12.9b. Each connected region has a unique label and hence unique color. Looking at the label values in this image, or by interactively probing the displayed label matrix using `imtool(label, [0 m])`, we see that the background has been labeled as 0, the leftmost blob is labeled 1, the rightmost blob is labeled 5, and so on.

To obtain an image containing just a particular blob is now easy. To select all pixels belonging to region 1 we create a logical image

```
>> reg1 = (label==1);
>> imshow(reg1)
```

which is shown in ▪ Fig. 12.9c. The total number of pixels in this blob is given by the total number of true-valued pixels in this logical image

```
>> sum(reg1(:))
ans =
     171060
```

In this example we have assumed 4-way connectivity, that is, pixels are connected within a region only through their north, south, east and west neighbors of the same class. The 8-way connectivity option allows connection via any of a pixel's eight neighbors of the same class. ▶

There is another function for connected component analysis called `bwconncomp`. It is more memory efficient and sometimes faster than `bwlabel`. `bwlabel` is slightly simpler to explain and therefore it is used in this section to illustrate basic concepts associated with connected component analysis.

To count the six black blobs (the background and the *holes*) we can first invert the input image `im`.

We have seen a label image previously. The output of the MSER function in ▪ Fig. 12.5 is a label image.

8-way connectivity can lead to surprising results. For example a black and white checkerboard would have just two regions; all white squares are one region and all the black squares another.

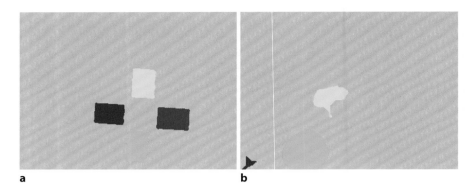

a   b

◻ **Fig. 12.10** Label images for the targets and garden examples in false color. The value of each pixel is the label of the spatially contiguous set to which the corresponding input pixel belongs

Returning now to the examples from the previous section. For the colored targets

```
>> targetsLabel = bwlabel(targetsBin);
>> imshow(label2rgb(targetsLabel));
```

and the garden image

```
>> tomatoesLabel = bwlabel(tomatoesBin);
>> imshow(label2rgb(tomatoesLabel));
```

the connected regions are shown in false color in ◻ Fig. 12.10. We are now starting to know something quantitative about these scenes: there are four yellow objects and three red objects respectively.

12.1.2.2 Graph-Based Segmentation

So far we have classified pixels based on some homogeneous characteristic of the object such as intensity or color. Consider now the complex scene

```
>> im = imread("58060.jpg");
```

The Berkeley segmentation site
► https://sn.pub/Mh5jb9 hosts these images plus a number of different human-made segmentations.

shown in ◻ Fig. 12.11a. The Gestalt principle of emergence says that we identify objects as a whole rather than as a collection of parts – we see a bowl of grain rather than deducing a bowl of grain by recognizing its individual components. However when it comes to a detailed pixel by pixel segmentation things become quite subjective – different people would perform the segmentation differently based on judgment calls about what is *important*. ◄ For example, should the colored stripes on the cloth be segmented? If segments represent real world objects, then the Gestalt view would be that the cloth should be just one segment. However the stripes are real, some effort was made to create them, so perhaps they should be segmented. This is why segmentation is a *hard* problem – humans cannot agree on what is correct. No computer algorithm could, or could be expected to, make this type of judgment.

Nevertheless, more sophisticated algorithms can do a very impressive job on complex real world scenes. The image can be represented as a graph (see Appendix I) where each pixel is a vertex and has 8 edges connecting it to its neighboring pixels. The weight of each edge is a nonnegative measure of the dissimilarity between the two pixels – the absolute value of the difference in color. The algorithm starts with every vertex assigned to its own set. At each iteration the edge weights are examined and if the vertices are in different sets but the edge weight is below a threshold the two vertex sets are merged.

For the image discussed above, to illustrate the power of the graph-based segmentation we start by converting the image to L*a*b* color space.

```
>> im_lab = rgb2lab(im);
```

◻ Fig. 12.11 Complex segmentation example. **a** Original color image (Image from the Berkeley Segmentation Dataset; Martin et al. 2001); **b** Yellow overlay of lazy snapping (graph-based) segmentation of the grains; **c** Image Segmenter app illustrating exemplar foreground (green scribbles) and background (red scribbles) selections

In the next step, we find subregions of the image with similar characteristics using a technique called superpixels. This step produces a familiar label matrix. The number of superpixels, 872, was chosen empirically.

```
>> L = superpixels(im_lab,872,IsInputLab=true);
```

The L*a*b* color space occupies ranges [0, 100] for L, [−86.18, 98.23] for a* and [−107.86, 94.48] for b*. To obtain more robust results we can rescale these channels to the range [0,1].

```
>> im_lab(:,:,1) = im_lab(:,:,1)/100;
>> im_lab(:,:,2) = (im_lab(:,:,2)+86.1827)/184.4170;
>> im_lab(:,:,3) = (im_lab(:,:,3)+107.8602)/202.3382;
```

Our objective is to segment the grains. Notice the many colors and shadows of the grains. To obtain reliable results, we guide the segmentation by loading manually selected representative foreground and background pixels. Finally, we apply a graph-based segmentation technique called lazy snapping (Li et al. 2004).

```
>> exemplars = load("graphSegMarkup.mat");
>> BW = lazysnapping(im_lab,L,exemplars.foregroundInd, ...
>>   exemplars.backgroundInd);
>> imshow(imoverlay(im, BW))
```

The resulting segmentation mask is shown overlayed in yellow on top of the original image in ◻ Fig. 12.11b. The easiest way to accomplish the entire segmentation is with the Image Segmenter (imageSegmenter) app shown in ◻ Fig. 12.11c. The exemplar foreground and background pixels can be easily selected using this tool.

12.1.3 **Description**

In the previous section, we learned how to find connected components in the image and how to isolate particular components such as shown in ◨ Fig. 12.9c. However, this representation of the component is still just an image with logical or labeled pixel values rather than a concise numeric description of object's size, position and shape.

12.1.3.1 **Bounding Boxes**

The simplest representation of size and shape is the bounding box – the smallest rectangle with sides parallel to the u- and v-axes that encloses the region. We will illustrate this with a simple binary image

```
>> sharks = imread("sharks.png");
```

which is shown in ◨ Fig. 12.12a. As described above we will label the pixels and select all those belonging to region 4

```
>> [label,m] = bwlabel(sharks);
>> blob = (label==4);
```

and the resulting logical image is shown in ◨ Fig. 12.12b. The number of pixels in this region is simply the sum

```
>> sum(blob(:))
ans =
    7728
```

The coordinates of all the nonzero (object) pixels are the corresponding elements of

```
>> [v,u] = find(blob);
```

where u and v are each vectors of size

```
>> size(u)
ans =
    7728          1
```

The bounds of the region ◄ are

The bounds and many other shape statistics can be obtained more simply using the regionprops function.

```
>> umin = min(u)
umin =
    443
>> umax = max(u)
umax =
    581
```

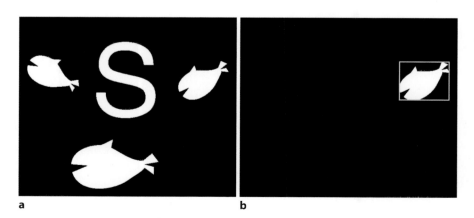

a   b

◨ **Fig. 12.12   a** Sharks image; **b** Region 4 with bounding box (yellow)

```
>> vmin = min(v)
vmin =
    125
>> vmax = max(v)
vmax =
    235
```

These bounds define a rectangle which we can superimpose on the image

```
>> imshow(insertShape(im2double(blob),"rectangle", ...
>>      [umin,vmin,umax-umin,vmax-vmin]))
```

as shown in ◘ Fig. 12.12b. The bounding box fits snugly around the blob and its center could be considered as the center of the blob. The sides of the bounding box are horizontal and vertical which means that as the blob rotates the size and shape of the bounding box would change, even though the size and shape of the blob does not.

12.1.3.2 Moments

Moments are a rich and computationally cheap class of image features that can describe region size and location as well as orientation and shape. The moment of an image $\mathbf{X}$ is a scalar

$$m_{pq}(\mathbf{X}) = \sum_{(u,v) \in \mathbf{X}} u^p v^q \, \mathbf{X}_{u,v} \tag{12.1}$$

where $(p + q)$ is the *order* of the moment. The zeroth moment $p = q = 0$ is

$$m_{00}(\mathbf{X}) = \sum_{(u,v) \in \mathbf{X}} \mathbf{X}_{u,v} \tag{12.2}$$

and for a binary image, where the background pixels are zero, this is simply the number of nonzero (white) pixels – the area of the region.

For the single shark, the zeroth moment is therefore

```
>> m00 = sum(blob(:))
m00 =
        7728
```

which is the area of the region in units of pixels.

Moments can be given a physical interpretation by regarding the image function as a mass distribution. Consider the region as being made out of thin plate where each pixel has one unit of area and one unit of mass. The total mass of the region is m_{00} and the center of mass or centroid of the region is

$$u_c = \frac{m_{10}}{m_{00}}, \quad v_c = \frac{m_{01}}{m_{00}} \tag{12.3}$$

where m_{10} and m_{01} are the first-order moments. For our example the centroid of the target region is

```
>> [v,u] = find(blob);
>> uc = sum(u(:))/m00
uc =
    503.4981
>> vc = sum(v(:))/m00
vc =
    184.7285
```

which we can display

```
>> hold on
>> plot(uc,vc,"gx",uc,vc,"go");
```

as shown in ◼ Fig. 12.12b.

The central moments μ_{pq} are computed with respect to the centroid

$$\mu_{pq}(\mathbf{X}) = \sum_{(u,v)\in\mathbf{X}} (u - u_c)^p (v - v_c)^q \mathbf{X}_{u,v} \qquad (12.4)$$

and are invariant to the position of the region. They are related to the moments m_{pq} by

$$\mu_{10} = 0, \qquad\qquad \mu_{01} = 0$$
$$\mu_{20} = m_{20} - \frac{m_{10}^2}{m_{00}}, \quad \mu_{02} = m_{02} - \frac{m_{01}^2}{m_{00}}, \quad \mu_{11} = m_{11} - \frac{m_{10}m_{01}}{m_{00}} \qquad (12.5)$$

Using the thin plate analogy again, the inertia tensor of the region about axes parallel to the u- and v-axes and intersecting at the centroid of the region is given by the symmetric matrix

$$\mathbf{J} = \begin{pmatrix} \mu_{20} & \mu_{11} \\ \mu_{11} & \mu_{02} \end{pmatrix} . \qquad (12.6)$$

The central second moments μ_{20}, μ_{02} are the moments of inertia and μ_{11} is the product of inertia. The product of inertia is nonzero if the shape is asymmetric with respect to the region's axes.

The *equivalent ellipse* is the ellipse that has the same inertia matrix as the region. For our example

```
>> u20 = sum(u(:).^2)-sum(u(:))^2/m00;
>> u02 = sum(v(:).^2)-sum(v(:))^2/m00;
>> u11 = sum(u(:).*v(:))-sum(u(:))*sum(v(:))/m00;
>> J = [u20 u11; u11 u02]
J =
   1.0e+06 *
    7.8299   -2.9169
   -2.9169    4.7328
```

The eigenvalues and eigenvectors of **J** are related to the radii of the ellipse and the orientation of its major and minor axes (see ▶ App. C.1.4). For this example, the eigenvalues

```
>> lambda = eig(J)
lambda =
   1.0e+06 *
    2.9788
    9.5838
```

are the principal moments of the region. The maximum and minimum radii of the equivalent ellipse are

MATLAB returns eigenvalues in increasing order: λ_1 then λ_2.

$$a = 2\sqrt{\frac{\lambda_2}{m_{00}}}, \ b = 2\sqrt{\frac{\lambda_1}{m_{00}}} \qquad (12.7)$$

respectively where $\lambda_2 \geq \lambda_1$. In MATLAB, this is ◀

```
>> a = 2*sqrt(lambda(2)/m00)
a =
   70.4313
>> b = 2*sqrt(lambda(1)/m00)
b =
   39.2663
```

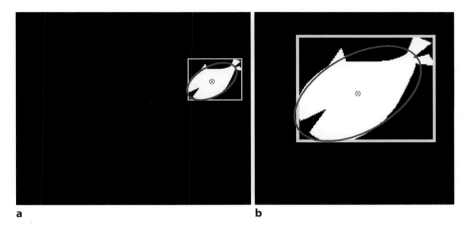

a   **b**

◻ Fig. 12.13 Sharks image. **a** Equivalent ellipse in red, centroid and bounding box in yellow for region 4 of the sharks image; **b** zoomed view

in units of pixels. These lengths are characteristic of this particular shape and are invariant to scale and rotation . The aspect ratio of the region

```
>> b/a
ans =
    0.5575
```

is a scalar that crudely characterizes the shape, and is invariant to position, scale and rotation of the region.

The eigenvectors of **J** are the principal axes of the ellipse – the directions of its major and minor axes. The major, or principal, axis is the eigenvector v corresponding to the maximum eigenvalue. For this example, the eigenvectors are the columns of

```
>> [x,lambda] = eig(J);
>> x
x =
   -0.5153   -0.8570
   -0.8570    0.5153
```

and v is always the *last* column of the returned eigenvector matrix

```
>> v = x(:,end);
```

since MATLAB returns eigenvalues in increasing order. The angle of this eigenvector, with respect to the horizontal axis, is

$$\theta = \tan^{-1} \frac{v_y}{v_x}$$

and for our example this is

```
>> orientation = atan(v(2)/v(1));
>> rad2deg(orientation)
ans =
  -31.0185
```

degrees which indicates that the major axis of the equivalent ellipse is approximately 30 degrees above horizontal. ▶

We can now superimpose the equivalent ellipse over the region

```
>> plotellipse([a b orientation],[uc vc],"r",LineWidth=2)
```

and the result is shown in ◻ Fig. 12.13.

With reference to ◻ Fig. C.2b the angle increases clockwise from the horizontal since the y-axis of the image is downward so the z-axis is into the page.

◘ Table 12.1   Region features and their invariance to camera motion: translation, rotation about the object's centroid and scale factor

	Translation	**Rotation**	**Scale**
Area	✓	✓	✗
Centroid	✗	✓	✓
Orientation θ	✓	✗	✓
Aspect ratio	✓	✓	✓
Circularity	✓	✓	✓

To summarize, we have created an image containing a spatially contiguous set of pixels corresponding to one of the objects in the scene that we segmented from the original color image. We have determined its area, a box that entirely contains it, its position (the location of its centroid), its orientation and its shape (aspect ratio). The equivalent ellipse is a crude indicator of the region's shape but it is invariant to changes in position, orientation and scale. The invariance of the different blob descriptors to camera motion is summarized in ◘ Tab. 12.1.

12.1.3.3 **Blob Descriptors**

The previous section illustrated the basic concepts about moments, but there is a simpler way to perform many of these computations

```
>> f = regionprops(blob,"Area","Centroid","Orientation", ...
>>  "MajorAxisLength","MinorAxisLength")
f =
  struct with fields:
              Area: 7728
          Centroid: [503.4981 184.7285]
    MajorAxisLength: 140.8673
    MinorAxisLength: 78.5410
        Orientation: 31.0185
```

The regionprops function can return many more statistics. You can easily see the complete list by invoking regionprops(blob,"all"). See the documentation for a thorough description of all the available region statistics.

which returns a struct that contains the specified features describing this region. ◄ These values can be easily accessed, for example

```
>> f.Area
ans =
  7728
>> f.Orientation
ans =
  31.0185
```

The regionprops function can also process labeled images. In this case, the returned statistics would be organized by the label number.

You can also compute features for *every* region in the image. ◄

```
>> fv = regionprops("table",sharks,"Area","BoundingBox", ...
>>              "Orientation")
fv =
  4×3 table
    Area              BoundingBox              Orientation
```

The bounding boxes are represented by a 4-vector containing coordinates of upper left corner of the box followed by width and height, [ulX, ulY, dX, dY]. The corner coordinates are fractional indicating that their locations fall on the pixel's upper left corner, rather than its center.

Area	BoundingBox				Orientation
7746	23.5	110.5	144	103	-28.871
18814	146.5	349.5	242	140	1.0701
14899	214.5	70.5	165	218	-84.56
7728	442.5	124.5	139	111	31.018

In this case, we requested the output in the table format, which is easy to visualize and convenient to manipulate. The table is a class and in our case, each row of the table holds statistics for individual blobs. For example, you can access all bounding boxes by ◄

```
>> bbox = fv.BoundingBox
bbox =
     23.5000   110.5000   144.0000   103.0000
    146.5000   349.5000   242.0000   140.0000
    214.5000    70.5000   165.0000   218.0000
    442.5000   124.5000   139.0000   111.0000
```

which returns a regular matrix. We can readily plot all the boxes using the showShape function

```
>> imshow(sharks), hold on
>> showShape("rectangle",bbox);
```

It is sometimes useful to exclude blobs touching the image border, for example, for the tomato image (▶ Sect. 12.1.1.2). This can be achieved using the imclearborder function

```
>> cleared_tomatoes = imclearborder(tomatoesBin);
```

This may be a good pre-processing step prior to running regionprops, in the cases where you do not want your statistics to be skewed by objects that are not fully visible.

The regionprops function is a powerful tool for clearing unwanted objects from your images. For example, to eliminate smaller items and those with negative orientation from the sharks image, we can use the ismember function

```
>> L = bwlabel(sharks);
>> stats = regionprops(L,"Area","Orientation");
>> idx = find([stats.Area] > 10000 & [stats.Orientation] > 0);
>> filtered_sharks = ismember(L,idx);
>> imshow(filtered_sharks)
```

The regionprops function provides a powerful way to analyze and manipulate binary images.

12.1.3.4 Shape from Object Boundary

The shape of a region is concisely described by its perimeter, boundary, contour or perimeter pixels – sometimes called edgels. ◘ Fig. 12.14 shows three common ways to represent the boundary of a region – each will give a slightly different estimate of the boundary length. A chain code is a list of the outermost pixels of the region whose center's are linked by short line segments. In the case of a 4-neighbor chain code, the successive pixels must be adjacent and the boundary segments have an orientation of $k \times 90°$, where $k \in \{0, 1, 2, 3\}$. With an 8-neighbor chain code, or Freeman chain code, the boundary segments have an orientation

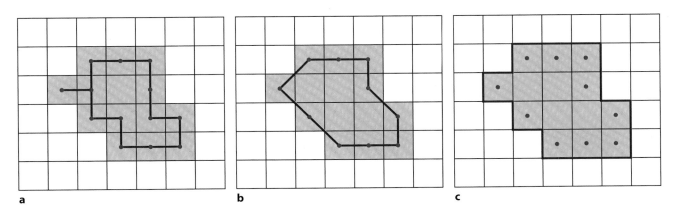

◘ **Fig. 12.14** Object boundary representations with region pixels shown in gray, boundary segments shown in blue and the center of boundary pixels marked by a red dot. **a** Chain code with 4 directions; **b** Freeman chain code with 8 directions; **c** crack code. The boundary lengths for this example are respectively 14, 12.2 and 18 pixels of $k \times 45°$, where $k \in \{0, \ldots, 7\}$. The crack code has its segments in the *cracks* between the pixels on the boundary of the region and the pixels outside the region. These have orientations of $k \times 90°$, where $k \in \{0, 1, 2, 3\}$.

The boundary can be encoded as a list of pixel coordinates (u_i, v_i) or very compactly as a bit string using just 2 or 3 bits to represent k for each segment. These various representations are equivalent and any representation can be transformed to another.

> 🛈 Note that for chain codes, the boundary follows a path that is on average half a pixel inside the true boundary and therefore underestimates the boundary length. The error is most significant for small regions.

There are two statistics computed by `regionprops` that rely on tracing around the boundary of the objects using 8-neighbor chain code and they include perimeter and circularity

```
>> sharks = imread("sharks.png");
>> fv = regionprops(sharks,"Perimeter","Circularity","Centroid");
>> fv(1)
ans =
  struct with fields:
       Centroid: [84.2352 160.6634]
    Circularity: 0.4074
      Perimeter: 488.7940
```

The boundary is a list of edge points represented as a matrix with one row per boundary point. To obtain boundaries for all regions, use the `bwtraceboundaries` function

```
>> b = bwboundaries(sharks);
>> size(b{1})
ans =
   445     2
```

In this case, there are 445 boundary points and the first five points are

```
>> b{1}(1:5,:)
ans =
   142     24
   141     25
   140     25
   139     25
   139     26
```

The displayed perimeter length of `488.7940` has had a heuristic correction applied to compensate for the underestimation due to chain coding.

The boundary and centroids can be overlaid on the image by

```
>> imshow(sharks); hold on
>> for k = 1:length(b)
>>     boundary = b{k};
>>     plot(boundary(:,2),boundary(:,1),"r",LineWidth=2)
>>     plot(fv(k).Centroid(1),fv(k).Centroid(2),"bx")
>> end
```

which is shown in ◘ Fig. 12.15.

Circularity is another commonly used and intuitive shape feature. It is defined as

$$\rho = \frac{4\pi m_{00}}{p^2} \tag{12.8}$$

where p is the region's perimeter length. Circularity has a maximum value of $\rho = 1$ for a circle, is $\rho = \frac{\pi}{4}$ for a square and zero for an infinitely long line. Circularity

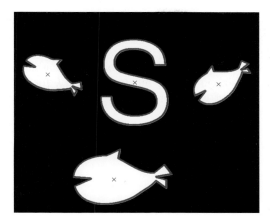

◻ Fig. 12.15 Boundaries (*red*) and centroids of four blobs

is also invariant to translation, rotation and scale. Circularity results for the `sharks` image

```
>> [fv.Circularity]
ans =
    0.4074    0.4004    0.1362    0.4092
```

show that the circularity is the same for blobs 1, 2 and 4, and much lower for blob 3 due to it being effectively a long line. ▶

For small blobs, quantization effects can lead to significant errors in computing circularity.

Every object has one external boundary, which may include a section of the image border if the object touches the border. An object with holes has one internal boundary per hole. The parent-child dependencies between boundaries and holes can be explored via an adjacency matrix returned by the `bwboundaries` function.

```
>> blobs = imread("multiblobs.png");
>> [B,L,N,A] = bwboundaries(blobs);
```

The returned values are: `B`, a cell array of boundaries specified as a list of perimeter points, `L` the label matrix, `N` the number of boundaries and `A`, the adjacency matrix which is square with side of length `max(L(:))`, in other words, the total number of objects and holes in the image. The row and column indicies of `A` correspond to the indicies of boundaries stored in `B`. If `A(i,j)` is one then object *i* is a child of object *j*. For example, using `A`, we can identify all top-level parent blobs and their hole boundaries, by

```
>> imshow(blobs);
>> hold on;
>> for k = 1:N
>>    if nnz(A(:,k)) > 0
>>      boundary = B{k};
>>      plot(boundary(:,2),boundary(:,1),"r",LineWidth=3);
>>      for l = find(A(:,k))'
>>        boundary = B{l};
>>        plot(boundary(:,2),boundary(:,1),"g",LineWidth=3);
>>      end
>>    end
>> end
```

which plots top level parent blob boundaries in red and their holes in green as shown in ◻ Fig. 12.16.

One way to analyze the rich shape information encoded in the boundary is to compute the distance and angle to every boundary point with respect to the object's centroid – this is computed by the `boundary2polar` function in the RVC Toolbox. Going back to the *sharks* image, we will process the first object.

```
>> [r,th] = boundary2polar(b(1),fv(1));
>> plot([r th])
```

◨ Fig. 12.16   First level parent boundaries in red and their holes in green

The result is shown in ◨ Fig. 12.17a. These are computed for 400 points (fixed) evenly spaced along the entire boundary of the object. Both, the radius and angle signatures describe the shape of the object. The angle signature is invariant to the scale of the object while the amplitude of the radius signature scales with object size. The radius signatures of all four blobs can be compared by

```
>> hold on
>> for i = 1:size(b,1)
>>    [r,t] = boundary2polar(b(i),fv(i));
>>    plot(r/max(r));
>> end
```

and are shown in ◨ Fig. 12.17b. We have normalized by the maximum value in order to remove the effect of object scale. The signatures are a function of normalized distance along the boundary. They all start at the left-most pixel on the object's boundary. Different objects of the same shape have identical signatures but possibly shifted horizontally and wrapped around – the first and last points in the horizontal direction are adjacent.

To compare the shape profile of objects requires us to compare the signature for all possible horizontal shifts. The radius signatures of all four blobs is

```
>> r_all = boundary2polar(b,fv);
```

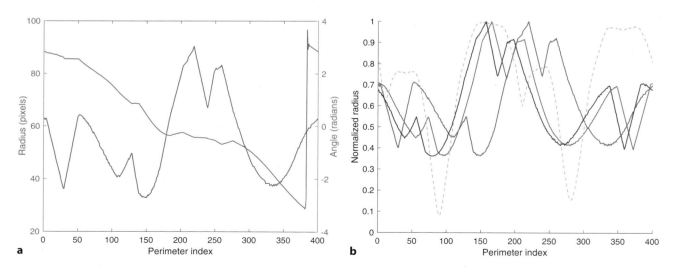

◨ Fig. 12.17   Radius signature matching. **a** Radius and angle signature for blob 1 (top left shark in ◨ Fig. 12.15), **b** normalized radius signatures for all blobs (letter S is shown dashed)

which is a 400×4 matrix with the radius signatures as columns. To compare the signature of blob 1 with all the signatures

```
>> boundmatch(r_all(:,1),r_all)
ans =
    1.0000    0.9886    0.6531    0.9831
```

which indicates that the shape of blob 1 closely matches the shape of itself and blobs 2 and 4, but not the shape of blob 3. The `boundmatch` function computes the 1-dimensional normalized cross correlation, see ◘ Tab. 12.1, for every possible rotation of one signature with respect to the other, and returns the highest value.

There are many variants to the approach described. The signature can be Fourier transformed and described more concisely in terms of a few Fourier coefficients. The boundary curvature can be computed which highlights corners, or the boundary can be segmented into straight lines and arcs.

12.1.4 Object Detection Using Deep Learning

Deep learning networks, introduced in ► Sect. 12.1.1.3, are capable of solving many tasks, including the detection of instances of objects in a scene and directly determining their class and location indicated by a bounding box. We will illustrate this with the YOLO multi-class object detector trained to detect seven indoor objects: exit sign, fire extinguisher, chair, clock, trash bin, screen and a printer based on the Indoor Object Detection dataset. ►

We start by loading a pretrained YOLO network

To learn about the Indoor Object Detection dataset, see ► https://zenodo.org/record/2654485. The YOLO training procedure is described in detail at ► https://sn.pub/LjtQ1T.

```
>> pretrainedURL = "https://www.mathworks.com/supportfiles/" + ...
>>     "vision/data/yolov2IndoorObjectDetector.zip";
>> pretrainedFolder = fullfile(tempdir,"pretrainedNetwork");
>> pretrainedNetworkZip = fullfile(pretrainedFolder, ...
>>     "yolov2IndoorObjectDetector.zip");
>> mkdir(pretrainedFolder);
>> disp("Downloading pretrained network (98 MB)...");
>> websave(pretrainedNetworkZip,pretrainedURL);
>> unzip(pretrainedNetworkZip, pretrainedFolder);
>> pretrainedNetwork = fullfile(pretrainedFolder, ...
>>     "yolov2IndoorObjectDetector.mat");
>> pretrained = load(pretrainedNetwork);
```

and then loading an image of an indoor scene

```
>> I = imread("doorScene.jpg");
>> imshow(I);
```

which is shown in ◘ Fig. 12.18a. We can now apply the network to recognize the objects in the scene

```
>> I = imresize(I,pretrained.detector.TrainingImageSize);
>> [bbox,score,label] = detect(pretrained.detector,I);
>> annotatedI = insertObjectAnnotation(I,"rectangle",bbox, ...
>>     "Label: "+string(label)+"; Score:"+string(score), ...
>>     LineWidth=3,FontSize=14,TextBoxOpacity=0.4);
>> imshow(annotatedI)
```

and the results are shown in ◘ Fig. 12.18b. You can imagine just how useful such a classification can be for a robot navigating through complex indoor environment.

We have shown pixel classification (semantic segmentation) and object detection using deep learning. Deep learning has a vast range of other applications, including pixel classification with simultaneous segmentation of individual objects (instance segmentation), activity recognition in videos, depth estimation from monocular images, and many more. We have only scratched the surface of these techniques. There is also a plethora of ever-improving network implementations

☐ **Fig. 12.18**   Deep neural network, YOLO v2, for detecting objects in indoor environment. **a** Input image; **b** The processed result which includes one misclassification of a fire extinguisher symbol as the real object, albeit with a lower classification score

Use of deep learning networks also requires Deep Learning Toolbox™.

for any one of these areas. For example, just in the area of multi-class object detection, there are networks such as YOLO, SSD, and Faster R-CNN which are also readily available in the Computer Vision Toolbox™. ◄

12.1.5   Summary

We have discussed the process of transforming an input image, gray scale or color, into concise descriptors of regions within the scene. The criteria for what constitutes a region is *application specific*. For a tomato picking robot it would be round red regions, for landing a drone it might be yellow targets on the ground.

The process outlined is the classical *bottom up* approach to machine vision applications and the key steps are:

— Classifying the pixels according to the application specific criterion, for example, "redness", "yellowness" or "person-ness". Each pixel is assigned to a class c.
— Grouping adjacent pixels of the same class into sets, and each pixel is assigned a label S indicating to which spatial set it has been assigned.
— Describing the sets in terms of features derived from their spatial extent, moments, equivalent ellipse and perimeter.

These steps are a progression from *low-level* to *high-level* representation. The low-level operations consider pixels in isolation, whereas the high-level is concerned with more abstract concepts such as size and shape. The MSER and graphcuts algorithms are powerful because they combine steps 1 and 2 and consider regions of pixels and localized differences in order to create a segmentation.

Importantly, none of these steps need to be perfect. Perhaps the first step has some false positives, isolated pixels misclassified as objects, that we can eliminate by morphological operations, or reject after connectivity analysis based on their small size. The first step may also have false negatives, for example specular reflection and occlusion may cause some object pixels to be classified incorrectly as non-object. In this case we need to develop some heuristics, for instance morphological processing, to fill in the gaps in the blob. Another option is to oversegment the scene – increase the number of regions and use some application-specific knowledge to merge adjacent regions. For example, a specular reflection colored region might be merged with surrounding regions to create a region corresponding to the whole fruit.

For some applications, it might be possible to engineer the camera position and illumination to obtain a high quality image but for a robot operating in the real world this luxury does not exist. A robot needs to glean as much useful information as it can from the image and move on.

Domain knowledge is always a powerful tool. Given that we know the scene contains tomatoes and plants, the observation of a large red region that is not circular allows us to infer that the tomato is occluded. The robot may choose to move to a location where the view of tomato is not occluded, or to seek a different tomato that is not occluded.

These techniques are suitable for relatively simple scenes and have the advantage of requiring only modest amounts of computation. However, in recent times they have been eclipsed by those based on deep learning which is a powerful and practical solution to complex image segmentation problems. A network transforms a color image, in a single step, to a pixel classification or object detections using human-meaningful semantic labels like "pedestrian" or "bicyclist" as shown in ▶ Sect. 12.1.1.3. MATLAB provides an extensive set of tools for deep learning and it is now relatively straightforward to train a network for the particular objects in your application. Training does however require a lot of labeled data and usually a lot of computation time.

12.2 Line Features

Lines are distinct visual features that are particularly common in human-made environments – for example the edges of roads, buildings and doorways. In ▶ Sect. 11.5.1.3 we discussed how image intensity gradients can be used to find edges within an image, and this section will be concerned with fitting line segments to such edges.

We will illustrate the principle using the very simple scene

```
>> im = imread("5points.png");
```

shown in ◻ Fig. 12.19a. Consider any one of these points – there are an infinite number of lines that pass through that point. If the point could vote for these lines,

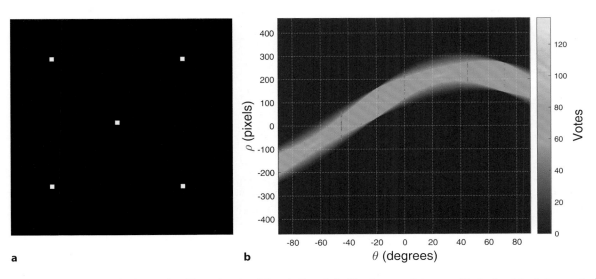

◻ **Fig. 12.19** Hough transform fundamentals. **a** Five points that define six lines; **b** the Hough accumulator array. The horizontal axis is an angle $\theta \in \mathbf{S}^1$ so we can imagine the graph wrapped around a cylinder and the left- and right-hand edges joined. The sign of ρ also changes at the join so the curve intersections on the left- and right-hand edges are equivalent

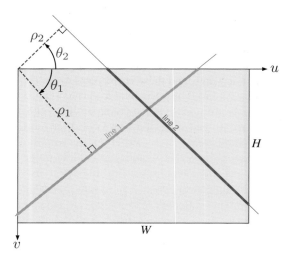

□ Fig. 12.20 (θ, ρ) parameterization for two line segments. Positive quantities are shown in blue, negative in red

then each possible line passing through the point would receive one vote. Now consider another point that does the same thing, casting a vote for all the possible lines that pass through it. One line (the line that both points lie on) will receive a vote from each point – a total of two votes – while all the other possible lines receive either zero or one vote.

We want to describe each line in terms of a minimum number of parameters but the standard form $v = mu + c$ is problematic for the case of vertical lines where $m = \infty$. Instead, it is common to represent lines using the (ρ, θ) parameterization shown in **□** Fig. 12.20

$$\rho = u \cos(\theta) + v \sin(\theta) \tag{12.9}$$

where $\theta \in [-90°, 90°)$ is the angle from the horizontal axis to the line, and $\rho \in [-\rho_{max}, \rho_{max}]$ is the perpendicular distance between the origin and the line. A horizontal line has $\theta = 90°$ and a vertical line has $\theta = 0$. Any line can therefore be considered as a point (θ, ρ) in the 2-dimensional space of all possible lines.

It is not practical to vote for one out of an infinite number of lines through each point, so we consider lines drawn from a finite set. The $\theta\rho$-space is quantized and a corresponding $N_\theta \times N_\rho$ array $\mathbf{A}$ is used to tally the votes – the *accumulator* array. For a $W \times H$ input image

$$\rho_{max} = -\rho_{min} = \sqrt{W^2 + H^2}$$

The array $\mathbf{A}$ has N_ρ elements spanning the interval $\rho \in [-\rho_{max}, \rho_{max}]$ and N_θ elements spanning the interval $\theta \in [-90°, 90°)$. The indices of the array are integers $(i, j) \subset \mathbb{N}^2$ such that

$$i \in [1, N_\theta] \mapsto \theta \in [-90°, 90°)$$
$$j \in [1, N_\rho] \mapsto \rho \in [-\rho_{max}, \rho_{max}]$$

An edge point (u, v) votes for *all* lines for which (i, j) pairs satisfy (12.9) and the elements $a_{i,j}$ are all incremented. For every i, the corresponding value of θ is computed, then ρ is computed according to (12.9) and mapped to a corresponding integer j. Every edge point adds a vote to N_θ elements of $\mathbf{A}$ that lie along a curve.

At the end of the process, those elements of $\mathbf{A}$ with the largest number of votes correspond to dominant lines in the scene. For the example of **□** Fig. 12.19a the resulting accumulator array is shown in ◨ Fig. 12.19b. Most of the array contains zero votes (dark blue) and the light curves are trails of single votes corresponding to each of the five input points. These curves intersect and those points correspond to lines with more than one vote. We see four locations where two curves intersect, resulting in cells with two votes, and these correspond to the lines joining the four outside points of ◨ Fig. 12.19a. The horizontal axis represents angle $\theta \in \mathbb{S}^1$ so the left- and right-hand ends are joined and ρ changes sign – the curve intersection points on the left- and right-hand sides of the array are equivalent. We also see two locations where three curves intersect, resulting in cells with three votes, and these correspond to the diagonal lines that include the middle point of ◨ Fig. 12.19a. This technique is known as the Hough (pronounced huff) transform.

Consider the more complex example of a solid square rotated counter-clockwise by 20°

```
>> im = zeros(256,"uint8");
>> im = insertShape(im,"filledrectangle",[64 64 128 128]);
>> im = imrotate(rgb2gray(im),20);
```

We compute the edge points

```
>> edges = edge(im,"canny");
```

which are shown in ◨ Fig. 12.21a. The Hough transform is computed by

```
>> [A,theta,rho] = hough(edges);
```

and returns, A, the two-dimensional vote accumulator array as well as theta and rho arrays of θ and ρ values used to create the accumulator. By default, the $\theta\rho$-plane is quantized into 179 columns ($-90°$ to $89°$) and $2D + 1$ rows, where D is the image diagonal in pixels. ▶ The accumulator array can be visualized as an image

> θ has a range of $[-90°, 90°)$, it is asymmetric about zero and has an even number of elements.

```
>> imagesc(theta,rho,A);
>> set(gca,YDir="normal");
```

which is shown in ◨ Fig. 12.21b. The four bright spots correspond to dominant edges in the input image. We can see that many other possible lines have received a small number of votes as well.

The next step is to find the peaks in the accumulator array and the corresponding lines

```
>> p = houghpeaks(A,4);
>> lines = houghlines(edges,theta,rho,p)
lines =
  1×4 struct array with fields:
    point1
    point2
    theta
    rho
```

which returns an array of structs holding line segments associated with the four strongest peaks that were selected using houghpeaks. Each of the four structs contains line segment end-points and a corresponding location of the peak θ and ρ. The line segments are found by examining the points in the original image that contributed to the particular peak in the accumulator image. ▶

Note that although the object has only four sides there are many more than four peaks in the accumulator array. We also note that there are clusters of peaks in the accumulator as shown in more detail in ◨ Fig. 12.21c. We see several bright spots (high numbers of votes) and this is due to quantization effects. Once again, we applied nonlocal maxima suppression to eliminate smaller peaks in the neighborhood of the maxima. The suppression is built into the houghpeaks function. Once a peak has been found all votes within the suppression distance are zeroed so as to eliminate any close maxima, and the process is repeated for all peaks in the voting array that exceed a specified fraction of the largest peak. ▶

> houghlines has two name-value pairs called FilledGap and MinLength. The first one controls merging of the line segments that might be close to each other and the other discards any line segments that are less than the specified minimum length. These options give you some control over the quality of returned results.

> With no argument all peaks greater than Threshold are returned by houghpeaks. This defaults to 0.5*max(A(:)), but can be set by the Threshold option to houghpeaks. The suppression region can be controlled by NHoodSize option specified as [numRows, numCols].

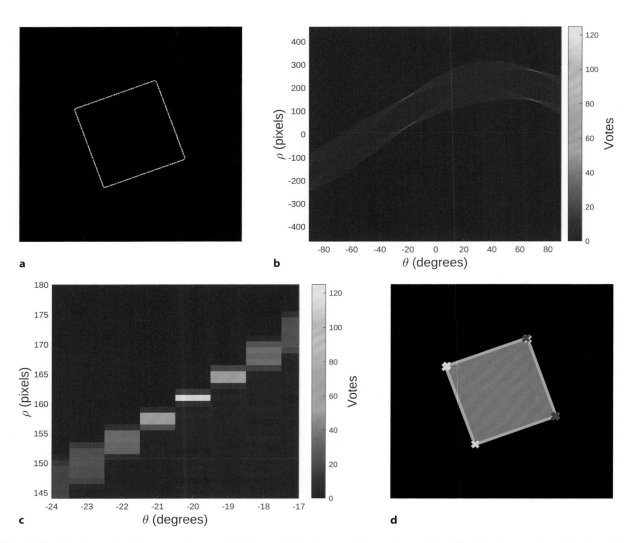

Fig. 12.21 Hough transform for a rotated square. **a** Edge image; **b** Hough accumulator; **c** closeup view of the Hough accumulator; **d** estimated lines overlaid on the original image with endpoints marked in red and yellow

The detected lines can be projected onto the original image

```
>> imshow(im), hold on
>> for k = 1:length(lines)
>>     xy = [lines(k).point1; lines(k).point2];
>>     plot(xy(:,1),xy(:,2),LineWidth=2,Color="green");
>>     plot(xy(1,1),xy(1,2),"xy",LineWidth=2);
>>     plot(xy(2,1),xy(2,2),"xr",LineWidth=2);
>> end
```

and the result is shown in ◻ Fig. 12.21d.

A real image example is

```
>> im = rgb2gray(imread("church.png"));
>> edges = edge(im,"sobel");
>> [A,theta,rho] = hough(edges);
>> p = houghpeaks(A,15);
>> lines = houghlines(edges,theta,rho,p);
```

12

◻ Fig. 12.22 Hough transform of a natural scene. The green line segments correspond to the fifteen strongest voting peaks

and the line segments corresponding to the strongest fifteen lines

```
>> imshow(im), hold on
>> for k = 1:length(lines)
>>     xy = [lines(k).point1; lines(k).point2];
>>     plot(xy(:,1),xy(:,2),LineWidth=2,Color="green");
>>     plot(xy(1,1),xy(1,2),"xy",LineWidth=2);
>>     plot(xy(2,1),xy(2,2),"xr",LineWidth=2);
>> end
```

are shown in ◻ Fig. 12.22. Many strong lines in the image have been found, and lines corresponding to the roof edges and building-ground line are correct. However, vertical line segments around the windows do not correspond precisely to lines in the image – they are the result of disjoint sections of high gradient *voting up* a line that passes through them. If we were to request more than fifteen peaks, the result would become even noisier. Note also that this time, we used the Sobel edge detector which in this case produced less noisy results as compared to Canny.

12.2.1 Summary

The Hough transform is elegant in principle and in practice it can work well or infuriatingly badly. It performs poorly when the scene contains a lot of texture or the edges are indistinct. Texture causes votes to be cast widely, but not uniformly, over the accumulator array which tends to mask the true peaks. Consequently, a lot of experimentation is required for the parameters of the edge detector and the Hough peak detector. The functions `hough`, `houghpeaks` and `houghlines` have many options which are described in the online documentation. Adjusting them can make a significant difference in the outcomes.

12.3 Point Features

The final class of features that we will discuss are point features. These are visually distinct points in the image that are also known as interest points, salient points, keypoints or corner points. We will first introduce some classical techniques for finding point features and then discuss scale-invariant techniques.

12.3.1 **Classical Corner Detectors**

We recall from ▶ Sect. 11.5.1.3 that a point on a line has a strong gradient in a direction normal to the line. However gradient *along* the line is low which means that a pixel on the line will look very much like its neighbors along the line. In contrast, an *interest point* is a point that has a high image gradient in orthogonal directions. It might be single pixel that has a significantly different intensity to all of its neighbors, or it might literally be a pixel on the corner of an object. Since interest points are quite distinct they have a much higher likelihood of being reliably detected in different views of the same scene. They are therefore key to multi-view techniques such as stereo and motion estimation which we will discuss in ▶ Chap. 14.

The earliest corner point detector was Moravec's *interest operator*, so called because it detected points in the scene that were *interesting* from a tracking perspective. It was based on the intuition that if a small image window $\mathcal{W}$ is to be unambiguously located in another image it must be quite different to the same-size window at any adjacent location. ▫ Fig. 12.23 shows, for each row, the contents of the red window as it is displaced horizontally and vertically. The first row is the case for a region of the image where all pixels have the same value, and we see that the displaced windows are all identical. The second row is the case for a vertical edge, and we see that the displaced windows have some variability but there are only three unique windows. The third and fourth rows show the case of an isolated point and a corner respectively – each of the displaced windows has a unique value. Clearly the isolated point and the corner point are the most distinctive.

Moravec defined the similarity between a region centered at (u, v) and an adjacent region, displaced by (δ_u, δ_v), as

$$s(u, v, \delta_u, \delta_v) = \sum_{(i,j) \in \mathcal{W}} \left(\mathbf{X}_{u+\delta_u+i, v+\delta_v+j} - \mathbf{X}_{u+i, v+j} \right)^2 \tag{12.10}$$

where $\mathcal{W}$ is some local image region and typically a $w \times w$ square window where w is odd. This is the sum of squared differences (SSD) similarity measure from ▫ Tab. 12.1 that we discussed previously – a low value means the regions are similar. Similarity is evaluated for displacements in eight cardinal ◀ directions

N, NE, E, ..., W, NW or $i, j \in \{-1, 0, 1\}$.

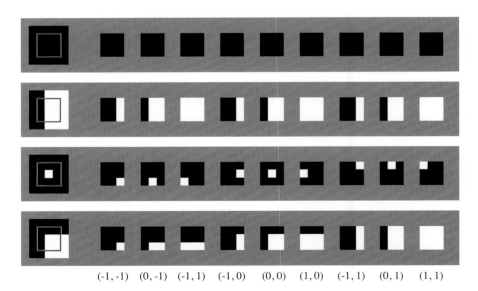

(-1, -1)   (0, -1)   (-1, 1)   (-1, 0)   (0, 0)   (1, 0)   (-1, 1)   (0, 1)   (1, 1)

▫ **Fig. 12.23** Principle behind the Morovec detector. At the left of each row is an image region with with a 3×3 window indicated in red. To the right are the contents of that red window for various displacements as shown across the bottom of the figure as (δ_u, δ_v)

$(\delta_u, \delta_v) \in \mathcal{D}$ and the minimum value is the interest measure or corner strength

$$C_M(u, v) = \min_{(\delta_u, \delta_v) \in \mathcal{D}} s(u, v, \delta_u, \delta_v) \tag{12.11}$$

which has a large value only if all the displaced patches are different to the original patch. The function $C_M(\cdot)$ is evaluated for every pixel in the image and interest points are those where C_M is *high*. The main limitation of the Moravec detector is that it is nonisotropic since it examines image change, essentially gradient, in a limited number of directions. Consequently the detector can give a strong output for a point on a line, which is not desirable.

We can generalize the approach by defining the similarity as the weighted sum of squared differences between the image region and the displaced region as

$$s(u, v, \delta_u, \delta_v) = \sum_{(i,j) \in \mathcal{W}} \mathbf{W}_{i,j} \Big(\underline{\mathbf{X}_{u+\delta_u+i,v+\delta_v+j}} - \mathbf{X}_{u+i,v+j} \Big)^2$$

where $\mathbf{W}$ is a weighting matrix, for example a 2D Gaussian, that emphasizes points closer to the center of the window $\mathcal{W}$. The underlined term can be approximated by a truncated Taylor series ▶

See ▶ App. E.

$$\mathbf{X}_{u+\delta_u,v+\delta_v} \approx \mathbf{X}_{u,v} + \delta_u \mathbf{X}_{u:u,v} + \delta_v \mathbf{X}_{v:u,v}$$

where $\mathbf{X}_{u:}$ and $\mathbf{X}_{v:}$ are the horizontal and vertical image gradients respectively. We can now write

$$\begin{aligned}
s(u, v, \delta_u, \delta_v) = {} & \delta_u^2 \sum_{(i,j) \in \mathcal{W}} \mathbf{W}_{i,j} \mathbf{X}_{u:u+i,v+j}^2 \\
& + \delta_v^2 \sum_{(i,j) \in \mathcal{W}} \mathbf{W}_{i,j} \mathbf{X}_{v:u+i,v+j}^2 \\
& + \delta_u \delta_v \sum_{(i,j) \in \mathcal{W}} \mathbf{W}_{i,j} \mathbf{X}_{u:u+i,v+j} \mathbf{X}_{v:u+i,v+j}
\end{aligned}$$

which can be written compactly in quadratic form as

$$s(u, v, \delta_u, \delta_v) = (\delta_u \; \delta_v) \mathbf{A} \begin{pmatrix} \delta_u \\ \delta_v \end{pmatrix}$$

where

$$\mathbf{A} = \begin{pmatrix} \sum \mathbf{W}_{i,j} \mathbf{X}_{u:u+i,v+j}^2 & \sum \mathbf{W}_{i,j} \mathbf{X}_{u:u+i,v+j} \mathbf{X}_{v:u+i,v+j} \\ \sum \mathbf{W}_{i,j} \mathbf{X}_{u:u+i,v+j} \mathbf{X}_{v:u+i,v+j} & \sum \mathbf{W}_{i,j} \mathbf{X}_{v:u+i,v+j}^2 \end{pmatrix} .$$

If the weighting matrix is a Gaussian kernel $\mathbf{W} = \mathbf{G}(\sigma_X)$ and we replace the summation by a convolution then

$$\mathbf{A} = \begin{pmatrix} \mathbf{G}(\sigma_X) * \mathbf{X}_u^2 & \mathbf{G}(\sigma_X) * \mathbf{X}_u \mathbf{X}_v \\ \mathbf{G}(\sigma_X) * \mathbf{X}_u \mathbf{X}_v & \mathbf{G}(\sigma_X) * \mathbf{X}_v^2 \end{pmatrix} \tag{12.12}$$

which is a symmetric 2×2 matrix referred to variously as the structure tensor, auto-correlation matrix or second moment matrix. It captures the intensity structure of the local neighborhood and its eigenvalues provide a rotationally invariant description of the neighborhood. The elements of the $\mathbf{A}$ matrix are computed from the image gradients, squared or multiplied element-wise, and then smoothed using a Gaussian kernel. The latter reduces noise and improves the stability and reliability of the detector. The gradient images $\mathbf{X}_u$ and $\mathbf{X}_v$ are typically calculated using a derivative of Gaussian kernel method (▶ Sect. 11.5.1.3) with a smoothing parameter σ_D.

An interest point (u, v) is one for which $s(u, v : \cdot)$ is high for *all* directions of the vector (δ_u, δ_v). That is, in whatever direction we move the window it rapidly becomes dissimilar to the original window. If we consider the original image $\mathbf{X}$ as a surface, the principal curvatures of the surface at (u, v) are λ_1 and λ_2 which are the eigenvalues of $\mathbf{A}$. If both eigenvalues are small then the surface is flat, that is the image region has approximately constant local intensity. If one eigenvalue is high and the other low, then the surface is ridge shaped which indicates an edge. If both eigenvalues are high the surface is sharply peaked which we consider to be a corner. ◀

The Shi-Tomasi detector considers the strength of the corner, or *cornerness*, as the minimum eigenvalue

$$C_{\mathrm{ST}}(u, v) = \min \lambda_1, \lambda_2 \tag{12.13}$$

Lossy image compression such as JPEG removes high-frequency detail from the image, and this is exactly what defines a corner. Ideally, corner detectors should be applied to images that have not been compressed and decompressed.

where λ_i are the eigenvalues of $\mathbf{A}$. Points in the image for which this measure is high are referred to as "*good features to track*". The Harris detector ◀ is based on this same insight but defines corner strength as

$$C_{\mathrm{H}}(u, v) = \det(\mathbf{A}) - k \operatorname{tr}(\mathbf{A}) \tag{12.14}$$

Sometimes referred to in the literature as the Plessey corner detector.

and again a large value represents a strong, distinct, corner. Since $\det(\mathbf{A}) = \lambda_1 \lambda_2$ and $\operatorname{tr}(\mathbf{A}) = \lambda_1 + \lambda_2$ the Harris detector responds when both eigenvalues are large and elegantly avoids computing the eigenvalues of $\mathbf{A}$ which has a somewhat higher computational cost. ◀ A commonly used value for k is 0.04. Another variant is the Noble detector

Evaluating eigenvalues for a 2×2 matrix involves solving a quadratic equation and therefore requires a square root operation.

$$C_{\mathrm{N}}(u, v) = \frac{\det(\mathbf{A})}{\operatorname{tr}(\mathbf{A})} \tag{12.15}$$

which is arithmetically simple but potentially singular.

The corner strength can be computed for every pixel and results in a corner-strength image. Then nonlocal maxima suppression is applied to only retain values that are greater than their immediate neighbors. A list of such points is created and sorted into descending corner strength. A threshold can be applied to only accept corners above a particular strength, or a particular fraction of the strongest corners, or simply the N strongest corners.

We will demonstrate the use of a Harris corner detector using a real image

```
>> b1 = imread("building2-1.png");
>> imshow(b1)
```

and the Harris features are computed by

```
>> C = detectHarrisFeatures(b1)
C =
  5891x1 cornerPoints array with properties:
    Location: [5891x2 single]
      Metric: [5891x1 single]
       Count: 5891
```

which returns a `cornerPoints` object. The detector found over 5000 corners that were local maxima of the corner strength image and these comprised 0.6% of all pixels in the image. The object contains a count of all corners, an array holding the location of the corners in the (u, v) image coordinates, and a metric indicating the strength of each corner.

The corners can be overlaid on the image as green markers

```
>> imshow(b1,[0 500]), hold on
>> plot(C.selectUniform(200,size(b1)))
```

Excurse 12.3: Determinant of the Hessian (DoH)

Another approach to determining image curvature is to take the determinant of the Hessian (DoH), and it is used by the SURF corner detector. The Hessian **H** is the matrix of second-order gradients at a point

$$h_{u,v} = \begin{pmatrix} x_{uu:u,v} & x_{uv:u,v} \\ x_{uv:u,v} & x_{vv:u,v} \end{pmatrix}$$

where $\mathbf{X}_{uu} = \partial^2 \mathbf{X}/\partial u^2$, $\mathbf{X}_{vv} = \partial^2 \mathbf{X}/\partial v^2$ and $\mathbf{X}_{uv} = \partial^2 \mathbf{X}^2/\partial u \partial v$. Thedeterminant det(**H**) has a large magnitude when there is gray-level variation in two directions. However, second derivatives accentuate image noise even more than first derivatives and the image must be smoothed first.

as shown in ◻ Fig. 12.24a. The `uint8` image values span the range from 0 to 255, but we requested scaling of the displayed pixels by `imshow` to span from 0 to 500 thus darkening the image to better accentuate the green markers. Additionally, rather than displaying all 5891 corners, we chose to select a subset of 200 corners.

◻ **Fig. 12.24** Harris corner detector applied to two views of the same building. **a** View one; **b** zoomed in view one; **c** view two; **d** zoomed in view two. Notice that a number of the detected corners are attached to the same world features in the two views

C.selectStrongest(200) can be used to select the strongest values.

Image registration is a process of aligning two images of the same scene that are taken by cameras placed at different locations.

These corners were chosen based on both their strength and their even distribution throughout the image. Normally, corners tend to cluster unevenly, with a greater density in regions of high contrast and texture. If we chose the 200 based only on their strength, ◀ they would mostly be located in the trees which may be unhelpful. Leaves can move with wind thus not providing stable features, and having all features localized in a small area can introduce a bias in certain computations such as estimating a geometric transform to register two images. ◀ A closeup view is shown in ◘ Fig. 12.24b and we see the features are indeed often located on the corners of objects.

We can apply standard vector operations and syntax to the cornerPoints object, for example

```
>> length(C)
ans =
      5891
```

and indexing

```
>> C(1:4)
ans =
   4×1 cornerPoints array with properties:
   Location: [4×2 single]
     Metric: [4×1 single]
      Count: 4
```

We can also create expressions such as

```
>> C(1:4).Metric' % transpose for display
ans =
  1×4 single row vector
    0.0005    0.0031    0.0037    0.0005
>> C(1).Location
ans =
  1×2 single row vector
    1.4843   267.3399
```

To plot the coordinate of every fifth feature in the first 800 features we can

```
>> imshow(b1, [0 500]), hold on
>> C(1:5:800).plot()
```

A cumulative histogram of the strength of the 200 selected corners is shown in ◘ Fig. 12.25. The strongest corner has $C_H \approx 0.04$ but most are much weaker than this, and only 10% of corners exceed $\frac{2}{3}$ of this value.

Consider another image of the same building taken from a different location

```
>> b2 = imread("building2-2.png");
```

and the detected corners

```
>> C2 = detectHarrisFeatures(b2);
>> imshow(b2,[0 500]),hold on
>> plot(C2.selectUniform(200,size(b2)));
```

are shown in ◘ Fig. 12.24c, d. For many applications in robotic vision – such as tracking, mosaicing and stereo vision that we will discuss in ▶ Chap. 14 – it is important that corner features are detected at the same world points irrespective of variation in illumination or changes in rotation and scale between the two views. From ◘ Fig. 12.24 we see that many, but not all, of the features are indeed *attached* to the same world feature in both views.

The Harris detector is computed from image gradients and is therefore robust to offsets in illumination. However, the detector is not invariant to changes in scale. As we *zoom in* the gradients around the corner points become lower – the same change in intensity is spread over a larger number of pixels. This reduces the image curvature and hence the corner strength. The next section discusses a remedy for this using *scale-invariant* feature detectors.

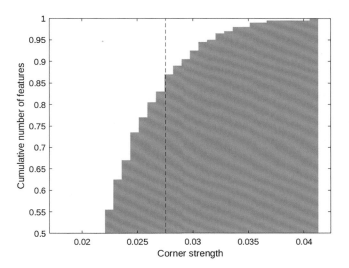

12.3.2 **Scale-Space Feature Detectors**

The Harris corner detector introduced in the previous section works very well in practice but responds poorly to changes in scale. Unfortunately, change in scale due to changing camera to scene distance or zoom, is common in many real applications. We also notice that the Harris detector responds strongly to fine texture, such as the leaves of the trees in ■ Fig. 12.24 but we would like to be able to detect features that are associated with larger-scale scene structures such as windows and balconies.

A wide variety of approaches have been proposed over the years to address the need for a reliable scale-space feature detector. One of the most effective feature detectors is Scale Invariant Feature Transform (SIFT). SIFT relies on Laplacian of Gaussian (LoG) computation to find features over different scales. Instead of computing LoG directly, it uses an approximation. Recalling the properties of a Gaussian from ▶ Sect. 11.5.1.1, a Gaussian convolved with a Gaussian is another wider Gaussian. Instead of convolving the original image with ever wider Gaussians, we can repeatedly apply the same Gaussian to the previous result. We also recall from ▶ Sect. 11.5.1.3 that the LoG kernel is approximated by the difference of two Gaussians. Using the properties of convolution we can write

$$(\mathbf{G}(\sigma_1) - \mathbf{G}(\sigma_2)) * \mathbf{X} = \mathbf{G}(\sigma_1) * \mathbf{X} - \mathbf{G}(\sigma_2) * \mathbf{X}$$

where $\sigma_1 > \sigma_2$. The difference of Gaussian (DoG) operator applied to the image is equivalent to the difference of the image at two different levels of smoothing. If we perform the smoothing by successive application of a Gaussian we have a sequence of images at increasing levels of smoothing. The difference between successive steps in the sequence is therefore an approximation to the Laplacian of Gaussian. ■ Fig. 12.26 shows this in diagrammatic form. At each stage of applying an increasingly larger Gaussian kernel, fine features begin to disappear and are replaced with larger blobs representing features at larger scales. In ■ Fig. 12.26, for example, we see the fine countours of Mona Lisa's eyes replaced with larger dark blobs. Note also that scale spaces are usually implemented as image pyramids. The images are not only repeatedly smoothed with a Gaussian but also sub-sampled to achieve a higher level of the pyramid. When using scale-space detectors, we encounter the concept of an *octave*. Typically, several differently sized filters are used per octave with filter size step and sub-sampling amount increasing between filter groups at each octave, thus catching ever coarser features. ▶

For example, the SURF feature detector uses filter sizes of 9×9, 15×15, 21×21 in the initial octave but then doubles the filter size increase in the following octaves thus jumping from 6 (15-9) to 12, 24, 48, and so on.

Gaussian sequence ⟶

Laplacian of Gaussian (LoG) sequence ⟶

■ **Fig. 12.26** Schematic for calculation of Gaussian and Laplacian of Gaussian scale-space sequence

These scale-space concepts underpin a number of popular feature detectors which find salient points within an image and determine their scale and also their orientation. SIFT is based on the maxima in a difference of Gaussian sequence as discussed earlier. Its close cousin, the Speeded Up Robust Feature (SURF) is based on the maxima in an approximate determinant of Hessian of a Gaussian sequence. Using the determinant of the Hessian rather than its trace (the Laplacian) activates the detector less on elongated, often less accurately localized structures. SURF is generally faster than SIFT due to use of even more approximate calculations. ◄

SURF uses simple box filters and integral images to speed up computation of the Hessian.

To illustrate use of the SURF detector, we will compute the SURF features for the building image used previously

```
>> sf1 = detectSURFFeatures(b1)
sf1 =
  4274×1 SURFPoints array with properties:
            Scale: [4274×1 single]
    SignOfLaplacian: [4274×1 int8]
      Orientation: [4274×1 single]
        Location: [4274×2 single]
          Metric: [4274×1 single]
           Count: 4274
```

which returns a SURFPoints object holding 4274 features. For example, the first feature is

```
>> sf1(1)
ans =
  SURFPoints with properties:
            Scale: 2.1333
    SignOfLaplacian: -1
      Orientation: 0
        Location: [579.6101 504.0464]
          Metric: 3.8738e+04
           Count: 1
```

The descriptor is a processed patch of data around a feature point that facilitates matching of corresponding features. The descriptor encodes the patch as a high-dimensional vector in a way which is invariant to brightness, scale and rotation. This enables feature descriptors to be unambiguously matched to a descriptor of the same world point in another image even if their scale and orientation are quite different.

Each feature includes the feature's location (estimated to subpixel accuracy), scale, sign of Laplacian, orientation, and metric indicating the strength of the feature. Orientation is defined by the dominant edge direction within the support region. It is initially set to zero until feature *descriptor* is computed. ◄ The sign of the Laplacian can be used to speed up matching of corresponding features by discarding the matches where the sign does not match without further computation involving SURF descriptors. SURF descriptors can be computed using extractFeatures and are vectors of length 64 or 128 which is selectable using FeatureSize parameter

Fig. 12.27   SURF descriptors showing the support region (scale)

Fig. 12.28   Histogram of feature scales shown with logarithmic vertical scale

to the `extractFeatures` function. We will cover the topic of using descriptors in more detail in ▸ Chap. 14.

This image contains more than 4000 SURF features but we can request the 200 strongest which we plot

```
>> imshow(b1,[0 500]), hold on
>> plot(sf1.selectStrongest(200))
```

and the result is shown in ◘ Fig. 12.27. The `plot` method draws a circle around the feature's location with a radius that indicates its scale – the size of the support region.

Feature scale varies widely and a histogram

```
>> hist(sf1.Scale,100);
```

shown in ◘ Fig. 12.28 indicates that there are many small features associated with fine image detail and texture. The bulk of the features have a scale less than 6 which corresponds to roughly a descriptor square patch with side of (20 × scale) 120 pixels, but some have scales over 10 which is roughly a square patch with side of 200 pixels. The computed responses are further weighted with a Gaussian thus emphasizing central part of the patch.

The SURF algorithm provides both a scale-invariant feature detector and a robust descriptor. Using the descriptors, we can match corresponding features between images taken from different perspectives. The difference in position, scale and orientation of the matched features gives some indication of the relative camera motion between the two views. Matching features between scenes is crucial to the problems that we will address in ▶ Chap. 14.

12.4 Applications

12.4.1 Character Recognition

An important class of visual objects are text characters. Our world is filled with informative text in the form of signs and labels that provide information about the names of places and directions to travel. We process much of this unconsciously, but this rich source of information is largely unavailable to robots. The Computer Vision Toolbox™ contains a deep learning detector for finding the text regions and an optical character recognition function.

The OCR function produces best results when it is given a reasonable estimate of the region in which the text is to be found. In the first step, we use the `detectTextCRAFT` function for a pretrained text detection deep learning network called Character Region Awareness for Text Detection (CRAFT). We start with the image of a sign we have used previously, locate bounding boxes of text regions as (u, v, w, h) where (u, v) are the coordinates of the top-left of the region and (w, h) are width and height

```
>> castle = imread("castle.png");
>> textBBoxes = detectTextCRAFT(castle)
textBBoxes =
    456    404    494    84
    554    310    302    82
    568    568    290    84
```

and then apply OCR to the three detected regions of interest

```
>> txt = ocr(castle,textBBoxes);
```

which returns a three-element vector of `ocrText` objects. The recognized words are

```
>> string([txt.Words])
ans =
  1×3 string array
    "information"    "Tourist"    "Castle"
```

and its confidence about those words is

```
>> [txt.WordConfidences]
ans =
  1×3 single row vector
    0.8930    0.8868    0.8671
```

We can highlight the location of the words and overlay the recognized text in the original image by

```
>> imshow(castle)
>> showShape("rectangle",textBBoxes,Color="y",Label=[txt.Words], ...
>>     LabelFontSize=18,LabelOpacity=0.8)
```

with the result shown in ◻ Fig. 12.29.

Fig. 12.29 Optical character recognition. The bounding boxes of detected words and recognized text are shown in the image

12.4.2 Image Retrieval

Given a set of images $\{\mathbf{X}_j,\, j = 1,\dots, N\}$ and a new image $\mathbf{X}'$ the image retrieval problem is to determine j such that $\mathbf{X}_j$ is the most similar to $\mathbf{X}'$. This is a difficult problem when we consider the effect of changes in viewpoint and exposure. Pixel-level similarity measures such as sum of absolute differences (SSD) or zero-mean normalized cross correlation (ZNCC) introduced in ▶ Sect. 11.5.2 are not suitable for this problem, since even small changes in viewpoint will result in almost zero similarity.

Image retrieval is useful to a robot to determine if it has visited a particular place before, or seen the same object before. If the previous images have some associated semantic data, such as the name of an object or the name of a place, then by inference that semantic data applies to the new image. For example, if a new image matches an existing image that has the semantic tag "kitchen" then it implies the robot is seeing the same scene and is therefore in or close to, the kitchen.

The particular technique that we will introduce is commonly referred to as "bag of words" and is useful for many robotic applications. It builds on techniques we have previously encountered such as SURF point features and k-means clustering.

We start by loading a set of twenty images

```
>> images = imageDatastore(fullfile(rvctoolboxroot,"images", ...
>>    "campus"));
```

into an image data store that lets us easily iterate over the data. We can easily display the images

```
>> montage(images)
```

which are shown in **Fig. 12.30.**

The bag of words technique was developed in the area of natural language processing. Since images do not contain discrete words, we first construct a *vocabulary* of SURF features representative of each image

```
>> rng(0) % set random seed for repeatable results
>> bag = bagOfFeatures(images,TreeProperties=[2 200])
Creating Bag-Of-Features.
------------------------
* Selecting feature point locations using the Grid method.
* Extracting SURF features from the selected feature point locations.
** The GridStep is [8 8] and the BlockWidth is [32 64 96 128].
* Extracting features from 20 images...done. Extracted 339200 features.
```

☐ **Fig. 12.30**   Twenty images from a university campus

```
*  Keeping 80 percent of the strongest features from each category.
*  Creating a 40000 word visual vocabulary.
*  Number of levels: 2
*  Branching factor: 200
*  Number of clustering steps: 201
...
*  [Step 201/201] Clustering vocabulary level 2, branch 200.
*  Number of features        : 1572
*  Number of clusters        : 200
*  Initializing cluster centers...100.00%.
*  Clustering completed 8/100 iterations (~0.02 seconds/iteration)
   converged in 8 iterations.
*  Finished creating Bag-Of-Features
```

To build a reliable visual vocabulary typically requires hundreds if not thousands of images and a vocabulary size that is also in the thousands. This example, with only 20 images, is meant to illustrate the basic concepts. Large vocabulary trees such as ours, with 40,000 words, require time to build, but that process only needs to happen once.

For SURF feature extractor of length 64, we are clustering feature points in 64-dimensional space.

where the `bagOfFeatures` function extracted SURF features from every image in our collection. We organized the vocabulary tree in a two level structure with 200 branches at each level resulting in $200 \times 200 = 40,000$ visual vocabulary words. This increase of vocabulary size over the default of 500 words produces more reliable outcomes for our data set. By default, the function extracts the features on a fixed grid which is faster, since it does not involve running the detector, and tends to introduce less bias into the visual vocabulary. ◀

The key insight behind the bag of words technique is that many of these features will describe visually similar scene elements, for example, windows in the buildings. Therefore, it is meaningful to then quantize the extracted features by constructing a more compact vocabulary using k-means clustering. We can think of the clouds of multi-dimensional feature vectors ◀ being represented by their centroids or cluster centers. It is these cluster centers that become the visual vocabulary that can then represent our data more compactly. The compact representation for any input image is obtained by computing image features and binning them into each visual word thus obtaining a histogram of word occurrences. It is this histogram that can then be used to rapidly and robustly compare images. In other words, the histogram of visual word occurrences becomes an image descriptor.

Our visual vocabulary comprises $k = 40{,}000$ visual words. We apply a technique from text document retrieval and describe *each* image by a weighted word-frequency vector

$$\boldsymbol{v}_i = (t_1, \ldots, t_K)^\top \in \mathbb{R}^k$$

which is a column vector whose elements describes the frequency of the corresponding visual words in an image.

$$t_j = \frac{n_{ij}}{n_i} \underbrace{\log \frac{N}{N_j}}_{\text{idf}} \tag{12.16}$$

where j is the visual word label, N is the total number of images in the database, N_j is the number of images which contain word j, n_i is the number of words in image i, and n_{ij} is the number of times word j appears in image i. The inverse document frequency (idf) term is a weighting that reduces the significance of words that are common across all images and which are thus less discriminatory. The weighted word frequency vector is obtained by the `encode` method of `bagOfFeatures` and for image 17 in the input set, it is

```
>> img17 = images.readimage(17);
>> v17 = bag.encode(img17);
>> size(v17)
ans =
     1    40000
```

which is a row vector that concisely describes the image in terms of its constituent visual words. ▶

The similarity between two images is the cosine of the angle between their corresponding word-frequency vectors

$$s(\boldsymbol{v}_1, \boldsymbol{v}_2) = \frac{\boldsymbol{v}_1^\top \boldsymbol{v}_2}{\|\boldsymbol{v}_1\| \, \|\boldsymbol{v}_2\|}$$

This is a large vector but it contains less than 5% of the number of elements of the original image. It is a very concise summary.

and is implemented by the `retrieveImages` function. A value of one indicates maximum similarity. There is one more step that we need to take prior to using `retrieveImages` and that is building the *inverted index* which efficiently maps the features from the input image to the location in the database of our twenty images ▶

```
>> invIdx = indexImages(images,bag);
```

and with that, the similarity of image 17 to all the images in the bag is simply

It is called inverted because it maps from content, visual words, to location in the document database (specific image) as opposed to the forward index which maps from the image to the content.

```
>> [simImgIdx,scores] = retrieveImages(img17,invIdx,NumResults=4);
>> simImgIdx' % transpose for display
ans =
  1x4 uint32 row vector
    17    16    15     8
>> scores'
ans =
    1.0000    0.4286    0.3786    0.3093
```

which returns a 4-vector of similarity values between the word frequency vector for image 17 and the others. The results are easily visualized

```
>> for i=1:4
>>    imgIdx = double(simImgIdx(i));
>>    similarImages(:,:,:,i) = ...
>>       insertText(images.readimage(imgIdx),[1 1], ...
>>       "image #" + imgIdx + ";similarity " + scores(i),FontSize=24);
>> end
>> montage(similarImages, BorderSize=6)
```

Fig. 12.31 Image retrieval. Image 17 is the query with a perfect score of 1, and in decreasing order of match quality we have retrieved images 16, 15 and 8

a   **b**

Fig. 12.32 Image retrieval for a new image. **a** The new query image 1; **b** the most similar image retrieved from the original database

and we see that image 17 is most similar to image 17 resulting in a perfect score of 1, as expected, and in decreasing order of similarity we have images 16, 15 and 8. These are shown in ◻ Fig. 12.31 along with the similarity scores, and we see that the algorithm has mostly recalled different views of the same building.

Now consider that we have some new images – our robot is out in the world – and we wish to retrieve previous images of the places it is seeing. The steps are broadly similar to the previous case. We load five new images

```
>> queryImages = imageDatastore(fullfile(rvctoolboxroot, ...
>>    "images","campus","holdout"));
```

but we do not need to perform clustering and instead use the existing visual vocabulary. For example, for image one in our new set, we would simply run

```
>> imgQuery = queryImages.readimage(1);
>> closestIdx = retrieveImages(imgQuery,invIdx,NumResults=1)
closestIdx =
  uint32
        3
>> montage(cat(4,imgQuery,images.readimage(double(closestIdx))), ...
>>    BorderSize=6);
```

to see the closest matching image which is shown in ◻ Fig. 12.32b

🔴 Note that the match would be just as strong if the query image was upside down, or even if it was cut up into lots of pieces and rearranged. Bag of words looks for the presence of visual words, not their spatial arrangement.

12.5 Wrapping Up

In this chapter we have discussed the extraction of features from an image. Instead of considering the image as millions of independent pixel values we succinctly describe regions within the image that correspond to distinct objects in the world. For instance we can find regions that are homogeneous with respect to intensity or color or semantic meaning and describe them in terms of features such as a bounding box, centroid, equivalent ellipse, aspect ratio, circularity and perimeter shape. Features have invariance properties with respect to translation, rotation about the optical axis and scale which are important for object recognition. Straight lines are common visual features in human-made environments, and we showed how to find and describe distinct straight lines in an image using the Hough transform.

We also showed how to find interest points that can reliably *associate* to particular points in the world irrespective of the camera view. These are key to techniques such as camera motion estimation, stereo vision, image retrieval, tracking and mosaicing that we will discuss in the next chapter.

12.5.1 MATLAB Notes

The Computer Vision Toolbox™ has a large number of additional feature detectors and descriptors, which also support code generation or operation with Simulink. Beyond Harris and Shi and Tomasi corner detectors, you will find functions for SURF, SIFT, FAST, MSER, BRISK, FREAK, KAZE, and ORB algorithms. Some of them are only descriptors, while others are both detectors and descriptors.

12.5.2 Further Reading

This chapter has presented a classical bottom up approach for feature extraction, starting with pixels and working our way up to higher level concepts such as regions and lines. Prince (2012) and Szeliski (2011) both provide a good introduction to high-level vision using probabilistic techniques that can be applied to problems such as object recognition, for example face recognition, and image retrieval. In the last few years computer vision, particularly object recognition, has undergone a revolution using deep convolutional neural networks. These have demonstrated very high levels of accuracy in locating and recognizing objects against complex background despite changes in viewpoint and illumination and resources are available at ▶ https://www.mathworks.com/solutions/deep-learning. html, ▶ https://pytorch.org/ and ▶ https://www.tensorflow.org/.

■ ■ Region features

Region-based image segmentation and blob analysis are classical techniques covered in many books and papers. Gonzalez and Woods (2018) and Szeliski (2022) provide a thorough treatment of the methods introduced in this chapter, in particular thresholding and boundary descriptors. Otsu's algorithm for threshold determination was introduced in Otsu (1975), and Bradley's algorithm for adaptive thresholding was introduced in Bradley and Roth (2007). The book by Nixon and Aguado (2012) expands on material covered in this chapter and introduces techniques such as deformable templates and boundary descriptors. The Freeman chain code was first described in Freeman (1974).

In addition to region homogeneity based on intensity and color it is also possible to describe the texture of regions – a spatial pattern of pixel intensities whose statistics can be described (Gonzalez and Woods 2018). Regions can then be segmented according to texture, for example a smooth road versus textured grass.

Clustering of data is an important topic in machine learning (Bishop 2006). In this chapter we have used a simple implementation of k-means, which is far from state-of-the-art in clustering, and requires the number of clusters to be known in advance. More advanced clustering algorithms are hierarchical and employ data structures such as kd-trees to speed the search for neighboring points. The initialization of the cluster centroids is also critical to performance. Szeliski (2011) introduces more general clustering methods as well as graph-based methods for computer vision. The graphcuts algorithm for segmentation was described by Felzenszwalb and Huttenlocher (2004). The maximally stable extremal region (MSER) algorithm is described by Matas et al. (2004). The Berkeley Segmentation Dataset at ▶ https://sn.pub/Mh5jb9 contains numerous complex real-world images each with several human-made segmentations.

Early work on using text recognition for robotics is described by Posner et al. (2010), while Lam et al. (2015) describe the application of OCR to parsing floor plans of buildings for robot navigation. A central challenge with OCR of real-world scenes is to determine which parts of the scene contain text and should be passed to the OCR engine. A powerful text detector is the stroke width transform described by Li et al. (2014). The MATLAB `ocr` function is based on the Tesseract open-source OCR engine which is available at ▶ https://github.com/tesseract-ocr and described by Smith (2007).

▪▪ Line features

The Hough transform was first first described in U.S. Patent 3,069,654 "Method and Means for Recognizing Complex Patterns" by Paul Hough, and its history is discussed in Hart (2009). The original application was automating the analysis of bubble chamber photographs and it used the problematic slope-intercept parameterization for lines. The currently known form with the (θ, ρ) parameterization was first described in Duda and Hart (1972) as a "generalized Hough transform" and is available at ▶ http://www.ai.sri.com/pubs/files/tn036-duda71.pdf. The Hough transform is covered in textbooks such as Szeliski (2022), Gonzalez and Woods (2018), Davies (2017) and Klette (2014). The latter has a good discussion on shape fitting in general and estimators that are robust with respect to outlier data points. The basic Hough transform has been extended in many ways and there is a large literature. A useful review of the transform and its variants is presented in Leavers (1993). The transform can be generalized to other shapes (Ballard 1981) such as circles of a fixed size where votes are cast for the coordinates of the circle's center. For circles of unknown size a three-dimensional voting array is required for the circle's center and radius.

▪▪ Point features

The literature on interest operators dates back to the early work of Moravec (1980) and Förstner (Förstner and Gülch 1987; Förstner 1994). The Harris corner detector (Harris and Stephens 1988) became very popular for robotic vision applications in the late 1980s since it was able to run in real-time on computers of the day, and the features were quite stable (Tissainayagam and Suter 2004) from image to image. The Noble detector is described in Noble (1988). The work of Shi, Tomasi, Lucas and Kanade (Shi and Tomasi 1994; Tomasi and Kanade 1991) led to the Shi-Tomasi detector and the Kanade-Lucas-Tomasi (KLT) tracker. Good surveys of the relative performance of many corner detectors include those by Deriche and Giraudon (1993) and Mikolajczyk and Schmid (2004).

Scale-space concepts have long been known in computer vision. Koenderink (1984), Lindeberg (1993) and ter Haar Romeny (1996) are a readable introduction to the topic. Scale-space was applied to classic corner detectors creating hybrid detectors such as scale-Harris (Mikolajczyk and Schmid 2004). An important development in scale-space feature detectors was the scale-invariant feature transform (SIFT) introduced in the early 2000s by Lowe (2004) and was a significant improvement for applications such as tracking and object recognition. A faster and nearly as robust alternative to SIFT was created later called Speeded Up Robust Features (SURF) (Bay et al. 2008). GPU-based parallel implementations have also been developed. The SIFT and SURF detectors do give different results and they are compared in Bauer et al. (2007).

Many other interest point detectors and features have been, and continue to be proposed. FAST by Rosten et al. (2010) has very low computational requirements and high repeatability ▶ http://www.edwardrosten.com/work/fast.html. CenSurE by Agrawal et al. (Agrawal et al. 2008) claims higher performance than SIFT, SURF and FAST at lower cost. BRIEF by Calonder et al. (2010) is not a feature detector but is a low cost and compact feature descriptor, requiring just 256 bits instead of 128 floating-point numbers per feature as required by SIFT. Other feature descriptors include histogram of oriented gradients (HOG), oriented FAST and rotated BRIEF (ORB), binary robust invariant scaleable keypoint (BRISK), fast retina keypoint (FREAK), aggregate channel features (ACF), vector of locally aggregated descriptors (VLAD), random ferns and many many more.

Local features have many advantages and are quite stable from frame to frame, but for outdoor applications the feature locations and the descriptors vary considerably with changes in lighting conditions, see for example Valgren and Lilienthal (2010). Night and day are obvious examples but even over a period of a few hours the descriptors change considerably. Over seasons the appearance change can be drastic: trees with or without leaves; the ground covered by grass or snow, wet or dry. Enabling robots to recognize places despite their changing appearance is the research field of robust place recognition which is introduced in Lowry et al. (2015).

The "bag of words" technique for image retrieval was first proposed by Sivic and Zisserman (2003) and has been used by many other researchers since. A notable extension for robotic applications is FABMAP (Cummins and Newman 2008) which explicitly accounts for the joint probability of feature occurrence and associates a probability with the image match.

12.5.3 Exercises

1. Grayscale classification
   a) Experiment with `graythresh` on the images `castle.png` and `castle2.png`.
   b) Experiment with Bradley's algorithm and vary the value of `sensitivity` and `NeighborhoodSize`.
   c) Apply `regionprops` to the output of the MSER segmentation. Develop an algorithm that uses the width and height of the bounding boxes to extract just those blobs that are letters.
   d) The function `detectMSERFeatures` has many parameters: `ThresholdDelta`, `RegionAreaRange`, `MaxAreaVariation`. Explore the effect of adjusting these.
   e) Use the `imageSegmenter` app to segment the `castle2.png` image. Understand and adjust the parameters to improve performance.
   f) Load the image `adelson.png` and attempt to segment the letters A and B.
2. Color classification
   a) Change k, the number of clusters, in the color classification examples. Is there a best value?

b) Write a function that determines which of the clusters represents the targets, that is, the yellow cluster or the red cluster.

c) Use imageSegmenter app on to the targets and garden image. How does it perform? Understand and adjust the parameters to improve performance. Generate MATLAB code from the app and explore it.

d) Experiment with the parameters of the morphological "cleanup" used for the targets and garden images.

e) Write code that loops over images captured from your computer's camera (Hint: use `webcam` function), applies a classification, and shows the result. The classification could be a grayscale threshold or color clustering.

3. Blobs. Create an image of an object with several holes in it. You could draw it and take a picture, export it from a drawing program, or write code to generate it.

a) Determine the outer, *inner* and total boundaries of the object.

b) Place small objects within the holes in the objects. Write code to display the topological hierarchy of the blobs in the scene.

c) For the same shape at different scales plot how the circularity changes as a function of scale. Explain the shape of this curve?

d) Create a square object and plot the estimated and true perimeter as a function of the square's side length. What happens when the square is small?

e) Create an image of a simple scene with a number of different shaped objects. Using the shape invariant features (aspect ratio, circularity) to create a simple shape classifier. How well does it perform?

f) Repeat the boundary matching example with some objects that you create. Modify the code to create a plot of edge-segment angle (k) versus θ and repeat the boundary matching example.

g) Another commonly used feature is the aligned rectangle. This is the smallest rectangle whose sides are aligned with the axes of the equivalent ellipse and which entirely contains the blob. The aspect ratio of this rectangle and the ratio of the blob's area to the rectangle's area are each scale and rotation invariant features. Write code to compute this rectangle, overlay the rectangle on the image, and compute the two features.

h) Write code to trace the perimeter of a blob.

4. Experiment with the `ocr` function.

a) What is the effect of a larger region of interest?

b) Capture your own image and attempt to read the text in it. How does accuracy vary with text size, contrast or orientation?

5. Hough transform

a) Experiment with varying the size of the Hough accumulator.

b) Explore the use of the Sobel edge detector versus Canny edge detector.

c) Experiment with varying the parameters `RhoResolution`, `Theta` in `hough` function, as well as parameters `FillGap`, `MinLength` in `houghlines` function.

d) Apply the Hough transform to one of your own images.

e) Write code that loops over images captured by your computer's camera, finding the two dominant lines and overlaying them on the image.

6. Corner and feature detectors

a) Experiment with the Harris detector. Vary the parameters `MinQuality` and `FilterSize`.

b) Compare the performance of the Harris and Shi-Tomasi corner detectors.

c) Implement the Moravec detector and compare to Harris detector.

d) Create a smoothed second derivative $\mathbf{X}_{uu}$, $\mathbf{X}_{vv}$ and $\mathbf{X}_{uv}$.

e) Detect SURF, SIFT and FAST features in your own image. Plot and compare the results.

Image Formation

» *Everything we see is a perspective,*
not the truth.
– Marcus Aurelius

Contents

13.1   Perspective Camera – 544

13.2   Camera Calibration – 558

13.3   Wide Field-of-View Cameras – 566

13.4   Unified Imaging Model – 574

13.5   Novel Cameras – 578

13.6   Applications – 581

13.7   Advanced Topics – 584

13.8   Wrapping Up – 587

© The Author(s), under exclusive license to Springer Nature Switzerland AG 2023
P. Corke, W. Jachimczyk, R. Pillat, *Robotics, Vision and Control*, Springer Tracts in Advanced Robotics 147,
https://doi.org/10.1007/978-3-031-07262-8_13

In this chapter we discuss how images, discussed in previous chapters, are formed and captured. It has long been known that a simple pinhole is able to create an inverted image on the wall of a darkened room. Some marine mollusks, for example the Nautilus, have pinhole camera eyes. The eyes of vertebrates use a lens to form an inverted image on the retina where the light-sensitive rod and cone cells, shown previously in ◘ Fig. 10.6, are arranged. A digital camera is similar in principle – a glass or plastic lens forms an image on the surface of a semiconductor chip where an array of light-sensitive devices converts the light to a digital image.

The process of image formation, in an eye or in a camera, involves a *projection* of the 3-dimensional world onto a 2-dimensional surface. The depth information is lost, and we can no longer tell from the image whether it is of a large object in the distance, or a smaller closer object. This transformation from 3 to 2 dimensions is known as perspective projection and is discussed in ▶ Sect. 13.1. ▶ Sect. 13.2 introduces the topic of camera calibration, the estimation of the parameters of the perspective transformation. ▶ Sects. 13.3 to 13.5 introduce alternative types of cameras capable of wide-angle, panoramic or light-field imaging. ▶ Sect. 13.6 introduces two applications: the use of artificial markers in scenes to determine the pose of objects, and the use of a planar homography for making image-based measurements of a planar surface. ▶ Sect. 13.7 introduces some advanced concepts such as projecting lines and conics, and non-perspective cameras.

13.1 Perspective Camera

13.1.1 Perspective Projection

A small hole in the wall of a darkened room will cast an inverted image of the outside world onto the opposite wall – a so-called pinhole camera. The pinhole camera produces a very dim image since, as shown in ◘ Fig. 13.1a, only a small fraction of the light leaving the object finds its way to the image. A pinhole camera has no focus adjustments – all objects are in focus irrespective of distance.

The key to brighter images is to use an objective lens, as shown in ◘ Fig. 13.1b, which collects light from the object over a larger area and directs it to the image. A convex lens can form an image just like a pinhole, and the fundamental geometry of image formation for a thin lens is shown in ◘ Fig. 13.2. The positive z-axis is the camera's optical axis.

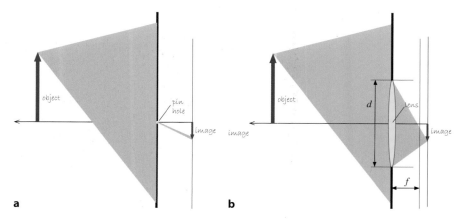

◘ **Fig. 13.1** Light gathering ability of **a** pinhole camera and **b** a lens

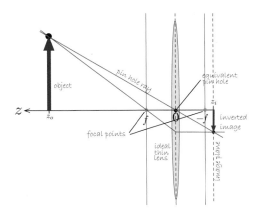

☐ **Fig. 13.2** Image formation geometry for a thin convex lens shown in 2-dimensional cross section. A lens has two focal points at a distance of f on each side of the lens. By convention the camera's optical axis is the z-axis

The z-coordinate of the object and its image, with respect to the lens center, are related by the thin lens equation

$$\frac{1}{|z_o|} + \frac{1}{|z_i|} = \frac{1}{f} \tag{13.1}$$

where z_o is z-coordinate of the object, z_i is the z-coordinate of the image, and f is the focal length of the lens. ▶ For $z_o > f$ an inverted image is formed on the image plane at $z_i < -f$.

The downside of using a lens is the need to focus. The image plane in a camera is the surface of the sensor chip, so the focus ring of the camera moves the lens along the optical axis so that it is a distance z_i from the image plane – for an object at infinity $z_i = f$. Our own eye has a single convex lens made from transparent crystalline proteins, and focus is achieved by muscles which change its shape – a process known as accommodation. A high-quality camera lens is a compound lens comprising multiple glass or plastic lenses.

In computer vision it is common to use the central perspective imaging model shown in ☐ Fig. 13.3. A ray from the world point P to the origin of the camera frame {C} intersects the image plane, located at $z = f$, at the projected point p. Unlike the pinhole or lens model, the projected image is non-inverted. The z-axis intersects the image plane at the principal point which is the origin of the 2D image coordinate frame. Using similar triangles we can show that the world point with a coordinate vector $\boldsymbol{P} = (X, Y, Z)$ is projected to a point on the image-plane with a

The inverse of focal length is known as diopter. For thin lenses placed close together their combined diopter is close to the sum of their individual diopters.

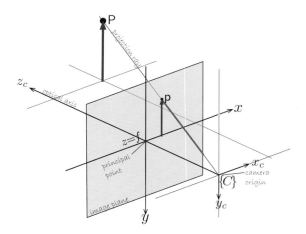

◘ Fig. 13.3 The central-projection model. The image plane is a distance f in front of the camera's origin, and a non-inverted image is formed. The camera's coordinate frame is right-handed with the z-axis defining the center of the field of view

coordinate vector $\boldsymbol{p} = (x, y)$ by

$$x = f\frac{X}{Z}, \quad y = f\frac{Y}{Z} \tag{13.2}$$

which are the *retinal* image-plane coordinates, and for the case where $f = 1$ the coordinates are referred to as the normalized or canonical image-plane coordinates. This is a projective transformation, or more specifically a perspective projection.

This mapping from the 3-dimensional world to a 2-dimensional image has consequences that we can see in ◘ Fig. 13.4 – parallel lines converge and circles become ellipses. More formally we can say that the transformation, from the world to the image plane, has the following characteristics:

— It performs a mapping from 3-dimensional space to the 2-dimensional image plane: $\mathcal{P} : \mathbb{R}^3 \mapsto \mathbb{R}^2$.

◘ Fig. 13.4 The effect of perspective transformation. **a** Parallel lines converge; **b** circles become ellipses

Excurse 13.2: Lens Aperture

The *f-number* of a lens, typically marked on the rim, is a dimensionless quantity $F = f/d$ where d is the diameter of the lens (often denoted ϕ on the lens rim). The f-number is *inversely* related to the light gathering ability of the lens. To reduce the amount of light falling on the image plane, the effective diameter is reduced by a mechanical aperture, or iris, which *increases* the f-number. Illuminance on the image plane is inversely proportional to F^2 since it depends on the light gathering area. Increasing the f-number by a factor of $\sqrt{2}$, or one "stop", will halve the illuminance at the sensor – photographers refer to this as "stopping down". The f-number graduations on the lens increase by $\sqrt{2}$ at each stop. An f-number is conventionally written in the form $f/1.4$ for $F = 1.4$.

Excurse 13.3: Focus and Depth of Field

Ideally a group of light rays from a point in the scene meet at a point on the image plane. With imperfect focus, the rays form a finite-sized spot called the circle of confusion which is the point-spread function of the optical system. By convention, if the size of the circle is around that of a pixel then the image is *acceptably* focused.

A pinhole camera has no focus control and always creates a focused image of objects irrespective of their distance. A lens does not have this property – the focus ring changes the distance between the lens and the image plane and must be adjusted so that the object of interest is *acceptably* focused. Photographers refer to depth of field, which is the range of object distances for which *acceptably* focused images are formed. Depth of field is high for small aperture settings where the lens is more like a pinhole, but this means less light and noisier images or longer exposure time and motion blur. This is the photographer's dilemma!

— Straight lines in the world are projected to straight lines on the image plane.
— Parallel lines in the world are projected to lines that intersect at a vanishing point as seen in ◧ Fig. 13.4a. In drawing, this effect is known as foreshortening. The exception is fronto-parallel lines – lines lying in a plane parallel to the image plane – which always remain parallel.
— Conics ▸ in the world are projected to conics on the image plane. For example, a circle is projected as a circle or an ellipse as shown in ◧ Fig. 13.4b.
— The size (area) of a shape is not preserved and depends on its distance from the camera.
— The mapping is not one-to-one and no unique inverse exists. That is, given (x, y) we cannot uniquely determine (X, Y, Z). All that can be said is that the world point P lies somewhere along the red projection ray shown in ◧ Fig. 13.3. This is an important topic that we will return to in ▸ Chap. 14.
— The transformation is not conformal – it does not preserve shape since internal angles are not preserved. Translation, rotation and scaling are examples of conformal transformations.

Conic sections, or conics, are a family of curves obtained by the intersection of a plane with a cone and are discussed in ▸ App. C.2.1.2. They include circles, ellipses, parabolas and hyperbolas.

13.1.2 Modeling a Perspective Camera

▶ App. C.2 provides a refresher on homogeneous coordinates.

We can write the coordinates of the image-plane point p as a homogeneous vector $\tilde{p} = (\tilde{x}, \tilde{y}, \tilde{z})^\top \in \mathbb{P}^2$ where

$$\tilde{x} = fX, \quad \tilde{y} = fY, \quad \tilde{z} = Z$$

and the tilde indicates homogeneous quantities. In compact matrix form, this becomes ◀

$$\tilde{p} = \begin{pmatrix} f & 0 & 0 \\ 0 & f & 0 \\ 0 & 0 & 1 \end{pmatrix} \begin{pmatrix} X \\ Y \\ Z \end{pmatrix} \tag{13.3}$$

where the nonhomogeneous image-plane coordinates are

$$x = \frac{\tilde{x}}{\tilde{z}}, \quad y = \frac{\tilde{y}}{\tilde{z}}$$

If we also write the coordinate of the world point P as a homogeneous vector ${}^C\tilde{P} = (X, Y, Z, 1)^\top \in \mathbb{P}^3$, then the perspective projection can be written in *linear* form as

$$\tilde{p} = \begin{pmatrix} f & 0 & 0 & 0 \\ 0 & f & 0 & 0 \\ 0 & 0 & 1 & 0 \end{pmatrix} {}^C\tilde{P} \tag{13.4}$$

or

$$\tilde{p} = \mathbf{C}\,{}^C\tilde{P} \tag{13.5}$$

where $\mathbf{C}$ is a 3×4 matrix known as the camera matrix and ${}^C\tilde{P}$ is the coordinate of the world point with respect to the camera frame {C}.

The camera matrix can be factored as

$$\tilde{p} = \begin{pmatrix} f & 0 & 0 \\ 0 & f & 0 \\ 0 & 0 & 1 \end{pmatrix} \underbrace{\begin{pmatrix} 1 & 0 & 0 & 0 \\ 0 & 1 & 0 & 0 \\ 0 & 0 & 1 & 0 \end{pmatrix}}_{\Pi} {}^C\tilde{P}$$

where the matrix $\mathbf{\Pi}$ is the projection matrix that maps three dimensions into two.

Using the RVC Toolbox, we can create a model of a central-perspective camera with a 15 mm lens

```
>> cam = CentralCamera(focal=0.015);
```

which returns an instance of a `CentralCamera` object with a 15 mm lens. By default, the camera is at the origin of the world frame with its optical axis pointing in the world z-direction as shown in ◘ Fig. 13.3. We define a world point

```
>> P = [0.3 0.4 3.0];
```

in units of meters and the corresponding retinal image-plane coordinates are

```
>> cam.project(P)
ans =
     0.0015     0.0020
```

Each projected point is in a row. Multiple world points are given as rows of the input array.

The point on the image plane is at (1.5, 2.0) mm with respect to the image center. ◀ This is a very small displacement but it is commensurate with the size of a typical image sensor.

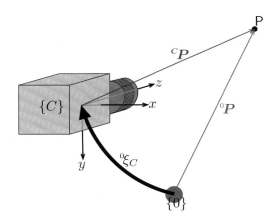

In general the camera will have an arbitrary pose ξ_C with respect to the world frame $\{0\}$ as shown in ■ Fig. 13.5. The coordinate vector of the world point P with respect to the camera frame $\{C\}$ is

$$^C\boldsymbol{P} = {}^C\xi_0 \cdot {}^0\boldsymbol{P} \tag{13.6}$$

or using homogeneous coordinates

$$^C\boldsymbol{P} = \left({}^0\mathbf{T}_C\right)^{-1} {}^0\boldsymbol{P}$$

where ${}^0\mathbf{T}_C \in \mathbf{SE}(3)$.

We can easily demonstrate this by moving our camera 0.5 m to the left

```
>> cam.project(P,pose=se3(eye(3),[-0.5 0 0]))
ans =
    0.0040    0.0020
```

where the pose of the camera ξ_C is provided as an `se3` object. We see that the x-coordinate has increased from 1.5 mm to 4.0 mm, that is, the image point has moved in the opposite direction – to the *right*.

13.1.3 **Discrete Image Plane**

In a digital camera, the image plane is a $W \times H$ grid of light-sensitive elements called photosites that correspond directly to the picture elements (or pixels) of the image as shown in ■ Fig. 13.6. The pixel coordinates are a 2-vector (u, v) of positive integers and by convention the origin is at the top-left hand corner of the image plane. In MATLAB® the top-left pixel is $(1, 1)$. The pixels are uniform in size and centered on a regular grid so the pixel coordinate is related to the image-plane coordinate (x, y) by

$$u = \frac{x}{\rho_w} + u_0, \quad v = \frac{y}{\rho_h} + v_0$$

where ρ_w and ρ_h are the width and height of each photosite respectively, and (u_0, v_0) is the principal point – the pixel coordinate of the point where the optical axis intersects the image plane with respect to the origin of the uv-coordinate frame. In matrix form, the homogeneous relationship between image plane coordinates and pixel coordinates is

$$\begin{pmatrix} \tilde{u} \\ \tilde{v} \\ \tilde{w} \end{pmatrix} = \begin{pmatrix} 1/\rho_{\mathrm{w}} & s & u_0 \\ 0 & 1/\rho_{\mathrm{h}} & v_0 \\ 0 & 0 & 1 \end{pmatrix} \begin{pmatrix} \tilde{x} \\ \tilde{y} \\ \tilde{z} \end{pmatrix}$$

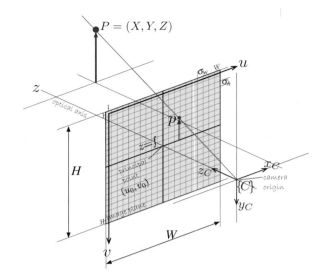

☐ Fig. 13.6 Central projection model showing image plane and discrete pixels

With precise semiconductor fabrication processes it is unlikely the axes on the sensor chip are not orthogonal, but this term may help model other imperfections in the optical system.

where $s = 0$ but is sometimes written with a nonzero value to account for skew, that is, the u- and v-axes being nonorthogonal. ◄ Now (13.4) can be written in terms of pixel coordinates

$$
\tilde{\boldsymbol{p}} = \underbrace{\begin{pmatrix} 1/\rho_{\mathrm{w}} & 0 & u_0 \\ 0 & 1/\rho_{\mathrm{h}} & v_0 \\ 0 & 0 & 1 \end{pmatrix} \begin{pmatrix} f & 0 & 0 \\ 0 & f & 0 \\ 0 & 0 & 1 \end{pmatrix}}_{\mathbf{K}} \underbrace{\begin{pmatrix} 1 & 0 & 0 & 0 \\ 0 & 1 & 0 & 0 \\ 0 & 0 & 1 & 0 \end{pmatrix}}_{\mathbf{\Pi}} {}^C\tilde{\boldsymbol{P}} \tag{13.7}
$$

where $\tilde{\boldsymbol{p}} = (\tilde{u}, \tilde{v}, \tilde{w})$ is the homogeneous coordinate, in units of pixels, of the projected world point P. The product of the first two terms is the 3×3 camera intrinsic matrix $\mathbf{K}$. The nonhomogeneous image-plane pixel coordinates are

$$
u = \frac{\tilde{u}}{\tilde{w}}, \quad v = \frac{\tilde{v}}{\tilde{w}} \; . \tag{13.8}
$$

Excurse 13.4: Image Sensors

The light-sensitive cells in a camera chip, the photosites (see ► Sect. 11.1.2), are arranged in a dense array. There is a 1:1 relationship between the photosites on the sensor and the pixels in the output image. Photosites are typically square with a side length in the range 1–10 μm. Professional cameras have large photosites for increased light sensitivity, whereas cellphone cameras have small sensors and therefore small less-sensitive photosites. The ratio of the number of horizontal to vertical pixels is the aspect ratio and is commonly 4 : 3 or 16 : 9 (see ► Exc. 11.6). The dimension of the sensor is measured diagonally across the array and is commonly expressed in inches, e.g. $\frac{1}{3}$, $\frac{1}{4}$ or $\frac{1}{2}$ inch. However, the active sensing area of the chip has a diagonal that is typically around $\frac{2}{3}$ of the given dimension.

For example, if the pixels are $10\,\mu$m square and the pixel array is 1280×1024 pixels with its principal point at image-plane coordinate $(640, 512)$ then

```
>> cam = CentralCamera(focal=0.015,pixel=10e-6, ...
>>      resolution=[1280 1024],center=[640 512],name="mycamera")
cam =
name: mycamera [central-perspective]
  focal length:    0.015
  pixel size:      (1e-05, 1e-05)
  principal pt:    (640, 512)
  number pixels:   1280 x 1024
  pose:            t = (0, 0, 0), RPY/zyx = (0, 0, 0) rad
```

which displays the parameters of the camera model including the camera pose. ▶ The nonhomogeneous image-plane coordinates of the previously defined world point are

```
>> cam.project(P)
ans =
   790    712
```

in units of pixels.

We have displayed a different roll-pitch-yaw rotation order YXZ. Given the way we have defined the camera axes, the camera orientation with respect to the world frame is a yaw about the vertical or y-axis, followed by a pitch about the x-axis followed by a roll about the optical axis or z-axis.

13.1.4 Camera Matrix

Combining (13.6) and (13.7) we can write the camera projection in general form as

$$\tilde{p} = \begin{pmatrix} f/\rho_w & 0 & u_0 \\ 0 & f/\rho_h & v_0 \\ 0 & 0 & 1 \end{pmatrix} \begin{pmatrix} 1 & 0 & 0 & 0 \\ 0 & 1 & 0 & 0 \\ 0 & 0 & 1 & 0 \end{pmatrix} (^0\mathbf{T}_C)^{-1} \tilde{P}$$

and substituting $^C\mathbf{T}_0$ for $(^0\mathbf{T}_C)^{-1}$ we can write

$$\tilde{p} = \begin{pmatrix} f/\rho_w & 0 & u_0 \\ 0 & f/\rho_h & v_0 \\ 0 & 0 & 1 \end{pmatrix} \begin{pmatrix} 1 & 0 & 0 & 0 \\ 0 & 1 & 0 & 0 \\ 0 & 0 & 1 & 0 \end{pmatrix} \begin{pmatrix} ^C\mathbf{R}_0 & ^C t_0 \\ \mathbf{0} & 1 \end{pmatrix} \tilde{P}$$

$$= \underbrace{\mathbf{K}}_{\text{intrinsic}} \underbrace{\begin{pmatrix} ^C\mathbf{R}_0 & ^C t_0 \end{pmatrix}}_{\text{extrinsic}} \tilde{P}$$

$$= \mathbf{C}\tilde{P} \qquad\qquad (13.9)$$

where all the terms are rolled up into the camera matrix $\mathbf{C}$. ▶ This is a 3×4 homogeneous transformation which performs scaling, translation and perspective projection. It is often also referred to as the projection matrix or the camera calibration matrix. It can be factored into a 3×3 camera intrinsic matrix $\mathbf{K}$, and a 3×4 camera extrinsic matrix.

The terms f/ρ_w and f/ρ_h are the focal length expressed in units of pixels.

🛑 It is important to note that the rotation and translation extrinsic parameters describe the world frame with respect to the camera.

The projection is often written in functional form as

$$p = \mathcal{P}(P, \mathbf{K}, \xi_C) \qquad\qquad (13.10)$$

where P is the coordinate vector of the point P in the world frame. $\mathbf{K}$ is the camera intrinsic matrix which comprises the intrinsic characteristics of the camera and

---

Excurse 13.5: Ambiguity of Perspective Projection

We have already mentioned the fundamental ambiguity with perspective projection, that we cannot distinguish between a large distant object and a smaller closer object. We can rewrite (13.9) as

$$\tilde{p} = \mathbf{C}\tilde{P} = \mathbf{C}(\mathbf{H}^{-1}\mathbf{H})\tilde{P} = (\mathbf{C}\mathbf{H}^{-1})(\mathbf{H}\tilde{P}) = \mathbf{C}'\tilde{P}'$$

where $\mathbf{H} \in \mathbb{R}^{3\times3}$ is an arbitrary nonsingular matrix. This implies that an infinite number of camera $\mathbf{C}'$ and world-point coordinate $\tilde{P}'$ combinations will result in the same image-plane projection $\tilde{p}$.

This illustrates the essential difficulty in determining 3-dimensional world coordinates from 2-dimensional projected coordinates. It can only be solved if we have information about the camera or the 3-dimensional object.

Excurse 13.6: Properties of the Camera Matrix

The camera matrix $\mathbf{C} \in \mathbb{R}^{3\times4}$ has some important structure and properties:

- It can be partitioned $\mathbf{C} = (\mathbf{M}|c_4)$ into a nonsingular matrix $\mathbf{M} \subset \mathbb{R}^{3\times3}$ and a vector $c_4 \in \mathbb{R}^3$, where $c_4 = -\mathbf{M}c$ and c is the world origin in the camera frame. We can recover this by $c = -\mathbf{M}^{-1}c_4$.
- The null space of $\mathbf{C}$ is $\tilde{c}$.
- A pixel at coordinate p corresponds to a ray in space parallel to the vector $\mathbf{M}^{-1}\tilde{p}$.
- The matrix $\mathbf{M} = \mathbf{K}\,^C\mathbf{R}_0$ is the product of the camera intrinsics and the camera inverse orientation. We can perform an RQ-decomposition of $\mathbf{M} = \bar{\mathbf{R}}\bar{\mathbf{Q}}$ where $\bar{\mathbf{R}}$ is an upper-triangular matrix (which is $\mathbf{K}$) and an orthogonal matrix $\bar{\mathbf{Q}}$ (which is $\,^C\mathbf{R}_0$). ◀
- The bottom row of $\mathbf{C}$ defines the principal plane, which is parallel to the image plane and contains the camera origin.
- If the rows of $\mathbf{M}$ are vectors m_i, $i \in \{1, 2, 3\}$ then:
  - $m_3^\top$ is a vector normal to the principal plane and parallel to the optical axis and $\mathbf{M}m_3^\top$ is the principal point in homogeneous form.
  - if the camera has zero skew, that is $\mathbf{K}_{1,2} = 0$, then
    $$(m_1 \times m_3) \cdot (m_2 \times m_3) = 0$$
  - and, if the camera has square pixels, that is $\rho_u = \rho_v = \rho$ then the focal length can be recovered by
    $$\|m_1 \times m_3\| = \|m_2 \times m_3\| = f/\rho$$

Confusingly, $\mathbf{R}$ has two different meanings here. MATLAB does not provide an RQ-decomposition but it can be determined by transforming the inputs to, and results of, the builtin MATLAB QR-decomposition function `qr`. There are many subtleties in doing this though: negative scale factors in the $\mathbf{K}$ matrix or det$\mathbf{R} = -1$, see Hartley and Zisserman (2003), or the implementation of `decomposeCam` for details.

sensor such as f, ρ_w, ρ_h, u_0 and v_0. ξ_C is the 3-dimensional pose of the camera and comprises a minimum of six parameters – the extrinsic parameters – that describe camera translation and orientation and is generally represented by an **SE**(3) matrix.

There are 5 intrinsic and 6 extrinsic parameters – a total of 11 independent parameters to describe a camera. The camera matrix has 12 elements so one degree of freedom, the overall scale factor, is unconstrained and can be arbitrarily chosen. Typically, the matrix is normalized such that the lower-right element is one.

The camera intrinsic parameter matrix $\mathbf{K}$ for this camera is

```
>> cam.K
ans =
   1.0e+03 *
   1.5000        0   0.6400
        0   1.5000   0.5120
        0        0   0.0010
```

> **Excurse 13.7: Field of View**
>
> The field of view of a camera is an open rectangular pyramid, a frustum, that subtends angles θ_h and θ_v in the horizontal and vertical planes respectively. A "normal" lens is one with a field of view around 50°, while a wide angle lens has a field of view >60°. Beyond 110° it is difficult to create a lens that maintains perspective projection, so nonperspective fisheye lenses are required.
>
> For very wide-angle lenses it is more common to describe the field of view as a solid angle which is measured in units of steradians (or sr). This is the area of the field of view projected onto the surface of a unit sphere. A hemispherical field of view is 2π sr and a full spherical view is 4π sr. If we approximate the camera's field of view by a cone with apex angle θ the corresponding solid angle is $2\pi(1 - \cos\frac{\theta}{2})$ sr. A camera with a field of view greater than a full hemisphere is termed omnidirectional or panoramic.

and the camera matrix **C** is

```
>> cam.C
ans =
   1.0e+03 *
   1.5000        0   0.6400        0
        0   1.5000   0.5120        0
        0        0   0.0010        0
```

and depends on the camera pose.

The field of view of a camera is a function of its focal length f and sensor size. A wide-angle lens has a small focal length, a telephoto lens has a large focal length, and a zoom lens has an adjustable focal length. The field of view can be determined from the geometry of ◘ Fig. 13.6. In the horizontal direction the half-angle of view is

$$\frac{\theta_h}{2} = \tan^{-1}\frac{W\rho_w}{2f}$$

where W is the number of pixels in the horizontal direction. We can then write

$$\theta_h = 2\tan^{-1}\frac{W\rho_w}{2f}, \quad \theta_v = 2\tan^{-1}\frac{H\rho_h}{2f} . \tag{13.11}$$

We note that the field of view is also a function of the dimensions of the camera chip which is $W\rho_w \times H\rho_h$. The field of view is computed by the `fov` method of the camera object

```
>> fov = cam.fov();
>> rad2deg(fov)
ans =
   46.2127   37.6930
```

in degrees in the horizontal and vertical directions respectively.

13.1.5 Projecting Points

The `CentralCamera` class is a subclass of the `Camera` class and inherits the ability to project multiple points or lines. Using the RVC Toolbox we create a 3×3 grid of points in the xy-plane with overall side length 0.2 m and centered at $(0, 0, 1)$

```
>> P = mkgrid(3,0.2,pose=se3(eye(3),[0 0 1.0]));
```

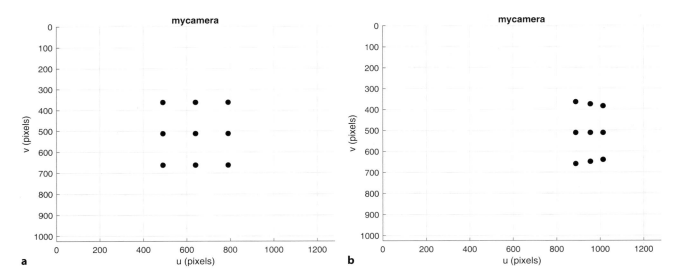

Fig. 13.7 Two views of a planar grid of points. **a** Frontal view, **b** oblique view

which returns a 9×3 matrix with one row per grid point where each row comprises the coordinates in X, Y, Z order. The first four rows are

```
>> P(1:4,:)
ans =
    -0.1000    -0.1000     1.0000
    -0.1000          0     1.0000
    -0.1000     0.1000     1.0000
          0    -0.1000     1.0000
```

By default `mkgrid` generates a grid in the xy-plane that is centered at the origin. The optional last argument is an **SE**(3) matrix that transforms the points and allows the plane to be arbitrarily positioned and oriented.

The image-plane coordinates of these grid points are

```
>> cam.project(P)
ans =
    490    362
    490    512
    490    662
    640    362
    640    512
    640    662
    790    362
    790    512
    790    662
```

which can be plotted

```
>> cam.plot(P)
```

to give the virtual camera view shown in ◩ Fig. 13.7a. The camera pose

```
>> Tcam = se3(eul2rotm([0 0.9 0]),[-1 0 0.5]);
```

results in an oblique view of the plane

```
>> cam.plot(P,pose=Tcam)
```

shown in ◩ Fig. 13.7b. We can clearly see the effect of perspective projection which has distorted the shape of the grid – the top and bottom edges, which are parallel lines, have been projected to lines that converge at a vanishing point.

The vanishing point for a line can be determined from the projection of its ideal line. The points in each row of the grid define lines that are parallel to the world

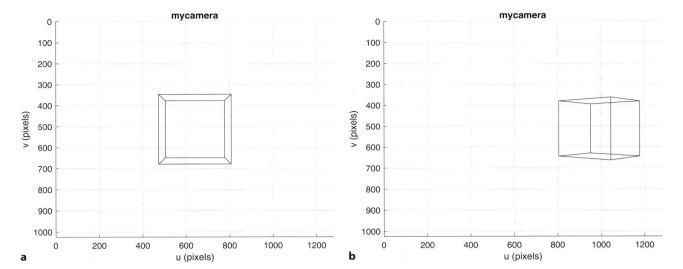

◘ Fig. 13.8   Line segment representation of a cube. **a** Frontal view, **b** oblique view

x-axis, or the vector $(1, 0, 0)$. The corresponding ideal line ▶ is the homogeneous vector $(1, 0, 0, 0)$ and the vanishing point is the projection of this vector

```
>> cam.project([1 0 0 0],pose=Tcam)
ans =
   1.0e+03 *
   1.8303    0.5120
```

which is $(1830, 512)$ and just to the right of the visible image plane.

The `plot` method can optionally return the image-plane coordinates

```
>> p = cam.plot(P,pose=Tcam);
```

just like the `project` method. For the oblique viewing case the image-plane coordinates

```
>> p(1:4,:)
ans =
   887.7638   364.3330
   887.7638   512.0000
   887.7638   659.6670
   955.2451   374.9050
```

have a fractional component which means that the point is not projected to the center of the pixel. However, a photosite responds to light equally ▶ over its surface area so the discrete pixel coordinate can be obtained by rounding.

A 3-dimensional object, a cube, can be defined and projected in a similar fashion. The vertices of a cube with side length 0.2 m and centered at $(0, 0, 1)$ can be defined by

```
>> cube = mkcube(0.2,pose=se3(eye(3),[0 0 1]));
```

which returns an 8×3 matrix with one row per vertex. The image-plane points can be plotted as before by

```
>> cam.plot(cube);
```

Alternatively, we can create an *edge* representation of the cube by

```
>> [X,Y,Z] = mkcube(0.2,"edge",pose=se3(eye(3),[0 0 1]));
```

and display it

```
>> cam.mesh(X, Y, Z)
```

as shown in ◘ Fig. 13.8 along with an oblique view generated by

```
>> Tcam = se3(eul2rotm([0 0.8 0]),[-1 0 0.4]);
>> cam.mesh(X,Y,Z,pose=Tcam);
```

"Ideal" in this context means special rather than perfect. This is a homogeneous line, see ▶ App. C.2.1, whose fourth element is zero, effectively a line at infinity. See also ideal point.

This is not strictly true for CMOS sensors where transistors reduce the light-sensitive area by the fill factor – the fraction of each photosite's area that is light sensitive.

The elements of the mesh (i, j) have coordinates $(x_{i,j}, y_{i,j}, z_{i,j})$.

The edges are in the same 3-dimensional mesh format ◄ as generated by MATLAB built-in functions such as `sphere`, `ellipsoid` and `cylinder`.

Successive calls to `plot` will redraw the points or line segments and provide a simple method of animation. The short piece of code

```
>> theta = [0:500]/100*2*pi;
>> [X,Y,Z] = mkcube(0.2,[],"edge");
>> for th=theta
>>    T_cube = se3(eul2rotm(th*[1.3 1.2 1.1]),[0 0 1.5]);
>>    cam.mesh(X,Y,Z,objpose=T_cube); drawnow
>> end
```

shows a cube tumbling in space. The cube is defined with its center at the origin and its vertices are transformed at each time step.

13.1.6 Lens Distortion

No lenses are perfect, and the low-cost lenses used in many webcams are far from perfect. Lens imperfections result in a variety of distortions including chromatic aberration (color fringing), spherical aberration or astigmatism (variation in focus across the scene), and geometric distortions where points on the image plane are displaced from where they should be according to (13.3). An example of geometric distortion is shown in ◘ Fig. 13.9a, and is generally the most problematic effect that we encounter for robotic applications. It has two components: radial distortion, and tangential distortion.

Radial distortion causes image points to be translated along radial lines from the principal point and is well approximated by a polynomial

$$\delta r = k_1 r^2 + k_2 r^4 + k_3 r^6 + \cdots \tag{13.12}$$

where r is the distance of the image point from the principal point. Barrel distortion occurs when magnification decreases with distance from the principal point which causes straight lines near the edge of the image to curve outward. Pincushion distortion occurs when magnification increases with distance from the principal point and causes straight lines near the edge of the image to curve inward. Tangential distortion occurs at right angles to the radii but is generally less significant than radial distortion. Tangential distortion takes place when the lens and imaging plane are not exactly parallel because of small errors in camera assembly. Examples of a distorted and undistorted image are shown in ◘ Fig. 13.9.

◘ **Fig. 13.9** Lens distortion. **a** Distorted image, the curvature of the top row of the squares is easily visible, **b** undistorted image. This calibration was done using the Camera Calibrator (`cameraCalibrator`) app (Images reprinted with permission of The MathWorks, Inc.)

Excurse 13.8: Understanding the Geometry of Vision

It has taken humankind a long time to understand light, color and human vision. The Ancient Greeks had two schools of thought. The emission theory, supported by Euclid and Ptolemy, held that sight worked by the eye emitting rays of light that interacted with the world somewhat like the sense of touch. The intromission theory, supported by Aristotle and his followers, had physical forms entering the eye from the object.

Euclid of Alexandria (325–265) arguably got the geometry of image formation correct, but his rays emanated from the eye, not the object. Claudius Ptolemy (100–170) wrote Optics and discussed reflection, refraction, and color but today there remains only a poor Arabic translation of his work.

The Arab philosopher Hasan Ibn al-Haytham (aka Alhazen, 965–1040) wrote a seven-volume treatise Kitab al-Manazir (Book of Optics) around 1020. He combined the mathematical rays of Euclid, the medical knowledge of Galen, and the intromission theories of Aristotle. He wrote that "from each point of every colored body, illuminated by any light, issue light and color along every straight line that can be drawn from that point". He understood refraction, but believed the eye's lens, not the retina, received the image – like many early thinkers he struggled with the idea of an inverted image on the retina. A Latin translation of his work was a great influence on later European scholars.

It was not until 1604 that geometric optics and human vision came together when the German astronomer and mathematician Johannes Kepler (1571–1630) published Astronomiae Pars Optica (The Optical Part of Astronomy). He was the first to recognize that images are projected inverted and reversed by the eye's lens onto the retina – the image being corrected later "in the hollows of the brain". (Image from Astronomiae Pars Optica, Johannes Kepler, 1604; reprinted from FelixPlatter, De corporis humani structura et usu. Libri III, König, 1583.)

A point (u_d, v_d) in the distorted image is related to its true coordinate (u, v) by

$$u_d = u + \delta_u, \quad v_d = v + \delta_v \tag{13.13}$$

where the displacement is

$$\begin{pmatrix} \delta_u \\ \delta_v \end{pmatrix} = \underbrace{\begin{pmatrix} u(k_1 r^2 + k_2 r^4 + k_3 r^6 + \cdots) \\ v(k_1 r^2 + k_2 r^4 + k_3 r^6 + \cdots) \end{pmatrix}}_{\text{radial}} + \underbrace{\begin{pmatrix} 2p_1 uv + p_2(r^2 + 2u^2) \\ p_1(r^2 + 2v^2) + 2p_2 uv \end{pmatrix}}_{\text{tangential}}. \tag{13.14}$$

In practice three coefficients are sufficient to describe the radial distortion and the distortion model is commonly parameterized by $(k_1, k_2, k_3, p_1, p_2)$ which are considered as additional intrinsic parameters. ▶ Distortion can be modeled by the `CentralCamera` class using the `distortion` option, for example

```
>> cam = CentralCamera(focal=0.015,pixel=10e-6, ...
>>    resolution=[1280 1024],center=[512 512], ...
>>    distortion=[k1 k2 k3 p1 p2]);
```

This popular camera model, introduced by Bouguet (Bouguet 2010) works well for cameras with field of view up to 95°. It is important to know that there are several camera models in existence and for cameras with wider field of view, a different model is necessary.

13.2 Camera Calibration

According to ANSI Standard PH3.13-1958 "Focal Length Marking of Lenses".

Changing the focus of a lens shifts the lens along the optical axis. In some designs, changing focus rotates the lens so if it is not perfectly symmetric this will move the distortions with respect to the image plane. Changing the aperture alters the parts of the lens that light rays pass through and hence the distortion that they incur.

The camera projection model (13.9) has a number of parameters that in practice are unknown. In general the principal point is *not* at the center of the photosite array. The focal length written on a lens is only accurate ◄ to 4% of what it is stated to be, and is only correct if the lens is focused at infinity. It is also common experience that the intrinsic parameters change if a lens is detached and reattached, or adjusted for focus or aperture. ◄ The only intrinsic parameters that it may be possible to obtain are the photosite dimensions ρ_w and ρ_h, from the sensor manufacturer's data sheet. The extrinsic parameters, the camera's pose, raises the question of which point in the camera should be considered its reference point?

Camera calibration is the process of determining the camera's intrinsic parameters, and the extrinsic parameters with respect to the world coordinate system. Calibration techniques rely on sets of world points whose relative coordinates are known and whose corresponding image-plane coordinates are also known. State-of-the-art techniques implemented in the Camera Calibrator (`cameraCalibrator`) app in Computer Vision Toolbox™ simply require a number of images of a planar checkerboard target such as shown in ◨ Fig. 13.12. From this, as discussed in ► Sect. 13.2.4, the intrinsic parameters (including distortion parameters) can be estimated as well as the relative pose of the checkerboard in each image. Classical calibration techniques require a single view of a 3-dimensional calibration target but are unable to estimate the distortion model. These methods are however easy to understand and they start our discussion in the next section.

13.2.1 Calibrating with a 3D Target

The homogeneous transform method allows direct estimation of the camera matrix $\mathbf{C}$ in (13.9). The elements of this matrix are functions of the intrinsic and extrinsic parameters. Setting $\tilde{p} = (u, v, 1)$, expanding (13.9) and substituting into (13.8) we can write

$$c_{1,1}X + c_{1,2}Y + c_{1,3}Z + c_{1,4} - c_{3,1}uX - c_{3,2}uY - c_{3,3}uZ - c_{3,4}u = 0$$
$$c_{2,1}X + c_{2,2}Y + c_{2,3}Z + c_{2,4} - c_{3,1}vX - c_{3,2}vY - c_{3,3}vZ - c_{3,4}v = 0$$

$$(13.15)$$

where (u, v) are the pixel coordinates corresponding to the world point (X, Y, Z) and $c_{i,j}$ are elements of the unknown camera matrix.

Calibration requires a 3-dimensional target such as shown in ◨ Fig. 13.10. The position of the center of each marker (X_i, Y_i, Z_i), $i = 1, \ldots, N$ with respect to the target frame {T} must be known, but {T} itself is not known. An image is captured and the *corresponding* image-plane coordinates (u_i, v_i) are determined. Assuming that $c_{3,4} = 1$ we stack the two equations of (13.15) for each of the N markers to

Fig. 13.10 A 3D calibration target where the circular markers are situated on three planes. The position of each circle with respect to {T} is accurately known. The image-plane coordinates of the markers are generally reconsidered to be the centroids of the circles in the image. (Image courtesy of Fabien Spindler)

form the matrix equation

$$
\begin{pmatrix}
X_1 & Y_1 & Z_1 & 1 & 0 & 0 & 0 & 0 & -u_1X_1 & -u_1Y_1 & -u_1Z_1 \\
0 & 0 & 0 & 0 & X_1 & Y_1 & Z_1 & 1 & -v_1X_1 & -v_1Y_1 & -v_1Z_1 \\
\vdots & \vdots & \vdots & \vdots & \vdots & \vdots & \vdots & \vdots & \vdots & \vdots & \vdots \\
X_N & Y_N & Z_N & 1 & 0 & 0 & 0 & 0 & -u_NX_N & -u_NY_N & -u_NZ_N \\
0 & 0 & 0 & 0 & X_N & Y_N & Z_N & 1 & -v_NX_N & -v_NY_N & -v_NZ_N
\end{pmatrix}
$$

$$
\times
\begin{pmatrix}
c_{1,1} \\
c_{1,2} \\
\vdots \\
c_{3,3}
\end{pmatrix}
=
\begin{pmatrix}
u_1 \\
v_1 \\
\vdots \\
u_N \\
v_N
\end{pmatrix}
\tag{13.16}
$$

which can be solved for the camera matrix elements $c_{1,1}, \ldots, c_{3,3}$. Equation (13.16) has 11 unknowns and for solution requires that $N \geq 6$. Often more than 6 points will be used leading to an over-determined set of equations which is solved using least squares.

If the points are coplanar then the left-hand matrix of (13.16) becomes rank deficient. This is why the calibration target must be 3-dimensional, typically an array of dots or squares on two or three planes as shown in ◘ Fig. 13.10.

We will illustrate this with an example where the calibration target is a cube. The calibration points are its vertices, its coordinate frame {T} is centered in the cube, and its axes are normal to the faces of the cube faces. The coordinates of the markers, with respect to {T}, are

```
>> P = mkcube(0.2);
```

The calibration target is at some "unknown pose" $^C\xi_T$ with respect to the camera which we choose to be

```
>> T_unknown = se3(eul2rotm([0.3 0.2 0.1]),[0.1 0.2 1.5]);
```

Next we create a perspective camera whose parameters we will attempt to estimate

```
>> rng(0) % set random seed for reproducibility of results
>> cam = CentralCamera(focal=0.015,pixel=10e-6, ...
>>    resolution=[1280 1024],noise=0.05);
```

We have also specified that zero-mean Gaussian noise with $\sigma = 0.05$ pix is added to the (u, v) coordinates to model camera noise and errors in the computer vision algorithms. The image-plane coordinates of the calibration target points at its "unknown" pose and observed by a camera with "unknown" parameters are

```
>> p = cam.project(P,objpose=T_unknown);
```

Now, using just the object model P and the observed image-plane points p we can estimate the camera matrix

```
>> C = camcald(P,p)
maxm residual 0.066733 pixels.
C =
   853.0895  -236.9378   634.2785   740.0438
   222.6439   986.6900   295.7327   712.0152
    -0.1304     0.0610     0.6495     1.0000
```

which is function of the intrinsic and extrinsic parameters as described by (13.9). Extracting these parameters from the matrix elements is covered in ▶ Sect. 13.2.2.

The residual, the worst-case error between the projection of a world point using the camera matrix **C** and the actual image-plane location is very small.

Linear techniques such as this cannot estimate lens distortion parameters. The distortion will introduce errors into the camera matrix elements but for many situations this might be acceptably low. Distortion parameters are often estimated using a nonlinear optimization over all parameters, typically 16 or more, and this approximate linear solution can be used as the initial parameter estimate.

13.2.2 Decomposing the Camera Calibration Matrix

The elements of the camera matrix are functions of the intrinsic and extrinsic parameters. However given a camera matrix most of the parameter values can be recovered.

The null space of **C** is the world origin in the camera frame. Using data from the example in ▶ Sect. 13.2.1, this is

```
>> wo = null(C)'  % transpose for display
wo =
    0.0809   -0.1709   -0.8138    0.5495
```

which is expressed in homogeneous coordinates that we can convert to Cartesian form

```
>> h2e(wo)
ans =
    0.1472   -0.3110   -1.4809
```

which is close to the true value

```
>> T_unknown.inv.trvec
ans =
    0.1464   -0.3105   -1.4772
```

To recover orientation as well as the intrinsic parameters we can *decompose* the previously estimated camera matrix

```
>> est = decomposeCam(C)
est =
name: decomposeCam [central-perspective]
  focal length:    1504
  pixel size:      (1, 0.9985)
  principal pt:    (646.8, 504.4)
  pose:            t = (0.147, -0.311, -1.48),
                   RPY/zyx = (-0.0327, -0.216, -0.286) rad
```

which returns a `CentralCamera` object with its parameters set to values that result in the same camera matrix. We note immediately that the focal length is very large compared to the true focal length of our lens which was 0.015 m, and that the pixel sizes are very large. From (13.9) we see that focal length and pixel dimensions always appear together as factors f/ρ_w and f/ρ_h. ▶ The function `decomposeCam` has set $\rho_w = 1$ but the ratios of the estimated parameters

```
>> est.f/est.rho(1)
ans =
    1.5040e+03
```

are very close to the ratio for the true parameters of the camera

```
>> cam.f/cam.rho(2)
ans =
    1.5000e+03
```

These quantities have units of pixels since ρ has units of m pixel^{-1}. It is quite common in the literature to consider $\rho = 1$ and the focal length is given in pixels. If the pixels are not square then different focal lengths f_u and f_v must be used for the horizontal and vertical directions respectively.

The small error in the estimated parameter values is due to the noisy image-plane coordinate values that we used in the calibration process.

The pose of the estimated camera is with respect to the calibration target {T} and is therefore $^T\hat{\xi}_C$. The true pose of the target with respect to the camera is $^C\xi_T$. If our estimation is accurate then $^C\xi_T \oplus {^T\hat{\xi}_C}$ will be ∅. We earlier set the variable `T_unknown` equal to $^C\xi_T$ and for our example we find that

```
>> printtform(T_unknown*est.T)
t = (4.13e-05, -4.4e-05, -0.00386),
RPY/zyx = (0.00516, 0.00441, -9.71e-05) rad
```

which is the relative pose between the true and estimated camera pose. The camera pose is estimated to better than 5 mm in position and a fraction of a degree in orientation.

We can plot the calibration markers as small red spheres

```
>> hold on; plotsphere(P,0.03,"r")
>> plottform(eye(4,4),frame="T",color="b",length=0.3)
```

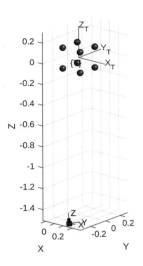

as well as frame {T} which we have set at the world origin. The estimated pose of the camera can be superimposed

```
>> est.plot_camera()
>> view([40 19])
```

and the result is shown in ☐ Fig. 13.11. The problem of determining the pose of a camera with respect to a calibration object is an important problem in photogrammetry known as the camera-location-determination problem.

13.2.3 Pose Estimation with a Calibrated Camera

The pose estimation problem is to determine the pose $^C\xi_T$ of a target's coordinate frame {T} with respect to the camera. The geometry of the target is known, that is, we know the position of a number of points $(X_i, Y_i, Z_i), i = 1, \ldots, N$ on the target with respect to {T}. The camera's intrinsic parameters are also known. An image is captured and the *corresponding* image-plane coordinates (u_i, v_i) are determined using computer vision algorithms.

Estimating the pose using $(u_i, v_i), (X_i, Y_i, Z_i)$ and camera intrinsic parameters is known as the Perspective-n-Point problem or PnP for short. It is a simpler problem than camera calibration and decomposition because there are fewer parameters to estimate. To illustrate pose estimation we will create a calibrated camera with known parameters

```
>> cam = CentralCamera(focal=0.015,pixel=10e-6, ...
>>       resolution=[1280 1024],center=[640 512]);
```

The object whose pose we wish to determine is a cube with side lengths of 0.2 m and the coordinates of the markers with respect to {T} are

```
>> P = mkcube(0.2);
```

which we can consider as a simple geometric model of the object. The object is at some arbitrary but unknown pose $^C\xi_T$ pose with respect to the camera

```
>> T_unknown = se3(eul2rotm([0.3 0.2 0.1]),[0.1 0.2 1.5]);
>> T_unknown.trvec
ans =
    0.1000    0.2000    1.5000
>> rad2deg(rotm2eul(T_unknown.rotm))
```

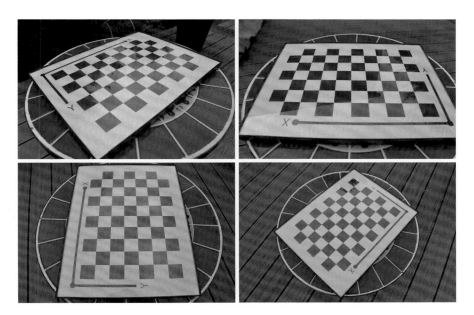

```
ans =
   17.1887   11.4592    5.7296
```

The image-plane coordinates of the object's points at its unknown pose are

```
>> p = cam.project(P,objpose=T_unknown);
```

Now using just the object model `P`, the observed image features `p` and the calibrated camera `cam` we estimate the relative pose $^C\xi_T$ of the object

```
>> T_est = cam.estpose(P,p);
>> T_est.trvec
ans =
    0.1000    0.2000    1.5000
>> rad2deg(rotm2eul(T_est.rotm))
ans =
   17.1887   11.4592    5.7296
```

which is the same (to four decimal places) as the unknown pose `T_unknown` of the object. ▶

In reality, the image features coordinates will be imperfectly estimated by the vision system and we have modeled this by adding zero-mean Gaussian noise to the image feature coordinates. It is also important that point correspondence is established, that is, each image plane point is associated with the correct 3D point in the object model.

For illustration purposes, we used the `estpose` method of the `CentralCamera` class. A more robust function that utilizes RANSAC to eliminate outliers from the PnP pose estimation problem is `estimateWorldCameraPose`.

13.2.4 Camera Calibration Tools

A powerful and practical tool for calibrating cameras using a planar checkerboard target or a grid of circles is the Camera Calibrator (`cameraCalibrator`) app that is part of the Computer Vision Toolbox™. A number of images, typically twenty, are taken of the target at different distances and orientations as shown in ■ Fig. 13.12.

The calibration tool is launched by

```
>> cameraCalibrator
```

Excurse 13.10: Camera Calibration Tips

There are a few things to keep in mind while calibrating a camera.

1. The calibration focuses on finding camera intrinsics. Extrinsics for each checkerboard pattern are computed during the calibration process, but are generally not useful afterwards beyond visualizing them to confirm accuracy of the calibration.

2. You should strive to fill as much of the field of view of the camera as possible. That should be done for individual images as well as across the entire set that you will collect. It is especially important to have the checkerboard placed as close as possible to the image boundaries where the effects of distortion are more significant. This assures accurate calibration for the entire lens as opposed to just its central region.

3. The distance to the checkerboard should be such that you are able to fill most of the field of view while not compromising the focus. That may require using a large checkerboard target.

4. It is key that the camera's parameters are not changed once it is calibrated. Otherwise, the calibration becomes invalid. That means that zoom and focus must be fixed and not allowed to adjust automatically.

5. Your calibration target should be of high quality to obtain accurate results. It must be perfectly planar.

6. It is critical to select the right camera model. Cameras with field of view less than 95° are well served by the pinhole camera model (Bouguet's model), but beyond 95° a different model must be used. The Camera Calibrator (`cameraCalibrator`) app offers Scaramuzza's model (2006) for the fisheye lenses.

The app is most convenient to use, but you can also calibrate a camera using MATLAB command line functionality. The easiest way to learn about the command line functions for calibration is to invoke the app, go through the calibration process and then use an option to export the steps as a MATLAB script.

and a graphical user interface (GUI) is displayed. ◄ The first step is to load the images using the Add Images button. Example checkerboard images can be found in `examples/calibrationImages` folder. While adding the images, you will be prompted to specify the size of the checkerboard square in world units, typically millimeters. Additionally, you can indicate if the image is highly distorted, which implies use of the fisheye lens. In that case, the checkerboard detection algorithm will make additional computations to compensate for the distortions. Cameras with the field of view beyond 95° will require this option. Once you specify these options during the image loading stage, and then push OK, the app automatically detects the keypoints in the checkerboard images.

At this point, we can inspect keypoint detections by clicking on individual images. The app's toolstrip provides a selection of the camera model, either standard (*pinhole*) or fisheye. Additionally, there is an Options button that can be used to select additional calibration parameters such as number of distortion coefficients for the standard model. Two coefficients are generally sufficient for most cameras but those with higher level of distortion may require three. Once these choices are made, the final step is to press the Calibrate button. When the calibration process completes, the app displays additional information about mean reprojection errors (bar plot in the upper right) and extrinsics estimated for each checkerboard (3D plot in lower right) as shown in ◨ Fig. 13.13

The camera pose relative to the target is estimated for each calibration image and the extrinsics are shown in a camera-centric view, as can be seen in ◨ Fig. 13.13, as well as checkerboard-centric view. The camera-centric view is more intuitive to inspect when the pattern is moved around while the checkerboard-centric view is more suitable for verifying the calibration results when the pattern is fixed in one location and the camera is moved around. It is important to note that despite having low reprojection errors, the calibration may still not be accurate if not enough high-quality target images were captured. One more way to verify the results is to use the Show Undistorted button which displays the undistorted images. We can verify these visually to gain further confidence in our calibration results. The details of the image undistortion transformation using image warping are discussed in ▶ Sect. 11.7.4. An example distorted and undistorted image are compared in ◨ Fig. 13.9.

◻ Fig. 13.13 Camera Calibrator app showing calibration results that include reprojection errors and extrinsics for each checkerboard.

Once we are satisfied with the calibration results, we can export them using the **Export Camera Parameters** button. The calibration parameters are returned to the MATLAB workspace as a `cameraParameters` object, by default named `cameraParams`

```
>> cameraParams
cameraParams =
  cameraParameters with properties:
  Camera Intrinsics
                        Intrinsics: [1×1 cameraIntrinsics]
  Camera Extrinsics
                 PatternExtrinsics: [16×1 rigidtform3d]
  Accuracy of Estimation
             MeanReprojectionError: 0.2956
                ReprojectionErrors: [54×2×16 double]
                 ReprojectedPoints: [54×2×16 double]
  Calibration Settings
                       NumPatterns: 16
                 DetectedKeypoints: [54×16 logical]
                       WorldPoints: [54×2 double]
                        WorldUnits: 'millimeters'
                      EstimateSkew: 0
  NumRadialDistortionCoefficients: 2
      EstimateTangentialDistortion: 0
```

The main output of the calibration is contained in the `cameraIntrinsics` object

```
>> cameraParams.Intrinsics
ans =
  cameraIntrinsics with properties:
            FocalLength: [1.2456e+03 1.2459e+03]
         PrincipalPoint: [813.2172 539.1277]
              ImageSize: [1080 1626]
        RadialDistortion: [-0.1788 0.1223]
    TangentialDistortion: [0 0]
                   Skew: 0
                      K: [3×3 double]
```

and contains core quantities including focal length in pixels as well as radial distortion coefficients. Camera intrinsics are a prerequisite for image undistortion, taking measurements in world units, camera pose estimation, SLAM and many other workflows. The camera intrinsics matrix is

```
>> cameraParams.Intrinsics.K
ans =
   1.0e+03 *
   1.2456        0    0.8132
        0   1.2459    0.5391
        0        0    0.0010
```

and it is in the form of matrix **K** as described by (13.7).

There is also an app for stereo calibration

```
>> stereoCameraCalibrator
```

It requires pairs of images as input and, in addition to computing camera intrinsics, it also estimates the stereo baseline, the translation and rotation between the two cameras forming the stereo pair.

13.3   Wide Field-of-View Cameras

We have discussed perspective imaging in quite some detail since it is the model of our own eyes and most cameras that we encounter. However, perspective cameras fundamentally constrain us to a limited field of view. The thin lens equation (13.1) is singular for points with $Z = f$ which limits the field of view to at most one hemisphere – real lenses achieve far less. As the focal length decreases, radial distortion is increasingly difficult to eliminate, and eventually a limit is reached, beyond which lenses cannot practically be built. The only way forward is to drop the constraint of perspective imaging. In ▶ Sect. 13.3.1 we describe the geometry of image formation with wide-angle lens systems.

An alternative to refractive optics is to use a reflective surface to form an image as shown in ◘ Fig. 13.14. Newtonian telescopes are based on reflection from concave mirrors rather than refraction by lenses. Mirrors are free of color fringing and are easier to scale up in size than a lens. Nature has also evolved reflective optics – the spookfish and some scallops (see ▶ Sect. 1.4) have eyes based on reflectors formed from guanine crystals. In ▶ Sect. 13.3.2 we describe the geometry of image formation with a combination of lenses and mirrors.

The cost of cameras is decreasing so an alternative approach is to combine the output of multiple cameras into a single image, and this is briefly described in ▶ Sect. 13.5.1.

13.3.1   Fisheye Lens Camera

A fisheye lens image is shown in ◘ Fig. 13.15 and we see that straight lines in the world are curved, and that the field of view is warped into a circle on the image

Fig. 13.14 Images formation by reflection from a curved surface (*Cloud Gate*, Chicago, Anish Kapoor, 2006). Note that straight lines have become curves

Fig. 13.15 Fisheye lens image. Note that straight lines in the world are no longer projected as straight lines. Note also that the field of view is mapped to a circular region on the image plane

plane. Image formation is modeled using the notation shown in ◧ Fig. 13.16 where the camera is positioned at the origin of the world frame O and its optical axis is the z-axis. The world point P is represented in spherical coordinates (R, θ, ϕ), where θ is the angle outward from the optical axis in the red plane, and ϕ is the angle of rotation of the red plane about the optical axis. We can write

$$R = \sqrt{X^2 + Y^2 + Z^2}, \quad \theta = \cos^{-1}\frac{R}{Z}, \quad \phi = \tan^{-1}\frac{Y}{X} .$$

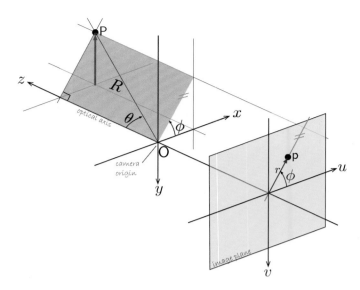

Fig. 13.16 Image formation for a fisheye lens camera. The world point P is represented in spherical coordinates (R, θ, ϕ) with respect to the camera's origin

Table 13.1 Fisheye lens projection models

Mapping	Equation
Equiangular	$r = k\theta$
Stereographic	$r = k\tan\left(\frac{\theta}{2}\right)$
Equisolid	$r = k\sin\left(\frac{\theta}{2}\right)$
Polynomial	$r = k_1\theta + k_2\theta^2 + \cdots$

On the image plane of the camera, we represent the projection p in polar coordinates (r, ϕ) with respect to the principal point, where $r = r(\theta)$. The Cartesian image-plane coordinates are

$$u = r(\theta)\cos\phi, \quad v = r(\theta)\sin\phi$$

and the exact nature of the function $r(\theta)$ depends on the type of fisheye lens. Some common projection models are listed in ◘ Tab. 13.1 and all have a scaling parameter k.

Using the RVC Toolbox we can create a fisheye camera model

```
>> cam = FishEyeCamera(name="fisheye",projection="equiangular", ...
>>    pixel=10e-6,resolution=[1280 1024]);
```

which is an instance of the `FishEyeCamera` class – a subclass of the `Camera` base class and polymorphic with the `CentralCamera` class discussed earlier. If k is not specified, as in this example, then it is computed such that a hemispheric field of view is projected into the maximal circle on the image plane. As is the case for perspective cameras, the parameters such as principal point and pixel dimensions are generally not known and must be estimated using a calibration procedure.

We create an edge-based model of a cube with side length 0.2 m

```
>> [X,Y,Z] = mkcube(0.2,"edge",center=[0.2 0 0.3]);
```

and project it to the fisheye camera's image plane

```
>> cam.mesh(X,Y,Z)
```

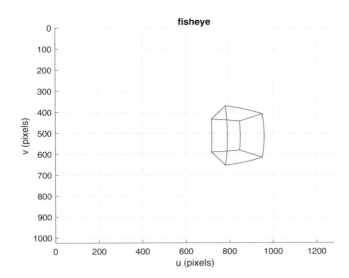

Fig. 13.17 A cube projected using the `FishEyeCamera` class. The straight edges of the cube are curves on the image plane

and the result is shown in ◘ Fig. 13.17. We see that straight lines in the world are no longer straight lines in the image.

Wide angle lenses are available with 180° and even 190° field of view, however they have some practical drawbacks. Firstly, the spatial resolution is lower since the camera's pixels are spread over a wider field of view. We also note from ◘ Fig. 13.15 that the field of view is a circular region which means that nearly 25% of the rectangular image plane is wasted. Secondly, outdoors images are more likely to include a lot of bright sky so the camera will automatically reduce its exposure. As a consequence, some non-sky parts of the scene could be underexposed.

13.3.2 Catadioptric Camera

A catadioptric imaging system comprises both reflective and refractive elements, ▶ a mirror and a lens, as shown in ◘ Fig. 13.18a. An example catadioptric image is shown in ◘ Fig. 13.18b.

From the Greek for curved mirrors (catoptrics) and lenses (dioptrics).

Image formation is modeled using the geometry shown in ◘ Fig. 13.19. A ray is constructed from the point P to the focal point of the mirror at O which is the origin of the camera system. The ray has an elevation angle of

$$\theta = \tan^{-1} \frac{Z}{X^2 + Y^2} + \frac{\pi}{2}$$

upward from the optical axis and intersects the mirror at the point M. The reflected ray makes an angle ψ with respect to the optical axis which is a function of the incoming ray angle, that is $\psi(\theta)$. The relationship between θ and ψ is determined by the tangent to the mirror at the point M and is a function of the shape of the mirror. Many different mirror shapes are used for catadioptric imaging including spherical, parabolic, elliptical and hyberbolic. In general the function $\psi(\theta)$ is nonlinear but an interesting class of mirror is the equiangular mirror for which

$$\theta = \alpha \psi .$$

The reflected ray enters the camera lens at angle ψ from the optical axis, and from the lens geometry we can write

$$r = \lambda \tan \psi$$

Fig. 13.18 Catadioptric camera. **a** Catadioptric camera system comprising a conventional perspective camera is looking upward at the mirror; **b** Catadioptric image. Note the dark spot in the center which is the support that holds the mirror above the lens. The floor is in the center of the image and the ceiling is at the edge (photos by Michael Milford)

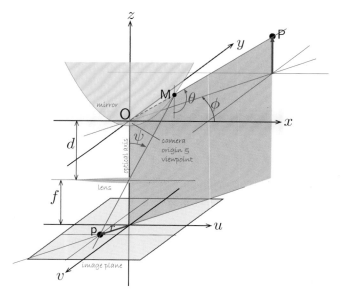

Fig. 13.19 Catadioptric image formation. A ray from point P at elevation angle θ and azimuth ϕ toward O is reflected from the mirror surface at M and is projected by the lens on to the image plane at p

which is the distance from the principal point. The image-plane point p can be represented in polar coordinates by $\boldsymbol{p} = (r, \phi)$ and the corresponding Cartesian coordinate is

$$u = r \cos \phi, \quad v = r \sin \phi,$$

where ϕ is the azimuth angle in the horizontal plane

$$\phi = \tan^{-1} \frac{Y}{X} .$$

In ■ Fig. 13.19 we have assumed that all rays pass through a single focal point or viewpoint – O in this case. This is referred to as central imaging and the resulting

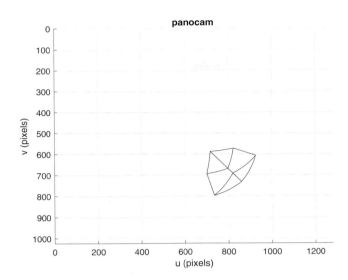

◘ Fig. 13.20 A cube projected using the `CatadioptricCamera` class

image can be transformed perfectly to a perspective image. The equiangular mirror does not meet this constraint and is therefore a noncentral imaging system – the focal point varies with the angle of the incoming ray and lies along a short locus within the mirror known as the caustic. Conical, spherical and equiangular mirrors are all noncentral. In practice the variation in the viewpoint is very small compared to the world scale and many such mirrors are well approximated by the central model.

Using the RVC Toolbox we can model a catadioptric camera, in this case for an equiangular mirror

```
>> cam = CatadioptricCamera(name="panocam", ...
>>    projection="equiangular", maxangle=pi/4,pixel=10e-6, ...
>>    resolution=[1280 1024]);
```

which is an instance of a `CatadioptricCamera` object. It is also a subclass of the RVC Toolbox's `Camera` object and polymorphic with the `CentralCamera` class discussed earlier. The option `maxangle` specifies the maximum elevation angle θ from which the parameters α and f are determined such that the maximum elevation angle corresponds to a circle that maximally fits the image plane. The parameters can be individually specified using the options `alpha` and `focal`. Other supported projection models include parabolic and spherical, and each camera type has different options as described in the online documentation.

We create an edge-based cube model

```
>> [X,Y,Z] = mkcube(1,"edge",center=[1 1 0.8]);
```

which we project onto the image plane

```
>> cam.mesh(X,Y,Z)
```

and the result is shown in ◘ Fig. 13.20.

Catadioptric cameras have the advantage that they can view 360° in azimuth, but they also have some practical drawbacks. They share many of the problems of fisheye lenses such as reduced spatial resolution, wasted image-plane pixels and exposure control. In some designs there is also a blind spot due to the mirror support, either a central stalk as seen in ◘ Fig. 13.18, or a number of side supports.

Excurse 13.11: Specular Reflection

Specular reflection occurs with a mirror-like surface. Incoming rays are reflected such that the angle of incidence equals the angle of reflection or $\theta_r = \theta_i$. This is in contrast to diffuse or Lambertian reflection which scatters incoming rays over a range of angles.

Speculum is Latin for mirror and speculum metal ($\frac{2}{3}$ copper, $\frac{1}{3}$ tin) is an alloy that can be highly polished. It was used by Newton and Herschel for the curved mirrors in their reflecting telescopes. The image is of the 48 inch speculum mirror from Herschel's 40 foot telescope, completed in 1789, which is now in the British Science Museum. (Image by Mike Peel (▶ https://www.mikepeel.net) licensed under CC-BY-SA)

13.3.3 Spherical Camera

The fisheye lens and catadioptric systems guide the light rays from a large field of view onto an image plane. Ultimately, the 2-dimensional image plane is a limiting factor and it is advantageous to consider instead an image *sphere* as shown in ◘ Fig. 13.21.

The world point P is projected by a ray to the origin of a unit sphere O. The projection is the point p where the ray intersects the surface of the sphere, and can be represented by the coordinate vector $\boldsymbol{p} = (x, y, z)$ where

$$x = \frac{X}{R}, \quad y = \frac{Y}{R}, \quad z = \frac{Z}{R} \tag{13.17}$$

and $R = \sqrt{X^2 + Y^2 + Z^2}$ is the radial distance to the world point. The surface of the sphere is defined by $x^2 + y^2 + z^2 = 1$ so one of the three Cartesian coordinates is redundant. A minimal two-parameter representation for a point on the surface of a sphere $\boldsymbol{p} = (\phi, \theta)$ comprises the angle of colatitude measured down from the North pole

$$\theta = \sin^{-1} r, \quad \theta \in [0, \pi] \tag{13.18}$$

where $r = \sqrt{x^2 + y^2}$, and the azimuth angle (or longitude)

$$\phi = \tan^{-1} \frac{y}{x}, \quad \phi \in [-\pi, \pi) . \tag{13.19}$$

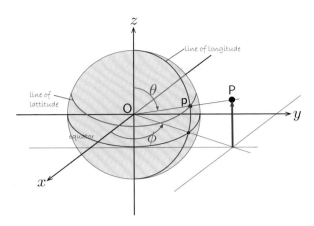

Fig. 13.21 Spherical image formation. The world point P is projected to p on the surface of the unit sphere and represented by the angles of colatitude θ and longitude ϕ

Fig. 13.22 A cube projected using the `SphericalCamera` class . The spherical image plane is represented in Cartesian coordinates

The polar and Cartesian coordinates of the point p are related by

$$x = \sin\theta\cos\phi, \quad y = \sin\theta\sin\phi, \quad z = \cos\theta \ . \tag{13.20}$$

Using the RVC Toolbox we can create a spherical camera

```
>> cam = SphericalCamera(name="spherical");
```

which is an instance of the `SphericalCamera` class – a subclass of the `Camera` base class and polymorphic with the `CentralCamera` class discussed earlier.

As in the previous examples, we can create an edge-based cube model

```
>> [X,Y,Z] = mkcube(1,"edge",center=[2 3 1]);
```

and project it onto the sphere

```
>> cam.mesh(X,Y,Z)
```

and this is shown in ▣ Fig. 13.22. To aid visualization, the spherical image plane has been unwrapped into a rectangle – lines of longitude and latitude are displayed as vertical and horizontal lines respectively. The top and bottom edges correspond to the north and south poles respectively.

It is not yet possible to buy a spherical camera, but prototypes have been demonstrated in several laboratories. The spherical camera is more useful as a conceptual construct to simplify the discussion of wide-angle imaging. As we show in the next section we can transform images from perspective, fisheye or catadioptric camera onto the sphere where we can treat them in a unified manner.

13.4 **Unified Imaging Model**

We have introduced a number of different imaging models in this chapter. Now we will discuss how to transform an image captured with one type of camera, to the image that would have been captured with a different type of camera. For example, given a fisheye lens projection we will generate the corresponding projection for a spherical camera or a perspective camera. The unified imaging model provides a powerful framework to consider very different types of cameras such as standard perspective, catadioptric and many types of fisheye lens.

The unified imaging model is a two-step process and the notation is shown in ◻ Fig. 13.23. The first step is spherical projection of the world point P to the surface of the unit sphere p′ as discussed in the previous section and described by (13.17) and (13.18). The view point O is the center of the sphere which is a distance m from the image plane along its normal z-axis. The single view point implies a *central* camera.

In the second step, the point p′ is reprojected to the image plane p using the view point F which is at a distance ℓ along the z-axis above O. The image-plane point p is described in polar coordinates as $\boldsymbol{p} = (r, \phi)$ where

$$r = \frac{(\ell + m)\sin\theta}{\ell - \cos\theta} \ . \tag{13.21}$$

The unified imaging model has only two parameters m and ℓ and these are a function of the type of camera as listed in ◻ Tab. 13.2. For a perspective camera, the two view points O and F are coincident and the geometry becomes the same as the central perspective model shown in ◻ Fig. 13.3.

For catadioptric cameras with mirrors that are conics, the focal point F lies between the center of the sphere and the north pole, that is, $0 < \ell < 1$. This projection model is somewhat simpler than the catadioptric camera geometry shown in ◻ Fig. 13.19. The imaging parameters are written in terms of the conic parameters eccentricity ε and latus rectum $4p$. ◄

The projection with F at the north pole is known as stereographic projection and is used in many fields to project thesurface of a sphere onto a plane. Many fisheyelenses are extremely well approximated by F above the north pole.

The length of a chord parallel to the directrix and passing through the focal point.

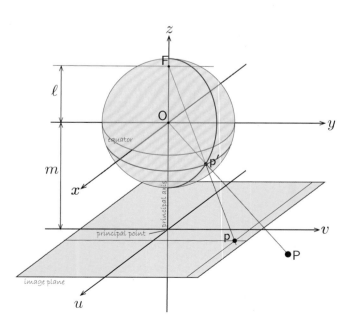

◻ **Fig. 13.23** Unified imaging model. The world point P is projected to p′ on the surface of the unit sphere using the viewpoint O, and then reprojected to point p on the image plane using the viewpoint F

Table 13.2 Unified imaging model parameters ℓ and m according to camera type. ε is the eccentricity of the conic and $4p$ is the latus rectum

Imaging	ℓ	m
Perspective	0	f
Stereographic	1	f
Fisheye	> 1 f	
Catadioptric (elliptical, $0 < \varepsilon < 1$)	$\frac{2\varepsilon}{1+\varepsilon^2}$	$\frac{2\varepsilon(2p-1)}{1+\varepsilon^2}$
Catadioptric (parabolic, $\varepsilon = 1$)	1	$2p-1$
Catadioptric (hyperbolic, $\varepsilon > 1$)	$\frac{2\varepsilon}{1+\varepsilon^2}$	$\frac{2\varepsilon(2p-1)}{1+\varepsilon^2}$

13.4.1 Mapping Wide-Angle Images to the Sphere

We can use the unified imaging model in reverse. Consider an image captured by a wide field of view camera, such as the fisheye image shown in ◘ Fig. 13.24a. If we know the location of F, then we can project each point p from the fisheye image onto the sphere at p′ to create a spherical image, even though we do not have a spherical camera.

In order to achieve this inverse mapping we need to know some parameters of the camera that captured the image. A common feature of images captured with a fisheye lens or catadioptric camera is that the outer bound of the image is a circle. This circle can be found and its center estimated quite precisely – this is the principal point. A variation of the camera calibration procedure of ▶ Sect. 13.2.4 is applied, which uses corresponding world and image-plane points from the planar calibration target shown in ◘ Fig. 13.24a. This particular camera has a field of view of 190° and its calibration parameters have been estimated to be: principal point $(528.1214, 384.0784)$, $\ell = 2.7899$ and $m = 996.4617$.

We will illustrate this using the image shown in ◘ Fig. 13.24a

```
>> fisheye = rgb2gray(imread("fisheye_target.png"));
>> fisheye = im2double(fisheye);
```

and we also define the domain of the input image

```
>> [Ui,Vi] = meshgrid(1:size(fisheye,2), 1:size(fisheye,1));
```

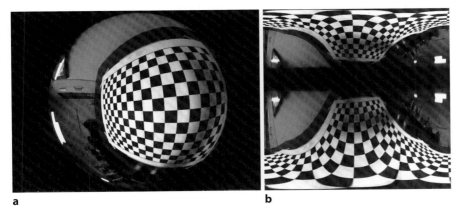

a   b

◘ **Fig. 13.24** Fisheye image of a planar calibration target. **a** Fisheye image (Image courtesy of Peter Hansen); **b** Image warped to (ϕ, θ) coordinates

We will use image warping from ▶ Sect. 11.7.4 to achieve this mapping. The output domain covers the entire sphere with longitude from $-\pi$ to $+\pi$ radians and colatitude from 0 to π radians with 500 steps in each direction

```
>> n = 500;
>> theta_range = linspace(0,pi,n);
>> phi_range = linspace(-pi,pi,n);
>> [Phi,Theta] = meshgrid(phi_range,theta_range);
```

For warping we require a function that returns the coordinates of a point in the input image given the coordinates of a point in the output spherical image. This function is the second step of the unified imaging model (13.21) which we implement as

```
>> l = 2.7899; m = 996.4617;
>> r = (l+m)*sin(Theta)./(l-cos(Theta));
```

from which the corresponding Cartesian coordinates in the input image are

```
>> u0 = 528.1214; v0 = 384.0784;
>> U = r.*cos(Phi) + u0;
>> V = r.*sin(Phi) + v0;
```

Using image warping from ▶ Sect. 11.7.4 we map the fisheye image to a spherical image

```
>> spherical = interp2(Ui,Vi,fisheye,U,V);
```

where the first three arguments define the input domain, and the last two arguments are the coordinates for which grayscale values will be interpolated from the input image and returned. We display the result

```
>> imshow(spherical)
```

which is shown in ▢ Fig. 13.24b. The image appears reflected about the equator and this is because the mapping from a point on the image plane to the sphere is double valued – F is above the north pole so the ray intersects the sphere twice. The top and bottom row of this image corresponds to the principal point, while the dark band above the equator corresponds to the circular outer edge of the input image.

The image is extremely distorted but this coordinate system is very convenient to texture map onto a sphere

```
>> sphere
>> h = findobj(Type="surface");
>> set(h,CData=flipud(spherical),FaceColor="texture");
>> colormap(gray); view(-53,-9)
```

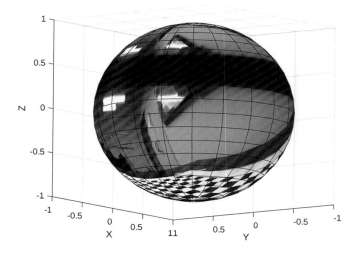

▢ Fig. 13.25 Fisheye image mapped to the unit sphere. We can see the planar grid lying on a table, the ceiling light, a door and a whiteboard

and this is shown in ◘ Fig. 13.25. Using the mouse, we can rotate the sphere and look at the image from different view points.

Any wide-angle image that can be expressed in terms of central imaging parameters can be similarly projected onto a sphere. So too can multiple perspective images obtained from a camera array, such as shown in ◘ Fig. 13.27.

13.4.2 Mapping from the Sphere to a Perspective Image

Given a spherical image we now want to reconstruct a perspective view in a particular direction. We can think of this as being at viewpoint O, inside the sphere, and looking outward at a small surface area which is close to flat and approximates a perspective camera view. This is the second step of the unified imaging model, but with F now at the center of the sphere – this makes the geometry of ◘ Fig. 13.23 similar to the central perspective geometry of ◘ Fig. 13.3. The perspective camera's optical axis is the negative z-axis of the sphere.

For this example we will use the spherical image created in the previous section. We wish to create a perspective image of 1000×1000 pixels and with a field-of-view of $45°$. The field of view can be written in terms of the image width W and the unified imaging parameter m as

$$\theta_{\text{FOV}} = 2 \tan^{-1} \frac{W}{2m}$$

For a $45°$ field-of-view, we require

```
>> W = 1000;
>> m = W/2/tand(45/2)
m =
   1.2071e+03
```

and for perspective projection we require

```
>> l = 0;
```

We also require the principal point to be in the center of the image

```
>> u0 = W/2; v0 = W/2;
```

The domain of the output image will be

```
>> [Uo,Vo] = meshgrid(0:W-1,0:W-1);
```

The polar coordinate (r, ϕ) of each point in the output image is

```
>> [phi,r] = cart2pol(Uo-u0,Vo-v0);
```

and the corresponding spherical coordinates (ϕ, θ) are

```
>> Phi_o = phi;
>> Theta_o = pi-atan(r/m);
```

We now warp from spherical coordinates to the perspective image plane

```
>> perspective = interp2(Phi,Theta,spherical,Phi_o,Theta_o);
```

and the result

```
>> imshow(perspective)
```

is shown in ◘ Fig. 13.26a. This is the view from a perspective camera at the center of the sphere looking down through the south pole. We see that the lines on the checkerboard calibration target are now straight as we would expect from a perspective image.

Of course we are not limited to just looking along the negative z-axis of the sphere. In ◘ Fig. 13.25 we can see some other features of the room such as a door,

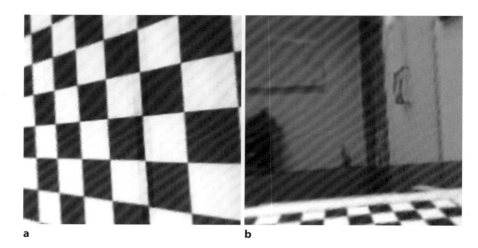

a   b

◻ Fig. 13.26   Perspective projection of spherical image ◻ Fig. 13.25 with a field of view of 45°. Note that the lines on the checkerboard are now straight. **a** looking down through the south pole; **b** looking toward the door and whiteboard

a whiteboard and some ceiling lights. We can point our virtual perspective camera in their direction by first rotating the spherical image

```
>> tform = se3(eul2rotm([0 0.9 -1.5],"XYZ"));
>> spherical = sphere_rotate(spherical,tform);
```

so that the negative z-axis now points toward the distant wall. Repeating the warp process

```
>> perspective = interp2(Phi,Theta,spherical,Phi_o,Theta_o);
>> imshow(perspective)
```

From a single wide-angle image we can create a perspective view in any direction without having any mechanical pan/tilt mechanism – it's just computation. In fact, multiple users could look in different directions simultaneously from a live feed of a single wide-angle camera.

we obtain the result shown in ◻ Fig. 13.26b in which we can clearly see a door and a whiteboard. ◄

The original wide-angle image contains a lot of detail though it can be hard to see because of the distortion. After mapping the image to the sphere we can create a virtual perspective camera view along any line of sight. This is only possible if the original image was taken with a central camera that has a single viewpoint. In theory we cannot create a perspective image from a noncentral wide-angle image but in practice, if the caustic is small, the parallax errors introduced into the perspective image will be negligible.

13.5 Novel Cameras

13.5.1 Multi-Camera Arrays

The cost of cameras and computation continues to fall making it feasible to warp and stitch images from multiple perspective cameras onto a cylindrical or spherical image plane. One such camera is shown in ◻ Fig. 13.27a and uses five cameras to capture a 360° panoramic view as shown in ◻ Fig. 13.27c. The camera in ◻ Fig. 13.27b uses six cameras to achieve an almost spherical field of view.

These camera arrays are not central cameras since light rays converge on the focal points of the individual cameras, not the center of the camera assembly. This can be problematic when imaging objects at short range but in typical use the distance between camera focal points, the caustic, is small compared to distances in the scene. The different viewpoints do have a real advantage however when it comes to capturing the light field.

■ Fig. 13.27 Omnidirectional camera array. **a** Five perspective cameras provide a 360° panorama with a 72° vertical field of view (camera by Occam Vision Group). **b** Panoramic camera array uses six perspective cameras to provide 90% of a spherical field of view. **c** A seamless panoramic image (3760 × 480 pixels) as output by the camera **a** (Images **a** and **c** by Edward Pepperell; image **b** courtesy of Point Grey Research)

13.5.2 Light-Field Cameras

As we discussed in the early part of this chapter a traditional perspective camera captures a representation of a scene using the two dimensions of the film or sensor. We can think of the captured image as a 2-dimensional function $\mathcal{L}(X, Y)$ that describes the light emitted by the 3D scene. The function is scalar-valued $\mathcal{L}(X, Y) \in \mathbb{R}$ for the monochrome case and vector-valued $\mathcal{L}(X, Y) \in \mathbb{R}^3$ for a tristimulus color representation. ▶

The pin-hole camera of ■ Fig. 13.1 allows only a very small number of light rays to pass through the aperture, yet space is filled with innumerable light rays that provide a richer and more complete description of the world. An objective lens directs multiple rays from a world point to a photosite – increasing the brightness – but the direction information associated with those rays is lost since the photosite responds to all light rays equally, irrespective of their angle to the surface.

The geometric description of *all* the light rays in the scene is called the plenoptic function. ▶ Each ray has a position and direction in 3-dimensional space and could be represented by $\mathcal{L}(X, Y, Z, \theta, \phi)$. ▶ However, a more convenient representation is $\mathcal{L}(s, t, u, v)$ using the 2-plane parameterization shown in ■ Fig. 13.29a which we can think of as an array of cameras, each with a 2-dimensional image plane. The conventional perspective camera image is the view from a single camera, or a 2-dimensional slice of the full plenoptic function.

The ideas behind the light field have been around for decades, but it is only in recent years that the technology to capture light fields has become widely available. Early light-field cameras were arrays of regular cameras arranged in a plane, such as shown in ■ Fig. 13.28, or on a sphere surrounding the scene, but these tended to be physically large, complex and expensive to construct. More recently, low-cost and compact light-field cameras based on microlens arrays have come on to the market. The selling point for early consumer light-field cameras was the ability

We could add an extra dimension to represent polarization of the light.

The word plenoptic comes from the Latin word plenus meaning full or complete.

Lines in 3D-space have four parameters, see Plücker lines in ▶ App. C.2.2.1.

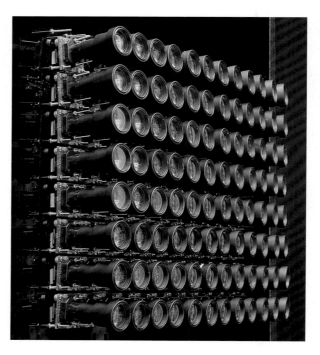

□ Fig. 13.28 An 8 × 12 camera array as described in Wilburn et al. (2005) (Image courtesy of Marc Levoy, Stanford University)

to refocus the image *after* taking the picture. However, the light-field image has many other virtues including synthesizing novel views, 3D reconstruction, low-light imaging and seeing through particulate obscurants.

The microlens or lenslet array is a regular grid of tiny lenses, typically comprising hundreds of thousands of lenses, which is placed a fraction of a millimeter above the surface of the camera's photosite array. The main objective lens focuses an image onto the surface of the microlens array as shown in □ Fig. 13.29b. The microlens directs incoming light to one of a small, perhaps 8×8, patch of photosites according to its direction. The resulting image captures information about both the origin of the ray (the lenslet) and its direction (the particular photosite beneath the lenslet). By contrast, in a standard perspective camera all the rays, irrespective of

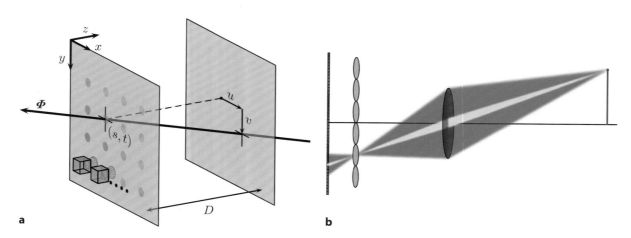

□ Fig. 13.29 **a** The light ray Φ passes through the image plane at point (u, v) and the center of the camera at (s, t). This is similar to the central projection model shown in □ Fig. 13.3. Any ray can be described by two points, in this case (u, v) and (s, t). **b** Path of light rays from object through main objective lens and lenslet array to the photosite array (Images courtesy Donald G. Dansereau)

◘ Fig. 13.30 **a** Closeup of image formed on the sensor by the lenslet array. The lenslets are circular but packed hexagonally, and they project hexagonal images because of the hexagonal iris which is designed to avoid overlap of the lenslet images. **b** array of images rendered from the light field for different camera view points (figures courtesy Donald G. Dansereau)

direction, contribute to the output of the photosite. The light-field camera pixels are sometimes referred to as *raxels* and the resolution of these cameras is typically expressed in megarays.

The raw image from the sensor array looks like ◘ Fig. 13.30a but can be *decoded* into a 4-dimensional light field, as shown in ◘ Fig. 13.30b, and used to render novel views.

13.6 Applications

13.6.1 Fiducial Markers

◘ Fig. 13.31 shows a scene that contains a number of artificial marker objects, often referred to as fiducial markers. In robotics, two families of fiducial markers are commonly used: AprilTags ► developed by the University of Michigan for tasks such as augmented reality, robotics and camera calibration, and ArUco markers, originally developed for augmented reality applications.

We will focus on AprilTags which have high-contrast patterns that make them easy to detect in a scene. The pattern of squares indicates the orientation of the marker but also encodes some binary data bits which can be used to uniquely identify the marker. If the camera intrinsics and the dimensions of the marker are known, then we can estimate the pose of the marker with respect to the camera.

Both types of markers are grouped into families. For example, the AprilTag markers contain several families with names such as 16H5, 21H7 (Circle), 25H9, ..., 52H13. The families labeled *Circle* are designed to fit roughly into a circular area. The first number in the name is the number of data bits corresponding to changeable blocks in the tag layout. The second number is the Hamming distance indicating tolerance to incorrect bits, for example caused by smudges or dirt. The larger the first number, the more data you can pack and more tags you can have in a family, but at a trade-off of space required to hold physically larger tags. ►

We begin by loading a scene with AprilTag markers

```
>> scene = imread("aprilTag36h11.jpg");
>> imshow(scene);
```

which is shown in ◘ Fig. 13.31.

See ► https://sn.pub/DdL5LA.

More about tag families ► https://sn.pub/Z2Mamv.

Fig. 13.31 Page with printed AprilTag markers and their corresponding numerical IDs. The *x*-, *y*- and *z*- axes are in red, green and blue respectively, with the *z*-axis pointing down into the page (Images reprinted with permission of The MathWorks, Inc.)

The intrinsic parameters of the camera were established earlier using the Camera Calibrator app and are simply loaded from a MAT file

```
>> data = load("camIntrinsicsAprilTag.mat");
>> intrinsics = data.intrinsics;
```

and the image is first undistorted to obtain precise readings

```
>> scene = undistortImage(scene,intrinsics,OutputView="same");
```

We are now ready to use the `readAprilTag` function

```
>> tagSize = 0.04; % in meters
>> [id,loc,pose] = readAprilTag(scene,"tag36h11",intrinsics,tagSize);
```

where we also pass in the name of the marker family, the camera intrinsic matrix and the side length of the marker's black square. The result is a list of marker IDs, their location and pose, which can be visualized by

```
>> % define axes points to draw
>> worldPoints = [0 0 0; tagSize /2 0 0;...
>>     0 tagSize /2 0; 0 0 tagSize /2];
>> for i = 1: length(pose)
>>     imagePoints = world2img(worldPoints ,pose(i),intrinsics);
>>     scene = insertShape(scene ,"Line" , ...
>>         [imagePoints (1,:) imagePoints (2,:); ...
>>         imagePoints (1,:) imagePoints (3,:); ...
>>         imagePoints (1,:) imagePoints (4,:)], ...
>>         Color =["red","green","blue"],LineWidth =12);
>>     scene = insertText(scene ,loc(1,:,i),id(i), ...
>>         BoxOpacity =1, FontSize =28);
>> end
>> imshow(scene)
```

and the result is shown in ☐ Fig. 13.31. These markers have a variety of applications, including assigning a unique identifier to a robot, creating a pattern for camera calibration, augmenting a scene with a projected 3D object based on the marker's extrinsic parameters (augmented reality) or simply establishing ground truth for verifying accuracy of your SLAM algorithm. Another interesting application of AprilTags is for landmark SLAM where robot odometry data can be combined with known fiducial marker locations to improve estimates of the robot trajectory. An extensive example related to this application can be found at ▶ https://sn.pub/Vqs8Co.

13.6.2 Planar Homography

Consider a coordinate frame attached to a planar surface such that the z-axis is normal to the plane and the x- and y-axes lie within the plane. We view the marker with a calibrated camera and recalling (13.9)

$$\begin{pmatrix} \tilde{u} \\ \tilde{v} \\ \tilde{w} \end{pmatrix} = \begin{pmatrix} c_{1,1} & c_{1,2} & c_{1,3} & c_{1,4} \\ c_{2,1} & c_{2,2} & c_{2,3} & c_{2,4} \\ c_{3,1} & c_{3,2} & c_{3,3} & 1 \end{pmatrix} \begin{pmatrix} X \\ Y \\ Z \\ 1 \end{pmatrix} .$$

For all points in the plane $Z = 0$, so we can remove Z from the equation

$$\begin{pmatrix} \tilde{u} \\ \tilde{v} \\ \tilde{w} \end{pmatrix} = \begin{pmatrix} c_{1,1} & c_{1,2} & \cancel{c_{1,3}} & c_{1,4} \\ c_{2,1} & c_{2,2} & \cancel{c_{2,3}} & c_{2,4} \\ c_{3,1} & c_{3,2} & \cancel{c_{3,3}} & 1 \end{pmatrix} \begin{pmatrix} X \\ Y \\ \cancel{Z} \\ 1 \end{pmatrix}$$

$$= \begin{pmatrix} c_{1,1} & c_{1,2} & c_{1,3} \\ c_{2,1} & c_{2,2} & c_{2,3} \\ c_{3,1} & c_{3,2} & 1 \end{pmatrix} \begin{pmatrix} X \\ Y \\ 1 \end{pmatrix} = \mathbf{H} \begin{pmatrix} X \\ Y \\ 1 \end{pmatrix}$$

resulting in an invertible linear relationship between homogeneous coordinate vectors in the world plane and in the image plane. This relationship is called a homography or a planar homography and the matrix $\mathbf{H} \in \mathbb{R}^{3 \times 3}$ is non singular.

We define a central perspective camera that is up high and looking obliquely downward

```
>> T_camera = se3(eul2rotm([0 0 -2.8]),[0 0 8]);
>> camera = CentralCamera("default",focal=0.012,pose=T_camera);
```

at a ground plane as shown in ◘ Fig. 13.32. A shape on the ground plane is defined by a set of 2-dimensional coordinates

```
>> P = [-1 1; -1 2; 2 2; 2 1];
```

where each row is the (X, Y) coordinate of a point.

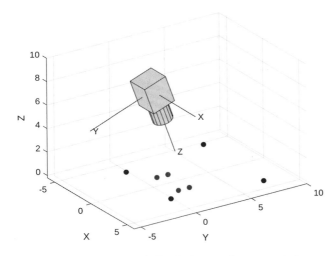

◘ **Fig. 13.32** An overhead camera looking obliquely downward to a ground plane on which a shape is defined by four red markers. The field of view of the camera at the groundplane is shown by the blue markers

We can project the ground plane points to the image plane in the familiar way using the `CentralCamera` object, after first adding $Z = 0$ for each world point

```
>> projP = camera.project(padarray(P,[0 1],0,"post"));
>> projP' % transpose for display
ans =
   359.5758   365.7968   804.4064   816.8484
   776.8649   628.2989   628.2989   776.8649
```

The homography is computed from the camera matrix by deleting column two

```
>> H = camera.C;
>> H(:,3) = []
H =
   1.0e+03 *
    1.2000     0.1715     3.8593
         0    -0.9592     7.0752
         0     0.0003     0.0075
```

and we can use that to transform the point coordinates from the ground plane directly to the image plane

```
>> Hobj = projtform2d(H);
>> projP = Hobj.transformPointsForward(P);
>> projP' % transpose for display
ans =
   359.5758   365.7968   804.4064   816.8484
   776.8649   628.2989   628.2989   776.8649
```

The `transformPointsForward` method of `projtform2d` class performs the appropriate conversions to and from homogeneous coordinates, and the results are the same as those obtained by perspective projection using the camera object.

Since **H** is invertible we can also perform the inverse mapping. The camera has a 1000×1000 image plane so the coordinates of its corners are

```
>> p = [0 0; 0 1000; 1000 1000; 1000 0];
```

and on the ground plane these are the points

```
>> Pi = Hobj.transformPointsInverse(p);
>> Pi' % transpose for display
ans =
   -4.2704    -3.1650     3.0167     4.0703
    7.3765    -0.3574    -0.3574     7.3765
```

which are shown in blue in ◘ Fig. 13.32. The figure was created by

```
>> camera.plot_camera(scale=2,color="y")
>> hold on
>> plotsphere(padarray(P,[0 1],0,"post"),0.2,"r")
>> plotsphere(padarray(Pi,[0 1],0,"post"),0.2,"b")
>> grid on, view([57 24])
```

A practical application of these ideas is captured in the example ► https://sn.pub/vn4Bc0 which uses a calibrated camera to measure a planar object.

13.7   Advanced Topics

13.7.1   Projecting 3D Lines and Quadrics

In ► Sect. 13.1 we projected 3D points to the image plane, and we projected 3D line segments by simply projecting their endpoints and joining them on the image plane. To project a continuous line in 3-dimensional space we must first decide how to represent it, and there are many possibilities which are discussed in ► App. C.1.2.2. One useful parameterization is Plücker coordinates – a 6-vector with many similarities to twists.

We can easily create a Plücker line using the RVC Toolbox. A line that passes through the points $(0, 0, 1)$ and $(1, 1, 1)$ would be

```
>> L = Plucker([0 0 1],[1 1 1])
L =
{ -1    1    0;  -1   -1    0}
```

which returns a `Plucker` object that is represented as a 6-vector with two components: a moment vector and a direction vector. Options can be used to specify a line using a point and a direction or the intersection of two planes. The direction of the Plücker line is the vector

```
>> L.w
ans =
   -1   -1    0
```

The `Plucker` object also has methods for plotting, as well as determining the intersection with planes or other Plücker lines. There are many representations of a Plücker line including the 6-vector used above, a minimal 4-vector, and a skew-symmetric 4×4 matrix computed using the `skew` method. The latter is used to project the line by

$$\ell = \bigvee\nolimits_x \left(\mathbf{C} \, [\mathbf{L}]_\times \mathbf{C}^\top \right) \in \mathbb{R}^3$$

where $\mathbf{C} \in \mathbb{R}^{3 \times 4}$ is the camera matrix, and results in a 2-dimensional line expressed in homogeneous coordinates. Observing this line with the default camera

```
>> cam = CentralCamera("default");
>> l = cam.project(L)
l =
    1   -1    0
```

results in a diagonal line across the image plane. We can plot this on the camera's virtual image plane by

```
>> cam.plot(L)
```

Quadrics, short for quadratic surfaces, are a rich family of 3-dimensional surfaces. There are 17 standard types including spheres, ellipsoids, hyperboloids, paraboloids, cylinders and cones all described by points $\tilde{x} \in \mathbb{P}^3$ such that

$$\tilde{x}^\top \mathbf{Q} \, \tilde{x} = 0$$

where $\mathbf{Q} \in \mathbb{R}^{4 \times 4}$ is symmetric. The *outline* of the quadric is projected to the image plane by

$$c^* = \mathbf{C} \mathbf{Q}^* \mathbf{C}^\top \in \mathbb{R}^{3 \times 3}$$

where $(\cdot)^*$ represents the adjugate operation, ▶ and c is a matrix representing a conic section on the image plane and the outline is the set of points p such that

$$\tilde{p}^\top c \, \tilde{p} = 0$$

and for $\tilde{p} = (u, v, 1)$ can be expanded as

$$A u^2 + B u v + C v^2 + D u + E v + F = 0$$

where

$$c = \left(\begin{array}{cc|c} A & B/2 & D/2 \\ B/2 & C & E/2 \\ \hline D/2 & E/2 & F \end{array} \right) .$$

The determinant of the top-left submatrix indicates the type of conic: negative for a hyperbola, 0 for a parabola and positive for an ellipse.

$\mathbf{A}^* = \det(\mathbf{A})\mathbf{A}^{-1}$ which is the transpose of the cofactor matrix. If $\mathbf{B} = \mathbf{A}^*$ then $\mathbf{A} = \mathbf{B}^*$. See ▶ App. B for more details.

To demonstrate this, we define a camera looking toward the origin

```
>> cam = CentralCamera("default", ...
>>    pose=se3eul2rotm([0 0 0.2]),[0.2 0.1 -5]));
```

and define a unit sphere at the origin

```
>> Q = diag([1 1 1 -1]);
```

then compute its projection to the image plane

```
>> Qs = inv(Q)*det(Q);  % adjugate
>> cs = cam.C*Qs*cam.C';
>> c = inv(cs)*det(cs); % adjugate
```

which is a 3×3 matrix describing a 2-dimensional conic. The determinant of the top-left submatrix

```
>> det(c(1:2,1:2))
ans =
    2.2862e+14
```

is positive indicating an ellipse, and a simple way to plot this is using the Symbolic Math Toolbox™

```
>> syms x y real
>> ezplot([x y 1]*c*[x y 1]',[0 1024 0 1024])
>> set(gca,Ydir="reverse")
```

13.7.2 Nonperspective Cameras

The camera matrix, introduced in ▶ Sect. 13.1.4, is a 3×4 matrix. Any 3×4 matrix corresponds to some type of camera, but most would result in wildly distorted images. The camera matrix in (13.9) has a special structure – it is a subset of all possible 3×4 matrices – and corresponds to a perspective camera. The camera projection matrix $\mathbf{C}$ from (13.9) can be written generally as

$$\mathbf{C} = \begin{pmatrix} c_{11} & c_{12} & c_{13} & c_{14} \\ c_{21} & c_{22} & c_{23} & c_{24} \\ c_{31} & c_{32} & c_{33} & 1 \end{pmatrix}$$

which has arbitrary scale so one element, typically c_{34} is set to one – this matrix has 11 unique elements or 11 DoF.

Orthographic or parallel projection is a simple perspective-free projection of 3D points onto a plane, like a "plan view". For small objects close to the camera this projection can be achieved using a telecentric lens. The apparent size of an object is independent of its distance.

For the case of an aerial robot flying high over relatively flat terrain the variation of depth, the vertical relief, Δ_Z is small compared to the average depth of the scene $\overline{Z}$, that is $\Delta_Z \ll \overline{Z}$. We can use a scaled-orthographic projection which is an orthographic projection followed by uniform scaling $m = f/\overline{Z}$.

These two nonperspective cameras are special cases of the more general affine camera model which is described by a matrix of the form

$$\mathbf{C} = \begin{pmatrix} c_{11} & c_{12} & c_{13} & c_{14} \\ c_{21} & c_{22} & c_{23} & c_{24} \\ 0 & 0 & 0 & 1 \end{pmatrix}$$

that can be factorized as

$$
\mathbf{C} = \underbrace{\begin{pmatrix} m_x & s & 0 \\ 0 & m_y & 0 \\ 0 & 0 & 1 \end{pmatrix}}_{\text{intrinsic}} \underbrace{\begin{pmatrix} 1 & 0 & 0 & 0 \\ 0 & 1 & 0 & 0 \\ 0 & 0 & 0 & 1 \end{pmatrix}}_{\text{projection}} \underbrace{\begin{pmatrix} \mathbf{R} & t \\ 0 & 1 \end{pmatrix}}_{\text{extrinsic}} .
$$

It can be shown that the principal point is undefined for such a camera model which simplifies the intrinsic matrix. A skew parameter s is commonly introduced to handle the case of nonorthogonal sensor axes. The projection matrix factor is different compared to the perspective case in (13.9) – the last two columns are swapped. We can delete the zero column of that matrix and compensate by deleting the third row of the extrinsic matrix resulting in

$$
\mathbf{C} = \underbrace{\begin{pmatrix} m_x & s & 0 \\ 0 & m_y & 0 \\ 0 & 0 & 1 \end{pmatrix}}_{\text{3 DoF}} \underbrace{\begin{pmatrix} r_{11} & r_{12} & r_{13} & t_x \\ r_{21} & r_{22} & r_{23} & t_y \\ 0 & 0 & 0 & 1 \end{pmatrix}}_{\text{5 DoF}}
$$

where $r_{ij} = \mathbf{R}_{i,j}$. This has at most 8 DoF ▶ and the independence from depth is very clear since t_z does not appear. The case where skew $s = 0$ and $m_x = m_y = 1$ is orthographic projection and has only 5 DoF, while the scaled-orthographic case when $s = 0$ and $m_x = m_y = m$ has 6 DoF. The case where $m_x \neq m_y$ is known as weak perspective projection, although this term is sometimes also used to describe scaled-orthographic projection.

The 2×3 submatrix of the rotation matrix has 6 elements but 3 constraints – the two rows have unit norms and are orthogonal – and therefore has 3 DoF.

13.8 Wrapping Up

This chapter has introduced the image formation process which connects the physical world of light rays to an array of pixels which comprise a digital image. The images that we are familiar with are perspective projections of the world in which 3 dimensions are compressed into 2 dimensions. This leads to ambiguity about object size – a large object in the distance looks the same as a small object that is close. Straight lines and conics are unchanged by this projection but shape distortion occurs – parallel lines can appear to converge and circles can appear as ellipses. We have modeled the perspective projection process and described it in terms of eleven parameters – intrinsic and extrinsic. Geometric lens distortion adds additional lens parameters. Camera calibration is the process of estimating these parameters and two approaches were introduced. We also discussed pose estimation where the pose of an object with known geometry can be estimated from a perspective projection obtained using a calibrated camera.

Perspective images are limited in their field of view and we discussed several wide-angle imaging systems based on the fisheye lens, catadioptrics and multiple cameras. We also discussed the ideal wide-angle camera, the spherical camera, which is currently still a theoretical construct. However, it can be used as an intermediate representation in the unified imaging model which provides one model for almost all camera geometries. We used the unified imaging model to convert a fisheye camera image to a spherical image and then to a perspective image along a specified view axis. We also covered some more recent camera developments such as panoramic camera arrays and light-field cameras. We used two application examples to introduce fiducial markers, and a linear mapping between a plane in the world and the image plane which is applicable to many problems. Finally, we covered some advanced topics such as projecting lines and quadrics from the world to the image plane, and some generalizations of the perspective camera matrix.

In this chapter we treated imaging as a problem of pure geometry with a small number of world points or line segments. In ► Chap. 14 we will discuss the geometric relationships between different images of the same scene.

13.8.1   Further Reading and Resources

Computer vision textbooks such as Davies, Klette (2014), Gonzalez and Woods (2018), Forsyth and Ponce (2011) and Szeliski at ► https://szeliski.org/Book (2022) all provide coverage of the topics introduced in this chapter. Hartley and Zisserman (2003) provide very detailed coverage of image formation using geometric and mathematical approaches, while Ma et al. (2003) provide a mathematical approach. Many topics in geometric computer vision have also been studied by the photogrammetric community, but different language is used. For example camera calibration is known as camera resectioning, and pose estimation is known as space resectioning. The updated Manual of Photogrammetry (McGlone 2013) provides comprehensive and definitive coverage of the field including history, theory and applications of aircraft and satellite imagery. The revised classic textbook by DeWitt and Wolf (2014) is a thorough and readable introduction to photogrammetry. Wade (2007) reviews the progression, over many centuries, of humankind's understanding of the visual process.

■ ■ **Camera calibration**

The homogeneous transformation calibration (Sutherland 1974) approach of ► Sect. 13.2.1 is also known as the direct linear transform (DLT) in the photogrammetric literature. The RVC Toolbox implementation `camcald` requires that the centroids of the calibration markers have already been determined which is a nontrivial problem (Corke 1996b, § 4.2). It also cannot estimate lens distortion. Wolf (1974) describes extensions to the linear camera calibration with models that include up to 18 parameters and suitable nonlinear optimization estimation techniques. A more concise description of nonlinear calibration is provided by Forsyth and Ponce (2011). Hartley and Zisserman (2003) describe how the linear calibration model can be obtained using features such as lines within the scene.

The Camera Calibrator (`cameraCalibrator`) and the Stereo Camera Calibrator (`stereoCameraCalibrator`) apps, in the Computer Vision Toolbox™, are tools that greatly simplify the camera calibration process. These apps are discussed in ► Sect. 13.2.4. There are also a number of good camera calibration toolboxes available on the web. The Camera Calibration Toolbox by Jean-Yves Bouguet (2010) is still available from ► http://www.vision.caltech.edu/bouguetj/calib_doc. Bouguet's model combines work by Zhang (calibration using flat checkerboard), Heikkila and Silven (tangential distortion). His toolbox has extensive online documentation and includes example calibration images. Several tools build on this and automatically find the checkerboard target which is otherwise tedious to locate in every image, for example the AMCC and RADOCC Toolboxes. The MATLAB Toolbox by Janne Heikkilä is available at ► http://www.ee.oulu.fi/~jth/calibr/ and works for planar or 3D targets with circular dot features and estimates lens distortion.

Pose estimation is a classic problem in computer vision and for which there exists a very large literature. The approaches can be broadly divided into analytic and iterative solutions. Assuming that lens distortion has been corrected the analytic solutions for three and four noncollinear points are given by Fischler and Bolles (1981), DeMenthon and Davis (1992) and Horaud et al. (1989). Typically multiple solutions exist, but for four coplanar points there is a unique solution. Six or more points always yield unique solutions, as well as the intrinsic camera calibration parameters. Iterative solutions were described by Rosenfield (1959) and

Excurse 13.12: Photogrammetry

Photogrammetry is the science of understanding the geometry of the world from images. The techniques were developed by the French engineer Aimé Laussedat (1819–1907) working for the Army Corps of Engineers in the 1850s. He produced the first measuring camera and developed a mathematical analysis of photographs as perspective projections. He pioneered the use of aerial photography as a surveying tool to map Paris – using rooftops as well as uncrewed balloons and kites.

Photogrammetry is normally concerned with making maps from images acquired at great distance, but the subfield of close-range or terrestrial photogrammetry is concerned with camera to object distances less than 100 m which is directly relevant to robotics. (Image from La Métrophotographie, Aimé Laussedat, 1899)

Lowe (1991). A more recent discussion based around the concept of bundle adjustment is provided by Triggs et al. (2000). Pose estimation requires a geometric model of the object and such computer vision approaches are known as *model-based vision*. An interesting historical perspective on model-based vision is the 1987 video by the late Joe Mundy which is available at ► http://www.archive.org/details/JosephMu1987.

■■ **Wide field-of-view cameras**

There is recent and growing interest in this type of camera and today, good quality lightweight fisheye lenses and catadioptric camera systems are available. Nayar (1997) provides an excellent motivation for, and introduction to, wide-angle imaging. A very useful online resource is the catadioptric sensor design page at ► http://www.math.drexel.edu/~ahicks/design and a page of links to research groups, companies and workshops at ► http://www.cis.upenn.edu/~kostas/omni.html. Equiangular mirror systems were described by Chahl and Srinivasan (1997) and Ollis et al. (1999). Nature's solution, the reflector-based scallop eye, is described in Colicchia et al. (2009). The book of Daniilidis and Klette (2006) is a collection of papers on nonperspective imaging and Benosman and Kang (2001) is another, earlier, published collection of papers. Some information is available through CVonline at ► http://homepages.inf.ed.ac.uk/rbf/CVonline in the section *Image Physics*.

The Camera Calibrator app from the Computer Vision Toolbox™ includes Davide Scaramuzza's model (Scaramuzza 2006) for wide-angle cameras. You can also find the original MATLAB toolbox by Davide Scaramuzza at ► https://sn.pub/SKZYOE. It is inspired by, and similar in usage, to Bouguet's Toolbox for perspective cameras. Another MATLAB Toolbox, by Juho Kannala, handles wide angle central cameras and is available at ► http://www.ee.oulu.fi/~jkannala/calibration.

The unified imaging model was introduced by Geyer and Daniilidis (2000) in the context of catadioptric cameras. Later it was shown (Ying and Hu 2004) that many fisheye cameras can also be described by this model. The fisheye calibration of ► Sect. 13.4.1 was described by Hansen et al. (2010) who estimates ℓ and m rather than a polynomial function $r(\theta)$ as does Scaramuzza's camera model.

There is a huge and growing literature on light-field imaging but as yet no textbook. A great introduction to light fields and its application to robotics is the thesis by Dansereau (2014). The same author has a MATLAB Toolbox available at ► https://sn.pub/ldLXpD. An interesting description of an early camera array is given by Wilburn et al. (2005) and the associated video demonstrates many capabilities. Light-field imaging is a subset of the larger, and growing, field of computational photography.

■■ **Fiducial markers**

AprilTags were described in Olson (2011) and there is an implementation at ► https://github.com/AprilRobotics/apriltag. ArUco markers are described in Romero-Ramirez et al. (2018) and also at ► https://www.uco.es/investiga/grupos/ava/node/26.

13.8.2 Exercises

1. Create a central camera and a cube target and visualize it for different camera and cube poses. Create and visualize different 3D mesh shapes such as created by the MATLAB functions `cylinder` and `sphere`.
2. Write code to fly the camera in an orbit around the cube, always facing toward the center of the cube.
3. Write code to fly the camera through the cube.
4. Create a central camera with lens distortion and which is viewing a 10×10 planar grid of points. Vary the distortion parameters and see the effect this has on the shape of the projected grid. Create pincushion and barrel distortion.
5. Repeat the homogeneous camera calibration exercise of ► Sect. 13.2.1 and the decomposition of ► Sect. 13.2.2. Investigate the effect of the number of calibration points, noise and camera distortion on the calibration residual and estimated target pose.
6. Determine the solid angle for a rectangular pyramidal field of view that subtends angles θ_h and θ_v.
7. Calibrate the camera on your computer using the Camera Calibrator app.
8. Derive (13.14).
9. For the camera calibration matrix decomposition example (► Sect. 13.2.2) determine the roll-pitch-yaw orientation error between the true and estimated camera pose.
10. Pose estimation (► Sect. 13.2.3)
    a) Repeat the pose estimation exercise for different object poses (closer, further away).
    b) Repeat for different levels of camera noise.
    c) What happens as the number of points is reduced?
    d) Does increasing the number of points counter the effects of increased noise?
    e) Change the intrinsic parameters of the camera `cam` before invoking the `estpose` method. What is the effect of changing the focal length and the principal point by say 5%.
11. Repeat exercises 2 and 3 for the fisheye camera and the spherical camera.
12. With reference to �«Fig. 13.19 derive the function $\psi(\theta)$ for a parabolic mirror.
13. With reference to �«Fig. 13.19 derive the equation of the equiangular mirror $z(x)$ in the xz-plane.
14. Quadrics
    a) Write a routine to plot a quadric given a 4×4 matrix. Hint use `meshgrid` and `isosurface`.
    b) Write code to compute the quadric matrix for a sphere at arbitrary location and of arbitrary radius.

c) Write code to compute the quadric matrix for an arbitrary circular cylinder.

d) Write numeric MATLAB code to plot the planar conic section described by a 3×3 matrix.

15. Project an ellipsoidal or spherical quadric to the image plane. The result will be the implicit equation for a conic – write code to plot the implicit equation.

Using Multiple Images

Contents

14.1   Point Feaature Correspondence – 595

14.2   Geometry of Multiple Views – 599

14.3   Sparse Stereo – 614

14.4   Dense Stereo – 623

14.5   Anaglyphs – 634

14.6   Other Depth Sensing Technologies – 635

14.7   Point Clouds – 637

14.8   Applications – 644

14.9   Wrapping Up – 654

© The Author(s), under exclusive license to Springer Nature Switzerland AG 2023
P. Corke, W. Jachimczyk, R. Pillat, *Robotics, Vision and Control*, Springer Tracts in Advanced Robotics 147,
https://doi.org/10.1007/978-3-031-07262-8_14

`chapter14.mlx`

▶ sn.pub/WUYryF

Almost! We can determine the translation of the camera only up to an unknown scale factor, that is, the translation is $\lambda \hat{t} \in \mathbb{R}^3$ where the direction t is known but λ is not. This ambiguity can be resolved by use of a stereo camera.

14

In ▶ Chap. 12 we learned about point features which are distinctive *points* in an image. They correspond to visually distinctive physical features in the world that can be reliably detected in different views of the same scene, irrespective of viewpoint or lighting conditions. They are characterized by high image gradients in orthogonal directions and frequently occur on the corners of objects. However, the 3-dimensional coordinate of the corresponding world point was lost in the perspective projection process which we discussed in ▶ Chap. 13 – we mapped a 3-dimensional world point to a 2-dimensional image coordinate. All we know is that the world point lies along some ray in space corresponding to the pixel coordinate, as shown in ◘ Fig. 13.6. To recover the missing third dimension we need additional information. In ▶ Sect. 13.2.3 the additional information was camera calibration parameters plus a geometric object model, and this allowed us to estimate the object's 3-dimensional pose from 2-dimensional image data.

In this chapter we consider an alternative approach in which the additional information comes from *multiple* views of the same scene. As already mentioned, the pixel coordinates from a single view constrains the world point to lie along some ray. If we can locate the same world point in another image, taken from a different but known pose, we can determine another ray along which that world point must lie. The world point lies at the intersection of these two rays – a process known as triangulation or 3D reconstruction. Even more powerfully, if we observe sufficient number of points, we can estimate the 3D motion of the camera between the views as well as the 3D structure of the world. ◄

The underlying challenge is to find the same world point in multiple images. This is the *correspondence problem*, an important but nontrivial problem that we will discuss in ▶ Sect. 14.1. In ▶ Sect. 14.2 we revisit the fundamental geometry of image formation developed in ▶ Chap. 13 for the case of a single camera. If you haven't yet read that chapter, or it has been a while since you read it, it would be helpful to reacquaint yourself with that material. We extend the geometry to encompass multiple image planes and show the geometric relationship between pairs of images. Stereo vision is an important technique for robotics where information from two images of a scene, taken from different viewpoints, is combined to determine the 3-dimensional structure of the world. We discuss sparse and dense approaches to stereo vision in ▶ Sects. 14.3 and 14.4 respectively. Bundle adjustment is an advanced concept, but a very general approach to combining information from many cameras, and is introduced in ▶ Sect. 14.3.2.

For some applications we might use RGBD cameras which return depth as well as color information and the underlying principles of such cameras are introduced in ▶ Sect. 14.6. The 3-dimensional information is typically represented as a *point cloud*, a set of 3D points, and techniques for plane fitting and alignment of such data are introduced in ▶ Sect. 14.7.

We finish this chapter, and this part of the book, with three application examples that put the concepts we have learned into practice. ▶ Sect. 14.8.1 describes how we can transform an image with obvious perspective distortion into one without it, effectively synthesizing the view from a virtual camera at a different location. ▶ Sect. 14.8.2 describes mosaicing which is the process of taking multiple overlapping images from a moving camera and *stitching* them together to form one large virtual image. ▶ Sect. 14.8.3 describes how we can process a sequence of images from a moving camera to locate consistent world points and to estimate the camera motion and 3-dimensional world structure.

14.1 Point Feaature Correspondence

Correspondence is the problem of finding the pixel coordinates in two different images that correspond to the same point in the world. ▶ Consider the pair of real images

```
>> im1 = rgb2gray(imread("eiffel2-1.jpg"));
>> im2 = rgb2gray(imread("eiffel2-2.jpg"));
```

This is another example of the data association problem that we have encountered several times in this book.

shown in color in ◘ Fig. 14.1. They show the same scene viewed from two different positions using two different cameras – the pixel size, focal length and number of pixels for each image are quite different. The scenes are complex and we see immediately that determining correspondence is not trivial. More than half the pixels in each scene correspond to blue sky and it is impossible to match a blue pixel in one image to the corresponding blue pixel in the other – these pixels are insufficiently distinct. This situation is common and can occur with homogeneous image regions such as dark shadows, expanses of water or snow, or smooth human-made objects such as walls or the bodies of cars.

The solution is to choose only those points that are distinctive. We can use the point feature detectors that were introduced in ▶ Chap. 12. Harris point features can be detected using

```
>> hf = detectHarrisFeatures(im1);
>> imshow(im1); hold on
>> plot(hf.selectStrongest(200))
```

and SURF features using

```
>> sf = detectSURFFeatures(im1);
>> imshow(im1); hold on
>> plot(sf.selectStrongest(200))
```

and the results are shown in ◘ Fig. 14.2. We simplified the problem – instead of millions of pixels we only have to focus on 200 distinctive points.

Consider the general case of two sets of features points: $\{{}^1\boldsymbol{p}_i \in \mathbb{R}^2, i = 1, \ldots, N_1\}$ in the first image and $\{{}^2\boldsymbol{p}_j \in \mathbb{R}^2, j = 1, \ldots, N_2\}$ in the second image. ▶ Since these are distinctive image points we would expect a significant number of points in image one would correspond to points found in image two. The problem is to determine which $({}^2u_j, {}^2v_j)$, if any, corresponds to each $({}^1u_i, {}^1v_i)$.

The feature coordinates are assumed to be real numbers, determined to subpixel precision.

We cannot use the feature coordinates alone to determine correspondence – the features will have different coordinates in each image. For example in ◘ Fig. 14.1 we see that most features are lower in the right-hand image. We cannot use the

a   b

◘ **Fig. 14.1** Two views of the Eiffel tower. The images were captured approximately simultaneously using two different handheld digital cameras. **a** 7 Megapixel camera with $f = 7.4$ mm; **b** 10 Megapixel camera with $f = 5.2$ mm (Image by Lucy Corke). The images have quite different scale and the tower is 700 and 600 pixels tall in **a** and **b** respectively. The camera that captured image **b** is held by the person in the bottom-right corner of **a** a   b

◻ Fig. 14.2   200 point features computed for ◻ Fig. 14.1a. **a** Harris point features; **b** SURF point features showing scale

intensity or color of the pixels either. Variations in white balance, illumination and exposure setting make it highly unlikely that corresponding pixels will have the same value. Even if intensity variation was eliminated there are likely to be tens of thousands of pixels in the other image with exactly the same intensity value – it is not sufficiently unique. We need some richer way of *describing* each feature.

In practice, we consider a region of pixels around the feature point called the *support region*. We describe this region with a *feature descriptor* which is a distinctive and unique numerical description of the point and its immediate surrounds. For corner feature points such as Harris, the `extractFeatures` function returns Fast Retina Keypoint (FREAK) descriptors by default.

```
>> hd = extractFeatures(im1,hf)
   hd =
      binaryFeatures with properties:
            Features: [1151×64 uint8]
              NumBits: 512
         NumFeatures: 1151
```

Feature descriptors are often referred to as feature vectors.

The FREAK feature descriptors are held in a `binaryFeatures` object that has a `Features` property which is an array of descriptors for each feature. The feature vectors ◄ are 512 bits long and are held in each row of the array in uint8 containers (`Features` property of `hd`). Binary descriptors such as FREAK use a compact binary string that encodes appearance of an image patch. These binary-encoded patches are then compared using the Hamming distance. ◄ Hamming distance provides a measure of similarity between the patches and thus is useful in solving the correspondence problem. The advantage of Hamming distance is that it is very fast to compute in contrast to a metric such as normalized cross correlation. Typically, we would compare feature $^1\!f_i \in \mathbb{Z}^M$ with all features in the other image $\{^2\!f_j \in \mathbb{Z}^M, j = 1, \ldots, N_2\}$ and choose the one that is most similar. ◄ The 512 bits of a FREAK descriptor generally hold enough information to enable accurate feature vector comparisons. For a 512 bit string, Hamming distance of 0 corresponds to the perfect match and Hamming distance of 512 to the worst match.

Hamming distance is a similarity measure computed by comparing two binary data strings. Hamming distance is the number of bit positions in which the two bits are different. For example, 111 and 111 would result in 0 while 101 and 011 would produce distance of 2.

If the world point is not visible in image two then the most similar feature will be an incorrect match.

The SURF algorithm computes a 64-element descriptor ◄ vector for the feature point in a way that is scale and rotationally invariant, and based on the pixels within the feature's support region. It is created from the image in the scale-space sequence corresponding to the feature's scale and rotated according to the feature's orientation. The vector is normalized to a unit vector to increase its invariance to changes in image intensity. Similarity between descriptors is based on sum of squared differences (SSD) distance. This descriptor is quite invariant to image intensity, scale and rotation. Additionally, SURF is both a feature detector and a

When using `extractFeatures` function, a 128-element SURF feature vector can be created by setting the parameter `FeatureSize` to a value of `128`. It makes the descriptor more discriminative hence generally producing fewer but stronger matches.

descriptor, whereas the Harris operator is only a corner detector which must be used with one of a number of different descriptors. ▶

For the remainder of this chapter we will use SURF features. They are computationally more expensive but pay for themselves in terms of the quality of matches between widely different views of the same scene. We compute SURF features for each image

```
>> sf1 = detectSURFFeatures(im1);
>> [sfd1,vsf1] = extractFeatures(im1,sf1);
>> size(sfd1)
  ans =
       917     64
>> sf2 = detectSURFFeatures(im2);
>> [sfd2,vsf2] = extractFeatures(im2,sf2);
>> size(sfd2)
  ans =
      1006     64
```

Note that you can even use the SURF descriptor with a Harris corner point. This can be accomplished by explicitly specifying `Method` parameter of the `extractFeatures` function.

which results in two arrays holding 917 and 1006 descriptors for the first and second image respectively. Additionally, `vsf1` and `vsf2` hold only valid SURF points that correspond exactly to the extracted, `sfd1` and `sfd2` descriptors. The number of valid points, `vsf1` and `vsf2`, can be smaller if some input points are eliminated due to being too close to the edge of the image or in the case of SIFT features, larger because more than one descriptor with different orientations can be extracted for a single SIFT point. In most cases, `vsf1=sf1` and `vsf2=sf2`.

Next, we match the two sets of SURF features based on the distance between the SURF descriptors

```
>> [idx,s] = matchFeatures(sfd1,sfd2)
idx =
  210×2 uint32 matrix
      8     29
     11     10
     14     10
  (listing suppressed)
s =
  210×1 single column vector
    0.0102
    0.0077
    0.0107
  (listing suppressed)
```

We refer to them as putatively matched because, although they are very likely to correspond, this has not yet been confirmed.

where `idx` is a 2-column matrix of indices of 210 putatively ▶ matched features and `s` holds the match score values. Column-one indices corresponds to the feature

◻ Fig. 14.3 Putatively matched features based on SURF descriptor similarity. A few matches represented by long lines are clearly incorrect. The two images were overlayed in red-cyan to assist in visualizing the matches

By default, `matchFeatures` uses a threshold that is set to the perfect score minus 1 % of the total score span.

vectors extracted from first image and column-two indices belong to feature vectors from the second image. The scores for the matches are returned as an optional second output. The returned matches had a threshold applied based on the scores. ◄

We can overlay these matches on the original image pair

```
>> matchedPts1 = vsf1(idx(:,1));
>> matchedPts2 = vsf2(idx(:,2));
>> showMatchedFeatures(im1,im2,matchedPts1,matchedPts2);
```

and the result is shown in ◻ Fig. 14.3. This display is a composite of the two images in red and cyan false colors to effectively visualize matching points. Yellow lines connect the matched features in each image showing a consistent pattern for points that match well. A few matches are obviously wrong and we will deal with these later. Note that it is also possible to obtain a side-by-side composite view of the two images with match lines crossing from one image to the other but with many features to display, it becomes too difficult to interpret. On the other hand, when you are matching much smaller image with a larger image, for example to locate an object using feature matching, the side-by-side display offers a better view. The `montage` option of `showMatchedFeatures` function creates a side-by-side view.

Feature matching is computationally expensive – it is an $O(N^2)$ problem since every feature descriptor in one image must be compared with every feature descriptor in the other image. More sophisticated algorithms store the descriptors in a data structure like a kd-tree so that similar descriptors, nearest neighbors in feature space, can be easily found. This option can be selected by using `method=approximate` setting of `matchFeatures` function.

Although the quality of matching shown in ◻ Fig. 14.3 looks quite good there are a few incorrect matches. We can discern a pattern in the lines joining the corresponding points. This pattern is a function of the relative pose between the two camera views, and understanding this is key to determining which of the candidate matches are correct. That is the topic of the next section.

14.2 Geometry of Multiple Views

We start by studying the geometric relationships between images of a single point P observed from two different viewpoints and this is shown in ◘ Fig. 14.4. This geometry could represent the case of two cameras simultaneously viewing the same scene, or one moving camera taking a picture from two different viewpoints. ▶ The center of each camera, the origins of {1} and {2}, plus the world point P defines a plane in space – the epipolar plane. The world point P is projected onto the image planes of the two cameras at points ^{1}p and ^{2}p respectively, and these points are known as conjugate points. The intersection of the epipolar plane and a camera's image plane is an epipolar line.

Consider image one. The image-plane point ^{1}p is a function of the world point P, and point ^{1}e is a function of the relative pose of camera two. These two points, plus the camera center define the epipolar plane, which in turn defines the epipolar line $^2\ell$ in image two. By definition, the conjugate point ^{2}p must lie on that line. Conversely, ^{1}p must lie along the epipolar line in image one $^1\ell$ that is defined by ^{2}p in image two.

This is a very fundamental and important geometric relationship – given a point in one image we know that its conjugate is constrained to lie along a line in the other image. We illustrate this with a simple example that mimics the geometry of ◘ Fig. 14.4

Assuming the world point does not move.

```
>> T1 = se3(eul2rotm([0 0.4 0]),[-0.1 0 0]);
>> cam1 = CentralCamera("default",name="camera 1", ...
>>   focal=0.002,pose=T1);
```

which returns an instance of the `CentralCamera` class as discussed previously in ▶ Sect. 13.1.2. Similarly for the second camera

```
>> T2 = se3(eul2rotm([0 -0.4 0]),[0.1 0 0]);
>> cam2 = CentralCamera("default",name="camera 2", ...
>>   focal=0.002,pose=T2);
```

and the pose of the two cameras is visualized by

```
>> axis([-0.4 0.6 -0.5 0.5 -0.2 1])
>> cam1.plot_camera("label",color="b")
>> cam2.plot_camera("label",color="r")
```

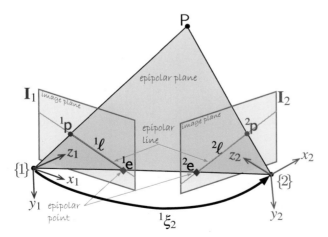

◘ **Fig. 14.4** Epipolar geometry showing the two cameras with associated coordinate frames {1} and {2} and image planes. Three black dots, the world point P and the two camera centers, define the epipolar plane. The intersection of this plane with the image-planes are the epipolar lines

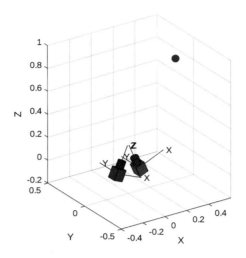

◘ Fig. 14.5   Visualization of two cameras and a target point. The origins of the two cameras are offset along the *x*-axis and the cameras are *verged*, that is, their optical axes intersect

which is shown in ◘ Fig. 14.5. We define an arbitrary world point

```
>> P=[0.5 0.1 0.8];
```

which we display as a small sphere

```
>> plotsphere(P,0.03,"b");
>> grid on, view([-34 26])
```

which is shown in ◘ Fig. 14.5. We project this point to both cameras

```
>> p1 = cam1.plot(P)
p1 =
   561.6861   532.6079
>> p2 = cam2.plot(P)
p2 =
   746.0323   546.4186
```

and this is shown in ◘ Fig. 14.6. The epipoles are computed by projecting the center of each camera to the other camera's image plane

```
>> cam1.hold
>> e1 = cam1.plot(cam2.center,Marker="d",MarkerFaceColor="k")
```

◘ Fig. 14.6   The virtual image planes of two `CentralCamera` objects

```
e1 =
   985.0445   512.0000
>> cam2.hold
>> e2 = cam2.plot(cam1.center,Marker="d",MarkerFaceColor="k")
e2 =
   38.9555   512.0000
```

and these are shown in ◨ Fig. 14.6 as a black ♦-marker.

14.2.1 The Fundamental Matrix

The epipolar relationship shown graphically in ◨ Fig. 14.4 can be expressed concisely and elegantly as

$$^2\tilde{\boldsymbol{p}}^\top \mathbf{F}\,^1\tilde{\boldsymbol{p}} = 0 \tag{14.1}$$

where $^1\tilde{\boldsymbol{p}}$ and $^2\tilde{\boldsymbol{p}}$ are homogeneous coordinate vectors describing the image points $^1\mathsf{p}$ and $^2\mathsf{p}$ respectively, and $\mathbf{F} \subset \mathbb{R}^{3\times3}$ is known as the fundamental matrix. We can rewrite this as

$$^2\tilde{\boldsymbol{p}}^\top\,^2\tilde{\boldsymbol{\ell}} = 0 \tag{14.2}$$

where

$$^2\tilde{\boldsymbol{\ell}} \simeq \mathbf{F}\,^1\tilde{\boldsymbol{p}} \tag{14.3}$$

describes a 2-dimensional line, the epipolar line, along which conjugate points in image two must lie. This line is a function of the point $^1\mathsf{p}$ in image one and (14.2) is a powerful test as to whether or not a point in image two is a possible conjugate.

Taking the transpose of both sides of (14.1) yields

$$^1\tilde{\boldsymbol{p}}^\top \mathbf{F}^\top\,^2\tilde{\boldsymbol{p}} = 0 \tag{14.4}$$

from which we can write the epipolar line for camera one

$$^1\tilde{\boldsymbol{\ell}} \simeq \mathbf{F}^\top\,^2\tilde{\boldsymbol{p}} \tag{14.5}$$

in terms of a point viewed by camera two.

Excurse 14.2: 2D Projective Geometry in Brief

The projective plane $\mathbb{P}^2$ is the set of all points $(x_1, x_2, x_3)^\top$, $x_i \in \mathbb{R}$ and x_i not all zero. Typically, the 3-tuple is considered a column vector. A point P with Euclidean coordinates (u, v) is represented in $\mathbb{P}^2$ by homogeneous coordinates $\tilde{\boldsymbol{p}} = (u, v, 1)^\top$. Scale is unimportant for homogeneous quantities and we express this as $\tilde{\boldsymbol{p}} \simeq \lambda\tilde{\boldsymbol{p}}$ where the operator $\simeq$ means equal up to a (possibly unknown) nonzero scale factor. A point in $\mathbb{P}^2$ can be represented in nonhomogeneous, or Euclidean, form $\boldsymbol{p} = (x_1/x_3, x_2/x_3)^\top$ in $\mathbb{R}^2$. The homogeneous vector $(u, v, f)^\top$, where f is the focal length in pixels, is a vector from the camera's origin that points toward the world point P. More details are given in ▶ App. C.2.

The RVC Toolbox functions e2h and h2e convert between Euclidean and homogeneous coordinates for points (a column vector) or sets of points (a 2D matrix with one column per point).

The fundamental matrix is a function of the camera parameters and the relative camera pose between the views

$$\mathbf{F} \simeq \mathbf{K}_2^{-\top} \left[{}^2\boldsymbol{t}_1 \right]_\times {}^2\mathbf{R}_1 \, \mathbf{K}_1^{-1} \tag{14.6}$$

If both images were captured with the same camera, then $\mathbf{K}_1 = \mathbf{K}_2$.

where $\mathbf{K}_1, \mathbf{K}_2 \in \mathbb{R}^{3\times3}$ are the camera intrinsic matrices defined in (13.7) ◀. ${}^2\mathbf{R}_1 \in \mathbf{SO}(3)$ and ${}^2\boldsymbol{t}_1 \in \mathbb{R}^3$ are the relative orientation and translation of camera one with respect to camera two or ${}^2\boldsymbol{\xi}_1$.

❗ This might be the inverse of what you expect, it is camera one with respect to camera two, but the mathematics are expressed more simply this way. RVC Toolbox functions always describe camera pose with respect to the world frame, or camera two with respect to camera one.

The fundamental matrix that relates the two views is returned by the method `F` of the `CentralCamera` class, for example

```
>> F = cam1.F(cam2)
F =
          0    -0.0000     0.0010
    -0.0000          0     0.0019
     0.0010     0.0001    -1.0208
```

and for the two image points computed earlier we can evaluate (14.1)

```
>> e2h(p2)*F*e2h(p1)'
ans =
    1.3878e-16
```

and we see that the result is zero within a tolerance.

The fundamental matrix has some important properties. It is singular with a rank of two

```
>> rank(F)
ans =
          2
```

The matrix $\mathbf{F} \subset \mathbb{R}^{3\times3}$ has seven underlying parameters so its nine elements are not independent. The overall scale is not defined, and it has a constraint that $\det(\mathbf{F}) = 0$.

and has seven degrees of freedom. ◀ The epipoles are *encoded* in the null space of the matrix. The epipole for camera one is the right null space of $\mathbf{F}$

```
>> null(F)'   % transpose for display
ans =
    -0.8873    -0.4612    -0.0009
```

in homogeneous coordinates or

```
>> e1 = h2e(ans)
e1 =
    985.0445 512.0000
```

This is the right null space of the matrix transpose. The MATLAB function `null` returns the right null space.

in Euclidean coordinates – as shown in ◘ Fig. 14.6. The epipole for camera two is the left null space ◀ of the fundamental matrix

```
>> null(F')';
>> e2 = h2e(ans)
e2 =
    38.9555 512.0000
```

The RVC Toolbox can display epipolar lines using the `plot_epiline` methods of the `CentralCamera` class

```
>> cam2.plot_epiline(F,p1,"r")
```

which is shown in ◘ Fig. 14.6 as a red line in the camera two image plane. We see, as expected, that the projection of P lies on this epipolar line. The epipolar line for camera one is

```
>> cam1.plot_epiline(F',p2,"r")
```

14.2.2 The Essential Matrix

The epipolar geometric constraint can also be expressed in terms of normalized image coordinates

$$^2\tilde{x}^\top \mathbf{E}\ ^1\tilde{x} = 0 \tag{14.7}$$

where $\mathbf{E} \subset \mathbb{R}^{3\times3}$ is the essential matrix and $^2\tilde{x}$ and $^1\tilde{x}$ are conjugate points in homogeneous normalized image coordinates. ▶ This matrix is a simple function of the relative camera pose

> For a camera with a focal length of 1 and the coordinate origin at the principal point, see ▶ Sect. 13.1.2.

$$\mathbf{E} \simeq [^2t_1]_\times\ ^2\mathbf{R}_1 \tag{14.8}$$

where $(^2\mathbf{R}_1, ^2t_1)$ represent $^2\boldsymbol{\xi}_1$, the relative pose of camera one with respect to camera two. The essential matrix is singular, has a rank of two, and has two equal nonzero singular values ▶ and one of zero. The essential matrix has only 5 degrees of freedom and is completely defined by 3 rotational and 2 translational ▶ parameters. For pure rotation, when $t = 0$, the essential matrix is not defined.

> See ▶ App. B.
>
> A 3-dimensional translation (x, y, z) with unknown scale can be considered as $(x', y', 1)$.

We recall from (13.7) that $\tilde{p} \simeq \mathbf{K}\tilde{x}$ and substituting into (14.7) we can write

$$^2\tilde{p}^\top \underbrace{\mathbf{K}_2^{-\top} \mathbf{E} \mathbf{K}_1^{-1}}_{\mathbf{F}}\ ^1\tilde{p} = 0\ . \tag{14.9}$$

Equating terms with (14.1) yields a relationship between the two matrices

$$\mathbf{E} \simeq \mathbf{K}_2^\top \mathbf{F} \mathbf{K}_1 \tag{14.10}$$

in terms of the intrinsic parameters of the two cameras involved. ▶ This is implemented by the E method of the `CentralCamera` class

> If both images were captured with the same camera then $\mathbf{K}_1 = \mathbf{K}_2$.

```
>> E = cam1.E(F)
E =
         0    -0.0779         0
   -0.0779         0    0.1842
         0    -0.1842    0.0000
```

where the intrinsic parameters of camera one (which is the same as camera two) are used. Like the camera matrix in ▶ Sect. 13.2.2, the essential matrix can be decomposed to yield the relative pose $^1\boldsymbol{\xi}_2$ as an **SE**(3) matrix. This is accomplished using the `estrelpose` function ▶

> Although (14.8) is written in terms of $^2\boldsymbol{\xi}_1 \sim (^2\mathbf{R}_1, ^2t_1)$, the `estrelpose` function returns $^1\boldsymbol{\xi}_2$.

```
>> pose = estrelpose(E,cam1.intrinsics(),p1,p2);
>> pose.A
ans =
    0.6967    0.0000   -0.7174    0.9211
    0.0000    1.0000    0.0000    0.0000
    0.7174   -0.0000    0.6967    0.3894
         0         0         0    1.0000
```

which returns a `rigidtform3d` object. The call to `cam1.intrinsics()` packs the camera intrinsics into a `cameraIntrinsics` object which is a required input to the `estrelpose` function. The true relative pose from camera one to camera two is

```
>> inv(cam1.T)*cam2.T
ans =
  se3
```

```
   0.6967          0    -0.7174     0.1842
        0     1.0000          0          0
   0.7174          0     0.6967     0.0779
        0          0          0     1.0000
```

As observed by Hartley and Zisserman (2003, p 259), not even the sign of *t* can be determined when decomposing an essential matrix.

The corresponding points are usually available as a byproduct of using RANSAC as shown in ▶ Sect. 14.3.1.

which matches perfectly with the exception of translation since $\mathbf{E} \simeq \lambda \mathbf{E}$ the translational part of the homogeneous transformation matrix has an unknown scale factor. ◀ By default, the `estrelpose` function sets the translation to be a unit vector. The direction of this vector is correct despite the unknown scale.

The inverse is not unique and in general there are two solutions. Since we do not know the pose of the two cameras, how did the `estrelpose` function determine the correct solution? It used the corresponding points `p1` and `p2` from the two cameras to determine whether the associated world points are visible. ◀ Internally, it tested each pose to determine if the points end up in front of the camera thus returning only the correct solution.

In summary, these 3×3 matrices, the fundamental and the essential matrix, encode the parameters and relative pose of the two cameras. The fundamental matrix and a point in one image defines an epipolar line in the other image along which its conjugate points must lie. The essential matrix encodes the relative pose of the two camera's centers and the pose can be extracted, with two possible values, and with translation scaled by an unknown factor. In this example, the fundamental matrix was computed from known camera motion and intrinsic parameters. The real world isn't like this – camera motion is difficult to measure and the camera may not be calibrated. Instead, we can estimate the fundamental matrix directly from corresponding image points.

14.2.3   Estimating the Fundamental Matrix from Real Image Data

Assume that we have N pairs of corresponding points in two views of the same scene ($^1\boldsymbol{p}_i, {}^2\boldsymbol{p}_i$), $i = 1, \dots, N$. To demonstrate this, we create a set of twenty random point features (within a $2 \times 2 \times 2$ m cube) whose center is located 3 m in front of the cameras

```
>> tform = rigidtform3d([0 0 0],[-1 -1 2]);
>> rng(0) % set random seed for reproducibility of results
>> P = tform.transformPointsForward(2*rand(20,3));
```

and project these points onto both camera image planes

```
>> p1 = cam1.project(P);
>> p2 = cam2.project(P);
```

If $N \geq 8$, the fundamental matrix can be estimated from these two sets of *corresponding* points

```
>> F = estimateFundamentalMatrix(p1,p2,Method="Norm8Point")
F =
    0.0000     0.0000    -0.0010
    0.0000    -0.0000    -0.0019
   -0.0010    -0.0001     1.0000
```

We estimated the fundamental matrix using the normalized 8-point algorithm which requires a minimum of eight matching points, hence the name. It is a direct computation which assumes that the points have correct correspondence. The estimated matrix has the required rank property

```
>> rank(F)
ans =
    2
```

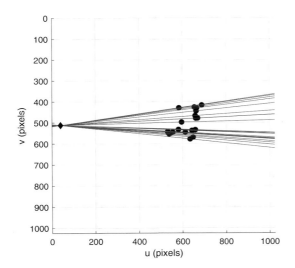

Fig. 14.7 A pencil of epipolar lines on the camera two image plane. Note how all epipolar lines pass through the epipole which is the projection of camera one's center

For camera two we can plot the projected points

```
>> cam2.plot(P);
>> cam2.hold
```

and overlay the epipolar lines generated by each point in image one

```
>> cam2.plot_epiline(F,p1,"r")
```

which is shown in ■ Fig. 14.7. We see a family or *pencil* of epipolar lines, and that every point in image two lies on an epipolar line. Note how the epipolar lines all converge on the epipole which is possible in this case ▶ because the two cameras are verged as shown in ■ Fig. 14.5.

One way to assess our solution of the fundamental matrix is to pick pairs of candidate corresponding points, presumed conjugates, and measure how far one is from the epipolar line defined by the other

```
>> epidist(F,p1(1,:),p2(1,:))
ans =
   2.8869e-13
>> epidist(F,p1(7,:),p2(7,:))
ans =
   0
```

We tried points 1 and 7 and clearly, other than the small numerical errors in the computation, we have a perfect fit.

To demonstrate the importance of correct point correspondence we will repeat the example above but introduce two *bad* data associations by swapping two elements in p2

```
>> p2([8 7],:) = p2([7 8],:);
```

The new fundamental matrix estimation

```
>> F1 = estimateFundamentalMatrix(p1,p2,Method="Norm8Point")
F1 =

    0.0000    0.0000   -0.0041
   -0.0000   -0.0000    0.0021
    0.0022   -0.0031    1.0000
```

can again be tested to see how far one point is from the epipolar line defined by the other.

The example has been contrived so that the epipoles lie within the images, that is, that each camera can see the center of the other camera. A common imaging geometry is for the optical axes to be parallel, such as shown in ■ Fig. 14.21 in which case the epipoles are at infinity (the third element of the homogeneous coordinate is zero) and all the epipolar lines are parallel.

```
>> epidist(F1,p1(1,:),p2(1,:))
ans =
    1.2119
>> epidist(F1,p1(7,:),p2(7,:))
ans =
    3.1116
```

The degradation in the estimate of the fundamental matrix is obvious with the distances significantly deviating from 0. This also means that the point correspondence can no longer be explained by the relationship (14.1). So how do we obtain an accurate estimate of the fundamental matrix when it is not guaranteed that we have perfectly matching points, as is the case with feature detection and matching techniques described in ▶ Chap. 12? We solve this problem with an ingenious algorithm called Random Sampling and Consensus or RANSAC.

The underlying principle is delightfully simple. Estimating a fundamental matrix requires eight points, so we randomly choose eight candidate corresponding points (the sample) from the set of twenty points, and estimate **F** to create a *model*. This model is tested against all the other candidate pairs and those that fit ◀ vote for this model. The process is repeated a number of times and the model that had the most supporters (the consensus) is returned. Since the sample is small, the chance that it contains all valid candidate pairs is high. The point pairs that support the model are termed inliers and those that do not are outliers.

RANSAC is remarkably effective and efficient at finding the inlier set, even in the presence of large numbers of outliers (more than 50%), and is applicable to a wide range of problems. By default, the `estimateFundamentalMatrix` function uses the RANSAC algorithm. There are a few variants of the RANSAC algorithm

> To within a defined threshold *t* that is computed similarly to what we have shown using the `epidist` function when we computed the distance between a point and an epipolar line.

Excurse 14.3: Robust Estimation with RANSAC

We demonstrate use of RANSAC for fitting a line to data with a few erroneous, or outlier, points. The blue dashed line is the least squares best fit and is clearly very different to the true line, due to the outlier data points. Despite 40% of the points not fitting the model, RANSAC finds the parameters of the consensus line, the line that the largest number of points fit, and the indices of the data points that support that model.

We begin this example by loading the points

```
>> load("pointsForLineFitting.mat");
>> plot(points(:,1),points(:,2),".r", ...
>>    MarkerSize=15)
>> hold on
```

The least squares fit is calculated using the `polyfit` function.

```
>> modelLeastSquares = polyfit( ...
>>    points(:,1),points(:,2),1);
>> x = [min(points(:,1)),max(points(:,1))];
>> y = modelLeastSquares(1)*x+ ...
>>    modelLeastSquares(2);
>> plot(x,y,"b--");
```

Next, we fit a line to the points using RANSAC

```
>> sampleSize = 2; maxDistance = 2;
>> fitLineFcn = @(points) polyfit( ...
>>    points(:,1),points(:,2),1);
>> evaluateFitFcn = ...
>>    @(model,points) sum((points(:,2)- ...
>>    polyval(model,points(:,1))).^2,2);
>> rng(0); % set random seed for
>>         % reproducibility
```

```
>> [ab,inliers] = ransac(points, ...
>>    fitLineFcn,evaluateFitFcn, ...
>>    sampleSize,maxDistance);
```

where `ab` holds the estimated parameters (a, b) of the line $ax + b = y$.

Finally, we can plot the line based on the $ax + b = y$ equation using the newly calculated line parameters.

```
>> y = ab(1)*x+ab(2);
>> plot(x,y,"b-");
```

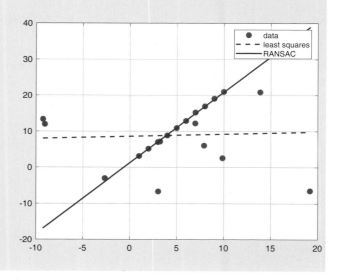

and for this step, we select the M-estimator SAmple Consensus (MSAC) version of RANSAC which will search for more inliers points. ▶

```
>> [F,in] = estimateFundamentalMatrix(p1,p2,Method="MSAC", ...
>>    DistanceThreshold=0.02);
```

We obtain the same fundamental matrix as for a perfect set of matching points despite the fact that we swapped two of them. A logical index to the set of inliers is also returned

```
>> in' % transpose for display
ans =
  1×20 logical array
   1 1 1 1 1 1 0 0 1 1 1 1 1 1 1 1 1 1 1 1
```

and the two incorrect associations, points 7 and 8, are notably false (0) in this list. The `DistanceThreshold` parameter to `estimateFundamentalMatrix` is the threshold which is used to determine whether or not a point pair supports the model. If it is chosen to be too small, RANSAC may require more trials than its default maximum, and this in turn requires adjustment of the `NumTrials` parameter. Keep in mind that the results of RANSAC will vary from run to run due to the random subsampling. Using RANSAC may involve some trial and error to choose efficient threshold based on the precision of the desired final result and the number of outliers. There are also a number of other options that are described in the online documentation of `estimateFundamentalMatrix`.

We return now to the pair of images of the Eiffel tower shown in ◼ Fig. 14.3. At the end of ▶ Sect. 14.1, we had found *putative* correspondences based on descriptor similarity but there were a number of clearly incorrect matches. We can now pass the putatively matched points directly to the `estimateFundamentalMatrix` function

```
>> rng(0) % set random seed for reproducibility of results
>> [F,in] = estimateFundamentalMatrix(matchedPts1,matchedPts2);
>> F
F =
    0.0000   -0.0000    0.0028
    0.0000    0.0000   -0.0041
   -0.0034    0.0036    1.0000
>> sum(in)
ans =
     105
```

Since we are working with a relatively small number of points, the resulting F matrix will vary slightly between runs. It is relatively stable using the default settings of the `estimateFundamentalMatrix` function. It is also unrealistic to expect a perfect estimate of the fundamental matrix since the real image data is subject to random error such as image sensor noise and systematic error such as lens distortion. ▶

RANSAC identified 105 inliers or correct data associations from the SURF feature matching stage which is approximately 50% of the *candidate* matches. ▶ We can plot the inliers

```
>> showMatchedFeatures(im1,im2,matchedPts1(in),matchedPts2(in));
```

and the outliers

```
showMatchedFeatures(im1,im2,matchedPts1(~in),matchedPts2(~in));
```

and these are shown in ◼ Fig. 14.8.

We can also overlay the epipolar lines computed from the corresponding points found in the second image

```
>> epiLines = epipolarLine(F',matchedPts2.Location(in,:));
>> pts = lineToBorderPoints(epiLines,size(im1));
>> imshow(im1), hold on
>> line(pts(:,[1 3])',pts(:,[2 4])');
```

The default Least Median of Squares RANSAC algorithm terminates after it finds the number of inliers equal to 50% of the total number of input points.

Lens distortion causes points to be displaced on the image plane and this violates the epipolar geometry. Images can be corrected by warping as discussed in ▶ Sect. 11.7.4 by using the efficient `undistortImage` function prior to feature detection.

When looking for inliers, similarity alone is not enough. The corresponding points in the two images must be *consistent* with the epipolar geometry as represented by the consensus fundamental matrix.

a   **b**

⬛ Fig. 14.8 Results of SURF feature matching after RANSAC. **a** Inlier matches; **b** Outlier matches, some are quite visibly incorrect while others are more subtly wrong

We only plot a small subset of the epipolar lines since they are too numerous and would obscure the image.

and the result is shown in ⬛ Fig. 14.9. The `epipolarLine` function returns $[A, B, C]$ parameters of an $A * x + B * y + C = 0$ line in each row of the $M \times 3$ `epiLines` matrix. The epipolar lines intersect at the epipolar point which we can clearly see is the projection of the second camera in the first image. ◄ The epipole

```
>> epole = h2e(null(F)');
>> plot(epole(1),epole(2),"bo");
```

is also superimposed on the plot. With two handheld cameras and a common view we have been able to pinpoint the second camera in the first image. The result is not quite perfect – there is a horizontal offset of about 50 pixels which is likely to be due to a small orientation error in one or both cameras which were handheld and only approximately synchronized.

⬛ Fig. 14.9 Image from ⬛ Fig. 14.1a showing epipolar lines converging on the projection of the second camera's center. In this case, the second camera is visible in the bottom right of the image

14.2.4 Planar Homography

In this section, we will consider two different cameras viewing a set of world points P_i that lie on a plane. The camera image plane projections ^1p_i and ^2p_i are related by

$$^2\tilde{p}_i \simeq \mathbf{H}\,^1\tilde{p}_i \tag{14.11}$$

where $\mathbf{H} \subset \mathbb{R}^{3\times3}$ is a nonsingular matrix known as a homography, a planar homography, or the homography *induced* by the plane. ▶

For example, consider again the pair of cameras from ▶ Sect. 14.2 repeated below

```
>> T1 = se3(eul2rotm([0 0.4 0]),[-0.1 0 0]);
>> cam1 = CentralCamera("default",name="camera 1", ...
>>    focal=0.002,pose=T1);
>> T2 = se3(eul2rotm([0 -0.4 0]),[0.1 0 0]);
>> cam2 = CentralCamera("default",name="camera 2", ...
>>    focal=0.002,pose=T2);
```

now observing a 3×3 grid of points

```
>> Tgrid = se3(eul2rotm([0.1 0.2 0],"XYZ"),[0 0 1]);
>> P = mkgrid(3,1.0,pose=Tgrid);
```

where `Tgrid` is the pose of the grid coordinate frame {G} and the grid points are centered in the frame's xy-plane. The points are projected to both cameras

```
>> p1 = cam1.plot(P,"o");
>> p2 = cam2.plot(P,"o");
```

and the projections are shown in ▫ Fig. 14.10a, b.

Just as we did for the fundamental matrix, we can estimate the transformation matrix from two sets of corresponding points

```
>> H = estgeotform2d(p1,p2,"projective");
>> H.A
```

A homography matrix has arbitrary scale and therefore 8 degrees of freedom. With respect to (14.13) the rotation, translation and normal have 3, 3 and 2 degrees of freedom respectively. Homographies form a group: the product of two homographies is another homography, the identity homography is a unit matrix and an inverse operation is the matrix inverse.

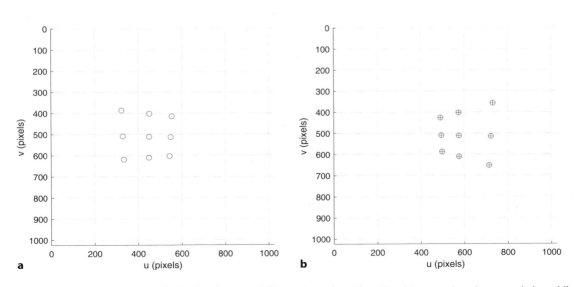

a   **b**

▫ **Fig. 14.10** Views of the oblique planar grid of points from two different view points. The grid points are projected as open circles, while the plus signs in **b** indicate points transformed from the camera one image plane by the homography

```
ans =
   -0.4282    -0.0006    408.0894
   -0.7030     0.3674    320.1340
   -0.0014    -0.0000      1.0000
```

where H is a `projtform2d` object holding the 3×3 transformation matrix. According to (14.11) we can predict the position of the grid points in image two from the corresponding image one coordinates

```
>> p2b = H.transformPointsForward(p1);
```

which we can can superimpose on image two as + symbols

```
>> cam2.hold()
>> cam2.plot(p2b,"+")
```

as shown in ◻ Fig. 14.10b. We see that the predicted points are perfectly aligned with the actual projection of the world points. The inverse of the homography matrix

$$ {}^1\tilde{\boldsymbol{p}}_i \simeq \mathbf{H}^{-1}\; {}^2\tilde{\boldsymbol{p}}_i \tag{14.12} $$

performs the inverse mapping, from image two coordinates to image one

```
>> p1b = H.transformPointsInverse(p2);
```

The fundamental matrix constrains the conjugate point to lie along a line but the homography tells us *exactly* where the conjugate point will be in the other image – provided that the points lie on a plane.

We can use this proviso to our advantage as a test for whether or not points lie on a plane. We will add some extra world points ◄ to our example

```
>> Q = [-0.2302    0.3287    0.4000
>>      -0.0545    0.4523    0.5000
>>       0.2537    0.6024    0.6000];
```

These points lie along the ray from the camera one center to an extra row of points in the grid plane. However, their z-coordinates have been chosen to be 0.4, 0.5 and 0.6 m respectively.

which we plot in 3D

```
>> axis([-1 1 -1 1 -0.2 1.8])
>> plotsphere(P,0.05,"b")
>> plotsphere(Q,0.05,"r")
>> cam1.plot_camera("label",color="b")
>> cam2.plot_camera("label",color="r")
>> grid on, view([-25 15])
```

and this is shown in ◻ Fig. 14.11. The new points, shown in red, are clearly not in the same plane as the original blue points. Viewed from camera one

```
>> p1 = cam1.plot([P;Q],"o");
```

as shown in ◻ Fig. 14.12a, these new points *appear* as an extra row in the grid of points we used above. However, in the second view

```
>> p2 = cam2.plot([P;Q],"o");
```

as shown in ◻ Fig. 14.12b these *out of plane* points no longer form a regular grid. If we apply the homography to the camera one image points

```
>> p2h = H.transformPointsForward(p1);
```

we find where they should be in the camera two image, *if* they belonged to the plane implicit in the homography

```
>> cam2.plot(p2h,"+");
```

We see that the original nine points overlap, but the three new points do not. We could make an automated test based on the prediction error

```
>> vecnorm(p2h'-p2')
ans =
  Columns 1 through 6
     0.0000     0.0000     0.0000     0.0000     0.0000     0.0000
```

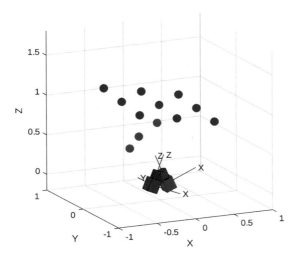

Fig. 14.11 World view of target points and two camera poses. Blue points lie in a planar grid, while the red points appear to lie in the grid from the viewpoint of camera one

```
Columns 7 through 12
   0.0000    0.0000    0.0000   50.5969   46.4423   45.3836
```

which is large for these last three points since they do not belong to the plane that induced the homography.

In this example, we estimated the homography based on two sets of corresponding points which were projections of known planar points. In practice, we do not know in advance which points belong to the plane so we can again use RANSAC, which is already built into the estgeotform2d function

```
>> [H,inlierIdx] = estgeotform2d(p1,p2,"projective");
>> H.A
ans =
   -0.4282   -0.0006  408.0894
   -0.7030    0.3674  320.1340
   -0.0014   -0.0000    1.0000
```

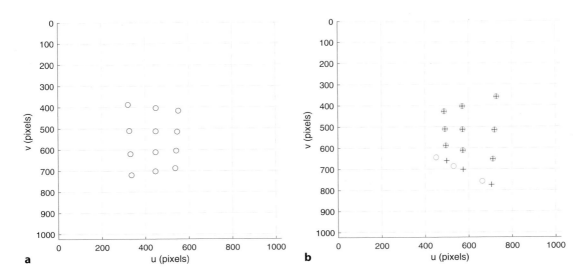

Fig. 14.12 The points observed from two different view points. The grid points are projected as open circles. Plus signs in **b** indicate points transformed from the camera one image plane by the homography. The bottom of row of points in each case are not coplanar with the other points

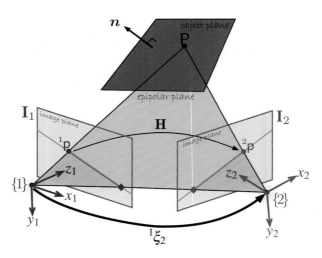

□ Fig. 14.13 Geometry of homography showing two cameras with associated coordinate frames {1} and {2} and image planes. The world point P belongs to a plane with surface normal n. $\mathbf{H}$ is the homography, a 3×3 matrix that maps ^{1}p to ^{2}p

```
>> inlierIdx'  % transpose for display
ans =
  1×12 logical array
   1  1  1  1  1  1  1  1  1  0  0  0
```

which finds the homography that best explains the relationship between the sets of image points. It has also identified those points which support the homography and the three *out of plane points*, points 10–12, are marked as zero and therefore they are not inliers.

See ▶ Sect. 13.1.2.

The geometry related to the homography is shown in □ Fig. 14.13. We can express the homography in normalized image coordinates ◄

$$^2\tilde{\boldsymbol{x}} \simeq \mathbf{H}_E\,{}^1\tilde{\boldsymbol{x}}$$

where $\mathbf{H}_E$ is the Euclidean homography which is written

$$\mathbf{H}_E \simeq {}^2\mathbf{R}_1 + \frac{{}^2\boldsymbol{t}_1}{d}\boldsymbol{n}^\top \tag{14.13}$$

in terms of the relative camera pose $^2\boldsymbol{\xi}_1$ and the plane $\boldsymbol{n}^\top \boldsymbol{P} + d = 0$ with respect to frame {1}. The Euclidean and projective homographies are related by

$$\mathbf{H}_E \simeq \mathbf{K}^{-1}\mathbf{H}\mathbf{K}$$

where $\mathbf{K}$ is the camera intrinsic parameter matrix.

Similarly to the essential matrix, the projective homography can be decomposed to yield the relative pose $^1\boldsymbol{\xi}_2$ in homogeneous transformation form ◄ as well as the normal to the plane. We create a `cameraIntrinsics` object from `cam1` and pass it to the `estrelpose` function

Although (14.13) is written in terms of $(^2\mathbf{R}_1, {}^2\boldsymbol{t}_1)$, the `estrelpose` function returns the inverse which is $^1\boldsymbol{\xi}_2$.

```
>> pose = estrelpose(H,cam1.intrinsics(),p1(inlierIdx,:), ...
>>    p2(inlierIdx,:));
>> pose.A
```

a   **b**

● **Fig. 14.14** Two pictures of a courtyard taken from different viewpoints. Image **b** was taken approximately 30 cm to the right of image **a**. Image **a** has superimposed features that fit a plane. The camera was handheld

```
ans =
     0.6967      0.0000     -0.7174      0.9211
    -0.0000      1.0000     -0.0000     -0.0000
     0.7174      0.0000      0.6967      0.3894
          0           0           0      1.0000
```

which returns a `rigidtform3d` object. Again, there are multiple solutions but with an input of observed inlier points to the `estrelpose` function, a correct, physically realizable solution is determined. As usual, the translational component of the transformation matrix has an unknown scale factor. We know from ● Fig. 14.11 that the camera motion is predominantly in the x-direction and that the plane normal is approximately parallel to the camera's optical- or z-axis and this matches our solution. The true relative pose from camera one to two is

```
>> inv(cam1.T)*cam2.T
ans =
  se3
     0.6967           0     -0.7174      0.1842
          0      1.0000           0           0
     0.7174           0      0.6967      0.0779
          0           0           0      1.0000
```

and further confirms the correct result. ▸ The pose of the grid with respect to camera one is

The translation scale factor is quite close to one in this example, but in general it must be considered unknown.

```
>> inv(cam1.T)*Tgrid
ans =
  se3
     0.9797     -0.0389     -0.1968     -0.2973
     0.0198      0.9950     -0.0978           0
     0.1996      0.0920      0.9756      0.9600
          0           0           0      1.0000
```

and the third column of the rotation submatrix is the grid's normal ▸ which matches the estimated normal of our solution.

Since the points are in the xy-plane of the grid frame {G} the normal is the z-axis.

We can apply this technique to a pair of real images

```
>> im1 = rgb2gray(imread("walls-l.jpg"));
>> im2 = rgb2gray(imread("walls-r.jpg"));
```

which are shown in ● Fig. 14.14. We start by finding the SURF features

```
>> pts1 = detectSURFFeatures(im1);
>> pts2 = detectSURFFeatures(im2);
>> [sf1,vpts1] = extractFeatures(im1,pts1);
>> [sf2,vpts2] = extractFeatures(im2,pts2);
```

We used the option `Unique` to avoid matching a single point to more than one point in the other set. It takes longer to process but it can produce more robust results, especialy with operations such as bundle adjustment discussed later where one-to-one matches are required to produce reliable point tracks.

For illustrative purposes, we initialized the random number generator using the `rng` function to obtain repeatable results. Because RANSAC uses a random number generator, the *dominant* wall could actually change.

and the putatively matched points ◄

```
>> idxPairs = matchFeatures(sf1,sf2,Unique=true);
>> matchedPts1 = vpts1(idxPairs(:,1));
>> matchedPts2 = vpts2(idxPairs(:,2));
```

then use RANSAC to find the set of corresponding points that best fits a plane in the world ◄

```
>> rng(0) % set random seed for reproducibility of results
>> [H,inliersWall] = estgeotform2d(matchedPts1, matchedPts2, ...
>>    "projective", MaxDistance=4);
>> wallPts = matchedPts1(inliersWall);
>> wallPts.Count
ans =
      1049
```

In this case, the majority of point pairs do not fit the model, that is they do not belong to the plane that induces the homography **H**. However, 1077 points out of 30,783 in image one *do* belong to the plane, and we can superimpose them on the figure

```
>> imshow("walls-1.jpg"), hold on
>> wallPts.plot()
```

as shown in ◘ Fig. 14.14a. RANSAC has found a consensus which is the plane containing the left-hand wall. The distance threshold was set to 4 to account for lens distortion and the planes being not perfectly smooth. If we remove the inlier points from the original input, keep the outliers

```
>> outliers = matchedPts1(~inliersWall);
```

and repeat the RANSAC homography estimation step, we will find the next most dominant plane in the scene, which turns out to be the right-hand wall. Planes are very common in man-made environments and we will revisit homographies and their decomposition in ▶ Sect. 14.8.1.

14.3 Sparse Stereo

Stereo vision is a technique for estimating the 3-dimensional structure of the world from two images taken from different viewpoints as, for example, shown in ◘ Fig. 14.14. Our eyes are separated by 50–80 mm and the difference between these two viewpoints is an important, but not the only, part of how we sense distance. This section introduces sparse stereo, a natural extension of what we have learned about feature matching, which recovers the world coordinate (X, Y, Z) for each corresponding point pair. Dense stereo, covered in ▶ Sect. 14.4 attempts to recover the world coordinate (X, Y, Z) for *every pixel* in the image.

14.3.1 3D Triangulation

To illustrate sparse stereo, we will return to the pair of images shown in ◘ Fig. 14.14. We have already found the SURF features and established candidate correspondences between them. Now, we estimate the fundamental matrix

```
>> rng(0) % set random seed for reproducibility of results
>> [F,in] = estimateFundamentalMatrix(matchedPts1,matchedPts2);
```

which captures the relative geometry of the two views. We can display the epipolar lines for a subset of right-hand image points overlaid on the left-hand image

```
>> epiLines = epipolarLine(F',matchedPts2(in));
>> pts = lineToBorderPoints(epiLines,size(im1));
```

Fig. 14.15 Image of ☐ Fig. 14.14a with epipolar lines for a subset of right image points superimposed

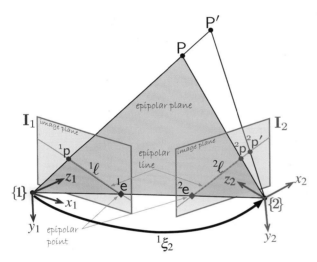

Fig. 14.16 Epipolar geometry for stereo vision. We can see clearly that as the depth of the world point increases, from P to P′, the projection moves rightward along the epipolar line in the second image plane

```
>> imshow(im1), hold on
>> line(pts(1:40,[1 3])',pts(1:40,[2 4])',color="y");
```

which is shown in ☐ Fig. 14.15. In this case, the epipolar lines are approximately horizontal and parallel which is expected for a camera motion that is a pure translation in the x-direction. ☐ Fig. 14.16 shows the epipolar geometry for stereo vision. It is clear that as points move away from the camera, P to P′ the conjugate points in the right-hand image move to the right along the epipolar line.

The origin of {1} and the image plane point $^1\mathbf{p}$ defines a ray in space, as does the origin of {2} and $^2\mathbf{p}$. These two rays intersect at the world point P – the process of triangulation – but to determine these rays we need to know the pose and intrinsic parameters of each camera. It is convenient to consider that the camera one frame {1} is the world frame {0}, and only the pose of camera two $\boldsymbol{\xi}_2$ is unknown. However, we can estimate the relative pose $^1\boldsymbol{\xi}_2$ by decomposing the essential matrix computed between the two views. We already have the fundamental matrix, but to determine the essential matrix according to (14.10) we need the camera's intrinsic parameters. With a little sleuthing we can find them!

The camera focal length is stored in the metadata of the image, as we discussed in ▶ Sect. 11.1.1 and can be examined

```
>> md = imfinfo("walls-1.jpg");
```

where `md` is a structure of text strings that contains various characteristics of the image – its metadata. The element `DigitalCamera` is a structure that describes the camera

```
>> f = md.DigitalCamera.FocalLength
f =
    4.1500
```

from which we determine that the focal length is 4.15 mm.

The dimensions of the pixels $\rho_w \times \rho_h$ are not included in the image header but some web-based research on this model camera

```
>> md.Model
ans =
    'iPhone 5s'
```

suggests that this camera has an image sensor with 1.5 μm pixels. We start by computing focal length in pixels and then use it with image size and principal point to construct the `cameraIntrinsics` object. Since we did not calibrate the camera, we will chose the center of the image plane as the approximate principal point. The 0.5 below is added to the `principalPoint` to land exactly in the middle of the image, between the square pixels for even-sized image or in the middle of the pixel for an odd-sized image ◀

Further explanation of pixel and spatial image coordinates used by the Image Processing Toolbox can be found at ▶ https://sn.pub/B62lt8.

```
>> flenInPix = (f/1000)/1.5e-6;
>> imSize = size(im1);
>> principalPoint = imSize/2+0.5;
>> camIntrinsics = cameraIntrinsics(flenInPix,principalPoint,imSize)
camIntrinsics =
  cameraIntrinsics with properties:
              FocalLength: [2.7667e+03 2.7667e+03]
           PrincipalPoint: [1.2245e+03 1.6325e+03]
                ImageSize: [2448 3264]
           RadialDistortion: [0 0]
       TangentialDistortion: [0 0]
                     Skew: 0
                        K: [3×3 double]
```

The essential matrix can be determined using the `estimateEssentialMatrix` function which, similarly to the `estimateFundamentalMatrix`, uses RANSAC to determine the essential matrix from putatively matched point pairs and the additional camera intrinsics

```
>> rng(0) % set random seed for reproducibility of results
>> [E,in] = estimateEssentialMatrix(matchedPts1,matchedPts2, ...
>>      camIntrinsics,MaxDistance=0.18,Confidence=95);
```

We can now decompose it to determine the camera motion

```
>> inlierPts1 = matchedPts1(in);
>> inlierPts2 = matchedPts2(in);
>> pose = estrelpose(E,camIntrinsics,inlierPts1,inlierPts2);
>> pose.R
ans =
    0.9999    0.0110   -0.0034
   -0.0111    0.9997   -0.0232
    0.0032    0.0232    0.9997
>> pose.Translation
ans =
    0.9708   -0.0745    0.2281
```

The inlier points passed into the `estrelpose` function were used to determine the correct solution for the relative camera motion. Since the camera orientation was kept fairly constant, the rotational part of the transformation is expected to be close to the identity matrix as we observe.

The estimated translation t from {1} to {2} has an unknown scale factor and is returned as a unit vector in `t`. Once again, we bring in an extra piece of information – when we took the images, the camera position changed by approximately 0.3 m to the right. The estimated translation has the correct direction, dominant x-axis motion, but the magnitude is set by the `estrelpose` function to a unit vector. We therefore scale the translation unit vector to match physical reality.

```
>> t = pose.Translation*0.3;
```

We now have an estimate of $^1\xi_2$ – the relative pose of camera two with respect to camera one represented as a rotation matrix `rot` and translation vector `t`.

Each point p in an image corresponds to a ray in space, sometimes called a raxel. We refer to the process of finding the world points where the rays intersect as triangulation. Before we can do that, we first need to set up camera projection matrices

```
>> tform1 = rigidtform3d;    % null tranform by default
>> camMatrix1 = cameraProjection(camIntrinsics,tform1);
>> cameraPose = rigidtform3d(pose.R,t);
>> tform2 = pose2extr(cameraPose);
>> camMatrix2 = cameraProjection(camIntrinsics,tform2);
```

and with that, we compute the 3D world point locations corresponding to image point pairs

```
>> P = triangulate(inlierPts1,inlierPts2,camMatrix1,camMatrix2);
```

where P holds all the world points, one per row. Note that due to errors in the estimate of camera two's pose the rays do not actually intersect, but their closest intersection points are returned by the `triangulate` function.

We use a subset of one hundred returned points and extract their last column

```
>> z = P(1:100,3);
```

which is the depth coordinate. We can superimpose the distance to each point on the image of the courtyard

```
>> circles = [inlierPts1.Location(1:100,:) repmat(15,[100 1])];
>> imAnnotated = insertObjectAnnotation(im1,"circle", ...
>>     circles,z,FontSize=50,LineWidth=4);
>> imshow(imAnnotated)
```

which is shown in ◼ Fig. 14.17 and the feature markers are annotated with the estimated depth in meters. The generated structure can also be easily viewed as a point cloud

```
>> pc = pointCloud(P);
>> pcshow(pcdenoise(pc),VerticalAxis="Y",VerticalAxisDir="Down");
```

and it is shown in ◼ Fig. 14.18. We first wrapped the 3D points in a convenient `pointCloud` object, then used `pcdenoise` function to clean up a few spurious points, and finally displayed it with the `pcshow` function. Considering the lack of rigor in this exercise, two handheld camera shots and only approximate knowledge of the magnitude of the camera displacement, the recovered depth information is quite remarkable.

This is an example of stereopsis, where we have used information from two overlapping images to infer the 3-dimensional position of points in the world. For obvious reasons, the approach used here is referred to as sparse stereo because we only compute distance at a tiny subset of pixels in the image. More commonly, the relative pose between the cameras would be known, as would the camera intrinsic parameters.

Fig. 14.17 Image of **□** Fig. 14.14a with depth of selected points indicated in meters

14

▶ sn.pub/2pj2MV

Fig. 14.18 Point cloud generated from a pair of images in **□** Fig. 14.14. Cooler colors represent point locations closer to the sensor

14.3.2 **Bundle Adjustment (advanced)**

In the previous section we used triangulation to estimate the 3D coordinates of a sparse set of landmark points in the world, but this was an approximation based on a guesstimate of the relative pose between the cameras. To assess the quality of our solution we can *reproject* the estimated 3D landmark points onto the image planes based on the estimated camera poses and the known camera model. The reprojection error is the image-plane distance between the back-projected landmark and its observed position on the image plane.

Continuing the example from the previous section, we can reproject the 3D triangulated points to each of the camera image planes. It turns out that our camera pose estimate based on the essential matrix from the previous section was already

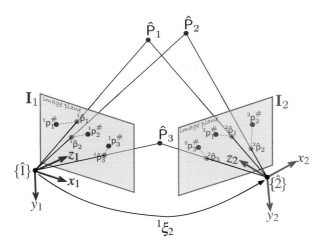

Fig. 14.19 Bundle adjustment notation illustrated with a simple problem comprising only two cameras and three landmark points P_i in the world. The estimated camera poses and point positions are indicated, as are the estimated and measured image-plane coordinates. The reprojection errors are shown as dashed gray lines. The problem is solved when the variables are adjusted so that the total reprojection error is as small as possible

very good with mean reprojection error less than one pixel. To illustrate the power of bundle adjustment, we will start by introducing an error in the translation vector.

```
>> T1 = tform1; T2 = tform2;
>> translationError = [0 0.01 -0.02];
>> T2.Translation = T2.Translation + translationError;
>> p1 = world2img(P,T1,camIntrinsics);
>> p2 = world2img(P,T2,camIntrinsics);
```

The distances between the back projections and observations across both cameras are

```
>> e = vecnorm([p1-inlierPts1.Location; p2-inlierPts2.Location]');
```

with statistics

```
>> mean(e)
ans =
  single
    6.7601
>> max(e)
ans =
  single
   46.8220
```

which clearly indicates the error in our solution – each back-projected point is in error by up to 46.8 pixels. Large reprojection errors imply that there can be an error in the estimate of the world point, the pose estimate for camera one or two, or both. However, we do know that a good estimate is one where this total back-projection error is low – ideally zero. To reduce this error, we can use a technique called bundle adjustment.

Bundle adjustment is an optimization process that simultaneously adjusts the camera poses and the landmark coordinates so as to minimize the total back-projection error. It uses 2D measurements from a set of images of the same scene to recover information related to the 3D geometry of the imaged scene, as well as the poses of the cameras. This is also called Structure from Motion (SfM) or Structure and Motion Estimation (SaM) – *structure* being the 3D landmarks in the world and *motion* being a sequence of camera poses. It is also called visual SLAM (VSLAM) since it is very similar to the pose-graph SLAM problem discussed in ▶ Sect. 6.4.

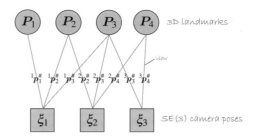

Fig. 14.20 A visibility graph showing camera vertices (red) and landmark vertices (blue). Lines connecting vertices represent a view of that landmark from that camera, and the edge value is the observed image-plane coordinate. Landmark P_1 is viewed by only one camera, P_2 and P_4 are viewed by two cameras, and P_3 is viewed by three cameras

That was a planar problem solved in the three dimensions of $\mathbf{SE}(2)$ whereas bundle adjustment involves camera poses in $\mathbf{SE}(3)$ and points in $\mathbb{R}^3$.

To formalize the problem, consider a camera with known intrinsic parameters at N different poses $\boldsymbol{\xi}_i \in \mathbf{SE}(3)$, $i = 1, \ldots, N$ and a set of M landmark points P_j represented by coordinate vectors $\boldsymbol{P}_j \in \mathbb{R}^3$, $j = 1, \ldots, M$. The camera with pose $\boldsymbol{\xi}_i$ observes P_j, and the image-plane projection p_j is represented by the coordinate vector $^i\boldsymbol{p}^{\#}[j] \in \mathbb{R}^2$. The notation is shown in ◨ Fig. 14.19 for two cameras.

In general, only a subset of landmarks is visible from any camera, and this visibility information can be represented elegantly using a graph as shown in ◨ Fig. 14.20, where each camera pose $\boldsymbol{\xi}_i$ and each landmark coordinate is a vertex. ◁ Edges between camera and landmark vertices represent observations, and the value of the edge is the observed image-plane coordinate. The estimated value of the image-plane projection of landmark j on the image plane of camera i is

The visibility information can also be represented by a visibility matrix, an $N \times M$ array where element (i, j) is 1 if landmark j is visible from camera i, otherwise 0.

$$^i\hat{\boldsymbol{p}}_j = \mathcal{P}\left(\hat{\boldsymbol{\xi}}_i,\ \boldsymbol{P}_j;\ \mathbf{K}\right)$$

and the back-projection error – the difference between the estimated and observed projection – is $^i\hat{\boldsymbol{p}}_j - {}^i\boldsymbol{p}^{\#}[j]$.

We begin by setting up a description of our problem in terms of camera views, corresponding feature points and what we already know about the camera poses. The `imageviewset` is designed to do all the necessary bookkeeping for us.

```
>> vSet = imageviewset;
```

We are only dealing with two views so it may feel like overkill, but when potentially hundreds of camera views are involved such a data structure is a necessity. We store the relative pose between the two cameras, and the locations of SURF point detections. Note that we wish to keep the first camera fixed thus defining the origin of our world coordinate system.

```
>> absPose1 = T1;
```

The second camera's relative pose is therefore the same as its absolute pose since we are only dealing with two views. It was derived earlier from the essential matrix. We add all the points

```
>> relPose2 = extr2pose(T2);
>> absPose2 = relPose2;
>> vSet = vSet.addView(1,absPose1,Points=pts1);
>> vSet = vSet.addView(2,absPose2,Points=pts2);
```

and next define connections between matching points in the two views. The points must also satisfy our epipolar constraint therefore they are a subset of the putatively matched points that we used to compute the essential matrix.

```
>> vSet = vSet.addConnection(1,2,relPose2,Matches=idxPairs(in,:));
```

We can now extract groups of these points that we refer to as *tracks*. For two views, these tracks are simple, but you can imagine a complex set of connections when multiple views are involved and which may include occlusion which would make the tracks have varying lengths. We also extract a table holding our absolute camera poses.

```
>> tracks = vSet.findTracks([1 2]);
>> camPoses = vSet.poses();
```

Now that the data is ready, let us discuss how to solve this optimization problem. We can put all the variables we wish to adjust into a single state vector. For bundle adjustment the state vector contains camera poses and landmark coordinates

$$x = \{\xi_1, \, \xi_2 \, \cdots \, \xi_N | \mathbf{P}_1, \, \mathbf{P}_2 \, \cdots \mathbf{P}_M \} \in \mathbb{R}^{6N+3M}$$

where the **SE**(3) camera pose is represented in a vector format $\xi_i \sim (t, r) \in \mathbb{R}^6$ comprising translation $t \in \mathbb{R}^3$ and rotation $r \in \mathbb{R}^3$; ▶ and $\mathbf{P}_j \in \mathbb{R}^3$.

The number of unknowns in this system is $6N + 3M$: 6 unknowns for each camera pose and 3 unknowns for the position of each landmark point. However, we have up to $2NM$ equations due to the measured projections of the points on the image planes. Typically, the pose of one camera is assumed to be the reference coordinate frame, and this reduces the number of unknowns to $6(N-1) + 3M$. In the problem we are discussing $N = 2$, but one camera is locked, and $M = 1152$ (`size(P,1)`) so we have $6 \times (2-1) + 3 \times 1152 = 3462$ unknowns and $2 \times 2 \times 1152 = 4608$ equations – an overdetermined set of equations for which a solution should be possible.

Bundle adjustment is a minimization problem – it finds the camera poses and landmark positions that minimize the total reprojection error across all the edges

$$x^* = \arg \min_x \sum_k F_k(x)$$

where $F_k(\cdot) > 0$ is a nonnegative scalar cost associated with the graph edge k from camera i to landmark j. The reprojection error of a landmark at P_j onto the camera at pose ξ_i is

$$f_k(x) = \mathcal{P}\left(\hat{\xi}_i, \, \hat{P}_j; \, \mathbf{K}\right) - {}^i p_j^\# \in \mathbb{R}^2$$

and the scalar cost is the squared Euclidean reprojection error

$$F_k(x) = f_k^\top(x) \, f_k(x) \,.$$

Although written as a function of the entire state vector, $F_k(x)$ depends on only two elements of that vector: ξ_i and P_j. The total error, the sum of the squared back-projection errors for all edges, can be computed for any value of the state vector. The bundle adjustment task is to adjust the camera and landmark parameters to reduce this value. We have framed bundle adjustment as a sparse nonlinear least squares problem, and this can be solved numerically if we have a sufficiently good initial estimate of x.

The first step in solving this problem is to linearize it. The reprojection error $f_k(x)$ can be linearized about the current state x_0 of the system

$$f_k'(\Delta) \approx f_{0,k} + \mathbf{J}_k \Delta$$

where $f_{0,k} = f_k(x_0)$ and

$$\mathbf{J}_k = \frac{\partial f_k(x)}{\partial x} \in \mathbb{R}^{2 \times (6N+3M)}$$

Possible representations of rotation include Euler angles, roll-pitch-yaw angles, angle-axis or exponential coordinate representations. For bundle adjustment it is common to use the vector component of a unit-quaternion which is singularity free and has only three parameters. The double cover property of unit-quaternions means that any unit-quaternion can be written with a nonnegative scalar component. By definition, the unit-quaternion has a unit norm, so the scalar component can be easily recovered

$$s = \sqrt{1 - v_x^2 - v_y^2 - v_z^2}$$ given the vector component.

Linearization and Jacobians are discussed in ▶ App. E, and solution of sparse nonlinear equations in ▶ App. F.

is a Jacobian matrix ◀ which depends only on the camera pose $\boldsymbol{\xi}_i$ and the landmark position $\mathbf{P}_j$ so is therefore mostly zeros

$$\mathbf{J}_k = \begin{pmatrix} \mathbf{0}_{2\times 6} & \cdots & \mathbf{A}_i & \cdots & \mathbf{B}_j & \cdots & \mathbf{0}_{2\times 3} \end{pmatrix}, \quad \text{where} \quad \mathbf{A}_i = \frac{\partial \boldsymbol{f}_k(\boldsymbol{x})}{\partial \boldsymbol{\xi}_i} \in \mathbb{R}^{2\times 6},$$

$$\mathbf{B}_j = \frac{\partial \boldsymbol{f}_k(\boldsymbol{x})}{\partial \mathbf{P}_j} \in \mathbb{R}^{2\times 3}.$$

The structure of the Jacobian matrix $\mathbf{A}_i$ is specific to the chosen representation of camera pose. The Jacobians, particularly $\mathbf{A}_i$, are quite complex to derive but can be automatically generated using the MATLAB Symbolic Math Toolbox™.

Since everything is in place and the approach described above is implemented in the efficient `bundleAdjustment` function, we can now solve our problem

```
>> [Pout,camPoses,e] = bundleAdjustment(P,tracks,camPoses, ...
>>    camIntrinsics,FixedViewId=1,PointsUndistorted=true, ...
>>    Verbose=true);
Initializing bundle adjustment solver ...
Starting Levenberg-Marquardt iterations:
Iteration 1:Mean squared reprojection error: 96.9447
Iteration 2:Mean squared reprojection error: 0.31999
bundle adjustment stopped because the mean squared reprojection
error was less than the specified absolute tolerance value.
Mean reprojection error before bundle adjustment: 6.7601.
Mean reprojection error after bundle adjustment: 0.41922.
```

and the displayed text shows how the total reprojection error decreases at each iteration, reducing by over an order of magnitude. The final result has an average reprojection error better than half a pixel and the maximum error is

```
>> max(e)
ans =
  single
    3.4504
```

which is significantly better than the initial maximum error of nearly 47 pixels.

While the overall error is low, we can look at the final reprojection errors in more detail. Each element of vector `e` is the average reprojection error in pixels for every landmark j across all cameras, in our case, two of them. The median of average errors is

```
>> median(e)
ans =
  single
    0.3067
```

which is a little over quarter of a pixel, but there are a handful of landmarks with a final average reprojection error that are greater than 1 pixel

```
>> size(find(e > 1),1)
ans =
    68
```

and the worst error

```
>> [mx,k] = max(e)
mx =
  single
    3.4504
k =
    1032
```

of 3.4504 pixels occurs for landmark 1032.

Bundle adjustment finds the optimal *relative* pose and positions – not absolute pose. For example, if all the cameras and landmarks moved 1 m in the x-direction, the total reprojection error would be the same. To remedy this we can fix or *anchor* one or more cameras or landmarks – in this example, we fixed the first camera. ▶ The values of the fixed poses and positions are kept in the state vector but they are not updated during the iterations – their Jacobians do not need to be computed and the Hessian matrix used to solve the update at each iteration is smaller since the rows and columns corresponding to those fixed parameters can be deleted.

The `FixedViewId` parameter of the `bundleAdjustment` function can be specified as a vector that permits fixing the camera for any of the views.

The fundamental issue of scale ambiguity with monocular cameras, as we discussed for the essential matrix in ▶ Sect. 14.2.2, applies here as well. A scaled model of the same world with a similarly scaled camera translation is indistinguishable from the real thing. More formally, if the whole problem was scaled so that $P'_j = \lambda P_j$, $[\xi'_i]_t = \lambda [\xi_i]_t$ and $\lambda \neq 0$, the total reprojection error would be the same. The solution we obtained above has an arbitrary scale or value of λ – changing the initial condition for the camera poses or landmark coordinates will lead to a solution with a different scale. We can remedy this by anchoring the pose of at least two cameras, one camera and one landmark, or two landmarks.

The bundle adjustment technique allows for constraints between cameras. For example, a multi-camera rig moving through space would use constraints to ensure the fixed relative pose of the cameras at each time step. Odometry from wheels or inertial sensing could be used to constrain the distance between camera coordinate frames to enforce the correct scale, or orientation from an IMU could be used to constrain the camera orientation. In the underlying graph representation of the problem, shown in ◩ Fig. 14.20, this would involve adding additional edges between the camera vertices. Constraints could also be added between landmarks that had a known relative position, for example the corners of a window – this would involve adding additional edges between the relevant landmark vertices. This is now very similar to the posegraph SLAM solution we introduced in ▶ Sect. 6.4.

The particular problem we studied is unusual in that every camera views every landmark. In a more common situation the camera might be moving in a very large environment so any one camera will only see a small subset of landmarks. In a real-time system, a limited bundle adjustment might be performed with respect to occasional frames known as keyframes, and a bundle adjustment over all frames, or all keyframes, performed at a lower rate in the background.

In this example, we have assumed the camera intrinsic parameters are known and constant. Theoretically, bundle adjustment can solve for intrinsic as well as extrinsic parameters. ▶ We simply add additional parameters for each camera in the state vector and adjust the Jacobian **A** accordingly. However, given the coupling between intrinsic and extrinsic parameters this may lead to poor performance. If we chose to estimate the elements of the camera matrix ($c_{1,1}, \ldots, c_{3,3}$) directly, then the state vector would contain 11 ▶ rather than 6 elements for each camera. If $\mathbf{C}_i$ is the true camera matrix, then bundle adjustment will estimate an arbitrary linear transformation $\mathbf{C}_i \mathbf{Q}$ where $\mathbf{Q} \in \mathbb{R}^{4 \times 4}$ is some nonsingular matrix. Correspondingly, the estimated world points will be $\mathbf{Q}^{-1} \tilde{P}_j$ where $\tilde{P}_j$ is their true value. Fortunately, projection matrices for realistic cameras have well defined structure (13.9) and properties as described in ▶ Sect. 13.1.4, and these provide constraints that allow us to estimate $\mathbf{Q}$. Estimating an arbitrary $\mathbf{C}_i$ is referred to as a projective reconstruction. This can be *upgraded* to an affine reconstruction (using an affine camera model) or a metric reconstruction (using a perspective camera model) by suitable choice of $\mathbf{Q}$.

The `bundleAdjustment` function does not provide this option.

The camera matrix has an arbitrary scale factor. Changes in focal length and z-axis translation have similar image-plane effects as do change in principal point and camera x- and y-axis translation.

14.4 Dense Stereo

A stereo image pair is commonly taken simultaneously using two cameras, generally with parallel optical axes, and separated by a known distance referred to as the camera baseline. ◩ Fig. 14.21 shows stereo camera systems which simultaneously

■ **Fig. 14.21** Stereo cameras that compute disparity onboard using FPGA hardware. Camera baselines of 100 and 250 mm are shown (Image courtesy of Nerian Vision GmbH)

capture images from both cameras, perform dense stereo matching and send the results to a host computer for action. Stereo cameras are a common sensor for mobile robotics, particularly outdoor robots, and can be seen in ■ Fig. 1.5a, ■ Fig. 1.9 and ■ Fig. 4.16.

To illustrate stereo image processing, we load the left and right images comprising a stereo pair

```
>> L = imread("rocks2-l.png");
>> R = imread("rocks2-r.png");
>> imshowpair(L,R,"montage")
>> L = rgb2gray(L); R = rgb2gray(R);
```

shown in ■ Fig. 14.22. We can interactively examine these two images by using the imfuse function and the imtool app, for example

```
>> imtool(imfuse(L,R))
```

as shown in ■ Fig. 14.23. We first zoom in on the rock that has the digit marks. Next, by selecting the measurement tool and placing it on the green-magenta false color composite of the two images we can easily explore distances between corresponding points. The measurement ruler, as shown in ■ Fig. 14.23, is set between points on the digit 5 written on one of the foreground rocks. Firstly, we observe that the spot has the same vertical coordinate in both images, and this implies that the epipolar lines are horizontal. Secondly, in the right-hand image, the spot has moved to the left by 142 pixels. If we probed more points we would see that lateral displacement decreases for conjugate points that are further from the camera.

■ **Fig. 14.22** Stereo pair of rock images. These images are from the Middlebury stereo database (Scharstein and Pal 2007)

Pixel info: (1253, 717) [111 149 111]

◘ Fig. 14.23 The `imtool` image exploration window showing a zoomed composite of stereo images overlayed in the green-magenta false colors. After selecting the measurement tool in the toolstrip, a ruler was placed in the left-hand image (green) at the top right of the digit 5 on a foreground rock. The other end of the ruler was placed in the corresponding location of the right-hand image (magenta). A label on top of the ruler indicates a horizontal shift of 142.0 pixels to the left. This stereo image pair is from the Middlebury stereo database (Scharstein and Pal 2007). The focal length f/ρ is 3740 pixels, and the baseline is 160 mm. The images have been cropped so that the actual disparity should be offset by 274 pixels

As shown in ◘ Fig. 14.16, the conjugate point in the right-hand image moves rightward along the epipolar line as the point depth increases. For the parallel-axis camera geometry the epipolar lines are parallel and horizontal, so conjugate points must have the same v-coordinate. If the coordinates of two corresponding points are $(^L u, {}^L v)$ and $(^R u, {}^R v)$ then $^R v = {}^L v$. The displacement along the horizontal epipolar line $d = {}^L u - {}^R u$, where $d \geq 0$, is called *disparity*.

The dense stereo process is illustrated in ◘ Fig. 14.24. For the pixel at $(^L u, {}^L v)$ in the left-hand image we know that its corresponding pixel is at some coordinate $(^L u - d, {}^L v)$ in the right-hand image where $d = d_{\min}, \ldots, d_{\max}$. To reliably find the corresponding point for a pixel in the left-hand image we create an $N \times N$ pixel *template* region **T** about that pixel – shown as a red square. We *slide* the template window horizontally leftward across the right-hand image. The position at which the template is most similar is considered to be the corresponding point from which disparity is calculated. Compared to the matching problem we discussed in ▶ Sect. 11.5.2, this one is much simpler because there is no change in relative scale or orientation between the two images.

The epipolar constraint means that we only need to perform a 1-dimensional search for the corresponding point. The template is moved in D steps of 1 pixel from $d_{\min}$ to $d_{\max}$. At each template position, we perform a template matching operation, as discussed in ▶ Sect. 11.5.2. For an $N \times N$ template these have a

◩ Fig. 14.24 Stereo matching. A search window in the right image, starting at $u = {}^L u$, is moved leftward along the horizontal epipolar line $v = {}^L v$ until it best matches the template window **T** from the left image

computational cost of $O(N^2)$, and for a $W \times H$ image the total cost of dense stereo matching is $O(DWHN^2)$. This is high, but feasible in real time using optimized algorithms, SIMD instructions sets, and possibly GPU hardware.

To perform stereo matching for the image pair in ◩ Fig. 14.23 is quite straightforward

```
>> D = disparityBM(L,R,DisparityRange=[80 208],BlockSize=13);
```

The result is a matrix of the same size as L and the value of each element $d_{u,v}$, or D(v,u) in MATLAB, is the disparity at that coordinate in the left image. The corresponding pixel in the right image would be at $(u - d_{u,v}, v)$. We can display the disparity as an image – a disparity image

```
>> imshow(D,[])
```

The disparityBM function requires that the difference between specified minimum and maximum disparity is divisible by 16 which lets the function take advantage of Intel® SIMD instruction to speed up the computation.

14

which is shown in ◩ Fig. 14.25. Disparity images have a distinctive ghostly appearance since all surface color and texture is absent. The third argument to disparityBM is the range of disparities to be searched, in this case from 80 to 208 pixels ◀ so the pixel values in the disparity image lie in the range $[80, 208]$. The disparity range was determined by examining some far and near points us-

◩ Fig. 14.25 Disparity image for the rock pile stereo pair, where brighter means higher disparity or shorter range. Red indicates values in the result where disparity could not be computed. These pixels were marked by a value of minimum disparity minus one ing `imfuse` with `imtool`. ▶ The fourth argument to `disparityBM` is the template size specified as an odd number. `disparityBM` uses the SAD similarity measure described earlier in ▶ Chap. 12.

In the disparity image, we can clearly see that the rocks at the bottom of the pile have a larger disparity and are closer to the camera than those at the top. There are also some errors, such as the anomalous bright values around the edges of some rocks. These pixels are indicated as being nearer than they really are. Around the edge of the image, where the similarity matching template falls off the edge of the image, and in places where the disparity could not be computed reliably, as described in the next section, the disparity is set to a value of minimum disparity minus one which is an out-of-range value. These values are displayed as red in ◘ Fig. 14.25.

Note that besides `disparityBM` ▶, there is another function, `disparitySGM`, that uses a more sophisticated algorithm called semi-global matching. It tends to compute smoother disparity output with fewer discontinuities in the disparity image but it is slower.

We could choose a range such as `[0,208]` but this increases the search time: 209 disparities would have to be evaluated instead of 128. It also increases the possibility of matching errors.

`BM` in `disparityBM` stands for block matching, which describes the main approach used by this function to compute the disparity image.

14.4.1 Stereo Failure Modes

Each element of the disparity image we saw earlier is simply the displacement of the template resulting in an optimal matching score somewhere along the epipolar line. In most cases there is a strong and unambiguous match, but not always. For example, it is not uncommon to find the template pattern more than once in the search region. This occurs when there are regular vertical features in the scene as is often the case in human-made scenes: brick walls, rows of windows, architectural features or a picket fence. The problem, illustrated in ◘ Fig. 14.26, is commonly known as the picket fence effect and more properly as spatial aliasing. There is no real cure for this problem ▶ but we can detect its presence. The ambiguity ratio is the ratio of the second extremum to the first extremum – a high-value indicates that the result is uncertain and should not be used. The chance of detecting an incorrect extremum can be reduced by ensuring that the disparity range used in `disparityBM` is as small as possible but this requires some knowledge of the expected range of objects.

Weak template matches typically occur when the corresponding scene point is not visible in the right-hand view due to occlusion – also known as the missing parts problem. Occlusion is illustrated in ◘ Fig. 14.27 and it is clear that point 3 is only visible to the left camera. The stereo matching algorithm will always return the best match so if the point is occluded it will return disparity to the most similar, but wrong, template or mark it as an undefined disparity value. Even though the figure is an exaggerated depiction, real images suffer this problem where the depth changes rapidly. In our example, this occurs at the edges of the rocks which in

Multi-camera stereo, using more than two cameras, is a powerful method to solve this ambiguity.

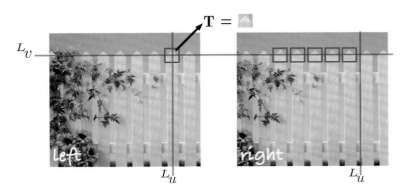

◘ **Fig. 14.26** Picket fence effect. The template will match well at a number of different disparities. This problem occurs in any scene with repeating patterns

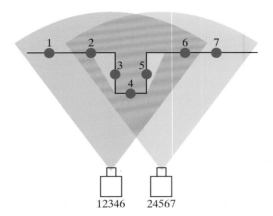

Fig. 14.27 Occlusion in stereo vision. The field of view of the two cameras are shown as colored sectors. Points 1 and 7 fall outside the overlapping view area and are seen by only one camera each. Point 5 is occluded from the left camera and point 3 is occluded from the right camera. The order of points seen by each camera is given underneath it

Fig. 14.25 is exactly where we observe the incorrect disparities or disparities marked red. The problem becomes more prevalent as the baseline increases. The problem also occurs when the corresponding point does not lie within the disparity search range, that is, the disparity search range is too small. This issue cannot be cured but it can be detected and disparity results where similarity is low can be discarded as is done by the `disparityBM` function.

A simple but effective way to test for occlusion is to perform the matching in two directions – left-right consistency checking. Starting with a pixel in the left-hand image, the strongest match in the right-image is found. Then the strongest match to that pixel is found in the left-hand image. If this is where we started, then the match is considered valid. However, if the corresponding point was occluded in the right image, the first match will be a weak one to a different feature, and there is a high probability that the second match will be to a different pixel in the left image.

From Fig. 14.27 it is clear that pixels on the left-side of the left-hand image may not overlap at all with the right-hand image – point 1, for example, is outside the field of view of the right-hand camera. To avoid returning large number of likely incorrect matches, it is common practice to discard the $d_{\max}$ left-most columns, 208 in this case, of the disparity image as shown in Fig. 14.25.

The final problem that can arise is a similarity function with a very broad extrema. The breadth makes it difficult to precisely estimate the maxima. This generally occurs when the template region has very low texture, for example corresponding to the sky, dark shadows, sheets of water, snow, ice or smooth human-made objects. Simply put, in a region that is all gray, a gray template matches equally well with any number of gray candidate regions. One approach to detect this is to look at the variability of pixel values in the template using measures such as the difference between the maximum and minimum value or the variance of the pixel values. If the template has too little variance, it is less likely to result in a strong extremum.

For the various problem cases just discussed, disparity cannot be determined, but the problem can be detected. This is important since it allows those pixels to be marked as having no known range and this allows a robot to be prudent with respect to regions whose 3-dimensional structure cannot be reliably determined.

14.4.1.1 Summary

The design of a stereo-vision system has three degrees of freedom. The first is the baseline distance between the cameras. As this increases, the disparities become larger making it possible to estimate depth to greater precision, but the occlusion problem becomes worse. Second, the disparity search range needs to be set carefully. If the maximum is too large the chance of spatial aliasing increases but if too small then points close to the camera will generate incorrect and weak matches. A large disparity range also increases the computation time. Third, template size involves a tradeoff between computation time and quality of the disparity image. A small template size can pick up fine depth structure but tends to give results that are much noisier since a small template is more susceptible to ambiguous matches. A large template gives a smoother disparity image, but requires greater computation. It also increases the chance that the template will contain pixels belonging to objects at different depths which is referred to as the mixed pixel problem. This can cause poor quality matching at the edges of objects, and the resulting disparity image appears blurred.

14.4.2 Refinement and Reconstruction

The result of stereo matching, such as shown in ◘ Fig. 14.25, have a number of imperfections for the reasons we have just described. For robotic applications such as path planning and obstacle avoidance it is important to know the 3-dimensional structure of the world, but it is also critically important to know what we don't know. Where reliable depth information from stereo vision is missing a robot should be prudent and treat it differently to free space. The `disparityBM` function automatically accounts for many of the stereo failure modes, including

- template extending past the edge of the image
- regions of stereo images that do not overlap
- weak template matches
- broad peaks resulting from patches lacking texture
- occlusion

The unreliable pixels are marked red in ◘ Fig. 14.25. The good news is that there are a lot of non-red pixels! In fact,

```
>> sum(D(:)~=79)/prod(size(L))*100
ans =
   77.6957
```

nearly 80% of disparity values pass our battery of quality tests. It is also important to know that disparity values were interpolated to better than a pixel location using techniques such as the one described in ▶ App. J that covers extremum refinement.

14.4.2.1 3D Reconstruction

The final step of the stereo vision process is to convert the disparity values, in units of pixels, to world coordinates in units of meters – a process known as 3D reconstruction. In the earlier discussion on sparse stereo, we determined the world point from the intersection of two rays in 3-dimensional space. For a parallel-axis stereo camera rig as shown in ◘ Fig. 14.21 the geometry is much simpler as illustrated in ◘ Fig. 14.28. ▶ For the red and blue triangles we can write

$$X = Z \tan \theta_1, \quad \text{and} \quad X - b = Z \tan \theta_2$$

where b is the baseline and the angles of the rays correspond to the horizontal image coordinate $^i u, i = \{L, R\}$

$$\tan \theta_i = \frac{\rho_u(^i u - u_0)}{f} .$$

The rock pile stereo pair was rectified to account for minor alignment errors in the stereo cameras. Rectification is discussed in ▶ Sect. 14.4.3.

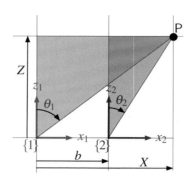

◻ Fig. 14.28 Stereo geometry for parallel camera axes. X and Z are measured with respect to camera one, b is the baseline

where ρ_u is the pixel width, u_0 is the horizontal coordinate of the principal point, and f is the focal length. Substituting and eliminating X gives

$$Z = \frac{fb}{\rho_u(^Lu - {}^Ru)} = \frac{fb}{\rho_u d}$$

which shows that depth is inversely proportional to disparity. We can also recover the X- and Y-coordinates so the 3D point coordinate is

$$\boldsymbol{P} = \frac{b}{d}\left({}^Lu - u_0, {}^Lv - v_0, \frac{f}{\rho_u}\right) \tag{14.14}$$

which can be computed for every pixel.

A good stereo system can estimate disparity with an accuracy of 0.2 pixels. Distant points have a small disparity and the error in the estimated 3D coordinate will be significant. A rule of thumb is that stereo systems typically have a maximum range of $50b$.

🛇 The images shown in ◻ Fig. 14.22, from the Middlebury dataset, were taken with a very wide camera baseline. The left edge of the left-image and the right edge of the right-image have no overlap and have been cropped. Cropping N pixels from the left of the left-hand image only, reduces the disparity by N. For this stereo pair the actual disparity must be increased by 274 to account for the cropping.

> The special floating-point value `NaN` (for not a number) has the useful property that the result of any arithmetic operation involving `NaN` is always `NaN`. Many MATLAB functions such as `max` or `min` ignore `NaN` values in the input matrix, and plotting and graphics functions do not display this value, leaving a hole in the graph or surface.

Earlier, we created a disparity image `D` and now, we set the disparity values that were identified as unreliable to `NaN`. ◄

```
>> D(D==79) = NaN;
```

The true disparity is

```
>> D = D+274;
```

> A process known as vectorizing. Using matrix and vector operations instead of `for` loops can increase the speed of MATLAB code execution. See ► https://sn.pub/zpw98n for details.

and we compute the X-, Y- and Z-coordinate of each pixel as separate matrices to exploit MATLAB's efficient matrix operations ◄

```
>> [U,V] = meshgrid(1:size(L,2),1:size(L,1));
>> u0 = size(L,2)/2; v0 = size(L,1)/2;
>> b = 0.160;
>> X = b*(U-u0)./D; Y = b*(V-v0)./D; Z = 3740*b./D;
```

▶ sn.pub/uk7vo6

◻ Fig. 14.29 3-dimensional reconstruction for parallel stereo cameras. Hotter colors indicate parts of the surface that are further from the camera

which can be displayed as a surface

```
>> Z = medfilt2(Z,[5 5]);
>> surf(Z)
>> shading interp; view(-74,44)
>> set(gca,ZDir="reverse"); set(gca,XDir="reverse")
>> colormap(parula)
```

as shown in ◻ Fig. 14.29. ▶ This is somewhat unimpressive in print but by using the mouse to rotate the image using the MATLAB figure toolbar *3D rotate* option the 3-dimensionality becomes quite clear. The axis reversals are required to have z increase from our viewpoint and to maintain a right-handed coordinate frame. There are *holes* in this surface which are the NaN values we inserted to indicate unreliable disparity values.

We used the medfilt2 function to eliminate some of the noisy Z values.

14.4.2.2 3D Texture Mapped Display

For human, rather than robot consumption, we can *drape* the left-hand image over the 3-dimensional surface using a process called texture mapping. We reload the left-hand image, this time in color

```
>> Lcolor = imread("rocks2-1.png");
```

and render the surface with the image texture mapped

```
>> surface(X,Y,Z,Lcolor,FaceColor="texturemap", ...
>>    EdgeColor="none",CDataMapping="direct")
>> set(gca,ZDir="reverse"); set(gca,XDir="reverse")
>> view(-74,44), grid on
```

which creates the image shown in ◻ Fig. 14.30. Once again, it is easier to get an impression of the 3-dimensionality by using the mouse to rotate the image using the MATLAB figure toolbar *3D rotate* option.

14.4.3 Stereo Image Rectification

The rock pile stereo pair of ◻ Fig. 14.23 has corresponding points on the same row in the left- and right-hand images – they are an epipolar-aligned image pair. Stereo cameras, such as shown in ◻ Fig. 14.21, are built with precision to ensure that the optical axes of the cameras are parallel and that the u- and v-axes of the two sensor

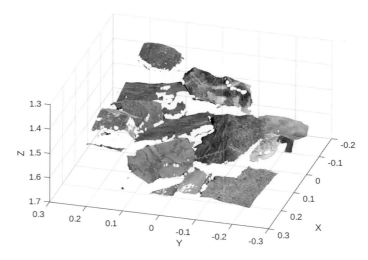

Fig. 14.30 3-dimensional reconstruction for parallel stereo cameras with image texture mapped onto the surface

chips are parallel. However, there are limits to the precision of mechanical alignment and additionally, lens distortion will introduce error. Typically, one or both images are warped to correct for these errors – a process known as rectification.

We will illustrate rectification using the courtyard stereo pair from ☐ Fig. 14.14

```
>> imL = imresize(rgb2gray(imread("walls-l.jpg")),0.25);
>> imR = imresize(rgb2gray(imread("walls-r.jpg")),0.25);
```

which we recall are far from being epipolar aligned. We first find the SURF features

```
>> ptsL = detectSURFFeatures(imL);
>> ptsR = detectSURFFeatures(imR);
>> [sfL,vptsL] = extractFeatures(imL,ptsL);
>> [sfR,vptsR] = extractFeatures(imR,ptsR);
```

and determine the candidate matches

```
>> idxPairs = matchFeatures(sfL,sfR);
>> matchedPtsL = vptsL(idxPairs(:,1));
>> matchedPtsR = vptsR(idxPairs(:,2));
```

then determine the epipolar relationship

```
>> [F,inlierIdx] = estimateFundamentalMatrix(matchedPtsL, ...
>>    matchedPtsR,Method="MSAC",NumTrials=6000, ...
>>    DistanceThreshold=0.04,Confidence=95);
```

The rectification step requires the fundamental matrix as well as the set of corresponding points

```
>> inlierPtsL = matchedPtsL(inlierIdx);
>> inlierPtsR = matchedPtsR(inlierIdx);
>> [tformL,tformR] = estimateStereoRectification(F, ...
>>    inlierPtsL,inlierPtsR,size(imL));
>> [rectL,rectR] = rectifyStereoImages(imL,imR,tformL,tformR);
```

and returns rectified versions of the two input images. We display these using `imshowpair`

```
>> imshowpair(rectL,rectR,"montage");
```

which is shown in ☐ Fig. 14.31. Corresponding points in the scene now have the same vertical coordinate. The function `estimateUncalibratedRectification` works by computing unique homographies to warp the left and right images. As

Fig. 14.31 Rectified images of the courtyard

Fig. 14.32 Dense stereo disparity image for the courtyard. The walls and ground show a clear depth gradient

we have observed previously, when warping images not all of the output pixels are mapped to the input images which results in undefined pixels. Instead of including those pixels, by default, `rectifyStereoImages` function crops the images to include only valid regions. ▶

We can think of these images as having come from a virtual stereo camera with parallel axes and aligned pixel rows, and they can now be used for dense stereo matching

```
>> d = disparitySGM(rectL,rectR,"DisparityRange",[0 64]);
>> imshow(d,[])
```

and the result is shown in ◘ Fig. 14.32. The disparity range parameters were determined interactively using `imtool(imfuse(imL, imR))` to check the disparity at near and far points in the rectified image pair. The noisy patches at the bottom and top right are due to occlusion – world points in one image are not visible in the other. Nevertheless, this is quite an impressive result – using only two images taken from a handheld camera we have been able to create a dense 3-dimensional representation of the scene.

When working with a calibrated stereo camera, the `rectifyStereoImages` function can also take calibration parameters as input. By taking advantage of the calibration information, a more reliable and accurate rectification can be produced.

14.5 Anaglyphs

Human stereo perception relies on each eye providing a different viewpoint of the scene. However even if we look at a 2D photograph of a 3D scene we still get some sense of depth, albeit reduced, because our brain uses many visual cues, in addition to stereo, to infer depth. Since the invention of photography in the 19th century people have been fascinated by 3D photographs and movies, and the enduring popularity of 3D movies is further evidence of this.

The key to most 3D display technologies is to take the image from two cameras, with a similar baseline to the human eyes (50–80 mm) and present those images again to the corresponding eyes. Old fashioned stereograms required a binocular viewing device or could, with difficulty, be viewed by squinting at the stereo pair and crossing your eyes. More modern and convenient means of viewing stereo pairs are LCD shutter (gaming) glasses or polarized glasses which allow full-color stereo movie viewing, or head mounted displays.

An old, but inexpensive, method of viewing and distributing stereo information is through anaglyph images where the left and right images are overlaid in different colors. Typically red is used for the left eye and cyan (greenish blue) for the right eye, but many other color combinations are commonly used. The anaglyph is viewed through glasses with different colored lenses. The red lens allows mainly the red part of the anaglyph image to enter the left eye, while the cyan lens allows mainly the cyan parts of the image to enter the right eye. The disadvantage is that the color is not reproduced accurately since the image colors of the stereo pair are manipulated to work with the glasses. The big advantage of anaglyphs is that they can be printed on paper or imaged onto ordinary movie film and viewed with simple and cheap glasses such as those shown in � Fig. 14.33a.

The rock pile stereo pair can be displayed as an anaglyph

```
>> Rcolor = imread("rocks2-r.png");
>> A = stereoAnaglyph(Lcolor,Rcolor);
>> imshow(A)
```

which is shown in ◻ Fig. 14.33b. If you have glasses other than the common red-cyan, different color combinations can be produced by using the `imfuse` function.

Excurse 14.4: Anaglyphs

The earliest developments occurred in France. In 1858 Joseph D'Almeida projected 3D magic lantern slide shows as red-blue anaglyphs and the audience wore red and blue goggles. Around the same time, Louis Du Hauron created the first printed anaglyphs. Later, around 1890, William Friese-Green created the first 3D anaglyphic motion pictures using a camera with two lenses. Anaglyphic films called plasticons or plastigrams were a craze in the 1920s.

Today, anaglyphs of Mars can be found at ▶ http://mars.nasa.gov/mars3d and a playful and interactive virtual reality-based anaglyphs at ▶ https://sn.pub/70G58P.

◘ Fig. 14.33 Anaglyphs for stereo viewing. **a** Anaglyph glasses shown with red and blue lenses, **b** anaglyph rendering of the rock scene from ◘ Fig. 14.23 with the left image in red and the right image in cyan

14.6 Other Depth Sensing Technologies

There are a number of alternative approaches to obtaining 3-dimensional information about the world around a robot.

14.6.1 Depth from Structured Light

A classic, yet effective, method of estimating the 3D structure of a scene is structured light. This is conceptually similar to stereo vision, but we replace the left camera with a projector that imposes a pattern of light on the scene.

The simplest example of structured light, shown in ◘ Fig. 14.34a, uses a vertical plane of light. ► This is equivalent, in a stereo system, to a left-hand image that is a vertical line. The image of the line projected onto the surface viewed from the right-hand camera will be a distorted version of the line, as shown in ◘ Fig. 14.34b. The disparity between the virtual left-hand image and the actual right-hand image is a function of the depth of points along the line.

Finding the light stripe on the scene is a relatively simple vision problem. In each image row we search for the pixel corresponding to the projected stripe, based on intensity or color. If the camera coordinate frames are parallel, then depth is computed by (14.14).

To achieve depth estimates over the whole scene, we need to move the light plane horizontally across the scene and there are many ways to achieve this: mechanically rotating the laser stripe projector, using a moving mirror to deflect the stripe or using a data projector and software to create the stripe image. However, sweeping the light plane across the scene is slow and fundamentally limited by the rate at which we can acquire successive images of the scene. One way to speed up the process is to project multiple lines on the scene, but then we have to solve the correspondence problem which is not simple if parts of some lines are occluded. Many solutions have been proposed but generally involve coding the lines in some way – using different colors or using a sequence of binary or gray-coded line patterns which can match 2^N lines in just N frames.

Laser-based line projectors, so called "laser stripers" or "line lasers", are available for just a few dollars. They comprise a low-power solid-state laser and a cylindrical lens or diffractive optical element.

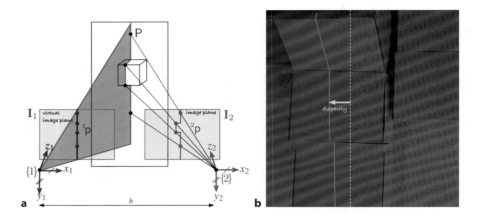

◘ Fig. 14.34 **a** Geometry of structured light showing a light projector on the left and a camera on the right; four conjugate points are marked with dots on the left and right images and the scene; **b** a real structured light scenario showing the light stripe falling on a stack of cardboard boxes. The superimposed dashed line represents the stripe position for a plane at infinity. Disparity, the left shift of the projected line relative to the dashed line, is inversely proportional to depth

This approach was used in the original Kinect for Xbox 360 introduced in 2010, and still available in the Asus Xtion sensor. It was developed by the Israeli company PrimeSense which was bought by Apple in 2013. The newer Azure Kinect uses per pixel time-of-flight measurement.

A related approach is to project a known but random pattern of *dots* onto the scene as shown in ◘ Fig. 14.35a. Each dot can be identified by the unique pattern of dots in its surrounding window. ◄ One lens projects an infrared dot pattern using a laser with a diffractive optical element which is viewed, see ◘ Fig. 14.35a, by an infrared sensitive camera behind another lens from which the depth image shown in ◘ Fig. 14.35c is computed. The shape of the dots also varies with distance, due to imperfect focus, and this provides additional cues about the distance of a point. A third lens is a regular color camera which provides the view shown in ◘ Fig. 14.35b. This is an example of an RGBD camera, returning RGB color values as well as depth (D) at every pixel.

Structured light approaches work well for ranges of a few meters indoors, for textureless surfaces, and in the dark. Outdoors, the projected pattern is overwhelmed by ambient illumination from the sun.

Some stereo systems, such as the Intel RealSense D400 series, also employ a dot pattern projector, sometimes known as a speckle or texture projector. This provides artificial texture that helps the stereo vision system when it is looking at textureless surfaces where matching is frequently weak and ambiguous, as discussed in ▶ Sect. 14.4.1. Such a sensor has the advantage of working on textureless surfaces which are common indoors where the sun is not a problem, and outdoors using pure stereo where scene texture is usually rich.

14.6.2 Depth from Time-Of-Flight

An alternative to stereo vision are cameras based on time-of-flight measurement. The camera emits a pulse of infrared light that illuminates the entire scene, and every pixel records the intensity and time of return of the reflected energy. The time measurement problem is challenging since a depth resolution of 1 cm requires timing precision of the order of 70 picoseconds and needs to be performed at every pixel.

This type of camera works well indoors and even in complete darkness, but outdoors under full sun the maximum range is limited just as it is for structured light. In fact, it is worse than structured light. The illumination energy is limited by eye-safety considerations and structured light concentrates that energy over a sparse dot field, whereas time-of-flight cameras spread it over an area.

◻ Fig. 14.35 3D imaging with the Kinect 360 sensor. **a** Random dot pattern as seen by the Kinect's infrared camera; **b** original scene captured with the Kinect's color camera; **c** computed depth image. *Red pixels* indicate NaN values where depth could not be computed due to occlusion or the maximum range being exceeded, as for example through the window on the left side of the scene (Images courtesy William Chamberlain)

14.7 Point Clouds

A point cloud is a set of coordinates $\{(X_i, Y_i, Z_i), i = 1, \ldots, N\}$ of points in the world. A colored point cloud is a set of tuples $\{(X_i, Y_i, Z_i, R_i, G_i, B_i), i = 1, \ldots, N\}$ comprising the coordinates of a world point and its tristimulus value. Point clouds can be derived from all sorts of 3-dimensional sensors including stereo vision, RGBD cameras and LiDARs. A point cloud is a useful way to combine 3-dimensional data from multiple sensors, of different types, or sensors moving over time.

For a robotics application, we need to extract some concise meaning from the thousands or millions of points in the point cloud. Common data reduction strategies include finding dominant planes, and registering points to known 3-dimensional models of objects.

The order of the points is generally not important, and a kd-tree is typically constructed to reduce the computation involved in finding the neighbors of a point. The exception is an organized point cloud where the points are arranged on a fixed rectangular grid typically determined by the scanning order of a particular sensor. For example, for the RGBD sensor, the point grid corresponds to pixels in the $M \times N$ RGB image, and holds the $[X, Y, Z]$ point locations in the $M \times N \times 3$ array. For LiDAR, the grid is organized by the azimuth and elevation of the laser beam. The key advantage of an organized point cloud is that it contains additional spatial information that can help with efficiency of processing the point cloud.

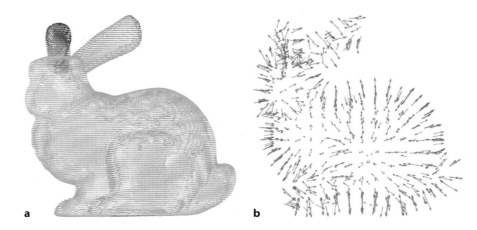

a   b

☐ **Fig. 14.36   a** Point cloud of the Stanford bunny with 35,947 points, nearer points have a warmer color; **b** A sparse needle plot of surface normals (Stanford bunny model courtesy Stanford Graphics Laboratory)

This model is well known in the computer graphics community. It was created by Greg Turk and Marc Levoy in 1994 at Stanford University, using a Cyberware 3030 MS scanner and a ceramic rabbit figurine.

The Polygon file format is a simple file format that contains (X, Y, Z) tuples for the points, as well as faces and color information. We can read PLY, PCD, STL file formats, as well as streaming data from Velodyne's PCAP files and others.

For example, finding nearest neighbors of a point is simplified as compared to randomly arranged points. When performing various point cloud operations such as downsampling, care must be taken not to disturb organization of the point cloud so that performance advantages can be maintained down the processing chain for as long as possible.

We will illustrate basic point cloud operations by loading a point cloud of the famous Stanford bunny. ◄ Just as for 2-dimensional images, there are many common file formats, and here we will load a PLY file ◄

```
>> bunny = pcread("bunny.ply")
bunny =
  pointCloud with properties:
    Location: [35947×3 single]
       Count: 35947
      XLimits: [-0.0947 0.0610]
      YLimits: [0.0330 0.1873]
      ZLimits: [-0.0619 0.0588]
       Color: []
      Normal: []
   Intensity: [35947×1 single]
>> pcshow(bunny, VerticalAxis="Y")
```

which is shown in ☐ Fig. 14.36a. The `pcread` function returns a `pointCloud` object which holds the point coordinates in the `Location` property. This object has many useful methods, including `findNearestNeighbors` which can be used to construct basic region processing functionality. It also builds and caches an internal kd-tree representation of the point cloud, whenever a function that requires region processing is invoked on this object. The `pcshow` function displays a single point cloud. It has a context menu that can be used to obtain various projections of the point cloud, or to set a colormap indicating depth in a selected axis direction. For streaming pointclouds obtained from an RGBD sensor or LiDAR, we can use the `pcplayer` function

```
>> veloReader = velodyneFileReader( ...
>>    "lidarData_ConstructionRoad.pcap","HDL32E");
>> xlimits = [-60 60]; ylimits = [-60 60]; zlimits = [-20 20];
>> player = pcplayer(xlimits, ylimits, zlimits);
>> xlabel(player.Axes,"X (m)"); ylabel(player.Axes,"Y (m)");
>> zlabel(player.Axes,"Z (m)");
>> while(hasFrame(veloReader) && player.isOpen())
>>    ptCloud = readFrame(veloReader);
>>    view(player,ptCloud);
>> end
```

which restricts the display dimensions to hand-picked limits so that the viewing area does not constantly resize. Unlike images, point clouds are not constrained by a fixed grid, but instead their limits correspond to physical dimensions of the scene and maximum range of the sensor.

Point clouds can be manipulated in various ways, such as by cropping or downsampling

```
>> bunny_d = pcdownsample(bunny,gridAverage=0.01);
>> bunny_d.Count
ans =
   755
```

which in this case downsamples the point cloud by computing an average value of point locations contained within each cube of side length 1 cm.

We can also compute the normal at each point, based on the surrounding points to which a local surface is fitted

```
>> normals = pcnormals(bunny_d);
>> pcshow(bunny_d,VerticalAxis="Y"); hold on
>> x = bunny_d.Location(:,1);
>> y = bunny_d.Location(:,2);
>> z = bunny_d.Location(:,3);
>> [u,v,w] = deal(normals(:,1),normals(:,2),normals(:,3));
>> quiver3(x,y,z,u,v,w);
```

and this is shown in ◻ Fig. 14.36b as a sparse needle map. Point clouds can also be triangulated into surface meshes for rendering as solid 3D objects.

14.7.1 Fitting Geometric Objects into a Point Cloud

Planes are common in human surroundings, and for robotics, planes can be used to model the ground and walls for wheeled mobile robot driving or for UAV landing. There are two common approaches to finding a plane of best fit in a point cloud. The first, and simplest, approach for finding the plane is to fit the data to an ellipsoid. The ellipsoid will have one very small radius in the direction normal to the plane – that is, it will be an elliptical plate. The inertia of the ellipsoid is equal to the moment matrix which is calculated directly from the points

$$\mathbf{J} = \sum_{i=1}^{N} x_i x_i^\top \in \mathbb{R}^{3\times3} \qquad (14.15)$$

where $x = P_i - \overline{P}$ are the coordinates of the points, as column vectors, with respect to the centroid of the points $\overline{P} = \frac{1}{N} \sum_{i=1}^{N} P_i$. The radii of the ellipsoid are the square root of the eigenvalues of $\mathbf{J}$, and the eigenvector corresponding to the smallest eigenvalue is the direction of the minimum radius which is the normal to the plane.

Outlier data points are problematic with this simple estimator since they significantly bias the solution. A number of approaches are commonly used but a simple one is to modify (14.15) to include a weight

$$\mathbf{J} = \sum_{i=1}^{N} w_i x_i x_i^\top$$

which is inversely related to the distance of x_i from the plane and solve iteratively. Initially, all weights are set to $w_i = 1$, and on subsequent iterations the weights ▶ are set according to the distance of point P_i from the plane estimated at the previous step. ▶ App. C.1.4 has more details about ellipses.

Alternatively a Cauchy-Lorentz function $w = \beta^2/(d^2 + \beta^2)$ is used where d is the distance of the point from the plane and β is the half-width. The function is smooth for $d = [0, \infty)$ and has a value of $\frac{1}{2}$ when $d = \beta$.

◻ Fig. 14.37 **a** Image of an office desk and **b** the corresponding point cloud (Images reprinted with permission of The MathWorks, Inc.)

▶ sn.pub/XNGAG7

Alternatively, we could apply RANSAC by taking samples of three points to estimate the parameters of a plane $aX + bY + cZ + d = 0$. We can illustrate this with an office scene captured using a Microsoft Kinect sensor

```
>> ptCloud = pcread("deskScene.pcd");
```

which can be displayed as both an image and a point cloud

```
>> imshow(ptCloud.Color), figure
>> pcshow(ptCloud)
```

as shown in ◻ Fig. 14.37. To isolate the plane, we set a tolerance in the form of a maximum distance from the plane (2 cm) and specify a reference normal vector $[0, 0, 1]$ to pick the plane that we wish to extract, which in our case happens to be in the xy-plane.

```
>> rng(0) % set random seed for reproducibility of results
>> [planeModel,inlierIdx,outlierIdx] = ...
>>    pcfitplane(ptCloud,0.2,[0,0,1]);
>> planeModel
planeModel =
  planeModel with properties:
    Parameters: [-0.0702 0.0425 0.9966 -0.1114]
        Normal: [-0.0702 0.0425 0.9966]
```

The plane can now be separated from the original scene

```
>> plane = select(ptCloud,inlierIdx);
>> remainingPoints = select(ptCloud,outlierIdx);
>> pcshow(plane), figure
>> pcshow(ptCloud), hold on
>> planeModel.plot()
```

and its isolated point cloud is shown in ◻ Fig. 14.38a while the estimated plane is overlaid in red in the original scene in ◻ Fig. 14.38b. To find the next plane, this process can be repeated using the `remainingPoints` point cloud.

Similarly, it is possible to fit other common geometric models to a point cloud

```
>> roi = [-inf 0.5 0.2 0.4 0.1 inf];
>> sampleIdx = findPointsInROI(ptCloud,roi);
>> rng(0) % set random seed for reproducibility of results
>> [model,inlierIdx] = ...
>>    pcfitsphere(ptCloud,0.01,SampleIndices=sampleIdx);
>> globe = select(ptCloud,inlierIdx);
>> pcshow(globe), figure
>> pcshow(ptCloud), hold on
>> plot(model)
```

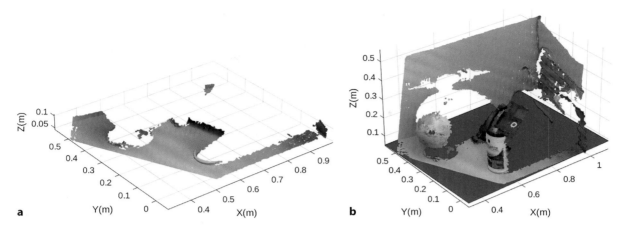

Fig. 14.38 **a** Horizontal plane (desk) isolated from the scene; **b** Fitted model of the plane displayed in red in the original scene

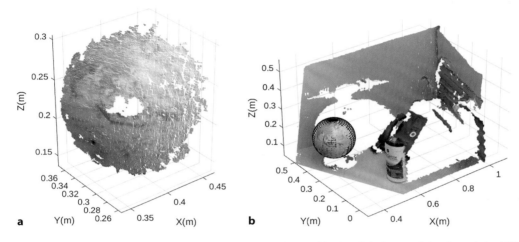

Fig. 14.39 **a** Globe isolated from the scene; **b** Fitted model of the globe displayed in the original scene. Sphere's mesh is visible where surface pixels are missing

and the result of fitting a sphere with 1 cm tolerance is shown in ■ Fig. 14.39. This time, to assist in the fitting, we first extracted a cuboid region from the scene, specified as [xmin,xmax,ymin,ymax,zmin,zmax], to separate the globe. Besides pcfitplane and pcfitsphere functions, the Computer Vision Toolbox™ also contains pcfitcylinder and a non-RANSAC segmentGroundFromLidarData function which requires an organized point cloud as input. It is a fast and robust means to find ground surfaces, which for real robotics problems cannot be assumed to be flat.

▶ sn.pub/TSKfK0

14.7.2 Matching Two Sets of Points

Consider the problem shown in ■ Fig. 14.40a where we have a model of the bunny at a known pose shown as a blue point cloud. The red point cloud is the result of either the bunny moving, or the 3D camera moving. By aligning the two point clouds we can determine the rigid-body motion of the bunny or the camera.

Formally, this is the point cloud registration problem. Given two sets of n-dimensional points: the model $\{M_i \in \mathbb{R}^n, i = 1, \ldots, N_M\}$ and some noisy observed data $\{D_j \in \mathbb{R}^n, j = 1, \ldots, N_D\}$ determine the rigid-body motion from the data

▶ sn.pub/A1yEMi

▶ sn.pub/EFaSyQ

☐ **Fig. 14.40** Iterative closest point (ICP) matching of two point clouds: model (magenta) and data (green) **a** before registration, **b** after registration; observed data points have been transformed to the model coordinate frame (Stanford bunny model courtesy Stanford Graphics Laboratory)

coordinate frame to the model frame

$$^{D}\hat{\boldsymbol{\xi}}_{M} = \arg\min_{\boldsymbol{\xi}} \sum_{i,j} \|\boldsymbol{D}_{j} - \boldsymbol{\xi} \cdot \boldsymbol{M}_{i}\| \ .$$

A well known solution to this problem is the iterative closest point algorithm or ICP. At each iteration, the first step is to compute a translation that makes the centroids of the two point clouds coincident ◀

We consider the general case where the two points clouds have different numbers of points, that is, $N_D \neq N_M$.

$$\overline{\boldsymbol{M}} = \frac{1}{N_M} \sum_{i=1}^{N_M} \boldsymbol{M}_i, \ \ \overline{\boldsymbol{D}} = \frac{1}{N_D} \sum_{j=1}^{N_D} \boldsymbol{D}_j$$

from which we compute a displacement

$$\boldsymbol{t} = \overline{\boldsymbol{D}} - \overline{\boldsymbol{M}} \ .$$

Next, we compute approximate correspondence. For each data point $\boldsymbol{D}_j$ we find the closest model point $\boldsymbol{M}_i$. ◀ Correspondence is not unique and quite commonly several points in one set can be associated with a single point in the other set, and consequently some points will be unpaired. Often, the sensor returns only a subset of points in the model, for instance a laser scanner can see the front but not the back of an object. This approach to correspondence is far from perfect but it is surprisingly effective in practice and *improves* the alignment of the point clouds so that in the next iteration the estimated correspondences will be a little more accurate.

This is computationally expensive, but organizing the points in a kd-tree improves the efficiency.

The corresponding points are used to compute the moment matrix

$$\mathbf{J} = \sum_{i,j} \left(\boldsymbol{M}_i - \overline{\boldsymbol{M}} \right) \left(\boldsymbol{D}_j - \overline{\boldsymbol{D}} \right)^{\top} \in \mathbb{R}^{n \times n}$$

which encodes the rotation between the two point sets. The singular value decomposition is $\mathbf{J} = \mathbf{U}\boldsymbol{\Sigma}\mathbf{V}^{\top}$ and the rotation matrix is ◀

It is possible to obtain a result where $\det(\mathbf{R}) = -1$, in which case $\mathbf{R}$ should be negated. See ▶ App. F.1.1.

$$\mathbf{R} = \mathbf{V}\mathbf{U}^{\top} \in \mathbf{SO}(3) \ .$$

The estimated relative pose between the two point clouds is $\boldsymbol{\xi}^\Delta \sim (\mathbf{R}, \boldsymbol{t})$ and the model points are transformed so that they are closer to the data points

$$\boldsymbol{M}_i \leftarrow \boldsymbol{\xi}^\Delta \cdot \boldsymbol{M}_i, \; i = 1, \ldots, N_M$$
$$\boldsymbol{\xi} \leftarrow \boldsymbol{\xi} \oplus \boldsymbol{\xi}^\Delta$$

and the process is repeated until it converges. The used correspondences are unlikely to have all been correct, and therefore the estimate of the relative orientation between the sets is only an approximation.

We will illustrate this with the bunny point cloud. The model will be a random subsampling of the bunny comprising 10% of the points

```
>> bunny = pcread("bunny.ply");
>> rng(0) % set random seed for reproducibility of results
>> model = pcdownsample(bunny,"random",0.1);
```

The data that we try to fit to the model is another random subsampling, this time only 5% of the points, and we apply a rigid-body transformation to those points ▶

```
>> data = pcdownsample(bunny, "random", 0.05);
>> tform = rigidtform3d([0 0 60], ...    % [rx, ry, rz] in degrees
>>                      [0.3 0.4 0.5]); % [tx, ty, tz] in meters
>> data = pctransform(data, tform);
```

Random sampling means that the chance of the same bunny points being in both point clouds is low. Despite this, ICP does a very good job of matching.

We can view the two pointclouds together in distinct colors to visualize their relative positions and orientations

```
>> pcshowpair(model,data,VerticalAxis="Y")
```

which is shown in ◘ Fig. 14.40a. The `pcshowpair` shows the two point clouds colored in magenta and green. This function is designed for comparing point clouds, typically in the context of registration problems.

We can now register the point clouds using the ICP algorithm with a tolerance of absolute difference between iterations of 1 cm and 0.1° in translation and rotation respectively.

```
>> [tformOut,dataReg] = pcregistericp(data,model, ...
>>    Tolerance=[0.01 0.1],MaxIterations=100);
>> tout = tformOut.invert
tout =
  rigidtform3d with properties:
    Dimensionality: 3
                 R: [3×3 double]
       Translation: [0.2997 0.3992 0.5007]
                 A: [4×4 double]
>> rad2deg(rotm2eul(tout.R)) % [rz ry rx] in degrees
ans =
    59.5367    0.2839   -0.5950
```

which is the "unknown" relative pose of the data point cloud that we chose above. The result of the registration can again be shown using the `pcshowpair` function

```
>> pcshowpair(model,dataReg,VerticalAxis="Y")
```

which is shown in ◘ Fig. 14.40b. We see that the magenta (`model`) and green (`dataReg`) point clouds are nearly perfectly aligned.

14.8 Applications

14.8.1 Perspective Correction

Consider the image

```
>> im = imread("notre-dame.jpg");
>> imshow(im)
```

shown in ◘ Fig. 14.41. The shape of the building is significantly distorted because the camera's optical axis was not normal to the plane of the building and we see evidence of perspective foreshortening or keystone distortion. We manually pick four points, clockwise from the bottom left, that are the corners of a large rectangle on the planar face of the building.

```
>> isInteractive = false; % set to true to select points manually
>> if isInteractive
>>    h = drawpolygon(Color="y");
>>    wait(h);
>> else % use preselected points for reproducibility
>>    p1 = [43 382;90 145;539 154;613 371];
>>    h = drawpolygon(Position=p1, Color="y");
>> end
```

To finalize the point selections, double-click in the middle of the polygon. The returned matrix

```
>> p1 = h.Position
p1 =
    43.0000   382.0000
    90.0000   145.0000
   539.0000   154.0000
   613.0000   371.0000
```

has one row per point that contains the u- and v-coordinate. The region is marked on the image of the cathedral with a translucent yellow keystone shape. We use the extrema of these points to define the vertices of a rectangle in the image

```
>> mn = min(p1);
>> mx = max(p1);
>> p2 = [mn(1) mx(2); mn(1) mn(2); mx(1) mn(2); mx(1) mx(2)];
```

which we overlay on the image in red

```
>> drawpolygon("Position",p2,Color="r",FaceAlpha=0);
```

The sets of points `p1` and `p2` are projections of world points that lie approximately in a plane, so we can compute a homography

```
>> H = fitgeotform2d(p1,p2,"projective");
>> H.A
ans =
    1.3712    0.3333  -120.9807
   -0.0629    1.7132   -71.6072
   -0.0002    0.0014    1.0000
```

that will transform the vertices of the yellow trapezoid to the vertices of the red rectangle ◄

A homography can also be computed from four lines in the plane, for example, the building's edges.

$$\tilde{p}_2 \simeq \mathbf{H}\tilde{p}_1 \ .$$

That is, the homography maps image coordinates from the distorted keystone shape to an undistorted rectangular shape.

We can apply this homography to the coordinate of every pixel in an output image to warp the input image. We use the generalized image warping function

```
>> imWarped = imwarp(im,H);
>> imshow(imWarped)
```

Fig. 14.41 Photograph taken from the ground shows the effect of foreshortening which gives the building a trapezoidal appearance (also known as keystone distortion). Four points on the approximately planar face of the building have been manually picked as indicated by the yellow shading (Notre Dame de Paris). The undistorted shape is indicated by the red rectangle

Fig. 14.42 A fronto-parallel view synthesized from ■ Fig. 14.41. The image has been transformed so that the marked points become the corners of a rectangle in the image

and the result shown in ■ Fig. 14.42 is a synthetic fronto-parallel view. This is equivalent to the view that would be seen by a camera high in the air with its optical axis normal to the face of the cathedral. However, points that are not in the plane, such as the left-hand side of the right bell tower have been distorted. The black pixels in the output image are due to the corresponding pixel coordinates not being present in the input image.

In addition to creating this synthetic view we can decompose the homography to recover the camera motion from the actual to the virtual viewpoint and also the surface normal of the cathedral. As we saw in ▶ Sect. 14.2.4, we need to determine the camera calibration matrix so that we can convert the projective homography

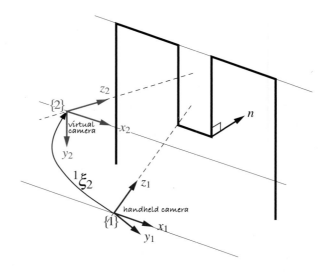

□ Fig. 14.43 Notre-Dame example showing the two camera coordinate frames. The blue frame {1} is that of the camera that took the image, and the red frame {2} is the viewpoint for the synthetic fronto-parallel view

to a Euclidean homography. We obtain the focal length from the metadata in the EXIF-format file that holds the image

```
>> md = imfinfo("notre-dame.jpg");
>> f = md.DigitalCamera.FocalLength
f =
    7.4000
```

which is in units of millimeters, and the sensor of this camera is known to be 7.18×5.32 mm. We create a camera intrinsics object

```
>> pixelDims = [7.18 5.32]./size(im,[2 1]); % in mm
>> focalLenInPix = f./pixelDims;
>> imSize = size(im,[1 2]);
>> principalPoint = imSize/2+0.5; % +0.5 for exact middle
>> camIntrinsics = cameraIntrinsics(focalLenInPix, ...
>>    principalPoint,imSize)
camIntrinsics =
  cameraIntrinsics with properties:
            FocalLength: [659.6100 592.5564]
         PrincipalPoint: [213.5000 320.5000]
              ImageSize: [426 640]
         RadialDistortion: [0 0]
    TangentialDistortion: [0 0]
                   Skew: 0
                      K: [3×3 double]
```

Now we use the camera model to decompose the Euclidean homography

```
>> pose = estrelpose(H,camIntrinsics,p1,p2)
pose =
  rigidtform3d with properties:
    Dimensionality: 3
                R: [3×3 double]
      Translation: [0.1928 -0.9221 0.3356]
                A: [4×4 double]
```

which returns a solution for ${}^1\xi_2$. The input points p1 and p2 to the estrelpose function are used to determine which of the multiple possible solutions is physically realizable. The coordinate frames for this example are sketched in □ Fig. 14.43 and show the actual and virtual camera poses. The camera translation vector, which is not to scale but has the correct sign, ▶ is dominantly in the negative y- and positive z-direction with respect to the frame {1}. The rotation in ZYX-angle form

See Malis and Vargas (2007).

```
>> rad2deg(tform2eul(pose.A))
ans =
    2.1269   -5.4791   -34.1974
```

indicates that the camera needs to be pitched downward (pitch is rotation about the camera's x-axis) by 34 degrees to achieve the attitude of the virtual camera. The normal to the frontal plane of the church n is defined with respect to {1} and is essentially in the camera z-direction as expected.

14.8.2 Image Mosaicing

Mosaicing or image stitching is the process of creating a large-scale composite image from a number of overlapping images. It is commonly applied to drone and satellite images to create a continuous single image of the Earth's surface. It can also be applied to images of the ocean floor captured from downward looking cameras on an underwater robot. The panorama generation software supplied with, or built into, digital cameras and smart phones is another example of mosaicing.

The input to the mosaicing process is a sequence of overlapping images. ▶ It is not necessary to know the camera calibration parameters or the pose of the camera where the images were taken – the camera can rotate arbitrarily between images and the scale can change.

As a rule of thumb images should overlap by 60% of area in the forward direction and 30% sideways.

We will illustrate the basic concepts of mosaicing with a real example using the pair of images

```
>> fetchExampleData("Mosaicing"); % Download example data
>> mosaicFolder = fullfile(rvctoolboxroot,"examples","mosaic");
>> im1 = imread(fullfile(mosaicFolder,"aerial2-01.png"));
>> im2 = imread(fullfile(mosaicFolder,"aerial2-02.png"));
```

which are each 1280×1024. We create an empty composite image that is 1200×1500

```
>> composite = zeros(1200,1500,"uint8");
```

that will hold the mosaic for this pair of images. The essentials of the mosaicing process are shown in ◻ Fig. 14.44.

The first image is easy and we simply paste it into the top left corner of the composite image as shown in red in ◻ Fig. 14.44. The next image, shown in blue,

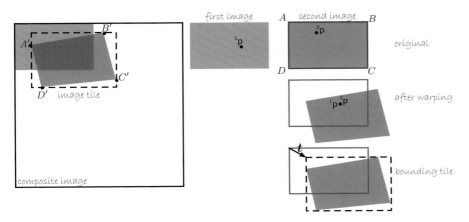

◻ **Fig. 14.44** The first image in the sequence is shown in red, the second in blue. The second image is warped into the image tile and then blended into the composite image

◻ Fig. 14.45 **a** Two mosaic images averaged together. The boundaries between tiles are clearly visible; **b** In this mosaic, the second image was carefully inserted to cleanly connect the tile boundaries

is more complex and needs to be rotated, scaled and translated so that it correctly overlays the red image.

For this problem, we assume that the scene is planar. This means that we can use a homography to relate the various camera views. The first step is to identify common feature points which are known as tie points, and we use now familiar tools

```
>> pts1 = detectSURFFeatures(im1);
>> [f1,pts1] = extractFeatures(im1,pts1);
>> pts2 = detectSURFFeatures(im2);
>> [f2,pts2] = extractFeatures(im2,pts2);
>> idxPairs = matchFeatures(f1,f2,Unique=true);
>> matchedPts1 = pts1(idxPairs(:,1));
>> matchedPts2 = pts2(idxPairs(:,2));
```

and then we use RANSAC to estimate the homography

```
>> rng(0) % set random seed for reproducibility of results
>> H = estgeotform2d(matchedPts2,matchedPts1, ...
>>     "projective",Confidence=99.9,MaxNumTrials=2000);
```

which maps ${}^{1}\tilde{\boldsymbol{p}}$ to ${}^{2}\tilde{\boldsymbol{p}}$. Now we wish to map ${}^{2}\tilde{\boldsymbol{p}}$ to its corresponding coordinates in the first image

$$
{}^{1}\tilde{\boldsymbol{p}} \simeq \mathbf{H}^{-1} \, {}^{2}\tilde{\boldsymbol{p}}
$$

We do this for every pixel in the new image by warping

```
>> refObj = imref2d(size(composite));
>> tile = imwarp(im2,H,OutputView=refObj);
```

The bounding box of the tile is computed by applying the homography to the image corners $A = (1, 1)$, $B = (W, 1)$, $C = (W, H)$ and $D = (1, H)$, where W and H are the width and height respectively, and finding the bounds in the u- and v-directions.

The fill values used by imwarp can be changed using FillValues parameter.

This time, we also used imref2d object to designate the size of our output space, otherwise imwarp would tightly clip around the rotated image. Instead, we wish to have the image placed within our composite canvas. As shown in ◻ Fig. 14.44 the warped blue image falls outside the bounds of the original blue image. ◄ The image shown with a dashed black line is referred to as a *tile*. In general, not every pixel in the tile has a corresponding point in the input image and those pixels are set to zero by default. ◄

Next, the tile has to be *blended* into the composite mosaic image

```
>> blended = imfuse(im1,tile,"blend");
>> imshow(blended)
```

and the result is shown in ◻ Fig. 14.45a. This averaged our tiles which were placed in the larger canvas. The overlapping area became brighter and for illustrative purposes, this shows exactly the boundaries of the overlap.

Fig. 14.46 Example image mosaic. At the bottom of the frame we can clearly see five overlapping views of the airport runway which shows accurate alignment between the frames

Instead of averaging the tiles together, this time we will insert the next tile into the canvas through a boolean mask that we warp to match the region of insertion

```
>> blender = vision.AlphaBlender(Operation="Binary mask", ...
>>     MaskSource="Input port");
>> mask1 = true(size(im1,[1 2]));
>> composite = blender.step(composite,im1,mask1);
>> tileMask = imwarp(true(size(im2,[1 2])),H,OutputView=refObj);
>> composite = blender.step(composite,tile,tileMask);
>> imshow(composite)
```

and the result is shown in ◘ Fig. 14.45b. This time, the transitions between the tiles are hardly visible. Generally, if the images were taken with the same exposure then the edges of the tiles would not be visible. If the exposures were different, the two sets of overlapping pixels have to be analyzed to determine the average intensity offset and scale factor which can be used to correct the tile before blending – a process known as tone matching.

The full code for this example is in `examples/mosaic.m`

```
>> mosaic
```

In addition to the steps described earlier, the example code shows how to compute the required canvas size to tightly fit the mosaic and it loops over the images, combining homography transforms to create cumulative transform needed to insert each tile. The output of running the entire loop over 10 images is shown in ◘ Fig. 14.46 and we can clearly see the images overlaid with accurate alignment.

We also need to consider the effect of points in the image that are not in the ground plane such as those on a tall building. An image taken from directly overhead will show just the roof of the building, but an image taken from further away will be an oblique view that shows the side of the building. In a mosaic, we want to create the illusion that we are directly above every point in the image so we should not see the sides of any building. This type of image is known as an orthophoto, and unlike a perspective view, where rays converge on the camera's focal point, the rays are all parallel which implies a viewpoint at infinity. ▶ At every pixel in the

Google Earth sometimes provides an imperfect orthophoto. When looking at cities we might see oblique views of buildings.

composite image we can choose a pixel from any of the overlapping tiles. To best approximate an orthophoto we should choose the pixel that is closest to overhead, that is, prior to warping the pixel that was closest to the principal point.

In photogrammetry, this type of mosaic is referred to as an uncontrolled digital mosaic since it does not use explicit control points – manually identified corresponding features in the images. ◄

Other related examples can be found online for panoramic image stitching at ► https://sn.pub/Bd0N3r and image registration at ► https://sn.pub/bEhvLD.

14.8.3 Visual Odometry

A common problem in robotics is to estimate the distance a robot has traveled, and this is a key input to all of the localization algorithms discussed in ► Chap. 6. For a wheeled robot we can use information from the wheel encoders, but these are subject to random errors (slippage) as well as systematic errors (imprecisely known wheel radius). However, for a flying or underwater robot the problem of odometry is much more difficult. Visual odometry (VO) is the process of using information from consecutive images to estimate the robot's relative motion from one camera image to the next.

We load a sequence of images taken from a car driving along a road

```
>> fetchExampleData("VisualOdometry"); % Download example data
>> visodomFolder = fullfile(rvctoolboxroot,"examples","visodom");
>> left = imageDatastore(fullfile(visodomFolder,"left"));
>> size(left.Files,1)
ans =
   251
```

These images were already rectified and contain a black border. ◄ They are also stored in a `uint16` container and do not span the full dynamic range. We adjust their contrast, convert them to `uint8` and crop out the black border. All of this is accomplished by transforming the original `ImageDatastore` object into a new one which will invoke our anonymous preprocessing function

```
>> tformFcn = @(in)imcrop(im2uint8(imadjust(in)),[17 17 730 460]);
>> left = left.transform(tformFcn);
```

on every `read()` operation. We can see that any transformation, however complex, can be applied to a set of images using this approach. The image sequence can be displayed as an animation.

```
>> while left.hasdata()
>>    imshow(left.read())
>> end
>> left.reset() % rewind to first image
```

For each frame we can detect point features and overlay these as an animation

```
>> while left.hasdata();
>>    frame = left.read();
>>    imshow(frame), hold on
>>    features = detectORBFeatures(frame);
>>    features.plot(ShowScale=false), hold off
>> end
>> left.reset();
```

and a single frame of this sequence is shown in ◼ Fig. 14.47. For variety, and computational efficiency, we use ORB features which are commonly used for this type of online application. We see that the point features *stick* reliably to points in the world over many frames, and show a preference for the corners of signs and cars, as well as the edges of trees. The motion of features in the image is known

The principles illustrated here can also be applied to the problem of image stabilization. The homography is used to map features in the new image to the location they had in the previous image.

This is the left image sequence from EISATS Bridge sequence in dataset 4 (Klette et al. 2011) located at ► https://sn.pub/Tf7f9F.

Frame 15

◻ Fig. 14.47 Frame number 15 from the `bridge-1` image sequence with overlaid features (Image from .enpeda.. project, Klette et al. 2011)

as optical flow, and is a function of the camera's motion through the world and the 3-dimensional structure of the world. ▶

To understand the 3-dimensional world structure we will use sparse stereo as discussed in ▶ Sect. 14.3. The vehicle in this example was fitted with a stereo camera and we load the corresponding right-camera images ▶

```
>> right = imageDatastore(fullfile(visodomFolder,"right"));
>> right = right.transform(tformFcn);
```

For each pair of left and right images we extract features, and determine correspondence by robustly matching features using descriptor similarity and the epipolar constraint implied by a fundamental matrix. Next, we compute horizontal disparity between corresponding features and, assuming the cameras are fully calibrated, we triangulate the image-plane coordinates to determine the world coordinates of the landmark points with respect to the left-hand camera on the vehicle.

We could match the 3D point clouds at the current and previous time step using a technique like iterative closest point (ICP) in order to determine the camera pose change. This is the so-called 3D-3D approach to visual odometry and, while the principle is sound, it works poorly in practice. Firstly, some of the 3D points may be on other moving objects and this violates the assumption of ICP that the sensor or the object moves, but not both. Secondly, the estimated range to distant points is quite inaccurate since errors in estimated disparity become significant when disparity is small.

An alternative approach, 3D-2D matching, projects the 3D points at the previous time step into the current image and finds the camera pose that minimizes the error with respect to the observed feature coordinates – this is bundle adjustment.

Typically, this is done for just one image and we will choose the left image. To establish correspondence of features over time we find correspondences between left-image features that had a match with the right image, and a match with features from the current left image – again enforcing an epipolar constraint. We now know the correspondence between points in the three views of the scene as shown in ◻ Fig. 14.48. The three views are current frame left and right images plus the previous frame left image.

At each time step we set up a bundle adjustment problem that has two cameras and a number of landmarks determined from stereo triangulation. The first camera

The magnitude of optical flow – the speed of a world point on the image plane – is proportional to camera velocity divided by the distance of the world point from the camera. Optical flow therefore has a scale ambiguity – a camera moving quickly through a world with distant points yields the same flow magnitude as a slower camera moving past closer points. To resolve this, we need to use additional information. For example, if we knew that the points were on the road surface, that the road was flat, and the height of the camera above the road then we can resolve this unknown scale. However, this assumption is quite strict and would not apply for something like a drone moving over unknown terrain.

Both sets of images are rectified, the image cropping is to exclude the warping artifacts.

Stereo matching   **Left frame temporal matching**

a   b

■ **Fig. 14.48**   Feature correspondence for visual odometry. **a** Matching features for a stereo pair at the current time step; **b** matching features for left frame's current and previous time step; both displays show epipolar consistent correspondences between the images (Images from .enpeda. project, Klette et al. 2011)

is associated with the previous time step and is fixed at the reference frame origin. The second camera is associated with the current time step and would be expected to have a translation in the positive z-axis direction. We could obtain an initial estimate of the second camera's pose by estimating and decomposing an essential matrix, but we will instead set it to the origin which we later expect to be moved by applying bundle adjustment.

The details can be found in the example script

```
>> visodom
```

which displays graphics like ■ Fig. 14.48 for every frame.

The final results for camera translation between frames are shown in ■ Fig. 14.49a, and we notice a value approaching 0.5 m at each time step. There are also some anomalous high values and these have two causes. Firstly, the bundle adjustment process has failed to converge properly as indicated by the per-frame final error shown in ■ Fig. 14.49b. The median error is around 0.76 pixels and 95% of the frames have a bundle-adjustment error of less than 1 pixels. However, the maximum is around 2 pixels and we could exclude such bad results and perhaps infer the translation from the previous value. The likely source of error are incorrect point correspondences. Bundle adjustment assumes that all points in the world are fixed but in this sequence there are numerous moving objects. We used the epipolar constraint between current and previous frame to ensure that only feature points consistent with a moving camera and a fixed world are in the inlier set. There is also the motion of cars on the overpass bridges that is not consistent with the motion induced by the car's own motion. A more sophisticated bundle adjustment algorithm would detect and reject such points. Finally, there is a preponderance of points in the top part of the scene which are quite distant from the cameras. A more sophisticated approach to feature detection would choose features more uniformly spread over the image.

The second cause of error is a common one when using video data for robots. The clue is that a number of camera displacements are suspiciously close to exactly twice the median value. Each image in the sequence was assigned a timestamp when it was received by the computer and those timestamps can be loaded from the left-image archive

```
>> ts = load(fullfile(visodomFolder,"timestamps.dat"));
```

and the difference between timestamps

```
>> plot(diff(ts))
```

Fig. 14.49 Visual odometry results. **a** Estimated norm of camera displacement; **b** bundle adjustment final error per frame; **c** image time stamp

is shown in ◘ Fig. 14.49c. We see that while the median time between images is 44.6 ms, there are many spikes where the interval is twice that. The computer logging the images has skipped a frame, perhaps it was unable to write image data to memory or disk as quickly as it was arriving. So the interval between the frames was twice as long, the vehicle traveled twice as far, and the spikes on our estimated displacement are in fact correct. This is not an uncommon situation – in a robot system all data should be timestamped and timestamps should be checked to detect problems like this. The displacements in ◘ Fig. 14.49a that have this timestamp issue are annotated with a red cross. The median velocity over these frames is $11.53 \, \text{m s}^{-1}$ or around $40 \, \text{km h}^{-1}$.

For a vehicle or ground robot the estimated displacements in 3-dimensional space over time are not independent – they are constrained by vehicle's kinodynamic model as discussed in ▶ Chap. 4. We could use this model to smooth the results and discount erroneous velocity estimates. If the bundle adjuster included constraints on camera pose we could set the weighting to penalize infeasible motion in the lateral and vertical directions as well as roll and pitch motion.

14.9 **Wrapping Up**

This chapter has covered many topics with the aim to demonstrate a multiplicity of concepts that are of use in real robotic vision systems. There have been two common threads through this chapter. The first was the use of point features which correspond to distinctive points in the world, and allow matching of these points between images taken from different viewpoints. The second thread was the loss of scale in the perspective projection process, and techniques based on additional sources of information to recover scale, for example stereo vision, structured light or bundle adjustment.

We extended the geometry of single camera imaging to the case of two cameras, and showed how corresponding points in the two images are constrained by the fundamental matrix. We showed how the fundamental matrix can be estimated from image data, the effect of incorrect data association, and how to overcome this using the RANSAC algorithm. Using camera intrinsic parameters the essential matrix can be computed and then decomposed to give the camera motion between the two views, but the translation has an unknown scale factor. With some extra information such as the magnitude of the translation, the camera motion can be estimated completely. Given the camera motion, then the 3-dimensional coordinates of points in the world can be estimated.

For the special case where world points lie on a plane they *induce* a homography that is a linear mapping of image points between images. The homography can be used to detect points that do not lie in the plane, and can be decomposed to give the camera motion between the two views (translation again has an unknown scale factor) and the normal to the plane.

If the fundamental matrix is known, then a pair of overlapping images can be rectified to create an epipolar-aligned stereo pair, and dense stereo matching can be used to recover the world coordinates for every point. Errors due to effects such as occlusion and lack of texture were discussed, as were techniques to detect these situations.

We used bundle adjustment to solve the structure and motion estimation problem – using 2D measurements from a set of images of the scene to recover information related to the 3D geometry of the scene as well as the locations of the cameras. Stereo vision is a simple case where the motion is known – fixed by the stereo baseline – and we are interested only in structure. The visual odometry problem is complementary and we were interested only in the motion of the camera, not the scene structure.

We used point clouds, including colored point clouds, to represent and visualize the 3-dimensional world. We also demonstrated some operations on point cloud data such as outlier removal, downsampling, normal estimation, plane fitting, cylinder fitting, and registration. We only scratched the surface of what can be done with point clouds using the Computer Vision Toolbox™.

Finally, these multi-view techniques were used for application examples such as perspective correction, mosaic creation, and visual odometry.

It is also worth noting that many Computer Vision Toolbox™ functions can be used inside Simulink® and support C as well as GPU automatic code generation for hardware.

14.9.1 **Further Reading**

3-dimensional reconstruction and camera pose estimation have been studied by the photogrammetry community since the mid nineteenth century, see ▶ Exc. 13.12. 3-dimensional computer vision or *robot vision* has been studied by the computer

◻ Table 14.1 Rosetta stone. Summary of notational differences between two other popular textbooks and this book

Object	Hartley & Zisserman 2003	Ma et al. 2003	This book
World point	$\mathbf{X}$	P	P
Image plane point	$\mathbf{x}, \mathbf{x}'$	x_1, x_2	$^1\mathrm{p}, ^2\mathrm{p}$
i^{th} image plane point	$\mathbf{x}_i, \mathbf{x}'_i$	x_1^i, x_2^i	$^1\mathrm{p}_i, ^2\mathrm{p}_i$
Camera motion	$\mathbf{R}, \mathbf{t}$	R, T	$\mathbf{R}, t$
Normalized coordinates	$\mathbf{x}, \mathbf{x}'$	x_1, x_2	$(\bar{u}, \bar{v})$
Camera matrix	P	Π	$\mathbf{C}$
Homogeneous quantities	$\mathbf{x}, \mathbf{X}$	x, P	$\tilde{p}, \tilde{P}$
Homogeneous equivalence	$\mathbf{x} = \mathbf{PX}$	$\lambda x = \Pi P$ $x \sim \Pi P$	$\tilde{p} \simeq \mathbf{C}\tilde{P}$

vision and artificial intelligence communities since the 1960s. This book follows the language and nomenclature associated with the computer vision literature, but the photogrammetric literature can be comprehended with only a little extra difficulty. The similarity of a stereo camera to our own two eyes is very striking, and while we do make strong use of stereo vision it is not the only technique we use to infer distance (Cutting 1997).

Significant early work on multi-view geometry was conducted at laboratories such as Stanford, SRI International, MIT AI laboratory, CMU, JPL, INRIA, Oxford and ETL Japan in the 1980s and 1990s and led to a number of text books being published in the early 2000s. The definitive references for multiple-view geometry are Hartley and Zisserman (2003) and Ma et al. (2003). These books present quite different approaches to the same body of material. The former takes a more geometric approach while the latter is more mathematical. Unfortunately, they use quite different notation, and each differs from the notation used in this book – a summary of the important notational elements is given in ◻ Tab. 14.1. These books all cover feature extraction (using Harris point features, since they were published before scale invariant feature detectors such as SIFT or SURF were developed); the geometry of one, two and N views; fundamental and essential matrices; homographies; and the recovery of 3-dimensional scene structure and camera motion through offline batch techniques. Both provide the key algorithms in pseudo-code and have some supporting MATLAB code on their associated web sites. The slightly earlier book by Faugeras et al. (2001) covers much of the same material using a fairly mathematical approach and with different notation again. The older book by Faugeras (1993) focuses on sparse stereo from line features. The book by Szeliski (2022 ▸ https://szeliski.org/Book/) provides a very readable and deeper discussion of the topics in this chapter.

SURF, SIFT and other feature detectors were previously discussed in ▸ Sects. 12.3.2 and 14.1. The performance of feature detectors and their matching performance is covered in Mikolajczyk and Schmid (2005) which reviews a number of different feature descriptors. Arandjelović and Zisserman (2012) discuss some important points when matching feature descriptors.

The RANSAC algorithm described by Fischler and Bolles (1981) is the workhorse of all the feature-based methods discussed in this chapter but fails with very small inlier ratios. Subsequent work includes vector field consensus (VFC) by Ma et al. (2014) and progressive sample consensus (PROSAC) by Chum and Matas (2005).

The term fundamental matrix was defined in the thesis of Luong (1992). The book by Xu and Zhang (1996) is a readable introduction to epipolar geometry.

Epipolar geometry can also be formulated for nonperspective cameras in which case the epipolar line becomes an epipolar curve (Mičušík and Pajdla 2003; Svoboda and Pajdla 2002). For three views the geometry is described by the trifocal tensor $\mathcal{T}$ which is a $3 \times 3 \times 3$ tensor with 18 degrees of freedom that relates a point in one image to epipolar lines in two other images (Hartley and Zisserman 2003; Ma et al. 2003). An important early paper on epipolar geometry for an image sequence is Bolles et al. (1987).

The essential matrix was first described a decade earlier in a letter to Nature (Longuet-Higgins 1981) by the theoretical chemist and cognitive scientist Christopher Longuet-Higgins (1923–2004). The paper describes a method of estimating the essential matrix from eight corresponding point pairs. The decomposition of the essential matrix was first described in Faugeras (1993, § 7.3.1) but is also covered in the texts by Hartley and Zisserman (2003) and Ma et al. (2003). In this chapter, we have estimated camera motion by first computing the essential matrix and then decomposing it. The first step requires at least eight pairs of corresponding points but algorithms such as Nistér (2003), Li and Hartley (2006) compute the motion directly from just five pairs of points. Decomposition of a homography is described by Faugeras and Lustman (1988), Hartley and Zisserman (2003), Ma et al. (2003), and the comprehensive technical report by Malis and Vargas (2007).

Stereo cameras and stereo matching software are available today from many sources and can provide high-resolution depth maps at more than 10 Hz on standard computers. In the 1990s this was challenging, and custom hardware including FPGAs was required to achieve real-time operation (Woodfill and Von Herzen 1997; Corke et al. 1999). The application of stereo vision for planetary rover navigation is discussed by Matthies (1992). More than two cameras can be used, and multi-camera stereo was introduced by Okutomi and Kanade (1993) and provides robustness to problems such as the picket fence effect.

Brown et al. (2003) provide a readable review of stereo vision techniques with a focus on real-time issues. An old, but clearly written, book on the principles of stereo vision is Shirai (1987). Scharstein and Szeliski (2002) consider the stereo process as four steps: matching, aggregation, disparity computation and refinement. The cost and performance of different algorithms for each step are compared. The dense stereo matching method presented in ▶ Sect. 14.4 is a very conventional correlation-based stereo algorithm, and would be described as: SSD matching, box filter aggregation, and winner takes all. The disparity is computed independently at each pixel, but for real scenes adjacent pixels belong to the same surface and disparity will be quite similar – this is referred to as the *smoothness constraint*. Ensuring smoothness can be achieved using Markov random fields (MRFs), total variation with regularizers (Pock 2008), or more efficient semi-global matching (SGM) algorithms (Hirschmüller 2008). The very popular library for efficient large-scale stereo matching (LIBELAS) by Geiger et al. (2010) uses an alternative to global optimization that provides fast and accurate results for a variety of indoor and outdoor scenes. Stereo vision involves a significant amount of computation but there is considerable scope for parallelization using multiple cores, MIMD instruction sets, GPUs, custom chips and FPGAs. The use of nonparametric local transforms is described by Zabih and Woodfill (1994) and Banks and Corke (2001). Additionally, deep learning can be applied to the stereo matching problem (2016). Deep networks can also estimate depth and visual odometry directly from monocular image sequences (Garg et al. 2016; Zhan et al. 2018).

The ICP algorithm (Besl and McKay 1992) is used for a wide range of applications from robotics to medical imaging. ICP is fast but determining the correspondences via nearest neighbors is an expensive $O(N^2)$ operation. This can be sped up by using a kd-tree to organize the data points. Many variations have been developed that make the approach robust to outlier data and to improve computational speed for large datasets. Salvi et al. (2007) provide a recent review and comparison of some different algorithms. Determining the relative orientation between two sets of points is a classical problem and the SVD approach used here is described by Arun et al. (1987). Solutions based on quaternions and orthonormal rotation matrices have been described by Horn (Horn et al. 1988; Horn 1987). Newer techniques include Normal Distribution Transform (NDT), generalized ICP, and Coherent Point Drift (CPD) algorithm which is particularly useful for registering deformable point clouds. All of these techniques are part of the Computer Vision Toolbox™.

Structure from motion (SfM), the simultaneous recovery of world structure and camera motion, is a classical problem in computer vision. Two useful review papers are by Huang and Netravali (1994) which provides a taxonomy of approaches, and Jebara et al. (1999). Broida et al. (1990) describe an early recursive SfM technique for a monocular camera sequence using an EKF where each world point is represented by its (X, Y, Z) coordinate. McLauchlan provides a detailed description of a variable-length state estimator for SfM (McLauchlan 1999). Azarbayejani and Pentland (1995) present a recursive approach where each world point is parameterized by a scalar, its depth with respect to the first image. A more recent algorithm with bounded estimation error is described by Chiuso et al. (2002) and also discusses the problem of scale variation. The MonoSlam system by Davison et al. (2007) is an impressive monocular SfM system that maintains a local map that includes features even when they are not currently in the field of view. A more recent extension by Newcombe et al. (2011) performs camera tracking and dense 3D reconstruction from a single moving RGB camera. The application of SfM to large-scale urban mapping is becoming increasing popular and Pollefeys et al. (2008) describe a system for offline processing of large image sets. The ORB-SLAM family of visual SLAM algorithms (Campos 2021) is a powerful and popular way to compute the 3D path of a camera given a sequence of RGB, RGBD or stereo images. An example implementation of ORB-SLAM is included in the Computer Vision Toolbox™.

Bundle adjustment or structure from motion (SfM) is a big field with a large literature that cover many variants of the problem, for example robustness to outliers, and specific applications and camera types. Classical introductions include Triggs et al. (2000) and Hartley and Zisserman (2003). Recent theses by Warren (2015), Sünderhauf (2012) and Strasdat (2012) are comprehensive and readable. Unfortunately, every reference uses different notation. Estimating the camera matrix for each view, computing a projective reconstruction, and then upgrading it to a Euclidean reconstruction is described by Hartley and Zisserman (2003) and Ma et al. (2003). A sparse bundle adjustment algorithm used in the Computer Vision Toolbox™ is described by Lourakis (2009).

The SfM problem can be simplified by using stereo rather than monocular image sequences (Molton and Brady 2000; Zhang et al. 1992), or by incorporating inertial data (Strelow and Singh 2004). A readable two-part tutorial introduction to visual odometry (VO) is Scaramuzza and Fraundorfer (2011) and Fraundorfer and Scaramuzza (2012). Visual odometry is discussed by Nistér et al. (2006) using point features and monocular or stereo vision. Maimone et al. (2007) describe experience with stereo-camera VO on the Mars rover and Corke et al. (2004) describe monocular catadioptric VO for a prototype planetary rover.

Mosaicing is a process as old as photography. In the past, it was highly skilled and labor intensive requiring photographs, scalpels and sandpaper. The surface of the Moon and nearby planets was mosaiced manually in the 1960s using imagery sent back by robotic spacecraft. High-quality offline mosaicing tools are available for creating panoramas, for example the Hugin open source project ▶ http://hugin. sourceforge.net and the proprietary AutoStitch, ▶ https://autostitch.en.softonic. com/. Mosaicing is included in nearly all high to mid-range smartphones.

Image sequence analysis is the core of many real-time robotic vision systems. Early work on real-time feature tracking across frames, such as Hager and Toyama (1998) and Lucas and Kanade (1981), was limited by available computation. It was typically based on the computationally cheaper Harris detectors or the pyramidal

Kanade-Lucas-Tomasi (KLT) tracker. Today, efficient implementations of complex feature detectors like SIFT and SURF, perhaps using GPU hardware, are capable of real-time performance.

14.9.2 Resources

The field of computer vision has progressed through the availability of standard datasets. These have enabled researchers to quantitatively compare the performance of different algorithms on the same data. One of the earliest collections of stereo image pairs was the JISCT dataset (Bolles et al. 1993). The more recent Middlebury dataset (Scharstein and Szeliski 2002) at ▶ http://vision.middlebury.edu/stereo provides an extensive collection of stereo images, at high resolution, taken at different exposure settings and including ground truth data. Stereo images from various NASA Mars rovers are available online as left+right pairs or encoded in anaglyphs. Motion datasets include people moving inside a building ▶ http://homepages.inf.ed.ac.uk/rbf/CAVIARDATA1, traffic scenes ▶ http://i21www.ira.uka.de/image_sequences, and from a moving vehicle ▶ http://www.mi.auckland.ac.nz/EISATS.

The popular LIBELAS library (▶ http://www.cvlibs.net/software/libelas) for large-scale stereo matching supports parallel processing using OpenMP and has MATLAB and ROS interfaces. Various stereo vision algorithms are compared for speed and accuracy at the KITTI (▶ http://www.cvlibs.net/datasets/kitti/eval_scene_flow.php) and Middlebury (▶ http://vision.middlebury.edu/stereo/eval3) benchmark sites.

The Epipolar Geometry Toolbox (Mariottini and Prattichizzo 2005) for MATLAB by Gian Luca Mariottini and Domenico Prattichizzo is available at ▶ http://egt.dii.unisi.it and handles perspective and catadioptric cameras. Andrew Davison's monocular visual SLAM system (MonoSLAM) for C and MATLAB is available at ▶ http://www.doc.ic.ac.uk/~ajd/software.html.

The sparse bundle adjustment software by Lourakis (▶ http://users.ics.forth.gr/~lourakis/sba) is an efficient C implementation that is widely used and has a MATLAB and OpenCV wrapper. Computer Vision Toolbox™ contains its own implementation of the sparse bundle adjustment algorithm by Lourakis and it also provides an option to use g^2o. One application is Bundler (▶ http://www.cs.cornell.edu/~snavely/bundler) which can perform matching of points from thousands of cameras over city scales and has enabled reconstruction of cities such as Rome (Agarwal et al. 2014), Venice and Dubrovnik. Some of these large-scale datasets are available from ▶ http://grail.cs.washington.edu/projects/bal and ▶ http://www.robots.ox.ac.uk/~vgg/data/data-mview.html. Other open source solvers that can be used for sparse bundle adjustment include g^2o, SSBA and CERES, all implemented in C++. g^2o by Kümmerle et al. (2011) (▶ https://github.com/RainerKuemmerle/g2o) can also be used to solve SLAM problems. SSBA by Christopher Zach is available at ▶ https://github.com/chzach/SSBA. The CERES solver from Google (▶ http://ceres-solver.org) is a library for modeling and solving large complex optimization problems on desktop and mobile platforms.

Open3D (▶ http://www.open3d.org) is a large-scale, open and standalone package for 2D/3D image and point cloud processing with support for feature detectors and descriptors, 3D registration, kd-trees, shape segmentation (Zhou et al. 2018). The Point Data Abstraction Library (PDAL) (▶ http://www.pdal.io) is a library and set of Unix command line tools for manipulating point cloud data.

Point clouds can be stored in a number of common formats that can include optional color information as well as point coordinates. Point Cloud Data (PCD) files were defined by the earlier Point Cloud library (PCL) project (▶ https://pointclouds.org). Polygon file format (PLY) files are designed to describe meshes but can be used to represent an unmeshed point cloud, and there are a number of great visualizers such as MeshLab and potree. Open3D and PDAL allow reading and writing many point cloud file formats from Python. LAS format is widely used for large point clouds and colored point clouds, is is a binary format and LAZ format is a compressed version. Cloud Compare (▶ https://www.danielgm.net/cc) is a cross-platform, open-source, GUI-based tool for displaying, comparing, and editing very large point clouds stored in a variety of formats.

A song about the fundamental matrix can be found at ▶ http://danielwedge.com/fmatrix/.

14.9.3 Exercises

1. Point features and matching (▶ Sect. 14.1). Examine the cumulative distribution of feature strength for Harris and SURF features. What is an appropriate way to choose strong features for feature matching?
2. Feature matching. We could define the quality of descriptor-based feature matching in terms of the percentage of inliers after applying RANSAC.
   a) Take any image. We will match this image against various transforms of itself to explore the robustness of SURF and Harris features. The transforms are: *(a)* scale the intensity by 70%; *(b)* add Gaussian noise with standard deviation of 0.05, 0.5 and 2 gray values; *(c)* scale the size of the image by 0.9, 0.8, 0.7, 0.6 and 0.5; *(d)* rotate by 5, 10, 15, 20, 30, 40 degrees.
   b) Is there any correlation between outlier matches and feature strength?
3. Write the equation for the epipolar line in image two, given a point in image one.
4. Show that the epipoles are the null space of the fundamental matrix.
5. Can you determine the camera matrix **C** for camera two given the fundamental matrix and the camera matrix for camera one?
6. Estimating the fundamental matrix (▶ Sect. 14.2.3)
   a) For the Eiffel tower example observe the effect of varying the parameters to RANSAC. Repeat this with just the top 50% of SURF features after sorting by feature strength.
   b) What is the probability of drawing 8 inlier points in a random sample (without replacement) from N inliers and M outliers?
7. Epipolar geometry (▶ Sect. 14.2)
   a) Create two central cameras, one at the origin and the other translated in the x-direction. For a sparse fronto-parallel grid of world points display the family of epipolar lines in image two that correspond to the projected points in image one. Describe these epipolar lines? Repeat for the case where camera two is translated in the y- and z-axes and rotated about the x-, y- and z-axes. Repeat this for combinations of motion such as x- and z-translation or x-translation and y-rotation.
   b) The example of ◼ Fig. 14.15 has epipolar lines that slope slightly upward. What does this indicate about the two camera viewpoints?
8. Essential matrix (▶ Sect. 14.2.2)
   a) Create a set of corresponding points for a camera undergoing pure rotational motion, and compute the fundamental and essential matrix. Can you recover the rotational motion?
   b) For a case of translational and rotational motion visualize both poses that result from decomposing the essential matrix. Sketch it or use `plotCamera`.
9. Homography (▶ Sect. 14.2.4)
   a) Compute Euclidean homographies for translation in the x-, y- and z-directions and for rotation about the x-, y- and z-axes. Convert these to projective homographies and apply to a fronto-parallel grid of points. Is the

resulting image motion what you would expect? Apply these homographies as a warp to a real image such as Mona Lisa.

b) Decompose the homography of ◨ Fig. 14.14, the courtyard image, to determine the plane of the wall with respect to the camera. You will need the camera intrinsic parameters.

c) Reverse the order of the points given to `estgeotform2d` and show that the result is the inverse homography.

10. Load a reference image of this book's cover from `rvc3_cover.png`. Next, capture an image that includes the book's front cover, compute SURF or SIFT features, match them and use RANSAC to estimate a homography between the two views of the book cover. Decompose the homography to estimate rotation and translation (Hint: use `estrelpose` function). Put all of this into a real-time loop and continually display the pose of the book relative to the camera.

11. Sparse stereo (▶ Sect. 14.3)

a) Introduce a small error in the computation of the fundamental matrix. What effect does it have on the quality of the outcomes?

b) The assumed camera translation magnitude was 30 cm. Repeat for 25 and 35 cm. Are the closing error statistics changed? Can you determine what translation magnitude minimizes this error?

12. Bundle adjustment (▶ Sect. 14.3.2)

a) Vary the initial condition for the second camera, for example, set it to the identity matrix.

b) Set the initial camera translation to 3 m in the x-direction, and scale the landmark coordinates by 10×. What is the final value of the back-projection error and the second camera pose.

c) Experiment with anchoring landmarks and cameras.

d) Derive the two Jacobians **A** (hard) and **B**.

13. Derive a relationship for depth in terms of disparity for the case of verged cameras. That is, cameras with their optical axes intersecting similar to the cameras shown in ◨ Fig. 14.5.

14. Stereo vision. Using the rock piles example (◨ Fig. 14.23)

a) Use `imtool` and `imfuse` to zoom in on the disparity image and examine pixel values on the boundaries of the image and around the edges of rocks.

b) Experiment with different window sizes. What effects do you observe in the disparity image and computation time?

c) Experiment with changing the disparity range. Try `[50,90]`, `[30,90]`, `[40,80]` and `[40,100]`. What happens to the disparity image and why?

d) Display the epipolar lines on image two for selected points in image one.

15. Anaglyphs (▶ Sect. 14.5)

a) Download an anaglyph image and convert it into a pair of grayscale images, then compute dense stereo.

b) Create your own anaglyph from a stereo pair.

c) Write code to plot shapes, like squares, that appear at different depths. Or, a single shape that moves in and out of the screen.

d) Write code to plot a 3-dimensional shape like a cube.

16. Stereo vision. For a pair of identical cameras with a focal length of 8 mm, 1000×1000 pixels that are 10 μm square on an 80 mm baseline and with parallel optical axes:

a) Sketch, or write code to display, the fields of views of the camera in a plan view. If the cameras are viewing a plane surface normal to the principal axes how wide is the horizontal overlapping field of view in units of pixels?

b) Assuming that disparity error is normally distributed with $\sigma = 0.1$ pixels compute and plot the distribution of error in the z-coordinate of the reconstructed 3D points which have a mean disparity of 0.5, 1, 2, 5, 10 and 20 pixels. Draw 1000 random values of disparity, convert these to Z and plot a histogram (distribution) of their values.

17. A da Vinci on your wall. Acquire an image of a room in your house and display it using `imshow`. Select four points, using `drawpolygon`, to define the corners of the virtual frame on your wall. Perhaps use the corners of an existing rectangular feature in your room such as a window, poster or picture. Estimate the appropriate homography, warp the Mona Lisa image and insert it into the original image of your room.

18. Geometric object fitting (▶ Sect. 14.7.1)
    a) Test the robustness of the plane fitting algorithm to additive noise and outlier points.
    b) Implement an iterative approach with weighting to minimize the effect of outliers.
    c) Create a RANSAC-based plane fit algorithm that takes random samples of three points.
    d) Try fitting a cylinder using data from (▶ Sect. 14.7.1). Hint: try using the function `pcfitcylinder`.

19. ICP (▶ Sect. 14.7.2)
    a) Change the initial relative pose between the point clouds. Try some very large rotations.
    b) Explore the options available for ICP.
    c) Explore the robustness of ICP by simulating some realistic sensor errors. For example, add Gaussian or impulse noise to the data points. Or add an increasing number of spurious data points.
    d) How does matching performance vary as the number of data points increases and decreases.
    e) Discover and explore all the ICP options provided by the `pcregistericp` function.
    f) Try to register point clouds using one of the other registration techniques: `pcregisterndt`, `pcregistercpd`. Explore their parameters.

20. Perspective correction (▶ Sect. 14.8.1)
    a) Create a virtual view looking downward at 45° to the front of the cathedral.
    b) Create a virtual view from the original camera viewpoint but with the camera rotated 20° to the left.
    c) Find another real picture with perspective distortion and correct it.

21. Mosaicing (▶ Sect. 14.8.2)
    a) Run the example file `mosaic` and watch the whole mosaic being assembled.
    b) Modify the way the tile is pasted into the composite image so that pixels closest to the principal point are used.
    c) Run the software on a set of your own overlapping images and create a panorama.

22. Image stabilization can be used to virtually stabilize an unsteady camera, perhaps one that is handheld, on a drone or on a mobile robot traversing rough terrain. Capture a short image sequence $\mathbf{I}_1, \mathbf{I}_2 \cdots \mathbf{I}_N$ from an unsteady camera. For frame i, $i \geq 2$ estimate a homography with respect to frame 1, warp the image appropriately, and store it in an array. Animate the stabilized image sequence.

23. Visual odometry (▶ Sect. 14.8.3). Modify the example script to
    a) use Harris, SURF or SIFT features instead of ORB. What happens to accuracy and execution time?
    b) track the features from frame to frame and display the displacement (optical flow) vectors at each frame
    c) using a scale-invariant point feature, track points across multiple frames. You will need to create a a data structure to manage the life cycle of each point: it is first seen, it is seen again, then it is lost.
    d) ensure that features are more uniformly spread over the scene, investigate the `selectUniform` method of objects holding point features, for example, SURFPoints.

    e) Try to implement a bundle adjuster and use it example data (hard).

    f) use a Kalman filter with simple vehicle dynamics to smooth the velocity estimates or optionally skip the update step if the bundle adjustment result is poor.

    g) explore the statistics of the bundle adjustment error over all the frames. Does increasing the number of iterations help?

    h) redo the bundle adjustment with three cameras: left and right cameras at the current time step, and the left camera at the previous time step.

24. Learn about kd-trees and create your own implementation.

25. Explore the monocular ORB-SLAM example included with the Computer Vision Toolbox™ (► https://www.mathworks.com/help/vision/ug/monocular-visual-simultaneous-localization-and-mapping.html).

Robotics, Vision & Control

Contents

Chapter 15   Vision-Based Control – 665

Chapter 16   Real-World Applications – 693

This part of the book brings together much that we have learned previously, but separately, regarding robotics, robot control and computer vision. Now we consider how we can use a camera and vision system as a sensor to guide a robot to perform a task. That task might be a robotic arm reaching towards some object, or it might be a self-driving car staying within lanes perceived using a camera, or a delivery drone avoiding obstacles on its way to a dropoff point.

The part comprises two chapters. ▶ Chap. 15 discusses two classical approaches to closed-loop vision-based control, commonly known as visual servoing. Such approaches involve continuous measurement using vision to create a feedback signal which drives the robot until the visually observed error between the robot and the target is zero. Vision-based control is quite different to taking an image, determining where the target is and then moving toward it. The advantage of continuous measurement and feedback is that it provides great robustness with respect to any errors in the system.

▶ Chap. 16 introduces several large-scale examples that leverage our knowledge from previous chapters, and the powerful tools available in the MATLAB® and Simulink® ecosystem. These show, for a self-driving road vehicle, how to find lane markings on the road, detect other cars, and safely change lanes; for a delivery drone how to navigate a collision free path using lidar; and for a pick-and-place robot how to grasp objects perceived using an RGBD camera.

Vision-Based Control

Contents

15.1   Position-Based Visual Servoing – 668

15.2   Image-Based Visual Servoing – 670

15.3   Wrapping Up – 686

It is common to talk about a robot moving to an object, but in reality the robot is only moving to a pose at which it expects the object to be. This is a subtle but deep distinction. A consequence of this is that the robot will fail to grasp the object if it is not at the expected pose. It will also fail if imperfections in the robot mechanism or controller result in the end-effector not actually achieving the end-effector pose that was specified. In order for this conventional approach to work successfully we need to solve two quite difficult problems: determining the pose of the object and ensuring the robot achieves that pose.

The first problem, determining the pose of an object, is typically avoided in manufacturing applications by ensuring that the object is always precisely placed. This requires mechanical jigs and fixtures which are expensive, and have to be built and set up for every different part the robot needs to interact with, somewhat negating the flexibility of robotic automation.

The second problem, ensuring the robot can achieve a desired pose, is also far from straightforward. As we discussed in ▶ Chap. 7 a robot end effector is moved to a pose by computing the required joint angles. This assumes that the kinematic model is accurate, which in turn necessitates high precision in the robot's manufacture: link lengths must be precise and axes must be exactly parallel or orthogonal. Further the links must be stiff so they do not deform under dynamic loading or gravity. It also assumes that the robot has accurate joint sensors and high-performance joint controllers that eliminate steady state errors due to friction or gravity loading. The nonlinear controllers we discussed in ▶ Sect. 9.4 are capable of this high performance, but they require an accurate dynamic model that includes the mass, center of gravity and inertia for every link, as well as the payload.

None of these problems are insurmountable but this approach has led us along a path toward high complexity. The result is a heavy and stiff robot that in turn needs powerful actuators to move it, as well as high quality sensors and a sophisticated controller – all this contributes to a high overall cost. However we should, whenever possible, avoid solving hard problems if we do not have to. Stepping back for a moment and looking at this problem it is clear that

❯❯ *the root cause of the problem is that the robot cannot see what it is doing.*

Consider if the robot could see the object and its end-effector, and could use that information to guide the end-effector toward the object. This is what humans call hand-eye coordination and what we will call vision-based control or visual servo control – the use of information from one or more cameras to guide a robot in order to achieve a task.

The pose of the target does not need to be known a priori; the robot moves toward the observed target wherever it might be in the workspace. There are numerous advantages of this approach: part position tolerance can be relaxed, the ability to deal with parts that are moving comes almost for free, and any errors in the robot's intrinsic accuracy will be compensated for.

The task in visual servoing is to control the pose of the robot's end effector, relative to some goal object, using visual features extracted from an image of the goal object. The features might be centroids of points, equations of lines or ellipses, or even brightness patterns of pixels. The image of the goal is a function of its camera-relative pose $^C\xi_G$, and so to are the features extracted from the image.

There are two possible configurations of camera and robot. The configuration of ◘ Fig. 15.1a has the camera mounted on the robot's end effector observing the goal, and is referred to as end-point closed-loop or eye-in-hand. The configuration of ◘ Fig. 15.1b has the camera at a fixed point in the world, observing both the goal and the robot's end effector, and is referred to as end-point open-loop. In the remainder of this book we will discuss only the eye-in-hand configuration. Vision-based control can work with other types of robots besides manipulator arms – the camera could be attached to a mobile ground robot or a drone.

Fig. 15.1 Visual servo configurations as a pose graph. The coordinate frames are: world {0}, end effector {E}, camera {C} and goal {G}. **a** End-point closed-loop configuration (eye-in-hand); **b** end-point open-loop configuration

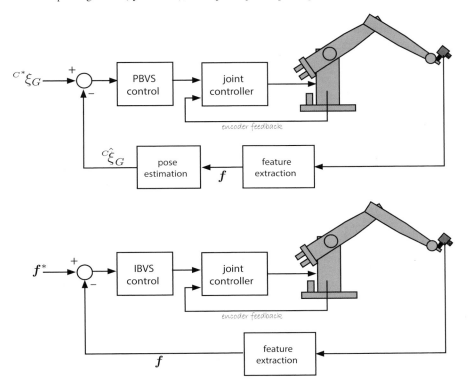

Fig. 15.2 The two distinct classes of visual servo system. **a** Position-based visual servo; **b** Image-based visual servo

For either configuration, there are two fundamentally different approaches to control: Position-Based Visual Servo (PBVS) and Image-Based Visual Servo (IBVS). Position-based visual servoing, shown in **Fig. 15.2a**, uses observed visual features, a calibrated camera and a known geometric model of the goal object to determine its pose with respect to the camera. The robot then moves toward the desired pose and the control is performed in task space. Good algorithms exist for pose estimation – see ▶ Sect. 13.2.3 – but it relies critically on the accuracy of the camera calibration and the model of the object's geometry. PBVS is discussed in ▶ Sect. 15.1.

Image-based visual servoing, shown in **Fig. 15.2b**, omits the pose estimation step, and uses the image features directly. The control is performed in image-coordinate space $\mathbb{R}^2$. The desired camera pose with respect to the goal is defined *implicitly* by the desired image feature values. IBVS is a challenging control problem since the image features are a highly nonlinear function of camera pose. IBVS is discussed in ▶ Sect. 15.2.

15.1 Position-Based Visual Servoing

For a PBVS system the relationships between the relevant poses is shown as a pose graph in ◘ Fig. 15.3. We wish to move the camera from its initial pose {C} to a desired pose {C*} which is defined with respect to the goal frame {G} by a relative pose $^{C^*}\xi_G$. The critical loop of the pose graph, indicated by the dashed-black arrow, is

$$\xi^\Delta \oplus {}^{C^*}\xi_G = {}^C\xi_G \tag{15.1}$$

where $^C\xi_G$ is the current relative pose of the goal with respect to the camera. However we cannot directly determine this since the pose of the goal in the world frame, ξ_G, shown by the gray arrow is unknown. Instead, we will use a vision-based estimate $^C\hat{\xi}_G$ of the goal pose relative to the camera. Pose estimation was discussed in ▶ Sect. 13.2.3 for a general object with known geometry, and in ▶ Sect. 13.6.1 for a fiducial marker. In both cases it requires a calibrated camera.

We rearrange (15.1) as

$$\xi^\Delta = {}^C\hat{\xi}_G \ominus {}^{C^*}\xi_G$$

which is the camera motion, in the camera frame, required to achieve the desired camera pose relative to the goal. The change in pose might be quite large so we do not attempt to make this movement in a single time step, rather we move to a point closer to {C*} by

$$\xi_{C\langle k+1\rangle} \leftarrow \xi_{C\langle k\rangle} \oplus \Lambda\big(\xi^\Delta_{\langle k\rangle}, \lambda\big)$$

which is a fraction $\lambda \in (0, 1)$ of the translation and rotation required – $\Lambda(\cdot)$ is an interpolation as discussed in ▶ Sect. 3.3.5.

Using the RVC Toolbox, we start by defining a camera with default intrinsic parameters

```
>> T_C = se3();  % identity pose
>> cam = CentralCamera("default",pose=T_C);
```

and an initial pose. The goal comprises four points that form a square of side length 0.5 m that lies in the *xy*-plane and is centered at the origin of the goal frame {G}

```
>> P = mkgrid(2,0.5);
```

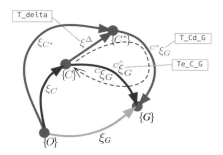

◘ **Fig. 15.3** Pose graph for PBVS example. Frame {C} is the current camera pose, frame {C*} is the desired camera pose, and frame {G} is the goal object. The Python variable names are shown in the gray boxes: an estimate (^) is indicated by the e suffix and a desired value (*) by the d suffix

We assume that {G} is unknown to the control system, and find the image-plane projections of the world points

```
>> p = cam.plot(P,objpose=se3([0 0 3],"trvec"));
>> p'  % transpose for display
p =
  445.3333   445.3333   578.6667   578.6667
  445.3333   578.6667   578.6667   445.3333
```

from which the pose of the goal with respect to the camera $^{C}\hat{\boldsymbol{\xi}}_{G}$ is estimated ▶

In code we represent $^{X}\boldsymbol{\xi}_{Y}$ as T_X_Y, an se3 instance, which we read as the transformation from X to Y.

```
>> Te_C_G = cam.estpose(P,p);
>> printtform(Te_C_G)
   Te_C_G: t = (0, 0, 3), RPY/zyx = (9.79e-13, -1.2e-12, 0) rad
```

indicating that the frame {G} is a distance of 3 m in front of the camera. The required relative pose is 1 m in front of the camera and fronto-parallel to it

```
>> T_Cd_G = se3([0 0 1],"trvec");
```

so the required change in camera pose $\boldsymbol{\xi}^{\Delta}$ is

```
>> T_delta = Te_C_G*T_Cd_G.inv();
```

and the incremental camera motion, given $\lambda = 0.05$, is

```
>> lambda=0.05;
>> T_inc = interp(se3,T_delta,lambda);
>> printtform(T_inc)
   T_inc: t = (0, 0, 0.1), RPY/zyx = (0, 0, 0) rad
```

which is a small move forward in the camera's z-direction – along its optical axis. The new value of camera pose is

```
>> cam.T = cam.T*T_inc;
```

We repeat the process, at each time step moving a fraction of the required relative pose until the goal is achieved. In this way, even if the goal moves or the robot has errors and does not move exactly as requested, the motion computed at the next time step will account for that error. ▶

The RVC Toolbox provides a class to simulate a PBVS system, display results graphically and also save the time history of the camera motion and related quantities. We start as before by defining a camera with some initial pose. For this example we choose the initial pose of the camera in world coordinates as

To ensure that the updated pose is a proper **SE**(3) matrix it should be normalized at each iteration using tformnorm.

```
>> T_C0 = se3(rotmz(0.6),[1 1 -3]);
```

and the desired pose of the goal with respect to the camera is

```
>> T_Cd_G = se3([0 0 1],"trvec");
```

which has the goal 1 m in front of the camera and fronto-parallel to it.

We create an instance of the PBVS class

```
>> pbvs = PBVS(cam,P=P,pose0=T_C0,posef=T_Cd_G, ...
>>    axis=[-1 2 -1 2 -3 0.5])
Visual servo object: camera=default
  200 iterations, 0 history
    P= -0.25        -0.25       0.25        0.25
       -0.25         0.25       0.25       -0.25
           0            0          0           0
    C_T0:   t = ( 1,  1,  -3), RPY/zyx = ( 0,  0,  0.6) rad
    C*_T_G: t = ( 0,  0,   1), RPY/zyx = ( 0,  0,  0) rad
```

which is a subclass of the VisualServo class and implements the controller outlined above. The arguments to the PBVS constructor are a CentralCamera object, the object points relative to frame {G}, the initial pose of the camera in the world frame, and the desired relative pose of the camera with respect to {G}. The run method implements the PBVS control algorithm that drives the camera and

```
>> pbvs.run(100);
```

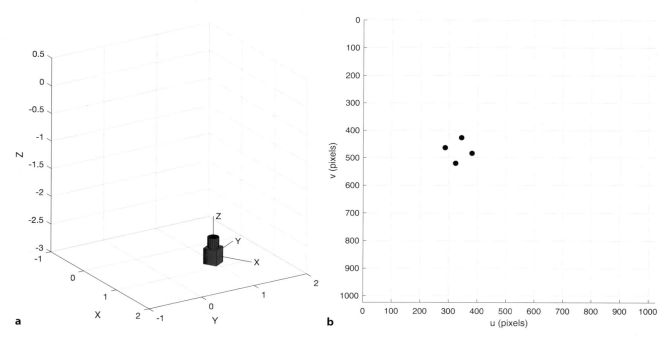

□ Fig. 15.4 Snapshop from the visual servo simulation. **a** An external view showing camera pose and features; **b** camera view showing current feature positions on the image plane

will run the simulation for 100 time steps – repeatedly calling the object's `step` method to simulate the motion for a single time step.

The simulation animates the features moving on the image plane of the camera and a 3-dimensional visualization of the camera and the world points – as shown in □ Fig. 15.4. The simulation completes after a defined number of iterations or when $\|\boldsymbol{\xi}^\Delta\|$ falls below some threshold.

The simulation results are stored within the object for later analysis. We can plot the path of the goal features in the image, the camera velocity versus time or camera pose versus time

```
>> pbvs.plot_p();
>> pbvs.plot_vel();
>> pbvs.plot_camera();
```

which are shown in □ Fig. 15.5. We see that the feature points have followed a curved path in the image, and that the camera's translation and orientation have converged smoothly on the desired values.

15.2 Image-Based Visual Servoing

IBVS differs fundamentally from PBVS by not estimating the relative pose of the goal. The relative pose is actually implicit in the values of the image features. □ Fig. 15.6 shows two views of a square object defined by its four corner points. The view from the initial camera pose is shown in red, and it is clear that the camera is viewing the object obliquely. The desired, or goal, view is shown in blue where the camera is further from the object, and its optical axis is normal to the plane of the object – a fronto-parallel view.

The control problem can be expressed in terms of image coordinates. The task is to move the feature points indicated by the round markers, to the points indicated by the star-shaped markers. The points may, but do not have to, follow the straight line paths indicated by the arrows. Moving the feature points in the image *implicitly*

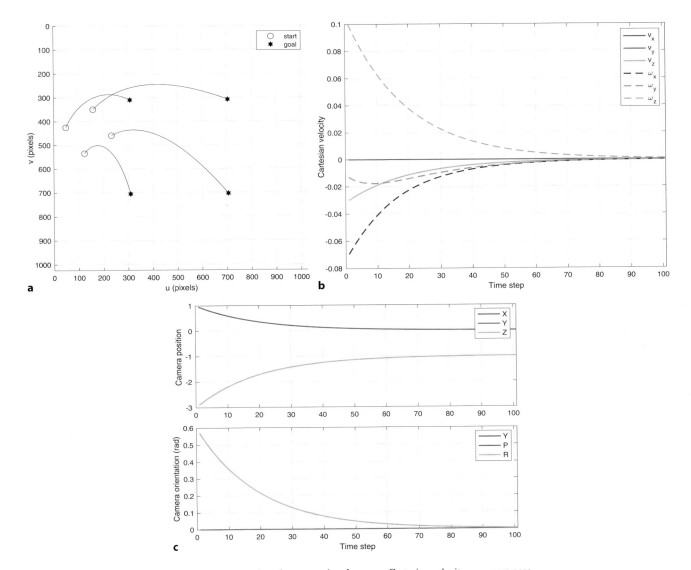

■ **Fig. 15.5** Results of PBVS simulation. **a** Image-plane feature motion; **b** camera Cartesian velocity; **c** camera pose

changes the camera pose – we have changed the problem from pose estimation to direct control of points on the image plane.

15.2.1 Camera and Image Motion

Consider the default camera

```
>> cam = CentralCamera("default");
```

and a world point at

```
>> P = [1 1 5];  % point as a row vector
```

which has image coordinates

```
>> p0 = cam.project(P)
p0 =
   672   672
```

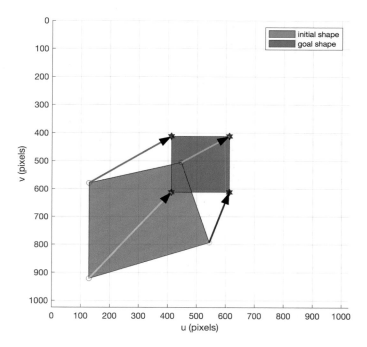

□ **Fig. 15.6**   Two different views of a square object: initial and goal

If we displace the camera slightly in the x-direction, the pixel coordinates will become

```
>> px = cam.project(P,pose=se3([0.1 0 0],"trvec"))
px =
    656    672
```

Using the camera coordinate conventions of □ Fig. 13.5, the camera has moved to the right so the image point has moved to the left. The *sensitivity* of image motion to camera motion $\Delta \boldsymbol{p}/\Delta x$ is

```
>> (px-p0)/0.1
ans =
  -160     0
```

which is an approximation to the derivative $\partial \boldsymbol{p}/\partial x$. We can repeat this for z-axis translation

```
>> (cam.project(P,pose=se3([0 0 0.1],"trvec"))-p0)/0.1
ans =
    32.6531    32.6531
```

which shows equal image-plane motion in the u- and v-directions. For x-axis rotation

```
>> (cam.project(P,pose=se3(rotmx(0.1)))-p0)/0.1
ans =
    40.9626    851.8791
```

the image motion is predominantly in the v-direction. It is clear that camera motion along, and about, the different degrees of freedom in 3-dimensional space causes quite different motion of image points. Earlier, in (13.10), we expressed perspective projection in functional form

$$\boldsymbol{p} = \mathcal{P}(\boldsymbol{P}, \mathbf{K}, \boldsymbol{\xi}_C) \tag{15.2}$$

and its derivative with respect to time is

$$\dot{\boldsymbol{p}} = \frac{\mathrm{d}\boldsymbol{p}}{\mathrm{d}t} = \frac{\partial \boldsymbol{p}}{\partial \boldsymbol{\xi}} \frac{\mathrm{d}\boldsymbol{\xi}}{\mathrm{d}t} = \mathbf{J}_p(\boldsymbol{P}, \mathbf{K}, \boldsymbol{\xi}_C)\, \boldsymbol{v} \tag{15.3}$$

where $\mathbf{J}_p$ is a Jacobian-like object, but because we have taken the derivative with respect to a pose $\boldsymbol{\xi} \in \mathbf{SE}(3)$, rather than a vector, it is technically called an *interaction matrix*. However, in the visual servoing world it is more commonly called an *image Jacobian* or a *feature sensitivity matrix*. This is a local linearization of the highly nonlinear function (15.2). ▶ The camera moves in 3-dimensional space and its velocity is a *spatial velocity* $\boldsymbol{v} = (\omega_x, \omega_y, \omega_z, v_x, v_y, v_z) \in \mathbb{R}^6$, a concept introduced in ▶ Sect. 3.1, and comprises rotational and translational velocity components.

See ▶ App. E.

Consider a camera moving with a body velocity $\boldsymbol{v} = (\boldsymbol{\omega}, \boldsymbol{v})$ in the world frame and observing a world point P described by a camera relative coordinate vector $^C\!\boldsymbol{P} = (X, Y, Z)$. The velocity of the point relative to the camera frame is

$$^C\!\dot{\boldsymbol{P}} = -\left(\boldsymbol{\omega} \times {}^C\!\boldsymbol{P} + \boldsymbol{v}\right) \tag{15.4}$$

which we can write in scalar form as

$$\dot{X} = Y\omega_z - Z\omega_y - v_x$$
$$\dot{Y} = Z\omega_x - X\omega_z - v_y$$
$$\dot{Z} = X\omega_y - Y\omega_x - v_z \ . \tag{15.5}$$

The perspective projection (11.2) for normalized image-plane coordinates is

$$x = \frac{X}{Z}, \quad y = \frac{Y}{Z}$$

and the temporal derivative, using the quotient rule, is

$$\dot{x} = \frac{\dot{X}Z - X\dot{Z}}{Z^2}, \quad \dot{y} = \frac{\dot{Y}Z - Y\dot{Z}}{Z^2} \ .$$

Substituting in (15.5), $X = xZ$ and $Y = yZ$, we can write these in matrix form as

$$\begin{pmatrix} \dot{x} \\ \dot{y} \end{pmatrix} = \begin{pmatrix} xy & -(1+x^2) & y & -\frac{1}{Z} & 0 & \frac{x}{Z} \\ 1+y^2 & -xy & -x & 0 & -\frac{1}{Z} & \frac{y}{Z} \end{pmatrix} \begin{pmatrix} \omega_x \\ \omega_y \\ \omega_z \\ v_x \\ v_y \\ v_z \end{pmatrix} \tag{15.6}$$

which relates camera spatial velocity to feature velocity in normalized image coordinates.

🚫 In much of the literature, including other editions of this book, the components of the spatial velocity are swapped and it is written as $\boldsymbol{v} = (\boldsymbol{v}, \boldsymbol{\omega})$. The first and last three columns of the visual Jacobian matrix are correspondingly swapped.

The normalized image-plane coordinates are related to the pixel coordinates by (13.7)

$$u = \frac{f}{\rho_u}x + u_0, \quad v = \frac{f}{\rho_v}y + v_0$$

which we rearrange as

$$x = \frac{\rho_u}{f}\bar{u}, \quad y = \frac{\rho_v}{f}\bar{v} \tag{15.7}$$

where $\overline{u} = u - u_0$ and $\overline{v} = v - v_0$ are the pixel coordinates relative to the principal point. The temporal derivative is

$$\dot{x} = \frac{\rho_u}{f}\dot{\overline{u}}, \quad \dot{y} = \frac{\rho_v}{f}\dot{\overline{v}} \tag{15.8}$$

and substituting (15.7) and (15.8) into (15.6) leads to

$$\begin{pmatrix} \dot{u} \\ \dot{v} \end{pmatrix} = \begin{pmatrix} \frac{\rho_v \overline{u}\,\overline{v}}{f} & -\frac{f^2 + \rho_u^2 \overline{u}^2}{\rho_u f} & \frac{\rho_v \overline{v}}{\rho_u} & -\frac{f}{\rho_u Z} & 0 & \frac{\overline{u}}{Z} \\ \frac{f^2 + \rho_v^2 \overline{v}^2}{\rho_v f} & -\frac{\rho_u \overline{u}\,\overline{v}}{f} & -\frac{\rho_u \overline{u}}{\rho_v} & 0 & -\frac{f}{\rho_v Z} & \frac{\overline{v}}{Z} \end{pmatrix} \begin{pmatrix} \omega_x \\ \omega_y \\ \omega_z \\ v_x \\ v_y \\ v_z \end{pmatrix}$$

noting that $\dot{u} = \dot{\overline{u}}$ and $\dot{v} = \dot{\overline{v}}$.

For the typical case where $\rho_u = \rho_v = \rho$ we can express the focal length in pixels $f' = f/\rho$ and write

$$\begin{pmatrix} \dot{u} \\ \dot{v} \end{pmatrix} = \underbrace{\begin{pmatrix} \frac{\overline{u}\,\overline{v}}{f'} & -\frac{f'^2 + \overline{u}^2}{f'} & \overline{v} & -\frac{f'}{Z} & 0 & \frac{\overline{u}}{Z} \\ \frac{f'^2 + \overline{v}^2}{f'} & -\frac{\overline{u}\,\overline{v}}{f'} & -\overline{u} & 0 & -\frac{f'}{Z} & \frac{\overline{v}}{Z} \end{pmatrix}}_{\mathbf{J}_p(\boldsymbol{p},Z)} \begin{pmatrix} \omega_x \\ \omega_y \\ \omega_z \\ v_x \\ v_y \\ v_z \end{pmatrix} \tag{15.9}$$

in terms of pixel coordinates *with respect to the principal point*. We can write this in concise matrix form as

$$\dot{\boldsymbol{p}} = \mathbf{J}_p(\boldsymbol{p}, Z)\,\boldsymbol{v} \tag{15.10}$$

where $\mathbf{J}_p$ is the 2×6 image Jacobian matrix for a point feature with coordinate vector $\boldsymbol{p} = (\overline{u}, \overline{v})$ ◄ and Z which is the z-coordinate of the point in the camera frame and often referred to as the point's depth.

The RVC Toolbox `CentralCamera` class provides the method `visjac_p` to compute the image Jacobian, and for the example above it is

```
>> J = cam.visjac_p([672 672],5)
J =
   32.0000 -832.0000  160.0000 -160.0000         0   32.0000
  832.0000  -32.0000 -160.0000         0 -160.0000   32.0000
```

where the first argument is the pixel coordinate of the point of interest, and the second argument is the depth of the point. The approximate numerical derivatives computed above correspond to the fourth, sixth and first columns respectively.

For a given camera velocity, the velocity of the point is a function of the point's coordinate, its depth and the camera's intrinsic parameters. Each column of the Jacobian indicates the velocity of an image feature point caused by one unit of the corresponding component of the velocity vector. The `flowfield` method of the `CentralCamera` class shows, for a particular camera velocity, the image-plane velocity for a grid of world points projected to the image plane. For camera translational velocity in the x-direction the flow field

```
>> cam.flowfield([0 0 0 1 0 0]);
```

This is commonly written in terms of u and v rather than $\overline{u}$ and $\overline{v}$ but we use the overbar notation to emphasize that the coordinates are with respect to the principal point, not the image origin which is typically in the top-left corner.

is shown in ◘ Fig. 15.7a. As expected, moving the camera to the right causes all the projected points to move to the left. The motion of points on the image plane is known as *optical flow* and can be computed from image sequences, using point feature tracking as discussed in ▶ Sect. 14.8.3. Equation (15.9) is often referred to as the optical flow equation.

For translation in the z-direction

```
>> cam.flowfield([0 0 0 0 0 1]);
```

the points radiate outward from the principal point – the Star Trek warp effect – as shown in ◘ Fig. 15.7b. Rotation about the z-axis

```
>> cam.flowfield([0 0 1 0 0 0]);
```

causes the points to rotate about the principal point as shown in ◘ Fig. 15.7c.

Rotational motion about the y-axis

```
>> cam.flowfield([0 1 0 0 0 0]);
```

is shown in ◘ Fig. 15.7d and is very similar to the case of x-axis translation, with some small curvature for points far from the principal point. This similarity is because the second and fourth column of the image Jacobian are approximately equal in this case.

Any point on the optical axis will project to the the principal point, and at a depth of 1 m, the image Jacobian is

```
>> cam.visjac_p(cam.pp,1)
ans =
          0  -800.0000         0  -800.0000         0         0
    800.0000         0         0         0  -800.0000         0
```

and we see that the second and fourth columns are equal. This implies that rotation about the y-axis causes the same image motion as translation in the x-direction. You can easily demonstrate this equivalence by watching how the world moves as you translate your head to the right or rotate your head to the right – in both cases the world appears to move to the left. As the focal length increases, the second column

$$\lim_{f' \to \infty} \begin{pmatrix} -\frac{f'^2 + \bar{u}^2}{f'} \\ -\frac{\bar{u}\,\bar{v}}{f'} \end{pmatrix} = \begin{pmatrix} -f' \\ 0 \end{pmatrix}$$

approaches a scalar multiple of the fourth column. Increasing the focal length to $f = 20\,\text{mm}$ (the default focal length is 8 mm) leads to the flow field

```
>> cam.f = 20e-3;
>> cam.flowfield([0 1 0 0 0 0]);
```

shown in ◘ Fig. 15.7e which is almost identical to that of ◘ Fig. 15.7a. Conversely, for small focal lengths (wide-angle cameras) the flow field

```
>> cam.f = 4e-3;
>> cam.flowfield([0 1 0 0 0 0]);
```

shown in ◘ Fig. 15.7f has much more pronounced curvature. The same principle applies for the first and fifth columns except for a difference of sign – there is an equivalence between rotation about the x-axis and translation in the $-y$-direction.

The curvature in the wide-angle flow field shown in ◘ Fig. 15.7f means that the ambiguity between translational and rotation motion, discussed earlier, is resolved. The edges of the field of view contain motion cues that differentiate these types of motion. As the field of view gets larger, this effect becomes more pronounced. For narrow field of view, such as for our own eyes, we need additional information from sensors to resolve this ambiguity and the angular rate sensors in our vestibular system serve that purpose.

15

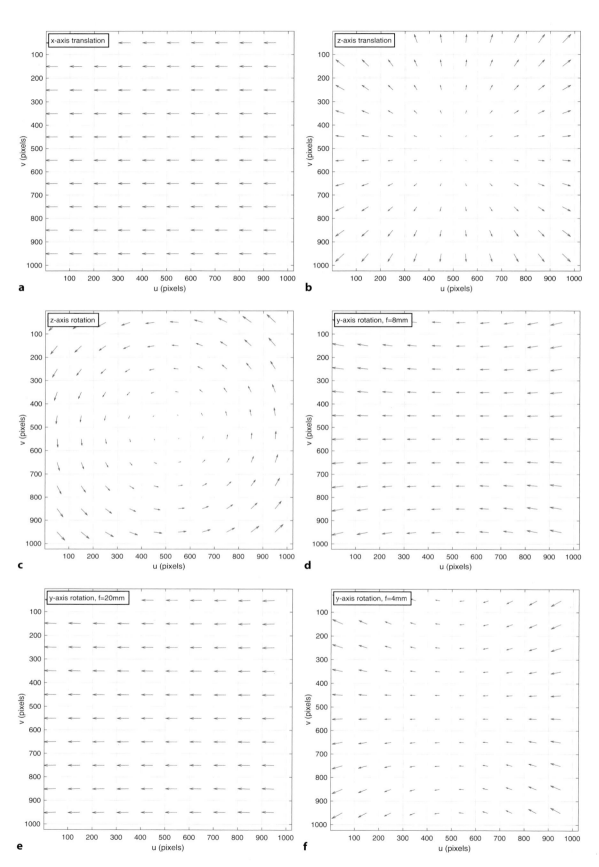

Fig. 15.7 Image-plane velocity vectors for canonic camera velocities where all corresponding world points lie in a fronto-parallel plane. The flow vectors are normalized – they are shown with correct relative scale within each plot, but not between plots

The Jacobian matrix of (15.9) has some interesting properties. It does not depend at all on the world coordinates X or Y, only on the image-plane coordinates $(\bar{u}, \bar{v})$. However, the last three columns depend on the point's depth Z and this reflects the fact that, for a translating camera, the image-plane velocity is inversely proportional to depth. Again, you can easily demonstrate this to yourself – translate your head sideways and observe that near objects move more in your field of view than distant objects. However, if you rotate your head all objects, near and far, move equally in your field of view.

The matrix has a rank of two, ► and therefore has a null space of dimension four. The null space comprises a set of spatial velocity vectors that individually, or in any linear combination, cause the world point to have *no motion* in the image. Consider the simple case of a world point lying on the optical axis which projects to the principal point

► The rank cannot be less than 2, even if $Z \to \infty$.

```
>> J = cam.visjac_p(cam.pp,1);
```

The null space of the Jacobian is

```
>> null(J)
ans =
         0    0.2357    0.6667         0
         0   -0.6667    0.2357         0
    1.0000         0         0         0
         0    0.6667   -0.2357         0
         0    0.2357    0.6667         0
         0         0         0    1.0000
```

The first column indicates that rotation about the z-axis, around the ray toward the point, results in no motion in the image. Nor does motion in the z-direction, as indicated by the last column. The second and third columns are more complex, combining rotation and translation. Essentially these exploit the image motion ambiguity mentioned above: x-axis rotation causes the same image motion as $-y$-axis translation, and y-axis rotation causes the same image motion as x-axis translation. If these rotational and translational motions cancel each other out then the resulting image motion will be zero – for the third column this is rotating up and to the right while translating to the right and down.

We can consider the motion of two points by stacking their Jacobians

$$\begin{pmatrix} \dot{u}_1 \\ \dot{v}_1 \\ \dot{u}_2 \\ \dot{v}_2 \end{pmatrix} = \begin{pmatrix} \mathbf{J}_p(\boldsymbol{p}_1, \ Z_1) \\ \mathbf{J}_p(\boldsymbol{p}_2, \ Z_2) \end{pmatrix} \boldsymbol{v}$$

to give a 4×6 matrix which will have a null space with just two columns. One of these null-space camera motions corresponds to rotation around a line joining the two points.

For three points

$$\begin{pmatrix} \dot{u}_1 \\ \dot{v}_1 \\ \dot{u}_2 \\ \dot{v}_2 \\ \dot{u}_3 \\ \dot{v}_3 \end{pmatrix} = \begin{pmatrix} \mathbf{J}_p(\boldsymbol{p}_1, \ Z_1) \\ \mathbf{J}_p(\boldsymbol{p}_2, \ Z_2) \\ \mathbf{J}_p(\boldsymbol{p}_3, \ Z_3) \end{pmatrix} \boldsymbol{v} \tag{15.11}$$

the matrix will be full rank, nonsingular, so long as the points are not coincident or collinear.

15.2.2 Controlling Feature Motion

So far we have shown how points move in the image plane as a consequence of camera motion. As is often the case, it is the inverse problem that is more useful – what camera motion is needed in order to move the image features at a desired velocity?

For the case of three points $\{(u_i, v_i), i = 1, 2, 3\}$ and corresponding velocities $\{(\dot{u}_i, \dot{v}_i)\}$, we can invert (15.11)

$$
\boldsymbol{v} = \begin{pmatrix} \mathbf{J}_p(\boldsymbol{p}_1, \, Z_1) \\ \mathbf{J}_p(\boldsymbol{p}_2, \, Z_2) \\ \mathbf{J}_p(\boldsymbol{p}_3, \, Z_3) \end{pmatrix}^{-1} \begin{pmatrix} \dot{u}_1 \\ \dot{v}_1 \\ \dot{u}_2 \\ \dot{v}_2 \\ \dot{u}_3 \\ \dot{v}_3 \end{pmatrix}
\tag{15.12}
$$

and solve for the required camera spatial velocity. The remaining question is how to determine the point velocity? Typically, we use a simple linear controller

$$
\dot{\boldsymbol{p}}^* = \lambda(\boldsymbol{p}^* - \boldsymbol{p})
\tag{15.13}
$$

that *drives* the image plane points toward their desired values $\boldsymbol{p}^*$. Combined with (15.12) we write ◀

$$
\boldsymbol{v} = \lambda \begin{pmatrix} \mathbf{J}_p(\boldsymbol{p}_1, \, Z_1) \\ \mathbf{J}_p(\boldsymbol{p}_2, \, Z_2) \\ \mathbf{J}_p(\boldsymbol{p}_3, \, Z_3) \end{pmatrix}^{-1} (\boldsymbol{p}^* - \boldsymbol{p})
$$

That's it! This controller computes a velocity that moves the camera in such a way that the feature points move toward their desired position in the image. It is important to note that nowhere have we required the pose of the camera or of the object– everything has been computed in terms of what can be measured on the image plane. ◀ As the controller runs, the image-plane points $\boldsymbol{p}_i$ move so $\mathbf{J}_p(\cdot)$ needs to be updated frequently, typically for every new camera image.

> ❶ **Correspondence between the observed and desired points is essential**
> It is critically important when computing (15.13) that the points $\boldsymbol{p}_i^*$ and $\boldsymbol{p}_i$ correspond. In a contrived scene, the points could be the centroids of shapes with uniques colors, sizes or shapes. In a more general case, correspondence can be solved if one point is unique and the ordering of the points, often with respect to the centroid of the points, is known.

For the general case where $N > 3$ points we can stack the Jacobians for all features and solve for camera velocity using the pseudoinverse

$$
\boldsymbol{v} = \lambda \begin{pmatrix} \mathbf{J}_p(\boldsymbol{p}_1, \, Z_1) \\ \vdots \\ \mathbf{J}_p(\boldsymbol{p}_N, \, Z_N) \end{pmatrix}^{+} (\boldsymbol{p}^* - \boldsymbol{p})
\tag{15.14}
$$

Note that it is possible to specify a set of feature point velocities which are inconsistent, that is, there is no possible camera motion that will result in the required

Note that papers based on the task function approach (Espiau et al. 1992) write this as actual position minus demanded position and use $-\lambda$ in (15.14) to ensure negative feedback. We use the common convention from control theory where the error signal is "demand minus actual".

We do require the point depth Z, but we will deal that issue shortly.

15

image motion. In such a case the pseudoinverse will find a solution that minimizes the norm of the feature velocity error.

The Jacobian is a first-order approximation of the relationship between camera motion and image-plane motion. Faster convergence is achieved by using a second-order approximation and it has been shown that this can be obtained very simply

$$
v = \frac{\lambda}{2} \left[\begin{pmatrix} \mathbf{J}_p(\boldsymbol{p}_1, \, Z_1) \\ \vdots \\ \mathbf{J}_p(\boldsymbol{p}_N, \, Z_N) \end{pmatrix}^+ + \begin{pmatrix} \mathbf{J}_p(\boldsymbol{p}_1^*, \, Z_1^*) \\ \vdots \\ \mathbf{J}_p(\boldsymbol{p}_N^*, \, Z_N^*) \end{pmatrix}^+ \right] (\boldsymbol{p}^* - \boldsymbol{p}) \qquad (15.15)
$$

by taking the mean of the pseudo inverse of the image Jacobians at the current and desired states.

For $N \geq 3$ the matrix can be poorly conditioned if the points are nearly co-incident or collinear. In practice, this means that some camera motions will cause very small image motions, that is, the motion has low perceptibility. There is strong similarity with the concept of manipulability that we discussed in ▶ Sect. 8.3.2 and we take a similar approach in formalizing perceptibility. Consider a camera spatial velocity of unit magnitude

$$
\boldsymbol{v}^\top \boldsymbol{v} = 1
$$

and from (15.10) we can write the camera velocity in terms of the pseudoinverse

$$
\boldsymbol{v} = \mathbf{J}^+ \dot{\boldsymbol{p}}
$$

where $\mathbf{J} \in \mathbb{R}^{2N \times 6}$ is the Jacobian stack and the point velocities are $\dot{\boldsymbol{p}} \in \mathbb{R}^{2N}$. Substituting yields

$$
\dot{\boldsymbol{p}}^\top \mathbf{J}^{+\top} \mathbf{J}^+ \dot{\boldsymbol{p}} = 1
$$
$$
\dot{\boldsymbol{p}}^\top \left(\mathbf{J}\mathbf{J}^\top \right)^{-1} \dot{\boldsymbol{p}} = 1
$$

which is the equation of an ellipse in the image-plane velocity space. The eigenvectors of $\mathbf{J}\mathbf{J}^\top$ define the principal axes of the ellipse and the singular values of $\mathbf{J}$ are the radii. The ratio of the maximum to minimum radius is given by the condition number of $\mathbf{J}\mathbf{J}^\top$ and indicates the anisotropy of the feature motion. A high value indicates that some of the points have low velocity in response to some camera motions. An alternative to stacking all the point feature Jacobians in (15.14) is to select just three that, when stacked, result in the best conditioned square matrix which can then be inverted.

Using the RVC Toolbox we start by defining a camera with default intrinsic parameters

```
>> T_C = se3();  % identity pose
>> cam = CentralCamera("default");
```

The goal comprises four points that form a square of side length 0.5 m that lies in the xy-plane and is centered at $(0, 0, 3)$

```
>> P = mkgrid(2,0.5,pose=se3([0 0 3],"trvec"));
```

The desired position of the goal features on the image plane are a 400×400 square centered on the principal point

```
>> pd = 200*[-1 -1;-1 1;1 1;1 -1] + cam.pp;
```

which implicitly has the square goal fronto-parallel to the camera. The camera's initial projection of the world points is

```
>> p = cam.plot(P,pose=T_C);
```

and p and pd each have one row per point.

We compute the image-plane error

```
>> e = pd-p;
```

and the stacked image Jacobian

```
>> J = cam.visjac_p(p,1);
```

Here we provide a single value which
is taken as the depth of all the points.
Alternatively we could provide a
vector to specify the depth of each
point individually.

is an 8×6 matrix in this case, since `p` contains four points. The Jacobian does require the point depth which we do not know, so for now we will just choose a constant value $Z = 1$. ◄ This is an important topic that we will address in ► Sect. 15.2.3.

The control law determines the required translational and angular velocity of the camera ◄

```
>> lambda = 0.05;
>> v = lambda*pinv(J)*reshape(e',[],1);
```

To create the error column vector as
described in (15.12) we need to flatten
the 2D array e in row-major order.

where `lambda` is the gain, a positive number, and we take the pseudoinverse of the nonsquare Jacobian to implement (15.14). `e` is a 4×2 matrix and we reshape it, rowwise, into a column vector. The computed camera velocity

```
>> v'   % transpose for display
ans =
   -0.0000   -0.0000    0.0000    0.0000   -0.0000    0.1000
```

is a pure translation in the camera's z-direction, driving the camera forward towards the target. Integrating this over a unit time step results in a spatial displacement of the same magnitude. The camera pose is updated by

$$\boldsymbol{\xi}_C \langle k+1 \rangle \leftarrow \boldsymbol{\xi}_C \langle k \rangle \oplus \Delta^{-1}(\boldsymbol{v}\langle k \rangle)$$

where $\Delta^{-1}(\cdot)$ is described in ► Sect. 3.1.5. This is implemented as

```
>> cam.T = cam.T*delta2se(v);
```

Similar to the PBVS example, we create an instance of the IBVS classRz

```
>> T_C0 = se3(rotmz(0.6),[1 1 -3]);
>> camera = CentralCamera("default",pose=T_C0);
>> ibvs = IBVS(camera,P=P,p_d=pd);
```

which is a subclass of the `VisualServo` class and which implements the controller outlined above. The IBVS constructor takes a `CentralCamera` object, with a specified initial pose, as its argument. The controller then drives the camera to achieve the desired image-plane point configuration specified by `pd`. The simulation is run for 25 time steps

```
>> ibvs.run(25);
```

which repeatedly calls the object's `step` method which simulates motion for a single time step. The simulation animates the image plane of the camera as well as a 3-dimensional visualization of the camera and the world points.

The simulation results are stored within the object for later analysis. We can plot the path of the goal features on the image plane, the camera velocity versus time or camera pose versus time

```
>> ibvs.plot_p();
>> ibvs.plot_vel();
>> ibvs.plot_camera();
```

which are shown in ■ Fig. 15.8. We see that the feature points have followed an approximately straight-line path in the image, and the camera pose has changed smoothly toward some final value. The condition number of the image Jacobian

```
>> ibvs.plot_jcond();
```

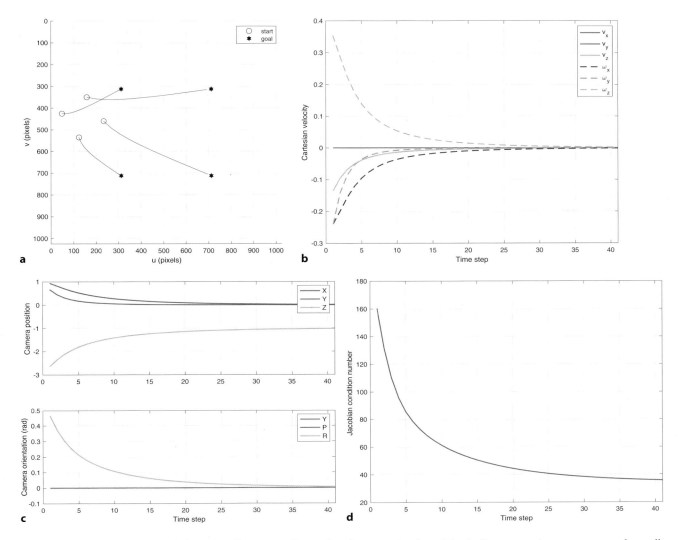

■ **Fig. 15.8** Results of IBVS simulation, created by `IBVS`. **a** Image-plane feature motion; **b** spatial velocity components; **c** camera pose where roll-pitch-yaw angles are in camera YXZ order; **d** image Jacobian condition number (lower is better)

decreases over the motion indicating that the Jacobian is becoming better conditioned, and this is a consequence of the features moving further apart – becoming less coincident.

How is p^* determined? The image points can be found by demonstration, by moving the camera to the desired pose and recording the observed image coordinates. Alternatively, if the camera calibration parameters and the goal geometry are known, the desired image coordinates can be computed for any specified goal pose. This calculation, world point point projection, is computationally cheap and performed only once before visual servoing commences.

15.2.3 Estimating Feature Depth

Computing the image Jacobian requires knowledge of the camera intrinsics, the principal point and focal length, but in practice it is quite tolerant to errors in these. The Jacobian also requires knowledge of Z_i, the z-coordinate, or the depth of, each point. In the simulations just discussed, we have assumed that depth is known – this is easy in simulation but not so in reality.

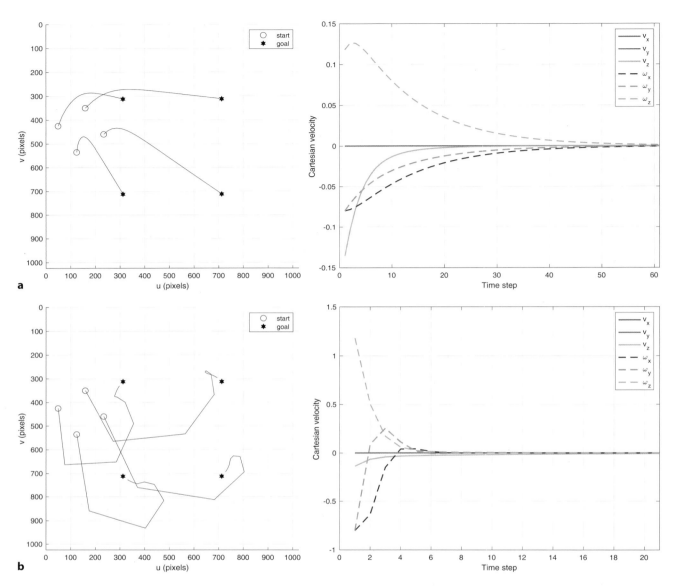

■ **Fig. 15.9** Image feature paths and camera pose for IBVS with different constant estimates of point depth: **a** For $Z = 1$; **b** For $Z = 10$

A number of approaches have been proposed to deal with the problem of unknown depth. The simplest is to just assume a constant value for the depth, and this is quite reasonable if the required camera velocity is approximately in a plane parallel to the plane of the object points. To evaluate the performance of different constant estimates of point depth, we can compare the effect of choosing $Z = 1$ and $Z = 10$ for the example above where the true depth is initially $Z = 3$

```
>> ibvs = IBVS(cam,pose0=T_C0,p_d=pd,depth=1)
>> ibvs.run(60)
>> ibvs = IBVS(cam,pose0=T_C0,p_d=pd,depth=10)
>> ibvs.run(20)
```

and the results are shown in ■ Fig. 15.9. We see that the image-plane paths are no longer straight, because the Jacobian is now a poor approximation of the relationship between the camera motion and image feature motion. We also see that for $Z = 1$ the convergence is much slower than for the $Z = 10$ case. The Jacobian for $Z = 1$ overestimates the optical flow, so the inverse Jacobian underestimates the required camera velocity. For the $Z = 10$ case, the camera displacement at each time step is large leading to a very jagged path. Nevertheless, for quite significant errors in depth, IBVS has converged, and it is widely observed in practice that IBVS is remarkably tolerant to errors in Z.

A second approach is to assume a calibrated camera and use standard computer vision techniques such as pose estimation or sparse stereo to estimate the value of Z. However, this does defeat an important advantage of IBVS which is to avoid explicit pose estimation and 3D models.

A third approach, which we will expand on here, is to estimate depth on-line using observed frame-to-frame feature motion, known camera intrinsics, and knowledge of the camera spatial velocity $(\boldsymbol{\omega}, \boldsymbol{v})$ which the controller computes. We can create a depth estimator by rewriting (15.9)

$$
\begin{pmatrix} \dot{u} \\ \dot{v} \end{pmatrix} = \left(\begin{array}{ccc|ccc} \frac{\rho_u \overline{uv}}{f} & -\frac{f^2 + \rho_u^2 \overline{u}^2}{\rho_u f} & \overline{v} & -\frac{f}{\rho_u Z} & 0 & \frac{\overline{u}}{Z} \\ \frac{f^2 + \rho_v^2 \overline{v}^2}{\rho_v f} & -\frac{\rho_v \overline{uv}}{f} & -\overline{u} & 0 & -\frac{f}{\rho_v Z} & \frac{\overline{v}}{Z} \end{array} \right) \begin{pmatrix} \boldsymbol{\omega} \\ \boldsymbol{v} \end{pmatrix}
$$

$$
= \left(\mathbf{J}_\omega \ \Big| \ \tfrac{1}{Z} \mathbf{J}_t \right) \begin{pmatrix} \boldsymbol{\omega} \\ \boldsymbol{v} \end{pmatrix}
$$

$$
= \mathbf{J}_\omega \boldsymbol{\omega} + \frac{1}{Z} \mathbf{J}_t \boldsymbol{v}
$$

and then rearranging into estimation form

$$
(\mathbf{J}_t \boldsymbol{v}) \frac{1}{Z} = \begin{pmatrix} \dot{u} \\ \dot{v} \end{pmatrix} - \mathbf{J}_\omega \boldsymbol{\omega} \tag{15.16}
$$

The right-hand side is the observed optical flow from which the expected optical flow, due to camera rotation, is subtracted – a process referred to as derotating optical flow. The remaining optical flow, after subtraction, is only due to translation. Writing (15.16) in compact form

$$
\mathbf{A}\boldsymbol{\theta} = \boldsymbol{b} \tag{15.17}
$$

we have a simple linear equation with one unknown parameter $\boldsymbol{\theta} = 1/Z$ which can be solved using least-squares.

In our example we can enable this by

```
>> ibvs = IBVS(cam,pose0=T_C0,p_d=pd,depthest=true);
>> ibvs.run()
>> ibvs.plot_z();
>> ibvs.plot_p();
```

and the result is shown in ◻ Fig. 15.10. ◻ Fig. 15.10b shows the estimated and true point depth versus time. The estimated depth was initially zero, a poor choice, but it has risen rapidly and then tracked the actual goal depth as the controller converges. ◻ Fig. 15.10a shows the feature motion, and we see that the features initially move in the wrong direction because the gross error in depth has led to an image Jacobian that predicts poorly how feature points will move.

15.2.4 Performance Issues

The control law for PBVS is defined in terms of the 3-dimensional pose so there is no mechanism by which the motion of the image features is directly regulated. For the PBVS example shown in ◻ Fig. 15.5, the feature points followed a curved path on the image plane, and therefore it is possible that they could leave the camera's

◘ Fig. 15.10 IBVS with online depth estimator. **a** Feature paths; **b** comparison of estimated and true depth for all four points

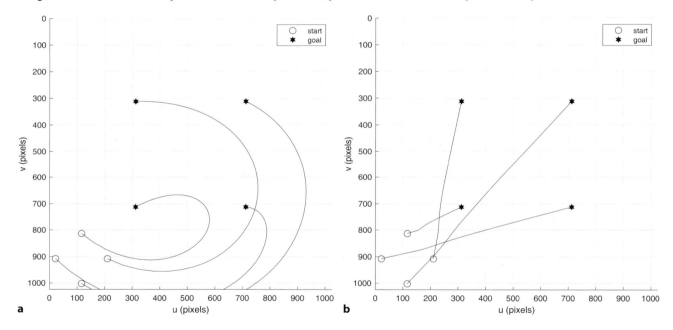

◘ Fig. 15.11 Image-plane feature paths for **a** PBVS and **b** IBVS

In this simulation the image plane coordinates are still computed and used, even though they fall outside the image bounds. Visibility checking can be enabled in the camera object by the `visibility` option to the constructor.

field of view. For a different initial camera pose

```
>> pbvs.T0 = se3(rotmz(5*pi/4),[-2.1 0 -3]);
>> pbvs.run()
```

the result is shown in ◘ Fig. 15.11a and we see that two of the points move beyond the image boundary which would cause the PBVS control to fail. ◄

By contrast, the IBVS control for the same initial camera pose

```
>> ibvs = IBVS(cam,pose0=pbvs.T0,p_d=pd,lambda=0.002, ...
>>    niter=Inf,eterm=0.5)
>> ibvs.run()
>> ibvs.plot_p();
```

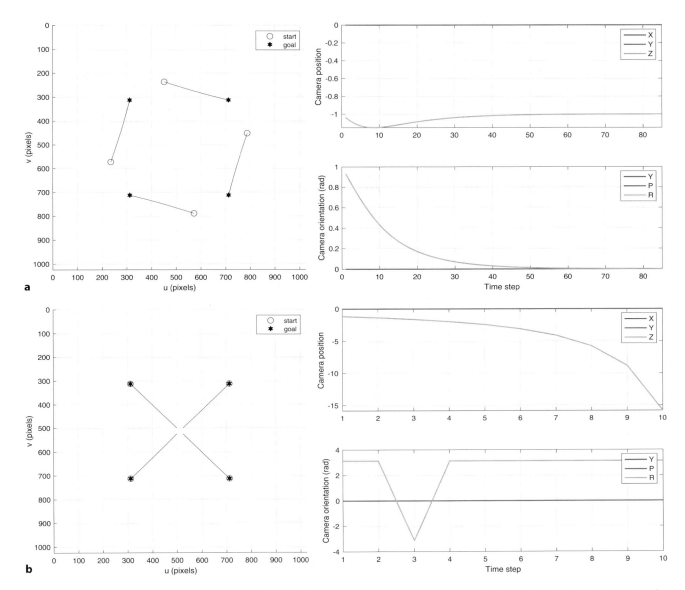

Fig. 15.12 Image feature paths and camera pose for IBVS with pure goal rotation about the optical axis. **a** for rotation of 1 rad; **b** for rotation of π rad, note that the roll angle at $t = 3$ is equivalent to the angle at $t = 2, 4$

gives the feature trajectories shown in ◘ Fig. 15.11b, but in this case there is no direct control over the task-space camera velocity. This can sometimes result in surprising motion, particularly when the goal is rotated about the optical axis

```
>> ibvs = IBVS(cam,pose0=se3(rotmz(1),[0 0 -1]),p_d=pd);
>> ibvs.run()
>> ibvs.plot_camera
```

where the image features and camera pose are shown in ◘ Fig. 15.12a. We see that the camera has performed an unnecessary translation along the z-axis – initially away from the goal, and then back again. This phenomenon is termed camera retreat. The resulting motion is not time optimal and can require large and possibly unachievable camera motion. An extreme example arises for a pure rotation about the optical axis by π rad

```
>> ibvs = IBVS(cam,pose0=se3(rotmz(pi),[0 0 -1]), ...
>>    p_d=pd, niter=10);
```

```
>> ibvs.run()
>> ibvs.plot_camera
```

As the image points converge on the principal point the Jacobian condition number increases and it will eventually become rank deficient.

which is shown in ◘ Fig. 15.12b. The feature points are, as usual, moving in a straight line toward their desired values, but for this problem the paths all pass through the principal point. The only way the goal feature points can be at the principal point is if the camera is at negative infinity, and that is where it is headed! ◄

A final consideration is that the image Jacobian is a linearization of a highly nonlinear system. If the motion at each time step is large, then the linearization is not valid and the features will follow curved rather than linear paths in the image, as we saw in ◘ Fig. 15.9. This can occur if the desired feature positions are a long way from the initial positions and/or the gain λ is too high. One solution is to limit the maximum norm of the commanded velocity

$$
v = \begin{cases} v_{max} \frac{v}{|v|} & \text{if } |v| > v_{max} \\ v & \text{if } |v| \leq v_{max} \end{cases}
$$

and this is enabled by the vmax option to the constructor. We can see from all the camera velocity plots that the highest velocity is commanded at the first time step when the feature error is largest. We can also use a technique called *smooth start* to modify the image-plane error $e = p^* - p$

$$
e' = e - e_0 e^{-\mu t}
$$

before it is used to compute the control, where e_0 is the value of e at time $t = 0$. The second term ensures that e' is initially zero and grows gradually with time. This can be enabled using the smoothstart=μ option to the constructor.

The feature paths do not have to be straight lines and nor do the features have to move with asymptotic velocity – we have used these only for simplicity. Using the trajectory planning methods of ► Sect. 3.3 the features could be made to follow any arbitrary trajectory in the image.

In summary, IBVS is a remarkably robust approach to vision-based control. We have seen that it is tolerant to errors in the depth of points. We have also shown that it can produce less than optimal Cartesian paths for the case of large rotations about the optical axis.

There are of course also some practical complexities. If the camera is on the end of the robot it might interfere with the task. Or, when the robot is close to the target the camera might be unable to focus, or the target might be obscured by the gripper.

15.3 Wrapping Up

In this chapter we have learned about the fundamentals of vision-based robot control, and the fundamental techniques developed over two decades up to the mid 1990s. There are two distinct configurations. The camera can be attached to the robot observing the goal (eye-in-hand) or fixed in the world observing both robot and goal. Another form of distinction is the control structure: Position-Based Visual Servo (PBVS) and Image-Based Visual Servo (IBVS). The former involves pose estimation based on a calibrated camera and a geometric model of the goal object, while the latter performs the control directly in the image plane. Each approach has certain advantages and disadvantages. PBVS performs efficient straight-line Cartesian camera motion in the world, but may cause image features to leave the image plane. IBVS always keeps features in the image plane, but may result in trajectories that exceed the reach of the robot, particularly if it requires a large amount of rotation about the camera's optical axis. IBVS also requires a touch of 3-dimensional information (the depth of the feature points) but is quite robust to errors in depth and it is quite feasible to estimate the depth as the camera moves.

15.3.1   Further Reading

The tutorial paper by Hutchinson et al. (1996) was the first comprehensive articulation and taxonomy of the field, and Chaumette and Hutchinson (2006 and 2007) provide a more recent tutorial introduction. Chapters on visual servoing are included in Siciliano and Khatib (2016, § 34) and Spong et al. (2006, § 12).

It is well known that IBVS is very tolerant to errors in depth and its effect on control performance is examined in detail in Marey and Chaumette (2008). Feddema and Mitchell (1989) performed a partial 3D reconstruction to determine point depth based on observed features and known goal geometry. Papanikolopoulos and Khosla (1993) described adaptive control techniques to estimate depth, as used in this chapter. Hosoda and Asada (1994), Jägersand et al. (1996) and Piepmeier et al. (1999) have shown how the image Jacobian matrix itself can be estimated online from measurements of robot and image motion. The second-order visual servoing technique was introduced by Malis (2004).

The most common image Jacobian is based on the motion of points in the image, but it can also be derived for the parameters of lines in the image plane (Chaumette 1990; Espiau et al. 1992) and the parameters of an ellipse in the image plane (Espiau et al. 1992). Mahoney et al. (2002) describe how the choice of features effects the closed-loop dynamics. Moments of binary regions have been proposed for visual servoing of planar scenes (Chaumette 2004; Tahri and Chaumette 2005). More recently the ability to servo directly from image pixel values, without segmentation or feature extraction, has been described by Collewet et al. (2008) and subsequent papers, and more recently by Bakthavatchalam et al. (2015) and Crombez et al. (2015). Visual servoing for line, ellipse and photometric features, as well as non-perspective cameras, is covered by Corke (2017) and includes MATLAB®implementations, albeit for an older and different MATLAB Toolbox. Deep learning has been applied to visual servoing by Bateux et al. (2018).

The problem of camera retreat was was first noted in Chaumette's paper (Chaumette 1998) which introduced the specific example given in ▶ Sect. 15.2.4. One of the first methods to address this problem was 2.5D visual servoing, proposed by Malis et al. (1999), which augments the image-based point features with a minimal Cartesian feature – the first so-called *hybrid* visual servo scheme. Other notable early hybrid methods were proposed by Morel et al. (2000) and Deguchi (1998) which partitioned the image Jacobian into a translational and rotational part. A homography is computed between the initial and final view (so the goal points must be planar) and then decomposed to determine a rotation and translation. Morel et al. combine this rotational information with translational control based on IBVS of the point features. Conversely, Deguchi et al. combine this translational information with rotational control based on IBVS. Since translation is only determined up to an unknown scale factor some additional means of determining scale is required.

Corke and Hutchinson (2001) presented an intuitive geometric explanation for the camera retreat problem and proposed a partitioning scheme split by axes: x- and y-translation and rotation in one group, and z-translation and rotation in the other. Another approach to hybrid visual servoing is to switch rapidly between IBVS and PBVS approaches (Gans et al. 2003). The performance of several partitioned schemes is compared by Gans et al. (2003).

The Jacobian for point features can also be written for points expressed in polar form, (r, θ) rather than (u, v), (Iwatsuki and Okiyama 2002a, 2002b; Chaumette and Hutchinson 2007). This handles the IBVS failure case nicely, but results in somewhat suboptimal camera translational motion (Corke et al. 2009) – the converse of what happens for the Cartesian formulation.

The Jacobian for a spherical camera is similar to the polar form. The two angle parameterization was first described in Corke (2010) and was used for control and structure-from-motion estimation. There has been relatively little work on spherical visual servoing. Fomena and Chaumette (2007) consider the case for a single spherical object from which they extract features derived from the projection to the spherical imaging "plane" such as the center of the circle and its apparent radius. Tahri et al. (2009) consider spherical image features such as lines and moments. Hamel and Mahony (2002) describe kino-dynamic control of an underactuated aerial robot using point features.

Visual servoing of nonholonomic robots is nontrivial since Brockett's theorem (1983) shows that no linear time-invariant controller can control it. IBVS approaches have been proposed (Tsakiris et al. 1998; Masutani et al. 1994) but require that the camera is attached to the mobile base by a robot with a small number of degrees of freedom. Mariottini et al. 2007) describe a two-step servoing approach where the camera is rigidly attached to the base and the epipoles of the geometry, defined by the current and desired camera views, are explicitly servoed. Usher (Usher et al. 2003; Usher 2005) describes a switching control law that takes the robot onto a line that passes through the desired pose, and then along the line to the pose – experimental results on an outdoor vehicle are presented. The similarity between mobile robot navigation and visual servoing problem is discussed in Corke (2001).

The literature on PBVS is much smaller, but the paper by Westmore and Wilson (1991) is a good introduction. They use an EKF to implicitly perform pose estimation – the goal pose is the filter state and the innovation between predicted and observed feature coordinates updates the goal pose state. Hashimoto et al. (1991) present simulations to compare position-based and image-based approaches.

▪▪ History and background

Visual servoing has a very long history – the earliest reference is by Shirai and Inoue (1973) who describe how a visual feedback loop can be used to correct the position of a robot to increase task accuracy. They demonstrated a system with a servo cycle time of 10 s, and this highlights the early challenges with real-time feature extraction. Up until the late 1990s this required bulky and expensive special-purpose hardware such as that shown in ❏ Fig. 15.13. Significant early work on industrial applications occurred at SRI International during the late 1970s (Hill and Park 1979; Makhlin 1985).

In the 1980s Weiss et al. (1987) introduced the classification of visual servo structures as either position-based or image-based. They also introduced a distinction between visual servo and dynamic look and move, the former uses only visual feedback whereas the latter uses joint feedback and visual feedback. This latter distinction is no longer in common usage and most visual servo systems today make use of joint-position and visual feedback, commonly encoder-based joint velocity loops as discussed in ► Chap. 9 with an outer vision-based position loop. Weiss (1984) applied adaptive control techniques for IBVS of a robot arm without joint-level feedback, but the results were limited to low degree of freedom arms due to the low-sample rate vision processing available at that time. Others have looked at incorporating the manipulator dynamics (9.11) into controllers that command motor torque directly (Kelly 1996; Kelly et al. 2002a, 2002b) but all still require joint angles in order to evaluate the manipulator Jacobian, and the joint rates to provide damping. Feddema (Feddema and Mitchell 1989; Feddema 1989) used closed-loop joint control to overcome problems due to low visual sampling rate and demonstrated IBVS for 4-DoF. Chaumette, Rives and Espiau (Chaumette et al. 1991; Rives et al. 1989) describe a similar approach using the task function

□ Fig. 15.13 In the early 1990s Corke (1994) used a 19-inch VMEbus rack of hardware image processing cards, capable of 10 Mpixs^{-1} throughput or 50 Hz framerate for 512×512 images, to do real-time visual servoing

method (Samson et al. 1990) and show experimental results for robot positioning using a goal object with four features. Feddema et al. (1991) describe an algorithm to select which subset of the available features give the best conditioned square Jacobian. Hashimoto et al. (1991) have shown that there are advantages in using a larger number of features and using a pseudoinverse to solve for velocity. Control and stability in closed-loop visual control systems was addressed by several researchers (Corke and Good 1992; Espiau et al. 1992; Papanikolopoulos et al. 1993) and feedforward predictive, rather than feedback, controllers were proposed by Corke (1994) and Corke and Good (1996).

The 1993 book edited by Hashimoto (1993) was the first collection of papers covering approaches and applications in visual servoing. The 1996 book by Corke (1996b) is now out of print but available free online and covers the fundamentals of robotics and vision for controlling the dynamics of an image-based visual servoing system. It contains an extensive, but dated, collection of references to visual servoing applications including industrial applications, camera control for tracking, high-speed planar micromanipulator, road vehicle guidance, aircraft refueling, and fruit picking. Another important collection of papers (Kriegman et al. 1998) stems from a 1998 workshop on the synergies between control and vision: how vision can be used for control and how control can be used for vision. Another workshop collection by Chesi and Hashimoto (2010) covers more recent algorithmic developments and application.

Visual servoing has been applied to a diverse range of problems that normally require human hand-eye skills such as ping-pong (Andersson 1989), juggling (Rizzi and Koditschek 1991) and inverted pendulum balancing (Dickmanns and Graefe 1988a; Andersen et al. 1993), catching (Sakaguchi et al. 1993; Buttazzo et al. 1993; Bukowski et al. 1991; Skofteland and Hirzinger 1991; Skaar et al. 1987; Lin et al. 1989), and controlling a labyrinth game (Andersen et al. 1993).

15.3.2 Exercises

1. Position-based visual servoing (▶ Sect. 15.1)
   a) Run the PBVS example. Experiment with varying parameters such as the initial camera pose, the path fraction λ and adding pixel noise to the output of the camera.
   b) Create a PBVS system that servos with respect to a fiducial marker from ▶ Sect. 13.6.1.
   c) Use a different camera model for the pose estimation (slightly different focal length or principal point) and observe the effect on final end-effector pose.
   d) Implement an EKF based PBVS system as described in Westmore and Wilson (1991).
2. Optical flow fields (▶ Sect. 15.2.1)
   a) Plot the optical flow fields for cameras with different focal lengths.
   b) Plot the flow field for some composite camera motions such as x- and y-translation, x- and z-translation, and x-translation and z-rotation.
3. For the case of two points the image Jacobian is 4×6 and the null space has two columns. What camera motions do they correspond to?
4. Image-based visual servoing (▶ Sect. 15.2.2)
   a) Run the IBVS example and experiment with varying the gain λ. Remember that λ can be a scalar or a diagonal matrix which allows different gain settings for each degree of freedom.
   b) Implement the function to limit the maximum norm of the commanded velocity.
   c) Experiment with adding pixel noise to the output of the camera.
   d) Experiment with different initial camera poses and desired image-plane coordinates.
   e) Experiment with different number of goal points, from three up to ten. For the cases where $N > 3$ compare the performance of the pseudoinverse with just selecting a subset of three points (first three or random three). Design an algorithm that chooses a subset of points which results in the stacked Jacobian with the best condition number?
   f) Create a set of desired image-plane points that form a rectangle rather than a square. There is no perspective viewpoint from which a square appears as a rectangle (why is this?). What does the IBVS system do?
   g) Create a set of desired image-plane points that cannot be reached, for example swap two adjacent world or image points. What does the IBVS system do?
   h) Use a different camera model for the image Jacobian (slightly different focal length or principal point) and observe the effect on final end-effector pose.
   i) Implement second-order IBVS using (15.15).
   j) For IBVS we generally force points to move in straight lines but this is just a convenience. Use a trajectory generator to move the points from initial to desired position with some sideways motion, perhaps a half or full cycle of a sine wave. What is the effect on camera Cartesian motion?
   k) Implement stereo IBVS. Hint: stack the point feature Jacobians for both cameras and determine the desired feature positions on each camera's image plane.
   l) Simulate the output of a camera at arbitrary pose observing a planar target. Use knowledge of homographies and image warping to achieve this.
   m) Implement an IBVS controller using SURF features. Use the SURF descriptors to ensure robust correspondence.

5. Derive the image Jacobian for the case where the camera is limited to just pan and tilt motion. What happens if the camera pans and tilts about a point that is not the camera center?

6. When discussing motion perceptibility we used the identity $\mathbf{J}_p^{+\top}\mathbf{J}_p^{+} = (\mathbf{J}_p\mathbf{J}_p^{\top})^{-1}$. Prove this. Hint, use the singular value decomposition $\mathbf{J} = \mathbf{U\Sigma V}^\top$ and remember that $\mathbf{U}$ and $\mathbf{V}$ are orthogonal matrices.

7. End-point open-loop visual servo systems have not been discussed in this book. Consider a group of goal points on the robot end effector as well as the those on the goal object, both being observed by a single camera (challenging).

   a) Create an end-point open-loop PBVS system.

   b) Use a different camera model for the pose estimation (slightly different focal length or principal point) and observe the effect on final end-effector relative pose.

   c) Create an end-point open-loop IBVS system.

   d) Use a different camera model for the image Jacobian (slightly different focal length or principal point) and observe the effect on final end-effector relative pose.

Real-World Applications

Contents

16.1   Lane and Vehicle Detection with a Monocular Camera – 694

16.2   Highway Lane Change Planner and Controller – 697

16.3   UAV Package Delivery – 699

16.4   Pick-and-Place Workflow in Gazebo Using Point-Cloud
Processing and RRT Path Planning – 701

© The Author(s), under exclusive license to Springer Nature Switzerland AG 2023
P. Corke, W. Jachimczyk, R. Pillat, *Robotics, Vision and Control*, Springer Tracts in Advanced Robotics 147,
https://doi.org/10.1007/978-3-031-07262-8_16

► sn.pub/LYTIWe

This chapter introduces four complex applications that bring together many tools from the MATLAB® and Simulink® ecosystem to illustrate real-world problems involving self-driving vehicles, autonomous drones, and pick-and-place robots.

This part of the book is aimed at a higher level than earlier parts. It assumes a good level of familiarity with the rest of the book, and the increasingly complex examples are sketched out rather than described in detail. The four selected examples are a small subset of the many complex examples that are provided with MATLAB, and which can be found on the MathWorks® website:

	Examples
Robotics System Toolbox™	► https://sn.pub/MKSjo2
Image Processing Toolbox™	► https://sn.pub/YqK8D6
Computer Vision Toolbox™	► https://sn.pub/96OIMj
Navigation Toolbox™	► https://sn.pub/wiTegx
Automated Driving Toolbox™	► https://sn.pub/4ptjoG
UAV Toolbox	► https://sn.pub/Tc6WKE

The examples are included when these MATLAB toolboxes are installed and it is easy to run them locally – some examples also run in MATLAB Online™.

16.1 Lane and Vehicle Detection with a Monocular Camera

Vehicles that contain advanced driver assistance system (ADAS) features, or which are designed to be fully autonomous, rely on multiple sensors to provide situational awareness. These sensors typically include sonar, radar, lidar and cameras. In this example, we illustrate how a monocular camera sensor can perform tasks such as:

- Lane boundary detection,
- Detection of vehicles, people, and other objects,
- Distance estimation from the ego vehicle ◄ to obstacles.

The ego vehicle is the reference vehicle that carries the camera.

These detections and distance estimates can be used to issue lane departure warnings, collision warnings, or to design a lane-keeping control system. In conjunction with other sensors, this can also be used to implement an emergency braking system and other safety-critical features.

This example is based on functionality included in the Automated Driving Toolbox, the Computer Vision Toolbox and the Image Processing Toolbox. The complete code example can be opened in MATLAB by

```
>> openExample("driving/VisualPerceptionUsingMonoCameraExample")
```

or viewed online.

► sn.pub/LYTIWe

▣ Fig. 16.1 Ego vehicle coordinate frame

■ **Fig. 16.2** Frame from the vehicle-mounted camera (Image from the Caltech Lanes Dataset, Aly et al. 2008)

■ **Fig. 16.3** Bird's eye view of the road generated from ■ Fig. 16.2. The area of the road furthest from the sensor, the right-hand side, is blurry due to the large magnification required and the effects of pixel interpolation

The ego vehicle's 3D coordinate system is shown in ■ Fig. 16.1 where the x-axis points in the ego vehicle's forward direction, the y-axis points to the left of the vehicle, and the z-axis points upward. The camera configuration is stored in a `monoCamera` object constructed from known camera height, pitch angle, and intrinsic parameters which can be determined using a checkerboard-based camera calibration procedure, as discussed in ▶ Sect. 13.1.4. For this camera, lens distortion is negligible and will be ignored.

The `monoCamera` object provides a method `imageToVehicle` for mapping image coordinates to the coordinates of points on the road in the vehicle coordinate frame, based on a homography matrix as discussed in ▶ Sect. 13.6.2. The inverse mapping is performed by `vehicleToImage`. ▶

An example image from the vehicle-mounted camera is shown in ■ Fig. 16.2. Finding and classifying painted lane markers, either solid lines or segments of a broken line, is a critical task. Here, we use a bird's-eye-view image transform, shown in ■ Fig. 16.3, computed by a `birdsEyeView` object. The lane markers in the bird's-eye view have uniform thickness which simplifies the segmentation process, and the lane markers belonging to the same lane boundary are parallel which simplifies later image analysis steps.

The segmented lane markers, shown in ■ Fig. 16.4, parallel to the vehicle's path, are used to fit a parabolic lane boundary. The segmented pixels contain many outliers, for example, from lane markers orthogonal to the path or signs painted on the roads. Therefore, we use robust curve fitting algorithm based on RANSAC, introduced in ▶ Sect. 14.2.3. Heuristics, such as curve length are used to reject undesired road markings such as crosswalk stripes.

Additional heuristics are used to classify lane markers as either solid or dashed which is critical for steering the vehicle automatically, for example, to avoid crossing a solid line. The detected lane markers, and their classification, in the bird's-eye and original camera view are shown in ■ Fig. 16.5.

The conversion between the coordinate systems assumes a flat road. Nonflat roads introduce errors in distance computations, especially at locations that are far from the vehicle.

◙ **Fig. 16.4** Segmented lane markers found by `segmentLaneMarkerRidge`

◙ **Fig. 16.5** Lane detections shown in the synthetic bird's eye view and the original camera image. The left boundary is red, and the right boundary is green. Overlaid lines are solid or dashed to reflect the type of land boundary on the road

◙ **Fig. 16.6** Vehicles and lane markers detected with a vehicle camera. The numbers in green boxes show the locations of cars in the ego vehicle's coordinate system (Image from the Caltech Lanes Dataset, Aly et al. 2008)

An important task for front collision warning (FCW) and autonomous emergency braking (AEB) systems is detecting and tracking other vehicles. We use an aggregate channel features (ACF) detector that is pretrained to detect the front and rear of vehicles, and configure it for the specific geometry of the scene to avoid running the detector above the horizon line.

We combine lane and vehicle detection in a tracking framework, to take advantage of temporal consistency in the video sequence, and a snapshot of the results is shown in ◼ Fig. 16.6. The classes used in this example allow most parameters to be specified in real world, rather than pixel units, and that makes it easy to adapt the code to different lane marker dimensions and cameras.

16.2　Highway Lane Change Planner and Controller

An automated lane change maneuver (LCM) system enables the ego vehicle to automatically move from one lane to another lane by controlling the longitudinal and lateral velocity. An LCM system senses the environment for most important objects (MIOs) using on-board sensors, identifies an optimal, smooth and safe trajectory that avoids these objects, and steers the ego vehicle along this trajectory.

This example simulates an LCM for highway driving scenarios. It uses functionality included in Simulink®, the Automated Driving Toolbox, Computer Vision Toolbox, Image Processing Toolbox, Model Predictive Control Toolbox™, and Navigation Toolbox. The complete code example can be opened in MATLAB by

```
>> openExample("autonomous_control/" + ...
>>    "HighwayLaneChangePlannerAndControllerExample")
```

▶ sn.pub/zSS9en

The example uses Simulink and comprises a number of subsystems:
- Scenario and Environment – specifies the scene, vehicles, and map data used for simulation. It uses the `Scenario Reader` block to provide road network and vehicle ground truth positions.
- Planner Configuration Parameters – specifies the configuration parameters required for the planner algorithm.
- Lane Change Controller – specifies the path-following controller that generates control commands to steer the ego vehicle along the generated trajectory. The longitudinal control maintains a user-set velocity, while the lateral control steers the ego vehicle along the center line of its lane.
- Vehicle Dynamics – specifies the dynamic model for the ego vehicle.
- Metrics Assessment – computes metrics to assess LCM behavior: collision of the ego vehicle with other vehicles, the headway between the ego and lead vehicles, the time gap based on the headway and the longitudinal velocity of the ego vehicle, and the longitudinal and lateral jerk for passenger comfort.

This example supports a number of scenarios which are listed in ◼ Tab. 16.1 and they are passed as a parameter within the example live script

```
>> helperSLHLCPlannerAndControllerSetup(scenarioFcnName = ...
>>    "scenario_LC_15_StopnGo_Curved")
```

These scenarios are created using the `Driving Scenario Designer` and are exported to a scenario file. The comments in these files provide details about the road and vehicles in each virtual scenario.

The Visualization block in the model creates a MATLAB figure that shows the chase view and top view of the scenario, and plots the ego vehicle, sampled trajectories, and other vehicles in the scenario.

The scenario can be simulated over different time intervals and some results are shown in ◼ Fig. 16.7. Over 8 seconds, shown in ◼ Fig. 16.7b, we see that the planner generates a trajectory to navigate around a slower lead vehicle. Over 18 seconds, shown in ◼ Fig. 16.7c the planner generates a trajectory to navigate the vehicle to the left lane and then to the right lane to avoid collision with the slow-moving lead vehicle. Observe that the ego vehicle performs a lane change twice to avoid collision, while maintaining a set velocity.

During simulation, the model logs signals to the base workspace. The example provides support functions for analysis of these simulation results with a plot that shows the chase view of the scenario, and a slider which selects the desired simulation step to display different parameters and performance measures.

◘ Fig. 16.7 Simulation of `scenario_LC_15_StopnGo_Curved` scenario over 5, 8 and 18 seconds

Table 16.1 Various highway driving scenarios that can be selected
scenario_LC_01_SlowMoving
scenario_LC_02_SlowMovingWithPassingCar
scenario_LC_03_DisabledCar
scenario_LC_04_CutInWithBrake
scenario_LC_05_SingleLaneChange
scenario_LC_06_DoubleLaneChange
scenario_LC_07_RightLaneChange
scenario_LC_09_CutInWithBrake_Curved
scenario_LC_10_SingleLaneChange_Curved
scenario_LC_11_MergingCar_HighwayEntry
scenario_LC_12_CutInCar_HighwayEntry
scenario_LC_13_DisabledCar_Ushape
scenario_LC_14_DoubleLaneChange_Ushape
scenario_LC_15_StopnGo_Curved [Default]

16.3 UAV Package Delivery

A UAV used for package delivery in a 3D city environment must be able to safely fly a path defined by waypoints while avoiding obstacles sensed by its onboard lidar. Additionally, the UAV should communicate with a ground control station (GCS) to report its status and accept new missions.

This example primarily uses tools from UAV Toolbox, Navigation Toolbox, and Aerospace Blockset™. It also requires that you install the QGroundControl Ground Control Station software (QGC) from ► http://qgroundcontrol.com. QGroundControl is a powerful open-source ground station that runs on Windows, MacOS, Linux, iOS, and Android.

We start by opening the example live script in MATLAB

```
>> openExample("uav/UAVPackageDeliveryExample");
```

This example is organized as a sequence of design iterations of increasing sophistication.

The UAV Toolbox supports the standard MAVLink protocol as used by popular autopilots such as PX4 and ArduPilot. This allows the GCS software to communicate with a simulated UAV implemented in Simulink which mimics a real drone – its mission can be modified by adding waypoints or moving those that are already in the mission. The figures in this example were created using QGC and Simulink, and a typical QGC view is shown in ◘ Fig. 16.8.

The UAV operates in a rich virtual 3D environment which simulates the lidar and camera sensors, providing point clouds and front-camera imagery respectively, as shown in ◘ Fig. 16.9. We can modify the mission using the GCS to add waypoints or move those that are already in the mission.

The simulated drone uses the simulated lidar data to avoid obstacles using "on-board" logic. At some point during the flight, we see the UAV pass through a narrow pass between two buildings as seen in ◘ Fig. 16.10.

The example supports different levels of fidelity for the aircraft's dynamics. The low-fidelity multirotor plant model is computationally efficient but only approximates the closed-loop dynamics. The high-fidelity model contains an inner-loop controller and a more sophisticated plant model based on blocks from the Aerospace

► sn.pub/eOfdmW

Fig. 16.8 QGroundControl display showing mission of the simulated UAV

Fig. 16.9 Flying full mission in a photorealistic simulation environment, showing the ground control station display (left), simulated 3D lidar (top right) and different camera views

Fig. 16.10 Obstacle avoidance guiding the UAV through a passage between buildings

Blockset. There are only minor differences in the UAV behavior due to the different models. The modular structure of the example makes it easy to replace the sensing modalities, the obstacle avoidance algorithm, the ground control station, or the UAV plant simulation, without affecting the other parts of the system.

16.4 Pick-and-Place Workflow in Gazebo Using Point-Cloud Processing and RRT Path Planning

Manipulator arms are often used to perform sorting tasks, moving objects to appropriate output bins according to their type. The robot is equipped with an RGBD camera which finds fixed obstacles and classifies the objects being sorted.

This example uses tools from Robotics System Toolbox, ROS Toolbox, Image Processing Toolbox, and Computer Vision Toolbox. It also requires that you install the Gazebo simulator, and ROS is used to connect the simulator to the rest of the application running in MATLAB. For convenience, a virtual machine (VM) containing ROS and Gazebo is available for download and can be run on Windows, Linux, and MacOS platforms.

We start by opening the example live script in MATLAB

```
>> openExample("robotics/"+ ...
>>    "PickandPlaceWorkflowInGazeboUsingPointCloudProcessingExample");
```

A flow chart of the workflow is shown in ◘ Fig. 16.11. First, the robot moves to five predefined scanning poses, visualized in ◘ Fig. 16.12, and captures a set of point clouds of the scene using an onboard depth sensor. At each of the scanning poses, the current camera pose is retrieved by reading the corresponding ROS transformation using `rostf` and `getTransform`.

The captured point clouds are transformed from camera to world frame using `pctransform` and merged to a single point cloud using `pcmerge`. The final point cloud is segmented based on Euclidean distance using `pcsegdist`. The re-

▶ sn.pub/298d4k

◘ **Fig. 16.11** Flowchart of the object sorting task

◘ **Fig. 16.12** Five scanning poses as seen in the Gazebo simulator

Fig. 16.13   Collision meshes built from the point clouds at each scanning pose

Fig. 16.14   Collision-free end-effector paths generated using RRT and shown using MATLAB graphics

sulting point cloud segments are then encoded as collision meshes as shown in ☑ Fig. 16.13 (see `collisionMesh`) which are used during RRT path planning.

Iterative closest point (ICP) registration implemented by `pcregistericp` identifies which of the segmented point clouds match the geometries of objects that should be picked.

The RRT algorithm is used to plan collision-free paths from an initial to a desired joint configuration as directed by the task planner `manipulatorRRT` object. The resulting path is shortened and then interpolated at a desired validation distance. To generate a trajectory, the `trapveltraj` function is used to assign time steps to each of the interpolated waypoints following a trapezoidal velocity profile.

The trajectory is packaged into joint-trajectory ROS messages and sent as an action request to the joint-trajectory controller implemented in the KINOVA ROS package. The planned paths are visualized in a MATLAB preview figure shown in ☑ Fig. 16.14 and are simulated, with dynamics, in Gazebo. The Gazebo world shows the robot moving in the simulated scene and sorting parts to the recycling bins. The robot continues working until all objects have been placed in the bins. Some frames from the simulation are shown in ☑ Fig. 16.15.

The particular example uses a KINOVA® Gen3 manipulator arm and uses RRT for planning. However, the example can be adapted to different scenarios, robots, planners, simulation platforms, and object detection options. A number of related and variant examples are provided as live scripts and are linked from this example live script.

◘ **Fig. 16.15** Four frames from the Gazebo simulation showing objects being sorted into the appropriate bins. Bottles are placed into the right-hand (green) bin, while cans are placed into the left-hand (blue) bin

Supplementary Information

A Software Installation – 706

B Linear Algebra – 708

C Geometry – 716

D Lie Groups and Algebras – 732

E Linearization, Jacobians, and Hessians – 737

F Solving Systems of Equations – 742

G Gaussian Random Variables – 751

H Kalman Filter – 754

I Graphs – 760

J Peak Finding – 763

References – 767

Index of People – 783

Index of Functions, Classes, and Methods – 785

Index of Apps – 793

Index of Models – 795

General Index – 797

appendices.mlx

▶ sn.pub/UP1TrC

GitHub

▶ sn.pub/fDK8WN

MATLAB Onramp

▶ sn.pub/VcQb6n

Simulink Onramp

▶ sn.pub/1A0zi4

A Software Installation

The most up-to-date instructions on installing the required software can always be found at

❯ https://github.com/petercorke/RVC3-MATLAB

The examples in this book are all based on MATLAB®, a programming platform for engineers and scientists, published by MathWorks®. The core MATLAB product is a powerful platform for linear algebra, graphics and programming and can extended by software toolboxes. There is a rich ecosystem of free toolboxes but this book depends on the toolboxes listed in ◘ Tab. A.1 which must be licensed from MathWorks. Not all products are required for every part of the book.

❗ This book requires that you must have at least MATLAB R2023a in order to access all the required MATLAB language and toolbox features. The code examples rely on recent MATLAB language extensions: strings which are delimited by double quotation marks (introduced in R 2016b); and `name=value` syntax for passing arguments to functions (introduced in R 2021a), for example, `plot(x,y,LineWidth=2)` instead of the old-style `plot(x,y,"LineWidth",2)`.

Another product, Simulink®, is used in the book to illustrate systems as block-diagram models. It is not essential to have a Simulink license unless you wish to experiment with these Simulink models, which are provided with the RVC Toolbox.

You may be able to access these products via your university or company. Many universities around the world have a license for the full suite of MathWorks' products. Personal licenses are also available from MathWorks. Time-limited free trials are available at the time of writing. Full and up-to-date license options are available at the GitHub repository.

If you are new to MATLAB or Simulink, you can take free (with registration) hands-on browser-based courses called Onramp.

In addition to the licensed toolboxes you will require the free open-source RVC Toolbox which can be obtained from the GitHub repository given above.

◘ **Table A.1** MathWorks software products required by this book. Those shown in bold are sufficient to run a large subset of the code examples. Those indicated with the † are only required for ▶ Chap. 16

MATLAB®	Optimization Toolbox™
Simulink®	**Robotics System Toolbox™**
Automated Driving Toolbox™ †	ROS Toolbox †
Computer Vision Toolbox™	Signal Processing Toolbox™
Deep Learning Toolbox™	Statistics and Machine Learning Toolbox™
Image Processing Toolbox™	Symbolic Math Toolbox™
Model Predictive Control Toolbox™ †	UAV Toolbox
Navigation Toolbox™	

A.1 Working in the Cloud with MATLAB®Online™

MATLAB Online allows users to run MATLAB in the cloud through only a web browser and no local software installation. It is available to many license categories, including campus wide, and provides access to the toolboxes that you are licensed to use.

A.2 Working with the Desktop Application

- Install MATLAB and the required toolboxes using your personal, university or company license.
- Install the RVC Toolbox which provides convenience functions and data files used in the code examples in this book, as well as the source code for all MATLAB generated figures. You can clone the RVC Toolbox from GitHub

```
% git clone https://github.com/petercorke/RVC3-MATLAB
```

to a suitable location in your file system. Alternatively, you can visit ► https://github.com/petercorke/RVC3-MATLAB and download a zip archive by clicking on "Latest release" in the right-hand side panel.
- Unzip this file and add its top-level folder to your MATLAB path using the `path` command, the interactive `pathtool` command or right clicking the folder in the MATLAB file browser pane. Alternatively, double click the project file `rvc3setup.prj`.

A.3 Getting Help

The GitHub repository hosts several other resources for users:
- *Wiki pages* provides answers to frequently asked questions.
- *Discussions* between users on aspects of the book and the core toolboxes.
- *Issues* can be used for reporting and discussing issues and bugs.

B Linear Algebra

B.1　Vectors

The term *vector* has multiple meanings which can lead to confusion:
- In computer science, a vector is an array of numbers or a tuple.
- In physics and engineering, a vector is a physical quantity like force or velocity which has a magnitude, direction and unit.
- In mathematics, a vector is an object that belong to a vector space.
- In linear algebra, a vector is a one-dimensional matrix organized as either a row or a column.

In this book, we use all these interpretations and rely on the context to disambiguate, but this section is concerned with the last two meanings. We consider only real vectors which are an ordered *n-tuple* of real numbers which is usually written as

$$\boldsymbol{v} = (v_1, v_2, \ldots, v_n) \in \mathbb{R}^n$$

where v_1, v_2, etc. are called the scalar components of $\boldsymbol{v}$, and v_i is called the ith component of $\boldsymbol{v}$. The symbol $\mathbb{R}^n = \mathbb{R} \times \mathbb{R} \times \cdots \times \mathbb{R}$ is a Cartesian product that denotes the set of ordered n-tuples of real numbers. For a 3-vector, we often write the elements as $\boldsymbol{v} = (v_x, v_y, v_z)$.

The coordinate of a point in an n-dimensional space is also represented by an *n-tuple* of real numbers. A coordinate vector is that same tuple but interpreted as a linear combination $\boldsymbol{v} = v_1 \boldsymbol{e}_1 + v_2 \boldsymbol{e}_2 + \cdots + v_n \boldsymbol{e}_n$ of the orthogonal basis vectors $\{\boldsymbol{e}_1, \boldsymbol{e}_2, \ldots, \boldsymbol{e}_n\}$ of the space – this is a vector from the origin to the point. In this book, the basis vectors are denoted by $\{\hat{\boldsymbol{x}}, \hat{\boldsymbol{y}}\}$ or $\{\hat{\boldsymbol{x}}, \hat{\boldsymbol{y}}, \hat{\boldsymbol{z}}\}$ for 2 or 3 dimensions respectively.

In mathematics, an n-dimensional *vector space* (also called a *linear space*) is a group-like object that contains a collection of objects called *vectors* $\boldsymbol{v} \in \mathbb{R}^n$. It supports the operations of vector addition $\boldsymbol{a} + \boldsymbol{b} = (a_1 + b_1, a_2 + b_2, \ldots, a_n + b_n)$, and multiplication ("scaling") by a number s called a scalar $s\boldsymbol{a} = (sa_1, sa_2, \ldots, sa_n)$. The negative of a vector, scaling by -1, is obtained by negating each element of the vector $-\boldsymbol{a} = (-a_1, -a_2, \ldots, -a_n)$.

The symbol $\mathbb{R}^n$ is used to denote a space of points or vectors, and context is needed to resolve the ambiguity. We need to be careful to distinguish points and vectors because the operations of addition and scalar multiplication, while valid for vectors, are meaningless for points. We can add a vector to the coordinate vector of a point to obtain the coordinate vector of another point, and we can subtract one coordinate vector from another, and the result is the displacement between the points.

For many operations in linear algebra, it is important to distinguish between column and row vectors

$$\boldsymbol{v} = \begin{pmatrix} v_1 \\ v_2 \\ \vdots \\ v_n \end{pmatrix} \quad \text{or} \quad \boldsymbol{v} = (v_1, v_2, \ldots, v_n)$$

which are equivalent to an $n \times 1$ and a $1 \times n$ matrix (see next section) respectively. Most vectors in this book are column vectors which are sometimes written compactly as $(v_1, v_2, \ldots, v_n)^\top$, where $\cdot^\top$ denotes a matrix transpose. Both can be denoted by $\mathbb{R}^n$, or distinguished by $\mathbb{R}^{n \times 1}$ or $\mathbb{R}^{1 \times n}$ for column and row vectors respectively.

The magnitude or length of a vector is a nonnegative scalar given by its p-norm

$$\|\boldsymbol{v}\|_p = \left(\sum_{i=1}^{n} |v_i|^p\right)^{1/p} \in \mathbb{R}_{\geq 0} .$$

norm(v)

The Euclidean length of a vector is given by $\|\boldsymbol{v}\|_2$ which is also referred to as the L_2 norm and is generally assumed when p is omitted, for example $\|\boldsymbol{v}\|$. A unit vector ▶ is one where $\|\boldsymbol{v}\|_2 = 1$ and is denoted as $\hat{\boldsymbol{v}}$. The L_1 norm is the sum of the absolute value of the elements of the vector, and is also known as the Manhattan distance, it is the distance traveled when confined to moving along the lines in a grid. The L_∞ norm is the maximum element of the vector.

The unit vector corresponding to the vector $\boldsymbol{v}$ is $\hat{\boldsymbol{v}} = \frac{1}{\|\boldsymbol{v}\|_2}\boldsymbol{v}$.

The dot or inner product of two vectors is a scalar

$$\boldsymbol{a} \cdot \boldsymbol{b} = \boldsymbol{b} \cdot \boldsymbol{a} = \sum_{i=1}^{n} a_i b_i = \|\boldsymbol{a}\| \|\boldsymbol{b}\| \cos\theta$$

dot(a,b)

where θ is the angle between the vectors. $\boldsymbol{a} \cdot \boldsymbol{b} = 0$ when the vectors are orthogonal. If $\boldsymbol{a}$ and $\boldsymbol{b}$ are column vectors, $\boldsymbol{a}, \boldsymbol{b} \in \mathbb{R}^{n \times 1}$, the dot product can be written as

$$\boldsymbol{a} \cdot \boldsymbol{b} = \boldsymbol{a}^\top \boldsymbol{b} = \boldsymbol{b}^\top \boldsymbol{a}$$

The outer product $\boldsymbol{a}\boldsymbol{b}^\top \in \mathbb{R}^{n \times n}$ has a maximum rank of one, and if $\boldsymbol{b} = \boldsymbol{a}$ is a symmetric matrix.

For 3-vectors, the cross product

$$\boldsymbol{a} \times \boldsymbol{b} = -\boldsymbol{b} \times \boldsymbol{a} = \det\left(\begin{pmatrix} \hat{\boldsymbol{x}} & \hat{\boldsymbol{y}} & \hat{\boldsymbol{z}} \\ a_1 & a_2 & a_3 \\ b_1 & b_2 & b_3 \end{pmatrix}\right) = \|\boldsymbol{a}\| \|\boldsymbol{b}\| \sin\theta \hat{\boldsymbol{n}}$$

cross(a,b)

where $\hat{\boldsymbol{x}}$ is a unit vector parallel to the x-axis, etc., and $\hat{\boldsymbol{n}}$ is a unit vector normal to the plane containing $\boldsymbol{a}$ and $\boldsymbol{b}$ whose direction is given by the right-hand rule. If the vectors are parallel, $\boldsymbol{a} \times \boldsymbol{b} = 0$.

B.2 Matrices

In this book we are concerned, almost exclusively, with real $m \times n$ matrices ▶

$$\mathbf{A} = \begin{pmatrix} a_{1,1} & a_{1,2} & \cdots & a_{1,n} \\ a_{2,1} & a_{2,2} & \cdots & a_{2,n} \\ \vdots & \vdots & \ddots & \vdots \\ a_{m,1} & a_{m,2} & \cdots & a_{m,n} \end{pmatrix} \in \mathbb{R}^{m \times n}$$

Real matrices are a subset of all matrices. For the general case of complex matrices, the term Hermitian is the analog of symmetric, and unitary is the analog of orthogonal. $\mathbf{A}^{\mathrm{H}}$ denotes the Hermitian transpose, the complex conjugate transpose of the complex matrix $\mathbf{A}$. Matrices are rank-2 tensors. The MATLAB postfix operator ' computes the Hermitian transpose, which for real matrices is the same as the transpose. The .' postfix operator computes the matrix transpose. This book uses ' throughout rather than the slightly more cumbersome .' operator.

with m rows and n columns. If $n = m$, the matrix is square.

Matrices of the same size $\mathbf{A} \in \mathbb{R}^{m \times n}$ and $\mathbf{B} \in \mathbb{R}^{m \times n}$ can be added

$$\mathbf{C} = \mathbf{A} + \mathbf{B}, \quad c_{i,j} = a_{i,j} + b_{i,j}, \quad i = 1, \ldots, m, \ j = 1, \ldots, n$$

A+B

or multiplied element-wise

$$\mathbf{C} = \mathbf{A} \circ \mathbf{B}, \quad c_{i,j} = a_{i,j} b_{i,j}, \quad i = 1, \ldots, m, \ j = 1, \ldots, n$$

A.*B

which is also called the Hadamard product. Two matrices with *conforming* dimensions $\mathbf{A} \in \mathbb{R}^{m \times n}$ and $\mathbf{B} \in \mathbb{R}^{n \times p}$ can be multiplied

$$\mathbf{C} = \mathbf{A}\mathbf{B}, \quad c_{i,j} = \sum_{k=1}^{n} a_{i,k} b_{k,j}, \quad i = 1, \ldots, m, \ j = 1, \ldots, p$$

A*B

which is matrix multiplication and $\mathbf{C} \in \mathbb{R}^{m \times p}$.

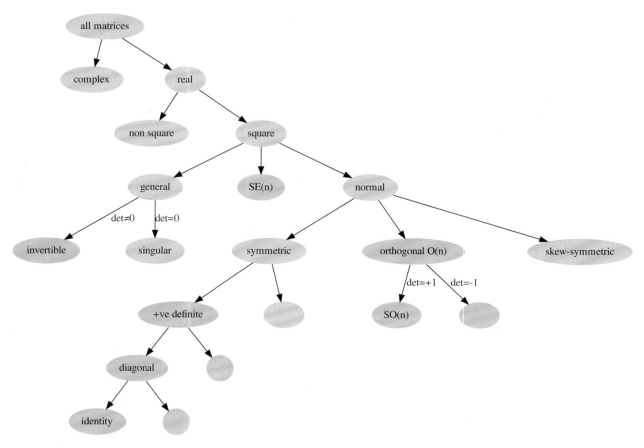

◘ Fig. B.1 Taxonomy of matrices. Matrices shown in blue are never singular

A'

The transpose of a matrix is

$$\mathbf{B} = \mathbf{A}^\top, \quad b_{i,j} = a_{j,i}, \quad \forall i, j$$

and it can be shown that

$$(\mathbf{AB})^\top = \mathbf{B}^\top \mathbf{A}^\top, \quad (\mathbf{ABC})^\top = \mathbf{C}^\top \mathbf{B}^\top \mathbf{A}^\top, \quad \text{etc.}$$

The elements of a matrix can also be matrices, creating a block or partitioned matrix

$$\mathbf{A} = \begin{pmatrix} \mathbf{B}_{1,1} & \mathbf{B}_{1,2} & \cdots & \mathbf{B}_{1,n} \\ \mathbf{B}_{2,1} & \mathbf{B}_{2,2} & \cdots & \mathbf{B}_{2,n} \\ \vdots & \vdots & \ddots & \vdots \\ \mathbf{B}_{m,1} & \mathbf{B}_{m,2} & \cdots & \mathbf{B}_{m,n} \end{pmatrix}$$

where all matrices in a column must have the same width, and all matrices in a row must have the same height. A block diagonal matrix has matrices, not necessarily of the same size, arranged along the diagonal

$$\mathbf{A} = \begin{pmatrix} \mathbf{B}_1 & & & \\ & \mathbf{B}_2 & & 0 \\ & & \ddots & \\ 0 & & & \mathbf{B}_p \end{pmatrix}$$

A taxonomy of matrices is shown in ◘ Fig. B.1.

B.2.1 Square Matrices

A square matrix may have an inverse $\mathbf{A}^{-1}$ in which case

inv(A)

$$\mathbf{A}\mathbf{A}^{-1} = \mathbf{A}^{-1}\mathbf{A} = \mathbf{1}_{n \times n}$$

where

$$\mathbf{1}_{n \times n} = \begin{pmatrix} 1 & & & 0 \\ & 1 & & \\ & & \ddots & \\ 0 & & & 1 \end{pmatrix} \in \mathbb{R}^{n \times n}$$

eye(n)

is the identity matrix, a unit diagonal matrix, sometimes written as $\mathbf{I}$. The inverse $\mathbf{A}^{-1}$ exists provided that the matrix $\mathbf{A}$ is non-singular, that is, its determinant $\det(\mathbf{A}) \neq 0$. The inverse can be computed from the matrix of cofactors. If $\mathbf{A}$ and $\mathbf{B}$ are square and non-singular, then

det(A)

$$(\mathbf{AB})^{-1} = \mathbf{B}^{-1}\mathbf{A}^{-1}, \quad (\mathbf{ABC})^{-1} = \mathbf{C}^{-1}\mathbf{B}^{-1}\mathbf{A}^{-1}, \quad \text{etc.}$$

and also

$$\left(\mathbf{A}^\top\right)^{-1} = \left(\mathbf{A}^{-1}\right)^\top = \mathbf{A}^{-\top} .$$

The inverse can be written as

$$\mathbf{A}^{-1} = \frac{1}{\det(\mathbf{A})} \text{adj}(\mathbf{A})$$

where $\text{adj}(\mathbf{A})$ is the transpose of the matrix of cofactors and known as the adjugate matrix and sometimes denoted by $\mathbf{A}^*$. ▶ If $\mathbf{A}$ is non-singular, the adjugate can be computed by

Confusingly, sometimes also referred to as the adjoint matrix, but in this book that term is reserved for the matrix introduced in ▶ Sect. 3.1.3.

$$\text{adj}(\mathbf{A}) = \det(\mathbf{A})\mathbf{A}^{-1} .$$

For a square $n \times n$ matrix, if:

$\mathbf{A} = \mathbf{A}^\top$ – the matrix is **symmetric**. The inverse of a symmetric matrix is also symmetric. Many matrices that we encounter in robotics are symmetric, for example, covariance matrices and manipulator inertia matrices.

$\mathbf{A}^{-1} = \mathbf{A}^\top$ – the matrix is **orthogonal**. The matrix is also known as **orthonormal** since its column (and row) vectors must be of unit length, and the columns (and rows) are orthogonal (normal) to each other, that is, their dot products are zero. The product of two orthogonal matrices is also an orthogonal matrix. The set of $n \times n$ orthogonal matrices forms a group $\mathbf{O}(n)$ under the operation of matrix multiplication known as the orthogonal group. The determinant of an orthogonal matrix is either $+1$ or -1. The subgroup of orthogonal matrices with determinant $+1$ is called the special orthogonal group denoted by $\mathbf{SO}(n)$.

$\mathbf{A} = -\mathbf{A}^\top$ – the matrix is **skew symmetric** or **anti symmetric**. Such a matrix has a zero diagonal, and is always singular if n is odd. Any matrix can be written as the sum of a symmetric matrix and a skew-symmetric matrix.

There is a mapping from a vector to a skew-symmetric matrix which, for the case $\boldsymbol{v} \in \mathbb{R}^3$, is

$$\mathbf{A} = [\boldsymbol{v}]_\times = \begin{pmatrix} 0 & -v_z & v_y \\ v_z & 0 & -v_x \\ -v_y & v_x & 0 \end{pmatrix} \tag{B.1}$$

vec2skew(v)

The inverse mapping is

$$\boldsymbol{v} = \vee_\times(\mathbf{A}) \ .$$

If $\boldsymbol{v} \in \mathbb{R}^n$ and $\mathbf{A} \in \mathbb{R}^{n \times n}$ then $[\mathbf{A}\boldsymbol{v}]_\times = \det(\mathbf{A})\mathbf{A}^{-\top}[\boldsymbol{v}]_\times \mathbf{A}^{-1}$.

For the 3-dimensional case, the cross product can be written as the matrix-vector product $\boldsymbol{v}_1 \times \boldsymbol{v}_2 = [\boldsymbol{v}_1]_\times \boldsymbol{v}_2$, and $\boldsymbol{v}^\top[\boldsymbol{v}]_\times = [\boldsymbol{v}]_\times \boldsymbol{v} = 0$, $\forall \boldsymbol{v}$. If $\mathbf{R} \in \mathbf{SO}(3)$ then $[\mathbf{R}\boldsymbol{v}]_\times = \mathbf{R}[\boldsymbol{v}]_\times \mathbf{R}^\top$.

$\mathbf{A}^\top \mathbf{A} = \mathbf{A}\mathbf{A}^\top$ – the matrix is **normal** and can be diagonalized by an orthogonal matrix $\mathbf{U}$ so that $\mathbf{U}^\top \mathbf{A}\mathbf{U}$ is a diagonal matrix. All symmetric, skew-symmetric and orthogonal matrices are normal matrices as are matrices of the form $\mathbf{A} = \mathbf{B}^\top \mathbf{B} = \mathbf{B}\mathbf{B}^\top$ where $\mathbf{B}$ is an arbitrary matrix.

The square matrix $\mathbf{A} \in \mathbb{R}^{n \times n}$ can be applied as a linear transformation to a vector $\boldsymbol{x} \in \mathbb{R}^n$

$$\boldsymbol{x}' = \mathbf{A}\boldsymbol{x}$$

which results in another vector, generally with a change in its length and direction. For example, in two dimensions, if $\boldsymbol{x}$ is the set of all points lying on a circle, then $\boldsymbol{x}'$ defines points that lie on some ellipse. However, there are some important special cases. If $\mathbf{A} \in \mathbf{SO}(n)$, the transformation is isometric and the vector's length is unchanged, $\|\boldsymbol{x}'\| = \|\boldsymbol{x}\|$. The (right) eigenvectors of the matrix are those vectors $\boldsymbol{x}$ such that

$$\mathbf{A}\boldsymbol{x} = \lambda_i \boldsymbol{x}, \quad i = 1, \dots, n \tag{B.2}$$

that is, their direction is unchanged when transformed by the matrix. They are simply scaled by λ_i, the corresponding eigenvalue. The matrix $\mathbf{A}$ has n eigenvalues (the *spectrum* of the matrix) which can be real or complex pairs. For an orthogonal matrix, the eigenvalues lie on a unit circle in the complex plane, $|\lambda_i| = 1$, and the eigenvectors are all orthogonal to one another.

If $\mathbf{A}$ is non-singular, then the eigenvalues of $\mathbf{A}^{-1}$ are the reciprocal of those of $\mathbf{A}$, and the eigenvectors of $\mathbf{A}^{-1}$ are parallel to those of $\mathbf{A}$. The eigenvalues of $\mathbf{A}^\top$ are the same as those of $\mathbf{A}$ but the eigenvectors are different.

The trace of a matrix is the sum of the diagonal elements

$$\text{tr}(\mathbf{A}) = \sum_{i=1}^n a_{i,i}$$

which is also the sum of the eigenvalues

$$\text{tr}(\mathbf{A}) = \sum_{i=1}^n \lambda_i \ .$$

For a rotation matrix, either 2D or 3D, the trace is related to the angle of rotation

$$\theta = \cos^{-1} \frac{\text{tr}(\mathbf{R}) - 1}{2}$$

about the rotation axis, but due to the limited range of $\cos^{-1}$, values of θ above π cannot be properly determined.

The determinant of the matrix is equal to the product of the eigenvalues

$$\det(\mathbf{A}) = \prod_{i=1}^n \lambda_i$$

thus a matrix with one or more zero eigenvalues will be singular.

The eigenvalues of a real **symmetric** matrix $\mathbf{A}$ are all real and we classify such a matrix according to the sign of its eigenvalues:

- $\lambda_i > 0$, $\forall i$, positive-definite, never singular, written as $\mathbf{A} \succ 0$
- $\lambda_i \geq 0$, $\forall i$, positive-semi-definite, possibly singular, written as $\mathbf{A} \succeq 0$
- $\lambda_i < 0$, $\forall i$, negative-definite, never singular, written as $\mathbf{A} \prec 0$
- otherwise, indefinite.

The diagonal elements of a positive-definite matrix are positive, $a_{i,i} > 0$, and its determinant is positive $\det(\mathbf{A}) > 0$. The inverse of a positive-definite matrix is also positive definite.

We will frequently encounter the matrix quadratic form

$$s = \mathbf{x}^\top \mathbf{A} \mathbf{x} \tag{B.3}$$

which is a scalar. If $\mathbf{A}$ is positive-definite, then $s > 0$, $\forall \mathbf{x} \neq 0$. For the case that $\mathbf{A}$ is diagonal this can be written

$$s = \sum_{i=1}^{n} a_{i,i} x_i^2$$

which is a weighted sum of squares. If $\mathbf{A} = \mathbf{1}$ then $s = (\|\mathbf{v}\|_2)^2$. If $\mathbf{A}$ is symmetric then

$$s = \sum_{i=1}^{n} a_{i,i} x_i^2 + 2 \sum_{i=1}^{n} \sum_{j=i+1}^{n} a_{i,j} x_i x_j$$

and the result also includes products or correlations between elements of $\mathbf{x}$.

The Mahalanobis distance is a weighted distance or norm

$$s = \sqrt{\mathbf{x}^\top \mathbf{P}^{-1} \mathbf{x}}$$

where $\mathbf{P} \in \mathbb{R}^{n \times n}$ is a covariance matrix which down-weights elements of $\mathbf{x}$ where uncertainty is high.

The matrices $\mathbf{A}^\top \mathbf{A}$ and $\mathbf{A} \mathbf{A}^\top$ are always symmetric and positive-semi-definite. This implies than any symmetric matrix $\mathbf{A}$ can be written as

$$\mathbf{A} = \mathbf{L} \mathbf{L}^\top$$

where $\mathbf{L}$ is the Cholesky decomposition of $\mathbf{A}$.   `chol(A)`

Other matrix factorizations of $\mathbf{A}$ include the matrix square root   `sqrtm(A)`

$$\mathbf{A} = \mathbf{S} \mathbf{S}$$

where $\mathbf{S}$ is the square root of $\mathbf{A}$ or $\mathbf{A}^{\frac{1}{2}}$ which is positive definite (and symmetric) if $\mathbf{A}$ is positive definite, and QR-decomposition   `qr(A)`

$$\mathbf{A} = \mathbf{Q} \mathbf{R}$$

where $\mathbf{Q}$ is an orthogonal matrix and $\mathbf{R}$ is an upper triangular matrix.

If $\mathbf{T}$ is any non-singular matrix, then

$$\mathbf{A} = \mathbf{T} \mathbf{B} \mathbf{T}^{-1}$$

is known as a similarity transformation or conjugation. $\mathbf{A}$ and $\mathbf{B}$ are said to be similar, and it can be shown that the eigenvalues are unchanged by this transformation.

The matrix form of (B.2) is

$$\mathbf{A}\mathbf{X} = \mathbf{X}\mathbf{\Lambda}$$

where $\mathbf{X} \in \mathbb{R}^{n \times n}$ is a matrix of eigenvectors of $\mathbf{A}$, arranged column-wise, and $\mathbf{\Lambda}$ is a diagonal matrix of corresponding eigenvalues. If $\mathbf{X}$ is non-singular, we can rearrange this as

$$\mathbf{A} = \mathbf{X}\mathbf{\Lambda}\mathbf{X}^{-1}$$

which is the eigen or spectral decomposition of the matrix. This implies that the matrix can be diagonalized by a similarity transformation

$$\mathbf{\Lambda} = \mathbf{X}^{-1}\mathbf{A}\mathbf{X} \,.$$

If $\mathbf{A}$ is symmetric, then $\mathbf{X}$ is orthogonal and we can instead write

$$\mathbf{A} = \mathbf{X}\mathbf{\Lambda}\mathbf{X}^{\top} \,. \tag{B.4}$$

The determinant of a square matrix $\mathbf{A} \in \mathbb{R}^{n \times n}$ is the factor by which the transformation scales volumes in an n-dimensional space. For two dimensions, imagine a shape defined by points x_i with an enclosed area Λ. The shape formed by the points $\mathbf{A}x_i$ would have an enclosed area of $\Lambda \det(\mathbf{A})$. If $\mathbf{A}$ is singular, the points $\mathbf{A}x_i$ would be coincident or collinear and have zero enclosed area. In a similar way for three dimensions, the determinant is a scale factor applied to the volume of a set of points transformed by $\mathbf{A}$.

rank(A)

The columns of $\mathbf{A} = (c_1, c_2, \ldots, c_n)$ can be considered as a set of vectors that define a space – the column space. Similarly, the rows of $\mathbf{A}$ can be considered as a set of vectors that define a space – the row space. The column rank of a matrix is the number of linearly independent columns of $\mathbf{A}$. Similarly, the row rank is the number of linearly independent rows of $\mathbf{A}$. The column rank and the row rank are always equal and are simply called the rank of $\mathbf{A}$, and this has an upper bound of n. A square matrix for which $\text{rank}(\mathbf{A}) < n$ is said to be rank deficient or not of full rank, and will be singular. The rank shortfall $n - \text{rank}(\mathbf{A})$ is the nullity of $\mathbf{A}$. In addition, $\text{rank}(\mathbf{A}\mathbf{B}) \leq \min(\text{rank}(\mathbf{A}), \text{rank}(\mathbf{B}))$ and $\text{rank}(\mathbf{A} + \mathbf{B}) \leq \text{rank}(\mathbf{A}) + \text{rank}(\mathbf{B})$. If x is a column vector, the matrix $xx^{\top}$, the outer product of x, has rank 1 for all $x \neq \mathbf{0}$.

B.2.2 Non-Square and Singular Matrices

Singular square matrices can be treated as a special case of a non-square matrix. For a non-square matrix $\mathbf{A} \in \mathbb{R}^{m \times n}$ the rank is the dimension of the largest non-singular square submatrix that can be formed from $\mathbf{A}$, and has an upper bound of $\min(m, n)$.

A non-square or singular matrix cannot be inverted, but we can determine the left generalized inverse or pseudoinverse or Moore-Penrose pseudoinverse

pinv(A)

$$\mathbf{A}^{+}\mathbf{A} = \mathbf{1}_{n \times n}$$

where $\mathbf{A}^{+} = (\mathbf{A}^{\top}\mathbf{A})^{-1}\mathbf{A}^{\top}$. The right generalized inverse is

$$\mathbf{A}\mathbf{A}^{+} = \mathbf{1}_{m \times m}$$

where $\mathbf{A}^{+} = \mathbf{A}^{\top}(\mathbf{A}\mathbf{A}^{\top})^{-1}$. The qualifier left or right denotes which side of $\mathbf{A}$ the pseudoinverse appears.

If the matrix $\mathbf{A}$ is not of full rank, then it has a finite null space or kernel. A vector x lies in the null space of the matrix if

$$\mathbf{A}x = \mathbf{0} \, .$$

null(A)

More precisely, this is the right-null space. A vector lies in the left-null space if

$$x^\top \mathbf{A} = \mathbf{0} \, .$$

The left-null space is equal to the right-null space of $\mathbf{A}^\top$.

The null space is defined by a set of orthogonal basis vectors whose cardinality is the nullity of $\mathbf{A}$. Any linear combination of these null-space basis vectors lies in the null space.

For a non-square matrix $\mathbf{A} \in \mathbb{R}^{m \times n}$, the analog to (B.2) is

$$\mathbf{A}v_i = \sigma_i u_i$$

where $u_i \in \mathbb{R}^m$ and $v_i \in \mathbb{R}^n$ are respectively the right and left singular vectors of $\mathbf{A}$, and σ_i its singular values. The singular values are nonnegative real numbers that are the square root of the eigenvalues of $\mathbf{A}\mathbf{A}^\top$ and u_i are the corresponding eigenvectors. v_i are the eigenvectors of $\mathbf{A}^\top \mathbf{A}$.

The singular value decomposition or SVD of the matrix $\mathbf{A}$ is

$$\mathbf{A} = \mathbf{U}\boldsymbol{\Sigma}\mathbf{V}^\top$$

svd(A)

where $\mathbf{U} \in \mathbb{R}^{m \times m}$ and $\mathbf{V} \in \mathbb{R}^{n \times n}$ are both orthogonal matrices comprising, as columns, the corresponding singular vectors u_i and v_i. $\boldsymbol{\Sigma} \in \mathbb{R}^{m \times n}$ is a diagonal matrix of the singular values σ_i in *decreasing* magnitude

$$\boldsymbol{\Sigma} = \begin{pmatrix} \sigma_1 & & & & & \\ & \ddots & & & 0 & \\ & & \sigma_r & & & \\ & & & 0 & & \\ & 0 & & & \ddots & \\ & & & & & 0 \end{pmatrix}$$

where $r = \operatorname{rank}(\mathbf{A})$ is the rank of $\mathbf{A}$ and $\sigma_{i+1} \leq \sigma_i$. For the case where $r < n$, the diagonal will have $n - r$ zero elements as shown. Columns of $\mathbf{V}^\top$ corresponding to the zero columns of $\boldsymbol{\Sigma}$ define the null space of $\mathbf{A}$. The condition number of a matrix $\mathbf{A}$ is $\max(\boldsymbol{\sigma})/\min(\boldsymbol{\sigma})$ and a high value means the matrix is close to singular or "poorly conditioned".

cond(A)

C Geometry

Geometric concepts such as points, lines, ellipses and planes are critical to the fields of robotics and robotic vision. We briefly summarize key representations in both Euclidean and projective (homogeneous coordinate) space.

C.1 Euclidean Geometry

C.1.1 Points

A point in an n-dimensional space is represented by an n-tuple, an ordered set of n numbers $(x_1, x_2, \ldots, x_n)^\top$ which define the coordinates of the point. The tuple can also be interpreted as a column vector – a coordinate vector – from the origin to the point. A point in 2-dimensions is written as $p = (x, y)^\top$, and in 3-dimensions is written as $p = (x, y, z)^\top$.

C.1.2 Lines

C.1.2.1 Lines in 2D

A line is defined by $\ell = (a, b, c)^\top$ such that

$$ax + by + c = 0 \tag{C.1}$$

which is a generalization of the line equation we learned in school $y = mx + c$, but which can easily represent a vertical line by setting $b = 0$. $v = (a, b)$ is a vector normal to the line, and $v = (-b, a)$ is a vector parallel to the line. The line that joins two points p_1 and p_2, $p_i = (x_i, y_i)$, is given by the solution to

$$\begin{pmatrix} x_1 & y_1 & 1 \\ x_2 & y_2 & 1 \end{pmatrix} \begin{pmatrix} a \\ b \\ c \end{pmatrix} = \mathbf{0}$$

which is found from the right null space of the leftmost term. The intersection point of two lines ℓ_1 and ℓ_2 is

$$\begin{pmatrix} a_1 & b_1 \\ a_2 & b_2 \end{pmatrix} \begin{pmatrix} x \\ y \end{pmatrix} = -\begin{pmatrix} c_1 \\ c_2 \end{pmatrix}$$

which has no solution if the lines are parallel – the leftmost term is singular.

We can also represent the line in polar form

$$x \cos \theta + y \sin \theta + \rho = 0$$

where θ is the angle from the x-axis to the line and ρ is the normal distance between the line and the origin, as shown in ◻ Fig. 12.20.

C.1.2.2 Lines in 3D and Plücker Coordinates

We can define a line by two points, $\mathbf{p}$ and $\mathbf{q}$, as shown in ◻ Fig. C.1, which would require a total of six parameters $\ell = (p_x, p_y, p_z, q_x, q_y, q_z)$. However, since these points can be arbitrarily chosen, there would be an infinite set of parameters that represent the same line making it hard to determine the equivalence of two lines.

There are advantages in representing a line as

$$\ell = (\boldsymbol{\omega} \times \boldsymbol{q}, \boldsymbol{p} - \boldsymbol{q}) = (\boldsymbol{v}, \boldsymbol{\omega}) \in \mathbb{R}^6$$

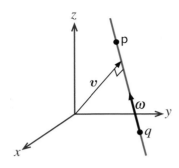

◻ Fig. C.1 Describing a line in three dimensions

where $\boldsymbol{\omega}$ is the direction of the line and $\boldsymbol{v}$ is the moment of the line – a vector from the origin to a point on the line and which is normal to the line. This is a Plücker coordinate vector – a 6-dimensional quantity subject to two constraints: the coordinates are homogeneous and thus invariant to overall scale factor; and $\boldsymbol{v} \cdot \boldsymbol{\omega} = 0$. Lines therefore have 4 degrees-of-freedom ► and the Plücker coordinates lie on a 4-dimensional manifold – the Klein quadric – in 6-dimensional space. Lines with $\boldsymbol{\omega} = \boldsymbol{0}$ lie at infinity and are known as ideal lines. ►

This is not intuitive but consider two parallel planes and an arbitrary 3D line passing through them. The line can be described by the 2-dimensional coordinates of its intersection point on each plane – a total of four coordinates.

Ideal as in imaginary, not as in perfect.

We will first define two points

```
>> P = [2 3 4]; Q = [3 5 7];
```

and then create a Plücker line object

```
>> L = Plucker(P,Q)
L =
{ 1  -2  1; -1  -2  -3}
```

which displays the $\boldsymbol{v}$ and $\boldsymbol{\omega}$ components. These can be accessed as properties

```
>> L.v
ans =
     1    -2     1
>> L.w
ans =
    -1    -2    -3
```

A Plücker line can also be represented as a skew-symmetric matrix

```
>> L.skew
ans =
     0     1     2    -1
    -1     0     1    -2
    -2    -1     0    -3
     1     2     3     0
```

To plot this line, we first define a region of 3D space ► then plot it in blue

Since lines have infinite length, we need to specify a finite volume in which to draw it.

```
>> axis([-5 5 -5 5 -5 5]);
>> L.plot("b");
```

The line is the set of all points

$$\boldsymbol{p}(\lambda) = \frac{\boldsymbol{v} \times \boldsymbol{\omega}}{\boldsymbol{\omega} \cdot \boldsymbol{\omega}} + \lambda\boldsymbol{\omega}, \quad \lambda \in \mathbb{R}$$

which can be generated parametrically in terms of the scalar parameter λ

```
>> L.point([0 1 2])
ans =
    0.5714    0.1429    -0.2857
    0.3042   -0.3917    -1.0875
    0.0369   -0.9262    -1.8893
```

where the rows are points on the line corresponding to $\lambda = 0, 1, 2$.

A point x is closest to the line when

$$\lambda = \frac{(x - q) \cdot \omega}{\omega \cdot \omega} \; .$$

For the point $(1, 2, 3)$, the closest point on the line, and its distance, is given by

```
>> [x,d] = L.closest([1 2 3])
x =
    1.5714    2.1429    2.7143
d =
    0.6547
```

The line intersects the plane $n^{\top} x + d = 0$ at the point coordinate

$$x = \frac{v \times n - d\,\omega}{\omega \cdot n} \; .$$

For the xy-plane, the line intersects at

```
>> L.intersect_plane([0 0 1 0])
ans =
    0.6667    0.3333         0
```

Two lines can be identical, coplanar or skewed. Identical lines have linearly dependent Plücker coordinates, that is, $\ell_1 = \lambda \ell_2$ or $\hat{\ell}_1 = \hat{\ell}_2$. If coplanar, they can be parallel or intersecting, and, if skewed they can be intersecting or not. If lines have $\omega_1 \times \omega_2 = 0$ they are parallel, otherwise, they are skewed.

The minimum distance between two lines is

$$d = \omega_1 \cdot v_2 + \omega_2 \cdot v_1$$

and is zero if they intersect.

For two lines ℓ^1 and ℓ^2, the side operator is a permuted dot product

$$\text{side}(\ell^1, \ell^2) = \ell_1^1 \ell_5^2 + \ell_2^1 \ell_6^2 + \ell_3^1 \ell_4^2 + \ell_4^1 \ell_3^2 + \ell_5^1 \ell_1^2 + \ell_6^1 \ell_2^2$$

which is zero if the lines intersect or are parallel, and is computed by the `side` method. For `Plucker` objects `L1` and `L2`, the operators `L1 ^ L2` and `L1 | L2` are true if the lines are respectively intersecting or parallel.

Excurse C.1: Julius Plücker

Plücker (1801–1868) was a German mathematician and physicist who made contributions to the study of cathode rays and analytical geometry. He was born at Elberfeld and studied at Düsseldorf, Bonn, Heidelberg and Berlin, and went to Paris in 1823 where he was influenced by the French geometry movement. In 1825, he returned to the University of Bonn, was made professor of mathematics in 1828 (at age 27), and professor of physics in 1836. In 1858, he proposed that the lines of the spectrum, discovered by his colleague Heinrich Geissler (of Geissler tube fame), were characteristic of the chemical substance which emitted them. In 1865, he returned to geometry and invented what was known as line geometry. He was the recipient of the Copley Medal from the Royal Society in 1866, and is buried in the Alter Friedhof (Old Cemetery) in Bonn.

We can transform a Plücker line between frames by

$$^B\boldsymbol{\ell}' = \mathrm{Ad}(\,^B\boldsymbol{\xi}_A)\,^A\boldsymbol{\ell}$$

where the adjoint of the rigid-body motion is described by (D.2).

C.1.3 Planes

A plane is defined by a 4-vector $\boldsymbol{\pi} = (a, b, c, d)^\top$ and is the set of all points $\boldsymbol{x} = (x, y, z)^\top$ such that

$$ax + by + cz + d = 0$$

which can be written in point-normal form as

$$\boldsymbol{n}^\top(\boldsymbol{x} - \boldsymbol{p}) = 0$$

where $\boldsymbol{n} = (a, b, c)$ is the normal to the plane, and $\boldsymbol{p} \in \mathbb{R}^3$ is a point in the plane.

A plane with the normal $\boldsymbol{n}$ and containing the point with coordinate vector $\boldsymbol{p}$ is $\boldsymbol{\pi} = (n_x, n_y, n_z, \boldsymbol{n} \cdot \boldsymbol{p})^\top$. A plane can also be defined by 3 points $\boldsymbol{p}_1$, $\boldsymbol{p}_2$ and $\boldsymbol{p}_3$, where $\boldsymbol{p}_i = (x_i, y_i, z_i)$

$$\begin{pmatrix} x_1 & y_1 & z_1 & 1 \\ x_2 & y_2 & z_2 & 1 \\ x_3 & y_3 & z_3 & 1 \end{pmatrix} \boldsymbol{\pi} = \boldsymbol{0}$$

and solved for using the right null space of the leftmost term, or by two nonparallel lines $\boldsymbol{\ell}_1$ and $\boldsymbol{\ell}_2$

$$\boldsymbol{\pi} = (\boldsymbol{\omega}_1 \times \boldsymbol{\omega}_2, \ \boldsymbol{v}_1 \cdot \boldsymbol{\omega}_2)$$

or by a Plücker line $(\boldsymbol{v}, \boldsymbol{w})$ and a point with coordinate vector $\boldsymbol{p}$

$$\boldsymbol{\pi} = (\boldsymbol{\omega} \times \boldsymbol{p} - \boldsymbol{v}, \ \boldsymbol{v} \cdot \boldsymbol{p}) \ .$$

A point can also be defined as the intersection point of three planes $\boldsymbol{\pi}_1$, $\boldsymbol{\pi}_2$ and $\boldsymbol{\pi}_3$

$$\begin{pmatrix} a_1 & b_1 & c_1 \\ a_2 & b_2 & c_2 \\ a_3 & b_3 & c_3 \end{pmatrix} \begin{pmatrix} x \\ y \\ z \end{pmatrix} = - \begin{pmatrix} d_1 \\ d_2 \\ d_3 \end{pmatrix} \ .$$

If the left-hand matrix is singular, then two or more planes are parallel and have either zero or infinitely many intersection points.

The Plücker line formed by the intersection of two planes $\boldsymbol{\pi}_1$ and $\boldsymbol{\pi}_2$ is

$$\boldsymbol{\ell} = (\boldsymbol{n}_1 \times \boldsymbol{n}_2, \ d_2\boldsymbol{n}_1 - d_1\boldsymbol{n}_2) \ .$$

C.1.4 Ellipses and Ellipsoids

An ellipse belongs to the family of planar curves known as conics. The simplest form of an ellipse, centered at $(0, 0)$, is defined implicitly by the points (x, y) such that

$$\frac{x^2}{a^2} + \frac{y^2}{b^2} = 1$$

A circle is a special case of an ellipse where $a = b = r$ and r is the radius.

and is shown in ◨ Fig. C.2a. ◄ This canonical ellipse is centered at the origin and has its major and minor axes aligned with the x- and y-axes. The radius in the x-direction is a and in the y-direction is b. The longer of the two radii is known as the semi-major axis length, and the other is the semi-minor axis length.

We can write the ellipse in matrix quadratic form (B.3) as

$$(x \quad y)\begin{pmatrix} 1/a^2 & 0 \\ 0 & 1/b^2 \end{pmatrix}\begin{pmatrix} x \\ y \end{pmatrix} = 1$$

and more compactly as

$$\boldsymbol{x}^\top \mathbf{E} \boldsymbol{x} = 1 \tag{C.2}$$

where $\boldsymbol{x} = (x, y)^\top$ and $\mathbf{E}$ is a symmetric matrix

$$\mathbf{E} = \begin{pmatrix} \alpha & \gamma \\ \gamma & \beta \end{pmatrix}. \tag{C.3}$$

Its determinant $\det(\mathbf{E}) = \alpha\beta - \gamma^2$ defines the type of conic

$$\det(\mathbf{E}) \begin{cases} > 0 & \text{ellipse} \\ = 0 & \text{parabola} \\ < 0 & \text{hyperbola} \end{cases}$$

An ellipse can therefore be represented by a positive-definite symmetric matrix $\mathbf{E}$. Conversely, any positive-definite symmetric 2×2 matrix, such as the inverse of an inertia tensor or covariance matrix, has an equivalent ellipse.

Nonzero values of γ change the orientation of the ellipse. The ellipse can be arbitrarily centered at $\boldsymbol{x}_c = (x_c, y_c)$ by writing it in the form

$$(\boldsymbol{x} - \boldsymbol{x}_c)^\top \mathbf{E}(\boldsymbol{x} - \boldsymbol{x}_c) = 1$$

which leads to the general ellipse shown in ◨ Fig. C.2b.

Since $\mathbf{E}$ is symmetric, it can be diagonalized by (B.4)

$$\mathbf{E} = \mathbf{X}\boldsymbol{\Lambda}\mathbf{X}^\top$$

where $\mathbf{X}$ is an orthogonal matrix comprising the eigenvectors of $\mathbf{E}$, and the diagonal elements of $\boldsymbol{\Lambda}$ are the eigenvalues of $\mathbf{E}$. The quadratic form (C.2) becomes

$$\boldsymbol{x}^\top \mathbf{X}\boldsymbol{\Lambda}\mathbf{X}^\top \boldsymbol{x} = 1$$
$$(\mathbf{X}^\top \boldsymbol{x})^\top \boldsymbol{\Lambda}(\mathbf{X}^\top \boldsymbol{x}) = 1$$
$$\boldsymbol{x}'^\top \boldsymbol{\Lambda}\boldsymbol{x}' = 1$$

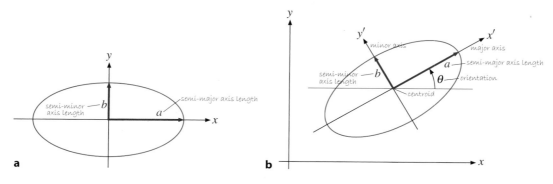

◨ **Fig. C.2** Ellipses. **a** Canonical ellipse centered at the origin and aligned with the x- and y-axes; **b** general form of ellipse which is similar to (C.2) but with the ellipse defined by the diagonal matrix $\mathbf{\Lambda}$ with respect to the rotated coordinated frame $\mathbf{x'} = \mathbf{X}^\top \mathbf{x}$. The major and minor ellipse axes are aligned with the eigenvectors of $\mathbf{E}$.

The radii of the ellipse are related to the inverse square root of the eigenvalues

$$r_i = \frac{1}{\sqrt{\lambda_i}} \tag{C.4}$$

so the major and minor radii of the ellipse, a and b respectively, are determined by the smallest and largest eigenvalues respectively. The area of an ellipse is

$$\Lambda = \pi r_1 r_2 = \frac{\pi}{\sqrt{\det(\mathbf{E})}} \tag{C.5}$$

since $\det(\mathbf{E}) = \prod \lambda_i$. The eccentricity is

$$\varepsilon = \frac{\sqrt{a^2 - b^2}}{a} \ .$$

Alternatively, the ellipse can be represented in polynomial form by writing as

$$(\mathbf{x} - \mathbf{x}_c)^\top \begin{pmatrix} \alpha & \gamma \\ \gamma & \beta \end{pmatrix} (\mathbf{x} - \mathbf{x}_c) = 1$$

and expanding to

$$e_1 x^2 + e_2 y^2 + e_3 xy + e_4 x + e_5 y + e_6 = 0$$

where $e_1 = \alpha$, $e_2 = \beta$, $e_3 = 2\gamma$, $e_4 = -2(\alpha x_c + \gamma y_c)$, $e_5 = -2(\beta y_c + \gamma x_c)$ and $e_6 = \alpha x_c^2 + \beta y_c^2 + 2\gamma x_c y_c - 1$. The ellipse has only five degrees of freedom, its center coordinate and the three unique elements in $\mathbf{E}$. For a nondegenerate ellipse where $e_1 \neq 0$, we can rewrite the polynomial in normalized form

$$x^2 + \epsilon_1 y^2 + \epsilon_2 xy + \epsilon_3 x + \epsilon_4 y + \epsilon_5 = 0 \tag{C.6}$$

with five unique parameters $\epsilon_i = e_{i+1}/e_1$, $i = 1, \ldots, 5$. Consider the ellipse

$$\mathbf{x}^\top \begin{pmatrix} 1 & 1 \\ 1 & 2 \end{pmatrix} \mathbf{x} = 1$$

which is represented in MATLAB by

```
>> E = [1 1;1 2];
```

We can plot this by

```
>> plotellipse(E)
```

which is shown in ◘ Fig. C.3. The eigenvectors and eigenvalues of $\mathbf{E}$ are

```
>> [x,e] = eig(E)
x =
   -0.8507    0.5257
    0.5257    0.8507
e =
    0.3820         0
         0    2.6180
```

◻ Fig. C.3 Ellipse corresponding to a symmetric 2×2 matrix. The arrows indicate the major and minor axes of the ellipse

where the eigenvalues are given in ascending order on the diagonal of e. The ellipse radii are

```
>> r = 1./sqrt(diag(e))
r =
    1.6180
    0.6180
```

If either radius is equal to zero the ellipse is degenerate and becomes a line. If both radii are zero the ellipse is a point.

The eigenvectors are unit vectors in the major and minor axis directions ◀ and we will scale them by the radii to yield radius vectors which we can plot

Since eig sorts the eigenvalues in ascending order.

```
>> p = x(:,1)*r(1); quiver(0,0,p(1),p(2),0,"k");
>> p = x(:,2)*r(2); quiver(0,0,p(1),p(2),0,"k");
```

The orientation of the ellipse is the angle of the major axis with respect to the horizontal axis and is

$$\theta = \tan^{-1} \frac{v_y}{v_x}$$

where $\boldsymbol{v} = (v_x, v_y)$ is the eigenvector corresponding to the smallest eigenvalue (the largest radius). For our example, this is

```
>> atan2d(x(2,1),x(1,1))
ans =
  148.2825
```

in units of degrees, counter-clockwise from the x-axis.

The RVC Toolbox function plotellipsoid will draw an ellipsoid, it is similar to plotellipse, but **E** is a 3×3 matrix.

C.1.4.1 Drawing an Ellipse

In order to draw an ellipse, we first define a set of points on the unit circle $y = (x, y)^\top$ such that

$$y^\top y = 1 .$$ (C.7)

We rewrite (C.2) as

$$x^\top \mathbf{E}^{\frac{1}{2}} \mathbf{E}^{\frac{1}{2}} x = 1$$ (C.8)

where $\mathbf{E}^{\frac{1}{2}}$ is the matrix square root, and equating (C.7) and (C.8), we can write

$$x^\top \mathbf{E}^{\frac{1}{2}} \mathbf{E}^{\frac{1}{2}} x = y^\top y .$$

which leads to

$$y = \mathbf{E}^{\frac{1}{2}} x$$

which we can rearrange as

$$x = \mathbf{E}^{-\frac{1}{2}} y$$

that transforms a point on the unit circle to a point on an ellipse, and $\mathbf{E}^{-\frac{1}{2}}$ is the inverse of the matrix square root. If the ellipse is centered at x_c, rather than the origin, we can perform a change of coordinates

$$(x - x_c)^\top \mathbf{E}^{\frac{1}{2}} \mathbf{E}^{\frac{1}{2}} (x - x_c) = 1$$

from which we write the transformation as

$$x = \mathbf{E}^{-\frac{1}{2}} y + x_c$$

Drawing an ellipsoid is tackled in an analogous fashion.

Continuing the code example above

```
>> E = [1 1; 1 2];
```

We define a set of points on the unit circle

```
>> th = linspace(0,2*pi,50);
>> y = [cos(th);sin(th)];
```

which we transform to points on the perimeter of the ellipse

```
>> x = inv(sqrtm(E))*y;
>> plot(x(:,1),x(:,2));
```

which is encapsulated in the RVC Toolbox function

```
>> plotellipse(E)
```

An optional second argument can be provided to specify the ellipse's center point which defaults to $(0,0)$. In many cases we wish to plot an ellipse given by $\mathbf{A}^{-1}$, for instance $\mathbf{A}$ is a covariance matrix, but computing the inverse of $\mathbf{A}$ is inefficient because `plotellipse` will invert it again – in this case we write `plotellipse(A,inverted=true)`.

C.1.4.2 Fitting an Ellipse to Data

A common problem is to find the equation of an ellipse that best fits a set of points that lie within the ellipse boundary, or that lie on the ellipse boundary.

From a Set of Interior Points

To fit an ellipse to a set of points within the ellipse boundary, we find the ellipse that has the same mass properties as the set of points. From the set of N points $\mathbf{x}_i = (x_i, y_i)^\top$, we can compute the moments

$$m_{00} = N, \quad m_{10} = \sum_{i=1}^{N} x_i, \quad m_{01} = \sum_{i=1}^{N} y_i .$$

The center of the ellipse is taken to be the centroid of the set of points

$$\mathbf{x}_c = (x_c, \ y_c)^\top = \left(\frac{m_{10}}{m_{00}}, \ \frac{m_{01}}{m_{00}} \right)^\top$$

which allows us to compute the central second moments

$$\mu_{20} = \sum_{i=1}^{N} (x_i - x_c)^2$$

$$\mu_{02} = \sum_{i=1}^{N} (y_i - y_c)^2$$

$$\mu_{11} = \sum_{i=1}^{N} (x_i - x_c)(y_i - y_c) .$$

The inertia tensor for a general ellipse is the symmetric matrix

$$\mathbf{J} = \begin{pmatrix} \mu_{20} & \mu_{11} \\ \mu_{11} & \mu_{02} \end{pmatrix}$$

where the diagonal terms are the moments of inertia and the off-diagonal terms are the products of inertia. Inertia can be computed more directly as the summation of N rank-1 matrices

$$\mathbf{J} = \sum_{i=1}^{N} (\mathbf{x} - \mathbf{x}_c)(\mathbf{x} - \mathbf{x}_c)^\top .$$

The inertia tensor and its equivalent ellipse are inversely related by

$$\mathbf{E} = \frac{m_{00}}{4} \mathbf{J}^{-1} .$$

To demonstrate this, we can create a set of points that lie within the ellipse used in the example above

```
>> % generate a set of points within the ellipse
>> rng(0);  % reset random number generator
>> x = [];  % empty point set
>> while size(x,2) < 500
>>      p = (rand(2,1)-0.5)*4;
>>      if norm(p'*E*p) <= 1
>>          x = [x p];
>>      end
>> end
>> plot(x(1,:),x(2,:),".")
>> % compute the moments
>> m00 = mpq_point(x,0,0);
>> m10 = mpq_point(x,1,0);
>> m01 = mpq_point(x,0,1);
```

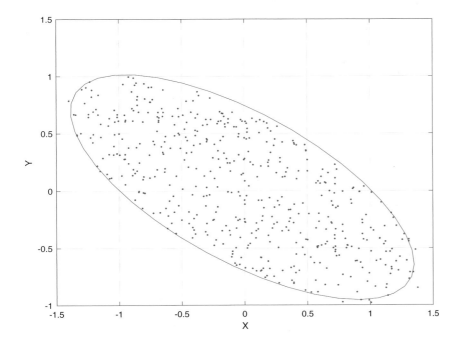

◘ Fig. C.4 500 random data points with a fitted ellipse

```
>> xc = m10/m00; yc = m01/m00;
>> % compute second moments relative to centroid
>> x0 = x - [xc; yc];
>> m20 = mpq_point(x0,2,0);
>> m02 = mpq_point(x0,0,2);
>> m11 = mpq_point(x0,1,1);
>> % compute the moments and ellipse matrix
>> J = [m20 m11;m11 m02];
>> E_est = m00*inv(J)/4;
```

which results in an estimate

```
>> E_est
E_est =
    0.9914    0.9338
    0.9338    1.9087
```

that is similar to the original value of E. The point data is shown in ◘ Fig. C.4. We can overlay the estimated ellipse on the point data

```
>> plotellipse(E_est,"r")
```

and the result is shown in red in ◘ Fig. C.4.

From a Set of Perimeter Points

Given a set of points (x_i, y_i) that lie on the perimeter of an ellipse, we use the polynomial form of the ellipse (C.6) for each point. We write this in matrix form with one row per point

$$
\begin{pmatrix}
y_1^2 & x_1 y_1 & x_1 & y_1 & 1 \\
y_2^2 & x_2 y_2 & x_2 & y_2 & 1 \\
\vdots & \vdots & \vdots & \vdots & \vdots \\
y_N^2 & x_N y_N & x_N & y_N & 1
\end{pmatrix}
\begin{pmatrix}
\epsilon_1 \\
\epsilon_2 \\
\epsilon_3 \\
\epsilon_4 \\
\epsilon_5
\end{pmatrix}
=
\begin{pmatrix}
-x_1^2 \\
-x_2^2 \\
\vdots \\
-x_N^2
\end{pmatrix}
$$

and for $N \geq 5$ we can solve for the ellipse parameter vector using least squares.

C.2 Homogeneous Coordinates

A point in homogeneous coordinates, or the projective space $\mathbb{P}^n$, is represented by a coordinate vector $\tilde{x} = (\tilde{x}_1, \tilde{x}_2, \ldots, \tilde{x}_{n+1})$ and the tilde is used to indicate that the quantity is homogeneous. The Euclidean coordinates are related to the projective or homogeneous coordinates by

$$x_i = \frac{\tilde{x}_i}{\tilde{x}_{n+1}}, \quad i = 1, \ldots, n$$

Conversely, a homogeneous coordinate vector can be constructed from a Euclidean coordinate vector by

$$\tilde{x} = (x_1, x_2, \ldots, x_n, 1) \ .$$

The extra *degree of freedom* offered by projective coordinates has several advantages. It allows points and lines at infinity, known as ideal points and lines, to be represented using only finite numbers. It also means that scale is unimportant, that is, $\tilde{x}$ and $\tilde{x}' = \lambda \tilde{x}$ both represent the same Euclidean point for all $\lambda \neq 0$. We express this as $\tilde{x} \simeq \tilde{x}'$. We can apply a rigid-body transformation to points $\tilde{x} \in \mathbb{P}^n$ by multiplying the homogeneous coordinate by an $(n+1) \times (n+1)$ homogeneous transformation matrix.

Homogeneous vectors are important in computer vision when we consider points and lines that exist in a plane, for example, a camera's image plane. We can also consider that the homogeneous form represents a ray in projective space as shown in ◻ Fig. C.5. Any point on the projective ray is equivalent to p and this relationship between points and rays is at the core of the projective transformation.

C.2.1 Two Dimensions

C.2.1.1 Points and Lines

In two dimensions, there is a duality between points and lines. In $\mathbb{P}^2$, a line is represented by a vector $\tilde{\ell} = (\ell_1, \ell_2, \ell_3)^\top$, not all zero, and the equation of the line is the set of all points $\tilde{x} \in \mathbb{P}^2$ such that

$$\tilde{\ell}^\top \tilde{x} = 0$$

which is the point equation of a line. This expands to $\ell_1 x + \ell_2 y + \ell_3 = 0$ and can be manipulated into the more familiar representation of a line $y = -\frac{\ell_1}{\ell_2} x - \frac{\ell_3}{\ell_2}$. Note that

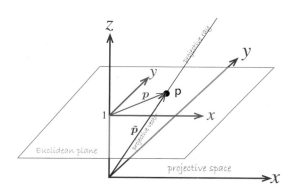

◻ **Fig. C.5** A point p on the Euclidean plane is described by a coordinate vector $p \in \mathbb{R}^2$ which is equivalent to the three-dimensional vector in the projective space $\tilde{p} \in \mathbb{P}^2$

this form can represent a vertical line which the familiar form $y = mx + c$ cannot. The nonhomogeneous vector $(\ell_1, \ell_2)^\top$ is normal to the line, ► and $(-\ell_2, \ell_1)^\top$ is parallel to the line.

Hence, this line representation is referred to as normal form.

A point is defined by the intersection of two lines. If we write the point equations for two homogeneous lines $\tilde{\ell}_1^\top \tilde{x} = 0$ and $\tilde{\ell}_2^\top \tilde{x} = 0$, their intersection is the point with homogeneous coordinates

$$\tilde{p} = \tilde{\ell}_1 \times \tilde{\ell}_2$$

and is known as the line equation of a point. Similarly, a line passing through two points $\tilde{p}_1, \tilde{p}_2 \in \mathbb{P}^2$, said to be *joining* the points, is given by the cross-product

$$\tilde{\ell}_{12} = \tilde{p}_1 \times \tilde{p}_2 \; .$$

Consider the case of two parallel lines at $45°$ to the horizontal axis

```
>> l1 = [1 -1 0];
>> l2 = [1 -1 -1];
```

which we can plot

```
>> plothomline(l1,"b")
>> plothomline(l2,"r")
```

The intersection point of these parallel lines is

```
>> cross(l1,l2)
ans =
     1     1     0
```

This is an *ideal point* since the third coordinate is zero – the equivalent Euclidean point would be at infinity. Projective coordinates allow points and lines at infinity to be simply represented and manipulated without special logic.

The distance from a line $\tilde{\ell}$ to a point $\tilde{p}$ is

$$d = \frac{\tilde{\ell}^\top \tilde{p}}{\tilde{p}_3 \sqrt{\tilde{\ell}_1^2 + \tilde{\ell}_2^2}} \; . \tag{C.9}$$

C.2.1.2 Conics

Conic sections are an important family of planar curves that includes circles, ellipses, parabolas and hyperbolas. They can be described generally as the set of points $\tilde{x} \in \mathbb{P}^2$ such that

$$\tilde{x}^\top \mathbf{\Omega} \tilde{x} = 0$$

where $\mathbf{\Omega}$ is a symmetric matrix

$$\mathbf{\Omega} = \begin{pmatrix} a & b/2 & d/2 \\ b/2 & c & e/2 \\ d/2 & e/2 & f \end{pmatrix} \; . \tag{C.10}$$

The determinant of the top-left submatrix indicates the type of conic: positive for an ellipse, zero for a parabola, and negative for a hyperbola.

C.2.2 Three Dimensions

In three dimensions, there is a duality between points and planes.

C.2.2.1 Lines

For two points $\tilde{p}, \tilde{q} \in \mathbb{P}^3$ in homogeneous form, the line that joins them is defined by a 4×4 skew-symmetric matrix

$$
\begin{aligned}
\mathbf{L} &= \tilde{q}\,\tilde{p}^\top - \tilde{p}\,\tilde{q}^\top \\
&= \begin{pmatrix}
0 & v_3 & -v_2 & -\omega_1 \\
-v_3 & 0 & v_1 & -\omega_2 \\
v_2 & -v_1 & 0 & -\omega_3 \\
\omega_1 & \omega_2 & \omega_3 & 0
\end{pmatrix}
\end{aligned}
\tag{C.11}
$$

whose six unique elements comprise the Plücker coordinate representation of the line. This matrix is rank 2 and the determinant is a quadratic in the Plücker coordinates – a 4-dimensional quadric hypersurface known as the Klein quadric. All points that lie on this manifold are valid lines. Many of the relationships in ▶ App. C.1.2.2 (between lines and points and planes) can be expressed in terms of this matrix. This matrix is returned by the `skew` method of the `Plucker` class.

For a perspective camera with a camera matrix $\mathbf{C} \in \mathbb{R}^{3\times4}$, the 3-dimensional Plücker line represented as a 4×4 skew-symmetric matrix $\mathbf{L}$ is projected onto the image plane as

$$
\tilde{\ell} = \mathbf{C}\mathbf{L}\mathbf{C}^\top \in \mathbb{P}^2
$$

which is a homogeneous 2-dimensional line. This is computed by the `project` method of the `CentralCamera` class.

C.2.2.2 Planes

A plane $\tilde{\pi} \in \mathbb{P}^3$ is described by the set of points $\tilde{x} \in \mathbb{P}^3$ such that $\tilde{\pi}^\top \tilde{x} = 0$. A plane can be defined by a line $\mathbf{L} \in \mathbb{R}^{4\times4}$ and a homogeneous point $\tilde{p}$

$$
\tilde{\pi} = \mathbf{L}\tilde{p}
$$

or three points

$$
\begin{pmatrix}
\tilde{p}_1^\top \\
\tilde{p}_2^\top \\
\tilde{p}_3^\top
\end{pmatrix} \tilde{\pi} = \mathbf{0}
$$

and the solution is found from the right-null space of the matrix.

The intersection, or incidence, of three planes is the dual

$$
\begin{pmatrix}
\tilde{\pi}_1^\top \\
\tilde{\pi}_2^\top \\
\tilde{\pi}_3^\top
\end{pmatrix} \tilde{p} = \mathbf{0}
$$

and is an ideal point, zero last component, if the planes do not intersect at a point.

C.2.2.3 Quadrics

Quadrics, short for quadratic surfaces, are a rich family of 3-dimensional *surfaces*. There are 17 standard types including spheres, ellipsoids, hyperboloids, paraboloids, cylinders and cones, all described by the set of points $\tilde{x} \in \mathbb{P}^3$ such that

$$\tilde{x}^\top \mathbf{Q} \tilde{x} = 0$$

and $\mathbf{Q} \in \mathbb{R}^{4\times4}$ is symmetric.

For a perspective camera with a camera matrix $\mathbf{C} \in \mathbb{R}^{3\times4}$, the *outline* of the 3-dimensional quadric is projected to the image plane by

$$\mathbf{\Omega}^* = \mathbf{C}\mathbf{Q}^*\mathbf{C}^\top$$

where $\mathbf{\Omega}$ is given by (C.10), and $(\cdot)^*$ represents the adjugate operation, see ▶ App. B.1.

C.3 Geometric Transformations

A linear transformation is

$$y = \mathbf{A}x \tag{C.12}$$

while an affine transformation

$$y = \mathbf{A}x + b \tag{C.13}$$

comprises a linear transformation *and* a change of origin. Examples of affine transformations include translation, scaling, homothety, ▶ reflection, rotation, shearing, and any arbitrary composition of these. Every linear transformation is affine, but not every affine transformation is linear.

Scaling about an arbitrary point.

In homogeneous coordinates, we can write (C.13) as

$$\tilde{y} = \mathbf{H}\tilde{x}, \quad \text{where } \mathbf{H} = \begin{pmatrix} \mathbf{A} & b \\ \mathbf{0} & 1 \end{pmatrix}$$

and the transformation operates on a point with homogeneous coordinates $\tilde{x}$.

Projective space is a generalization of affine space which, in turn, is a generalization of Euclidean space. An affine space has no distinguished point that serves as an origin and hence no vector can be uniquely associated to a point. An affine space has only displacement vectors between two points in the space. Subtracting two points results in a displacement vector, and adding a displacement vector to a point results in a new point. If a displacement vector is defined as the difference between two homogeneous points $\tilde{p}$ and $\tilde{q}$, then the difference $\tilde{p} - \tilde{q}$ is a 4-vector whose last element will be zero, distinguishing a point from a displacement vector.

In two dimensions, the most general transformation is a projective transformation, also known as a collineation

$$\mathbf{H} = \begin{pmatrix} h_{1,1} & h_{1,2} & h_{1,3} \\ h_{2,1} & h_{2,2} & h_{2,3} \\ h_{3,1} & h_{3,2} & 1 \end{pmatrix}$$

which is unique up to scale and one element has been normalized to one. It has 8 degrees of freedom.

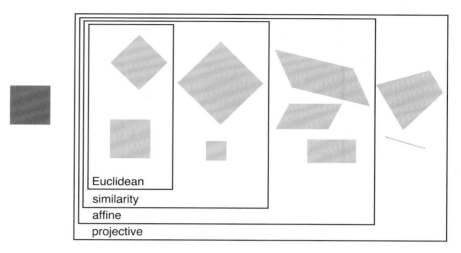

◘ Fig. C.6 A 2-dimensional dark gray square with various transformations applied: from the most limited (Euclidean) to the most general (projective)

The affine transformation is a subset where the elements of the last row are fixed

$$\mathbf{H} = \begin{pmatrix} a_{1,1} & a_{1,2} & t_x \\ a_{2,1} & a_{2,2} & t_y \\ 0 & 0 & 1 \end{pmatrix}$$

and has 6 degrees of freedom, two of which are the translation (t_x, t_y).

The similarity transformation is a more restrictive subset

$$\mathbf{H} = \begin{pmatrix} s\mathbf{R} & t \\ \mathbf{0} & 1 \end{pmatrix} = \begin{pmatrix} sr_{1,1} & sr_{1,2} & t_x \\ sr_{2,1} & sr_{2,2} & t_y \\ 0 & 0 & 1 \end{pmatrix}$$

where $\mathbf{R} \in \mathbf{SO}(2)$ resulting in only 4 degrees of freedom, and $s < 0$ causes a reflection. Similarity transformations, without reflection, are sometimes referred to as a Procrustes transformation.

Finally, the Euclidean or rigid-body transformation

$$\mathbf{H} = \begin{pmatrix} \mathbf{R} & t \\ \mathbf{0} & 1 \end{pmatrix} = \begin{pmatrix} r_{1,1} & r_{1,2} & t_x \\ r_{2,1} & r_{2,2} & t_y \\ 0 & 0 & 1 \end{pmatrix} \in \mathbf{SE}(2)$$

is the most restrictive and has only 3 degrees of freedom. Some graphical examples of the effect of the various transformations on a square are shown in ◘ Fig. C.6. The possible geometric transformations for each type of transformation are summarized in ◘ Tab. C.2 along with the geometric properties which are unchanged, or invariant, under that transformation. We see that while Euclidean is most restrictive in terms of the geometric transformations it can perform, it is able to preserve important properties such as length and angle.

◨ Table C.2 For various planar transformation families, the possible geometric transformations and the geometric properties which are preserved are listed

	Euclidean	Similarity	Affine	Projective
Geometric transformation				
Rotation	✓	✓	✓	✓
Translation	✓	✓	✓	✓
Reflection		✓	✓	✓
Uniform scaling		✓	✓	✓
Nonuniform scaling			✓	✓
Shear			✓	✓
Perspective projection				✓
	Euclidean	Similarity	Affine	Projective
Preserved geometric properties (invariants)				
Length	✓			
Angle	✓	✓		
Ratio of lengths	✓	✓		
Parallelism	✓	✓	✓	
Incidence	✓	✓	✓	✓
Cross ratio	✓	✓	✓	✓

D Lie Groups and Algebras

We cannot go very far in the study of rotations or rigid-body motion without coming across the terms Lie groups, Lie algebras or Lie brackets – all named in honor of the Norwegian mathematician Sophus Lie. Rotations and rigid-body motion in two and three dimensions can be represented by matrices that have special structure, they form Lie groups, and they have Lie algebras.

We will start simply by considering the set of all real 2×2 matrices $\mathbf{A} \in \mathbb{R}^{2 \times 2}$

$$\mathbf{A} = \begin{pmatrix} a_{1,1} & a_{1,2} \\ a_{2,1} & a_{2,2} \end{pmatrix}$$

which we could write as a linear combination of basis matrices

$$\mathbf{A} = a_{1,1} \begin{pmatrix} 1 & 0 \\ 0 & 0 \end{pmatrix} + a_{1,2} \begin{pmatrix} 0 & 1 \\ 0 & 0 \end{pmatrix} + a_{2,1} \begin{pmatrix} 0 & 0 \\ 1 & 0 \end{pmatrix} + a_{2,2} \begin{pmatrix} 0 & 0 \\ 0 & 1 \end{pmatrix}$$

where each basis matrix represents a *direction* in a 4-dimensional space of 2×2 matrices. That is, the four axes of this space are *parallel* with each of these basis matrices. Any 2×2 matrix can be represented by a point in this space – this particular matrix is a point with the coordinates $(a_{1,1}, a_{1,2}, a_{2,1}, a_{2,2})$.

All proper rotation matrices, those belonging to **SO**(2), are a *subset* of points within the space of all 2×2 matrices. For this example, the points lie in a 1-dimensional subset, a closed curve, in the 4-dimensional space. This is an instance of a manifold, a lower-dimensional smooth *surface* embedded within a space.

The notion of a curve in the 4-dimensional space makes sense when we consider that the **SO**(2) rotation matrix

$$\mathbf{A} = \begin{pmatrix} \cos\theta & \sin\theta \\ -\sin\theta & \cos\theta \end{pmatrix}$$

has only one free parameter, and varying that parameter moves the point along the manifold.

Invoking mathematical formalism, we say that rotations **SO**(2) and **SO**(3), and rigid-body motions **SE**(2) and **SE**(3) are matrix Lie groups and this has two implications. Firstly, they are an *algebraic group*, a mathematical structure comprising

Excurse D.2: Sophus Lie

Lie, pronounced lee, (1842–1899) was a Norwegian mathematician who obtained his Ph.D. from the University of Christiania in Oslo in 1871. He spent time in Berlin working with Felix Klein, and later contributed to Klein's Erlangen program to characterize geometries based on group theory and projective geometry. On a visit to Milan during the Franco-Prussian war, he was arrested as a German spy and spent one month in prison. He is best known for his discovery that continuous transformation groups (now called Lie groups) can be understood by linearizing them and studying their generating vector spaces. He is buried in the Vår Frelsers gravlund in Oslo. (Image by Ludwik Szacinski)

elements and a single operator. In simple terms, a group $\mathbf{G}$ has the following properties:

1. If g_1 and g_2 are elements of the group, that is, $g_1, g_2 \in \mathbf{G}$, then the result of the group's operator $\diamond$ is also an element of the group: $g_1 \diamond g_2 \in \mathbf{G}$. In general, groups are not commutative, so $g_1 \diamond g_2 \neq g_2 \diamond g_1$. For rotations and rigid-body motions, the group operator $\diamond$ represents composition. ▶
2. The group operator is associative, that is, $(g_1 \diamond g_2) \diamond g_3 = g_1 \diamond (g_2 \diamond g_3)$.
3. There is an identity element $I \in \mathbf{G}$ such, for every $g \in \mathbf{G}$, $g \diamond I = I \diamond g = g$. ▶
4. For every $g \in \mathbf{G}$, there is a unique inverse $h \in \mathbf{G}$ such that $g \diamond h = h \diamond g = I$. ▶

In this book's notation, $\oplus$ is the group operator for relative pose.

In this book's notation, the identity for relative pose (implying null motion) is denoted by $\varnothing$ so we can say that $\xi \oplus \varnothing = \varnothing \oplus \xi = \xi$.

The second implication of being a Lie group is that there is a smooth (differentiable) manifold structure. At any point on the manifold, we can construct tangent vectors. The set of all tangent vectors at that point form a vector space – the tangent space. This is the multidimensional equivalent to a tangent line on a curve, or a tangent plane on a solid. We can think of this as the set of all possible derivatives of the manifold at that point.

In this book's notation, we use the operator $\ominus \xi$ to form the inverse of a relative pose.

The tangent space *at the identity* is described by the Lie algebra of the group, and the basis directions of the tangent space are called the generators of the group. Points in this tangent space map to elements of the group via the exponential function. If $\mathbf{g}$ is the Lie algebra for group $\mathbf{G}$, then

$$e^{\mathbf{X}} \in \mathbf{G}, \quad \forall X \in \mathbf{g}$$

where the elements of $\mathbf{g}$ and $\mathbf{G}$ are matrices of the same size and each of them has a specific structure.

The surface of a sphere is a manifold in a 3-dimensional space and at any point on that surface we can create a tangent vector. In fact, we can create an infinite number of them and they lie within a plane which is a 2-dimensional vector space – the tangent space. We can choose a set of basis directions and establish a 2-dimensional coordinate system and we can map points on the plane to points on the sphere's surface.

Now, consider an arbitrary real 3×3 matrix $\mathbf{A} \in \mathbb{R}^{3 \times 3}$

$$\mathbf{A} = \begin{pmatrix} a_{1,1} & a_{1,2} & a_{1,3} \\ a_{2,1} & a_{2,2} & a_{2,3} \\ a_{3,1} & a_{3,2} & a_{3,3} \end{pmatrix}$$

we could write as a linear combination of basis matrices

$$a_{1,1} \begin{pmatrix} 1 & 0 & 0 \\ 0 & 0 & 0 \\ 0 & 0 & 0 \end{pmatrix} + a_{1,2} \begin{pmatrix} 0 & 1 & 0 \\ 0 & 0 & 0 \\ 0 & 0 & 0 \end{pmatrix} + \cdots + a_{3,3} \begin{pmatrix} 0 & 0 & 0 \\ 0 & 0 & 0 \\ 0 & 0 & 1 \end{pmatrix}$$

basis matrix represents a *direction* in a 9-dimensional space of $3 \times$ very possible 3×3 matrix is represented by a point in this space. trices in this space are proper rotation matrices belonging to $\mathbf{SO}(3)$, will lie on a manifold since $\mathbf{SO}(3)$ is a Lie group. The null rotation, he identity matrix, is one point in this space. At that point, we can nt space which has only 3 dimensions. Every point in the tangent essed as a linear combination of basis matrices

$$\underbrace{\begin{pmatrix} 0 \\ -1 \\ 0 \end{pmatrix}}_{} + \omega_2 \underbrace{\begin{pmatrix} 0 & 0 & 1 \\ 0 & 0 & 0 \\ -1 & 0 & 0 \end{pmatrix}}_{\mathbf{G}_2} + \omega_3 \underbrace{\begin{pmatrix} 0 & -1 & 0 \\ 1 & 0 & 0 \\ 0 & 0 & 0 \end{pmatrix}}_{\mathbf{G}_3} \qquad (\text{D.1})$$

of the $\mathbf{SO}(3)$ group. The bases of this space: $\mathbf{G}_1$, $\mathbf{G}_2$ and rs of $\mathbf{SO}(3)$ and belong to $\mathbf{so}(3)$. ▶

The equivalent algebra is denoted using lowercase letters and is a vector space of matrices.

Equation (D.1) can be written as a skew-symmetric matrix

$$\mathbf{\Omega} = [\boldsymbol{\omega}]_\times = \begin{pmatrix} 0 & -\omega_3 & \omega_2 \\ \omega_3 & 0 & -\omega_1 \\ -\omega_2 & \omega_1 & 0 \end{pmatrix} \in \mathbf{so}(3)$$

parameterized by the vector $\boldsymbol{\omega} = (\omega_1, \omega_2, \omega_3) \in \mathbb{R}^3$ which is known as the Euler vector or exponential coordinates. This reflects the 3 degrees of freedom of the **SO**(3) group embedded in the space of all 3×3 matrices. The 3DOF is consistent with our intuition about rotations in 3D space and also Euler's rotation theorem.

Mapping between vectors and skew-symmetric matrices is frequently required and the following shorthand notation will be used

$$[\cdot]_\times : \mathbb{R} \mapsto \mathbf{so}(2), \ \mathbb{R}^3 \mapsto \mathbf{so}(3),$$

$$\vee_\times(\cdot) : \mathbf{so}(2) \mapsto \mathbb{R}, \ \mathbf{so}(3) \mapsto \mathbb{R}^3 .$$

The first mapping is performed by the RVC Toolbox function `vec2skew` and the second by `skew2vec`.

The exponential of *any* matrix in **so**(3) is a valid member of **SO**(3)

$$\mathbf{R}(\theta\hat{\boldsymbol{\omega}}) = \mathrm{e}^{[\theta\hat{\boldsymbol{\omega}}]_\times} \in \mathbf{SO}(3)$$

and an efficient closed-form solution is given by Rodrigues' rotation formula

$$\mathbf{R}(\theta\hat{\boldsymbol{\omega}}) = \mathbf{1} + \sin\theta[\hat{\boldsymbol{\omega}}]_\times + (1 - \cos\theta)[\hat{\boldsymbol{\omega}}]_\times^2$$

Finally, consider an arbitrary real 4×4 matrix $\mathbf{A} \in \mathbb{R}^{4\times4}$

$$\mathbf{A} = \begin{pmatrix} a_{1,1} & a_{1,2} & a_{1,3} & a_{1,4} \\ a_{2,1} & a_{2,2} & a_{2,3} & a_{2,4} \\ a_{3,1} & a_{3,2} & a_{3,3} & a_{3,4} \\ a_{4,1} & a_{4,2} & a_{4,3} & a_{4,4} \end{pmatrix}$$

which we could write as a linear combination of basis matrices

$$\mathbf{A} = a_{1,1}\begin{pmatrix} 1 & 0 & 0 & 0 \\ 0 & 0 & 0 & 0 \\ 0 & 0 & 0 & 0 \\ 0 & 0 & 0 & 0 \end{pmatrix} + a_{1,2}\begin{pmatrix} 0 & 1 & 0 & 0 \\ 0 & 0 & 0 & 0 \\ 0 & 0 & 0 & 0 \\ 0 & 0 & 0 & 0 \end{pmatrix} + \cdots + a$$

where each basis matrix represents a *direction* in a 16-d possible 4×4 matrices. Every 4×4 matrix is represente

Not all matrices in this space are proper rigid-bo belonging to **SE**(3), but those that do lie on a smoot (zero rotation and translation), which is represented point in this space. At that point, we can constr 6 dimensions in this case, and points in the tan linear combination of basis matrices

$$\mathbf{\Sigma} = \omega_1 \begin{pmatrix} 0 & 0 & 0 & 0 \\ 0 & 0 & -1 & 0 \\ 0 & 1 & 0 & 0 \\ 0 & 0 & 0 & 0 \end{pmatrix} + \omega_2$$

$$+ v_1 \begin{pmatrix} 0 & 0 & 0 & 1 \\ 0 & 0 & 0 & 0 \\ 0 & 0 & 0 & 0 \\ 0 & 0 & 0 & 0 \end{pmatrix} +$$

and these generator matrices belong to the Lie algebra of the group $\mathbf{SE}(3)$ which is denoted by $\mathbf{se}(3)$. This can be written in general form as

$$\mathbf{\Sigma} = [S] = \left(\begin{array}{ccc|c} 0 & -\omega_3 & \omega_2 & v_1 \\ \omega_3 & 0 & -\omega_1 & v_2 \\ -\omega_2 & \omega_1 & 0 & v_3 \\ \hline 0 & 0 & 0 & 0 \end{array} \right) \in \mathbf{se}(3)$$

which is an augmented skew-symmetric matrix parameterized by $S = (\omega, v) \in \mathbb{R}^6$ which is referred to as a twist and has physical interpretation in terms of a screw axis direction and position. The sparse matrix structure and this concise parameterization reflects the 6 degrees of freedom of the $\mathbf{SE}(3)$ group embedded in the space of all 4×4 matrices. We extend our earlier shorthand notation

$$[\cdot] \colon \mathbb{R}^3 \mapsto \mathbf{se}(2), \ \mathbb{R}^6 \mapsto \mathbf{se}(3),$$
$$\vee(\cdot) \colon \mathbf{se}(2) \mapsto \mathbb{R}^3, \ \mathbf{se}(3) \mapsto \mathbb{R}^6$$

and we can use these operators to convert between a twist representation which is a 6-vector and a Lie algebra representation which is a 4×4 augmented skew-symmetric matrix. The first mapping is performed by the RVC Toolbox function `vec2skewa` and the second by `skewa2vec`.

We convert the Lie algebra to the Lie group representation using

$$\mathbf{T}(\theta \hat{S}) = e^{[\theta \hat{S}]} \in \mathbf{SE}(3)$$

or the inverse using the matrix logarithm. The exponential and the logarithm each have an efficient closed-form solution but the builtin general-purpose MATLAB exponential and the logarithm functions are actually faster in practice.

Transforming a Twist – the Adjoint Representation

We have seen in ▶ Sect. 2.4.7 that rigid-body motions can be described by a twist which represents motion in terms of a screw axis direction and position. For example, in ◘ Fig. D.1, the twist $^A S$ can be used to transform points on the body. If the screw is rigidly attached to the body which undergoes some rigid-body motion $^A \xi_B$ the new twist is

$$^B S = \mathrm{Ad}\left(^B \xi_A \right) {}^A S$$

where

$$\mathrm{Ad}\left(^A \xi_B \right) = \left(\begin{array}{cc} ^A \mathbf{R}_B & \mathbf{0} \\ \left[^A t_B \right]_\times {}^A \mathbf{R}_B & ^A \mathbf{R}_B \end{array} \right) \in \mathbb{R}^{6 \times 6} \tag{D.2}$$

◘ **Fig. D.1** Points in the body (gray cloud) can be transformed by the twist $^A S$. If the body and the screw axis undergo a rigid-body transformation $^A \xi_B$, the new twist is $^B S$

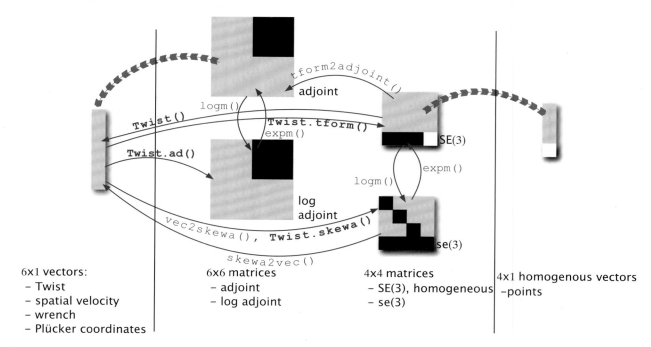

◘ Fig. D.2 The menagerie of **SE**(3) related quantities. Matrix values are coded as: 0 (black), 1 (white), other values (gray). Transformations between types are indicated by blue arrows with the relevant class plus method name. Operations are indicated by red arrows: the tail-end object operates on the head-end object and results in another object of the head-end type

is the adjoint representation of the rigid-body motion. Alternatively, we can write

$$\mathrm{Ad}(e^{[S]}) = e^{\mathrm{ad}(S)}$$

where $\mathrm{ad}(S)$ is the logarithm of the adjoint and defined in terms of the twist parameters as

$$\mathrm{ad}(S) = \begin{pmatrix} [\boldsymbol{\omega}]_\times & \mathbf{0} \\ [\boldsymbol{v}]_\times & [\boldsymbol{\omega}]_\times \end{pmatrix} \in \mathbb{R}^{6\times 6} .$$

The relationship between the various mathematical objects discussed are shown in ◘ Fig. D.2.

E Linearization, Jacobians, and Hessians

In robotics and computer vision, the equations we encounter are often nonlinear. To apply familiar and powerful analytic techniques, we must work with linear or quadratic approximations to these equations. The principle is illustrated in ■ Fig. E.1 for the 1-dimensional case, and the analytical approximations shown in red are made at $x = x_0$. The approximation equals the nonlinear function at x_0 but is increasingly inaccurate as we move away from that point. We call this a *local approximation* since it is valid in a region local to x_0 – the size of the valid region depends on the severity of the nonlinearity. This approach can be extended to an arbitrary number of dimensions.

E.1 Scalar Function of a Scalar

The function $f \colon \mathbb{R} \mapsto \mathbb{R}$ can be expressed as a Taylor series

$$f(x_0 + \Delta) = f(x_0) + \frac{\mathrm{d}f}{\mathrm{d}x}\Delta + \frac{1}{2}\frac{\mathrm{d}^2 f}{\mathrm{d}x^2}\Delta^2 + \cdots$$

which we truncate to form a first-order or linear approximation

$$f'(\Delta) \approx f(x_0) + J(x_0)\Delta$$

or a second-order approximation

$$f'(\Delta) \approx f(x_0) + J(x_0)\Delta + \frac{1}{2}H(x_0)\Delta^2$$

where $\Delta \in \mathbb{R}$ is an infinitesimal change in x relative to the linearization point x_0, and the first and second derivatives are given by

$$J(x_0) = \left.\frac{\mathrm{d}f}{\mathrm{d}x}\right|_{x_0}, \quad H(x_0) = \left.\frac{\mathrm{d}^2 f}{\mathrm{d}x^2}\right|_{x_0}$$

respectively.

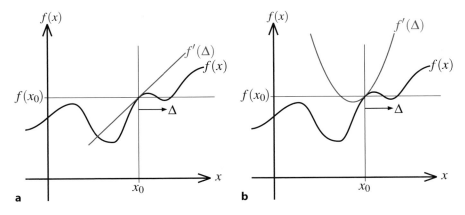

a   **b**

■ **Fig. E.1** The nonlinear function $f(x)$ is approximated at the point $x = x_0$ by **a** the red line – a linear or first-order approximation, **b** the red parabola – a second-order approximation. At the linearization point, both curves are equal and tangent to the function while for **b** the second derivatives also match

Excurse E.3: Ludwig Otto Hesse

Hesse (1811–1874) was a German mathematician, born in Königsberg, Prussia, who studied under Jacobi and Bessel at the University of Königsberg. He taught at Königsberg, Halle, Heidelberg and finally at the newly established Polytechnic School in Munich. In 1869, he joined the Bavarian Academy of Sciences. He died in Munich, but was buried in Heidelberg which he considered his second home.

E.2   Scalar Function of a Vector

The scalar field $f(\boldsymbol{x}): \mathbb{R}^n \mapsto \mathbb{R}$ can be expressed as a Taylor series

$$f(\boldsymbol{x}_0 + \boldsymbol{\Delta}) = f(\boldsymbol{x}_0) + \mathbf{J}(\boldsymbol{x}_0)\boldsymbol{\Delta} + \frac{1}{2}\boldsymbol{\Delta}^\top \mathbf{H}(\boldsymbol{x}_0)\boldsymbol{\Delta} + \cdots$$

which we can truncate to form a first-order or linear approximation

$$f'(\boldsymbol{\Delta}) \approx f(\boldsymbol{x}_0) + \mathbf{J}(\boldsymbol{x}_0)\boldsymbol{\Delta}$$

or a second-order approximation

$$f'(\boldsymbol{\Delta}) \approx f(\boldsymbol{x}_0) + \mathbf{J}(\boldsymbol{x}_0)\boldsymbol{\Delta} + \frac{1}{2}\boldsymbol{\Delta}^\top \mathbf{H}(\boldsymbol{x}_0)\boldsymbol{\Delta}$$

where $\boldsymbol{\Delta} \in \mathbb{R}^n$ is an infinitesimal change in $\boldsymbol{x} \in \mathbb{R}^n$ relative to the linearization point $\boldsymbol{x}_0$, $\mathbf{J} \in \mathbb{R}^{1 \times n}$ is the vector version of the first derivative, and $\mathbf{H} \in \mathbb{R}^{n \times n}$ is the Hessian – the matrix version of the second derivative.

The derivative of the function $f(\cdot)$ with respect to the vector $\boldsymbol{x}$ is

$$\mathbf{J}(\boldsymbol{x}) = \nabla f(\boldsymbol{x}) = \begin{pmatrix} \frac{\partial f}{\partial x_1} \\ \frac{\partial f}{\partial x_2} \\ \vdots \\ \frac{\partial f}{\partial x_n} \end{pmatrix} \in \mathbb{R}^n$$

and is a column vector that points in the direction at which the function $f(\boldsymbol{x})$ has maximal increase. It is often written as $\nabla_x f$ to make explicit that the differentiation is with respect to $\boldsymbol{x}$.

The Hessian is an $n \times n$ symmetric matrix of second derivatives

$$\mathbf{H}(\boldsymbol{x}) = \begin{pmatrix} \frac{\partial^2 f}{\partial x_1^2} & \frac{\partial^2 f}{\partial x_1 \partial x_2} & \cdots & \frac{\partial^2 f}{\partial x_1 \partial x_n} \\ \frac{\partial^2 f}{\partial x_1 \partial x_2} & \frac{\partial^2 f}{\partial x_2^2} & \cdots & \frac{\partial^2 f}{\partial x_2 \partial x_n} \\ \vdots & \vdots & \ddots & \vdots \\ \frac{\partial^2 f}{\partial x_1 \partial x_n} & \frac{\partial^2 f}{\partial x_2 \partial x_n} & \cdots & \frac{\partial^2 f}{\partial x_n^2} \end{pmatrix} \in \mathbb{R}^{n \times n} .$$

The function is at a critical point when $\|\mathbf{J}(\boldsymbol{x})\| = 0$. If $\mathbf{H}(\boldsymbol{x})$ is positive-definite then the function is at a local minimum, if negative-definite then a local maximum, otherwise the result is inconclusive.

For functions which are quadratic in $\boldsymbol{x}$, as is the case for least-squares problems, it can be shown that the Hessian is

$$\mathbf{H}(\boldsymbol{x}) = \mathbf{J}^{\top}(\boldsymbol{x})\,\mathbf{J}(\boldsymbol{x}) + \sum_{i=1}^{m} f_i(\boldsymbol{x})\frac{\partial^2 f_i}{\partial \boldsymbol{x}^2} \approx \mathbf{J}^{\top}(\boldsymbol{x})\,\mathbf{J}(\boldsymbol{x})$$

which is frequently approximated by just the first term and this is key to Gauss-Newton least-squares optimization discussed in ▶ App. F.2.3.

E.3   Vector Function of a Vector

The vector field $\boldsymbol{f}(\boldsymbol{x})\colon \mathbb{R}^n \mapsto \mathbb{R}^m$ can also be written as

$$\boldsymbol{f}(\boldsymbol{x}) = \begin{pmatrix} f_1(\boldsymbol{x}) \\ f_2(\boldsymbol{x}) \\ \vdots \\ f_m(\boldsymbol{x}) \end{pmatrix} \in \mathbb{R}^m$$

where $f_i\colon \mathbb{R}^m \to \mathbb{R}$, $i = 1,\ldots,m$. The derivative of $\boldsymbol{f}$ with respect to the vector $\boldsymbol{x}$ can be expressed in matrix form as a Jacobian matrix

$$\mathbf{J}(\boldsymbol{x}) = \begin{pmatrix} \frac{\partial f_1}{\partial x_1} & \cdots & \frac{\partial f_1}{\partial x_n} \\ \vdots & \ddots & \vdots \\ \frac{\partial f_m}{\partial x_1} & \cdots & \frac{\partial f_m}{\partial x_n} \end{pmatrix} \in \mathbb{R}^{m \times n}$$

which can also be written as

$$\mathbf{J}(\boldsymbol{x}) = \begin{pmatrix} \nabla f_1^{\top} \\ \nabla f_2^{\top} \\ \vdots \\ \nabla f_m^{\top} \end{pmatrix}$$

This derivative is also known as the tangent map of $\boldsymbol{f}$, denoted by $\mathrm{T}\boldsymbol{f}$, or the differential of $\boldsymbol{f}$ denoted by $\mathrm{D}\boldsymbol{f}$. To make explicit that the differentiation is with respect to $\boldsymbol{x}$ this can be denoted as $\mathbf{J}_x$, $\mathrm{T}_x\boldsymbol{f}$, $\mathrm{D}_x\boldsymbol{f}$ or even $\partial\boldsymbol{f}/\partial\boldsymbol{x}$.

The Hessian in this case is a rank-3 tensor which we can think of as a list, of length m, of $n \times n$ Hessians $\mathbf{H} = \big(\mathbf{H}_1(\boldsymbol{x}),\ldots,\mathbf{H}_m(\boldsymbol{x})\big)$. In MATLAB® this could be represented by an $n \times n \times m$ matrix.

E.4   Deriving Jacobians

Jacobians of functions are required for many optimization algorithms, for example pose-graph optimization, as well as for the extended Kalman filter. They can be computed either numerically or symbolically.

Consider (6.9) for the range and bearing angle of a landmark given the pose of the vehicle and the position of the landmark. We can express this as a MATLAB anonymous function

```
>> zrange = @(xi,xv,w) ...
>>        [sqrt((xi(1)-xv(1))^2 + (xi(2)-xv(2))^2) + w(1);
>>        atan((xi(2)-xv(2))/(xi(1)-xv(1)))-xv(3) + w(2)];
```

To estimate the Jacobian $\mathbf{H}_{\boldsymbol{x}_v} = \partial \boldsymbol{h} / \partial \boldsymbol{x}_v$ for $\boldsymbol{x}_v = (1, 2, \frac{\pi}{3})$ and $\boldsymbol{x}_i = (10, 8)$, we can compute a first-order numerical difference

```
>> xv = [1 2 pi/3]; xi = [10 8]; w= [0,0];
>> h0 = zrange(xi,xv,w)
h0 =
    10.8167
    -0.4592
>> d = 0.001;
>> J = [zrange(xi,xv+[1 0 0]*d,w)-h0 ...
>>      zrange(xi,xv+[0 1 0]*d,w)-h0 ...
>>      zrange(xi,xv+[0 0 1]*d,w)-h0]/d
J =
    -0.8320   -0.5547        0
     0.0513   -0.0769  -1.0000
```

which shares the characteristic last column with the Jacobian shown in (6.14). Note that in computing this Jacobian we have set the measurement noise w to zero. The principal difficulty with this approach is choosing d, the difference used to compute the finite-difference approximation to the derivative. Too large and the results will be quite inaccurate if the function is nonlinear, too small and numerical problems will lead to reduced accuracy.

Alternatively, we can perform the differentiation symbolically. This particular function is relatively simple and the derivatives can be determined easily by hand using differential calculus. The numerical derivative can be used as a quick check for correctness. To avoid the possibility of error, or for more complex functions, we can perform the symbolic differentiation using any of a large number of computer algebra packages. Using the MATLAB Symbolic Math Toolbox™, we can declare some symbolic variables

```
>> syms xi yi xv yv thetav wr wb
```

and then evaluate the same function as above

```
>> z = zrange([xi yi],[xv yv thetav],[wr wb])
z =
        wr + ((xi - xv)/(yi - yv)^2)^(1/2)
 wb - thetav + atan((yi - yv)/(xi - xv))
```

which is simply (6.9) in MATLAB symbolic form. The Jacobian is computed by a Symbolic Math Toolbox function

```
>> J = jacobian(z,[xv yv thetav])
J =
[ -(2*xi - 2*xv)/(2*((xi - xv)^2 + (yi - yv)^2)^(1/2)),
   -(2*yi - 2*yv)/(2*((xi - xv)^2 + (yi - yv)^2)^(1/2)), 0]
[ (yi - yv)/((xi - xv)^2*((yi - yv)^2/(xi - xv)^2 + 1)),
   -1/((xi - xv)*((yi - yv)^2/(xi - xv)^2 + 1)), -1]
```

which has the required dimensions

```
>> whos J
  Name        Size            Bytes  Class    Attributes
  J           2x3                 8  sym
```

and the characteristic last column. We could cut and paste this code into our program or automatically create a MATLAB callable function

```
>> Jf = matlabFunction(J);
```

where Jf is a MATLAB function handle. We can evaluate the Jacobian at the operating point given above

```
>> xv = [1 2 pi/3]; xi = [10 8]; w = [0 0];
>> Jf(xi(1),xv(1),xi(2),xv(2))
ans =
   -0.8321 -0.5547 0
    0.0513 -0.0769 -1.0000
```

which is similar to the numerical approximation obtained above. The function `matlabFunction` can also write the function to a MATLAB file. The function `ccode` will generate efficient C code to compute the Jacobian.

Another approach is automatic differentiation (AD) of code. The package ADOL-C is an open-source tool that can differentiate C and C++ programs, that is, given a function written in C, it will return a Jacobian or higher-order derivative function written in C. It is available at ▶ https://github.com/coin-or/ADOL-C. For MATLAB code, there is an open-source tool called ADiGator available at ▶ https://github.com/matt-weinstein/adigator.

F Solving Systems of Equations

Solving systems of linear and nonlinear equations, particularly over-constrained systems, is a common problem in robotics and computer vision.

F.1 Linear Problems

F.1.1 Nonhomogeneous Systems

These are equations of the form

$$\mathbf{A}x = b$$

where we wish to solve for the unknown vector $x \in \mathbb{R}^n$ where $\mathbf{A} \in \mathbb{R}^{m \times n}$ and $b \in \mathbb{R}^m$ are known constants.

If $n = m$ then $\mathbf{A}$ is square, and if $\mathbf{A}$ is non-singular then the solution is

$$x = \mathbf{A}^{-1}b .$$

In practice, we often encounter systems where $m > n$, that is, there are more equations than unknowns. In general, there will not be an exact solution, but we can attempt to find the *best* solution, in a least-squares sense, which is

$$x^* = \arg\min_x ||\mathbf{A}x - b|| .$$

That solution is given by

$$x^* = \left(\mathbf{A}^\top \mathbf{A}\right)^{-1} \mathbf{A}^\top b = \mathbf{A}^+ b$$

which is known as the pseudoinverse or more formally the left-generalized inverse. Using SVD where $\mathbf{A} = \mathbf{U\Sigma V}^\top$, this is

$$x = \mathbf{V\Sigma}^{-1}\mathbf{U}^\top b$$

where $\mathbf{\Sigma}^{-1}$ is simply the element-wise inverse of the diagonal elements of $\mathbf{\Sigma}$.

If the matrix is singular, or the system is under constrained $n < m$, then there are infinitely many solutions. We can again use the SVD approach

$$x = \mathbf{V\Sigma}^{-1}\mathbf{U}^\top b$$

where this time $\mathbf{\Sigma}^{-1}$ is the element-wise inverse of the *nonzero* diagonal elements of $\mathbf{\Sigma}$, all other zeros are left in place.

Using MATLAB®, all these problems can be solved using the backslash operator

```
>> x = A\b
```

F.1.2 Homogeneous Systems

These are equations of the form

$$\mathbf{A}x = 0 \tag{F.1}$$

and always have the trivial solution $x = 0$. If $\mathbf{A}$ is square and non-singular, that is the only solution. Otherwise, if $\mathbf{A}$ is not of full rank, that is, the matrix is non-square, or square and singular, then there are an infinite number of solutions which are linear combinations of vectors in the right null space of $\mathbf{A}$ which is computed by the function `null`.

F.1.3 Finding a Rotation Matrix

Consider two sets of points in $\mathbb{R}^n$ related by an unknown rotation $\mathbf{R} \in \mathbf{SO}(\mathrm{n})$. The points are arranged column-wise as $\mathbf{P} = \{\boldsymbol{p}_1, \ldots, \boldsymbol{p}_m\} \in \mathbb{R}^{n \times m}$ and $\mathbf{Q} = \{\boldsymbol{q}_1, \ldots, \boldsymbol{q}_m\} \in \mathbb{R}^{n \times m}$ such that

$$\mathbf{RP} = \mathbf{Q} \ .$$

To solve for $\mathbf{R}$, we first compute the moment matrix

$$\mathbf{M} = \sum_{i=1}^{m} \boldsymbol{q}_i \, \boldsymbol{p}_i^\top$$

and then take the SVD such that $\mathbf{M} = \mathbf{U\Sigma V}^\top$. The least squares estimate of the rotation matrix is

$$\mathbf{R} = \mathbf{U} \begin{pmatrix} 1 & 0 & 0 \\ 0 & 1 & 0 \\ 0 & 0 & s \end{pmatrix} \mathbf{V}^\top, \quad s = \det(\mathbf{U}) \det(\mathbf{V}^\top) \tag{F.2}$$

and is guaranteed to be an $\mathbf{SO}(3)$ matrix (Horn 1987; Umeyama 1991).

F.2 Nonlinear Problems

Many problems in robotics and computer vision involve sets of nonlinear equations. Solution of these problems requires linearizing the equations about an estimated solution, solving for an improved solution and iterating. Linearization is discussed in ▶ App. E.

F.2.1 Finding Roots

Consider a set of equations expressed in the form

$$\boldsymbol{f}(\boldsymbol{x}) = \mathbf{0} \tag{F.3}$$

where $\boldsymbol{f}: \mathbb{R}^n \mapsto \mathbb{R}^m$. This is a nonlinear version of (F.1), and we first linearize the equations about our best estimate of the solution $\boldsymbol{x}_0$

$$\boldsymbol{f}(\boldsymbol{x}_0 + \boldsymbol{\Delta}) = \boldsymbol{f}_0 + \mathbf{J}(\boldsymbol{x}_0)\boldsymbol{\Delta} + \frac{1}{2}\boldsymbol{\Delta}^\top \mathbf{H}(\boldsymbol{x}_0)\boldsymbol{\Delta} + \text{H.O.T.} \tag{F.4}$$

where $\boldsymbol{f}_0 = \boldsymbol{f}(\boldsymbol{x}_0) \in \mathbb{R}^{m \times 1}$ is the function value and $\mathbf{J} = \mathbf{J}(\boldsymbol{x}_0) \in \mathbb{R}^{m \times n}$ the Jacobian, both evaluated at the linearization point, $\boldsymbol{\Delta} \in \mathbb{R}^{n \times 1}$ is an infinitesimal change in $\boldsymbol{x}$ relative to $\boldsymbol{x}_0$ and H.O.T. denotes higher-order terms. $\mathbf{H}(\boldsymbol{x}_0)$ is the $n \times n \times m$ Hessian tensor. ▶

We take the first two terms of (F.4) to form a linear approximation

To evaluate $\boldsymbol{\Delta}^\top \mathbf{H}(\boldsymbol{x}_0)\boldsymbol{\Delta}$ we write it as $\boldsymbol{\Delta}^\top (\mathbf{H}(\boldsymbol{x}_0)\bar{\times}_2 \boldsymbol{\Delta})$ which uses a mode-2 tensor-vector product.

$$\boldsymbol{f}'(\boldsymbol{\Delta}) \approx \boldsymbol{f}_0 + \mathbf{J}(\boldsymbol{x}_0)\boldsymbol{\Delta} \tag{F.5}$$

and solve an approximation of the original problem $\boldsymbol{f}'(\boldsymbol{\Delta}) = \mathbf{0}$

$$\boldsymbol{f}_0 + \mathbf{J}(\boldsymbol{x}_0)\boldsymbol{\Delta} = \mathbf{0} \ \Rightarrow \ \boldsymbol{\Delta} = -\mathbf{J}^{-1}(\boldsymbol{x}_0)\boldsymbol{f}_0 \ .$$

If $n \neq m$, then $\mathbf{J}$ is non-square and we can use the pseudoinverse or the MATLAB backslash operator `-J\f0`. The computed step $\mathbf{\Delta}$ is based on an approximation to the original nonlinear function so $x_0 + \mathbf{\Delta}$ will generally not be the solution but it will be closer. This leads to an iterative solution – the Newton-Raphson method:

repeat
> compute $f_0 = f(x_0)$, $\mathbf{J} = \mathbf{J}(x_0)$
> $\mathbf{\Delta} = -\mathbf{J}^{-1} f_0$
> $x_0 \leftarrow x_0 + \mathbf{\Delta}$

until $\|f_0\| < \epsilon$

F.2.2 Nonlinear Minimization

A very common class of problems involves finding the *minimum* of a scalar function $f(x) \colon \mathbb{R}^n \mapsto \mathbb{R}$ which can be expressed as

$$x^* = \arg \min_x f(x) \ .$$

The derivative of the linearized system of equations (F.5) is

$$\frac{\mathrm{d} f'}{\mathrm{d}\mathbf{\Delta}} = \mathbf{J}(x_0)$$

and if we consider the function to be a multi-dimensional surface, then $\mathbf{J}(x_0) \in \mathbb{R}^{n \times 1}$ is a vector indicating the direction and magnitude of the *slope* at $x = x_0$ so an update of

$$\mathbf{\Delta} = -\beta \mathbf{J}(x_0)$$

will move the estimate *down hill* toward the minimum. This leads to an iterative solution called gradient descent:

repeat
> compute $\mathbf{J} = \mathbf{J}(x_0)$
> $\mathbf{\Delta} = -\beta \mathbf{J}$
> $x_0 \leftarrow x_0 + \mathbf{\Delta}$

until $\|\mathbf{\Delta}\| < \epsilon$

and the challenge is to choose the appropriate value of β which controls the step size. If too small, then many iterations will be required. If too large, the solution may be unstable and not converge.

If we include the second-order term from (F.4), the approximation becomes

$$f'(\mathbf{\Delta}) \approx f_0 + \mathbf{J}(x_0)\mathbf{\Delta} + \frac{1}{2}\mathbf{\Delta}^\top \mathbf{H}(x_0)\mathbf{\Delta}$$

and includes the $n \times n$ Hessian matrix. To find its minima we take the derivative and set it to zero

$$\frac{\mathrm{d} f'}{\mathrm{d}\mathbf{\Delta}} = \mathbf{0} \ \Rightarrow \ \mathbf{J}(x_0) + \mathbf{H}(x_0)\mathbf{\Delta} = \mathbf{0}$$

and the update is

$$\mathbf{\Delta} = -\mathbf{H}^{-1}(x_0)\mathbf{J}(x_0) \ .$$

This leads to another iterative solution – Newton's method. The challenge is determining the Hessian of the nonlinear system, either by numerical approximation or symbolic manipulation.

F.2.3 **Nonlinear Least-Squares Minimization**

Very commonly, the scalar function we wish to optimize is a quadratic cost function

$$F(x) = ||f(x)||^2 = f(x)^\top f(x)$$

where $f(x): \mathbb{R}^n \mapsto \mathbb{R}^m$ is some vector-valued nonlinear function which we can linearize as

$$f'(\Delta) \approx f_0 + \mathbf{J}\Delta$$

and the scalar cost is

$$
\begin{aligned}
F(\Delta) &\approx (f_0 + \mathbf{J}\Delta)^\top (f_0 + \mathbf{J}\Delta) \\
&\approx f_0^\top f_0 + \underline{f_0^\top \mathbf{J}\Delta} + \underline{\Delta^\top \mathbf{J}^\top f_0} + \Delta^\top \mathbf{J}^\top \mathbf{J}\Delta \\
&\approx f_0^\top f_0 + 2 f_0^\top \mathbf{J}\Delta + \Delta^\top \mathbf{J}^\top \mathbf{J}\Delta
\end{aligned}
\tag{F.6}
$$

where $\mathbf{J}^\top \mathbf{J} \in \mathbb{R}^{n \times n}$ is the *approximate* Hessian from ▶ App. E.2. ▶

To minimize the error of this linearized least squares system, we take the derivative with respect to Δ and set it to zero

$$\frac{\mathrm{d}F}{\mathrm{d}\Delta} = \mathbf{0} \ \Rightarrow \ 2 f_0^\top \mathbf{J} + \Delta^\top \mathbf{J}^\top \mathbf{J} = \mathbf{0}$$

which we can solve for the locally optimal update

$$\Delta = -\left(\mathbf{J}^\top \mathbf{J}\right)^{-1} \mathbf{J}^\top f_0 \tag{F.7}$$

which is the pseudoinverse or left generalized-inverse of $\mathbf{J}$. Once again, we iterate to find the solution – a Gauss-Newton iteration.

One of the underlined terms is the transpose of the other, but since both result in a scalar, the transposition doesn't matter.

Numerical Issues

When solving (F.7), we may find that the Hessian $\mathbf{J}^\top \mathbf{J}$ is poorly conditioned or singular and we can add a damping term λ

$$\Delta = -\left(\mathbf{J}^\top \mathbf{J} + \lambda \mathbf{1}_{n \times n}\right)^{-1} \mathbf{J}^\top f_0$$

which makes the system more positive-definite. Since $\mathbf{J}^\top \mathbf{J} + \mathbf{1}_{n \times n}$ is effectively in the denominator, increasing λ will decrease $\|\Delta\|$ and slow convergence.

How do we choose λ? We can experiment with different values, but a better way is the Levenberg-Marquardt algorithm (■ Fig. F.1) which adjusts λ to ensure convergence. If the error increases compared to the last step, then the step is repeated with increased λ to reduce the step size. If the error decreases, then λ is reduced to increase the convergence rate. The updates vary continuously between Gauss-Newton (low λ) and gradient descent (high λ).

For problems where n is large, inverting the $n \times n$ approximate Hessian is expensive. Typically, $m < n$ which means the Jacobian is not square and (F.7) can be rewritten as

$$\Delta = -\mathbf{J}^\top \left(\mathbf{J}\mathbf{J}^\top\right)^{-1} f_0$$

which is the right pseudoinverse and involves inverting a smaller matrix. We can reintroduce a damping term

$$\Delta = -\mathbf{J}^\top \left(\mathbf{J}\mathbf{J}^\top + \lambda \mathbf{1}_{m \times m}\right)^{-1} f_0$$

```
1  initialize λ
2  repeat
3  |   compute f₀ = f(x₀), J = J(x₀), H = JᵀJ
4  |   Δ = −(H + λ1ₙ×ₙ)⁻¹Jᵀf₀
5  |   if  f(x₀ + Δ) < f(x₀) then
6  |   |   – error decreased: reduce damping
7  |   |   x₀ ← x₀ + Δ
8  |   |   λ ← λ/c
9  |   else
10 |   |   – error increased: discard and raise damping
11 |   |   λ ← cλ
12 |   end
13 until  ‖Δ‖ < ε
```

□ **Fig. F.1** Levenberg-Marquadt algorithm, c is typically chosen in the range 2 to 10

and if λ is large this becomes simply

$$\boldsymbol{\Delta} \approx -\beta \mathbf{J}^\top \boldsymbol{f}_0$$

but exhibits very slow convergence.

If $\boldsymbol{f}_k(\cdot)$ has additive noise that is zero mean, normally distributed and time invariant, we have a maximum likelihood estimator of $\boldsymbol{x}$. Outlier data has a significant impact on the result since errors are squared. Robust estimators minimize the effect of outlier data and in an M-estimator

$$F(\boldsymbol{x}) = \rho(\boldsymbol{f}_k(\boldsymbol{x}))$$

the squared norm is replaced by a loss function $\rho(\cdot)$ which models the likelihood of its argument. Unlike the squared norm, these functions flatten off for large values, and some common examples include the Huber loss function and the Tukey biweight function.

F.2.4 Sparse Nonlinear Least Squares

For a large class of problems, the overall cost is the sum of quadratic costs

$$F(\boldsymbol{x}) = \sum_k \| \boldsymbol{f}_k(\boldsymbol{x}) \|^2 = \sum_k \boldsymbol{f}_k(\boldsymbol{x})^\top \boldsymbol{f}_k(\boldsymbol{x}) \ . \tag{F.8}$$

Consider the problem of fitting a model $\boldsymbol{z} = \phi(\boldsymbol{w}; \boldsymbol{x})$ where $\phi \colon \mathbb{R}^p \mapsto \mathbb{R}^m$ with parameters $\boldsymbol{x} \in \mathbb{R}^n$ to a set of data points $(\boldsymbol{w}_k, \boldsymbol{z}_k)$. The error vector associated with the kth data point is

$$\boldsymbol{f}_k(\boldsymbol{x}) = \boldsymbol{z}_k - \phi(\boldsymbol{w}_k; \boldsymbol{x}) \in \mathbb{R}^m$$

and minimizing (F.8) gives the optimal model parameters $\boldsymbol{x}$.

Another example is pose-graph optimization as used for pose-graph SLAM and bundle adjustment. Edge k in the graph connects nodes i and j and has an associated cost $\boldsymbol{f}_k(\cdot) \colon \mathbb{R}^n \mapsto \mathbb{R}^m$

$$\boldsymbol{f}_k(\boldsymbol{x}) = \hat{\boldsymbol{e}}_k(\boldsymbol{x}) - \boldsymbol{e}_k^{\#} \tag{F.9}$$

where $\boldsymbol{e}_k^{\#}$ is the observed value of the edge parameter and $\hat{\boldsymbol{e}}_k(\boldsymbol{x})$ is the estimate based on the state $\boldsymbol{x}$ of the pose graph. This is linearized

$$\boldsymbol{f}_k'(\boldsymbol{\Delta}) \approx \boldsymbol{f}_{0,k} + \mathbf{J}_k \boldsymbol{\Delta}$$

and the squared error for the edge is

$$F_k(x) = f_k^\top(x)\Omega_k f_k(x)$$

where $\Omega_k \in \mathbb{R}^{m \times m}$ is a positive-definite constant matrix ▶ which we combine as

This can be used to specify the significance of the edge $\det(\Omega_k)$ with respect to other edges, as well as the relative significance of the elements of $f_k(\cdot)$.

$$
\begin{aligned}
F_k(\Delta) &\approx (f_{0,k} + J_k\Delta)^\top \Omega_k (f_{0,k} + J_k\Delta) \\
&\approx f_{0,k}^\top \Omega_k f_{0,k} + f_{0,k}^\top \Omega_k J_k \Delta + \Delta^\top J_k^\top \Omega_k f_{0,k} + \Delta^\top J_k^\top \Omega_k J_k \Delta \\
&\approx c_k + 2b_k^\top \Delta + \Delta^\top H_k \Delta
\end{aligned}
$$

where $c_k = f_{0,k}^\top \Omega_k f_{0,k}$, $b_k^\top = f_{0,k}^\top \Omega_k J_k$ and $H_k = J_k^\top \Omega_k J_k$. The total cost is the sum of all edge costs

$$
\begin{aligned}
F(\Delta) &= \sum_k F_k(\Delta) \\
&\approx \sum_k \left(c_k + 2b_k^\top \Delta + \Delta^\top H_k \Delta\right) \\
&\approx \sum_k c_k + 2\left(\sum_k b_k^\top\right)\Delta + \Delta^\top \left(\sum_k H_k\right)\Delta \\
&\approx c + 2b^\top \Delta + \Delta^\top H\Delta
\end{aligned}
$$

where

$$b^\top = \sum_k f_{0,k}^\top \Omega_k J_k, \quad H = \sum_k J_k^\top \Omega_k J_k$$

are summations over the edges of the graph. Once they are computed, we proceed as previously, taking the derivative with respect to Δ and setting it to zero, solving for the update Δ and iterating using ◻ Fig. F.1.

State Vector

The state vector is a concatenation of all poses and coordinates in the optimization problem. For pose-graph SLAM, it takes the form

$$x = \{\xi_1, \xi_2, \ldots, \xi_N\} \in \mathbb{R}^{N_x}.$$

Poses must be represented in a vector form and preferably one that is compact and singularity free. For **SE**(2), this is quite straightforward and we use $\xi \sim (x, y, \theta) \in \mathbb{R}^3$, and $N_\xi = 3$. For **SE**(3), we will use $\xi \sim (t, r) \in \mathbb{R}^6$ which comprises translation $t \in \mathbb{R}^3$ and rotation $r \in \mathbb{R}^3$, and $N_\xi = 6$. Rotation r can be triple angles (Euler or roll-pitch-yaw), axis-angle (Euler vector), exponential coordinates, or the vector part of a unit quaternion as discussed in ▶ Sect. 2.3.1.7. The state vector x has length $N_x = NN_\xi$, and comprises a sequence of subvectors one per pose – the ith subvector is $x_i \in \mathbb{R}^{N_\xi}$.

For pose-graph SLAM with landmarks, or bundle adjustment, the state vector comprises poses and coordinate vectors

$$x = \{\xi_1, \xi_2, \ldots, \xi_N | P_1, P_2, \ldots, P_M\} \in \mathbb{R}^{N_x}$$

and we denote the ith and j^{th} subvectors of x as $x_i \in \mathbb{R}^{N_\xi}$ and $x_j \in \mathbb{R}^{N_P}$ that correspond to ξ_i and P_j respectively, where $N_P = 2$ for **SE**(2) and $N_P = 3$ for **SE**(3). The state vector x length is $N_x = NN_\xi + MN_P$.

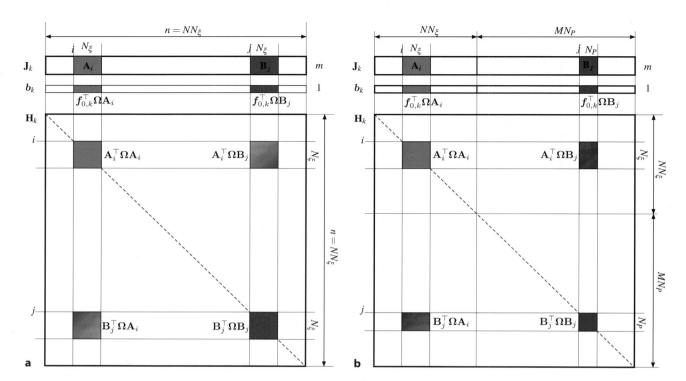

◘ Fig. F.2 Inherent structure of the error vector, Jacobian and Hessian matrices for graph-based least-squares problems. **a** Pose-graph SLAM with N nodes representing robot pose as $\mathbb{R}^{N_\xi}$; **b** bundle adjustment with N nodes representing camera pose as $\mathbb{R}^{N_\xi}$ and M nodes representing landmark position as $\mathbb{R}^{N_P}$. The indices i and j denote the ith and j^{th} block – not the ith and j^{th} row or column. Zero values are indicated by white

Inherent Structure

A key observation is that the error vector $\boldsymbol{f}_k(\boldsymbol{x})$ for edge k depends only on the associated nodes i and j, and this means that the Jacobian

$$\mathbf{J}_k = \frac{\partial \boldsymbol{f}_k(\boldsymbol{x})}{\partial \boldsymbol{x}} \in \mathbb{R}^{m \times N_x}$$

is mostly zeros

$$\mathbf{J}_k = \begin{pmatrix} \mathbf{0} & \cdots & \mathbf{A}_i & \cdots & \mathbf{B}_j & \cdots & \mathbf{0} \end{pmatrix}, \quad \mathbf{A}_i = \frac{\partial \boldsymbol{f}_k(\boldsymbol{x})}{\partial \boldsymbol{x}_i}, \quad \mathbf{B}_j = \frac{\partial \boldsymbol{f}_k(\boldsymbol{x})}{\partial \boldsymbol{x}_j}$$

where $\mathbf{A}_i \in \mathbb{R}^{m \times N_\xi}$ and $\mathbf{B}_j \in \mathbb{R}^{m \times N_\xi}$ or $\mathbf{B}_j \in \mathbb{R}^{m \times N_P}$ according to the state vector structure.

This sparse block structure means that the vector $\boldsymbol{b}_k$ and the Hessian $\mathbf{J}_k^\top \boldsymbol{\Omega}_k \mathbf{J}_k$ also have a sparse block structure as shown in ◘ Fig. F.2. The Hessian has just four small nonzero blocks, so rather than compute the product $\mathbf{J}_k^\top \boldsymbol{\Omega}_k \mathbf{J}_k$, which involves many multiplications by zero, we can just compute the four nonzero blocks and add them into the Hessian for the least-squares system. All blocks in a row have the same height, and in a column have the same width. For pose-graph SLAM with landmarks, or bundle adjustment, the blocks are of different sizes as shown in ◘ Fig. F.2b.

If the value of an edge represents pose, then (F.9) must be replaced with $\boldsymbol{f}_k(\boldsymbol{x}) = \hat{\boldsymbol{e}}_k(\boldsymbol{x}) \ominus \boldsymbol{e}_k^{\#}$. We generalize this with the $\boxminus$ operator to indicate the use of $-$ or $\ominus$, subtraction or relative pose, as appropriate. Similarly, when updating the state vector at the end of an iteration, the poses must be compounded $\boldsymbol{x}_0 \leftarrow \boldsymbol{x}_0 \oplus \boldsymbol{\Delta}$ and positions incremented by $\boldsymbol{x}_0 \leftarrow \boldsymbol{x}_0 + \boldsymbol{\Delta}$ which we generalize with the $\boxplus$ operator. The pose-graph optimization is solved by the iteration in ◘ Fig. F.3.

```
 1  repeat
 2  │  H ← 0, b ← 0
 3  │  foreach k do
 4  │  │   f₀,ₖ(x₀) = êₖ(x) ⊟ eₖ#
 5  │  │   (i, j) = vertices(k)
 6  │  │   compute Aᵢ(xᵢ), Bⱼ(xⱼ)
 7  │  │   bᵢ ← bᵢ + f₀,ₖᵀ Ωₖ Aᵢ
 8  │  │   bⱼ ← bⱼ + f₀,ₖᵀ Ωₖ Bⱼ
 9  │  │   Hᵢ,ᵢ ← Hᵢ,ᵢ + Aᵢᵀ Ωₖ Aᵢ
10  │  │   Hᵢ,ⱼ ← Hᵢ,ⱼ + Aᵢᵀ Ωₖ Bⱼ
11  │  │   Hⱼ,ᵢ ← Hⱼ,ᵢ + Bⱼᵀ Ωₖ Aᵢ
12  │  │   Hⱼ,ⱼ ← Hⱼ,ⱼ + Bⱼᵀ Ωₖ Bⱼ
13  │  end
14  │  Δ = −H⁻¹b
15  │  x₀ ← x₀ ⊞ Δ
16  until ‖Δ‖ < ε
```

Fig. F.3 Pose graph optimization. For Levenberg-Marquardt optimization, replace line 14 with lines 4–12 from ◻ Fig. F.1

F.2.4.1 Large Scale Problems

For pose-graph SLAM with thousands of poses or bundle adjustment with thousands of cameras and millions of landmarks, the Hessian matrix will be massive leading to computation and storage challenges. The overall Hessian is the summation of many edge Hessians structured as shown in ◻ Fig. F.2, and the total Hessian for two problems we have discussed are shown in ◻ Fig. F.4. They have clear structure which we can exploit.

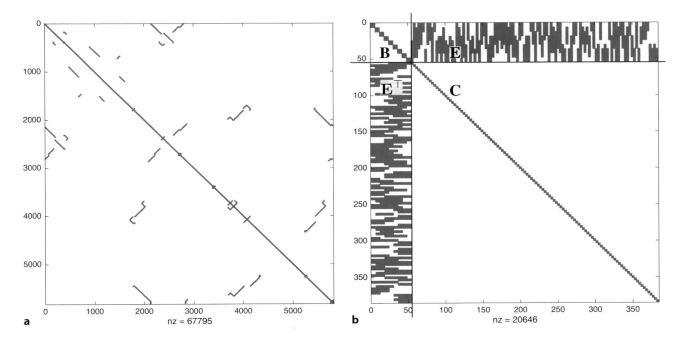

Fig. F.4 Hessian sparsity maps produced using the MATLAB `spy` function, the number of nonzero elements is shown beneath the plot. **a** Hessian for the pose-graph SLAM problem of ◻ Fig. 6.32, the diagonal elements represent pose constraints between successive nodes due to odometry, the off-diagonal terms represent constraints due to revisiting locations (loop closures); **b** Hessian for a bundle adjustment problem with 10 cameras and 110 landmarks (`toolbox/examples/bundleAdjDemo.m`)

Firstly, in both cases the Hessian is sparse – that is, it contains mostly zeros. MATLAB has built-in support for such matrices and instead of storing all those zeros (at 8 bytes each), it simply keeps a list of the nonzero elements. All the standard matrix operations employ efficient algorithms for manipulating sparse matrices.

Secondly, for the bundle adjustment case in ◨ Fig. F.2b, we see that the Hessian has a block structure, so we partition the system as

$$
\begin{pmatrix} \mathbf{B} & \mathbf{E} \\ \mathbf{E}^\top & \mathbf{C} \end{pmatrix} \begin{pmatrix} \mathbf{\Delta}_\xi \\ \mathbf{\Delta}_P \end{pmatrix} = \begin{pmatrix} \boldsymbol{b}_\xi \\ \boldsymbol{b}_P \end{pmatrix}
$$

A block diagonal matrix is inverted by simply inverting each of the nonzero blocks along its diagonal.

where $\mathbf{B}$ and $\mathbf{C}$ are block diagonal. ◀ The subscripts ξ and P denote the blocks of $\mathbf{\Delta}$ and $\boldsymbol{b}$ associated with camera poses and landmark positions respectively. We solve first for the camera pose updates $\mathbf{\Delta}_\xi$

$$
\mathbf{S}\mathbf{\Delta}_\xi = \boldsymbol{b}_\xi - \mathbf{E}\mathbf{C}^{-1}\boldsymbol{b}_P
$$

where $\mathbf{S} = \mathbf{B} - \mathbf{E}\mathbf{C}^{-1}\mathbf{E}^\top$ is the Schur complement of $\mathbf{C}$ and is a symmetric positive-definite matrix that is also block diagonal. Then we solve for the update to landmark positions

$$
\mathbf{\Delta}_P = \mathbf{C}^{-1}\big(\boldsymbol{b}_P - \mathbf{E}^\top\mathbf{\Delta}_\xi\big) \ .
$$

More sophisticated techniques exploit the fine-scale block structure to further reduce computational time, for example GTSAM (▶ https://github.com/borglab/gtsam) and SLAM++ (▶ https://sourceforge.net/projects/slam-plus-plus).

F.2.4.2 Anchoring

Optimization provides a solution where the *relative* poses and positions give the lowest overall cost, and the solution will have an arbitrary transformation with respect to a global reference frame. To obtain absolute poses and positions, we must anchor or fix some nodes – assign them values with respect to the global frame and prevent the optimization from adjusting them. The appropriate way to achieve this is to remove from $\mathbf{H}$ and $\boldsymbol{b}$ the rows and columns corresponding to the anchored poses and positions. We then solve a lower dimensional problem for $\mathbf{\Delta}'$ which will be shorter than $\boldsymbol{x}$. Careful bookkeeping is required to add the subvectors of $\mathbf{\Delta}'$ into $\boldsymbol{x}$ for the update.

G Gaussian Random Variables

The 1-dimensional Gaussian function

$$g(x) = \frac{1}{\sqrt{\sigma^2 2\pi}} e^{-\frac{1}{2\sigma^2}(x-\mu)^2} \tag{G.1}$$

is completely described by the position of its peak μ and its width σ. The total area under the curve is unity and $g(x) > 0, \forall x$. The function can be plotted using the function `gaussfunc`

```
>> x = linspace(-6,6,500);
>> plot(x,gaussfunc(0,1,x),"r")
>> hold on
>> plot(x,gaussfunc(0,2^2,x),"b")
```

and � Fig. G.1 shows two Gaussians with zero mean and $\sigma = 1$ and $\sigma = 2$. Note that the second argument to `gaussfunc` is σ^2.

If the Gaussian is considered to be a probability density function (PDF), then this is the well known normal distribution and the peak position μ is the mean value and the width σ is the standard deviation. A random variable drawn from a normal distribution is often written as $X \sim N(\mu, \sigma^2)$, and σ^2 is the variance. $N(0, 1)$ is referred to as the standard normal distribution – the MATLAB® function `randn` draws random numbers from this distribution. To draw one hundred Gaussian random numbers with mean `mu` and standard deviation `sigma` is

```
>> g = sigma*randn(100,1) + mu;
```

The probability that a random value falls within an interval $x \in [x_1, x_2]$ is obtained by integration

$$P = \int_{x_1}^{x_2} g(x)\,dx = \Phi(x_2) - \Phi(x_1), \text{ where } \Phi(x) = \int_{-\infty}^{x} g(x)\,dx$$

or evaluation of the cumulative normal distribution function $\Phi(x)$ returned by `norminv`. The marked points in ◆ Fig. G.1 at $\mu \pm 1\sigma$ delimit the 1σ confidence interval. The area under the curve over this interval is 0.68, so the probability of a random value being drawn from this interval is 68%.

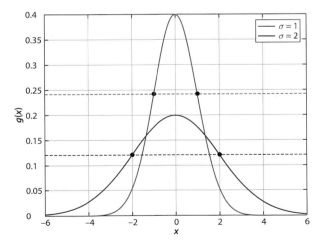

◆ **Fig. G.1** Two Gaussian functions, both with mean $\mu = 0$, and with standard deviation $\sigma = 1$, and $\sigma = 2$. The markers indicate the points $x = \mu \pm 1\sigma$. The area under both curves is unity

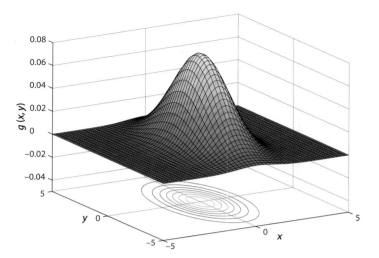

■ **Fig. G.2** The 2-dimensional Gaussian with covariance $\mathbf{P} = \mathrm{diag}(1^2, 2^2)$. Contour lines of constant probability density are shown beneath

The Gaussian can be extended to an arbitrary number of dimensions. The n-dimensional Gaussian, or multivariate normal distribution, is

$$g(x) = \frac{1}{\sqrt{\det(\mathbf{P})(2\pi)^n}} e^{-\frac{1}{2}(x-\mu)^\top \mathbf{P}^{-1}(x-\mu)} \tag{G.2}$$

and compared to the scalar case of (G.1) $x \in \mathbb{R}^n$ and $\mu \in \mathbb{R}^n$ have become vectors, the squared term in the exponent has been replaced by a matrix quadratic form, and σ^2, the variance, has become a symmetric positive-definite matrix $\mathbf{P} \in \mathbb{R}^{n \times n}$. We can plot a 2-dimensional Gaussian

```
>> [x,y] = meshgrid(-5:0.1:5,-5:0.1:5);
>> P = diag([1 2]).^2;
>> surfc(x,y,gaussfunc([0 0],P,x,y))
```

as a surface which is shown in ■ Fig. G.2. The contour lines are ellipses and in this example, the radii in the y- and x-directions are in the ratio $2 : 1$ as defined by the ratio of the standard deviations. If the covariance matrix is diagonal, as in this case, then the ellipses are aligned with the x- and y-axes as we saw in ▶ App. C.1.4, otherwise they are rotated.

If the 2-dimensional Gaussian is considered to be a PDF, then $\mathbf{P}$ is the covariance matrix where the diagonal elements $p_{i,i}$ represent the variance of x_i and the off-diagonal elements $p_{i,j}$ are the correlationss between x_i and x_j. The inverse of the covariance matrix is known as the *information matrix*. If the variables are independent or uncorrelated, the matrix $\mathbf{P}$ would be diagonal. In this case, $\mu = (0, 0)$ and $\mathbf{P} = \mathrm{diag}(1^2, 2^2)$ which corresponds to uncorrelated variables with standard deviation of 1 and 2 respectively. If the PDF represented the position of a vehicle, it is most likely to be at coordinates $(0, 0)$. The probability of being within any region of the xy-plane is the volume under the surface of that region, and involves a non-trivial integration to determine.

The connection between Gaussian probability density functions and ellipses can be found in the quadratic exponent of (G.2) which is the equation of an ellipse or ellipsoid. ◀ All the points that satisfy

It is also the definition of Mahalanobis distance, the covariance weighted distance between x and μ.

$$(x - \mu)^\top \mathbf{P}^{-1}(x - \mu) = s$$

result in a constant probability density value, that is, a contour of the 2-dimensional Gaussian. If p is the probability that the point x lies inside the ellipse, then s is given by the inverse-cumulative χ^2 distribution function ▶ with n degrees of freedom, 2 in this case. For example, the 50% confidence interval is

```
>> s = chi2inv(0.5,2)
s =
    1.3863
```

where the first argument is the probability and the second is the number of degrees of freedom ▶.

To draw a 2-dimensional covariance ellipse, we use the general approach for ellipses outlined in ▶ App. C.1.4, but the right-hand side of the ellipse equation is s not 1, and $\mathbf{E} = \mathbf{P}^{-1}$. For example, to plot the 95% confidence interval for a covariance matrix P we could write `plotellipse(P,confidence=0.95,inverted=true)`.

If we draw a vector of length n from the multivariate Gaussian, each element is normally distributed. The sum of squares of independent normally distributed values has a χ^2 (chi-squared) distribution with n degrees of freedom.

This function requires the MATLAB Statistics and Machine Learning Toolbox™.

H Kalman Filter

» *All models are wrong. Some models are useful.*
– George Box

Consider the system shown in ◘ Fig. H.1. The physical robot is a "black box" which has a true state or pose x that evolves over time according to the applied inputs. We cannot directly measure the state, but sensors on the robot have outputs which are a function of that true state. Our challenge is: given the system inputs and sensor outputs, *estimate* the unknown true state x and how certain we are of that estimate.

At face value, this might seem hard or even impossible, but we will assume that the system has the model shown inside the black box, and we know how this system behaves. Firstly, we know how the state evolves over time as a function of the inputs – this is the state transition ◄ model $f(\cdot)$, and we know the inputs u that are applied to the system. Our model is unlikely to be perfect ◄ and it is common to represent this uncertainty by an imaginary random number generator which is corrupting the system state – process noise. Secondly, we know how the sensor output depends on the state – this is the sensor model $h(\cdot)$ and its uncertainty is also modeled by an imaginary random number generator – sensor noise.

The imaginary random number sources v and w are *inside* the black box so the random numbers are also unknowable. However, we can describe the characteristics of these random numbers – their *distribution* which tells us how likely it is that we will *draw* a random number with a particular value. A lot of noise in physical systems can be modeled well by the Gaussian (aka normal) distribution $N(\mu, \sigma^2)$ which is characterized by a mean μ and a standard deviation σ, and the Gaussian distribution has some nice mathematical properties that we will rely on. ◄ Typically, we assume that the noise has zero mean, and that the covariance of the process and sensor noise are $\mathbf{V}$ and $\mathbf{W}$ respectively.

Often called the process or motion model.

For example, wheel slippage on a mobile ground robot or wind gusts for a UAV.

There are infinitely many possible distributions that these noise sources could have, which could be asymmetrical or have multiple peaks. We should never assume that noise is Gaussian – we should attempt to determine the distribution by understanding the physics of the process and the sensor, or from careful measurement and analysis.

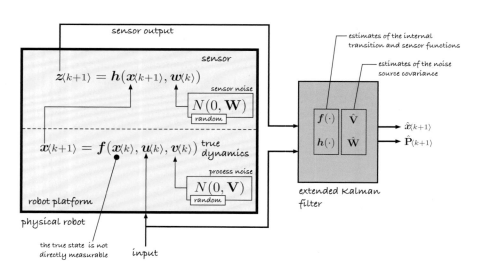

◘ **Fig. H.1** The physical robot on the left is a "black box" whose true state cannot be directly measured. However, it can be inferred from sensor output which is a function of the unknown true state, and the assumed model shown inside the black box

In general terms, the problem we wish to solve is:

> Given a model of the system $f(\cdot)$, $h(\cdot)$, estimates of $\mathbf{V}$ and $\mathbf{W}$; the known inputs applied to the system u; and some noisy sensor measurements z, find an estimate $\hat{x}$ of the system state and the uncertainty $\hat{\mathbf{P}}$ in that estimate.

In a robotic localization context, x is the unknown position or pose of the robot, u is the commands sent to the motors, and z is the output of various sensors on the robot. For a ground robot, x would be the pose in $\mathbf{SE}(2)$, u would be the motor commands, and z might be the measured odometry or range and bearing to landmarks. For an aerial robot, x would be the pose in $\mathbf{SE}(3)$, u are the known forces applied to the airframe by the propellers, and z might be the measured accelerations and angular velocities. ▶

The state is a vector and there are many approaches to mapping pose to a vector, especially the rotational component – Euler angles, quaternions, and exponential coordinates are commonly used, see ▶ App. F.2.4.

H.1 Linear Systems – Kalman Filter

Consider the transition model described as a discrete-time linear time-invariant system

$$x_{\langle k+1\rangle} = \mathbf{F}x_{\langle k\rangle} + \mathbf{G}u_{\langle k\rangle} + v_{\langle k\rangle} \tag{H.1}$$

$$z_{\langle k\rangle} = \mathbf{H}x_{\langle k\rangle} + w_{\langle k\rangle} \tag{H.2}$$

where k is the time step, $x \in \mathbb{R}^n$ is the state vector, and $u \in \mathbb{R}^m$ is a vector of inputs to the system at time k, for example, a velocity command, or applied forces and torques. The matrix $\mathbf{F} \in \mathbb{R}^{n \times n}$ describes the dynamics of the system, that is, how the states evolve with time. The matrix $\mathbf{G} \in \mathbb{R}^{n \times m}$ describes how the inputs are coupled to the system states. The vector $z \in \mathbb{R}^p$ represents the outputs of the system as measured by sensors. The matrix $\mathbf{H} \in \mathbb{R}^{p \times n}$ describes how the system states are mapped to the system outputs which we can observe.

To account for errors in the motion model ($\mathbf{F}$ and $\mathbf{G}$) or unmodeled disturbances, we introduce a Gaussian random variable $v \in \mathbb{R}^n$ termed the process noise. $v_{\langle k\rangle} \sim N(0, \mathbf{V})$, that is, it has zero mean and covariance $\mathbf{V} \in \mathbb{R}^{n \times n}$. Covariance is a matrix quantity which is the variance for a multi-dimensional distribution – it is a symmetric positive-definite matrix. The sensor measurement model $\mathbf{H}$ is not perfect either, and this is modeled by sensor measurement noise, a Gaussian random variable $w \in \mathbb{R}^p$, $w_{\langle k\rangle} \sim N(0, \mathbf{W})$ and covariance $\mathbf{W} \in \mathbb{R}^{p \times p}$.

The Kalman filter is an optimal estimator for the case where the process and measurement noise are both zero-mean Gaussian noise. The filter has two steps: prediction and update. The prediction is based on the previous state and the inputs that were applied

$$\hat{x}^+_{\langle k+1\rangle} = \mathbf{F}\hat{x}_{\langle k\rangle} + \mathbf{G}u_{\langle k\rangle} \tag{H.3}$$

$$\hat{\mathbf{P}}^+_{\langle k+1\rangle} = \underline{\mathbf{F}\hat{\mathbf{P}}_{\langle k\rangle}\mathbf{F}^\top} + \hat{\mathbf{V}} \tag{H.4}$$

where $\hat{x}^+$ is the predicted estimate of the state and $\hat{\mathbf{P}} \in \mathbb{R}^{n \times n}$ is the predicted estimate of the covariance, or uncertainty, in $\hat{x}^+$. The notation $^+$ makes explicit that the left-hand side is a prediction at time $k + 1$ based on information from time k. $\hat{\mathbf{V}}$ is a constant and our best estimate of the covariance of the process noise.

The underlined term in (H.4) *projects* the estimated covariance from the current time step to the next. Consider a one dimensional example where F is a scalar and the state estimate $\hat{x}_{\langle k\rangle}$ has a PDF which is Gaussian with a mean $\mu_{\langle k\rangle}$ and a variance $\sigma^2\langle k\rangle$. The prediction equation maps the state and its Gaussian distribution to a new Gaussian distribution with a mean $F\mu\langle k\rangle$ and a variance $F^2\sigma^2\langle k\rangle$. The term $\mathbf{F}\langle k\rangle\mathbf{P}\langle k\rangle\mathbf{F}\langle k\rangle^\top$ is the matrix form of this since

$$\text{cov}(\mathbf{F}x) = \mathbf{F}\,\text{cov}(x)\mathbf{F}^\top \tag{H.5}$$

which scales the covariance appropriately.

The prediction of $\hat{\mathbf{P}}$ in (H.4) involves the addition of two positive-definite matrices so the uncertainty will increase – this is to be expected since we have used an uncertain model to predict the future value of an already uncertain estimate. $\hat{\mathbf{V}}$ must be a reasonable estimate of the covariance of the actual process noise. If we overestimate it, that is our estimate of process noise is larger than it really is, then we will have a larger than necessary increase in uncertainty at this step, leading to a pessimistic estimate of our certainty.

To counter this growth in uncertainty, we need to introduce new information such as measurements made by the sensors since they depend on the state. The difference between what the sensors measure and what the sensors are predicted to measure is

$$\boldsymbol{v} = \boldsymbol{z}^{\#}\langle k+1\rangle - \mathbf{H}\hat{\boldsymbol{x}}^{+}\langle k+1\rangle \in \mathbb{R}^{p} \ .$$

Some of this difference is due to noise in the sensor, the measurement noise, but the remainder provides valuable information related to the error between the actual and the predicted value of the state. Rather than considering this as error, we refer to it more positively as *innovation* – new information.

The second step of the Kalman filter, the *update* step, maps the innovation into a correction for the predicted state, optimally tweaking the estimate based on what the sensors observed

$$\hat{\boldsymbol{x}}\langle k+1\rangle = \hat{\boldsymbol{x}}^{+}\langle k+1\rangle + \mathbf{K}\boldsymbol{v}, \tag{H.6}$$

$$\hat{\mathbf{P}}\langle k+1\rangle = \hat{\mathbf{P}}^{+}\langle k+1\rangle - \mathbf{K}\mathbf{H}\hat{\mathbf{P}}^{+}\langle k+1\rangle . \tag{H.7}$$

Uncertainty is now *decreased* or *deflated*, since new information, from the sensors, is being incorporated. The matrix

$$\mathbf{K} = \mathbf{P}^{+}\langle k+1\rangle \mathbf{H}^\top \left(\underline{\mathbf{H}\mathbf{P}^{+}\langle k+1\rangle\mathbf{H}^\top} + \hat{\mathbf{W}}\right)^{-1} \in \mathbb{R}^{n\times p} \tag{H.8}$$

is known as the Kalman gain. The term in parentheses is the estimated covariance of the innovation, and comprises the uncertainty in the state and the estimated measurement noise covariance. If the innovation has high uncertainty in some dimensions, then the Kalman gain will be correspondingly small, that is, if the new information is uncertain, then only small changes are made to the state vector. The underlined term *projects* the covariance of the state estimate into the space of sensor values. $\hat{\mathbf{W}}$ is a constant and our best estimate of the covariance of the sensor noise.

The covariance matrix must be symmetric, but after many updates the accumulated numerical errors may result in the matrix no longer being symmetric. The symmetric structure can be enforced by using the Joseph form of (H.7)

$$\hat{\mathbf{P}}\langle k+1\rangle = (\mathbf{1}_{n\times n} - \mathbf{K}\mathbf{H})\hat{\mathbf{P}}^{+}\langle k+1\rangle(\mathbf{1}_{n\times n} - \mathbf{K}\mathbf{H})^\top + \mathbf{K}\hat{\mathbf{V}}\mathbf{K}^\top$$

but this is computationally more costly.

The equations above constitute the classical Kalman filter which is widely used in robotics, aerospace and econometric applications. The filter has a number of important characteristics. Firstly, it is optimal, but only if the noise is truly Gaussian with zero mean and time-invariant parameters. This is often a good assumption but not always. Secondly, it is recursive, the output of one iteration is the input to the next. Thirdly, it is asynchronous. At a particular iteration, if no sensor information is available, we only perform the prediction step and not the update. In the case that there are different sensors, each with their own $\mathbf{H}$, and different sample rates, we just apply the update with the appropriate z and $\mathbf{H}$ whenever sensor data becomes available.

The filter must be initialized with some reasonable value of $\hat{x}$ and $\hat{\mathbf{P}}$, as well as good choices of the estimated covariance matrices $\hat{\mathbf{V}}$ and $\hat{\mathbf{W}}$. As the filter runs, the estimated covariance $\|\hat{\mathbf{P}}\|$ decreases but never reaches zero – the minimum value can be shown to be a function of $\hat{\mathbf{V}}$ and $\hat{\mathbf{W}}$. The Kalman-Bucy filter is a continuous-time version of this filter.

The covariance matrix $\hat{\mathbf{P}}$ is rich in information. The diagonal elements $\hat{p}_{i,i}$ are the variance, or uncertainty, in the state x_i. The off-diagonal elements $\hat{p}_{i,j}$ are the correlations between states x_i and x_j and indicate that the errors in the states are not independent. The correlations are critical in allowing any piece of new information to *flow through* to adjust all the states that affect a particular process output.

If elements of the state vector represent 2-dimensional position, the corresponding 2×2 submatrix of $\hat{\mathbf{P}}$ is the inverse of the equivalent uncertainty ellipse $\mathbf{E}^{-1}$, see ▶ App. C.1.4.

H.2 Nonlinear Systems – Extended Kalman Filter

For the case where the system is not linear, it can be described generally by two functions: the state transition (the motion model in robotics) and the sensor model

$$x_{\langle k+1 \rangle} = f(x_{\langle k \rangle},\ u_{\langle k \rangle},\ v_{\langle k \rangle}) \tag{H.9}$$

$$z_{\langle k \rangle} = h(x_{\langle k \rangle},\ w_{\langle k \rangle}) \tag{H.10}$$

and as before we represent model uncertainty, external disturbances and sensor noise by Gaussian random variables v and w.

We linearize the state transition function about the current state estimate $\hat{x}_{\langle k \rangle}$ as shown in ◼ Fig. H.2 resulting in

$$x'_{\langle k+1 \rangle} \approx \mathbf{F}_x\, x'_{\langle k \rangle} + \mathbf{F}_u\, u_{\langle k \rangle} + \mathbf{F}_v\, v_{\langle k \rangle} \tag{H.11}$$

$$z'_{\langle k \rangle} \approx \mathbf{H}_x\, x'_{\langle k \rangle} + \mathbf{H}_w\, w_{\langle k \rangle} \tag{H.12}$$

where $\mathbf{F}_x = \partial f / \partial x \in \mathbb{R}^{n \times n}$, $\mathbf{F}_u = \partial f / \partial u \in \mathbb{R}^{n \times m}$, $\mathbf{F}_v = \partial f / \partial v \in \mathbb{R}^{n \times n}$, $\mathbf{H}_x = \partial h / \partial x \in \mathbb{R}^{p \times n}$ and $\mathbf{H}_w = \partial h / \partial w \in \mathbb{R}^{p \times p}$ are Jacobians, see ▶ App. E, of the functions $f(\cdot)$ and $h(\cdot)$. Equating coefficients between (H.1) and (H.11) gives $\mathbf{F} \sim \mathbf{F}_x$, $\mathbf{G} \sim \mathbf{F}_u$ and $v_{\langle k \rangle} \sim \mathbf{F}_v v_{\langle k \rangle}$; and between (H.2) and (H.12) gives $\mathbf{H} \sim \mathbf{H}_x$ and $w_{\langle k \rangle} \sim \mathbf{H}_w w_{\langle k \rangle}$.

Taking the prediction equation (H.9) with $v_{\langle k \rangle} = 0$, and the covariance equation (H.4) with the linearized terms substituted, we can write the prediction step as

$$\hat{x}^{+}_{\langle k+1 \rangle} = f(\hat{x}_{\langle k \rangle},\ u_{\langle k \rangle})$$

$$\hat{\mathbf{P}}^{+}_{\langle k+1 \rangle} = \mathbf{F}_x\hat{\mathbf{P}}_{\langle k \rangle}\mathbf{F}_x^{\top} + \mathbf{F}_v\hat{\mathbf{V}}\mathbf{F}_v^{\top}$$

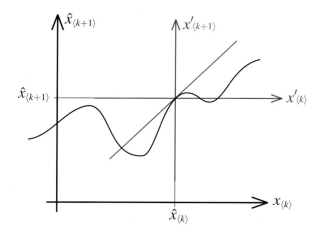

◻ Fig. H.2 One dimensional example illustrating how the nonlinear state transition function $f : x_k \mapsto x_{k+1}$ shown in black is linearized about the point $(\hat{x}_{\langle k \rangle}, \hat{x}_{\langle k+1 \rangle})$ shown in red

and the update step as

$$\hat{x}_{\langle k+1 \rangle} = \hat{x}^+_{\langle k+1 \rangle} + \mathbf{K}\nu$$
$$\hat{\mathbf{P}}_{\langle k+1 \rangle} = \hat{\mathbf{P}}^+_{\langle k+1 \rangle} - \mathbf{KH}_x \hat{\mathbf{P}}^+_{\langle k+1 \rangle}$$

where the Kalman gain is now

$$\mathbf{K} = \mathbf{P}^+_{\langle k+1 \rangle} \mathbf{H}_x^\top \left(\mathbf{H}_x \mathbf{P}^+_{\langle k+1 \rangle} \mathbf{H}_x^\top + \mathbf{H}_w \hat{\mathbf{W}} \mathbf{H}_w^\top \right)^{-1} . \tag{H.13}$$

Properly, these matrices should be denoted as depending on the time step, i.e. $\mathbf{F}_{x\langle k \rangle}$, but this has been dropped in the interest of readability.

These equations are only valid at the linearization point $\hat{x}_{\langle k \rangle}$ – the Jacobians $\mathbf{F}_x$, $\mathbf{F}_v$, $\mathbf{H}_x$, $\mathbf{H}_w$ must be computed at every iteration. ◄ The full procedure is summarized in ◻ Fig. H.3.

A fundamental problem with the extended Kalman filter is that PDFs of the random variables are no longer Gaussian after being operated on by the nonlinear functions $f(\cdot)$ and $h(\cdot)$. We can illustrate this by considering a nonlinear scalar function $y = (x + 2)^2/4$. We will draw a million Gaussian random numbers from the normal distribution $N(5, 4)$ which has a mean of 5 and a standard deviation of 2

```
>> x = 2*randn(1000000,1) + 5;
```

then apply the function

```
>> y = (x+2).^2/4;
```

and plot the probability density function of y

```
>> histogram(y,Normalization="pdf");
```

which is shown in ◻ Fig. H.4. We see that the PDF of y is substantially changed and no longer Gaussian. It has lost its symmetry, so the mean value is greater than the mode. The Jacobians that appear in the EKF equations appropriately scale the covariance but the resulting non-Gaussian distribution breaks the assumptions which guarantee that the Kalman filter is an optimal estimator. Alternative approaches to dealing with system nonlinearity include the iterated EKF described

Input: $\hat{x}_{\langle k \rangle} \in \mathbb{R}^n$, $\hat{P}_{\langle k \rangle} \in \mathbb{R}^{n \times n}$, $u_{\langle k \rangle} \in \mathbb{R}^m$, $z_{\langle k+1 \rangle} \in \mathbb{R}^p$; $\hat{V} \in \mathbb{R}^{n \times n}$,
  $\hat{W} \in \mathbb{R}^{p \times p}$

Output: $\hat{x}_{\langle k+1 \rangle} \in \mathbb{R}^n$, $\hat{P}_{\langle k+1 \rangle} \in \mathbb{R}^{n \times n}$

– linearize about $x = \hat{x}_{\langle k \rangle}$

  compute Jacobians: $\mathbf{F}_x \in \mathbb{R}^{n \times n}$, $\mathbf{F}_v \in \mathbb{R}^{n \times n}$, $\mathbf{H}_x \in \mathbb{R}^{p \times n}$, $\mathbf{H}_w \in \mathbb{R}^{p \times p}$

– the prediction step

$$\hat{x}^+_{\langle k+1 \rangle} = f(\hat{x}_{\langle k \rangle}, u_{\langle k \rangle}) \quad \textit{// predict state at next time step}$$

$$\hat{P}^+_{\langle k+1 \rangle} = \mathbf{F}_x \hat{P}_{\langle k \rangle} \mathbf{F}_x^\top + \mathbf{F}_v \hat{V} \mathbf{F}_v^\top \quad \textit{// predict covariance at next time step}$$

– the update step

$$\nu = z_{\langle k+1 \rangle} - \mathbf{h}(\hat{x}^+_{\langle k+1 \rangle}) \quad \textit{// innovation: measured - predicted sensor value}$$

$$\mathbf{K} = \mathbf{P}^+_{\langle k+1 \rangle} \mathbf{H}_x^\top \left(\mathbf{H}_x \, \mathbf{P}^+_{\langle k+1 \rangle} \mathbf{H}_x^\top + \mathbf{H}_w \hat{W} \mathbf{H}_w^\top \right)^{-1} \quad \textit{// Kalman gain}$$

$$\hat{x}_{\langle k+1 \rangle} = \hat{x}^+_{\langle k+1 \rangle} + \mathbf{K}\nu \quad \textit{// update state estimate}$$

$$\hat{P}_{\langle k+1 \rangle} = \hat{P}^+_{\langle k+1 \rangle} - \mathbf{K}\mathbf{H}_x \hat{P}^+_{\langle k+1 \rangle} \quad \textit{// update covariance estimate}$$

◘ **Fig. H.3** Procedure EKF

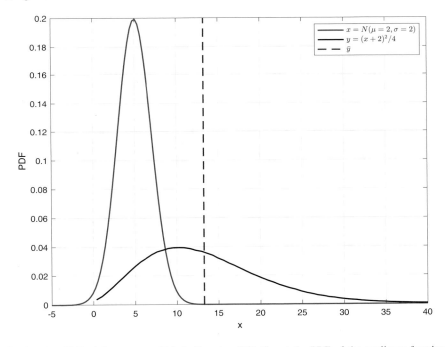

◘ **Fig. H.4** PDF of the state x which is Gaussian $N(5, 4)$ and the PDF of the nonlinear function $y = (x + 2)^2/4$

by Jazwinski (2007), the Unscented Kalman Filter (UKF) (Julier and Uhlmann 2004), or the sigma-point filter which uses discrete sample points (sigma points) to approximate the PDF.

I Graphs

A graph is an abstract representation of a set of objects connected by links and depicted visually as shown in ◼ Fig. I.1. Mathematically, a graph is denoted by $G(V, E)$ where V are the nodes or vertices, and E are the links that connect pairs of nodes and are called edges or arcs. Edges can be directed (shown as arrows) or undirected (shown as line segments) as in this case. Edges can have an associated weight or cost associated with moving from one node to another. A sequence of edges from one node to another is a path, and a sequence that starts and ends at the same node is a cycle. An edge from a node to itself is a loop. Graphs can be used to represent transport, communications or social networks, and this branch of mathematics is graph theory.

The RVC Toolbox provides a graph class called `UGraph` that supports undirected embedded graphs where the nodes are associated with a point in an n-dimensional space. ◀ To create a new undirected graph

This class wraps the native MATLAB `graph` class and is used for the mobile robot graph-based path planning algorithms in ▶ Sect. 5.3.

```
>> g = UGraph
g =
   2 dimensions
   0 nodes
   0 edges
   0 components
```

and, by default, the nodes of the graph exist in 2-dimensional space. We can add nodes to the graph

```
>> rng(10)
>> for i = 1:5
>>     g.add_node(rand(2,1));
>> end
```

and each has a random coordinate. The nodes are numbered 1 to 5, and we add edges between pairs of nodes

```
>> g.add_edge(1,2);
```

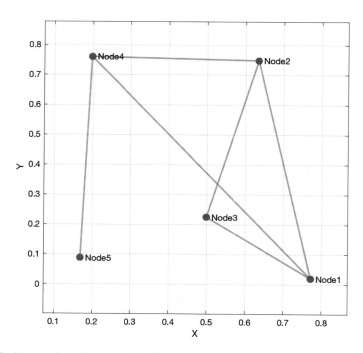

◼ **Fig. I.1** An example graph generated by the `UGraph` class

```
>> g.add_edge(1,3);
>> g.add_edge(1,4);
>> g.add_edge(2,3);
>> g.add_edge(2,4);
>> g.add_edge(4,5);
>> g
g =
  2 dimensions
  5 nodes
  6 edges
  1 components
```

By default, the edge cost is the Euclidean distance between the nodes, but this can be overridden by a third argument to add_edge. The methods add_node and add_edge return an integer that uniquely identifies the node or edge just created. The graph has one component, that is, all the nodes are connected into one network. The graph can be plotted by

```
>> g.plot(labels=true);
```

as shown in ◘ Fig. I.1. The nodes are shown as blue circles, and the option labels displays the node index next to the circle. Edges are shown as blue lines joining nodes. Many options exist to change default plotting behavior. Note that only graphs embedded in 2- and 3-dimensional space can be plotted.

The neighbors of node 2 are

```
>> g.neighbors(2)
ans =
     1     3     4
```

which are nodes connected to node 2 by edges. Each edge has a unique index and the edges connecting to node 2 are

```
>> e = g.edges(2)
e =
     1     4     5
```

The cost or length of an edge is

```
>> g.cost(e)
ans =
    0.7410    0.5412    0.4357
```

and clearly edge 5 has a lower cost than edges 4 and 1. Edge 5

```
>> g.nodes(5)'  % transpose for display
ans =
     2     4
```

joins nodes 2 and 4, and node 4 is clearly the closest neighbor of node 2. Frequently, we wish to obtain a node's neighboring nodes and their distances at the same time, and this can be achieved conveniently by

```
>> [n,c] = g.neighbors(2)
n =
     1     3     4
c =
    0.7410    0.5412    0.4357
```

Concise information about a node can be obtained by

```
>> g.about(1)
Node 1 #1@ (0.771321 0.0207519 )
  neighbours: 2 3 4
  edges: 1 2 3
```

Arbitrary data can be attached to any node or edge by the methods setndata and setedata respectively, and retrieved by the methods ndata and edata respectively.

The node closest to the coordinate $(0.5, 0.5)$ is

```
>> g.closest([0.5 0.5])
ans =
     3
```

and the node closest to an interactively selected point is given by `g.pick`.

The minimum cost path between any two nodes in the graph can be computed using well-known algorithms such as A* (Hart 1968)

```
>> g.path_Astar(3, 5)
ans =
     3     2     4     5
```

which is a sequence of node indices.

Methods exist to compute various other representations of the graph such as adjacency, incidence, degree and Laplacian matrices.

J Peak Finding

A commonly encountered problem is finding, or refining an estimate of, the position of the peak of some discrete 1- or 2-dimensional signal.

J.1 1D Signal

Consider the discrete 1-dimensional signal $y(k)$, $k \in \mathbb{N}$

```
>> load peakfit1
>> plot(y,"-o")
```

shown in ◻ Fig. J.1a.

Finding the position of the peak to the nearest integer k is straightforward using the MATLAB® function `max`

```
>> [ypk,k] = max(y)
ypk =
    0.9905
k =
    8
```

which indicates the peak occurs at the eighth element and has a value of 0.9905. In this case, there is more than one peak and we can use the Signal Processing Toolbox function `findpeaks` instead

```
>> [ypk,k] = findpeaks(y,SortStr="descend");
>> ypk'  % transpose for display
ans =
    0.9905    0.6718    -0.5799
>> k'  % transpose for display
ans =
    8    25    16
```

which has returned three maxima ordered by descending value. A common test of the quality of a peak is its magnitude, and the ratio of the height of the second peak to the first peak

```
>> ypk(2)/ypk(1)
ans =
    0.6783
```

which is called the ambiguity ratio and is ideally small.

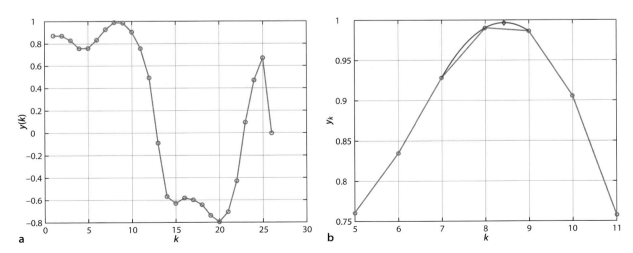

◻ **Fig. J.1** Peak fitting. **a** A signal with several local maxima; **b** close-up view of the first maximum with the fitted curve (red) and the estimated peak

The signal $y(k)$ in ▪ Fig. J.1a is a sampled representation of a continuous underlying signal $y(x)$ and the real peak might actually lie between the samples. If we look at a zoomed version of the signal, ▪ Fig. J.1b, we can see that although the maximum is at $k = 8$, the value at $k = 9$ is only slightly lower, so the peak lies somewhere between these two points. A common approach is to fit a parabola

$$y = a\delta_x^2 + b\delta_x + c, \; \delta_x \in \mathbb{R} \tag{J.1}$$

to the points surrounding the peak: $(-1, y_{k-1})$, $(0, y_k)$ and $(1, y_{k+1})$ where $\delta_x = 0$ when $k = 8$. Substituting these into (J.1), we can write three equations

$$y_{k-1} = a - b + c$$
$$y_k = c$$
$$y_{k+1} = a + b + c$$

or in compact matrix form as

$$\begin{pmatrix} y_{k-1} \\ y_k \\ y_{k+1} \end{pmatrix} = \begin{pmatrix} 1 & -1 & 1 \\ 0 & 0 & 1 \\ 1 & 1 & 1 \end{pmatrix} \begin{pmatrix} a \\ b \\ c \end{pmatrix}$$

and then solve for the parabolic coefficients

$$\begin{pmatrix} a \\ b \\ c \end{pmatrix} = \begin{pmatrix} 1 & -1 & 1 \\ 0 & 0 & 1 \\ 1 & 1 & 1 \end{pmatrix}^{-1} \begin{pmatrix} y_{k-1} \\ y_k \\ y_{k+1} \end{pmatrix} = \frac{1}{2} \begin{pmatrix} 1 & -2 & 1 \\ -1 & 0 & 1 \\ 0 & 2 & 0 \end{pmatrix} \begin{pmatrix} y_{k-1} \\ y_k \\ y_{k+1} \end{pmatrix}. \tag{J.2}$$

The maximum of the parabola occurs when its derivative is zero

$$2a\delta_x + b = 0$$

and substituting the values of a and b from (J.2), we find the displacement of the peak of the fitted parabola with respect to the discrete maximum

$$\delta_x = \frac{1}{2} \frac{y_{k-1} - y_{k+1}}{y_{k-1} - 2y_k + y_{k+1}}, \quad \delta_x \in (-1, 1)$$

so the refined, or interpolated, position of the maximum is at

$$\hat{x} = k + \delta_x \in \mathbb{R}$$

and the estimated value of the maximum is obtained by substituting δ_x into (J.1).

The coefficient a will be negative for a maximum. If $|a|$ is large, it indicates a well-defined sharp peak, whereas a low value indicates a very broad peak for which estimation of a refined peak may not be so accurate.

Continuing the earlier example, we can use MATLAB's polynomial functions to estimate the refined position of the first peak

```
>> range=k(1)-1:k(1)+1
range =
      7     8     9
>> p = polyfit(range,y(range),2)   % fit second-order polynomial
p =
   -0.0331    0.5582   -1.3585
>> pd = polyder(p)   % derivative of fitted polynomial
pd =
   -0.0661    0.5582
>> roots(pd)   % zero value of the derivative
ans =
    8.4394
```

The fitted parabola is overlaid on the data points in red in ▪ Fig. J.1b.

If the signal has superimposed noise, then there are likely to be multiple peaks, many of which are quite minor, and this can be overcome by specifying the *scale* of the peak. For example, the peaks that are greater than all other values within ± 5 values in the horizontal direction are

```
>> ypk = findpeaks(y,MinPeakDistance=5)'  % transpose for display
ypk =
    0.9905   -0.5799    0.6718
```

This technique is called nonlocal maxima suppression and requires that an appropriate scale is chosen. In this case, the result is unchanged since the signal is fairly smooth.

J.2 2D Signal

A 2-dimensional discrete signal was loaded from `peakfit1` earlier

```
>> z
z =
   -0.0696    0.0348    0.1394    0.2436    0.3480
    0.0800    0.2000    0.3202    0.4400    0.5600
    0.0400    0.1717    0.3662    0.4117    0.5200
    0.0002    0.2062    0.8766    0.4462    0.4802
   -0.0400    0.0917    0.2862    0.3317    0.4400
   -0.0800    0.0400    0.1602    0.2800    0.4000
```

and the maximum value of 0.8766 is at element $(3, 4)$ using the image coordinate indexing conventions. We can find this by

```
>> [zmax,i] = max(z(:))
zmax =
    0.8766
i =
    16
```

and the maximum is at the sixteenth element in column-major order ▶ which we convert to array subscripts

Counting the elements, starting with 1 at the top-left, down each column then back to the top of the next right-most column.

```
>> [y,x] = ind2sub(size(z),i)
y =
    4
x =
    3
```

We can find the local peaks more conveniently using the Computer Vision Toolbox

```
>> LMaxFinder = vision.LocalMaximaFinder(MaximumNumLocalMaxima=3, ...
>>    NeighborhoodSize=[3 3],Threshold=0);
```

which is a MATLAB system object that we pass the 2-dimensional data to

```
>> LMaxFinder(z)
ans =
  3×2 uint32 matrix
   3    4
   5    2
   5    4
```

and it has returned three local maxima, one per row, using the image coordinate indexing convention. This function will return all nonlocal maxima where the size of the local region is given by the `NeighborhoodSize` option.

To refine the position of the peak we follow a similar procedure to the 1D-case, but instead fit a *paraboloid*

$$z = ax^2 + by^2 + cx + dy + e \tag{J.3}$$

which has five coefficients that can be calculated from the center value (the discrete maximum) and its four neighbors (north, south, east and west) using a similar procedure to above. ◄ The displacement of the estimated peak with respect to the central point is

A paraboloid normally has an xy term giving a total of 6 parameters. This would require using one additional point in order to solve the equation, which breaks the simplicity and symmetry of this scheme. All neighbors could be used, requiring a least-squares solution.

$$\delta_x = \frac{1}{2} \frac{z_e - z_w}{2z_c - z_e - z_w}, \quad \delta_x \in (-1, 1)$$

$$\delta_y = \frac{1}{2} \frac{z_n - z_a}{2z_c - z_n - z_s}, \quad \delta_y \in (-1, 1)$$

In this case, the coefficients a and b represent the sharpness of the peak in the x- and y-directions, and the quality of the peak can be considered as being $\min(|a|, |b|)$.

References

Achtelik MW (2014) Advanced closed loop visual navigation for micro aerial vehicles. Ph.D. thesis, ETH Zürich

Adorno B, Marinho M (2021) DQ Robotics: A Library for Robot Modeling and Control. IEEE Robotics Automation Magazine, vol. 28, 102–116

Agarwal S, Furukawa Y, Snavely N, Simon I, Curless B, Seitz SM, Szeliski R (2011) Building Rome in a day. Commun ACM 54(10):105–112

Agarwal P, Burgard W, Stachniss C (2014) Survey of geodetic mapping methods: Geodetic approaches to mapping and the relationship to graph-based SLAM. IEEE Robot Autom Mag 21(3):63–80

Agrawal M, Konolige K, Blas M (2008) CenSurE: Center surround extremas for realtime feature detection and matching. In: Forsyth D, Torr P, Zisserman A (eds) Lecture notes in computer science. Computer Vision – ECCV 2008, vol 5305. Springer-Verlag, Berlin Heidelberg, pp 102–115

Albertos P, Mareels I (2010) Feedback and control for everyone. Springer-Verlag, Berlin Heidelberg

Altmann SL (1989) Hamilton, Rodrigues, and the quaternion scandal. Math Mag 62(5):291–308

Alton K, Mitchell IM (2006) Optimal path planning under different norms in continuous state spaces. In: Proceedings of the IEEE International Conference on Robotics and Automation (ICRA). pp 866–872

Aly, Mohamed (2008) Real time Detection of Lane Markers in Urban Streets, IEEE Intelligent Vehicles Symposium, Eindhoven, The Netherlands

Andersen N, Ravn O, Sørensen A (1993) Real-time vision based control of servomechanical systems. In: Chatila R, Hirzinger G (eds) Lecture notes in control and information sciences. Experimental Robotics II, vol 190. Springer-Verlag, Berlin Heidelberg, pp 388–402

Andersson RL (1989) Dynamic sensing in a ping-pong playing robot. IEEE T Robotic Autom 5(6):728–739

Antonelli G (2014) Underwater robots: Motion and force control of vehicle-manipulator systems, 3rd ed. Springer Tracts in Advanced Robotics, vol 2. Springer-Verlag, Berlin Heidelberg

Arandjelović R, Zisserman A (2012) Three things everyone should know to improve object retrieval. In: IEEE Conference on Computer Vision and Pattern Recognition (CVPR). pp 2911–2918

Arkin RC (1999) Behavior-based robotics. MIT Press, Cambridge, Massachusetts

Armstrong WW (1979) Recursive solution to the equations of motion of an N-link manipulator. In: Proceedings of the 5th World Congress on Theory of Machines and Mechanisms, Montreal, Jul, pp 1343–1346

Armstrong BS (1988) Dynamics for robot control: Friction modelling and ensuring excitation during parameter identification. Stanford University

Armstrong B (1989) On finding exciting trajectories for identification experiments involving systems with nonlinear dynamics. Int J Robot Res 8(6):28

Armstrong B, Khatib O, Burdick J (1986) The explicit dynamic model and inertial parameters of the Puma 560 Arm. In: Proceedings of the IEEE International Conference on Robotics and Automation (ICRA), vol 3. pp 510–518

Armstrong-Hélouvry B, Dupont P, De Wit CC (1994) A survey of models, analysis tools and compensation methods for the control of machines with friction. Automatica 30(7):1083–1138

Arun KS, Huang TS, Blostein SD (1987) Least-squares fitting of 2 3-D point sets. IEEE T Pattern Anal 9(5):699–700

Asada H (1983) A geometrical representation of manipulator dynamics and its application to arm design. J Dyn Syst-T ASME 105:131

Astolfi A (1999) Exponential stabilization of a wheeled mobile robot via discontinuous control. J Dyn Syst-T ASME 121(1):121–126

Azarbayejani A, Pentland AP (1995) Recursive estimation of motion, structure, and focal length. IEEE T Pattern Anal 17(6):562–575

Bailey T, Durrant-Whyte H (2006) Simultaneous localization and mapping: Part II. IEEE Robot Autom Mag 13(3):108–117

Bakthavatchalam M, Chaumette F, Tahri O (2015) An improved modelling scheme for photometric moments with inclusion of spatial weights for visual servoing with partial appearance/disappearance. In: Proceedings of the IEEE International Conference on Robotics and Automation (ICRA). pp 6037–6043

Baldridge AM, Hook SJ, Grove CI, Rivera G (2009) The ASTER spectral library version 2.0. Remote Sens Environ 113(4):711–715

Ball RS (1876) The theory of screws: A study in the dynamics of a rigid body. Hodges, Foster & Co., Dublin

Ball RS (1908) A treatise on spherical astronomy. Cambridge University Press, New York

Ballard DH (1981) Generalizing the Hough transform to detect arbitrary shapes. Pattern Recogn 13(2):111–122

Banks J, Corke PI (2001) Quantitative evaluation of matching methods and validity measures for stereo vision. Int J Robot Res 20(7):512–532

Barfoot T (2017) State Estimation for Robotics. Cambridge University Press

Bar-Shalom Y, Fortmann T (1988) Tracking and data association. Mathematics in science and engineering, vol 182. Academic Press, London Oxford

Bar-Shalom Y, Li XR, Kirubarajan T (2001) Estimation with applications to tracking and navigation. John Wiley & Sons, Inc., Chichester

Bateux Q, Marchand E, Leitner J, Chaumette F, Corke P (2018) Training Deep Neural Networks for Visual Servoing. IEEE International Conference On Robotics And Automation (ICRA). pp 3307–3314

Bauer J, Sünderhauf N, Protzel P (2007) Comparing several implementations of two recently published feature detectors. In: IFAC Symposium on Intelligent Autonomous Vehicles (IAV). Toulouse

Bay H, Ess A, Tuytelaars T, Van Gool L (2008) Speeded-up robust features (SURF). Comput Vis Image Und 110(3):346–359

Beard RW, McLain TW (2012) Small Unmanned Aircraft - Theory and Practice. Princeton University Press, Princeton, New Jersey

Benosman R, Kang SB (2001) Panoramic vision: Sensors, theory, and applications. Springer-Verlag, Berlin Heidelberg

Benson KB (ed) (1986) Television engineering handbook. McGraw-Hill, New York

Berns RS (2019) Billmeyer and Saltzman's Principles of Color Technology, 4th ed. Wiley

Bertolazzi E, Frego M (2015) G1 fitting with clothoids. Mathematical Methods in the Applied Sciences 38(5):881–897

Bertozzi M, Broggi A, Cardarelli E, Fedriga R, Mazzei L, Porta P (2011) VIAC expedition: Toward autonomous mobility. IEEE Robot Autom Mag 18(3):120–124

Besl PJ, McKay HD (1992) A method for registration of 3-D shapes. IEEE T Pattern Anal 14(2): 239–256

Bhat DN, Nayar SK (2002) Ordinal measures for image correspondence. IEEE T Pattern Anal 20(4): 415–423

Biber P, Straßer W (2003) The normal distributions transform: A new approach to laser scan matching. In: Proceedings of the IEEE/RSJ International Conference on intelligent robots and systems (IROS), vol 3. pp 2743–2748

Bishop CM (2006) Pattern recognition and machine learning. Information science and statistics. Springer-Verlag, New York

Blewitt M (2011) Celestial navigation for yachtsmen. Adlard Coles Nautical, London

Bolles RC, Baker HH, Marimont DH (1987) Epipolar-plane image analysis: An approach to determining structure from motion. Int J Comput Vision 1(1):7–55, Mar

Bolles RC, Baker HH, Hannah MJ (1993) The JISCT stereo evaluation. In: Image Understanding Workshop: proceedings of a workshop held in Washington, DC apr 18–21, 1993. Morgan Kaufmann, San Francisco, pp 263

Bolton W (2015) Mechatronics: Electronic control systems in mechanical and electrical engineering, 6th ed. Pearson, Harlow

Borenstein J, Everett HR, Feng L (1996) Navigating mobile robots: Systems and techniques. AK Peters, Ltd. Natick, MA, USA, Out of print and available at http://www-personal.umich.edu/~johannb/Papers/pos96rep.pdf

Borgefors G (1986) Distance transformations in digital images. Comput Vision Graph 34(3):344–371

Bostrom N (2016) Superintelligence: Paths, dangers, strategies. Oxford University Press, Oxford, 432 p

Bouguet J-Y (2010) Camera calibration toolbox for MATLAB®. http://www.vision.caltech.edu/bouguetj/calib_doc

Brady M, Hollerbach JM, Johnson TL, Lozano-Pérez T, Mason MT (eds) (1982) Robot motion: Planning and control. MIT Press, Cambridge, Massachusetts

Bradley D, Roth G, Adapting Thresholding Using the Integral Image, Journal of Graphics Tools. Vol. 12, No. 2, 2007, pp.13–21.

Braitenberg V (1986) Vehicles: Experiments in synthetic psychology. MIT Press, Cambridge, Massachusetts

Bray H (2014) You are here: From the compass to GPS, the history and future of how we find ourselves. Basic Books, New York

Brockett RW (1983) Asymptotic stability and feedback stabilization. In: Brockett RW, Millmann RS, Sussmann HJ (eds) Progress in mathematics. Differential geometric control theory, vol 27. pp 181–191

Broida TJ, Chandrashekhar S, Chellappa R (1990) Recursive 3-D motion estimation from a monocular image sequence. IEEE T Aero Elec Sys 26(4):639–656

Brooks RA (1986) A robust layered control system for a mobile robot. IEEE T Robotic Autom 2(1):14–23

Brooks RA (1989) A robot that walks: Emergent behaviors from a carefully evolved network. MIT AI Lab, Memo 1091

Brown MZ, Burschka D, Hager GD (2003) Advances in computational stereo. IEEE T Pattern Anal 25(8):993–1008

Brynjolfsson E, McAfee A (2014) The second machine age: Work, progress, and prosperity in a time of brilliant technologies. W.W. Norton & Co., New York

Buehler M, Iagnemma K, Singh S (eds) (2007) The 2005 DARPA Grand Challenge: The great robot race. Springer Tracts in Advanced Robotics, vol 36. Springer-Verlag, Berlin Heidelberg

Buehler M, Iagnemma K, Singh S (eds) (2010) The DARPA Urban Challenge. Tracts in Advanced Robotics, vol 56. Springer-Verlag, Berlin Heidelberg

Bukowski R, Haynes LS, Geng Z, Coleman N, Santucci A, Lam K, Paz A, May R, DeVito M (1991) Robot hand-eye coordination rapid prototyping environment. In: Proc ISIR, pp 16.15–16.28

Buttazzo GC, Allotta B, Fanizza FP (1993) Mousebuster: A robot system for catching fast moving objects by vision. In: Proceedings of the IEEE International Conference on Robotics and Automation (ICRA). Atlanta, pp 932–937

Calonder M, Lepetit V, Strecha C, Fua P (2010) BRIEF: Binary robust independent elementary features. In: Daniilidis K, Maragos P, Paragios N (eds) Lecture notes in computer science. Computer Vision – ECCV 2010, vol 6311. Springer-Verlag, Berlin Heidelberg, pp 778–792

Campos C, Elvira R, Rodríguez JJG, Montiel JMM, Tardós JD (2021) ORB-SLAM3: An Accurate Open-Source Library for Visual, Visual–Inertial, and Multimap SLAM. IEEE Trans on Robotics, 37(6), pp 1874–1890

Canny JF (1983) Finding edges and lines in images. MIT, Artificial Intelligence Laboratory, AI-TR-720. Cambridge, MA

Canny J (1987) A computational approach to edge detection. In: Fischler MA, Firschein O (eds) Readings in computer vision: Issues, problems, principles, and paradigms. Morgan Kaufmann, San Francisco, pp 184–203

Censi A (2008) An ICP variant using a point-to-line metric. In: Proceedings of the IEEE International Conference on Robotics and Automation (ICRA). pp 19–25

Chahl JS, Srinivasan MV (1997) Reflective surfaces for panoramic imaging. Appl Optics 31(36):8275–8285

Chaumette F (1990) La relation vision-commande: Théorie et application et des tâches robotiques. Ph.D. thesis, Université de Rennes 1

Chaumette F (1998) Potential problems of stability and convergence in image-based and position-based visual servoing. In: Kriegman DJ, Hager GD, Morse AS (eds) Lecture notes in control and information sciences. The confluence of vision and control, vol 237. Springer-Verlag, Berlin Heidelberg, pp 66–78

Chaumette F (2004) Image moments: A general and useful set of features for visual servoing. IEEE T Robotic Autom 20(4):713–723

Chaumette F, Hutchinson S (2006) Visual servo control 1: Basic approaches. IEEE Robot Autom Mag 13(4):82–90

Chaumette F, Hutchinson S (2007) Visual servo control 2: Advanced approaches. IEEE Robot Autom Mag 14(1):109–118

Chaumette F, Rives P, Espiau B (1991) Positioning of a robot with respect to an object, tracking it and estimating its velocity by visual servoing. In: Proceedings of the IEEE International Conference on Robotics and Automation (ICRA). Seoul, pp 2248–2253

Chesi G, Hashimoto K (eds) (2010) Visual servoing via advanced numerical methods. Lecture notes in computer science, vol 401. Springer-Verlag, Berlin Heidelberg

Chiuso A, Favaro P, Jin H, Soatto S (2002) Structure from motion causally integrated over time. IEEE T Pattern Anal 24(4):523–535

Choset HM, Lynch KM, Hutchinson S, Kantor G, Burgard W, Kavraki LE, Thrun S (2005) Principles of robot motion. MIT Press, Cambridge, Massachusetts

Chum O, Matas J (2005) Matching with PROSAC - progressive sample consensus. IEEE Computer Society Conference On Computer Vision And Pattern Recognition (CVPR), pp 220–226

Colicchia G, Waltner C, Hopf M, Wiesner H (2009) The scallop's eye – A concave mirror in the context of biology. Physics Education 44(2):175–179

Collewet C, Marchand E, Chaumette F (2008) Visual servoing set free from image processing. In: Proceedings of IEEE International Conference on Robotics and Automation (ICRA). pp 81–86

Commission Internationale de L'Éclairage (1987) Colorimetry, 2nd ed. Commission Internationale de L'Eclairage, CIE No 15.2

Corke PI (1994) High-performance visual closed-loop robot control. University of Melbourne, Dept. Mechanical and Manufacturing Engineering. https://hdl.handle.net/11343/38847

Corke PI (1996a) In situ measurement of robot motor electrical constants. Robotica 14(4):433–436

Corke PI (1996b) Visual control of robots: High-performance visual servoing. Mechatronics, vol 2. Research Studies Press (John Wiley). Out of print and available at http://www.petercorke.com/bluebook

Corke PI (2001) Mobile robot navigation as a planar visual servoing problem. In: Jarvis RA, Zelinsky A (eds) Springer tracts in advanced robotics. Robotics Research: The 10th International Symposium, vol 6. IFRR, Lorne, pp 361–372

Corke PI (2007) A simple and systematic approach to assigning Denavit-Hartenberg parameters. IEEE T Robotic Autom 23(3):590–594

Corke PI (2010) Spherical image-based visual servo and structure estimation. In: Proceedings of the IEEE International Conference on Robotics and Automation (ICRA). Anchorage, pp 5550–5555

Corke PI, Armstrong-Hélouvry BS (1994) A search for consensus among model parameters reported for the PUMA 560 robot. In: Proceedings of the IEEE International Conference on Robotics and Automation (ICRA). San Diego, pp 1608–1613

Corke PI, Armstrong-Hélouvry BS (1994) A meta-study of PUMA 560 dynamics: A critical appraisal of literature data. Robotica 13(3):253–258

Corke PI, Good MC (1992) Dynamic effects in high-performance visual servoing. In: Proceedings of the IEEE International Conference on Robotics and Automation (ICRA). Nice, pp 1838–1843

Corke PI, Good MC (1996) Dynamic effects in visual closed-loop systems. IEEE T Robotic Autom 12(5):671–683

Corke PI, Haviland J (2021) Not your grandmother's toolbox – the Robotics Toolbox reinvented for Python. IEEE International Conference on Robotics and Automation (ICRA). pp 11357–11363

Corke PI, Hutchinson SA (2001) A new partitioned approach to image-based visual servo control. IEEE T Robotic Autom 17(4):507–515

Corke PI, Dunn PA, Banks JE (1999) Frame-rate stereopsis using non-parametric transforms and programmable logic. In: Proceedings of the IEEE International Conference on Robotics and Automation (ICRA). Detroit, pp 1928–1933

Corke PI, Lobo J, Dias J (2007) An introduction to inertial and visual sensing. The International Journal of Robotics Research, 26(6). pp 519–536

Corke PI, Strelow D, Singh S (2004) Omnidirectional visual odometry for a planetary rover. In: Proceedings of the International Conference on Intelligent Robots and Systems (IROS). Sendai, pp 4007–4012

Corke PI, Spindler F, Chaumette F (2009) Combining Cartesian and polar coordinates in IBVS. In: Proceedings of the International Conference on Intelligent Robots and Systems (IROS). St. Louis, pp 5962–5967

Corke PI, Paul R, Churchill W, Newman P (2013) Dealing with shadows: Capturing intrinsic scene appearance for image-based outdoor localisation. In: Proceedings of the IEEE/RSJ International Conference on Intelligent Robots and Systems (IROS). pp 2085–2092

Corke PI (2017) Robotics, Vision and Control , 2nd ed. STAR Series, vol 118. Springer

Craig JJ (1987) Adaptive control of mechanical manipulators. Addison-Wesley Longman Publishing Co., Inc. Boston

Craig JJ (2005) Introduction to robotics: Mechanics and control, 3rd ed. Pearson/Prentice Hall

Craig JJ (1986) Introduction to robotics: Mechanics and control, 1st ed. Addison-Wesley

Craig JJ, Hsu P, Sastry SS (1987) Adaptive control of mechanical manipulators. Int J Robot Res 6(2):16–28

Crombez N, Caron G, Mouaddib EM (2015) Photometric Gaussian mixtures based visual servoing. In: Proceedings of the IEEE/RSJ International Conference on Intelligent Robots and Systems (IROS). pp 5486–5491

Crone RA (1999) A history of color: The evolution of theories of light and color. Kluwer Academic, Dordrecht

Cummins M, Newman P (2008) FAB-MAP: Probabilistic localization and mapping in the space of appearance. Int J Robot Res 27(6):647

Cutting JE (1997) How the eye measures reality and virtual reality. Behav Res Meth Ins C 29(1):27–36

Daniilidis K, Klette R (eds) (2006) Imaging beyond the pinhole camera. Computational Imaging, vol 33. Springer-Verlag, Berlin Heidelberg

Dansereau DG (2014) Plenoptic signal processing for robust vision in field robotics. Ph.D. thesis, The University of Sydney

Davies ER (2017) Computer Vision: Principles, Algorithms, Applications, Learning 5th ed. Academic Press

Davison AJ, Reid ID, Molton ND, Stasse O (2007) MonoSLAM: Real-time single camera SLAM. IEEE T Pattern Anal 29(6):1052–1067

Deguchi K (1998) Optimal motion control for image-based visual servoing by decoupling translation and rotation. In: Proceedings of the International Conference on Intelligent Robots and Systems (IROS). Victoria, Canada, pp 705–711

Dellaert F, Fox D, Burgard W, Thrun S (1999) Monte Carlo Localization for Mobile Robots. In: Proceedings of the IEEE International Conference on Robotics and Automation (ICRA)

Dellaert F (2012) Factor graphs and GTSAM: A hands-on introduction. Technical Report (GT-RIM-CP&R-2012-002), Georgia Institute of Technology

Dellaert F, Kaess M (2017) Factor Graphs for Robot Perception. Foundations and Trends in Robotics, Vol. 6, No. 1-2, pp 1–139

Dellaert F, Kaess M (2006) Square root SAM: Simultaneous localization and mapping via square root information smoothing. Int J Robot Res 25(12):1181–1203

DeMenthon D, Davis LS (1992) Exact and approximate solutions of the perspective-three-point problem. IEEE T Pattern Anal 14(11):1100–1105

Denavit J, Hartenberg RS (1955) A kinematic notation for lower-pair mechanisms based on matrices. J Appl Mech-T ASME 22(1):215–221

Deo AS, Walker ID (1995) Overview of damped least-squares methods for inverse kinematics of robot manipulators. J Intell Robot Syst 14(1):43–68

Deriche R, Giraudon G (1993) A computational approach for corner and vertex detection. Int J Comput Vision 10(2):101–124

DeWitt BA, Wolf PR (2000) Elements of photogrammetry (with applications in GIS). McGraw-Hill, New York

Dickmanns ED (2007) Dynamic vision for perception and control of motion. Springer-Verlag, London

Dickmanns ED, Graefe V (1988a) Applications of dynamic monocular machine vision. Mach Vision Appl 1:241–261

Dickmanns ED, Graefe V (1988b) Dynamic monocular machine vision. Mach Vision Appl 1(4):223–240

Dickmanns ED, Zapp A (1987) Autonomous high speed road vehicle guidance by computer vision. In: Tenth Triennial World Congress of the International Federation of Automatic Control, vol 4. Munich, pp 221–226

Dijkstra EW (1959) A note on two problems in connexion with graphs. Numer Math 1(1):269–271

Dougherty ER, Lotufo RA (2003) Hands-on morphological image processing. Society of Photo-Optical Instrumentation Engineers (SPIE)

Dubins LE (1957). On Curves of Minimal Length with a Constraint on Average Curvature, and with Prescribed Initial and Terminal Positions and Tangents. American Journal of Mathematics, 79(3), 497–516

Duda RO, Hart PE (1972) Use of the Hough transformation to detect lines and curves in pictures. Commun ACM 15(1):11–15

Durrant-Whyte H, Bailey T (2006) Simultaneous localization and mapping: Part I. IEEE Robot Autom Mag 13(2):99–110

Espiau B, Chaumette F, Rives P (1992) A new approach to visual servoing in robotics. IEEE T Robotic Autom 8(3):313–326

Everett HR (1995) Sensors for mobile robots: Theory and application. AK Peters Ltd., Wellesley

Faugeras OD (1993) Three-dimensional computer vision: A geometric viewpoint. MIT Press, Cambridge, Massachusetts

Faugeras OD, Lustman F (1988) Motion and structure from motion in a piecewise planar environment. Int J Pattern Recogn 2(3):485–508

Faugeras O, Luong QT, Papadopoulou T (2001) The geometry of multiple images: The laws that govern the formation of images of a scene and some of their applications. MIT Press, Cambridge, Massachusetts

Featherstone R (1987) Robot dynamics algorithms. Springer

Featherstone R (2010a) A Beginner's Guide to 6-D Vectors (Part 1). IEEE Robotics Automation Magazine, vol. 17, 83–94

Featherstone R (2010b) A Beginner's Guide to 6-D Vectors (Part 2). IEEE Robotics Automation Magazine, vol. 17, 88–99

Feddema JT (1989) Real time visual feedback control for hand-eye coordinated robotic systems. Purdue University

Feddema JT, Mitchell OR (1989) Vision-guided servoing with feature-based trajectory generation. IEEE T Robotic Autom 5(5):691–700

Feddema JT, Lee CSG, Mitchell OR (1991) Weighted selection of image features for resolved rate visual feedback control. IEEE T Robotic Autom 7(1):31–47

Felzenszwalb PF, Huttenlocher DP (2004) Efficient graph-based image segmentation. Int J Comput Vision 59(2):167–181

Ferguson D, Stentz A (2006) Using interpolation to improve path planning: The Field D* algorithm. J Field Robotics 23(2):79–101

Fischler MA, Bolles RC (1981) Random sample consensus: A paradigm for model fitting with applications to image analysis and automated cartography. Commun ACM 24(6):381–395

Flusser J (2000) On the independence of rotation moment invariants. Pattern Recogn 33(9):1405–1410

Fomena R, Chaumette F (2007) Visual servoing from spheres using a spherical projection model. In: Proceedings of the IEEE International Conference on Robotics and Automation (ICRA). Rome, pp 2080–2085

Ford M (2015) Rise of the robots: Technology and the threat of a jobless future. Basic Books, New York

Förstner W (1994) A framework for low level feature extraction. In: Ecklundh J-O (ed) Lecture notes in computer science. Computer Vision – ECCV 1994, vol 800. Springer-Verlag, Berlin Heidelberg, pp 383–394

Förstner W, Gülch E (1987) A fast operator for detection and precise location of distinct points, corners and centres of circular features. In: ISPRS Intercommission Workshop. Interlaken, pp 149–155

Forsyth DA, Ponce J (2012) Computer vision: A modern approach, 2nd ed. Pearson, London

Fraundorfer F, Scaramuzza D (2012) Visual odometry: Part II – Matching, robustness, optimization, and applications. IEEE Robot Autom Mag 19(2):78–90

Freeman H (1974) Computer processing of line-drawing images. ACM Comput Surv 6(1):57–97

Friedman DP, Felleisen M, Bibby D (1987) The little LISPer. MIT Press, Cambridge, Massachusetts

Frisby JP, Stone JV (2010) Seeing: The Computational Approach to Biological Vision. MIT Press

Fu KS, Gonzalez RC, Lee CSG (1987) Robotics: Control, Sensing, Vision, and Intelligence. McGraw-Hill

Funda J, Taylor RH, Paul RP (1990) On homogeneous transforms, quaternions, and computational efficiency. IEEE T Robotic Autom 6(3):382–388

Gálvez-López D, Tardós JD (2012) Bags of binary words for fast place recognition in image sequences. IEEE T Robotic Autom 28(5):1188–1197

Gans NR, Hutchinson SA, Corke PI (2003) Performance tests for visual servo control systems, with application to partitioned approaches to visual servo control. Int J Robot Res 22(10–11):955

Garg R, Kumar B, Carneiro G, Reid I (2016) Unsupervised CNN for Single View Depth Estimation: Geometry to the Rescue. ECCV

Gautier M, Khalil W (1992) Exciting trajectories for the identification of base inertial parameters of robots. Int J Robot Res 11(4):362

Geiger A, Roser M, Urtasun R (2010) Efficient large-scale stereo matching. In: Kimmel R, Klette R, Sugimoto A (eds) Computer vision – ACCV 2010: 10th Asian Conference on Computer Vision, Queenstown, New Zealand, November 8–12, 2010, revised selected papers, part I. Springer-Verlag, Berlin Heidelberg, pp 25–38

Geraerts R, Overmars MH (2004) A comparative study of probabilistic roadmap planners. In: Boissonnat J-D, Burdick J, Goldberg K, Hutchinson S (eds) Springer tracts in advanced robotics. Algorithmic Foundations of Robotics V, vol 7. Springer-Verlag, Berlin Heidelberg, pp 43–58

Gevers T, Gijsenij A, van de Weijer J, Geusebroek J-M (2012) Color in computer vision: Fundamentals and applications. John Wiley & Sons, Inc., Chichester

Geyer C, Daniilidis K (2000) A unifying theory for central panoramic systems and practical implications. In: Vernon D (ed) Lecture notes in computer science. Computer vision – ECCV 2000, vol 1843. Springer-Verlag, Berlin Heidelberg, pp 445–461

Glover A, Maddern W, Warren M, Reid S, Milford M, Wyeth G (2012) OpenFABMAP: An open source toolbox for appearance-based loop closure detection. In: Proceedings of the IEEE International Conference on Robotics and Automation (ICRA). pp 4730–4735

Gonzalez RC, Woods RE (2018) Digital image processing, 4th ed. Pearson

Gonzalez R, Woods R, Eddins S (2020) Digital image processing using MATLAB, 3rd ed. Gatesmark Publishing

Grassia FS (1998) Practical parameterization of rotations using the exponential map. Journal of Graphics Tools 3(3):29–48

Gregory RL (1997) Eye and brain: The psychology of seeing. Princeton University Press, Princeton, New Jersey

Grey CGP (2014) Humans need not apply. YouTube video, www.youtube.com/watch?v=7Pq-S557XQU

Grisetti G, Kümmerle R, Stachniss C, Burgard W (2010) A Tutorial on Graph-Based SLAM. IEEE Intelligent Transportation Systems Magazine 2(4). pp 31–43

Groves PD (2013) Principles of GNSS, inertial, and multisensor integrated navigation systems, 2nd ed. Artech House, Norwood, USA

Hager GD, Toyama K (1998) X Vision: A portable substrate for real-time vision applications. Comput Vis Image Und 69(1):23–37

Hamel T, Mahony R (2002) Visual servoing of an under-actuated dynamic rigid-body system: An image based approach. IEEE T Robotic Autom 18(2):187–198

Hamel T, Mahony R, Lozano R, Ostrowski J (2002) Dynamic modelling and configuration stabilization for an X4-flyer. IFAC World Congress 1(2), p 3

Hansen P, Corke PI, Boles W (2010) Wide-angle visual feature matching for outdoor localization. Int J Robot Res 29(1–2):267–297

Harris CG, Stephens MJ (1988) A combined corner and edge detector. In: Proceedings of the Fourth Alvey Vision Conference. Manchester, pp 147–151

Hart PE (2009) How the Hough transform was invented [DSP history]. IEEE Signal Proc Mag 26(6):18–22

Hart PE, Nilsson NJ, Raphael B (1968) A formal basis for the heuristic determination of minimum cost paths. IEEE Trans Systems Science and Cybernetics, 4(2):100–107

Hartenberg RS, Denavit J (1964) Kinematic synthesis of linkages. McGraw-Hill, New York

Hartley R, Zisserman A (2003) Multiple view geometry in computer vision. Cambridge University Press, New York

Hashimoto K (ed) (1993) Visual servoing. In: Robotics and automated systems, vol 7. World Scientific, Singapore

Hashimoto K, Kimoto T, Ebine T, Kimura H (1991) Manipulator control with image-based visual servo. In: Proceedings of the IEEE International Conference on Robotics and Automation (ICRA). Seoul, pp 2267–2272

Heikkila J, and Silven O (1997) A four-step camera calibration procedure with implicit image correction, Proc IEEE Computer Society Conference on Computer Vision and Pattern Recognition (CVPR), pp 1106–1112

Herschel W (1800) Experiments on the refrangibility of the invisible rays of the sun. Phil Trans R Soc Lond 90:284–292

Hess W, Kohler D, Rapp H, Andor D (2016) Real-Time Loop Closure in 2D LIDAR SLAM. In: Proceedings of the IEEE International Conference on Robotics and Automation (ICRA)

Hill J, Park WT (1979) Real time control of a robot with a mobile camera. In: Proceedings of the 9th ISIR, SME. Washington, DC. Mar, pp 233–246

Hirata T (1996) A unified linear-time algorithm for computing distance maps. Inform Process Lett 58(3):129–133

Hirschmüller H (2008) Stereo processing by semiglobal matching and mutual information. IEEE Transactions on Pattern Analysis and Machine Intelligence 30(2):328–341

Hirt C, Claessens S, Fecher T, Kuhn M, Pail R, Rexer M (2013) New ultrahigh-resolution picture of Earth's gravity field. Geophys Res Lett 40:4279–4283

Hoag D (1963) Consideration of Apollo IMU gimbal lock. MIT Instrumentation Laboratory, E–1344, https://www.hq.nasa.gov/alsj/e-1344.htm

Holland O (2003) Exploration and high adventure: the legacy of Grey Walter. Philosophical Trans of the Royal Society of London. Series A: Mathematical, Physical and Engineering Sciences, 361(1811)

Hollerbach JM (1980) A recursive Lagrangian formulation of manipulator dynamics and a comparative study of dynamics formulation complexity. IEEE T Syst Man Cyb 10(11):730–736, Nov

Hollerbach JM (1982) Dynamics. In: Brady M, Hollerbach JM, Johnson TL, Lozano-Pérez T, Mason MT (eds) Robot motion – Planning and control. MIT Press, Cambridge, Massachusetts, pp 51–71

Horaud R, Canio B, Leboullenx O (1989) An analytic solution for the perspective 4-point problem. Comput Vision Graph 47(1):33–44

Horn BKP (1987) Closed-form solution of absolute orientation using unit quaternions. J Opt Soc Am A 4(4):629–642

Horn BKP, Hilden HM, Negahdaripour S (1988) Closed-form solution of absolute orientation using orthonormal matrices. J Opt Soc Am A 5(7):1127–1135

Hosoda K, Asada M (1994) Versatile visual servoing without knowledge of true Jacobian. In: Proceedings of the International Conference on Intelligent Robots and Systems (IROS). Munich, pp 186–193

Howard TM, Green CJ, Kelly A, Ferguson D (2008) State space sampling of feasible motions for high-performance mobile robot navigation in complex environments. J Field Robotics 25(6–7):325–345

Hu MK (1962) Visual pattern recognition by moment invariants. IRE T Inform Theor 8:179–187

Hua M-D, Ducard G, Hamel T, Mahony R, Rudin K (2014) Implementation of a nonlinear attitude estimator for aerial robotic vehicles. IEEE T Contr Syst T 22(1):201–213

Huang TS, Netravali AN (1994) Motion and structure from feature correspondences: A review. P IEEE 82(2):252–268

Humenberger M, Zinner C, Kubinger W (2009) Performance evaluation of a census-based stereo matching algorithm on embedded and multi-core hardware. In: Proceedings of the 19th International Symposium on Image and Signal Processing and Analysis (ISPA). pp 388–393

Hunt RWG (1987) The reproduction of colour, 4th ed. Fountain Press, Tolworth

Hunter RS, Harold RW (1987) The measurement of appearance. John Wiley & Sons, Inc., Chichester

Hutchinson S, Hager G, Corke PI (1996) A tutorial on visual servo control. IEEE T Robotic Autom 12(5):651–670

Ings S (2008) The Eye: A Natural History. Bloomsbury Publishing

Huynh DQ (2009) Metrics for 3D Rotations: Comparison and Analysis. J Math Imaging Vis 35, pp 155–164

Iwatsuki M, Okiyama N (2002a) A new formulation of visual servoing based on cylindrical coordinate system with shiftable origin. In: Proceedings of the International Conference on Intelligent Robots and Systems (IROS). Lausanne, pp 354–359

Iwatsuki M, Okiyama N (2002b) Rotation-oriented visual servoing based on cylindrical coordinates. In: Proceedings of the IEEE International Conference on Robotics and Automation (ICRA). Washington, DC, May, pp 4198–4203

Izaguirre A, Paul RP (1985) Computation of the inertial and gravitational coefficients of the dynamics equations for a robot manipulator with a load. In: Proceedings of the IEEE International Conference on Robotics and Automation (ICRA). Mar, pp 1024–1032

Jägersand M, Fuentes O, Nelson R (1996) Experimental evaluation of uncalibrated visual servoing for precision manipulation. In: Proceedings of the IEEE International Conference on Robotics and Automation (ICRA). Albuquerque, NM, pp 2874–2880

Jarvis RA, Byrne JC (1988) An automated guided vehicle with map building and path finding capabilities. In: Robotics Research: The Fourth international symposium. MIT Press, Cambridge, Massachusetts, pp 497–504

Jazwinski AH (2007) Stochastic processes and filtering theory. Dover Publications, Mineola

Jebara T, Azarbayejani A, Pentland A (1999) 3D structure from 2D motion. IEEE Signal Proc Mag 16(3):66–84

Julier SJ, Uhlmann JK (2004) Unscented filtering and nonlinear estimation. P IEEE 92(3):401–422

Kaehler A, Bradski G (2017) Learning OpenCV 3: Computer vision in C++ with the OpenCV library. O'Reilly & Associates, Köln

Kaess M, Ranganathan A, Dellaert F (2007) iSAM: Fast incremental smoothing and mapping with efficient data association. In: Proceedings of the IEEE International Conference on Robotics and Automation (ICRA). pp 1670–1677

Kahn ME (1969) The near-minimum time control of open-loop articulated kinematic linkages. Stanford University, AIM-106

Kálmán RE (1960) A new approach to linear filtering and prediction problems. J Basic Eng-T Asme 82(1):35–45

Kane TR, Levinson DA (1983) The use of Kane's dynamical equations in robotics. Int J Robot Res 2(3):3–21

Karaman S, Walter MR, Perez A, Frazzoli E, Teller S (2011) Anytime motion planning using the RRT*. In: Proceedings of the IEEE International Conference on Robotics and Automation (ICRA). pp 1478–1483

Kavraki LE, Svestka P, Latombe JC, Overmars MH (1996) Probabilistic roadmaps for path planning in high-dimensional configuration spaces. IEEE T Robotic Autom 12(4):566–580

Kelly R (1996) Robust asymptotically stable visual servoing of planar robots. IEEE T Robotic Autom 12(5):759–766

Kelly A (2013) Mobile robotics: Mathematics, models, and methods. Cambridge University Press, New York

Kelly R, Carelli R, Nasisi O, Kuchen B, Reyes F (2002a) Stable visual servoing of camera-in-hand robotic systems. IEEE-ASME T Mech 5(1):39–48

Kelly R, Shirkey P, Spong MW (2002b) Fixed-camera visual servo control for planar robots. In: Proceedings of the IEEE International Conference on Robotics and Automation (ICRA). Washington, DC, pp 2643–2649

Kenwright K (2012) Dual-Quaternions: From Classical Mechanics to Computer Graphics and Beyond. https://xbdev.net/misc_demos/demos/dual_quaternions_beyond/paper.pdf

Khalil W, Creusot D (1997) SYMORO+: A system for the symbolic modelling of robots. Robotica 15(2):153–161

Khalil W, Dombre E (2002) Modeling, identification and control of robots. Kogan Page Science, London

Khatib O (1987) A unified approach for motion and force control of robot manipulators: The operational space formulation. IEEE T Robotic Autom 3(1):43–53

King-Hele D (2002) Erasmus Darwin's improved design for steering carriages and cars. Notes and Records of the Royal Society of London 56(1):41–62

Klafter RD, Chmielewski TA, Negin M (1989) Robotic engineering – An integrated approach. Prentice Hall, Upper Saddle River, New Jersey

Klein CA, Huang CH (1983) Review of pseudoinverse control for use with kinematically redundant manipulators. IEEE T Syst Man Cyb 13:245–250

Klein G, Murray D (2007) Parallel tracking and mapping for small AR workspaces. In: Sixth IEEE and ACM International Symposium on Mixed and Augmented Reality (ISMAR 2007). pp 225–234

Klette R, Kruger N, Vaudrey T, Pauwels K, van Hulle M, Morales S, Kandil F, Haeusler R, Pugeault N, Rabe C (2011) Performance of correspondence algorithms in vision-based driver assistance using an online image sequence database. IEEE T Veh Technol 60(5):2012–2026

Klette R (2014) Concise Computer Vision: An Introduction into Theory and Algorithms. Springer

Koenderink JJ (1984) The structure of images. Biol Cybern 50(5):363–370

Koenderink JJ (2010) Color for the sciences. MIT Press, Cambridge, Massachusetts

Koenig S, Likhachev M (2005) Fast replanning for navigation in unknown terrain. IEEE T Robotic Autom 21(3):354–363

Krajník T, Cristóforis P, Kusumam K, Neubert P, Duckett T (2017) Image features for visual teach-and-repeat navigation in changing environments. Robotics And Autonomous Systems. **88** pp 127–141

Kriegman DJ, Hager GD, Morse AS (eds) (1998) The confluence of vision and control. Lecture notes in control and information sciences, vol 237. Springer-Verlag, Berlin Heidelberg

Kuipers JB (1999) Quaternions and rotation sequences: A primer with applications to orbits, aerospace and virtual reality. Princeton University Press, Princeton, New Jersey

Kümmerle R, Grisetti G, Strasdat H, Konolige K, Burgard W (2011) g^2o: A general framework for graph optimization. In: Proceedings of the IEEE International Conference on Robotics and Automation (ICRA). pp 3607–3613

Lam O, Dayoub F, Schulz R, Corke P (2015) Automated topometric graph generation from floor plan analysis. In: Proceedings of the Australasian Conference on Robotics and Automation. Australasian Robotics and Automation Association (ARAA)

Lamport L (1994) LATEX: A document preparation system. User's guide and reference manual. Addison-Wesley Publishing Company, Reading

Land EH, McCann J (1971) Lightness and retinex theory. J Opt Soc Am A 61(1):1–11

Land MF, Nilsson D-E (2002) Animal eyes. Oxford University Press, Oxford

LaValle SM (1998) Rapidly-exploring random trees: A new tool for path planning. Computer Science Dept., Iowa State University, TR 98–11

LaValle SM (2006) Planning algorithms. Cambridge University Press, New York

LaValle SM (2011a) Motion planning: The essentials. IEEE Robot Autom Mag 18(1):79–89

LaValle SM (2011b) Motion planning: Wild frontiers. IEEE Robot Autom Mag 18(2):108–118

LaValle SM, Kuffner JJ (2001) Randomized kinodynamic planning. Int J Robot Res 20(5):378–400

Laussedat A (1899) La métrophotographie. Enseignement supérieur de la photographie. Gauthier-Villars, 52 p

Leavers VF (1993) Which Hough transform? Comput Vis Image Und 58(2):250–264

Lee CSG, Lee BH, Nigham R (1983) Development of the generalized D'Alembert equations of motion for mechanical manipulators. In: Proceedings of the 22nd CDC, San Antonio, Texas. pp 1205–1210

Lepetit V, Moreno-Noguer F, Fua P (2009) EPnP: An accurate $O(n)$ solution to the PnP problem. Int J Comput Vision 81(2):155–166

Li H, Hartley R (2006) Five-point motion estimation made easy. In: 18th International Conference on Pattern Recognition ICPR 2006. Hong Kong, pp 630–633

Li Y, Jia W, Shen C, van den Hengel A (2014) Characterness: An indicator of text in the wild. IEEE T Image Process 23(4):1666–1677

Li Y, Sun J, Tang C, Shum H (2004), Lazy Snapping, ACM Transactions on Graphics, http://home.cse. ust.hk/~cktang/sample_pub/lazy_snapping.pdf

Li T, Bolic M, Djuric P (2015) Resampling methods for particle filtering: Classification, implementation, and strategies. IEEE Signal Proc Mag 32(3):70–86

Lin Z, Zeman V, Patel RV (1989) On-line robot trajectory planning for catching a moving object. In: Proceedings of the IEEE International Conference on Robotics and Automation (ICRA). pp 1726–1731

Lindeberg T (1993) Scale-space theory in computer vision. Springer-Verlag, Berlin Heidelberg

Lipkin H (2005) A Note on Denavit-Hartenberg Notation in Robotics. Proceedings of the 29th ASME Mechanisms and Robotics Conference, pp 921–926

Lloyd J, Hayward V (1991) Real-time trajectory generation using blend functions. In: Proceedings of the IEEE International Conference on Robotics and Automation (ICRA). Seoul, pp 784–789

Longuet-Higgins H (1981) A computer algorithm for reconstruction of a scene from two projections. Nature 293:133–135

Lovell J, Kluger J (1994) Apollo 13. Coronet Books

Lowe DG (1991) Fitting parametrized three-dimensional models to images. IEEE T Pattern Anal 13(5): 441–450

Lowe DG (2004) Distinctive image features from scale-invariant keypoints. Int J Comput Vision 60(2):91–110

Lowry S, Sunderhauf N, Newman P, Leonard J, Cox D, Corke P, Milford M (2015) Visual place recognition: A survey. Robotics, IEEE Transactions on (99):1–19

Lu F, Milios E (1997) Globally consistent range scan alignment for environment mapping. Auton Robot 4:333–349

Lucas SM (2005) ICDAR 2005 text locating competition results. In: Proceedings of the Eighth International Conference on Document Analysis and Recognition, ICDAR05. pp 80–84

Lucas BD, Kanade T (1981) An iterative image registration technique with an application to stereo vision. In: International joint conference on artificial intelligence (IJCAI), Vancouver, vol 2. https:// www.ijcai.org/Past/%20Proceedings/IJCAI-81-VOL-2/PDF/017.pdf, pp 674–679

Luh JYS, Walker MW, Paul RPC (1980) On-line computational scheme for mechanical manipulators. J Dyn Syst-T ASME 102(2):69–76

Lumelsky V, Stepanov A (1986) Dynamic path planning for a mobile automaton with limited information on the environment. IEEE T Automat Contr 31(11):1058–1063

Luo W, Schwing A, Urtasun R (2016) Efficient Deep Learning for Stereo Matching. IEEE Conference On Computer Vision And Pattern Recognition (CVPR).

Luong QT (1992) Matrice fondamentale et autocalibration en vision par ordinateur. Ph.D. thesis, Université de Paris-Sud, Orsay, France

Lynch KM, Park FC (2017) Modern robotics: Mechanics, planning, and control. Cambridge University Press, New York

Ma Y, Kosecka J, Soatto S, Sastry S (2003) An invitation to 3D. Springer-Verlag, Berlin Heidelberg

Ma J, Zhao J, Tian J, Yuille A, Tu Z (2014) Robust Point Matching via Vector Field Consensus, IEEE Trans on Image Processing, 23(4), pp 1706–1721

Maddern W, Pascoe G, Linegar C, Newman P (2016) 1 Year, 1000km: The Oxford RobotCar Dataset. The International Journal of Robotics Research 36(1). pp 3–15

Magnusson M, Lilienthal A, Duckett T (2007) Scan registration for autonomous mining vehicles using 3D-NDT. J Field Robotics 24(10):803–827

Magnusson M, Nuchter A, Lorken C, Lilienthal AJ, Hertzberg J (2009) Evaluation of 3D registration reliability and speed – A comparison of ICP and NDT. In: Proceedings of the IEEE International Conference on Robotics and Automation (ICRA). pp 3907–3912

Mahony R, Corke P, Chaumette F (2002) Choice of image features for depth-axis control in image based visual servo control. IEEE/RSJ International Conference On Intelligent Robots And Systems, pp 390–395 vol.1

Mahony R, Kumar V, Corke P (2012) Multirotor aerial vehicles: Modeling, estimation, and control of quadrotor. IEEE Robot Autom Mag (19):20–32

Maimone M, Cheng Y, Matthies L (2007) Two years of visual odometry on the Mars exploration rovers. J Field Robotics 24(3):169–186

Makhlin AG (1985) Stability and sensitivity of servo vision systems. In: Proc 5th International Conference on Robot Vision and Sensory Controls – RoViSeC 5. IFS (Publications), Amsterdam, pp 79–89

Malis E (2004) Improving vision-based control using efficient second-order minimization techniques. In: Proceedings of the IEEE International Conference on Robotics and Automation (ICRA). pp 1843–1848

Malis E, Vargas M (2007) Deeper understanding of the homography decomposition for vision-based control. Research Report, RR-6303, Institut National de Recherche en Informatique et en Automatique (INRIA), 90 p, https://hal.inria.fr/inria-00174036v3/document

Malis E, Chaumette F, Boudet S (1999) 2-1/2D visual servoing. IEEE T Robotic Autom 15(2):238–250

Lourakis M, Argyros A (2009), SBA: A Software Package for Generic Sparse Bundle Adjustment, ACM Transactions on Mathematical Software 36, no. 1

Marey M, Chaumette F (2008) Analysis of classical and new visual servoing control laws. In: Proceedings of the IEEE International Conference on Robotics and Automation (ICRA). Pasadena, pp 3244–3249

Mariottini GL, Prattichizzo D (2005) EGT for multiple view geometry and visual servoing: Robotics vision with pinhole and panoramic cameras. IEEE T Robotic Autom 12(4):26–39

Mariottini GL, Oriolo G, Prattichizzo D (2007) Image-based visual servoing for nonholonomic mobile robots using epipolar geometry. IEEE T Robotic Autom 23(1):87–100

Marr D (2010) Vision: A computational investigation into the human representation and processing of visual information. MIT Press, Cambridge, Massachusetts

Martin D, Fowlkes C, Tal D, Malik J (2001) A database of human segmented natural images and its application to evaluating segmentation algorithms and measuring ecological statistics. Proceedings of the 8th International Conference on Computer Vision, vol 2. pp 416–423

Martins FN, Celeste WC, Carelli R, Sarcinelli-Filho M, Bastos-Filho TF (2008) An adaptive dynamic controller for autonomous mobile robot trajectory tracking. Control Eng Pract 16(11):1354–1363

Masutani Y, Mikawa M, Maru N, Miyazaki F (1994) Visual servoing for non-holonomic mobile robots. In: Proceedings of the International Conference on Intelligent Robots and Systems (IROS). Munich, pp 1133–1140

Matarić MJ (2007) The robotics primer. MIT Press, Cambridge, Massachusetts

Matas J, Chum O, Urban M, Pajdla T (2004) Robust wide-baseline stereo from maximally stable extremal regions. Image Vision Comput 22(10):761–767

Matthews ND, An PE, Harris CJ (1995) Vehicle detection and recognition for autonomous intelligent cruise control. Technical Report, University of Southampton

Matthies L (1992) Stereo vision for planetary rovers: Stochastic modeling to near real-time implementation. Int J Comput Vision 8(1):71–91

Mayeda H, Yoshida K, Osuka K (1990) Base parameters of manipulator dynamic models. IEEE T Robotic Autom 6(3):312–321

McGee LA, Schmidt SF (1985) Discovery of the Kalman filter as a Practical Tool for Aerospace and Industry. NASA-TM-86847

McGlone C (ed) (2013) Manual of photogrammetry, 6th ed. American Society of Photogrammetry

McLauchlan PF (1999) The variable state dimension filter applied to surface-based structure from motion. University of Surrey, VSSP-TR-4/99

Mellinger D, Michael N (2012) Trajectory Generation and Control for Precise Aggressive Maneuvers with Quadrotors. Int J Robot Res 31(5):664-674

Merlet JP (2006) Parallel robots. Kluwer Academic, Dordrecht

Mettler B (2003) Identification modeling and characteristics of miniature rotorcraft. Kluwer Academic, Dordrecht

Mičušík B, Pajdla T (2003) Estimation of omnidirectional camera model from epipolar geometry. In: IEEE Conference on Computer Vision and Pattern Recognition, vol 1. Madison, pp 485–490

Middleton RH, Goodwin GC (1988) Adaptive computed torque control for rigid link manipulations. Syst Control Lett 10(1):9–16

Mikolajczyk K, Schmid C (2004) Scale and affine invariant interest point detectors. Int J Comput Vision 60(1):63–86

Mikolajczyk K, Schmid C (2005) A performance evaluation of local descriptors. IEEE T Pattern Anal 27(10):1615–1630

Mindell DA (2008) Digital Apollo. MIT Press, Cambridge, Massachusetts

Molton N, Brady M (2000) Practical structure and motion from stereo when motion is unconstrained. Int J Comput Vision 39(1):5–23

Montemerlo M, Thrun S (2007) FastSLAM: A scalable method for the simultaneous localization and mapping problem in robotics, vol 27. Springer-Verlag, Berlin Heidelberg

Montemerlo M, Thrun S, Koller D, Wegbreit B (2003) FastSLAM 2.0: An improved particle filtering algorithm for simultaneous localization and mapping that provably converges. In: Proceedings of the 18th International Joint Conference on Artificial Intelligence. Morgan Kaufmann, San Francisco, pp 1151–1156

Moravec H (1980) Obstacle avoidance and navigation in the real world by a seeing robot rover. Ph.D. thesis, Stanford University

Morel G, Liebezeit T, Szewczyk J, Boudet S, Pot J (2000) Explicit incorporation of 2D constraints in vision based control of robot manipulators. In: Corke PI, Trevelyan J (eds) Lecture notes in control and information sciences. Experimental robotics VI, vol 250. Springer-Verlag, Berlin Heidelberg, pp 99–108

Muja M, Lowe DG (2009) Fast approximate nearest neighbors with automatic algorithm configuration. International Conference on Computer Vision Theory and Applications (VISAPP), Lisbon, Portugal (Feb 2009), pp 331–340

Murray RM, Sastry SS, Zexiang L (1994) A mathematical introduction to robotic manipulation. CRC Press, Inc., Boca Raton

NASA (1970) Apollo 13: Technical air-to-ground voice transcription. Test Division, Apollo Spacecraft Program Office, https://www.hq.nasa.gov/alsj/a13/AS13_TEC.PDF

Nayar SK (1997) Catadioptric omnidirectional camera. In: Proceedings of the IEEE Conference on Computer Vision and Pattern Recognition. Los Alamitos, CA, pp 482–488

Neira J, Tardós JD (2001) Data association in stochastic mapping using the joint compatibility test. IEEE T Robotic Autom 17(6):890–897

Neira J, Davison A, Leonard J (2008) Guest editorial special issue on Visual SLAM. IEEE T Robotic Autom 24(5):929–931

Newcombe RA, Lovegrove SJ, Davison AJ (2011) DTAM: Dense tracking and mapping in real-time. In: Proceedings of the International Conference on Computer Vision, pp 2320–2327

Ng J, Bräunl T (2007) Performance comparison of bug navigation algorithms. J Intell Robot Syst 50(1):73–84

Niblack W (1985) An introduction to digital image processing. Strandberg Publishing Company Birkeroed, Denmark

Nistér D (2003) An efficient solution to the five-point relative pose problem. In: IEEE Conference on Computer Vision and Pattern Recognition, vol 2. Madison, pp 195–202

Nistér D, Naroditsky O, Bergen J (2006) Visual odometry for ground vehicle applications. J Field Robotics 23(1):3–20

Nixon MS, Aguado AS (2019) Feature extraction and image processing for Computer Vision, 4th ed. Academic Press, London Oxford

Noble JA (1988) Finding corners. Image Vision Comput 6(2):121–128

Okutomi M, Kanade T (1993) A multiple-baseline stereo. IEEE T Pattern Anal 15(4):353–363

Ollis M, Herman H, Singh S (1999) Analysis and design of panoramic stereo vision using equi-angular pixel cameras. Robotics Institute, Carnegie Mellon University, CMU-RI-TR-99-04, Pittsburgh, PA

Olson E (2011) AprilTag: A robust and flexible visual fiducial system. In: Proceedings of the IEEE International Conference on Robotics and Automation (ICRA). pp 3400–3407

Orin DE, McGhee RB, Vukobratovic M, Hartoch G (1979) Kinematics and kinetic analysis of open-chain linkages utilizing Newton-Euler methods. Math Biosci 43(1/2):107–130

Ortega R, Spong MW (1989) Adaptive motion control of rigid robots: A tutorial. Automatica 25(6):877–888

Otsu N (1975) A threshold selection method from gray-level histograms. Automatica 11:285–296

Owen M, Beard RW, McLain TW (2015) Implementing Dubins Airplane Paths on Fixed-Wing UAVs. Handbook of Unmanned Aerial Vehicles, pp 1677–1701.

Papanikolopoulos NP, Khosla PK (1993) Adaptive robot visual tracking: Theory and experiments. IEEE T Automat Contr 38(3):429–445

Papanikolopoulos NP, Khosla PK, Kanade T (1993) Visual tracking of a moving target by a camera mounted on a robot: A combination of vision and control. IEEE T Robotic Autom 9(1):14–35

Patel S, Sobh T (2015) Manipulator performance measures-a comprehensive literature survey. Journal Of Intelligent and Robotic Systems, vol. 77, pp 547–570

Paul R (1972) Modelling, trajectory calculation and servoing of a computer controlled arm. Ph.D. thesis, technical report AIM-177, Stanford University

Paul R (1979) Manipulator Cartesian path control. IEEE T Syst Man Cyb 9:702–711

Paul RP (1981) Robot manipulators: Mathematics, programming, and control. MIT Press, Cambridge, Massachusetts

Paul RP, Shimano B (1978) Kinematic control equations for simple manipulators. In: IEEE Conference on Decision and Control, vol 17. pp 1398–1406

Paul RP, Zhang H (1986) Computationally efficient kinematics for manipulators with spherical wrists based on the homogeneous transformation representation. Int J Robot Res 5(2):32–44

Pieper DL (1968) The Kinematics Of Manipulators Under Computer Control. Ph.D. thesis, Stanford University

Piepmeier JA, McMurray G, Lipkin H (1999) A dynamic quasi-Newton method for uncalibrated visual servoing. In: Proceedings of the IEEE International Conference on Robotics and Automation (ICRA). Detroit, pp 1595–1600

Pivtoraiko M, Knepper RA, Kelly A (2009) Differentially constrained mobile robot motion planning in state lattices. J Field Robotics 26(3):308–333

Pock T (2008) Fast total variation for computer vision. Ph.D. thesis, Graz University of Technology

Pollefeys M, Nistér D, Frahm JM, Akbarzadeh A, Mordohai P, Clipp B, Engels C, Gallup D, Kim SJ, Merrell P, et al. (2008) Detailed real-time urban 3D reconstruction from video. Int J Comput Vision 78(2):143–167, Jul

Pomerleau D, Jochem T (1995) No hands across America Journal. https://www.cs.cmu.edu/~tjochem/nhaa/Journal.html

Pomerleau D, Jochem T (1996) Rapidly adapting machine vision for automated vehicle steering. IEEE Expert 11(1):19–27

Posner I, Corke P, Newman P (2010) Using text-spotting to query the world. In: IEEE/RSJ International Conference on Intelligent Robots and Systems (IROS). IEEE, pp 3181–3186

Pounds P (2007) Design, construction and control of a large quadrotor micro air vehicle. Ph.D. thesis, Australian National University

Pounds P, Mahony R, Gresham J, Corke PI, Roberts J (2004) Towards dynamically-favourable quadrotor aerial robots. In: Proceedings of the Australasian Conference on Robotics and Automation. Canberra

Pounds P, Mahony R, Corke PI (2006) A practical quad-rotor robot. In: Proceedings of the Australasian Conference on Robotics and Automation. Auckland

Poynton CA (2012) Digital video and HD algorithms and interfaces. Morgan Kaufmann, Burlington

Press WH, Teukolsky SA, Vetterling WT, Flannery BP (2007) Numerical recipes, 3rd ed. Cambridge University Press, New York

Prince SJ (2012) Computer vision: Models, learning, and inference. Cambridge University Press, New York

Prouty RW (2002) Helicopter performance, stability, and control. Krieger, Malabar FL

Pujol J (2012) Hamilton, Rodrigues, Gauss, Quaternions, and Rotations: A Historical Reassessment. Communications In Mathematical Analysis, vol. 13

Pynchon T (2006) Against the day. Jonathan Cape, London

Reeds J, Shepp L (1990) Optimal paths for a car that goes both forwards and backwards. Pacific Journal Of Mathematics, vol. 145, pp 367–393

Rekleitis IM (2004) A particle filter tutorial for mobile robot localization. Technical report (TR-CIM-04-02), Centre for Intelligent Machines, McGill University

Rives P, Chaumette F, Espiau B (1989) Positioning of a robot with respect to an object, tracking it and estimating its velocity by visual servoing. In: Hayward V, Khatib O (eds) Lecture notes in control and information sciences. Experimental robotics I, vol 139. Springer-Verlag, Berlin Heidelberg, pp 412–428

Rizzi AA, Koditschek DE (1991) Preliminary experiments in spatial robot juggling. In: Chatila R, Hirzinger G (eds) Lecture notes in control and information sciences. Experimental robotics II, vol 190. Springer-Verlag, Berlin Heidelberg, pp 282–298

Roberts LG (1963) Machine perception of three-dimensional solids. MIT Lincoln Laboratory, TR 315, https://dspace.mit.edu/handle/1721.1/11589

Romero-Ramirez FJ, Muñoz-Salinas R, Medina-Carnicer R (2018) Speeded up detection of squared fiducial markers. Image and Vision Computing, vol 76, pp 38–47

Rosenfield GH (1959) The problem of exterior orientation in photogrammetry. Photogramm Eng 25(4):536–553

Rosten E, Porter R, Drummond T (2010) FASTER and better: A machine learning approach to corner detection. IEEE T Pattern Anal 32:105–119

Raghavan M, Roth B (1990). Kinematic analysis of the 6R manipulator of general geometry. International Symposium on Robotics Research, pp. 314–320

Russell S, Norvig P (2020) Artificial intelligence: A modern approach, 4th ed. Pearson

Rusu RB, Blodow N, Beetz M (2009) Fast point feature histograms (FPFH) for 3D registration. In: Proceedings of the IEEE International Conference on Robotics and Automation (ICRA), pp. 3212–3217

Sakaguchi T, Fujita M, Watanabe H, Miyazaki F (1993) Motion planning and control for a robot performer. In: Proceedings of the IEEE International Conference on Robotics and Automation (ICRA). Atlanta, May, pp 925–931

Salvi J, Matabosch C, Fofi D, Forest J (2007) A review of recent range image registration methods with accuracy evaluation. Image Vision Comput 25(5):578–596

Samson C, Espiau B, Le Borgne M (1990) Robot control: The task function approach. Oxford University Press, Oxford

Scaramuzza D, Martinelli A, Siegwart R (2006) A Toolbox for Easy Calibrating Omnidirectional Cameras. Proc IEEE/RSJ Int Conf on Intelligent Robots and Systems, pp 5695–5701

Scaramuzza D, Fraundorfer F (2011) Visual odometry [tutorial]. IEEE Robot Autom Mag 18(4):80–92

Scharstein D, Pal C (2007) Learning conditional random fields for stereo. In: IEEE Computer Society Conference on Computer Vision and Pattern Recognition (CVPR). Minneapolis, MN

Scharstein D, Szeliski R (2002) A taxonomy and evaluation of dense two-frame stereo correspondence algorithms. Int J Comput Vision 47(1):7–42

Selig JM (2005) Geometric fundamentals of robotics. Springer-Verlag, Berlin Heidelberg

Sharf I, Nahon M, Harmat A, Khan W, Michini M, Speal N, Trentini M, Tsadok T, Wang T (2014) Ground effect experiments and model validation with Draganflyer X8 rotorcraft. Int Conf Unmanned Aircraft Systems (ICUAS). pp 1158–1166

Sharp A (1896) Bicycles & tricycles: An elementary treatise on their design an construction; With examples and tables. Longmans, Green and Co., London New York Bombay

Sheridan TB (2003) Telerobotics, automation, and human supervisory control. MIT Press, Cambridge, Massachusetts, 415 p

Shi J, Tomasi C (1994) Good features to track. In: Proceedings of the Computer Vision and Pattern Recognition. IEEE Computer Society, Seattle, pp 593–593

Shih FY (2009) Image processing and mathematical morphology: Fundamentals and applications, CRC Press, Boca Raton

Shirai Y (1987) Three-dimensional computer vision. Springer-Verlag, New York

Shirai Y, Inoue H (1973) Guiding a robot by visual feedback in assembling tasks. Pattern Recogn 5(2):99–106

Shoemake K (1985) Animating rotation with quaternion curves. In: Proceedings of ACM SIGGRAPH, San Francisco, pp 245–254

Siciliano B, Khatib O (eds) (2016) Springer handbook of robotics, 2nd ed. Springer-Verlag, New York

Siciliano B, Sciavicco L, Villani L, Oriolo G (2009) Robotics: Modelling, planning and control. Springer-Verlag, Berlin Heidelberg

Siegwart R, Nourbakhsh IR, Scaramuzza D (2011) Introduction to autonomous mobile robots, 2nd ed. MIT Press, Cambridge, Massachusetts

Silver WM (1982) On the equivalence of Lagrangian and Newton-Euler dynamics for manipulators. Int J Robot Res 1(2):60–70

Sivic J, Zisserman A (2003) Video Google: A text retrieval approach to object matching in videos. In: Proceedings of the Ninth IEEE International Conference on Computer Vision. pp 1470–1477

Skaar SB, Brockman WH, Hanson R (1987) Camera-space manipulation. Int J Robot Res 6(4):20–32

Skofteland G, Hirzinger G (1991) Computing position and orientation of a freeflying polyhedron from 3D data. In: Proceedings of the IEEE International Conference on Robotics and Automation (ICRA). Seoul, pp 150–155

Smith R (2007) An overview of the Tesseract OCR engine. In: 9th International Conference on Document Analysis and Recognition (ICDAR). pp 629–633

Sobel D (1996) Longitude: The true story of a lone genius who solved the greatest scientific problem of his time. Fourth Estate, London

Soille P (2003) Morphological image analysis: Principles and applications. Springer-Verlag, Berlin Heidelberg

Solà J (2017) Quaternion kinematics for the error-state Kalman filter. arXiv 1711.02508

Solà J, Deray J, Atchuthan D (2018) A micro Lie theory for state estimation in robotics. arXiv 1812.01537

Spong MW (1989) Adaptive control of flexible joint manipulators. Syst Control Lett 13(1):15–21

Spong MW, Hutchinson S, Vidyasagar M (2006) Robot modeling and control, 2nd ed. John Wiley & Sons, Inc., Chichester

Srinivasan VV, Venkatesh S (1997) From living eyes to seeing machines. Oxford University Press, Oxford

Stachniss C, Burgard W (2014) Particle filters for robot navigation. Foundations and Trends in Robotics 3(4):211–282

Steinvall A (2002) English colour terms in context. Ph.D. thesis, Ume Universitet

Stentz A (1994) The D* algorithm for real-time planning of optimal traverses. The Robotics Institute, Carnegie-Mellon University, CMU-RI-TR-94-37

Stone JV (2012) Vision and brain: How we perceive the world. MIT Press, Cambridge, Massachusetts

Strasdat H (2012) Local accuracy and global consistency for efficient visual SLAM. Ph.D. thesis, Imperial College London

Strelow D, Singh S (2004) Motion estimation from image and inertial measurements. Int J Robot Res 23(12):1157–1195

Sünderhauf N (2012) Robust optimization for simultaneous localization and mapping. Ph.D. thesis, Technische Universität Chemnitz

Sussman GJ, Wisdom J, Mayer ME (2001) Structure and interpretation of classical mechanics. MIT Press, Cambridge, Massachusetts

Sutherland IE (1974) Three-dimensional data input by tablet. P IEEE 62(4):453–461

Svoboda T, Pajdla T (2002) Epipolar geometry for central catadioptric cameras. Int J Comput Vision 49(1):23–37

Szeliski R (2022) Computer vision: Algorithms and applications. Springer-Verlag, Berlin Heidelberg

Tahri O, Chaumette F (2005) Point-based and region-based image moments for visual servoing of planar objects. IEEE T Robotic Autom 21(6):1116–1127

Tahri O, Mezouar Y, Chaumette F, Corke PI (2009) Generic decoupled image-based visual servoing for cameras obeying the unified projection model. In: Proceedings of the IEEE International Conference on Robotics and Automation (ICRA). Kobe, pp 1116–1121

Taylor RA (1979) Planning and execution of straight line manipulator trajectories. IBM J Res Dev 23(4):424–436

Terzakis G, Lourakis MIA, Ait-Boudaoud D (2018) Modified Rodrigues Parameters: An Efficient Representation of Orientation in 3D Vision and Graphics. J Math Imaging Vis. 60:422–442

ter Haar Romeny BM (1996) Introduction to scale-space theory: Multiscale geometric image analysis. Utrecht University

Thrun S, Burgard W, Fox D (2005) Probabilistic robotics. MIT Press, Cambridge, Massachusetts

Tissainayagam P, Suter D (2004) Assessing the performance of corner detectors for point feature tracking applications. Image Vision Comput 22(8):663–679

Titterton DH, Weston JL (2005) Strapdown inertial navigation technology. IEE Radar, Sonar, Navigation and Avionics Series, vol 17, The Institution of Engineering and Technology (IET), 576 p

Tomasi C, Kanade T (1991) Detection and tracking of point features. Carnegie Mellon University, CMU-CS-91-132

Triggs B, McLauchlan P, Hartley R, Fitzgibbon A (2000) Bundle adjustment – A modern synthesis. Lecture notes in computer science. Vision algorithms: theory and practice, vol 1883. Springer-Verlag, Berlin Heidelberg, pp 153–177

Tsai RY (1986) An Efficient and Accurate Camera Calibration Technique for 3D Machine Vision. Proc IEEE Conference on Computer Vision and Pattern Recognition (CVPR), pp 364–374

Tsakiris D, Rives P, Samson C (1998) Extending visual servoing techniques to nonholonomic mobile robots. In: Kriegman DJ, Hager GD, Morse AS (eds) Lecture notes in control and information sciences. The confluence of vision and control, vol 237. Springer-Verlag, Berlin Heidelberg, pp 106–117

Uicker JJ (1965) On the dynamic analysis of spatial linkages using 4 by 4 matrices. Dept. Mechanical Engineering and Astronautical Sciences, NorthWestern University

Umeyama, S (1991) Least-Squares Estimation of Transformation Parameters Between Two Point Patterns. IEEE Trans Pattern Analysis and Machine Intellence 13(4). pp 376–380

Usher K (2005) Visual homing for a car-like vehicle. Ph.D. thesis, Queensland University of Technology

Usher K, Ridley P, Corke PI (2003) Visual servoing of a car-like vehicle – An application of omnidirectional vision. In: Proceedings of the IEEE International Conference on Robotics and Automation (ICRA). Taipei, Sep, pp 4288–4293

Valgren C, Lilienthal AJ (2010) SIFT, SURF & seasons: Appearance-based long-term localization in outdoor environments. Robot Auton Syst 58(2):149–156

Vanderborght B, Sugar T, Lefeber D (2008) Adaptable compliance or variable stiffness for robotic applications. IEEE Robot Autom Mag 15(3):8–9

Vince J (2011) Quaternions for Computer Graphics. Springer

Wade NJ (2007) Image, eye, and retina. J Opt Soc Am A 24(5):1229–1249

Walker MW, Orin DE (1982) Efficient dynamic computer simulation of robotic mechanisms. J Dyn Syst-T ASME 104(3):205–211

Walter WG (1950) An imitation of life. Sci Am 182(5):42–45

Walter WG (1951) A machine that learns. Sci Am 185(2):60–63

Walter WG (1953) The living brain. Duckworth, London

Warren M (2015) Long-range stereo visual odometry for unmanned aerial vehicles. Ph.D. thesis, Queensland University of Technology

Weinmann M, Jutzi B, Mallet C (2014) Long-range stereo visual Semantic 3D Scene Interpretation: A Framework Combining Optimal Neighborhood Size Selection with Relevant Features. ISPRS Annals of Photogrammetry, Remote Sensing and Spatial Information Sciences II–3, pp. 181–188

Weiss LE (1984) Dynamic visual servo control of robots: An adaptive image-based approach. Ph.D. thesis, technical report CMU-RI-TR-84-16, Carnegie-Mellon University

Weiss L, Sanderson AC, Neuman CP (1987) Dynamic sensor-based control of robots with visual feedback. IEEE T Robotic Autom 3(1):404–417

Westmore DB, Wilson WJ (1991) Direct dynamic control of a robot using an end-point mounted camera and Kalman filter position estimation. In: Proceedings of the IEEE International Conference on Robotics and Automation (ICRA). Seoul, Apr, pp 2376–2384

Whitney DE (1969) Resolved motion rate control of manipulators and human prostheses. IEEE T Man Machine 10(2):47–53

Wiener N (1965) Cybernetics or control and communication in the animal and the machine. MIT Press, Cambridge, Massachusetts

Wilburn B, Joshi N, Vaish V, Talvala E-V, Antunez E, Barth A, Adams A, Horowitz M, Levoy M (2005) High performance imaging using large camera arrays. ACM Transactions on Graphics (TOG) – Proceedings of ACM SIGGRAPH 2005 24(3):765–776

Wolf PR, DeWitt BA (2014) Elements of photogrammetry, 4th ed. McGraw-Hill, New York

Woodfill J, Von Herzen B (1997) Real-time stereo vision on the PARTS reconfigurable computer. In: Proceedings of the IEEE Symposium on FPGAs for Custom Computing Machines, Grenoble. pp 201–210

Xu G, Zhang Z (1996) Epipolar geometry in stereo, motion, and object recognition: A unified approach. Springer-Verlag, Berlin Heidelberg

Yang H, Antonante P, Tzoumas V, Carlone L (2020) Graduated Non-Convexity for Robust Spatial Perception: From Non-Minimal Solvers to Global Outlier Rejection. IEEE Robotics and Automation Letters 5(2):1127–1134

Yi K, Trulls E, Lepetit V, Fua P (2016) LIFT: Learned Invariant Feature Transform. ECCV 2016. pp 467–483

Ying X, Hu Z (2004) Can we consider central catadioptric cameras and fisheye cameras within a unified imaging model. In: Pajdla T, Matas J (eds) Lecture notes in computer science. Computer vision – ECCV 2004, vol 3021. Springer-Verlag, Berlin Heidelberg, pp 442–455

Yoshikawa T (1984) Analysis and control of robot manipulators with redundancy. In: Brady M, Paul R (eds) Robotics research: The first international symposium. MIT Press, Cambridge, Massachusetts, pp 735–747

Zabih R, Woodfill J (1994) Non-parametric local transforms for computing visual correspondence. In: Ecklundh J-O (ed) Lecture notes in computer science. Computer Vision – ECCV 1994, vol 800. Springer-Verlag, Berlin Heidelberg, pp 151–158

Zarchan P, Musoff H (2005) Fundamentals of Kalman filtering: A practical approach. Progress in Astronautics and Aeronautics, vol 208. American Institute of Aeronautics and Astronautics

Zhan H, Garg R, Weerasekera C, Li K, Agarwal H, Reid I (2018) Unsupervised Learning of Monocular Depth Estimation and Visual Odometry With Deep Feature Reconstruction. IEEE Conference On Computer Vision And Pattern Recognition (CVPR).

Zhang Z, Faugeras O, Kohonen T, Hunag TS, Schroeder MR (1992) Three D-dynamic scene analysis: A stereo based approach. Springer-Verlag, New York

Zhou Q, Park J, Koltun V (2018) Open3D: A Modern Library for 3D Data Processing. ArXiv:1801.09847

Ziegler J, Bender P, Schreiber M, Lategahn H, Strauss T, Stiller C, Thao Dang, Franke U, Appenrodt N, Keller CG, Kaus E, Herrtwich RG, Rabe C, Pfeiffer D, Lindner F, Stein F, Erbs F, Enzweiler M, Knöppel C, Hipp J, Haueis M, Trepte M, Brenk C, Tamke A, Ghanaat M, Braun M, Joos A, Fritz H, Mock H, Hein M, Zeeb E (2014) Making Bertha drive – An autonomous journey on a historic route. IEEE Intelligent Transportation Systems Magazine 6(2):8–20

Index of People

A

Ackermann, Rudolph **131**
Asimov, Isaac   3

B

Babbage, Sir Charles   2
Ball, Sir Robert   73, **78**
Bayer, Bryce E.   405
Bayes, Reverend Thomas   **221**
Beer, August   399
Black, Harold   8
Bode, Henrik   8
Boltzman, Ludwig   397
Braitenberg, Valentino   **165**
Brunel, Isambard Kingdom   357
Bryan, George   **49**

C

Čapek, Karel   2, 6
Cardano, Gerolamo   **50**
Chasles, Michel   **75**
Chrétien, Henri   **445**
Clifford, William   66
Cook, Captain James   216, 236
Coriolis, Gaspard-Gustave de   **98**
Coulomb, Charles-Augustin de   **359**

D

Dalton, John   407
Davy, Sir Humphry   398
Delaunay, Boris   189
Denavit, Jacques   **297**
Descartes, René   **30**
Devol, George C. Jr.   **3**, 4
Dijkstra, Edsger   **208**
Draper, Charles Stark (Doc)   110, **111**, 222

E

Edison, Thomas Alva   398
Einstein, Albert   97
Engelberger, Joseph F.   **4**, 8
Euclid of Alexandria   **29**, **557**
Euler, Leonhard   **48**, 95, 170, 373

G

Gauss, Carl Friedrich   84, **464**
Gibbs, Josiah   84
Goethe, Johann Wolfgang von   404
Goetz, Raymond   7
Gray, John McFarlane   357

H

Hall, Edwin   **116**
Hamilton, Sir William Rowan   **59**, 84
Harrison, John   216
Hartenberg, Richard   **297**

H

Helmholtz, Hermann von   404
Hering, Karl Ewald   404
Herschel, William   398, **572**
Hesse, Ludwig Otto   **738**
Hough, Paul   **540**

I

Ibn al-Haytham, Hasan   **557**
Ilon, Bengt   145

J

Jacobi, Carl Gustav Jacob   331, **333**

K

Kálmán, Rudolf   221, **222**
Kepler, Johannes   **557**

L

Lagrange, Joseph-Louis   **373**
Lambert, Johann Heinrich   424
Land, Edwin   421
Laplace, Pierre-Simon   464, **465**
Laussedat, Aimé   **589**
Lazzarini, Mario   240
Leclerc, Georges-Louis   **240**
Lie, Sophus   **732**
Longuet-Higgins, Christopher   **656**
Lovelace, Countess Ada   2

M

Markov, Andrey   189
Marr, David   **470**
Maxwell, James Clerk   404
McCarthy, John   8
McCulloch, Warren   8
Metropolis, Nicholas   240
Minsky, Marvin   8
Moler, Cleve   **13**

N

Newell, Allen   8
Newton, Sir Isaac   **94**, 95, 391, 396, **572**
Nyquist, Harold   8, **483**

P

Pitts, Walter   8
Planck, Max   397
Plücker, Julius   **718**
Price, Richard   221
Ptolemy, Claudius   **557**

R

Rodrigues, Olinde   **55**, 84

S

Scheinman, Victor   282
Schmidt, Stanley F.   **222**
Shannon, Claude   8, **483**
Simon, Herbert   8
Sobel, Irwin   **489**
Stefan, Jozef   397
Swan, Sir Joseph   398

T

Tait, Peter   **49**, 84
Tesla, Nikola   7
Turing, Alan   8

U

Ulam, Stanislaw   240

V

von Kármán, Theodore   297
von Neumann, John   240
Voronoi, Georgy Feodosevich   189

W

Wald, George   403
Walter, William Grey   8, 162, **163**
Wien, Wilhelm   397
Wiener, Norbert   8

Y

Young, Thomas   404

Index of Functions, Classes, and Methods[1]

A

adaptthresh 499
anaglyph 634
analyticalInverseKinematics 302, 304, 306, 326, 327
– .generateIKFunction 302
– .showdetails 302, 304
angdiff 134, 159, 205
angle 55
animtform 45, 104, 106, 110
atan 136
atan2 133, 136, 451
atan2d 722
axang2rotm 55

B

bagOfFeatures 535
– .encode 537
bicycleKinematics 165
BicycleVehicle 158, 220, 225, 229, 234, 237, 241
– .addDriver 220, 229, 234, 237, 241
– .derivative 158
– .f 220
– .Fv 223
– .Fx 223
– .plotxy 223, 230, 234, 237, 243
– .q 158, 220
– .qhist 220
– .run 220
– .step 158, 220
binaryOccupancyMap 166, 181, 199, 205
– .copy 205
– .grid2world 181
– .GridOriginInLocal 199
– .inflate 205
– .setOccupancy 181, 199
– .show 166, 181
– .world2grid 181
birdsEyeView 695
blackbody 397, 399, 413, 419, 421
boundary2polar 517
boundmatch 519
bsplinepolytraj 202, 311
Bug2 167
– .plot 167
– .run 167, 168
buildMap 250, 265
bundleAdjustment 622, 623
bwboundaries 517
bwconncomp 507
bwlabel 507
bwmorph 189, 480
bwtraceboundaries 516

C

camcald 588
camera 553, 568, 571, 573
cameraCalibrator 556, 558, 563, 564, 588
cameraIntrinsics 566, 603, 612, 616

cameraParameters 565
cast 452
CatadioptricCamera 571
ccode 741
cellfun 292, 376
CentralCamera 548, 551, 557, 560, 568, 571, 573, 586, 599, 668, 671, 674, 679, 680
– .C 553
– .E 603
– .estpose 669
– .F 602
– .f 675
– .flowfield 674, 675
– .fov 553
– .intrinsics 603, 612
– .K 552
– .mesh 555, 556, 568, 571
– .plot 554–556, 605, 609, 610, 669, 679
– .plot_camera 584
– .plot_epiline 602, 605
– .pp 677, 679
– .project 548, 549, 551, 554, 555, 560, 563, 584, 604, 671, 672, 728
– .T 669, 680
– .visjac_p 674, 675, 677, 680
chi2inv 225, 753
clf 223
cmfrgb 410, 411
collisionBox 324
– .Pose 324, 325
– .show 325
collisionCylinder 325
collisionMesh 325, 702
– .Pose 325
collisionSphere 325
colon, : 194
colorname 414, 419, 502, 503, 505
colorThresholder 449, 452
cond 339
controllerPurePursuit 139
conv2 460
coriolisMatrix 375
cornerPoints 528, 530, 595
– .plot 528, 530, 595
cross 95, 120, 727
cubicpolytraj 105, 311
cylinder 590

D

decomposeCam 552, 561
deg2rad 36, 102, 220, 234, 237, 280, 281, 286–288, 306, 338, 340, 348
delta2se 680
delta2tform 95
det 33, 70, 71, 93, 338, 339
detectHarrisFeatures 528, 530, 595
detectMSERFeatures 500
detectORBFeatures 650
detectSURFFeatures 532, 595, 597, 613, 632, 648
detectTextCRAFT 534

---

[1] Classes are shown in **bold**, and methods are prefixed by a dot. All others are functions.

DGraph 179
diag 220, 223, 228, 229, 234, 237, 241, 242, 340, 341
DigitalCamera 616
digraph 175, 177
– .Nodes 177
– .plot 175, 177
disparityBM 626, 628, 629
disparitySGM 627, 633
DistanceTransformPlanner 181–183, 185
– .plan 182, 183, 185
– .plot 182, 183, 185
– .plot3d 183
– .query 182, 183, 185
drawpolygon 644
DStarPlanner 186
– .costmap 186
– .modifyCost 186
– .niter 186, 187
– .plan 186, 187
– .query 186, 187
dubinsConnection 193
– .connect 193
– .MinTurningRadius 193
dubinsPathSegment 193
– .interpolate 194
– .MotionLengths 193
– .show 193

E

e2h 601, 602
edge 465, 467, 523, 524
eig 54, 340, 341, 377, 512, 721
EKF 223–225, 229, 234, 237
– .landmarks 234
– .P_est 234
– .plotellipse 224, 230
– .plotmap 234, 237
– .plotP 225
– .plotxy 224, 230, 237
– .run 223, 230, 234, 237
– .transform 238
– .x_est 234
ellipsoid 556
epidist 605, 606
epipolarLine 607, 614
estgeotform2d 609, 611, 614, 648
estimateEssentialMatrix 616
estimateFundamentalMatrix 604–606, 632
estimateStereoRectification 632
estimateWorldCameraPose 563
estrelpose 603, 612, 616, 646
esttheta 430
ETS2 278–283, 299, 300, 330, 340
– * 278–280, 282, 299, 300, 330, 340
– .fkine 279, 280, 282, 299, 301, 327, 330
– .jacob0 343
– .njoints 281
– .param 281
– .plot 281, 282
– .Rz 278–280, 282, 299, 300, 330, 340
– .structure 281, 282
– .T 281
– .teach 280, 282, 327, 340, 348, 352
– .Tx 278–280, 282, 299, 300, 330, 340
ets2rbt 288, 321

ETS3 282, 288, 317, 321
– * 282, 288, 317, 321
– .fkine 283
– .njoints 283
– .Rx 317
– .Ry 282, 288, 321
– .Rz 282, 288, 317, 321
– .structure 283
– .teach 283
– .Tx 321
– .Ty 317, 321
– .Tz 282, 288, 317, 321
eul2jac 334
eul2rotm 48–50, 54, 70, 90, 104, 105, 559
eul2tform 65, 76, 302
eval 300
expm 34, 40, 56, 57, 62, 63, 75, 92
externalForce 379
extractFeatures 532, 596, 597, 613, 632, 648

F

find 196, 315, 511
findpeaks 448, 763, 765
FishEyeCamera 568, 569
fitgeotform2d 644
flipud 166
fminsearch 301
fspecial 457, 460, 463, 464, 467

G

g2oread 259
gait 319
gaussfunc 751, 752
geobasemap 171
geoplot 171
graythresh 498
groundTruthLabeler 505

H

h2e 38, 601, 602, 608
histeq 449, 451
histogram 452, 758
homtrans 38
hough 523–525
houghlines 523–525
houghpeaks 523–525

I

IBVS 680, 682–685
– .plot_camera 680, 685
– .plot_jcond 680
– .plot_p 680, 683, 684
– .plot_vel 680
– .plot_z 683
– .run 680, 682–685
– .step 680
ikineTraj 311, 313, 314, 316
ikineTrajNum 314, 318
im2double 438, 439, 449, 452, 486
im2uint8 449
imadjust 449, 451, 650
ImageDatastore
– .read 650

– .readimage 538
– .transform 650, 651
imageDatastore 535, 538, 650, 651
imageLabeler 505
imagesc 523
imageSegmenter 498, 509, 541
imageviewset 620
imclearborder 515
imclose 478, 479
imcrop 482, 490, 650
imdilate 476
imerode 475, 479
imeshgrid 485, 487
imfilter 456, 457, 460, 462–464, 467
imfinfo 439, 440, 616, 646
imformats 439
imfuse 624, 627, 634, 648
imgaussfilt 457
imhist 498
imopen 478, 479, 502
impixelinfo 438
importrobot 293
impyramid 484
imread 430, 436, 439, 441, 444, 452, 457, 471, 477, 482, 483, 487,
    496, 498, 500, 505, 506, 508, 521, 524, 528, 530, 534, 575, 581, 595,
    613, 624, 631, 644, 647
imref2d 648
imresize 452, 483
imrotate 487, 523
imsegkmeans 502
imshow 426, 430, 438, 439, 447, 451–453, 455, 457, 459, 462,
    471–474, 479, 482, 490, 497, 499, 500, 502, 506–508, 511, 524, 525,
    528, 530, 533, 576, 577, 595, 614, 626, 644, 650
imshowpair 488, 632
imtool 438, 490, 499, 624, 625, 627, 633
imudata 110, 119
imuSensor 119
imwarp 487, 644, 648
ind2sub 765
indexImages 537
insertObjectAnnotation 472, 617
insertShape 446, 511, 523, 582
insertText 582
interactiveRigidBodyTree 290, 292, 327, 333, 352
– .addConfiguration 290
– .Configuration 290
– .StoredConfigurations 290
interp2 486, 488
inv 38, 95, 259, 334, 339–341, 344, 377, 723, 724
inverseKinematics 304, 306, 308, 327, 345, 350
– .SolverAlgorithm 350
– .SolverParameters 306
– .step 304–308, 345
inverseKinematicsDesigner 298
ismember 515
isnan 315

J

jacobian 259, 331, 740
jsondecode 170

K

kmeans 504

L

label2rgb 500, 507
labelmatrix 500
lambda2rg 410, 412
lambda2xy 413, 419–421, 424
LandmarkMap 227, 229, 230, 234, 237, 240, 241
– .landmark 228
– .plot 227, 230, 234, 237, 243
LandmarkSensor 228, 229, 232–234, 237, 241, 255, 271
– .h 229
– .Hp 233
– .Hw 229
– .Hx 229
– .reading 228
LatticePlanner 197, 199
– .plan 197, 199
– .plot 197–199
– .query 197–199
lidarScan 246, 247, 256
– .Angles 247
– .plot 256
– .Ranges 247
lidarSLAM 263
– .addScan 264
– .LoopClosureSearchRadius 264
– .LoopClosureThreshold 264
– .PoseGraph 265
– .scansAndPoses 265
– .show 264
likelihoodFieldSensorModel 252
– .ExpectedMeasurementWeight 253
– .Map 252
– .MeasurementNoise 252
– .RandomMeasurementWeight 253
lin2rgb 451
line 614
lineToBorderPoints 614
linspace 99, 104, 105, 194, 287, 370, 723, 751
load 166, 246, 263, 315
loadrobot 288, 289, 292, 293, 301, 304, 308, 324, 327, 345, 370,
    379
loadrvcrobot 306, 332, 352, 353, 370, 384
loadspectrum 399–401, 403, 421, 423, 427, 432
logm 34, 39, 56, 62
logspace 194
lscov 352
luminos 401, 402

M

manipulability 314, 341, 342, 378
manipulatorRRT 702
mapClutter 181
– .show 181
mapMaze 181, 211
matchFeatures 597, 598, 614, 632, 648
matchScans 247–249
matchScansGrid 247–249
matlabFunction 740, 741
mclPlot 253
– .plot 253
mdl_quadrotor 152
mean 172
medfilt2 473, 631
meshgrid 372, 374, 486, 576, 577, 590, 630, 752
minjerkpolytraj 311

mkcube   555, 556, 559, 562, 568, 571, 573
mkgrid   609, 668, 679
mobileRobotPRM   189
– .findpath   190, 192
– .NumNodes   190
– .show   190
mobileRobotPropagator   204, 205
– .ControlStepSize   205
– .KinematicModel   204, 205
– .MaxControlSteps   205
– .SpeedLimit   204
– .StateBounds   205
– .SteerLimit   204
– .WheelBase   205
monoCamera   695
monteCarloLocalization   251, 253
– .GlobalLocalization   253, 254
– .InitialCovariance   254
– .InitialPose   253
– .MotionModel   251
– .ParticleLimits   253
– .SensorModel   252
– .step   253
mpq_point   724
mstraj   160, 315, 318

N

navPathControl   206
– .copy   206
– .interpolate   206
ncc   470
nlfilter   477
norm   301, 344, 346
norminv   751
normxcorr2   471, 472, 490
null   346, 602, 677, 742

O

oa2rotm   54
oa2tform   308, 316
occupancyMap   249, 250, 265
– .insertRay   249
– .LocalOriginInWorld   249
– .show   249, 250, 265
ocr   534, 540
ocrText   534
odometryMotionModel   251, 252
– .Noise   252
– .showNoiseDistribution   252
openExample   139, 202, 207, 262, 304, 312, 694, 697, 699, 701
optimizePoseGraph   259, 261, 265
ORBPoints
– .plot   650
ordfilt2   473, 477

P

ParticleFilter   242, 244
– .plotpdf   243
– .plotxy   243
– .run   242
– .std   243
PBVS   669
– .plot_p   670
– .plot_pose   670

– .plot_vel   670
– .run   669, 684
pcdenoise   617
pcdownsample   639, 643
pcfitcylinder   641
pcfitplane   640, 641
pcfitsphere   640, 641
pcmerge   701
pcplayer   638
pcread   638, 643
pcregistericp   643, 702
pcsegdist   701
pcshow   617, 638, 639, 643
pcshowpair   643
pctransform   643, 701
pinv   342–346, 352, 680
plannerControlRRT   204, 205, 207
– .GoalReachedFcn   205, 206
– .NumGoalExtension   205
– .plan   204, 206
plannerRRT   203
plannerRRTStar   203
plot3   316
plotCamera   659
plotChromaticity   413
plotellipse   721, 723, 725
plotellipsoid   340, 341, 377
plothomline   727
plotpoint   38, 39
plotsphere   561, 584, 600, 610
plottform   45, 46, 61, 64, 81, 95
plottform2d   32, 36–39
plotTransforms   81, 307
plotvehicle   206
Plucker   76, 585, 717, 728
– ^   718
– .closest   718
– .intersect_plane   718
– .plot   64, 76, 717
– .point   717
– .side   718
– .skew   717, 728
– .v   717
– .w   717
pointCloud   617, 638
– .findNearestNeighbors   638
– .normals   639
polyder   764
polyfit   606, 764
polyshape   207
poseGraph   259–262, 265
– .addRelativePose   265
– .copy   265
– .edgeResidualErrors   265
– .nodeEstimates   259
– .show   259–261, 265
poseGraph3D   259
poseGraphSolverOptions   266
printtform   81, 286, 289, 293, 302–305, 307, 308, 669
printtform2d   280, 301
projtform2d   584
– .transformPointsForward   584, 610
– .transformPointsInverse   584, 610

Q

qr 552
qtdecomp 180
quaternion 58, 59, 70, 82, 95, 104, 110, 119, 120
– * 58, 59, 120
– .compact 59
– .conj 59, 82, 120
– .dist 70, 119, 121
– .euler 110
– .normalize 71
– .rotatepoint 59, 120
– .rotmat 59
– .slerp 104
quinticpolytraj 99, 104, 309, 370

R

rad2deg 280, 300
rand 189, 191, 724
randn 191, 751, 758
RandomDriver 220, 229, 234, 241
rangeSensor 255, 256
– .HorizontalAngle 256
– .HorizontalAngleNoise 256
– .HorizontalAngleResolution 256
– .Range 256
– .RangeNoise 256
– .step 256
rank 338, 345, 602
ransac 606, 611
rateControl 287, 309, 316, 318, 319, 379
– .waitfor 287, 309, 316, 318, 319, 379
rbtTform 319
readAprilTag 231, 582
rectifyStereoImages 632
reedsSheppConnection 194, 196
– .connect 194, 196
– .ForwardCost 196
– .MinTurningRadius 194
– .ReverseCost 195, 196
reedsSheppPathSegment 194, 196
– .interpolate 195
– .Length 196
– .MotionDirections 195
– .show 195
referencePathFrenet 201, 202
– .show 201, 202
regionprops 510, 514–516
retrieveImages 537
rgb2gray 449, 595
rgb2hsv 415, 416
rgb2lab 427, 502
rgb2xyz 427
rigidBody 284–286, 296, 322, 371, 376
– .CenterOfMass 293, 294, 371, 376
– .Children 288, 292
– .Inertia 293, 294, 371
– .Joint 286, 288, 296, 302, 322
– .Mass 293, 294, 371, 376
– .Name 292
– .Parent 288
rigidBodyJoint 285, 286, 296, 322, 327
– .HomePosition 286
– .JointToParentTransform 286, 302
– .PositionLimits 288, 327
– .setFixedTransform 286, 296, 322

– .Type 288, 302
rigidBodyTree 284–286, 288, 289, 292, 317, 319, 353, 370–372, 379
– .addBody 286
– .BaseName 286
– .Bodies 287, 371, 376
– .centerOfMass 377
– .checkCollision 324, 325
– .copy 370
– .DataFormat 286
– .forwardDynamics 379, 390
– .geometricJacobian 332, 333, 338–341, 345, 347, 348, 377
– .getBody 287, 292, 294, 302, 376
– .getTransform 286, 288, 289, 293, 303–305, 307–309, 313, 317, 333, 336
– .Gravity 370, 372, 376
– .gravityTorque 372, 376
– .homeConfiguration 286, 288, 293, 304, 306, 308, 345
– .inverseDynamics 370–372
– .massMatrix 373, 374, 376, 377
– .removeBody 345
– .show 287–290, 292, 294, 301, 302, 308, 309, 316–319, 324, 379
– .showdetails 286, 289
– .subtree 292
– .velocityProduct 375
rigidtform3d 81, 617, 643
– .A 81
– .invert 643
– .R 81
– .Translation 81
rng 181, 189–191, 204, 206, 220, 223, 229, 234, 237, 240, 253, 306, 308
roots 764
rostf 701
rotm2axang 54, 93
rotm2d 32, 33, 36, 487
rotm2eul 48–50, 102, 106
rotmx 45, 46, 56–58, 70, 80–82, 92, 672
rotmy 45–47
rotmz 45, 47, 70, 669, 680, 684, 685
rottraj 104, 105
rpy2jac 334

S

sad 470
se2 279, 280, 301, 323
– * 279, 280
– .trvec 301
se3 80, 90, 105, 286, 289, 293, 302–311, 313, 316–319, 324, 333, 336, 345, 549, 559, 583, 599, 609, 668, 669, 672, 679, 680, 684, 685
– * 306, 308, 313, 316, 319, 345
– .eul 311
– .interp 105, 669
– .inv 97
– .rotm 80, 102, 106
– .tform 80, 286, 306–308, 324, 345
– .transform 81
– .trvec 80, 102, 106, 310, 311, 317, 336
segmentGroundFromLidar 641
segmentLaneMarkerRidge 696
Sensor 228
sensorfield 165, 211
setdiff 194
shadowRemoval 430
showcolorspace 428

showControlRRT   204, 206
showMatchedFeatures   598, 607
showShape   515, 534
sim   132, 134, 135, 137, 138, 142, 152, 165, 336, 338, 363, 367, 379, 381, 382, 388
simplify   33, 259
Simulink.SimulationData.Dataset   134
Simulink.SimulationOutput   132, 134, 152, 336, 379, 381, 382, 388
− .find   134, 135, 142, 379
− .get   133
− .plot   134
skew2vec   33, 34, 56, 57, 734
skewa2vec   40, 62, 63, 735
slamMapBuilder   266
so2
− .transform   102
so3   104
solve   299
sphere   556
SphericalCamera   573
− .mesh   573
spy   749
sqrtm   723
ssd   470
stackedplot   99, 100, 110, 133, 336
stateEstimatorPF   244
stereoCameraCalibrator   566, 588
stlread   325
strel   478, 479, 502, 503
subs   299, 300
superpixels   509
surfc   752
surfl   374
SURFPoints   532, 595, 632
− .plot   533, 595, 614
syms   33, 259, 299, 330, 740

T

table   514
testpattern   446
tform2adjoint   91, 97
tform2d   259
tform2delta   95
tform2eul   647
tform2rotm   61
tform2trvec   61, 313
tformnorm   70, 93, 669
tformr2d   36, 38, 41
tformrx   61, 62, 79
tic   490
timeit   249, 306
toc   490
tororead   260
transformScan   248
transformtraj   105, 107, 311, 313
trapveltraj   100–102, 107, 309, 702
triangulation   325
− .Points   325
trimpLoopClosures   266
tristim2cc   412, 419
trvec2tform   36–39, 41, 61, 62, 64, 65, 76, 79, 302, 304, 308, 313, 316, 325
Twist   65, 76
− *   76

− .compact   75
− .exp   64
− .line   64, 76
− .pitch   65
− .pole   65
− .printline   76
− .skewa   63, 75
− .tform   63, 76, 77
− .theta   65, 76
− .unit   76
− .UnitPrismatic   64
− .UnitRevolute   63, 64, 75
− .v   75
− .w   65, 75
Twist2d   41, 323
− *   323
− .compact   41
− .exp   323
− .pole   41, 42
− .tform   41, 42
− .UnitPrismatic   41
− .UnitRevolute   40, 323
− .w   41
twoJointRigidBodyTree   286

U

uavDubinsConnection   193
UGraph   171, 179, 208, 760
− .about   761
− .add_edge   171, 179, 208, 760
− .add_node   171, 179, 208, 760
− .adjacency   208
− .closest   762
− .cost   176, 761
− .degree   172
− .edgeinfo   172
− .edges   172, 761
− .highlight_node   175, 177
− .highlight_path   173
− .lookup   172
− .n   172
− .nc   172
− .ne   172
− .neighbors   172, 761
− .nodes   761
− .path_Astar   177, 179, 762
− .path_BFS   173
− .path_UCS   174–176
− .plot   172, 173, 761
uint8   436, 437, 439
undistortImage   488, 582, 607
unicycleKinematics   144

V

vec2skew   33, 34, 57, 92, 734
vec2skewa   40, 41, 62, 63, 735
vecnorm   179, 619
Vehicle   220, 223
vellipse   341
velodyneFileReader   638
velxform   90, 333
videoLabeler   505
VideoReader   443, 454
− .read   444
− .readFrame   444, 454, 455

view 64, 249, 287
vision.AlphaBlender 649
vision.LocalMaximaFinder 471, 765
VisualServo 669, 680
voronoi 189

W

webcam 441, 542
– .preview 441
– .snapshot 441
webcamlist 441
whos 79–81, 104, 106, 110, 166, 168, 194, 246, 287, 315, 436, 483, 740

wordcloud 414, 437, 443
world2img 582
worldToImage 618
wrapToPi 134

X

xplot 309, 313

Z

zncc 471

Index of Apps

C

Camera Calibrator (cameraCalibrator)   556, 558, 563, 564, 582, 588
Color Thresholder (colorThresholder)   449, 452

D

Driving Scenario Designer (drivingScenarioDesigner)   697

G

Ground Truth Labeler (groundTruthLabeler)   505

I

Image Labeler (imageLabeler)   505

Image Segmenter (imageSegmenter)   497, 509
Inverse Kinematics Designer (inverseKinematicsDesigner)   298

S

SLAM Map Builder (slamMapBuilder)   266
Stereo Camera Calibrator (stereoCameraCalibrator)   588

T

Triple Angle Sequence (tripleangle)   51

V

Video Labeler (videoLabeler)   505

Index of Models

D

drivingscenarioandsensors
– /Scenario Reader   697

R

roblocks   134
– /angdiff   135, 137, 138
– /Control Mixer 4   151
– /Joint vloop   363, 364, 368
– /Quadrotor   151
– /tform2delta   337, 338
– /wrapToPi   379
robotcorelib
– /Coordinate Transformation Conversion   151, 153, 313, 337
– /Polynomial Trajectory   380–382
– /Trapezoidal Velocity Profile Trajectory   312, 313, 368
robotmaniplib
– /Forward Dynamics   379–381
– /Get Jacobian   335, 337
– /Get Transform   337
– /GetTransform   313
– /Inverse Dynamics   380, 381
robotmobilelib   **144**
– /Bicycle Kinematic Model   132, 135, 137, 138, 141, 142, 164, 165
– /Differential Drive Kinematic Model   144
– /Unicycle Kinematic Model   144

S

simulink/SignalRouting
– /From   151
– /Goto   151
simulink/Sources
– /Clock   313
sl_braitenberg   **164**, 165, 166
sl_computed_torque   **381**, 382
sl_driveline   **137**, 138
sl_drivepoint   134, **135**
sl_drivepose   **141**, 142
sl_feedforward   **380**, 381
sl_jointspace   312, **313**
sl_lanechange   **132**
sl_ploop_test   367, **368**, 391
sl_pursuit   **138**, 139
sl_quadrotor   **151**, 152, 153
sl_quadrotor_guidance   **153**
sl_rrmc   **335**, 336
sl_rrmc2   **337**, 338
sl_sea   **388**
sl_vloop_test   363, **364**, 391
sl_zerotorque   **379**
slpidlib
– /PID Controller   138

U

uavalgslib
– /Guidance Model   153, 154
uavutilslib
– /UAV Animation   151, 153

General Index

ξ (relative pose) **22**
3D reconstruction 594, **629**

A

A* search **177**
abelian group 26
absorption
– coefficient 399
– color change 423
– spectrum 399, 423
abstract pose ξ **22**
acceleration **95**, 111, 112, 118
– angular **95**
– centripetal **98**
– Coriolis **98**, 122
– Euler **98**
– gravitational **97**, 111–113
– inertial **112**
– proper **112**
– sensor 113, **118**
– specific **112**
accelerometer 52, **111**, 112, 118
– triaxial **113**, 118
accommodation **545**
Ackermann steering 132
active transformation **67**
actuator 356
– brushed motor 356
– brushless motor 356
– electro-hydraulic 356
– pneumatic 356
– series-elastic (SEA) **387**, **388**
– stepper motor 356
added mass 147
addition
– Minkowski **476**
adjoint
– logarithm **736**
– matrix 91, 96, 323, 350, 719, **736**
adjugate **711**
aerial robot **147**
affine
– camera **586**
– reconstruction 623
– space 729
– transformation **729**
AGAST detector 597
AHRS (Attitude and Heading Reference System) **118**
aircraft
– configuration space **155**
– fixed-wing 156
– quadrotor 147
Airy pattern **457**
albedo 400
algebraic topology 71
algorithm
– A* **177**
– Adaptive Monte-Carlo Localization (AMCL) 251
– Bradley 499
– Breadth-First Search (BFS) **173**
– Bresenham 250
– bug2 **167**
– D* **185**
– Dijkstra search **207**
– FastSLAM **269**
– graphcuts 520
– ICP 269
– k-means **502**
– Levenberg-Marquardt (LM) **350**, 622, **745**
– localization 650
– maximally stable extremal region (MSER) 500, 520
– Newton-Euler **370**
– Probabilistic Roadmaps (PRM) **189**
– Random Sampling and Consensus (RANSAC) 606
– Rapidly-Exploring Random Tree (RRT) 204
– resolved-rate motion control 335
– skeletonization **189**
– stereo matching 625, 627
– SURF 596
– thinning 189
– Uniform-Cost Search (UCS) **174**
alias transformation **67**
aliasing
– spatial 483, 627, 629
alibi transformation **67**
ambiguity ratio **627**, 763
AMCL (Adaptive Monte-Carlo Localization) **251**
ampullae **113**
anaglyph 45, 634
analytical Jacobian **334**, 384
anamorphic lens **445**
angle
– azimuth **572**
– colatitude **572**
– declination **116**
– difference **134**
– dip **116**
– elevation 216
– heading **117**
– inclination **116**
– solid 397, 406, 553
– steered wheel **131**, 132
– steering wheel 131
– subtraction **134**
– yaw **117**
angle-axis representation 54, 58, 92, 621
angles
– bank, attitude, heading **49**
– Cardan 43, **49**
– Euler 43, **47**, **51**, 52, 101
  – rate of change **334**
– nautical **49**
– roll-pitch-yaw 101, 106, 107, 334
  – rate of change 334
  – XYZ **49**, **50**, 311, 312
  – YXZ 681
  – ZYX 49
– rotation **52**, 57
– Tait-Bryan **49**
angular
– acceleration **95**, 375, 383
– momentum **95**, 108
– uncertainty 225

– velocity **88**, 95, 97, 107, **109**, 120, 218, 334, 755
– velocity, of a point **89**
anthropomorphic 210, 284, **285**
anti-aliasing
– filter 483, 486, 488
– graphics **447**
anti-symmetric matrix, see skew-symmetric matrix **711**
aperture
– lens **547**, 558
– pinhole **544**, 579
– stop **547**
Apollo
– 13 51, 53
– Lunar Module **52**, 110
application
– character recognition **534**
– fiducial marker **581**
– highway lane changing **697**
– image retrieval **535**
– inertial navigation **107**
– lane and vehicle detection **694**
– lidar-based localization **251**
– lidar-based map building **249**
– lidar-based odometry **246**
– mosaicing **647**
– motion detection **453**
– perspective correction **644**
– pick and place with point cloud **701**
– planar homography **583**
– quadraped walking robot **316**
– resolved-rate motion control 335
– series-elastic actuator **387**
– UAV package delivery **699**
– visual odometry **650**
– writing on a surface **315**
approach vector **53**, 306, 308
April tag **581**
arithmetic
– error 70
– precision 71
array
– camera **578**
– lenslet **580**
– microlens 579, 580
– photosite **442**, **550**, 580
artificial intelligence 8, 11, 18, 655
ArUco marker **581**
aspect ratio **445**, **513**, 514, **550**
assorted pixel array 405
astigmatism **556**
Asus Xtion 636
ASV (Autonomous Surface Vehicle) **128**
Attitude and Heading Reference System (AHRS) **118**
augmented skew-symmetric matrix **40**, 63, 72, 74
AUS (Autonomous Uncrewed System) **128**
autocorrelation matrix **527**
automata 166
AUV (Autonomous Underwater Vehicle) **128**
axis
– of motion **101**
– of rotation **88**, 95
– optical 54, 631, 644, 645, 670, 685
– screw **73**
azimuth angle **572**

B

back
– EMF **367**
– end, SLAM **263**
– projection 619, 620
bag of words **536**
barometric pressure 118
barrel distortion 556
base wrench 376
basis vector **30**, 31, 32, 42–44, **708**
Bayer
– filtering **405**
– pattern **405**
Beer's law **399**
behavior-based robot **166**
Beidou (satellite navigation system) **218**
Bézier spline 202
bicycle motion model **130**, 140
bimodal distribution 498
binarization **496**
binary
– classification **496**
– descriptor 596
– image **449**
bi-quaternion (see dual quaternion) **66**
blackbody
– radiator 429
blackbody radiator **397**, 418
blade flapping 149
blend **101**
blob (see region feature) **507**
body-fixed frame 51, 77, **98**, 108
Boltzmann constant **397**
bootstrap filter 240
boundary
– blob **515**
– detection 479
– gamut 412
– processing 460
bounding box **510**
box filter **457**
Bradley's method for thresholding **499**
Braitenberg vehicle 164, **165**
Bresenham algorithm 250
BRISK descriptor 597
Brockett's theorem 156, **688**
brushfire planner 184
B-spline 202
Buffon's needle problem 240
bug algorithm **166**
bundle adjustment 271, **589**, **618**, **619**, 621, **622**, 623, 651
buoyancy force 147

C

calibration
– camera 551, **558**
– target 422
– technique 558
camera 256
– affine **586**, 623
– aperture 443
– array **578**, 579, 580
– baseline **623**, 630
– black level 470
– calibration **558**, 560, 563, 645, 668, 695

– catadioptric **569**, 571, 575, 589
– CCD   405
– center   558, 580
– center point   **560**
– central   545, **570**, 577, 578
– central perspective   **545**, 548
– depth of field   443
– digital   404, 426
– DSLR   443
– exposure   596
  – control   571
  – time   443, 547
  – value   **442**
– extreme low light   442
– extrinsic parameters   **552**, 558, 587, 623
– field of view   **553**
– fisheye   **568**, 575
– f-number   **442**
– focal length   442
– frame   **545**, 546, 548
– global shutter   442
– high dynamic range   405
– hyperspectral   **431**
– image plane   **545**
– image plane, discrete   **549**
– infrared   **431**, 636
– intrinsic parameters   **551**, **552**, 558, 582, 602, 603, 615, 623, 668, 674, 679, 681
– lens   545
– light field   **579**, 580, 590
– location determination problem   **562**
– low-light   580
– matrix   **548**, **551**, 552, 553, 558, 560
– motion   615, 616, 645, 651
– multispectral   405
– noncentral   571
– nonperspective   566, **586**
– omnidirectional   **553**
– optical axis   **544**, 644, 685
– orientation   552
– panoramic   422, **553**
– parameter   557, 560, 612, 623
– parameters   **552**
– perspective   **544**, 548, 566, **570**, 574, 579, 623
– pinhole   **544**, 579
– plenoptic   **579**
– pose   269, 552, 615, 651, 668, 669
– pose estimation   669
– principal point   674, 675, 679
– quantum efficiency   **442**
– resectioning   588
– retreat   **685**
– RGBD   **636**
– sensor   402, 429, 430
– SLR   445
– spherical   **572**, 573, 688
– stereo   10, 623, 629, 631
– SWIR   **431**
– thermal   **431**
– thermographic   **431**
– ultraviolet   **431**
– unified   **574**, 575
– velocity   673, 674, 677, 678, 682, 685
– verged   605
– video   426
– wide angle   **566**, 589

– wide-angle   675
CamVid dataset   **505**
Canny edge operator   **465**
canonical image coordinate   **546**
car   156
Cartesian
– coordinate system   **30**
– geometry   **29**
– motion   **105**, 308, **311**
– plane   **29**
– product   708
caustic   **571**, 578
CCD sensor   442
celestial navigation   216
CenSurE descriptor   597
center
– of gravity   **359**
– of gravity law (color)   411
– of mass   **359**, 371, 376, 377
central
– imaging   **570**, 578
– moments   **512**
– perspective model   **545**
centrifugal
– acceleration   111
– force   **98**
centripetal
– acceleration   **98**
chamfer matching   482
character recognition   499, **534**
Character Region Awareness for Text Detection (CRAFT)   534
charge well   **442**, 443
Chasles' theorem   **73**
Cholesky decomposition   **713**
chroma keying   **452**
chromatic aberration   **556**
chromaticity   **410**, 418, 427
– coordinate   **410**, 413
– diagram   **410**, 413
– plane   411
chromophore   **403**
CIE
– 1931 standard primary colors   **407**
– CIELUV color space   417
– color space   415
– standard primary colors   411
– XYZ color space   **411**
circle   727
– of confusion   **547**
circularity   **516**, 539
city-block distance   180
classification   **496**
– binary   **496**
– color   **500**
– grayscale   **496**, 497
– pixel   498, 502
clothoid   131, **200**
clustering   540
– k-means   502
CMOS sensor   **442**, 555
coarse-to-fine strategy   **484**
coefficient
– filter   456
colatitude   **572**
collineation   **729**
color   **401**

– apparent change   **423**
– blindness   **407**
– constancy   396
– filter   405
– filtering   407
– gamut   412
– image   439
– matching   408, 409, 427, 432
– matching function   410, **413**
– measuring   405
– plane   **439**, 456
– primary   **407**, 408
– reproduction   **407**, 410
– saturation   410
– space   **415**, 417, 427
  – L*u*v*   **417**
  – CIELUV   **417**
  – HSV   426
  – perceptually uniform   **417**
  – rgY   **415**
  – xyY   **415**, 426
  – XYZ   **411**, 413
  – YC_BC_R **426**
  – YUV   **426**
– spectral   410
– temperature   **420**, 429
Color Checker   428
colored point cloud   **637**
colorimetry   411, 431
column space   **714**
CoM (see also center of mass)   359
Commission Internationale de L'Éclairage (CIE)   **411**
compass   **116**, 141, 218, 232
– heading   214
compound lens   **545**
compression
– format   440
– gamma   **425**
– image   440, 443, 528
– lossless   436
– lossy   **528**
computed torque control   **382**
Concurrent Mapping and Localization (CML)   **235**
condition number (see matrix condition number)   339
cone   729
cone cell (see human eye)   **402**, 403
confidence
– ellipse   **224**
– test   231
configuration
– kinematic   336
– space   77, **78**, **130**, 154, 156, 281, 306, 308, 331
  – aircraft   147, **154**
  – car   **155**
  – helicopter   **155**
  – hovercraft   **154**
  – train   **154**
  – underwater robot   **154**
– zero-angle   295
conformal transformation   **547**
conic   **586**
conics   **547**, 574, **727**
– eccentricity   574
– latus rectum   **574**
conjugate point   **599**, 601, 603–605, 610, 615, 625
conjugation   **39**, **713**

connected components   **507**
– analysis   **507**
– graph   761
– image   507
connectivity analysis   **507**
consistency, left-right check   628
constant
– Boltzmann   **397**
– Planck   **397**
constraint
– epipolar   603, 625, 651, 652
– holonomic   78, 79, **156**
– nonholonomic   132, 144, **156**, 193
– nonintegrable   **156**
– rolling   **156**
continuity   **194**
– C^0 **194**
– C^1 **193**
– C^2 **194**
contour (see boundary)   515
control
– admittance   **387**
– altitude   152
– application
  – car-like vehicle   133
  – computed torque   **382**
  – quadrotor   **151**
  – robot joint   **356**
  – series-elastic actuator   **387**
  – torque feedforward   **381**
– attitude   **152**
– compliance   **387**
– computed-torque   **382**
– feedback linearization   **385**
– feedforward   152, **367**, 368
– hierarchical   **151**
– impedance   **385**
– integral   **365**, 391
– inverse-dynamic   **382**
– loop, nested   **356**
– mobile robot   **133**
– proportional (P)   133, 136, 151, **363**
– proportional-derivative (PD)   150
– proportional-integral (PI)   **365**
– proportional-integral-derivative (PID)   152, 367
– resolved-rate motion   **335**, 352
– shared   **7**
– traded   **7**
– velocity   133, **362**, 367
– vision-based   **666**
convolution   **456**, 463
– kernel   456
– property   **457**
– window   **455**
coordinate frame   **30**, 42, 249
– acceleration   95
– axes   **24**, 42
– basis vectors   **25**, 30, 42
– body-fixed   **25**, 51, **98**, 108, 148
– inertial   95, **97**, 108
– link   **283**, 295
– noninertial   97
– orientation   **24**
– origin   **24**
– right-handed   **24**, **42**
– translation   24

– world reference   **24**
coordinate vector   **25**, 26, 30, 43
coordinates
– Cartesian   **29**, **30**, **43**
– Euclidean   **726**
– exponential   **56**, 69, 101, 334, **734**
– generalized   **77**, 281, 296
– homogeneous   **35**, **726**
– image-plane   673
– joint   330
– nonhomogeneous   **35**
– normalized   673
– Plücker   **717**
Coriolis
– acceleration   **98**, 122
– coupling   370, 375
– matrix   **375**
corner detector (see point feature)   531
corner point (see point feature)   525
Cornu spiral   **200**
correction, perspective   654
correlation   **456**
correspondence   595, 635, 643, 651
– inliers   **606**
– outliers   **606**
– point   605, 625, 652
– point feature   **595**, 678
– problem   594
– putative   607
cost map   **185**, 186
Coulomb friction   **358**, 361
covariance matrix   **217**, 220, 225, 228, 233, 234, 244, **752**
– correlation   217, 752, 757
– ellipse   **225**
– extending   **233**
– transformation   **756**, 758
cover, double   621
crack code   **516**
crawl gait   **317**
cropping   482
cross product   57, 71, **709**
CRT (cathode ray tube)   417, 425
C-space (see configuration space)   78
curvature   200, 530
– boundary   519
– image   **528**
– maximum   **192**
– path   200
– principal   **528**
cybernetics   4, **8**, 163, 164, 210
cylinder   556, 729

D

D* planning   **185**
D_{65}
– chromaticity   419
– white   418, 419, 427
d'Alembert, force   **97**
Daltonism   407
damped inverse   **342**, 745
data association   215, 216, **231**, 595, 605, 607
DCM (Direction Cosine Matrix)   **68**
dead reckoning   **214**, 218
declination
– angle   115, **116**

– magnetic   **116**
decomposition   614
– camera   562
– essential matrix   **603**
– homography matrix   **612**
– image   484
– plane   614
– QR   **713**
– RQ-   552
– singular value   642, **715**, 742
– spectral   **714**
deep learning   394, **505**
– CamVid dataset   **505**
– Faster R-CNN   520
– Indoor Object Detection dataset   **519**
– object detection   **519**
– ResNet-18   505
– ResNet-50   **505**
– SSD   520
– YOLO   519, 520
degrees of freedom   51, **78**, 101, 147, **154**, 156, 281, 304, 313, 332, 338, 341, 343, 672, 690
Denavit-Hartenberg
– constraints   295, 320
– modified notation   **321**
– notation   **295**, 322
– parameters   291, 296, 320
– standard vs modified   **322**
dense stereo, see stereo   **623**
depth of field   **547**
derivative
– SO(3)   **88**, **89**
– Euler angles   **334**
– exponential coordinates   **335**
– of Gaussian   465, 528
– orientation   **88**
– pose   **89**
– quaternion   **89**
– roll-pitch-yaw angles   152
Derivative of Gaussian kernel   463
derotation, optical flow   683
descriptor   **532**
detector
– Harris   595
– SIFT   597
– SURF   595
– zero crossing   467
determinant   529, **714**
– of Hessian   529
dichromat   **403**
dichromatic reflection   **424**
difference of Gaussian (DoG)   **467**
differential
– kinematics   **330**
– steering   **142**
diopter (see also focal length)   **545**
dip angle   **116**
Dirac function   396, 429
direct linear transform   588
direction cosine matrix (DCM)   **68**
disparity   **625**
– image   626
displacement, spatial   **94**
distance
– city-block   **180**
– Euclidean   **180**, 181

– Hamming **596**
– L_1 **180**, **709**
– L_2 **180**, **709**
– Mahalanobis **231**, **713**, 752
– Manhattan **180**, 181, **709**
– orientation **69**
– p-norm **709**
– transform 180, 186, 189, 481, **482**
– triangle inequality **70**
distortion
– barrel 556
– hard-iron **118**
– keystone 644
– lens 487, **556**, 632
– perspective 594, 644
– pincushion 556
– radial 556
– soft-iron **118**
– tangential 556
distribution
– bimodal 498
– Gaussian **751**, 754, 755, 757
– non-Gaussian 758
– Poisson 442
– von Mises **219**
– χ^2 753
DoG (difference of Gaussian) **467**
dot
– pattern 637
– product 462, **709**
double cover 621
double mapping **54**, 58, 70
drag, see also friction
– aerodynamic 149
– hydrodynamic 147
drone **147**
dual
– number **66**
– quaternion **66**
Dubins path **193**
dynamic range **443**
dynamics
– error 382, 385
– forward 150, 356, **378**
– inverse **370**
– quadrotor **152**
– rigid-body **95**, **370**, **380**

E

Earth
– diameter **111**
– gravity **111**
– magnetic field **115**
– shape **111**
east-north-up (ENU) **108**
eccentricity 574, **721**
edge
– detector 465, 473, 489
– graph, of 746, **760**
– preserving filter **473**
edgels **515**
EGNOS 218
eigenvalue 54, 225, 342, 377, 378, 512, 527, 528, 639, **712**
eigenvector 54, 512, 639, **712**
EISPACK 13

EKF SLAM **234**
elasticity
– joint **387**
– link 388
elementary transformation sequence **278**, 296, 320, 322
ellipse 546, 585, **719**, 727, 752
– area **225**
– axes **721**
– – principal 679
– canonical 720
– confidence 234, 237
– covariance 234, 753
– drawing **723**
– eccentricity **721**
– equivalent **512**, 514, 720
– error **225**, 230
– fitting to
– – perimeter **725**
– – points **723**
– force 348
– inertia of **724**
– major axis **720**, 722
– minor axis **720**
– orientation **722**
– polynomial form **721**
– radii **720**
– uncertainty 224
ellipsoid 585, **719**, 729, 752
– acceleration 377
– equivalent **639**
– force **348**
– principal axes 340
– velocity **340**
– volume 342, 351
– wrench **348**
encoder
– joint 361, 362
– wheel 246
end effector **276**
– coordinate frame 333
– force 348
– torque 348
– velocity 331
– wrench 347
ENU (east-north-up) **108**
Eötvös, effect **122**
ephemeris **216**
epipolar
– -aligned image **631**
– constraint 625, 651, 652
– geometry **599**
– line **599**, 601, 604, 605, 608, 614, 624
– plane **599**
epipole 600, 602
equal-energy white 420
equation
– differential 72
– ellipse 679, **719**
– motion, of
– – bicycle **132**
– – quadrotor 150
– – rigid-body **369**, 378
– – unicycle **143**
– optical flow **675**
– Planck radiation 397
– prediction 233

– Rodrigues' **55**, **57**, 74
– rotational motion (Euler's) **95**, 150
– sparse nonlinear 622
– warp 486
equations
– linear
  – homogeneous **742**
  – nonhomogeneous **742**
– nonlinear 743
– over determined **343**
– sparse nonlinear **746**
– under determined **343**
equiangular mirror **569**
equivalence principle **97**
equivalent
– ellipse 720
– ellipsoid **639**
error 263
– cumulative 257
– ellipse **225**
– landmark 244
– reprojection 622
essential matrix **603**, 604, 612, 615, 616
estimation
– camera pose 654, 669
– fundamental matrix **604**, 605
– homography 614
– Monte-Carlo 221, **239**, 268
– pose **562**
– Structure and Motion (SaM) **619**, 688
– vehicle pose **217**
estimator
– maximum likelihood 746
– robust 746
Euclidean
– coordinates **726**
– distance **29**, 181, **761**
– geometry **29**, **716**
– homography 646
– length **709**
– plane 29, **726**
– point **726**
– space 77, **729**
– transformation **730**
Euler
– acceleration **98**
– angles **47**, 334
  – rate of change **334**
  – singularity 49
– parameters 56, **58**
– -Rodrigues formula **56**
– rotation theorem 47
– rotational motion equation 95, 150, 370
– spiral **200**
– vector **56**, 96, **734**
EV (exposure value) **442**
EXIF file format **440**, 646
explicit complementary filter **120**
exponential
– coordinates 57, 69, 334, 335
  – rate of change **335**
– mapping **71**, 73
– matrix **33**, **56**, **62**, 71, 74
– product of (PoE) 323
exposure 470
– interval **442**

– time 441, 442
– value **442**
– value (EV) 443
extended Kalman filter (EKF), see Kalman Filter 221
exteroception **8**
extromission theory **396**
eye
– evolution of **11**
– -in-hand visual servo, see visual servo **666**
– spots **11**
– tristimulus values 427

F

FAST detector 597
Faster R-CNN network **520**
FastSLAM **244**
feature **494**
– blob 514
– depth 681
– detector 480, 532
– Harris corner 528
– image 563
– line (see line feature) **521**
– map 229, 237
– point (see point feature) **525**, 531
– region (see region feature) **496**
– sensitivity matrix, see Jacobian, image 673
– vector **496**
feedforward control 152, 368, 380
fibre-optic gyroscope (FOG) 109
fictitious force **97**, 112
fiducial marker 126, 231, **581**
field
– geomagnetic 114
– magnetic, flux density 117
– of view **553**, 566, 568, 569, 577, 578
– robot **5**, **128**
– vector **739**
file type
– Collada 294
– g2o 259
– image 437, 440
– JFIF 426
– PLY 638
– STL 294
– toro 260
– URDF 293
– video 443
fill factor **442**, 555
filter
– anti-aliasing **483**, 486, 488
– bandpass **464**
– Bayer 405
– box **457**
– coefficient 456
– color 405
– complementary **120**
– edge preserving 473
– extended Kalman (EKF) 120, **757**
– high-pass 464
– Kalman 121, 221, 228, 232, 238, 239, 244, 267, 270, **755**
– low-pass 464, 483, 488
– median **473**, 489
– particle **239**, 240, 241, 243, 244
– separable **458**

– spatial **456**
fisheye lens **566**
flux
– density, magnetic **116**
– luminous 402, 406
– visual 396
f-number **442**, **547**
focal
– length **545**
– point **545**
FOG (fibre-optic gyroscope) 109
follow the carrot **138**
force 95, **346**
– apparant **97**
– d'Alembert **97**
– ellipsoid **348**, 377
– fictitious **97**, 112
– inertial **97**
– pseudo **97**
foreshortening **547**, 644, 645
forward
– dynamics **378**
– kinematics **277**, 280, 323, 332
Fourier transform 519
fovea **404**
frame (see coordinate frame) **24**
FREAK descriptor **596**, 597
Freeman chain code **515**, 540
Fresnel
– integrals 201
– reflection **424**
friction **358**
– aerodynamic 149
– Coulomb **358**, 361
– hydrodynamic 147
– Stribeck **358**
– viscous **358**, 361
front end, SLAM **263**
fronto-parallel 547
– view 645
fronto-parallel view 670
frustum **553**
function
– Cauchy-Lorentz 639
– Dirac 429
– observation 232
– plenoptic **579**
– point spread 547
– probability density (PDF) **216**, 239, **751**
– signed distance 482
fundamental matrix **601**, 604
– estimation **604**

G

G protein-coupled receptors (GPCRs) 403
gait
– crawl 317
– wave 317
Galileo (satellite navigation system) **218**
gamma **425**
– compression **425**
– correction **425**
– decoding **425**, 426, 451
– decompression **425**
– encoding **425**, 451, 488

– sRGB 426, 451
gamma encoding 427, 451
ganglion layer 404
gantry robot **276**
Gaussian
– distribution 754
– function **457**, 458, 464, **751**
– kernel 483, 527
– multivariate **752**
– noise 560, 563, 755, 757
– probability 752
– properties 458
– random variables **751**
– smoothing 527
gearbox 360–362
– ratio 360
generalized
– coordinates **77**, 281, 296
– force **347**, 348, 349, **370**
– matrix inverse **714**
– velocity 339
– Voronoi diagram 480
generator matrix **733**, 735
geomagnetic field 114
geometric
– distortion **556**
– invariant 730
– Jacobian **332**
– transformation **731**
  – affine 729
  – homothety **729**
  – reflection 729
  – rotation 729
  – scaling 729
  – shear mapping 729
  – translation 729
geometry
– 3D reconstruction 594, **629**
– analytic **29**
– Cartesian **29**
– epipolar **599**
– Euclidean **29**, **716**
– steering **131**
Gestalt principle **508**
gimbal **50**, 51, 109, 118, 300
– lock **51**, 304, 313, 338
Global Positioning System (GPS) 151, 214, **218**, 232
– differential 218
– multi-pathing 218
– RTK **218**
– selective availability **218**
global shutter camera 442
GLONASS (satellite navigation system) **218**
GNSS, see Global Positioning System **218**
goal seeking behavior 166
good features to track **528**
GPCRs (G protein-coupled receptors) 403
GPS, see Global Positioning System **218**
GPU 626
gradient 462, 597
– descent **744**, 745
– edge 463
– image 527, 594
– intensity 499, 521
graph **169**, 508, 620, **760**
– average degree **172**

– components **172**, **761**
– connected 172
– directed 27, **168**, **760**
– edge 168, **760**
– embedded **168**, **760**
– node 168, **760**
– node degree 172
– non-embedded **168**
– path finding **762**
– pose **27**
– search
　– A* **177**
　– admissible heuristic **178**
　– breadth-first (BFS) **173**
　– closed set **173**
　– cost to come **174**
　– cost to go **174**, 178
　– expansion **174**
　– explored set **173**
　– frontier set **173**
　– heuristic distance **177**
　– open set **173**
　– uniform cost (UCS) **174**
– theory **170**
– topological **168**
– undirected **168**, **760**
Graphical Processing Unit (GPU) 505, vii
grassfire planner 184
Grassmann's laws **409**
gravity
– acceleration **111**, 114, 149, 359
– compensation 152, **367**, **384**
– load 360, 367, 370, **372**, 379
– Newtonian mechanics, in 97
gray value 436, **438**, 447–450
great circle **104**
ground effect **149**
group **27**, 708, 711, 732
– abelian 27
– homographies **609**
– identity element **733**
– inverse **733**
– Lie **733**
– operator 93, **733**
– orthogonal 31, **711**
– **SE**(2) **36**
– **SE**(3) **61**
– **SO**(2) **31**
– **SO**(3) **45**
– special orthogonal **711**
gyroscope 51, **108**, 118, 132, 218
– fibre-optic (FOG) 109
– ring-laser (RLG) 109
– strapdown **109**
– triaxial **109**, 118

H

Hall effect **116**
– sensor **116**
Hamming distance **596**
hard-iron distortion **118**
Harris corner feature **528**, 531, 597
heading
– angle **117**, **130**, 136, 152
– control 136, 152

– rate (see yaw rate) 132
helicoidal
– interpolation **77**
– motion 63
helicopter 156
Hessian 737, **738**, 744
– approximate 739, 745
– determinant 529
– matrix **529**, 623
– tensor **739**, 743
heuristic distance **177**
higher-pair joint **277**
histogram 452, 498, 530, 597
– cumulative 240
– equalization **449**
– image **447**, 449
– normalization **449**, 488
– peak finding **448**
hit and miss transform **479**
HOG descriptor 597
holonomic constraint 78, 79
homogeneous
– conic **727**, 729
– coordinates 35, 61, 72, 717, **726**
– line, 2D **726**, 728
– plane **728**
– point **726**, 728
– quadric **729**
– transformation 612, **726**
– transformation, 2D **35**, **36**
– transformation, 3D **60**, 61
homography **609**, 610, 612, 614, 632, 644, 645, 648
– estimation 614
– Euclidean **612**, 646
– matrix 610
– planar **583**, 609
– plane-induced 609
– projective **612**, 645
homothety **729**
Hough transform **523**
hovercraft 154–156
Huber loss function 746
hue 410, **415**
human eye 396, **402**
– accommodation **545**
– cone cell **402**, 403, 404
　– spectral response **403**
– dynamic range **443**
– fovea **404**
– opponent color model **404**
– photopic response **401**
– retina 402
– rod cell **402**, 443
– scotopic response **401**
– theory of vision 557
– trichromatic color model **404**
humanoid robot **5**, 6
hybrid trajectory **100**
hydrodynamic friction 147
hyperbola 727
hyperboloid 569, 585, 729
hysteresis threshold **465**

I

IBVS, see visual servo **670**

ICP (iterated closest point)   269
– algorithm   269
ICP (iterative closest point)   651
ICR (Instantaneous Center of Rotation)   143
ideal
– line   **554**, **726**
– point   **726**
illuminance   **406**, 421
illumination, infrared   636
image
– anaglyph   **634**
– binary   **449**
– color   440
– compression   436, 440, 528
– connected components   **507**
– coordinate
  – canonical   **546**
  – normalized   **546**
  – retinal   **546**
– cropping   482
– curvature   **528**, 529
– data type
  – double   438
  – uint8   436
– decimation   **483**
– disparity   626
– distance transform   180
– edge detector   465
– epipolar-aligned   631
– feature (see feature)   **494**
– file format   **437**, 440
– from
  – camera   **441**
  – code   **445**
  – video   **441**
  – webcam   **444**
– gradient   526, 527
– histogram   447
– Jacobian   **673**, 674, 681
– matching   **535**
– moments   **511**
– Mona Lisa   473
– monochromatic   **438**
– mosaicing   647
– noise   **442**
  – impulse   473
  – salt and pepper noise   473
  – shot   442
  – thermal   **442**
– perspective   451
– plane   **545**, 549, 726
– processing   16, 181, 188, **436**
– pyramid   **484**
– rectification   **631**
– region   507
– resizing   483
– retrieval   **535**
– segmentation   **496**
– similarity   469, 472, 526
– skeletonization   189
– smoothing   **457**, 464
– sphere   **572**
– stabilization   650
– stitching   647
– subsampling   **483**
– thinning   189

– thresholding   **449**, 455, 465, **496**
– warping   485, 564, 576, 633, 644, 648
Image-Based Visual Servo, see visual servo   **670**
impulse
– noise   473
– response   456
IMU (inertial measurement unit)   52, 118
– strapdown   **118**
incandescence   **396**
inclination angle   115, **116**
Indoor Object Detection dataset   **519**
inertia
– effective   **361**, 362
– link   359
– load   361
– matrix   360, **373**, 385, 639
– moments of   **96**
– motor   361
– product of   **96**, 374, **724**
– reduction   **360**
– rotational   108, **357**
– tensor   95, **96**, 150, 371
inertia tensor   512
inertial
– force   **97**
– measurement unit (IMU)   52, **118**
– navigation system (INS)   107, 118, 150, 151
– parameters   371
– reference frame   95, **97**, 108
inflation
– obstacles   **185**
– ratio   **185**
information matrix   258, **752**
infrared
– camera   **431**, 636
– near (NIR)   **431**
– pattern   636
– radiation   **396**, 398, 399
– short-wavelength (SWIR)   **431**
inner product   58, 69, **709**
innovation   120, **228**, 229, 231, 236, 257, **756**
INS (inertial navigation system)   107
Instantaneous Center of Rotation (ICR)   **131**, 143
integral
– action   **365**
– control   **369**
– dynamics   **378**
– windup   **366**
intensity   415
– dimension   415
– edge   461, 467
– gamma encoded   426
– illuminance   421
– light   404
– linear wedge   426
– luminous   406
– ramp   **446**
– sinusoid   **446**
– surface   468
interaction matrix, see Jacobian, image   **673**
interest point (see point feature)   **525**
International Telecommunication Union (ITU)   **411**
interpolation
– helicoidal   **77**
– linear   105
– orientation   **104**

– pixel value   486
– **SE**(3)   **105**
– unit quaternion   **104**, 105
inter-reflection   425
interval   **479**
intromission theory   **396**
invariance   532
– brightness   532
– geometric   **730**
– rotation   527
– rotational   596
– scale   513, 532, 534
inverse
– dynamics   **370**
– kinematics   **297**
inverted index   **537**
ISO setting   442
isosurface   590
isotropic
– velocity   **340**, 351
– wrench   **348**
iterated
– extended Kalman filter (IEKF)   **270**
iterative closest point (ICP)   **642**, 651

J

Jacobian   **331**, 737
– analytical   **333**, 334, 384
– damped inverse   342
– derivative   **383**
– end-effector coordinate frame   **333**
– geometric   **332**
– image   **673**, 675
  – point feature   **673**, 674, 679, 680
– insertion   233, 236
– manipulability   **339**
– manipulator   **330**, 332, 370
– matrix   223, 258, 315, 330–332, **739**
– numerical   **739**
– overactuated robot   **345**
– pseudoinverse   343, 345
– singularity   **338**, **342**
– symbolic   **739**
– transpose   349
– underactuated robot   **343**
jello effect   **442**
jerk   **98**, **99**
JFIF file format   426
joint
– actuator   **357**
– angles   **281**
– configuration   **281**
– control, independent   **356**
– coordinates   **281**
– dynamics   **362**
– elasticity   **387**
– higher pair   **277**
– lower pair   **277**
– modeling   **361**
– prismatic   **277**, 281, 282
– revolute   **277**, 281
– space   309
  – force   **347**
  – generalized torque   **347**, 384
  – velocity   **330**, 331

– structure   **281**
– transmission   356, 360, **387**
– velocity, see joint space velocity   **330**
Joseph form   **756**
JPEG file format   **440**

K

Kalman filter   121, **221**, 222, 230, 232, 244, 755
– extended (EKF)   120, 221, 222, 238, **757**
– gain   **229**, **236**, **756**
– innovation   **756**
– iterated extended   **270**, 759
– Joseph form   **756**
– motion model   754
– prediction   222, **755**
– process
  – model   754
  – noise   **754**
– sensor
  – model   754
  – noise   **755**
– unscented   270, 759
– update   228, **756**
kd-tree   598, 637
kernel   456
– convolution   456, 462, 468, 474
– density function   **268**
– difference of Gaussian (DoG)   **467**
– Gaussian   **457**, 467, 483, 527
– Laplacian   466
  – of Gaussian (LoG)   467, 531
– Marr-Hildreth   **467**
– Mexican hat   **467**
– separable   **458**
– smoothing   457
– Sobel   462, 489
– top hat   **460**
key point (see point feature)   **525**
keyframe   **623**
keystone distortion   **644**, 645
kidnapped robot   243
Kinect
– 360 camera   637
– Azure   636
kinematic chain   **277**
kinematics
– differential   **330**
– forward   **277**, 323, 330
– inverse   **297**
  – analytical   301
  – closed-form   298, 301
  – numerical   300, 304
– manipulator   330
– velocity   **330**
Klein quadric   717, **728**
k-means
– algorithm   **502**
– clustering   504
k-means clustering   535
Königsberg bridge problem   **170**

L

L*a*b* color space   427
L*u*v* color space   **417**
Lambertian reflection   424, 425, 572

landmark   **215**, 244, 251, 620
– ambiguity   231
– artificial   231
– association   215
– identity   231
lane marker   **695**
LAPACK   13
Laplacian
– kernel   **466**
– of Gaussian (LoG)   **467**
laser rangefinder   **245**
lateral motion   **131**, 132
latus rectum   **574**
law
– Asimov's   4
– Beer's   **399**, 423
– center of gravity (color)   411, 433
– Moore's   **vii**
– Newton's
    – first   97
    – second   97, 112, 149, 370, 385, 391
– power   426
– Stefan-Boltzman   397
– Wien displacement   **397**
LCD (liquid crystal display)   407
least squares problem
– linear   725
– non-linear   258, 621, **745**
left generalized inverse   742
left-right consistency check   **628**
lens   **544**, 545
– aberration
    – chromatic   **556**
    – spherical   556
– anamorphic   **445**
– aperture   **547**
– cardinal points   **560**
– compound   **545**, 560
– diopter   **545**
– distortion   **556**, 557, 587, 607
    – barrel   556
    – geometric   556
    – parameters   558, 560
    – pincushion   556
    – radial   **556**, 566
    – tangential   556
– entrance pupil   **560**
– equation   **545**
– fisheye   **566**, 568, 571, 574
– f-number   **547**
– focal length   **545**
– focal point   **545**, **578**
– objective   **544**
– simple   544
– stop   **547**
– telecentric   586
– thin   **544**
– wide angle   **566**
– zoom   **553**
lenslet array   **580**, 581
Levenberg-Marquardt algorithm   350, **745**
lidar   **245**, 256
– map building   **249**
– noise   249
– odometry   **246**
– remission   **246**

Lie
– algebra   33, 40, 57, 63, 74, 75, 732, **733**
– group   34, 71, **732**
    – generator   **74**
light
– absorption   **423**
– monochromatic   **396**, **407**
– solar spectrum   **399**
– structured   635, 636
– visible   **396**
light field camera   **579**
line
– $\rho - \theta$ form   **522**
– feature   **521**
– fronto-parallel   547
– ideal   **554**
– in 2D   **716**
    – direction   **716**, 727
    – distance to point   **727**
    – equation of a point   **727**
    – intersection point   **727**
    – joining two points   **716**, 727
    – normal   **716**, 727
    – polar form   **716**
– in 3D   **717**
    – direction   **717**
    – distance between   **718**
    – distance to point   **718**
    – intersection of planes   **719**
    – intersection point   **718**
    – joining two points   **717**
    – moment   **717**
    – normal   **717**
    – skew-symmetric matrix   **717**
    – transformation   **719**
– of no motion   **131**
linearization   270, 381, 382, 385, 673, 686, **737**, 758
link   **283**, 358
– collision shape   **294**
– inertia   **362**
– inertial parameters   **371**, 384
– shape   **294**
LINPACK   13
liquid crystal display (LCD)   407
localization   **214**, **234**, 249, 650
– CML (concurrent mapping and localization)   **235**
– error   216
– lidar-based   **251**
– Monte-Carlo, sequential   **239**
– problem   **216**, 217
– SLAM (simultaneous localization and mapping)   **235**
LoG kernel   **467**, 531
logarithm
– matrix   **34**, **39**, 56, 62
– principal   **34**, 56
longitude
– problem   **216**, 271
longitudinal motion   **131**
lookup table   **451**
loop closure   **257**
LORAN   **218**
lower-pair joint   **277**
lumen   401
luminance   **406**, 410, 413, 425
luminous
– flux   **406**

– intensity **406**
LUT **451**
LWIR **431**

M

magnetic
– compass **116**
– declination **116**
– dip angle **116**
– distortion
  – hard-iron **118**
  – soft-iron **118**
– flux
  – density **115**, **116**, 117
  – line **114**
– inclination **116**
– north **116**
– pole **114**
magnetometer **116**, 118
– triaxial **117**, 118
Mahalanobis distance 231, **713**, 752
maneuver 129, **153**, 155, 156
Manhattan distance 180, 181, 709
manifold 717, 728, **732**, 733, 734
manipulability **314**, **339**, 341, 342, 351, 679
– dynamic **377**, 378
manipulator **276**
– arm 282
– dexterity **314**
– dynamics **356**
– gantry **276**, 282
– high-speed 387
– Jacobian **332**, 347, 370
– joint limit **290**, 303, 304, 306, 308
– joint, see joint 276
– kinematics 330
– manipulability **314**, 341
– maximum payload 375
– number of joints **282**, **295**
– number of links **282**, **295**
– overactuated 79, 308
– Panda 308
– parallel link **276**
– planar 298
– polar-coordinate 281
– pose 249
– PUMA 560 **284**, 321, 362
– redundant 308, 326, **345**, 346
– SCARA **276**, 306
– serial-link 277
– singularity 304, 313
– Stanford arm **282**
– trajectory 244
– underactuated 281
– wrist singularity 313
– zero joint angle configuration 295
manufacturing robot 3, **6**
map 231, 244
– graph-based **170**
– grid-based **170**
– landmark-based **226**
– localization, using **226**
mapping **234**
– concurrent mapping and localization (CML) 235
– exponential 73

– lidar-based **249**
– simultaneous localization and mapping (SLAM) 235
Marr-Hildreth operator **467**
mass
– center of 359
– matrix **373**
matching
– 3D-2D **651**
– 3D-3D 651
– function, color 408, 427
mathematical
– graph **168**
– morphology 189, **474**
  – closing **477**, 503
  – dilation 185, **476**
  – erosion **476**
  – hit and miss **479**, 480
  – opening **476**, 502
MATLAB **13**
matrix
– addition **709**
– adjoint 91, 96, 323, 350, 711, 719, **736**
– adjugate **711**, 729
– anti-symmetric **33**, **57**, 711
– augmented skew-symmetric, 2D **40**
– augmented skew-symmetric, 3D **63**, 72, 74
– block **710**
– camera **548**, 551, 552, 558, 560, 586, 603, 623
– column space **714**
– condition number 339, 679, 680, **715**
– covariance 217, 220, 223, 225, 228, 230, 232–234, 240, 244, **752**
– damped inverse **342**, 745
– decomposition
  – Cholesky **713**
  – eigen **714**
  – QR **713**
  – singular value **715**, 742
  – spectral **714**
  – square root **713**
– determinant 70, 339, 342, 712, **714**
– diagonalization **714**
– eigenvalue **712**
– eigenvector **712**
– element-wise multiplication **709**
– essential **603**, 604, 612, 615, 652
– exponential **33**, 34, **40**, **56**, **62**, 71, 72, 74
– feature sensitivity **673**
– generalized inverse 342, **714**
– generator **74**, **733**, 735
– Hadamard product **709**
– Hessian **529**, 623, **738**
– homogeneous transformation **36**, 71
– identity 92, **711**
– information 258, **752**
– interaction **673**
– inverse **711**
– Jacobian 258, 330, **331**, 673, **739**
– kernel **715**
– linear transformation **712**
– logarithm **34**
– Moore-Penrose pseudoinverse 342, 343, 345, 678–680, **714**, **742**
– multiplication **709**
– negative-definite **713**, 739
– normal **712**
– normalization **70**
– null space 602, 677, **715**

– null-space 346
– orthogonal 31, 44, 45, **711**
– orthonormal **711**
– positive-definite **713**, 739, 747
– positive-semi-definite **713**
– product **709**
– quadratic form **713**
– rank **714**
– rotation 32, 33, 47, 56, 59, 70, 71, 334, 743
  – 2D **31**
  – 3D **45**
  – derivative **88**
  – reading 47
– row space **714**
– Schur complement **750**
– similar **713**
– skew-symmetric 55, 88, 92, **711**, 734, 735
  – 2D **33**, 40
  – 3D 56, **57**, 71
– sparse **750**
– square root **713**, 723
– symmetric **711**
– trace **712**
– transpose **710**
MAV (Micro Air Vehicle) **147**
maximally stable extremal regions (MSER) **500**, 501
maximum likelihood estimator **746**
mecanum wheel **145**
mechanical pole **362**
median filter **473**
MEMS (micro-electro-mechanical system) **109**
metamers **406**
Mexican hat kernel **467**
micro air vehicle (MAV) **147**
micro-electro-mechanical system (MEMS) **109**
microlens array **580**
minimization, see optimization 744
minimum
– jerk trajectory **103**
– norm solution 308, 314, 345
Minkowski
– addition **476**
– subtraction **476**
mirror
– catadioptric camera 569
– concave 566
– conical 571
– elliptical 569
– equiangular **569**
– hyberbolic 569
– parabolic 569
– spherical 569, 571
missing parts problem **627**
mixed pixel problem **472**, **629**
mobile robot **4**
– aerial **147**
– car-like **130**
– control **133**
– differentially-steered **142**
– localization and mapping **214**
– navigation **162**
– omnidirectional **144**
– SLAM **214**
model
– bicycle **130**, 139, 144
– central camera 545

– imaging 576
– kinematic 139, 147, 200
– motion 192, 203
– quadrotor **148**
– rigid-body dynamics 147
– unicycle 139, **142**
model-based control 380
modified Denavit-Hartenberg **321**
moment
– central **512**, 724
– of inertia **96**, **512**, 724
– of Plücker line **717**
– of twist 41
– principal **512**
moments **511**
momentum, angular **95**, 108
monochromatic
– image **438**
– light **396**, **407**
Monte-Carlo
– estimation **239**, 268
– Localization **251**
– localization **239**
Moore-Penrose pseudoinverse **714**
Moore's law **vii**
Moravec interest operator **526**
morphology (see mathematical morphology) 474
mosaicing **647**
motion
– ξ **22**
– algebra **26**
– axis 101
– blur 443
– Cartesian **105**, 308
– control, resolved-rate 335, 337
– detection **453**, 454
– helicoidal 63
– joint-space **308**
– lateral **131**, 144
– longitudinal **131**
– model
  – bicycle **130**, 192
  – unicycle **142**
– multi-axis **101**
– null **23**, 26
– omnidirectional **144**, 167, 180, 192
– perceptibility **679**
– planning (see also path planning) **163**
– rigid-body 94, 732
– rotational 72, 95
– screw 64
– segment 102
– sickness 113
– straight-line **311**
– task-space 336
– translational 41, 64, 72, 95
motor 361, 362
– limit 365
  – power 365
  – speed 100, 365
  – torque **365**
– servo 356
– stepper 356
– torque 357, 360
– torque constant **357**, 365, 367
MSER

– descriptor   597
MSER (maximally stable extremal regions)   501
multi-path effect   **218**
multirotor, see quadrotor   **147**

N

NaN   630
nautical
– angles   **49**
– chronometer   **216**
navigation   157, **162**
– chart   215
– dead reckoning   **214**
– inertial   88, 94, **107**, 118, 151
– landmark-based   214, **215**, **226**
– map-based   163, **168**
– marine   234
– principles   214
– radio   107, **218**
– reactive   162, **163**
– spacecraft   51, 109
NCC   **469**
NDT (see Normal Distributions Transform)   248
near infrared (NIR)   **431**
NED (north-east-down)   **108**
nested control loop   **356**
Newton
– -Euler method   370, 389, 390
– first law   **97**
– method   **744**
– -Raphson method   **744**
– second law   **95**, 97, 112, 149, 370, 385, 391
Newtonian telescope   566
NIR   **431**
node   263
noise   463
– dark current   **442**
– Gaussian   221, **225**, 232, 560, 563
– high-frequency   463
– image   **442**, 488
– impulse   473, 489
– odometry   219, 222
– pixel   478
– process   **754**
– reduction   463, 527
– removal   **477**
– salt and pepper   473
– sensor   229, 239, **754**
– shot   442
– thermal   **442**
nomenclature   **xix**
noncentral imaging   **571**
nonholonomic   129
– constraint   132, 144, **156**
– system   **156**
nonholonomic constraint   653
nonintegrable constraint   **156**
nonlocal maxima suppression   **465**, **474**, 523, 528, 765
Normal Distributions Transform   248
normal matrix   **712**
normalization   447
– **SE**(3) matrix   70
– histogram   449, 488
– homogeneous transformation   669
– rotation matrix   **70**

– unit dual quaternion   66
– unit quaternion   **71**
normalized
– cross correlation (see NCC)   **469**
– image coordinate   487, **546**, 603, 612
– image coordinates   673
north
– magnetic   **116**
– true   **116**
north-east-down (NED)   **108**
notation   13, **xix**
null space   **715**, 742
nullity   **715**
number
– dual   **66**
– hypercomplex   **58**
numerical optimization   300

O

object detection
– deep learning   **519**
– segmentation   **506**
objective lens   **544**
obstacle inflation   **185**
occlusion   621, 627–629, 633, 654
occupancy grid   166, **180**, **181**, 249, 250
– binary   166, 170, 180
– probabilistic   180, 244
OCR (see optical character recognition)   **534**
octave   **531**
odometer   **218**
odometry
– lidar   246
– visual   218, **650**
– wheel   **218**, 219, 256
OLED   407
omnidirectional
– camera   **553**, 579
– motion   130, **144**, 167, 180, 192
– wheel   145
OmniSTAR   218
operational space   **77**
– dynamics   **382**, 384
operator
– ∘   **58**
– ⊖   **134**
– ⊖   **23**, 25
– ⊕   **22**, 25
– ·   **25**, **58**
– \   742, 744
– binary arithmetic   **451**
– Canny edge   465, **466**
– closing   **477**
– correlation   **456**
– Difference of Gaussian   466
– dilation   **476**
– edge   466
– Gaussian   465, 531
– group   **733**
– Harris   597
– interest point   526, 540
– Laplacian   465, **466**
– Marr-Hildreth   **467**
– MATLAB, overloaded   76
– monadic   451, 496

– Sobel edge   542
– spatial   455, 474
– spatial displacement   94
opponent color
– space   417
– theory   **404**
opsins   **403**
optical
– axis   54, **544**, 549
– character recognition (OCR)   **534**
– flow   651, **675**, 682, 683
   – derotation   683
   – equation   **675**
   – field   **676**
optimization
– algorithm   350
– bundle adjustment   619
– Gauss-Newton   **745**
– gradient descent   **744**
– Levenberg-Marquardt (LM)   **350**
– Newton-Raphson   **744**
– Newton's method   **744**
– nonlinear   588, **744**
   – least squares   745
   – sparse least squares   746
– numerical   300
– pose graph   258, **259**, 261, 268, 269
ORB, descriptor   597
order
– filter   **473**
orientation   44
– camera   552, 616
– derivative   **88**
– error   118, 120
– estimation   **109**, 114, 120
– feature   596
– in 3D   **43**
– interpolation   **104**
– rate of change   152
– vector   **53**
– vehicle   **130**, 141
orthogonal
– group   **711**
– matrix   **711**, 715
– special group in 2D   **31**
– special group, 3D   **45**
orthonormal matrix (see orthogonal matrix)   **711**
orthophoto   **649**
Otsu threshold   **498**
outer product   **709**
overactuated manipulator   **308**, **345**

P

panoramic camera   **553**
parabola   727
paraboloid   569, 585, 729, 765
parallel-link manipulator   **276**
particle filter   **239**, 244
– resampling   240
passive transformation   **67**
passive, alias or extrinsic transformation   **67**
path
– admissible   **180**
– curvature   **192**, 200
– feasible   **192**

– following   **137**, 191, 192
– planning   137, **163**
   – complete   **180**
   – complexity   **180**
   – D*   **185**
   – distance transform   **180**, 207
   – Dubins   **193**
   – global   **208**
   – graph-based   **170**
   – grid-based   **180**
   – lattice   **197**
   – local   **208**
   – map-based   **168**
   – optimal   **180**
   – phase   **187**
   – PRM   **189**
   – query phase   **187**
   – Reeds-Shepp   194
   – roadmap   **187**, 188
   – RRT   **203**
payload   356
– effect of   **375**
PBVS, see visual servo   **668**
peak
– finding   498
   – 1D   **763**
   – 2D   **765**
– refinement   **764**, **766**
pencil of lines   605
perceptibility of motion   **679**
perceptually uniform color space   417
perimeter
– circularity   516, 539
– length   515
– shape   **515**
perspective
– camera   **544**, 568, 573, 574, 578, 580
– correction   **644**
– distortion   472, 546, 587, 594, **644**
– foreshortening   **547**, 644
– image   **545**, 566, 571
– projection   **546**, 548, 551, 554, 577, 587, 594, 672, 673
– weak   **587**
Perspective-n-Point (PnP) problem   **562**
photogrammetry   589, 654
photometric unit   402
photometry   **431**
photopic response   **401**
photopsin   403
photoreceptor   11, 403
photosite   **404**, **442**, 443, 550
– array   **442**, 580, 581
phototaxis   **164**
picket fence effect   **627**
pincushion distortion   **556**
pinhole camera   **544**, 547
pin-hole camera   **11**
pixel   **436**
– array, assorted   405
– classification   496
– color   **439**
– data type   **436**
– grayscale   **436**
– noise   463
– value, distribution   447
planar

– homography **609**
– transformation **731**
Planck
– constant **397**
– radiation formula **397**
Planckian source **397**
plane **719**, 728
– point-normal form **719**
– principal 552
planner, see path planning **163**
plenoptic
– camera **579**
– function **579**
Plücker
– coordinates 64, 76, 584
– line, see line in 3D 585, **716**
PnP (Perspective-n-Point) **562**
point 494
– 2D **30**
– 3D 43
– boundary **516**
– cloud 251, 270, **637**, 643
　　– colored **637**
　　– normal estimation **639**
　　– organized **637**
　　– plane fitting **639**
　　– registration **641**
– conjugate 599, 601, 603, 604, 610, 615
– coordinate 72
– coordinate vector **25**, 35
– corresponding 605, 607, 609, 611, 614, 625, 628, 632
– depth **674**
– epipolar 608
– equation of line, 2D **726**
– Euclidean 38, **727**
– feature 532
– focal **578**, 649
– homogeneous **38**
– ideal **727**
– image plane **545**
– iterative closest (ICP) **642**
– landmark 618, 621
– line intersection in 2D **727**
– principal 487, **545**, 616
– scale-space feature **531**
– spread function 547
– task space 78
– vanishing 547
– world **545**, 547, 548, 550, 552, 558, 560, 594
point feature 525, **527**, 531, 532, 595–597, 607, 613, 614, 632
– correspondence **595**
– descriptor 532
　　– ACF 541
　　– BRIEF 541
　　– BRISK 541, 597
　　– CenSurE 597
　　– FREAK **596**, 597
　　– HOG 541, 597
　　– ORB 541, 597
　　– SURF 532
　　– VLAD 541
– detector 526, 530, 540, 594
　　– AGAST 597
　　– corner 531
　　– FAST 541, 597
　　– Harris **528**, 541, 597

– Morovec **526**
– Noble **528**
– Plessey **528**
– Shi-Tomasi **528**, 597
– SIFT 541
– SURF 532, 541
– Harris 595
　　– detector 597
– orientation 596
– strength 527, 528, 595
– support region 596
– SURF 535
Poisson distribution 442
pole
– magnetic **115**
– mechanical **362**
– rotational **41**
– twist **65**
polynomial
– ellipse 721
– quintic **99**
pose **22**
– abstact ξ **22**
– concrete
　　– as a 2D twist **42**
　　– as a 3D twist **65**
　　– as a unit dual quaternion **66**
　　– as a unit quaternion **60**
　　– as a vector-quaternion pair **66**
　　– as an **SO**(2) matrix **32**, 36
　　– as an **SO**(3) matrix **46**, 62
– end effector 276, 330
– estimation 113, **562**, 581, 588
– graph **27**, 67, 83, 256, 257, 260, 263, 278
– in 2D **34**
– in 3D **60**
– interpolation **105**
– rate of change 88, **89**
– trajectory 98
Position-Based Visual Servoing (PBVS) **668**
positive-definite matrix **713**
posterior probability **221**
posterization **451**
power
– distribution, spectral (SPD) 432
– motor limit **365**
primal sketch 470
primary
– CIE **407**, 410, 418
– color 408
principal
– axis **513**
– curvature **528**
– moment **512**
– plane 552
– point **549**, 556, 558, 568, 570, 575, 577, 616, 650
principle
– equivalence **97**
– Gestalt 508
prior probability **221**
prismatic joint **277**, 281
Probabilistic Roadmaps (PRM), see path planning **189**
probability 217, 221, 240
– conditional 221
– density 225, 239, 751, 752
– density function (PDF) **216**

– Gaussian   225, 231, **751**
– posterior   221
– prior   221
problem
– correspondence   594, 635
– missing part   **627**
– mixed pixel   629
process noise   219, **755**
Procrustes transformation   **730**
product
– of exponentials (PoE)   323
– of inertia   512
projection
– catadioptric   **569**
– fisheye   **566**, 568
– matrix   **548**
– orthographic   **586**
– parallel   **586**
– perspective   544
– spherical   **572**
– stereographic   **574**
– weak perspective   **587**
projective
– homography   612
– plane   **601**
– reconstruction   623
– space   **35**
– transformation   **546**, 729
proof mass   **112**
proper acceleration   **112**
proprioception   **8**
pseudo
– force   **97**
– inverse   **714**, **742**
– random number   191
PUMA 560 robot   **284**, 321, 362
pure pursuit   **137**
purple boundary   411
pyramidal decomposition   484

Q

quadrants   55, **136**
quadratic
– form   **713**
– surface   **585**
quadric   584, **585**, **729**
quadrotor   79, **147**, 154
– control   151
– dynamics   **150**
– model   **148**
quantum efficiency   **442**
quaternion   **58**
– conjugate   **58**, **60**
– dual   **66**
– dual unit   **66**
– Hamilton product   **58**
– identity   59
– inner product   **58**
– JPL   **60**, 84
– matrix form   **58**
– multiplication   **58**
– pure   **58**, 89, 93
– unit   *see* unit, quaternion
quintic polynomial   **99**
quiver plot   464

R

radial distortion   **556**
radiation
– absorption   399
– electromagnetic   396
– infrared   **396**, 398, 402
– Planck formula   397
– ultraviolet   **396**
radio navigation   107, **218**
radiometric unit   402
radius, turning   197
random
– coordinate   760
– dot pattern   637
– number   191, 240, 606, 754
  – generator   **191**
  – seed   **191**, 643
– sampling   189, 203, 240
– variable   751, 755, 757, 758
range
– measurement   245
– shadow   **246**
rank
– deficiency   338, 559, 686, **714**
– of matrix   **714**
RANSAC   695
RANSAC (Random Sampling and Consensus) algorithm   **606**, 611, 614, 640
Rao-Blackwellized SLAM   **244**
Rapidly-Exploring Random Tree (RRT)   **203**
rate of change   *see* derivative
ratio
– ambiguity   627
– aspect   494, 513
– test   **473**
raw image file   405
raxel   **581**
recognition, characters   **534**
reconstruction   629
– affine   **623**
– projective   **623**
rectification   **632**
recursive Newton-Euler   **370**
redundant manipulator   **308**, **345**
reflectance   **400**, 421, 422
reflection
– dichromatic   **424**, 431
– Fresnel   **424**
– Lambertian   424, 425, 572
– specular   249, 424, 503, **572**
reflectivity   246, 249, **400**
reflector, diffuse   424
region feature
– area   **511**
– aspect ratio   **513**
– boundary   **515**
– bounding box   **510**
– centroid   **511**
– contour   **515**
– edge   **515**
– equivalent ellipse   **512**
– inertia matrix   **512**
– invariance   **514**
– maximally stable extrema (MSER)   500, 597
– moments   **511**
– orientation   **513**

– perimeter **515**
– shape 516
region of interest 482
relative pose **22**, **24**, 27
remission **246**
replanning **185**
reprojection **618**
– error 621, 622
resampling **240**, 244, 251
resectioning **215**
ResNet-18 network **505**
resolved-rate motion control **335**
retina 402–404
– ganglion layer 404
retinal
– image coordinates **546**
– molecule 403
retinex theory **421**, 431
revolute joint **277**, 278, 280
RGBD camera **636**
rhodopsin 403
right-hand rule 30, **42**, 709
rigid body
– chain **277**
– displacement 73
– dynamics **370**
– motion **22**
– transformation **22**
ring-laser gyroscope (RLG) 109
RLG (ring-laser gyroscope) 109
RNG (see also random number generator) 191
roadmap path planning **187**, 188
robot
– aerial **147**
– definition **8**, 164, 168
– ethics **11**
– field **5**
– humanoid **5**
– manipulator *see* manipulator
– manufacturing **4**, **6**
– mobile *see* mobile robot
– service **4**
– surgical 8
– telerobot **7**
robotic vision **9**
rod cell (see human eye) **402**
Rodrigues'
– equation 74
– rotation formula **55**, 57
– vector **56**
rolling
– constraint **156**
– shutter **442**
roll-pitch-yaw angles **49**, 51, 334
– rate of change 152, 334
– singularity **51**
– XYZ order **50**
– ZYX order **49**
root finding **743**
Rossum's Universal Robots (RUR) **6**
rotation 71
– active **67**
– angle 57
– axis 41, 54, 57, **88**
– distance **69**
– Euler vector 54

– extrinsic **67**
– incremental **92**
– intrinsic **67**
– matrix 33, 487, 647
  – in 2D **31**
  – in 3D **44**eigenvalues 55
  – normalization 70
– passive **67**
– shortest distance 105
rotational pole **41**
rotor disk **149**
rotorcraft, see quadrotor **147**
row space **714**
RRT (Rapidly exploring Random Tree) **203**, 702
rule
– anti-cyclic rotation **51**
– cyclic rotation **51**
– right-hand **42**
– rotation about a vector **44**

S

saccule 113
SAD **469**, 627
salient point (see point feature) **525**
salt and pepper noise **473**
SaM (Structure and Motion Estimation) **619**
sampling
– artifacts 483
– importance **240**
– Random Sampling and Consensus (RANSAC) **606**
– Shannon-Nyquist theorem 483
– spatial 483
satellite navigation system **218**
saturation **410**, **415**
scalar field **738**
scale
– space 464, 484, 596
– spatial 464
SCARA robot **276**, 306
scene luminance 442
Schur complement **750**
scotopic response **401**
screw **63**, 73
– axis 63, **73**, 74
– motion 63, 64
– pitch 74
– theory 73
SE(2) matrix **36**
se(2) matrix **40**
SE(3) matrix **61**, 549
se(3) matrix **63**, **74**, 75
SEA (series-elastic actuator) **388**
segmentation 478
– binary 502
– color image **500**
– graph-based **508**
– grayscale image **496**
– semantic **505**
selective availability **218**
semantic
– segmentation **505**
– tag 535
semi-global matching 627
sensor
– acceleration **111**, 118

– angular velocity **108**, **109**
– bias **118**, 119, 121
– calibration 118
– CCD 442
– CMOS 442
– drift **118**
– exteroceptive **8**
– fusion **118**, 120, **232**
– gyroscopic **109**
– Hall effect **116**
– magnetic field **114**
– noise **118**, 120
– observation **227**
– proprioceptive **8**
– range and bearing **227**
– scale factor **118**
serial-link manipulator *see* manipulator
series-elastic actuator (SEA) **388**
service robots **4**
servo-mechanism **357**
SfM (Structure from Motion) **619**
shadow 429
– range **246**
– removal 428
Shannon-Nyquist theorem 483
shape 503
– circularity 516, 539
– descriptor **515**
– equivalent ellipse **512**
– fitting 540
– from boundary **515**
– object 518
– perimeter 539
shared control **7**
shift invariance **457**
Shi-Tomasi detector 597
short-wavelength infrared (SWIR) 431
shutter
– glasses 634
– global 442
– rolling **442**
SIFT **531**, 597
sigma-point filter 759
signed distance function 482
similarity
– image 472
– transformation **713**
similarity transformation 223
simultaneous localization and mapping (SLAM) **235**
singleton dimension **439**
singular
– configuration 338
– value 679, **715**
– value decomposition (SVD) **715**, 742, 743
– vector **715**
singularity **51**
– Euler angles **49**, 52
– Jacobian **338**, **342**
– motion through **313**
– representational **334**
– roll-pitch-yaw angles **51**
– wrist **304**, 313
skeletonization 189
skew-symmetric matrix 88, 585, **711**, 728, 734, 735
skid steering **144**
SLAM **235**

– back end **263**
– EKF **234**
– FastSLAM **244**
– front end **263**
– pose graph **256**, 257, 263
– Rao-Blackwellized **244**
– visual (vSLAM) **619**
slerp **104**
smooth
– function **98**
– start **686**
smoothness constraint **656**
SO(2) matrix **31**
so(2) matrix **33**, **34**
SO(3) matrix **45**
so(3) matrix **57**
Sobel kernel **462**
soft-iron distortion **118**
solar spectrum **399**
solid angle 406, **553**
SOS (standard output sensitivity) **442**
space
– affine 729
– configuration 77, **78**, 154, 308
– Euclidean 77, 729
– linear **708**
– operational **77**
– projective **35**
– resectioning 588
– task **77**, 308
– vector **708**
sparse
– matrix 750
– stereo 614, **617**, 629
spatial
– acceleration **377**
– aliasing 483, 627
– displacement **94**
– sampling rate 483
– velocity **89**, **331**, 333, 384, 673, 677
special
– Euclidean group in 2D **36**
– Euclidean group in 3D **61**
– orthogonal group in 3D **45**
– orthogonal group, 2D **31**
speckle projector **636**
spectral
– decomposition **714**
– locus 410–413, 415
– power distribution (SPD) 432
– response
  – camera 429
  – human eye 397, 403, 405, 408, 429, 430photopic 401scotopic 401
spectrum
– absorption 399, 423
– D65 standard white 427
– illumination 421
– infrared 402
– luminance 400, 405, 410, 427
– reflection 400
– solar 399
– visible 399
specular reflection 249, 424, 503, **572**
speculum, metal 572
sphere 585, 590, 729

spherical
– aberration **556**
– linear interpolation **104**
– wrist   282, 285, 291, **300**, **301**
spring
– SEA   387
– torsional   387
– virtual   **386**
SSD   **469**, 526, 535
– deep learning network   **520**
stabilization, image   650
standard output sensitivity (SOS)   442
Stanford robot arm   **282**
steering
– Ackermann   **132**
– differential   **144**
– ratio   **131**
– skid   **144**
Stefan-Boltzman law   **397**
steradian   **397**, **553**
stereo
– anaglyph   634
– baseline   **623**, 634, 654
– camera   629, 631
– dense   **623**
– display   634
– glasses
    – anaglyph   45
– lack of texture   636
– matching   627, 629, 633
– pair   623, 630, 631, 634
– perception, human   634
– sparse   **614**, 651, 655
– system   635
– triangulation   651
– vision   614, 628, 629, 635, 636, 654
stereographic projection   **574**
stereopsis   **617**
stiction   **358**
straight-line motion   **311**
strapdown
– gyroscope   **109**
– IMU   **118**
– sensor configuration   **109**
Stribeck friction   **358**
Structure
– and Motion (SaM)   688
– and Motion Estimation (SaM)   **619**
– from Motion (SfM)   **619**
structure tensor   **527**
structured light   **635**
structuring element   **474**
subpixel precision   595
subsumption architecture   166
subtraction, Minkowski   **476**
sum of
– absolute differences (SAD)   469
– squared differences(SAD)   469
support region   596
suppression, nonlocal maxima   465, 466, **474**, 523, 528
SURF feature detecto   534
surface
– geometry   425
– normal   638
– quadratic   585, **729**
– sphere   572, 574

SVD (see singular value decomposition)   **715**
Swedish wheel   145
SWIR   **431**
symbolic computation   259, 298, 299, 330
symmetric matrix   527, **711**
system
– configuration   **77**
– navigation, inertial (INS)   **107**, 118
– nonholonomic   **156**
– nonlinear   757
– reference, attitude and heading (AHRS)   **118**
– Type 0   **364**, 365
– Type 1   **365**, **368**
– Type 2   **368**
– underactuated   79, **154**
– vestibular   **109**, 113

T

Tait-Bryan angles   **49**
tangent space   **733**
tangential distortion   556
task space   **77**, 281, 308, **338**, 340, 667
– acceleration   **384**
– Coriolis matrix   **384**
– dimension   **332**, 343
– dynamics   **382**
– gravity   **384**
– inertia matrix   **384**
– pose   **384**
– velocity   **331**, **384**
taxis   **164**
Taylor series   **737**
TCP (see also tool center point)   **291**
tearing   442
telecentric lens   **586**
telerobot   **7**
temperature
– color   420, 429
– Sun   **399**
– tungsten lamp   **399**
template matching   625
tensor
– inertia   95, **96**
– structure   **527**
tetrachromat   **403**
texture
– lack of   626, 628, 629
– mapping   **576**, 631
– projector   **636**
theorem
– Brockett's   156, **688**
– Chasles'   **73**
– Euler's rotation   **47**
theory
– extromission   **396**
– intromission   **396**
– Lie group   34, **732**
– opponent color   **404**
– retinex   **421**, 431
– screw   **73**
– trichromatic   **404**
thin lens   **544**
thinning   189
thresholding   **449**, 455, 488, **496**, 498
– adaptive   539

– Bradley's method   539
– hysteresis   465
– Otsu's method   **498**
tie point   **648**
time
– of flight   636
– series data   xxiv
– varying pose   88, **98**
tone matching   **649**
tool center point   **291**
top-hat kernel   **460**
topological
– graph   **168**
– skeleton   189
torque
– disturbance   356
– feedforward control   **367**, 381
– gravity   359, **372**
– maximum   **365**
– motor   357
trace of matrix   712
traded control   **7**
train   156
trajectory   **98**, 104, 244, 308, 356, 370
– Cartesian   121, **311**
– hybrid   **100**
– image plane   686
– joint-space   309, 311
– lane-changing   133
– multi-axis   **101**
– multi-segment   **102**
– piecewise-linear   102
– polynomial   **99**
– trapezoidal   **100**
transconductance   357
transform   **61**
– distance   180, 481, 482
– post-multiplication   **67**
– pre-multiplication   **67**
transformation
– affine   **729, 730**
– conformal   **547**
– Euclidean   **730**
– geometric   **729**, 731
– homogeneous, 2D   **35**
– homogeneous, 3D   **60**
– linear   **712**
– normalization   **70**
– planar   **731**
– Procrustes   **730**
– projective   **729**
– similarity   223, **713, 730**
transmission, light   423
trapezoidal trajectory   **100**, 368
traversability   **168**
triangle inequality   70, 96, 714
triangulation   215, **594**, 615, 617, 618, 651
triaxial
– accelerometer   **113**
– gyroscope   **109**
– magnetometer   **117**
trichromat   **403**
trichromatic theory   **404**
tri-level hypothesis   **470**
tristimulus   439
tristimulus values   **405**, 406, 408, 410, 411, 413–419, 422, 426, 430

true north   **116**
Tukey biweight function   746
tumbling motion   95, 110
turning radius   **131**, 192, 197
twist   40, 69, **73**, 323, 584, 735
– axis   **63**, 73, 74
– moment   **41**, 73
– nonunit   65
– pitch   **63**, 74
– pole   **41**
– transforming   **735**
– unit   **41, 63**, 74
– – rotation, 2D   **40**
– – rotation, 3D   **63**
– – translation, 2D   **41**
– – translation, 3D   **64**
– vector   **41, 63**
– velocity   **350**

U

UAV (uncrewed aerial vehicle)   **147**
ultrasonic rangefinder   245
ultraviolet radiation   **396**, 399
uncertainty   226, 228, 230
uncrewed aerial vehicle   147
underactuated
– manipulator   **343**
– robot   **343**
– system   79, 153, **154**, 155, 156, 342
underwater imaging   424
unicycle motion model   **142**
unified imaging model   **574**
Unified Robot Description Format (URDF)   **293**
unit
– dual quaternion   **66**, 69
– photometric   402
– quaternion   **58**, 69, 621
– – computational efficiency   **58**
– – derivative   **89**
– – double mapping   **54**
– – interpolation   **104**
– – to **SO**(3) matrix   59
– – vector   **60**
– – vector part   **56**
– twist   41, **63**, 74
units
– candela   **406**
– lux   **406**
– nautical mile   **214**
– nit   **406**
– radiometric   402
– steradian   **397**
Universal Transverse Mercator (UTM)   **170**
Unscented Kalman Filter (UKF)   **270**
URDF (see also Unified Robot Description Format)   293
UTM, see also Universal Transverse Mercator   170
utricle   **113**

V

vanishing point   547, **554**
Vaucanson's duck   2
vector   **708**
– addition   **708**
– approach   **53**
– cross product   **709**, 727

– dot product **709**
– Euler **56**, **734**
– field **739**
– inner product **709**
– normal 54
– orientation **53**
– outer product **709**
– Rodrigues' **56**
– rotation 41
– scalar multiplication **708**
– space **708**
vectorizing **630**
vehicle
– aerial **129**, **147**, 156
– Braitenberg 164
– car-like **130**, 142
– configuration **130**, 192
– coordinate system **131**
– differentially-steered 130, 142
– heading **130**, 141
– omnidirectional **144**
– path 136, 137, 142
– underwater **129**, 156
– wheeled **129**
velocity
– angular **88**, 92, 95, 109
– control 362, 367, 368
– ellipsoid **340**
– end-effector **330**, 331, **332**
– joint **330**, 331, **332**
– kinematics **330**
– rotational 88, 91
– spatial **89**, 90, 677
– transformation **90**, 96
– translational 88, 91
– twist 91, **350**
verged cameras 605
vertical relief 586
vestibular system **109**, 113, 675
via point **102**
view
– field of 628
– fronto-parallel 645, 669, 670
vignetting 442
visibility matrix **620**
visual
– odometry (VO) **650**
– servo 664, **666**
    – camera retreat **685**
    – end-point closed-loop **666**
    – end-point open-loop **666**
    – eye in hand **666**
    – image based (IBVS) **670**
    – position based (PBVS) **668**
– SLAM (VSLAM) **619**
von Mises distribution **219**
Voronoi
– cell **189**
– diagram 189, 480
– roadmap **188**

– tessellation **189**
VSLAM (visual SLAM) **619**
V-SLAM (visual SLAM) 271

W

WAAS (Wide Area Augmentation System) **218**
wave gait **317**
wavefront planner 184
waypoint **102**, 220
web camera (webcam) **444**
wheel
– base **131**
– encoder 246
– mecanum **145**
– separation 143
– slip 227
white **419**
– D_{65} **418**
– balance **422**, 423
– equal energy **418**
– point 415
Wide Area Augmentation System (WAAS) **218**
Wien's displacement law **397**
world reference frame **24**
wrench **96**, 346, 349
– base **376**
– ellipsoid 348
– end effector 348, 383
– transformation **96**, 347
wrist
– singularity **304**
– spherical **300**

X

XYZ
– color space 413, 427
– roll-pitch-yaw angles 50, 311, 334
– tristimulus values 418

Y

yaw
– angle **130**
– rate **132**, 232
YOLO network **519**, 520
Yoshikawa's manipulability measure **341**
YUV color space 426

Z

zero crossing detector 467
ZNCC **469**, 471, 535
– similarity measure 535
zoom lens **553**
ZSAD **469**
ZSSD **469**
ZYX roll-pitch-yaw angles **49**
ZYZ Euler angles **47**, 282

Printed by Wilco bv, the Netherlands